*Chemical Reactor
Design and Operation*

Chemical Reactor Design and Operation

K. R. WESTERTERP
W. P. M. VAN SWAAIJ
A. A. C. M. BEENACKERS
Chemical Reaction Engineering Laboratories,
Twente University of Technology, Enschede, The Netherlands

JOHN WILEY & SONS
Chichester · New York · Brisbane · Toronto · Singapore

1st edition Elements of Chemical Reactor Design and Operation by H. Kramers and K. R. Westerterp
© 1963 Netherlands University Press, Amsterdam.

Copyright © 1984 by John Wiley & Sons Ltd.

All rights reserved.

No part of this book may be reproduced by any means, nor
transmitted, nor translated into a machine language
without the written permission of the publisher.

Library of Congress Cataloguing in Publication Data:

Westerterp, K. R.
   Chemical reactor design and operation.
   Rev. ed. of: Elements of chemical reactor design
and operation / H. Kramers. 1963.
   Includes bibliographies and index.
   1. Chemical reactors. I. Swaaij, W. P. M. van.
II. Beenackers, A. A. C. M. III. Kramers, H. Elements of
chemical reactor design and operation. IV. Title.
TP157.W43   1983   660.2′81   83-5769

ISBN 0 471 90183 0

British Library Cataloguing in Publication Data:

Westerterp, K. R.
   Chemical reactor design and operations—2nd ed.
   1. Chemical reactors
   I. Title   II. Swaaij, W. P. M. van   III. Beenackers,
A. A. C. M.
   IV. Kramers, H.—Elements of chemical reactor
design and operation
   660.2′844   TP157

ISBN 0 471 90183 0

Typeset in Great Britain by Speedlith Photo Litho Ltd., Stretford, Manchester.

Printed by Page Bros. (Norwich) Ltd.

Preface to the First Edition

The chemical reactor is a vital element in every chemical manufacturing process. Chemical as well as physical phenomena take place in it and its final design and construction are determined by mechanical factors. The art of the design, construction and economic operation of a reactor can, therefore, be seen as a synthesis of the principles of chemistry, physics, mechanics and economics.

The design and operation of reactors has received much attention in the past decade and it is rapidly developing from an empirical art towards a synthetic and rational activity. In chemical engineering physical operations have already gone through such a development. The systematics of physical transport phenomena and of physical separation processes have reached a state of maturity by way of 'unit operations'. Compared to this development the chemical reactor aspects of a manufacturing process have lagged behind. The concept of 'unit processes' has had certain advantages in bringing some system into technical organic chemistry. It does not, however, provide a basis for a systematic treatment of chemical reactors since other aspects, such as heat effects and states of aggregation and of dispersion, may equally be determining factors. Therefore, the approach adopted in this book is rather based on physical considerations, as will be explained below.

For a process which has to be developed commercially or which has to be improved in actual operation, it must be assumed that, as a result of chemical research, information regarding the chemistry of the process will be available. This information is specific to the process and relates to equilibria, heat effects, conversion rates of desired and undesired reactions under various conditions, the influence of impurities, the behaviour of catalysts, etc. When such data have been supplied in a more or less complete form, the chemical engineer has to consider the following questions:

—what manner of operation has to be adopted?

—what reactor type is the most suitable?

—what reactor dimensions are required?

An answer to these questions should ultimately lead to a reactor design or a strategy of operation by means of which the desired materials can be manufactured at the desired rate and at the lowest cost.

The underlying philosophy of this book is that the above questions can be treated by methods which are not specific, but general. Accordingly, the text is limited to a discussion of the non-specific aspects of chemical reactors. As a starting point, the isothermal operation of a few model reactor types has been selected. Subsequently, the various complications are discussed which are connected with flow phenomena in actual reactors, the heat effect of the reaction, the state of dispersion of the reaction mixture and the economic requirements of a manufacturing process, respectively.

This limitation to the engineering principles of chemical reactors implies that for a complete development and design of an industrial reactor supplementary information of a more specific nature will be necessary. Although this kind of information is not treated in this book, its importance cannot be overemphasized. This applies not only to chemical and physico-chemical data, the principles of which have been described in many excellent textbooks on chemical thermodynamics and kinetics, but also to data regarding materials of construction, corrosion, strength of material and costs. Evidently, in acquiring such information the chemical engineer has to rely on the work of a number of specialists.

Since this book has evolved from a lecture course for senior chemical engineering students at the Technical University of Delft, it has the character of a textbook. Accordingly, many illustrative examples have been included. At the same time, the authors address themselves to chemical engineers and chemists in practice who, in the authors' experience, sometimes get wrapped up in specific reactor problems to such an extent that some reminder as to the reactor engineering principles may be useful. No attempt was made to make a complete survey of the existing literature in the field. In elucidating the principles, a number of papers have been referred to which may be consulted for further study. Throughout the book, calculations have been made in MKS (meter, kilogramme, second) units, with the degree Kelvin as a unit of temperature difference and the kilomol as a unit of the amount of substance. This system has been recommended by the *Conférence Générale des Poids et Mésures* in 1960; conversion tables to other systems are readily available. It has been attempted to use throughout the book a consistent set of symbols as indicated in the list of symbols on page xxiii.

Summarizing the contents of the book, it may be said that in Chapter I some of the basic data needed for our purposes are reviewed. In particular, the formalism of conversion rates and degrees of conversion is discussed at some length; in our opinion much confusion still prevails with respect to the definitions used for these important properties of a reaction mixture.

In Chapter II the material balance is applied to various basic reactor types working under isothermal conditions. The reactor types discussed are 'model' reactors which, however, are often closely approached in practice. The special properties of these reactors are demonstrated with respect to the degree of conversion and the capacity, and to the selectivity and the yield.

The residence time distribution and the degree of mixing in continuous flow reactors are discussed in Chapter III. On the basis of these flow phenomena a comparison can be made between the performance of actual reactors and that of the model reactors.

In Chapter IV the heat balance is taken into account along with the material balance for the treatment of non-isothermal reactor operation. The autothermal operation of reactor systems is discussed at some length because of its importance for the heat economy of a plant. The occurrence of 'hot spots' in tubular reactors is described and the relativity of the concept of 'maximum allowable temperature' is demonstrated.

Chapter V has been devoted to the various ways in which diffusion and mass transfer may influence the conversion rate, the selectivity and the yield of reactions carried out in heterogeneous systems. Since this field of combined chemical reaction and physical transport phenomena is extremely extensive, the text has been limited to several demonstrative subjects which can be used as a basis for further study.

The subject of Chapter VI, the optimization of chemical reactors, is at present rapidly expanding. We have limited ourselves to the economic and technical principles and to the description of a few practical results. Some of the mathematical methods of optimization have been mentioned at the end of this chapter.

In many places the economic background of reactor design and operation has been emphasized. Two subjects which could not well be fitted into the systematics given above have been briefly discussed in the two appendices. The first contains recommendations for obtaining dependable conversion rate data in bench-scale reactors. The dynamic behaviour of cooled tank reactors is considered in Appendix 2.

Many persons have contributed to this book either directly or indirectly. Only a few of them can be mentioned by name. The well-known text *Chemical Process Principles* by O. A. Hougen and K. M. Watson and the pioneering works by G. Damköhler and by D. A. Frank-Kamenetski have greatly influenced our thinking. Professor K. G. Denbigh's early contributions from the U.K. in the field of chemical reaction engineering were very stimulating. We have profited much from the material which was brought together at the first and the second international symposium on Chemical Reaction Engineering in Amsterdam in 1957 and 1960, respectively, which were held under the auspices of the European Federation of Chemical Engineering. We had the pleasure of having frequent contacts with the main initiators of these symposia, Professors D. W. van Krevelen and J. C. Vlugter. Of the many people with whom we had fruitful discussions on subjects contained in this book we also wish to mention Professor R. B. Bird, Professor P. V. Danckwerts, Professor P. Le Goff, Dr. C. van Heerden, Dr. J. W. Hiby, Dr. H. Hofmann, Mr. P. J. Hoftijzer, Dr. F. Horn, Professor M. Letort, Dr. B. H. Messikommer, Professor R. L. Pigford, Professor K. Rietema, Professor K. Schoenemann and Professor E. Wicke. In particular, we gratefully acknowledge the assistance of Dr. W. J. Beek in the preparation of Chapter VI and Appendix 2.

Finally, we wish to express our gratitude to our wives who witnessed the writing of this book with much patience and understanding.

December 1962

H. KRAMERS
Technical University, Delft, The Netherlands

K. R. WESTERTERP
N. V. Petrochemie AKU-Amoco, Arnhem, The Netherlands

Preface to the Second Edition

The first edition of this book, which appeared in 1963, was sold out within two years. The Russians published a translation in 1967. In view of the vast development of chemical reaction engineering we had to write a completely new book retaining the major elements of the first edition, especially the systematic construction starting with simple problems and gradually building up the complexity. This implies that in this second edition the scope of the book as outlined in the first preface has not been changed.

The field of chemical reaction engineering has reached full maturity in the past twenty years; the advent of the computer in particular has enabled the chemical reaction engineer to make his mathematical models much more realistic and to explore much better all the aspects of the chemical reactor. Also many more data have been obtained, especially in multi-phase reactors, and experimental techniques have been refined. We have included in this new edition what we regard as the important developments of the last two decades, aiming at practical usefulness and understanding of the underlying theories; we hope the reader agrees with our selection. Even with this restriction we could not avoid an increase in volume by more than a factor two. It reflects the significant progress in reaction engineering over the past two decades.

Nowadays, *yield and selectivity* are much better understood and two special chapters (III and VIII) are dedicated to their study in homogeneous and heterogeneous reaction systems respectively. Determination of *residence time distributions* in chemical reactors is daily practice now, but the development of the theory on *micromixing* and also a realistic application in practice is still in its infancy, mainly due to the complicated experimental techniques and the complexity of the phenomenon. Nevertheless, sufficient progress has been made in recent years to justify a separate Chapter V, especially because we feel the chemical reaction engineer should be at least fully aware of some of its inherent pitfalls. Great progress has been made in the field of *heterogeneous reactors*. The original single chapter has now been replaced by three new ones. In Chapter VII we treat the vastly expanded theory of single reactions in heterogeneous gas–liquid and gas–solid systems simultaneously—this is against common practice but we feel there are no principal differences in these systems, while their analogy is striking and convenient in its application—and also discuss experimental techniques for the measurement of kinetic data; in Chapter VIII selectivity and yield problems in heterogeneous reactors are treated. In Chapter IX we combine the mainly isothermal thinking of the two preceding chapters with the *heat effect* in order to come to complete heterogeneous reactor designs. As all our efforts are still judged on their impact on plant economics we retained Chapter X on *optimization*, which is necessarily only an introduction to a vast specialized field. The assistance of Dr. Jan Fontein on computing techniques in this chapter is gratefully acknowledged. *Concepts* and *model reactors* are treated in Chapters I and II and heat effects in *homogeneous reactors* extended with tank reactor dynamics are treated in Chapter V.

Many new worked out *examples* have been included in this edition. The majority of them were taken from our industrial and development practice. The reader must be aware of the many pitfalls that besiege the practising engineer, who constantly must check the validity of the assumptions made to tackle a problem. Our example problems, therefore, cannot be copied as such, although we vouch for their elaboration and conclusions.

We kept struggling with *symbols*. The American Chemical Society and the American Institute of Chemical Engineers produced a list of recommended symbols, the Working Party on Chemical Reaction Engineering of the European Federation of Chemical Engineering produced a different one. We think we found a happy compromise by taking from both and adding a few of our own, meanwhile keeping consistency of style with the previous edition.

We wish to *acknowledge* the moral support of H. Kramers. He retired a few years ago and preferred not to be further involved in this second edition. He very generously waived his author rights. We thank him for all the years of friendship and for what we and so many other chemical engineers learnt from him, both at the university and in industry.

Finally, we wish to thank Lucy, Joke and Bernardine for seeing us through the writing of this book in so many evening and weekend hours.

December 1982

K. R. WESTERTERP, W. P. M. VAN SWAAIJ, A. A. C. M. BEENACKERS
*Chemical Reaction Engineering Laboratories,
Twente University of Technology, The Netherlands.*

Contents

Preface to the First Edition		v
Preface to the Second Edition		ix
List of Symbols		xxiii

CHAPTER I		Fundamentals of chemical reactor calculations .	1
I.1		Introduction	1
I.2		The material, energy and economic balance	3
		—Material balance	3
		—Energy balance	4
		—Economic balance	5
I.3		Thermodynamic data: heat of reaction and chemical equilibrium	6
		—Heat of reaction	6
		—Chemical equilibrium	10
I.4		Conversion rate, chemical reaction rate and chemical reaction rate equations	14
		—Influence of temperature on kinetics	16
		—Influence of concentration on kinetics	19
I.5		The degree of conversion	24
		—Relation between conversion and concentration expressions	26
I.6		Selectivity and yield	29
		—Selectivity and yield in a reactor section with recycle of non-converted reactant	30
I.7		Classification of chemical reactors	32
		References	36
CHAPTER II		Model reactors: single reactions, isothermal single phase reactor calculations	38
II.1		The well-mixed batch reactor	39
II.2		The continuously operated ideal tubular reactor . . .	43
II.3		The continuously operated ideal tank reactor	49

xi

II.4	The cascade of tank reactors	54
II.5	The semi-continuous tank reactor	61
II.6	The recycle reactor	68
II.7	A comparison between the different model reactors	72
	—Batch versus continuous operation	72
	—Tubular reactor versus tank reactor	74
II.8	Some examples of the influence of reactor design and operation on the economics of the process	77
	—The use of one of the reactants in excess	77
	—Recirculation of unconverted reactant	78
	—Maximum production rate and optimum load with intermittent operation	79
References		82

CHAPTER III	Model reactors: multiple reactions, isothermal single phase reactors	83
III.1	Fundamental concepts	84
	—Differential selectivity and selectivity ratio	85
	—The reaction path	87
III.2	Parallel reactions	93
	—Parallel reactions with equal order rate equations	93
	—Parallel reactions with differing reaction order rate equations	95
	—A cascade of tank reactors	99
III.3	The continuous cross flow reactor system	101
III.4	Consecutive reactions	108
	—First order consecutive reactions in a plug flow reactor	109
	—First order consecutive reactions in a tank reactor	113
	—General discussion	114
III.5	Combination reactions	124
	—Graphical methods	126
	—Optimum yield in a cascade of tank reactors	131
	—Algebraic methods	135
III.6	Autocatalytic reactions	148
	—Single biochemical reactions	150
	—Multiple autocatalytic reactions	155
References		158

CHAPTER IV	Residence time distribution and mixing in continuous flow reactors	160
IV.1	The residence time distribution (RTD)	160
	—The E and the F diagram	160
	—The application of the RTD in practice	172
IV.2	Experimental determination of the residence time distribution	173
	—Input functions	173

IV.3	Residence time distribution in a continuous plug flow and in a continuous ideally stirred tank reactor	177
IV.4	Models for intermediate mixing	182
	—Model of a cascade of N equal ideally mixed tanks	183
	—The axially dispersed plug flow model	185
IV.5	Conversion in reactors with intermediate mixing	199
IV.6	Some data on the longitudinal dispersion in continuous flow systems	207
	—Flow through empty tubes	207
	—Packed beds	212
	—Fluidized beds	214
	—Mixing in gas–liquid reactors	216
References		225

CHAPTER V Influence of micromixing on chemical reactions . . 227

V.1	Nature of the micromixing phenomena	228
	—Macro or gross overall mixing as characterized by the residence time distribution	228
	—The state of aggregation of the reacting fluid	229
	—The earliness of the mixing	229
V.2	Boundaries to micromixing phenomena	232
	—The model tubular and tank reactors	232
	—Boundaries for micromixing for reactors with arbitrary RTDs	239
V.3	Intermediate degree of micromixing in continuous stirred tank reactors	243
	—Formal models	244
	—Agglomeration models	
	—Model for micromixing via exchange of mass between agglomerates and their 'average' environment, the IEM model	250
V.4	Experimental results on micromixing in stirred vessels	254
V.5	Concluding remarks on micromixing	258
References		259

CHAPTER VI The role of the heat effect in model reactors . . 260

VI.1	The energy balance and heat of reaction	261
VI.2	The well-mixed batch reactor	264
	—Batch versus semi-batch operation	273
VI.3	The tubular reactor with external heat exchange	281
	—Maximum temperature with exothermic reactions; parametric sensitivity	289
VI.4	The continuous tank reactor with heat exchange	302
VI.5	Autothermal reactor operation	310
	—The tank reactor	312

	—An adiabatic tubular reactor with heat exchange between reactants and products	320
	—A multi-tube reactor with internal heat exchange between the reaction mixture and the feed	323
	—Determination of safe operating conditions	327
VI.6	Maximum permissible reaction temperatures	331
VI.7	The dynamic behaviour of model reactors	339
	—The autothermal tank reactor	340
	—Tubular reactor	354
References		355

CHAPTER VII   Multiphase reactors, single reactions 357

VII.1	The role of mass transfer	357
VII.2	A qualitative discussion on mass transfer with homogeneous reaction	359
	—Concentration distribution in the reaction phase	360
VII.3	General material balance for mass transfer with reaction	362
VII.4	Mass transfer without reaction	363
	—Stagnant film model	367
	—Penetration models of Higbie and Danckwerts	369
VII.5	Mass transfer with homogeneous irreversible first order reaction	371
	—Penetration models	371
	—Stagnant film model	377
	—General conclusion on mass transfer with homogeneous irreversible first order reaction	380
	—Applications	383
VII.6	Mass transfer with homogeneous irreversible reaction of complex kinetics	390
VII.7	Mass transfer with homogeneous irreversible reaction of order $(1,1)$ with $Al \gg 1$	393
	—Slow reaction	395
	—Fast reaction	396
	—Instantaneous reaction	398
	—General approximated solution	400
VII.8	Mass transfer with irreversible homogeneous reaction of arbitrary kinetics with $Al \gg 1$	409
VII.9	Mass transfer with irreversible reaction of order $(1,1)$ for a small Hinterland coefficient	412
VII.10	Mass transfer with reversible homogeneous reactions	412
VII.11	Reaction in a fluid–fluid system with simultaneous mass transfer to the non-reaction phase (desorption)	420
VII.12	The influence of mass transfer on heterogeneous reactions	424
	—Heterogeneous reaction at an external surface	426
	—Reactions in porous solids	433
VII.13	General criterion for absence of mass transport limitation	448

VII.14	Heat effects in mass transfer with reaction.	452
	—Mass transfer with reaction in series.	452
	—Mass transfer with simultaneous reaction in a gas–liquid system	454
	—Mass transfer with simultaneous reaction in a porous pellet	455
VII.15	Model reactors for studying mass transfer with chemical reaction in heterogeneous systems	458
	—Model reactors for gas–liquid reactions.	459
	—Model reactors for liquid–liquid reactions	466
	—Model reactors for fluid–solid reactions.	466
VII.16	Measurement techniques for mass transfer coefficients and specific contact areas in multi-phase reactors.	471
	—Measurement of the specific contact area a.	472
	—Measurement of the product $k_L a$.	474
	—Measurement of the product $k_G a$.	475
	—Measurement of mass transfer coefficients k_L, k_G	476
VII.17	Numerical values of mass transfer coefficients and specific contact areas in multi-phase reactors	479
	—Fluid–solid reactors	479
	—Fluid–fluid (–solid) reactors	487
References		490

CHAPTER VIII Multi-phase reactors, multiple reactions . . . 495

VIII.1	Introduction.	495
VIII.2	Simultaneous mass transfer of two reactants A and A' with independent parallel reactions A → P and A' → X (Type I Selectivity).	497
	—Mass transfer and reaction in series.	498
	—Mass transfer and reaction in parallel	500
VIII.3	Mass transfer of one reactant (A) followed by two dependent parallel reactions $$A(+B) \to P \quad A(+B, B') \to X$$ (Type II Selectivity).	503
	—Mass transfer and reaction in series.	504
	—Mass transfer and reaction in parallel	506
VIII.4	Simultaneous mass transfer of two reactants (A and A') followed by dependent parallel reactions with a third reactant: A + B → P, A' + B → X.	518
	—Complete mass transfer limitation in non-reaction phase	522
	—One reactant mass transfer limited in non-reaction phase	522
	—One reaction instantaneous	526
	—Both reactions instantaneous.	531
	—No diffusion limitation of reactant originally present in reaction phase	531
	—More complex systems	531
VIII.5	Simultaneous mass transfer of two reactants (A and A') which	

	react with each other	536
VIII.6	Mass transfer with consecutive reactions A → P → X (Type III Selectivity)	541
	—Mass transfer and reaction in series	541
	—Mass transfer and reaction in parallel	545
VIII.7	Mass transfer with mixed consecutive parallel reactions	557
	—The system: A(1) → A(2); A(2) + B(2) → P(2); P(2) + B(2) → X(2)	558
	—The system: A(1) → A(2); A(2) + B(2) → P(2); A(2) + P(2) → X(2)	560
	—Complex systems	565
	References	567

CHAPTER IX   Heat effects in multi-phase reactors — 571

IX.1	Gas–liquid reactors	575
	—General	577
	—Column reactors	580
	—Bubble column reactors	590
	—Agitated gas–liquid reactors	596
IX.2	Gas–solid reactors	612
	—Single particle behaviour	613
	—Catalytic gas–solid reactors	627
	—The moving bed gas–solid reactor	639
	—Thermal stability and dynamic behaviour of gas–solid reactors	648
IX.3	Gas–liquid–solid reactors	663
	References	669

CHAPTER X   The optimization of chemical reactors — 674

X.1	The object and means of optimization	675
	—The objective function	678
	—The optimization variables	680
	—Relation between technical and economic optima	680
X.2	Optimization by means of temperature	688
	—The optimization of exothermic equilibrium reactions	689
	—Temperature optimization with complex reaction systems	705
X.3	Some mathematical methods of optimization	710
	—Geometric programming	712
	—The Lagrange multiplier technique	716
	—Numerical search routines	724
	—Dynamic programming	730
	—Pontryagin's maximum principle	738
	References	745

Author Index 747

Subject Index 756

List of Illustrative Examples

Chapter I *Page*

I.2	Material balance for a reactant in a closed system . . .	4
I.3.a	Calculation of a heat of reaction	8
I.3.b	Equilibrium constant in the synthesis of ammonia . . .	11
I.5.a	Relation between conversion and concentration expressions	26
I.5.b	Conversion rate as a function of the degree of conversion, liquid reaction	27
I.5.c	Conversion rate as a function of the degree of conversion, gas reaction	28
I.6.a	Selectivity and yield in a reactor section with recycle of non-converted reactant	30
I.6.b	Influence of reactor yield on plant profitability	31

Chapter II *Page*

II.1	Production of ethyl acetate in a batch reactor	41
II.2a	Production of ethyl acetate in the liquid phase in a tubular reactor	45
II.2.b	Reaction at varying density	46
II.2.c	Catalysed gas reaction in a tubular reactor	48
II.3.a	Check on ideal mixing	52
II.3.b	Production of ethyl acetate in a continuous tank reactor .	52
II.3.c	Start-up of a CISTR	52
II.4a	Production of ethyl acetate in a cascade of three tank reactors	58
II.4.b	Polymerization in a cascade	59
II.5.a	A semi-continuous (or semi-batch) experiment	62
II.5.b	Slow addition of reactant to reduce heat evolution, first order reaction	63
II.5.c	Production of ethyl acetate in a batch reactor with removal of product	64
II.5.d	Slow addition of reactant to reduce heat evolution, second order reaction	66
II.6	Autocatalytic reaction in various model reactors . . .	70

xviii

| II.7 | Comparison of various model reactors | 76 |
| II.8 | Maximum production and optimum load in a batch process | 81 |

Chapter III *Page*

III.1	Determination of selectivity and yield data in a PFR from laboratory experiments	90
III.2	Parallel reactions in a cascade	100
III.3	Parallel reactions in an idealized cross flow reactor	102
III.4.a	Relation between reactor selectivity and relative conversion	112
III.4.b	Instruction on reactor operation	115
III.4.c	The chlorination of benzene in various reactor systems	116
III.4.d	A chain of consecutive reactions, the chlorination of methane	120
III.5.a	Alkylation of isobutane with proplene	133
III.5.b	Yield improvement with increase of reactant concentration in the feed	142
III.5.c	Lumping of kinetics for the desulphurization of gas-oil	146

Chapter IV *Page*

IV.1.a	Some numerical examples of E and F diagrams and properties	167
IV.1.b	Applications of the convolution integral to flow regions in series	170
IV.2	Injection and sampling of tracer within a flow section	176
IV.3.a	E-diagram and moments derived from Laplace transformation	180
IV.3.b	Modelling a circulation reactor	181
IV.4.a	E-diagram and the moments of a cascade of mixers from Laplace transformation	185
IV.4.b	Moments of the RTD of the axially dispersed plug flow model by Laplace transformation	189
IV.4.c	Different boundary conditions for the axially dispersed plug flow model	190
IV.4.d	Dead zones, two region models and combined models	191
IV.5.a	Prediction of the conversion for a first order chemical reaction	203
IV.5.b	Consecutive reactions in a reactor with axially dispersed plug flow	205
IV.6.a	Blending in pipe-line transport	211
IV.6.b	Determination of parameters from pulse response tests	217

Chapter V *Page*

V.1	Emulsion polymerization in a continuous tank reactor	237
V.3	Application of the IEM model to the conversion calculation for a second order reaction (after Villermaux)	251
V.4	Macro and micromixing in a stirred tank reactor	256

Chapter VI

		Page
VI.2.a	Influence of the policy of operation on the reaction time	269
VI.2.b	An explosion	271
VI.2.c	Synthesis of a maleic acid mono-ester in a BR	277
VI.2.d	The synthesis of a maleic acid mono-ester in a SBR	278
VI.3.a	Tube diameters required in an isothermal laboratory plug flow reactor	283
VI.3.b	A catalytic dehydrogenation reaction in a tubular reactor	284
VI.3.c	First order reaction in an adiabatic PFR	287
VI.3.d	Parametric sensitivity	291
VI.3.e	Maximum reaction temperature	300
VI.3.f	Preliminary reactor design to prevent runaway	301
VI.4	Heat transfer requirements in a cascade of tank reactors	303
VI.5.a	Comparison of a non-authothermal and an autothermal reactor	310
VI.5.b	Static stability of tank reactors in a cascade	317
VI.5.c	Design of a CISTR	319
VI.6	The oxidation of naphthalene in a fluidized bed	335
VI.7.a	Will a tank reactor be stable?	351
VI.7.b	Stability of a liquid phase reactor	354

Chapter VII

		Page
VII.4	Chemical absorption of NH_3 from air in a packed column	366
VII.5.1	Laminar jet model reactor	376
VII.5.3	Influence of the gas phase resistance on the flux through an interface; first order reaction	382
VII.7.4.a	The influence of a gas phase resistance on the flux through an interface for a reaction of order (1,1)	403
VII.7.4.b	Soap Ltd, soap boilers since 1775	406
VII.7.4.c	Chemical absorption of CO_2 from air in a packed column	407
VII.12.1.a	Hydrogenation of an aldehyde in a gas–liquid slurry continuous tank reactor	430
VII.12.1.b	Dimerization in a tube reactor	432
VII.12.2.a	Catalytic equilibrium reaction in a spherical porous particle	436
VII.12.2.b	Packed bed reactor versus riser	443
VII.12.2.c	Conversion rate of a solid according to a sharp interface shrinking unreacted core model	445
VII.13	Gasification of wood char according to a grain model	450
VII.15.1.a	Kinetic data from absorption experiments in a stirred cell model reactor	461
VII.15.1.b	Kinetic data from experiments in a wetted wall reactor	463
VII.15.3	The measurement of reaction kinetics and internal pellet diffusivity with a single pellet diffusion reactor	471
VII.16.4	Simultaneous measurements of k_L and a in a cyclone reactor	477
VII.17.1	Design of a fluid bed reactor	485

Chapter VIII   *Page*

VIII.3.1   Mass transfer with parallel first order reactions. . . .   505
VIII.3.2   The absorption of CO_2 in amines aided by a shuttle . .   516
VIII.4.6.a The influence of mass transfer on the selective absorption of
           H_2S from a mixture of CO_2 and H_2S   534
VIII.4.6.b The negative enhancement factor.   535
VIII.6.2   Mass transfer with a slow consecutive reaction in a stirred tank
           reactor   552
VIII.7.2   Selectivity for a consecutive reaction with the controlling
           resistance completely in the non-reaction phase . . .   564

Chapter IX   *Page*

IX.1.a   Temperature difference over the liquid film, chlorination of
         toluene   579
IX.1.b   Check for isothermicity of a column reactor   581
IX.1.c   Chemical absorption of H_2S from a natural gas stream .   584
IX.1.d   A bubble column reactor with incomplete mixing in the liquid
         phase   592
IX.1.e   Oxidation of a dissolved hydrocarbon in an agitated gas–liquid
         reactor under pressure   609
IX.2.a   Catalytic combustion of NH_3 with atmospheric air . .   617
IX.2.b   Check for intra- and interparticle resistances   621
IX.2.c   Temperature and concentration distribution in an experimental
         catalytic reactor   637
IX.2.d   Reduction of iron ore in a semi-technical reactor . . .   647
IX.2.e   Decoking of a catalyst   658
IX.3   Sour gas absorption in a co-current trickle flow column .   665

Chapter X   *Page*

X.1.a   Influence of economics on optimal reactor design and operation   684
X.1.b   Optimal reactant concentration in the reactor feed . . .   686
X.2.a   Optimization of an ammonia cold shot converter with
        temperature constraints   701
X.2.b   Denbigh's problem   709
X.3.a   Minimum outer surface area of a cylindrical storage tank .   713
X.3.b   Minimum total volume in an isothermal cascade . . .   714
X.3.c   The maximum yield for first order consecutive reactions in one
        tank reactor   718
X.3.d   Temperature optimization of three parallel reactions . .   723
X.3.e   Optimal design of an autothermic reactor   726
X.3.f   The minimum reaction volume of an isothermal cascade .   731
X.3.g   The minimum reaction volume of a non-isothermal cascade   732

		Page
X.3.h	Cost minimization in a cascade	735
X.3.i	Optimum temperature profiles in several types of model reactors	740
X.3.j	Parallel reactions in a batch reactor	743

List of Symbols

Symbol	Quantity	Units
a	activity	—
a	thermal diffusivity	m^2/s
a	interfacial area per unit volume	m^2/m^3
a_p	external interfacial area per unit volume	m^2/m^3
a_s	internal surface area per unit porous particle volume	m^2/m^3
A	heat or mass transfer area	m^2
Al	Hinterland coefficient (equation (VIII.71))	—
$\bar{c}$	flow average concentration $(\iint_S cv\,dS / \iint_S v\,dS)$	$kmol/m^3$
$\bar{\bar{c}}$	area average concentration $(\iint_S c\,dS / \iint_S dS)$	$kmol/m^3$
c_J	molar concentration of species J	$kmol/m^3$
$\bar{c}_J$	molar concentration of species J in the bulk or, in stagnant phases, at $x = \delta$ (film theory)	$kmol/m^3$
c_p, c_v	heat capacity per unit mass, at constant pressure and volume, respectively	$J/kg\,°C$
C	reduced concentration, c/c_0	—
C_p, C_v	molar heat capacity, at constant pressure and volume, respectively	$J/kmol\,°C$
d	characteristic dimension, diameter	m
d_b	bubble diameter	m
d_p	particle diameter	m
d_t	tube diameter	m

Symbol	Quantity	Units
D	coefficient of molecular diffusion	m²/s
D_i	coefficient of internal diffusion in a porous particle	m²/s
D_t	coefficient of transverse dispersion	m²/s
D_l	coefficient of longitudinal dispersion	m²/s
e	relative excess of one reactant with respect to the other reactant	—
E	activation energy	J/kmol
E_J	extraction factor $= m_J \Phi_{vG}/\Phi_{vL}$	—
$E_J, E_{J'}$	enhancement factor of J and J' respectively; rate of mass transfer with reaction divided by the rate of mass transfer without reaction	—
E_{J0}	enhancement factor of J according to equation (VII.132)	—
$E_{J\infty}$	enhancement factor of J for instantaneous reactions	—
$E(t)$	residence time distribution function or frequency function	s⁻¹
$E(\theta)$	residence time distribution function for the dimensionless time	—
f	Fanning friction factor	—
$F(t)$	cumulative residence time distribution function	—
g	acceleration by gravity	m/s²
G	free energy, $H - TS$	J
G_J	molar free energy of species J	J/kmol
h	enthalpy per unit mass	J/kg
H	enthalpy	J
ΔH_a	heat of absorption	J/kmol
H_J	molar enthalpy of species J	J/kmol
ΔH_r	heat of reaction at constant pressure as associated with the stoichiometric equation	J
$(\Delta H_r)_J$	heat of reaction at constant pressure for the conversion of or to one molar unit of J	J/kmol

Symbol	Quantity	Units
J	molar flux	kmol/m²s
J_h	heat flux	W/m²
$\bar{J}_{J\tau}, \bar{J}_J, J_J$	average molar flux of species J during the contact time τ according to the penetration theory	kmol/m²s
k	homogeneous reaction velocity constant, dimension depends on the kinetics; for nth order:	$m^{3(n-1)}/kmol^{(n-1)}s$
k''	reaction velocity constant of surface reactions, dimension depends on kinetics; for nth order reaction:	$m^{3(n-1)+1}/kmol^{(n-1)}s$
k_P, k_X	homogeneous reaction velocity constants of reactions resulting in P and X respectively; dimensions depend on kinetics	—
k_P'', k_X''	the same, for surface reactions	—
$k_{n,m}$	reactions velocity constant of reaction of order n, m as in $R_A = -kc_A^n c_B^m$	$\dfrac{m^{3(n+m-1)}}{kmol^{(n+m-1)}s}$
k_G, k_L, k_S, k_F	mass transfer coefficients	m/s
k_{LS}	the same, from liquid to solid	m/s
k_∞, k_∞''	pre-exponential factor in the Arrhenius equation for homogeneous reaction and surface reaction respectively, dimension depends on the kinetics; see $k_{n,m}$:	$m^{3(n+m-1)}/kmol^{(n+m-1)}s$
K	chemical equilibrium constant, dimension depends on the kinetics	—
K'	dimensionless chemical equilibrium constant, e.g. equation (VII.186)	—
L	length	m
m, m_J	distribution coefficient (for species J), e.g. $c_L = mc_G$	—
m	total mass of a system	kg
m_J	mass of species J in a system	kg
m_s	mass of solid or catalyst in a system	kg

Symbol	Quantity	Units
$\bar{M}$	average molar mass of a reaction mixture	kg/kmol
M_J	molar mass of species J	kg/kmol
MO_k	kth moment of $E(t)$	s^k
MR_k	kth moment of $E(\theta)$	—
n	sequence number of a tank reactor in a cascade or of a bed in a multibed reactor	—
n_{inj}	number of kmoles injected	kmol
N	total number of tank reactors in a cascade or beds in a multibed reactor	—
p	total pressure	$N/m^2 = Pa$
p_J	partial pressure of species J	$N/m^2 = Pa$
Q	amount of heat	J
$\dot{Q}$	heat flow	W
r	cylindrical or spherical coordinate	m
r_p	radius of spherical particle	m
r_h	hydraulic radius	m
R_J	molar rate of production of J per unit volume of the reaction phase	$kmol/m^3 s$
R_{SJ}	the same, per unit mass of solids	$kmol/kg\,s$
R_J''	molar rate of production of J per unit area	$kmol/m^2 s$
$\langle R_J \rangle$	average molar rate of production of J per unit volume of a multiphase reactor	$kmol/m^3 s$
$\|R\|_J$	molar rate of conversion per unit volume of the reaction phase	$kmol/m^3 s$
$\|R\|_{SJ}$	molar rate of conversion per unit mass of solid	$kmol/kg\,s$
R	gas constant, 8315	$J/kmol\,K$
R	recycle ratio	—
s	Laplace transform variable	—
S	cross-sectional area	m^2
S	entropy	J/K
t	time	s

Symbol	Quantity	Units
t	age, residence time	s
$\bar{t}$	first moment of $E(t)$	s
T	temperature	K or °C
ΔT_{ad}	adiabatic temperature rise of a reaction mixture after complete conversion	°C
TR	transfer function	—
u	internal energy per unit mass	J/kg
u	superficial fluid velocity (Φ_v/S)	m/s
U	internal energy	J
U	overall heat transfer coefficient	W/m² °C
U_J	molar internal energy of species J	J/kmol
ΔU_r	heat of reaction at constant volume as associated with a stoichiometric equation	J
$(\Delta U_r)_J$	heat of reaction at constant reaction volume for the conversion of or to one molar unit of J	J/kmol
$U(t)$	unit step function	—
v	velocity	m/s
v_r	relative velocity; velocity of one phase relative to another phase	m/s
$\langle v \rangle$	average actual fluid velocity (in a packed bed = u/ε)	m/s
V	volume	m³
V_r	volume of reaction mixture	m³
V_{bL}	bulk volume of liquid phase	m³
w_J	mass fraction of species J	—
W	amount of work	N m = J
$\dot{W}$	rate of work done on the surroundings	N m/s = W
x_J	molar fraction of species J in any phase	—
x	coordinate perpendicular to interface	m
y_J	molar fraction of species J in the gas phase only used in connection with x_J for the liquid phase	—

Symbol	Quantity	Units
z	coordinate in the direction of the flow	m
Z	dimensionless coordinate in the direction of flow, z/L	—
α	heat transfer coefficient	W/m² °C
α_m	volumetric rate coefficient for mass exchange	s⁻¹
β	temperature factor E/RT_a; index a varies, depending on the situation	—
β_{dyn}	dynamic hold-up fraction	—
β_{stat}	static hold-up fraction	—
γ	fugacity coefficient	—
$-\gamma$	concentration gradient at $x = \delta$ (film theory)	kmol/m⁴
$\underline{\gamma}$	dimensionless adiabatic temperature rise or ratio of reaction heat liberation rate to heat conduction rate (see text)	—
δ	film thickness or characteristic length	m
δ_h	film thickness for heat transfer	m
δ_r	thickness of reaction zone or distance of reaction plane from interface (instantaneous reaction, film theory)	m
$\delta(t)$	Dirac function	s⁻¹
$\delta(\theta)$	Dirac function with dimensionless time	—
ε	fractional volume of dispersed phase in fluid–fluid systems; fraction of fluid phase in fluid–solid systems	—
ϑ	dimensionless temperature, e.g. T/T_0 or $(T - T_0)/T_0$, see text	—
θ	dimensionless residence time, t/τ	—
θ_t	dimensionless time, $t/\bar{t}$	—
ζ	relative degree of conversion, fraction of reactant converted	
κ	ratio between reaction rate constants	—

Symbol	Quantity	Units
κ_R	dimensionless reaction velocity constant relative to its value at a reference temperature	—
λ	thermal conductivity	W/m °C
λ_t	effective thermal conductivity transverse to main flow	W/m °C
λ_0	thermal conductivity of packed bed without flow	W/m °C
μ	dynamic viscosity	N s/m^2
ν	kinematic viscosity	m^2/s
ν_J	stoichiometric coefficient of species J	—
ξ_J	degree of conversion of species J	—
η	effectiveness factor (equation (VII.96)) or degree of utilization, equation (VII.86)	—
η''	effectiveness factor for external surface reaction, equation (VII.205)	—
η_J	yield of species J; fraction of reactant fed which is converted to product J	—
ρ	density, specific mass	kg/m^3
ρ_B	bulk density of the bed	kg/m^3
ρ_s	apparent density of solid (catalyst) particles	kg/m^3
σ	surface tension	N/m
σ_J	selectivity with respect to J, amount of a reactant converted to J divided by the total amount converted of the same reactant	—
σ'_J	differential selectivity with respect to J; conversion rate of a reactant to J divided by the total conversion rate of that reactant	—
σ_t	standard deviation of the residence time	s
σ_θ	standard deviation of the dimensionless residence time	—

Symbol	Quantity	Units
τ	average residence time in reactor system	s
τ_L	residence time in a tubular reactor of length L	s
τ_L	residence time of the liquid	s
ϕ	reaction of Thiele modulus	—
ϕ_P	reaction modulus for a consecutive reaction	—
ϕ'	modified reaction modulus, equation (VII.99)	—
ϕ''	reaction modulus expressed in observables, Section VII.13	—
ϕ_β	dynamic hold-up divided by the total hold-up	—
Φ_J	production rate of J	kg/s; kmol/s
$\Phi_{m'}$ (Φ_{mJ})	mass flow rate; the same for species J	kg/s
Φ_{mr}	mass flow rate of recycle flow	kg/s
Φ_v	volumetric flow rate	m^3/s
$\Delta\Phi$	net inflow (in–out)	m^3/s

Subscripts

A, B, A'	reactants A, B, A'
J	arbitrary species J
L, G, S, F	liquid, gas, solid, fluid phases
L	at the outlet of a tubular reactor of length L
N	output of a cascade of N tank reactors or of a multibed reactor with N beds
P, Q	desired products
R	at reference temperature T_R
X, Y, X'	undesired products
o	initial or feed conditions
1	output of one tank reactor
b	bubble
c	coolant or continuous phase
d	dispersed phase; in fluidized bed; dense phase
e	equilibrium

i	interface	
l	longitudinal	
mf	minimum fluidization	
n	output of the nth tank reactor in a cascade, or of a segment n	
r	reaction	
t	transverse	
v	volumetric	
w	wall	
$\bar{M}$	average M	
$\bar{c}$	c in bulk	
$\underline{E}$	Laplace transform of E	

Dimensionless groups

Bi	Biot number	$\alpha d/\lambda_S$ or $\alpha_w d_t/\lambda_t$
Bo	Bodenstein number	$\langle v \rangle d/D_1$
Fo	Fourier number	dt/D^2
Ha	Hatta number; form depends on kinetics	
Le	Lewis number	$\lambda/c_p \rho D$
Nu	Nusselt number	$\alpha d/\lambda$
Pe	Péclet number for longitudinal dispersion	$\langle v \rangle L/D_1$
Pe_h	Péclet number for heat dispersion	$\langle v \rangle L \rho c_p/\lambda$
Pe_t	Péclet number for transverse dispersion	$\dfrac{\langle v \rangle d_t}{D_t}$
Pr	Prandtl number	$\mu c_p/\lambda$
Re	Reynolds number	vd/ν
Re_p	Reynolds number in packed bed related to particle diameter	$\langle v \rangle d_p/\nu$
Sc	Schmidt number	ν/D
Sh	Sherwood number	$k_G d/D$; $k_L d/D$
St	Stanton number, for heat transfer	$\alpha/\rho c_p \langle v \rangle$
St	Stanton number, for mass transfer	$k_G/\langle v \rangle$ or $k_L/\langle v \rangle$
St'	modified Stanton number	$UA/\rho c_p \Phi_v$

Comments on the list of symbols

We tried to use the available standards for our symbols. Regretfully there does not exist yet a fully accepted set of symbols. Most of our symbols agree with common practice. There are a few exceptions and therefore, for readers' convenience, we summarize here deviations from either the European or the American practice. A European list has been published by U. Hoffmann, *Chem. Ing. Techn.* **51**, 1128 (1979) and an American list by E. Buck, *Chem. Eng. Prog.* **74**(10), 73 (1978). The following list shows only deviations between our symbols and theirs.

	Our	*Hoffmann*	*Buck*
mass of catalyst	m_{cat}	W	m_{cat}
thermal diffusivity $\lambda/\rho c_p$	a	α	α
distribution coefficient	m	κ	K^*
volumetric flow rate	Φ_v	Q	q
mass flow rate	Φ_m	—	F, L, B, V, N
molar flow rate	Φ_{mol}	F	F, L, B, V, N
mass flow rate per unit area	Φ_m''	G, L	G, L
volumetric rate per unit area	u	u	V
heat flow rate per unit area	$\dot{Q}/S$	J_w	q/S
heat transfer coefficient	α	h	h
pre-exponential factor	k_∞	A	A
relative degree of conversion	ζ_A	X	X
yield	η_p	θ_p, Φ_p	—
selectivity	σ_p	S_{12}	—
observable Thiele modulus	ϕ''	Φ	—
heat	Q	—	Q
diameter	d	d	D
distance along path	z	z	s, x, X
height above plane	h	h	Z
lateral distance from plane	y	y	Y
thickness	d	d	B
area per unit mass	a_s	a_s	A_w, s
area per unit volume	a	a_v	A_s, a
molecular diffusivity	D	D	δ_E, D_v, δ
latent heat at a phase change	ΔH	ΔH	λ
thermal conductivity	λ	λ	k

Chapter I
Fundamentals of chemical reactor calculations

I.1 INTRODUCTION

A rational approach to all problems relating to a physical or chemical change of matter must be based on the elementary physical conservation laws. The formulation of these laws is, therefore, the point of departure in any chemical engineering textbook, and it has been handled most systematically in, for example, the books by Bird *et al.* [1], and by Beek and Muttzall [2]. For the treatment of chemical reactors, the laws of conservation of matter and of energy are of primary importance; they will be consistently applied in this book. Flow phenomena, governed by the principle of conservation of momentum, naturally are equally important for chemical reactors; however, the reader is assumed to have a working knowledge of engineering fluid dynamics, so this will be applied where necessary without further explanation.

The principle of conservation of energy is also expressed in the first law of thermodynamics, whereas the second law of thermodynamics is concerned with the degradation of energy associated with irreversible processes. The application of the first and second laws to the chemical transformation of matter, i.e. chemical thermodynamics, provides a consistent framework for the calculation of heats of transformation and for the relationships describing chemical equilibrium. Of the numerous textbooks in which this material is treated, we only mention those by Hougen *et al.* [3], by Modell and Reid [4] and by Denbigh [5].

The conservation laws and the relevant relationships supplied by chemical thermodynamics will be briefly reviewed in Sections I.2 and I.3. Also an economical balance is treated in Section I.2 to remind the reader of the fact that the ultimate judgment on the performance of a chemical reactor or a reactor section is based on economics. Although no physical conservation principle is here involved, we may speak of an economic balance, which has to be taken into account for obtaining maximum profitability of any industrial operation. However, economics have to be judged on the scale of the whole plant; at the same time safety, pollution of the environment, raw materials availability and social factors, which all have a strong influence on the chemical reactor design and operation, are equally important factors in plant performance.

The second kind of information needed for a quantitative treatment of chemical reactor problems is concerned with *rates*. The rate at which a reaction proceeds may not only be determined by the chemical kinetics of the reaction proper but also by physical transport phenomena (Section I.4). The interaction between chemical kinetics and physical transport rates has received considerable attention during the past decades. Damköhler [6] and particularly Frank-Kamenetski [7] have systematically developed this field. The latter author distinguished in this respect 'microkinetics' (i.e. chemical kinetics) and 'macrokinetics' (i.e. physical rates). Many scientists, e.g. van Krevelen [8], Danckwerts [9] and Astarita [10] have stressed the great importance of the scale of the scrutiny, at which we consider the phenomena in a reacting system. In principle we have to take into account three different ones:

(a) the scale of the molecular mean free path (diffusion), for example in the pore of a catalyst pellet,
(b) the scale of an eddy, or dispersion in heterogeneous systems (transfer), where we take a catalyst particle or one gas bubble as the basis for deriving the appropriate equations, and
(c) the scale of the reactor as a whole (convection transport), where we are concerned with the integration of small scale phenomena over the entire reactor and where macroscopic mixing is most important.

Whereas for obtaining chemical kinetic data in general we still must entirely rely on experimental results relative to the particular reaction under consideration, it is almost always possible to estimate physical transport and transfer rates on the basis of fluid properties, flow circumstances and the geometry of the system. Since the reader is assumed to be familiar with the concepts of diffusion, mass transfer and heat transfer, these will be used where necessary without prior discussion. Therefore, in Section I.4 of this chapter we shall only briefly mention a few facts about chemical kinetics and indicate which formulation will be used for the chemical conversion rate and the degree of conversion.

In Section I.5 the concept of degree of conversion, which plays such an important role in all chemical reactor calculations, is introduced. We will define the degree of conversion as fundamentally as possible in order to avoid pitfalls; after that we introduce the less fundamental approach, which is in more common use by chemical reaction engineers. As the driving force for a chemical reaction to take place is provided by molar concentrations (moles per unit volume) rather than mass fractions, we can run into translation troubles, introducing reaction rates dependent on molar concentrations into the material balance, if the number of moles or the density of a reacting system are not constant.

In Section I.6 the concepts of yield and selectivity are introduced, because they will appear repeatedly in this book. In industrial chemical operations generally not only desired products will be obtained in the reactor, but undesired products as well. The relationship between the productions of desired and undesired products can be described by the yield and selectivity in reactor operations. They generally dominate the final economic performance of the whole chemical plant to a far greater extent than the reactor volume itself, so that we wish to introduce these concepts right at the start.

Finally, for the reader's orientation, some remarks are made in Section I.7 on the great variety of chemical reactors. This leads to the three basic reactor types to be discussed further: the mixed batch reactor, and, for continuous operation, the stirred tank reactor and the tubular reactor.

I.2 THE MATERIAL, ENERGY AND ECONOMIC BALANCE

I.2.1 Material balance

For any reactor calculation the application of the principle of conservation of matter is indispensable. We will follow the systematic formulation of balances used by Bird et al. [1]. A material balance may be set up for any molecular species taking part in the reaction. In words, such a balance over a system, e.g. for a component J, may be formulated in the following manner:

$$\begin{Bmatrix} \text{accumulation of} \\ \text{mass of J} \\ \text{in the system} \end{Bmatrix} = \begin{Bmatrix} \text{mass of J} \\ \text{into the} \\ \text{system} \end{Bmatrix} - \begin{Bmatrix} \text{mass of J} \\ \text{out of the} \\ \text{system} \end{Bmatrix} + \begin{Bmatrix} \text{mass of J} \\ \text{produced by} \\ \text{reaction} \end{Bmatrix}.$$

This equation can be set up for a certain time interval in which case the terms are expressed in units of mass; in view of the applications, however, it is preferable to use units of mass per unit time, so that the terms represent mass rates of J.

The mathematical formulation of the material balance greatly depends on the nature of the system under consideration. The scale at which we set up the mass balance should be carefully chosen. In non-homogeneous systems or in reactors with reaction conditions varying over volume of the entire reactor, we must generally start with a differential volume dV. As a basis for further calculations we select as our system a volume V for which a material balance may be put into the form:

$$\frac{dm_J}{dt} = -\Delta\Phi_{mJ} + \langle M_J R_J \rangle V \text{ [kg/s]}, \tag{I.1}$$

where

m_J = mass of J in the system,
$-\Delta\Phi_{mJ}$ = net inflow of J by convection and possibly also by a diffusional process, in which $\Delta\Phi_{mJ} = \Phi_{mJout} - \Phi_{mJin}$,
$M_J R_J$ = mass rate of production of J by chemical reaction per unit volume,
$\langle \; \rangle$ = space average of $M_J R_J$.

The sum of the material balances for each chemical species gives the total material balance:

$$\frac{dm}{dt} = -\Delta\Phi_m \text{ [kg/s]}. \tag{I.2}$$

Since we have expressed the amounts of the various components in units of mass, the production terms in equation (I-1) cancel each other in this summation. This would, of course, not be generally so if other units were used (mols or units of volume). The necessity of making use of units of mass in the material balance, together with the theoretically well-founded practice of using molar quantities for chemical reactions, calls for much care (also see Section I.5).

Example I.2 *Material balance for a reactant in a closed system*
It is required to derive a material balance for a reactant in a closed system $(-\Delta\Phi_{mJ} = 0)$ based on molar concentrations. By dividing equation (I.1) by the molecular weight M_J (kg J/kmol J) of J, with $m_J = M_J c_J V$, if c_J is expressed in kmoles of J per unit of volume of the reacting system and with $V = m/\rho$ (the total mass m in the closed system remains constant!) we find

$$m \, d\left(\frac{c_J}{\rho}\right)/dt = \langle R_J \rangle \frac{m}{\rho}$$

and after rearranging we can derive:

$$\langle R_J \rangle = \frac{dc_J}{dt} - \frac{c_J}{\rho}\frac{d\rho}{dt}$$

From this equation we see that the chemical conversion rate is equal to the change of the concentration with time of the component under consideration only in case the density of the reaction mixture remains constant.

I.2.2 Energy balance

The application of the principle of conservation of energy leads to an energy balance which in general states that

$$\left\{\begin{array}{l}\text{rate of}\\ \text{accumulation}\\ \text{energy}\end{array}\right\} = \left\{\begin{array}{l}\text{rate of}\\ \text{energy}\\ \text{in}\end{array}\right\} - \left\{\begin{array}{l}\text{rate of}\\ \text{energy}\\ \text{out}\end{array}\right\} + \left\{\begin{array}{l}\text{rate of}\\ \text{energy}\\ \text{production}\end{array}\right\}.$$

Strictly speaking, all forms of energy must be taken into account: heat, kinetic energy, potential energy in a gravitational, electrical and magnetic field. For a complete formulation of the energy balance, the reader is referred to e.g. [1], [2] and [11]. In most reactor calculations the terms with thermal energy and work done on the surroundings are of main importance. Leaving out the other effects, the energy balance for a system in which reaction takes place can easily be derived from the first law of thermodynamics. For a closed system this law states that the increase of internal energy ΔU of the system in a certain time interval equals the heat Q supplied to the system minus the work W done by the system on its surroundings in the same time interval:

$$\Delta U = Q - W.$$

In an open flow system there is an inflow of mass Φ_m with an internal energy u per unit of mass and work must be done to press this mass into the system, being

$p_{in}\Phi_{Vin}$ per unit of volume or $(p\Phi_m/\rho)_{in}$. The same holds for the outflow of mass out of the system. From thermodynamics we know that $H = U + pV$, so that

$$\Phi_m(u + p/\rho) = \Phi_m h.$$

For an open system the first law of thermodynamics therefore becomes

$$\frac{d(\langle u \rangle m)}{dt} = -\Delta(h\Phi_m) + \dot{Q} - \dot{W} \;[\text{J/s} = \text{W}] \tag{I.3}$$

here

$\langle u \rangle m$ = total internal energy of the system,
$-\Delta(h\Phi_m)$ = net inflow of enthalpy,
$\dot{Q}$ = rate of heat supply from the surroundings,
$\dot{W}$ = rate of work done on the surroundings.

Note that for a closed system (i.e., no material flowing in or out) the second term in equation (I.3) disappears. The term $\dot{W}$ accounts for the rate of work done by a device for delivering work to the surroundings and forming an integral part of the reacting system, e.g. an expansion turbine or a piston in a combustion engine. Effects of reaction heat in the energy balance will be further discussed in Chapters VI and IX.

In principle both the material and the energy balance have to be considered on the three different scales, as mentioned in Section I.1. Whether the smaller scales can be neglected and whether we may start directly with a macroscopic balance over the entire reactor depends on the state of mixing in the reactor; this will be elucidated further in this book.

I.2.3 Economic balance

It is not only required from a design of a reactor system that the specifications demanded of the process can actually be realized; it is also important to know the production costs, the pay-out time and the return on investment for the installation considered. In general, therefore, each project is evaluated economically in an early stage of its development. In the economic evaluation all costs involved in the production of a certain product are balanced against the total expected income (see, e.g., Schweyer [12] and Peters and Timmerhaus [13]). When the income exceeds the production costs, a profit will result. The best (or optimum) design for a certain production will show the greatest profit *per unit time*. It will be clear that both an increase in income and a reduction of the production costs may cause the profit to rise. With much imagination, by analogy with the material and energy balance we can define an economic balance for a system such as an entire chemical plant:

$$\left\{\begin{array}{l}\text{rate of}\\\text{accumulation}\\\text{of money}\end{array}\right\} = \left\{\begin{array}{l}\text{rate of}\\\text{money}\\\text{in}\end{array}\right\} - \left\{\begin{array}{l}\text{rate of}\\\text{money}\\\text{out}\end{array}\right\} + \left\{\begin{array}{l}\text{rate of}\\\text{money}\\\text{production}\end{array}\right\}$$

The first term represents the profit made in the chemical operation and is positive or negative. The second term reflects the income from sales of products and the third all expenses for operating the plant: raw materials, catalysts and solvents;

labour for operators, maintenance and supervision; power and utilities; overhead, administrative, distribution and marketing expenses; insurances, rent, taxes, etc. The last term reflects the effects of wear and tear, depreciation, depletion, amortization and economic obsolescence, and is always negative. In optimization studies often only the last two terms are taken into account and minimized, as for example when only a part of a plant, such as a reactor or a reactor section, are studied, in which case as reactor size increases (because of increasing depreciation costs) the raw materials consumption is diminished (because of decreasing raw material costs). An optimum is reached at the lowest total costs. Inherent to these studies is the assumption that if the costs of all parts of a plant are minimized, the economic results of the entire chemical production unit are maximized. However, this is not always true.

We will not endeavour to elaborate the given economic balance equation further, but will make a frequent use throughout this book of economic data, because the results of all engineering efforts in the end will be judged by the economic results of the entire plant combined with a sound solution of all safety, pollution and social problems connected with the particular chemical operation.

The main factors which determine the economics of a *reactor* are: composition of the feed, pressure, temperature, reaction time and the materials of construction to be used. For the economics of a *production unit* as a whole, the plant yield, the costs of auxiliary equipment, e.g. for conditioning the feed and for isolating the desired product, are equally very important.

Since these many variables are to a great extent mutually dependent and since their influence on the performance of the plant is not generally known sufficiently beforehand, an exact prediction of the optimum solution is not feasible. In practice, near-to-optimum designs are arrived at from a combination of technical data obtained from the laboratory and pilot-plant units, of experience obtained with similar production units and of engineering common sense of the designers.

I.3 THERMODYNAMIC DATA: HEAT OF REACTION AND CHEMICAL EQUILIBRIUM

I.3.1 Heat of reaction

Generally there is a difference between the sum of absolute enthalpies of the reaction products and that of the reactants. For a reaction

$$\nu_A A + \nu_B B + \cdots \rightarrow \nu_P P + \nu_Q Q + \cdots,$$

the corresponding heat of reaction, ΔH_r, is defined as the heat absorbed by the system when the reaction proceeds completely in the direction indicated by the arrow, at constant temperature and pressure. Hence

$$\Delta H_r \equiv (\nu_P H_P + \nu_Q H_Q + \cdots) - (\nu_A H_A + \nu_B H_B + \cdots)$$
$$= + \sum (\nu_J H_J)_{\text{prod}} - \sum (\nu_J H_J)_{\text{react}}. \tag{I.4}$$

This result follows from the energy balance for a closed system which is identical with the first law of thermodynamics. ΔH_r is positive for an endothermic reaction and negative for an exothermic reaction.

For a certain reaction, ΔH_r can be calculated from the heat effects of other reactions, e.g. the heats of formation of the species involved, or the heats of combustion of these compounds. For obtaining the numerical value of ΔH_r, the stoichiometric equation, the direction of the reaction, the temperature, the pressure and the physical state of the components must be specified. When the reactants and the products are at standard conditions and the reaction proceeds under standard conditions, the standard heat of reaction is ΔH_{rs}^o. The index s refers to the standard temperature and the superscript o to the standard pressure and standard physical state. Its value can then be derived from the standard heats of formation $(\Delta H_{fs}^o)_J$ according to

$$\Delta H_{rs}^o = \sum [v_J(\Delta H_{fs}^o)_J]_{prod} - \sum [v_J(\Delta H_{fs}^o)_J]_{react}, \tag{I.5}$$

and from the standard heats of combustion $(\Delta H_{cs}^o)_J$ from

$$\Delta H_{rs}^o = \sum [v_J(\Delta H_{cs}^o)_J]_{react} - \sum [v_J(\Delta H_{cs}^o)_J]_{prod}. \tag{I.6}$$

The standard pressure is generally 1 bar and the standard temperature 18 or 25°C. Values of ΔH_{fs}^o and ΔH_{cs}^o can be found in the book by Hougen et al. [11] and other sources.

The influence of pressure on the heat of reaction can be neglected for solids, liquids and ideal gases, so that $\Delta H_{rs}^o = \Delta H_{rs}$. Corrections to be used for non-ideal gases can be found in the references [3], [14] and [15]; also see Example 1.3a. The heat of reaction generally varies with temperature. The value of ΔH_r^o at a temperature T follows from ΔH_{rs}^o according to

$$\Delta H_r = \Delta H_{rs}^o + \int_{T_s}^{T} \Delta C_p \, dT \tag{I.7}$$

where T_s is the standard temperature selected, and

$$\Delta C_p = \sum (v_J C_{pJ})_{prod} - \sum (v_J C_{pJ})_{react}.$$

The molar specific heats C_{pJ} and hence ΔC_p are in general weak functions of temperature. Therefore the average value $\Delta \bar{C}_p$ over the temperature interval under consideration is often used. Equation (I.7) is illustrated in figure I.1.

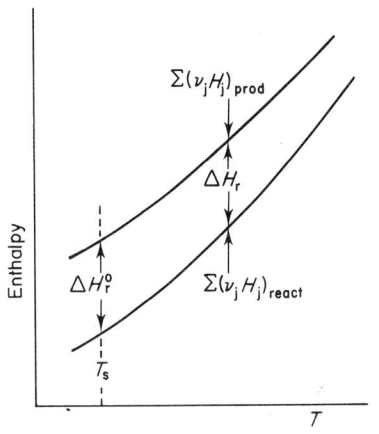

Fig. I.1 Enthalpy–temperature diagram for an endothermic reaction

Example I.3.a *Calculation of a heat of reaction*
It is required to calculate ΔH_r for the reaction $\frac{1}{3}N_2 + H_2 \rightarrow \frac{2}{3}NH_3$, at 400°C and for 1 and for 200 bars of pressure.

ΔH_r at 1 bar

The data given by Hougen *et al.* [11] are used to calculate ΔH_r° from the standard heats of formation and ΔC_p from the specific heats of the three components. Both quantities refer to the amount reacting according to the above reaction formula, i.e. 1 kmol of H_2 converted. The heats of formation and the specific heats are given in Table I.1.

Table I.1

Component	$(\Delta H_f^\circ)_J$ kcal/kmol (18°C)	C_{pJ} kcal/kmol K (T in K)
P = NH_3(g)	−11 000	$5.92 + 8.963 \times 10^{-3} T - 1.764 \times 10^{-6} T^2$
A = N_2(g)	0	$6.46 + 1.389 \times 10^{-3} T - 0.069 \times 10^{-6} T^2$
B = H_2(g)	0	$6.95 - 0.196 \times 10^{-3} T + 0.476 \times 10^{-6} T^2$

$$\Delta C_p = \tfrac{2}{3}C_{pP} - \tfrac{1}{3}C_{pA} - C_{pB}$$
$$= -5.15 + 5.71 \times 10^{-3} T - 1.63 \times 10^{-6} T^2 \text{ kcal/K}$$
$$= -21.6 + 23.9 \times 10^{-3} T - 6.83 \times 10^{-6} T^2 \text{ kJ/K}$$
$$\Delta H_{rs}^\circ = -\tfrac{2}{3} \times 11\,000 = -7333 \text{ kcal}$$
$$= -30.73 \times 10^3 \text{ kJ}$$

At 400 °C and 1 bar we find for ΔH_r,

$$\Delta H_r^\circ = -30.73 \times 10^3 + \int_{291.1}^{673.1} (-21.6 + 23.9 \times 10^{-3} T - 6.83 \times 10^{-6} T^2) \, dT$$
$$= (-30.73 - 8.49 + 4.40 - 0.63) \times 10^3 = -35.45 \times 10^3 \text{ kJ}.$$

It is seen that in this case the influence of temperature on ΔH_r is quite large.

ΔH_r at 200 bars

In order to calculate ΔH_r at 200 bars and 400 °C with the use of the known value of ΔH_r at 1 bar and 400 °C, the following procedure is followed:

(i) We imagine the reactant mixture ($\frac{1}{3}$ kmol N_2 + 1 kmol H_2) to expand isothermally at 400 °C from 200 to 1 bar;
(ii) We let the reaction proceed at 400 °C and 1 bar;
(iii) The product ($\frac{2}{3}$ kmol NH_3) is isothermally compressed at 400 °C from 1 to 200 bars.

If we call the enthalpy changes involved in the first and third processes ΔH_1 and ΔH_3 respectively, we have at 400 °C and 200 bars:

$$\Delta H_r = \Delta H_1 + \Delta H_r^\circ + \Delta H_3.$$

The values of ΔH_1 and ΔH_3 can be found from a generalized graph where the enthalpy correction per kmol of a gas is given as a function of the reduced pressure, p_r, and the reduced temperature, T_r (see e.g. Hougen et al. [3], Fig. 141). The critical pressure, p_c, and the critical temperature T_c, of the reactant mixture are estimated as follows:

$$(p_c)_{react} = x_A p_{cA} + x_B p_{cB} = 0.25 \times 33.5 + 0.75 \times 12.8 = 18.0 \text{ bar}$$
$$(T_c)_{react} = x_A T_{cA} + x_B T_{cB} = 0.25 \times 126 + 0.75 \times 33.2 = 56.4 \text{ K}.$$

From this we find for the reactant mixture at 200 bars and 400 °C:

$$(p_r)_{react} = \frac{200}{18.0} = 11.1 \text{ and } (T_r)_{react} = \frac{673.1}{56.4} = 11.9.$$

For these values Fig. 105 in [16] gives

$$H(1 \text{ bar}) - H(200 \text{ bars}) = -3.2(T_c)_{react} = -3.2 \times 56.4$$
$$= -181 \text{ kcal/mol}$$
$$= -756 \text{ kJ/kmol of reactant}.$$

Hence:

$$\Delta H_1 = -\frac{4}{3} \times 756 = -1010 \text{ kJ}.$$

Similarly, we find for the product ($NH_3 = P$), with $p_{rP} = 200/111.5 = 1.79$ and $T_{rp} = 673.1/405.5 = 1.66$:

$$H(200 \text{ bar}) - H(1 \text{ bar}) = -1.55 \, T_{cp} = -1.55 \times 404.4$$
$$= -629 \text{ kcal/mol}$$
$$= -2630 \text{ kJ/kmol}.$$

Accordingly,

$$\Delta H_3 = -\tfrac{2}{3} \times 2630 = -1750 \text{ kJ}.$$

As a result, the heat of reaction at 400 °C and 200 bars is found to be

$$\Delta H_r = -1010 - 35450 - 1750 = -38.2 \times 10^3 \text{ kJ}.$$

Instead of applying the correction to the mixture the correction can also be applied to the individual components according to

$$\Delta H_r = \Delta H_r^\circ - \left\{ \sum [v_J(H^\circ - H)_{T,J}]_{products} - \sum [v_J(H^\circ - H)_{T,J}]_{reactants} \right\}.$$

We then find the data for the corrections shown in Table I.2. Hence, if we apply the correction to the individual components, we find for the reaction heat at 400 °C and 200 bars,

$$\Delta H_r = -35\,450 - \{\tfrac{2}{3} \times 2630 - (\tfrac{1}{3} \times 111 - 167)\}$$
$$- 37.3 \times 10^3 \text{ kJ}.$$

Table I.2

Component	T_c(K)	p_c(bar)	T_r	p_r	$(H^\circ - H)/T_c \dfrac{\text{kcal}}{\text{kmol K}}$	$(H^\circ - H)$kJ/kmol
N_2	126.2	33.5	5.33	5.97	0.21	111
H_2	33.3	12.8	16.1	9.62	-4.0	-167
NH_3	405.5	111.3	1.66	1.80	1.55	2630

I.3.2 Chemical equilibrium

A chemical reaction by itself proceeds in the direction in which the Gibbs free energy (or the free enthalpy), G, of the reaction mixture diminishes. When equilibrium is reached, this quantity has a minimum value. Hence, from the value of G as a function of the extent of the reaction it can be predicted whether the reaction will proceed in a certain direction, and what will be the composition of the reaction mixture at chemical equilibrium. This follows from the second law of thermodynamics, which also furnishes a relationship between the equilibrium constant K and the difference between the free enthalpies of the product mixture and of the reactant mixture, ΔG. When the reaction proceeds at a constant temperature T and the reactants and products remain at the standard state (denoted by superscript o), this relationship is:

$$-RT\ln K = \Delta G^\circ = \Delta H_r^\circ - T\Delta S^\circ. \tag{I.8}$$

ΔG° can be calculated in various ways, see e.g. van Krevelen and Chermin [17], who directly calculate ΔG° for organic chemicals from group contributions. However, in general, more accurate values of ΔG° can be derived by estimating ΔH_r° and ΔS° separately [15]. Also we have:

$$\Delta H_r^\circ = \Delta H_{rs}^\circ + \int_{T_s}^{T} \Delta C_p^\circ \, dT \tag{I.9}$$

and

$$\Delta S^\circ = \Delta S_s^\circ + \int_{T_s}^{T} \frac{\Delta C_p}{T} \, dT \tag{I.10}$$

where the subscript s refers to the standard temperature. K is the *true* equilibrium constant for the reaction

$$\nu_A A + \nu_B B + \cdots \rightarrow \nu_P P + \nu_Q Q + \cdots$$

In chemistry the symbol ⇌ is used for the indication of an equilibrium reaction. However, we shall in general use only one arrow, which indicates in which direction the reaction proceeds. If the reaction is an equilibrium reaction, this will be automatically clear from the chemical reaction rate equation. K now is defined as

$$K = a_P^{v_P} a_Q^{v_Q} \cdots / a_A^{v_A} a_B^{v_B} \cdots . \tag{I.11}$$

For ideal gases the activities a_j are proportional to the partial pressures of the corresponding species; in other cases, fugacities have to be used, see, e.g. Hougen et al. [3]. The true equilibrium constants are frequently used, and they may depend on pressures and composition as well. Therefore, in using such equilibrium constants it is necessary to define them carefully.

The variation of K with temperature follows from equations (I.8–I.10):

$$\frac{d \ln K}{dT} = \frac{\Delta H_{rs}^\circ}{RT^2} + \frac{1}{RT^2} \int_{T_s}^{T} \Delta C_P^\circ \, dT = \frac{\Delta H_r^\circ}{RT^2}. \tag{I.12}$$

Accordingly, e.g. with an exothermic reaction, K decreases with increasing temperature.

Knowledge of the equilibrium composition of a reaction mixture makes it possible to determine whether a reaction can proceed in the desired direction; also, the circumstances can be predicted under which a desired product yield is obtainable. Not only the reaction temperature and pressure, but also the composition of the reaction mixture can be selected with the purpose of obtaining a favourable yield of products; examples of this are the use of high pressure in NH_3 synthesis, of excess air in the catalytic oxidation of SO_2 to SO_3, and of excess steam in the water–gas reaction ($H_2O + CO \rightarrow CO_2 + H_2$).

However, the degree to which the equilibrium is approached depends on the conversion *rate* and the time during which the volume elements of the reacting mixture are exposed to the reaction conditions. These factors determine to a great extent the reactor type and conditions needed for successful operation on an industrial scale. The degree of success is then mainly judged on the basis of the economics of the whole manufacturing unit, of which the reactor forms only a part.

Example I.3.b *Equilibrium constant in the synthesis of ammonia*
For the reaction

$$\tfrac{1}{3}N_2 + H_2 \rightarrow \tfrac{2}{3}NH_3$$
$$\text{(A)} \quad \text{(B)} \quad \text{(P)}$$

it is required to calculate, at 400 °C,

(i) the true equilibrium constant K;
(ii) the equilibrium composition as a function of pressure for an initial composition with a molar ratio $N_2:H_2 = 1:3$;
(iii) the equilibrium composition at 200 bars as a function of the initial composition.

(i) *The true equilibrium constant*
The standard molal entropies at 298.1 K are [3]:

$$N_2: 191.9 \text{ kJ/kmol K};$$
$$H_2: 130.9 \text{ kJ/kmol K};$$
$$NH_3: 192.9 \text{ kJ/kmol K}.$$

Hence

$$\Delta S_s^\circ = (\tfrac{2}{3} \times 192.9 - \tfrac{1}{3} \times 191.9 - 130.9) = -66.2 \text{ kJ/K};$$

this is the entropy change for a complete conversion to NH_3 according to the above reaction equation, i.e. for the case where 1 kmol of H_2 is converted. With equation (I.10) and with the C_p values given in Example I.3a, we find for the entropy change at 400 °C

$$\Delta S^\circ = -66.2 + [-21.6 \ln T + 23.9 \times 10^{-3} T - 3.41 \times 10^{-6} T^2]_{298.1}^{673.1}$$
$$= -76.1 \text{ kJ/K}.$$

With equation (I.8) and the ΔH_r° value from Example I.3a, we have

$$\ln K = \left(\frac{35.45 \times 10^3}{673.1} - 76.1 \right) \frac{1}{8.315} = -2.79,$$

and

$$K = \frac{a_P^{2/3}}{a_A^{1/3} a_B} = 0.062$$

Note that the value of K is very sensitive to small errors in ΔH_r° and ΔS° and that this equilibrium constant belongs to the reaction equation given at the beginning of this illustration. If, however, we wish to write the reaction equation, for example, in the form of:

$$\tfrac{1}{2} N_2 + \tfrac{3}{2} H_2 \xrightarrow{K'} NH_3$$

then we find as the new equilibrium constant $K' = K^{3/2} = 0.0154$. In general for the equation

$$v_A A + v_B B + \cdots \to v_P P + v_Q Q + \cdots$$

we find

$$K_{v_i = 1} = [K_{v_j = 1}]^{v_i/v_j}$$

(ii) *The equilibrium composition as a function of pressure* (400°C)
The activity of a component of a gas mixture is equal to the product of its fugacity coefficient γ, its mole fraction x and the total pressure p, here expressed in the number of bars. Thus for component A,

$$a_A = \gamma_A x_A p.$$

Accordingly, the true equilibrium constant K for NH_3 synthesis can be written as

$$K = \frac{\gamma_P^{2/3}}{\gamma_A^{1/3} \gamma_B} \frac{x_P^{2/3}}{x_A^{1/3} x_B} \frac{1}{p^{2/3}} = \frac{1}{p^{2/3}} K_\gamma K_x$$

K_γ, which is unity for ideal gases, can be derived from the generalized relationship between γ, the reduced pressure p_r and the reduced temperature T_r (see, for example, Fig. 142 in [3]). K_x can be expressed in terms of the ratio of the reactants and the degree of conversion of one of them; if we call the relative conversion of $H_2(B) \zeta_B$, we have for the stoichiometric ratio between N_2 and H_2

$$x_A = \frac{0.25(1 - \zeta_B)}{1 - \frac{1}{2}\zeta_B}, \quad x_B = \frac{0.75(1 - \zeta_B)}{1 - \frac{1}{2}\zeta_B}, \quad x_P = \frac{\frac{1}{2}\zeta_B}{1 - \frac{1}{2}\zeta_B}$$

and

$$K_x = \frac{4}{3} \frac{\zeta_B^{2/3} (1 - \frac{1}{2}\zeta_B)^{2/3}}{(1 - \zeta_B)^{4/3}}.$$

Table I.3

Bars		10	50	100	200	400	600
N_2, $T_r = 5.34$							
	p_r	0.30	1.50	2.99	5.97	11.94	17.91
	γ_A	1.00	1.00	1.02	1.08	1.20	1.30
H_2, $T_r = 20.3$							
	p_r	0.78	3.91	7.81	15.63	31.25	46.88
	γ_B	1.00	1.00	1.04	1.10	1.15	1.35
NH_3, $T_r = 1.66$							
	p_r	0.09	0.45	0.90	1.79	3.59	5.39
	γ_P	1.00	0.98	0.95	0.90	0.80	0.70
	K_γ	1.00	0.98	0.93	0.81	0.66	0.49
	$p^{2/3}$	4.65	13.6	23.0	34.2	54.2	71.0
$Kp^{2/3}/K_\gamma = K_x$		0.288	0.827	1.52	2.63	5.10	9.00
equilibrium value of:							
	ζ_B	0.09	0.29	0.46	0.61	0.75	0.83
	x_P	0.047	0.17	0.30	0.44	0.60	0.71

The results shown in Table I.3 are obtained. The equilibrium mole fraction of NH_3, x_P, at 400 °C is shown as a function of pressure in figure I.2.

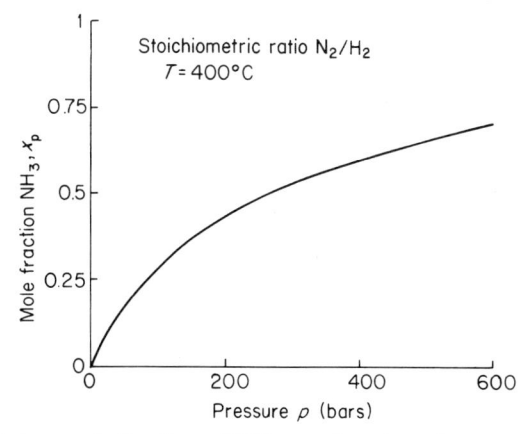

Fig. I.2 Equilibrium mole fraction of NH_3 as a function of pressure; Example I.3.b

(iii) *The equilibrium composition as a function of the initial composition* (400 °C, 200 bars)

From the Table I.3 it is seen that at 400 °C and 200 bars, $K_x = 2.63$. If the initial mole fractions of H_2 and N_2 are x_{Bo} and $(1 - x_{Bo})$, respectively, and the relative degree of conversion of H_2 is ζ_B, we have for the reaction mixture:

$$x_A = \frac{1 - x_{Bo} - \tfrac{1}{3}x_{Bo}\zeta_B}{1 - \tfrac{2}{3}x_{Bo}\zeta_B}, \quad x_B = \frac{x_{Bo} - x_{Bo}\zeta_B}{1 - \tfrac{2}{3}x_{Bo}\zeta_B}, \quad x_P = \frac{\tfrac{2}{3}x_{Bo}\zeta_B}{1 - \tfrac{2}{3}x_{Bo}\zeta_B}.$$

From the last expression we find

$$x_{Bo}\zeta_B = \frac{3}{2}\frac{x_P}{1 + x_P}$$

so that K_x can be expressed in terms of x_{Bo} and x_P. The result is that, for equilibrium at 400 °C and 200 bars, the following relation must apply:

$$K_x = 2.63 = \frac{\left(\dfrac{x_P}{1+x_P}\right)^{2/3}\left(1 - \dfrac{x_P}{1+x_P}\right)^{2/3}}{\left(1 - x_{Bo} - \dfrac{1}{2}\dfrac{x_P}{1+x_P}\right)^{1/3}\left(x_{Bo} - \dfrac{3}{2}\dfrac{x_P}{1+x_P}\right)}$$

The resulting dependence of the equilibrium value of x_P on x_B is shown in figure I.3. The maximum of x_P is found at a stoichiometric composition of the initial reaction mixture.

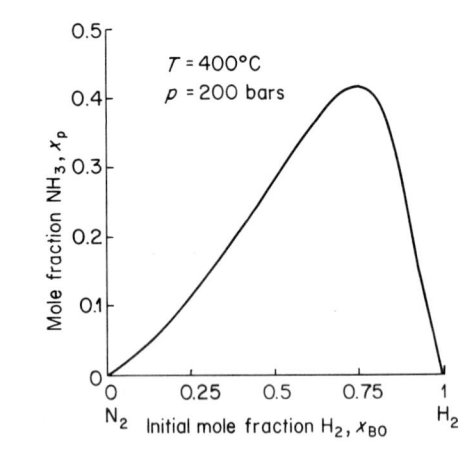

Fig. I.3 Equilibrium mole fraction of NH_3 as a function of initial composition; Example I.3.b

I.4 CONVERSION RATE, CHEMICAL REACTION RATE AND CHEMICAL REACTION RATE EQUATIONS

For the application of the material balance to reactor problems (see e.g. equation (I.1)) an expression must be available for the *chemical production rate* of the species for which the balance has been written down. For the production rate of a component J, the symbol R_J is used if it is expressed in molar units per unit time and volume; the production rate in mass units is then found by multiplying by the molecular weight M_J. When we have a reaction proceeding according to the reaction formula

$$v_A A + v_B B \to v_P P + v_Q Q, \tag{I.13}$$

it is often convenient to use the concept of the *chemical conversion rate*, which is always positive when the reaction proceeds in the direction of the arrow. If we denote by $|R|$ the number of moles converted according to the stoichiometric formula, and per unit time and volume, we can write

$$|R| = \frac{|R|_A}{v_A} = \frac{|R|_B}{v_B} = \frac{|R|_P}{v_P} = \frac{|R|_Q}{v_Q} \tag{I.14}$$

Thus the molar conversion rates $|R|_J$ are the absolute values of the molar production rates, R_J; similarly, the mass conversion rates are the absolute values of the mass production rates. If instead of conversion rates production rates are used, it will be clear that for the above reaction the production rates of the reactants A and B are negative, because they are consumed. For the production rates we would find:

$$\text{molar production rate} = -\frac{R_A}{v_A} = -\frac{R_B}{v_B} = \frac{R_P}{v_P} = \frac{R_Q}{v_Q}$$

In this book in general we will use in our equations the conversion rate in order to avoid difficulties in applying the correct sign for the reaction rate in the material balance. If the conversion of any reactant is also positive as long as the reaction proceeds in the direction of the arrow, the material balance (I.1) based on a conversion ξ_J of the component J changes into

$$\frac{dm\xi_J}{dt} = -\Delta\Phi_m \xi_J + M_J |R|_J V \tag{I.15}$$

The conversion ξ will be further discussed in the next section. The conversion rate R is to be considered as a phenomenological property of the reaction mixture under its operating conditions. It will generally depend on the composition, temperature and pressure, on the properties of a catalyst which may be involved and in principle also on the conditions of flow, mixing, mass transfer and heat transfer in the reaction system. If one of these variables changes, it is manifested in a change of the conversion rate. Since some of these variables will change from place to place in nearly all reactors, a proper knowledge of the relationship between the conversion rate and the pertinent variables is indispensable for integrating the material balance.

In this connection, it is important to note that in principle the *conversion rate is not identical with the chemical reaction rate*. The latter quantity only reflects the *chemical kinetics* of the system, i.e. the conversion rate measured under such conditions that it is not influenced by physical transport (diffusion and mass transfer) of reactants towards the reaction site or of products away from it. This situation is often encountered with homogeneous reactions in fluids where the reactants are well mixed on a molecular scale, and with reactions in heterogeneous systems which proceed slowly with respect to the potential physical transport. On the other hand, with homogeneous reactions in a poorly mixed fluid and with relatively rapid reactions in heterogeneous systems, the physical transport phenomena may reduce the conversion rate, so that here in general the conversion rate is lower than the chemical reaction rate under circumstances of no physical transport limitations. The latter aspect will be treated extensively in Chapters VII and VIII, so that a few remarks on chemical kinetics properly speaking will suffice here. For further pertinent information, the reader is referred to the large number of available texts, of which we mention the books by Hougen and Watson [19], Smith [14], Gardiner [20], Ashmore [21], Benson [22], Boudart [23], Frost and Pearson [24], Laidler [25] and Walas [26].

Although the reader should always be aware of the difference between the real conversion rate and the chemical reaction rate, we assume it to be clear from the case under consideration whether the conversion or chemical reaction rate are dealt with and we will not use different symbols to distinguish between them in this book. In many cases the results of measurements of the chemical reaction rate can be described by means of properly chosen functions of temperature and composition in terms of concentration of the reaction mixture. Such expressions do not necessarily reflect the mechanism of the reaction proper, although their form often suggests, rightly or wrongly, a certain mechanism.

I.4.1 Influence of temperature on kinetics

In the reaction rate expressions, the influence of temperature and composition can usually be represented separately. Thus, the influence of temperature is accounted for in the reaction velocity constant k which has the form:

$$k = k_\infty \exp(-E/RT), \qquad (I.16)$$

where E is the activation energy of the reaction. The fact that k is currently called a 'constant' only stems from the circumstance that it is not supposed to be a function of composition. For most reactions the activation energy lies in the range of 40–300 MJ/kmol (10 to 60 × 10^3 kcal/kmol)—see Table I.4 for some examples—resulting in an increase of k by a factor of 2 to 50 for a temperature rise of 10 °C at room temperature and by a factor of 1.1 to 1.6 at a reaction temperature of 600 °C. The temperature rise required for doubling the reaction rate constant as a function of the temperature and the activation energy is given in figure I.4. We see that the higher the activation energy the more temperature sensitive the reactions are and that the reaction velocity constant becomes less sensitive to temperature at higher temperatures, although the absolute value, of

Table I.4 Activation energies for some chemical reactions

Reaction	Activation energy in MJ/kmol
gaseous decomposition of trimethylacetic acid [26]	274
decomposition of dipropyl ether, uncatalyzed [26]	253
decomposition of diethyl ether, uncatalyzed [26]	224
decomposition of methylethyl ether, uncatalyzed [26]	197
decomposition of acetaldehyde, uncatalysed [26]	189
second order gas reaction between $C_2H_2 + H_2$ [26]	181
first order gaseous decomposition of methyl iodide [26]	180
first order gaseous decomposition of ethyl nitrite [26]	158
second order gas reaction between i-C_4H_8 + HCl [26]	121
second order gas reaction between i-C_4H_8 + HBr [26]	94
decomposition of H_2O_2 uncatalysed [26]	75
first order gaseous decomposition of nitrogen tetroxide [26]	58
decomposition of H_2O_2, catalysed by iodide ion [26]	57
by colloidal platinum	48
by liver catalase [26]	23
$CO_2 + OH^-$ [27]	55
SO_3 + p-dichlorobenzene (in trichlorofluoromethane) [28]	54
CO_3 + ammonia [29]	49
$COCl_2 + OH^-$ [30]	46
$C_2H_4 + Cl_2 + H_2O \rightarrow CH_2ClCH_2OH + HCl$ [31]	46
CO_2 + monoisopropanolamine [32]	44
$C_6H_5Cl + SO_3$ (in nitromethane) [28]	31

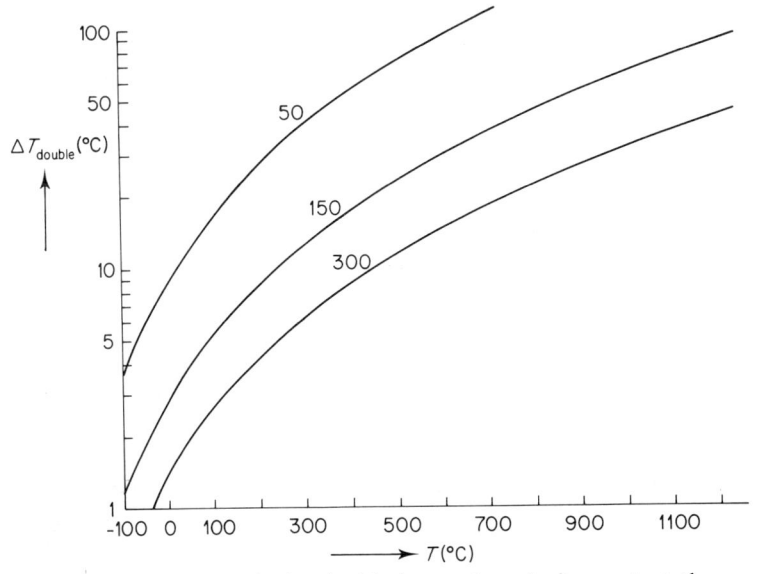

Fig. I.4 Temperature rise required to double the reaction velocity constant; the parameter is E in MJ/kmole

course, increases. This is demonstrated in figure I.5, where the reaction velocity constant is shown as a function of temperature, both in dimensionless form. As E/R ranges from 5000 to 35 000 K, in industrial practice only the range of $RT/E < 0.6$ is of interest: this region is shown enlarged in part b of the figure. We see that, although at extreme temperatures k approaches the value k_∞, in practice it is several orders of magnitude lower.

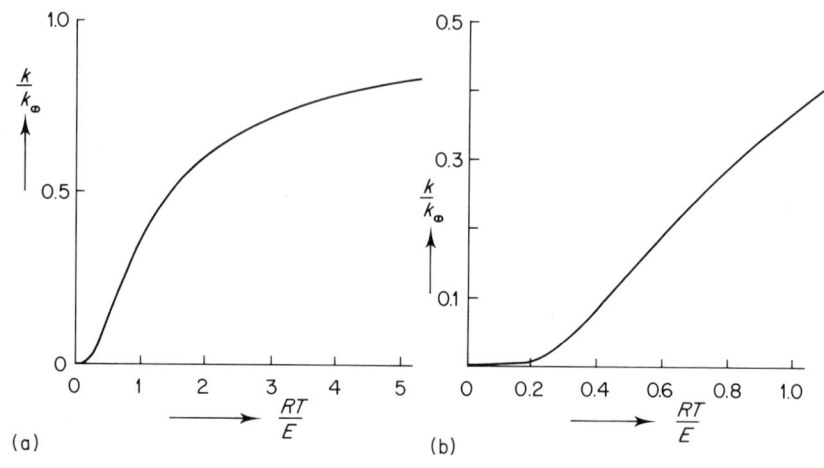

Fig. I.5 Temperature dependence of the reaction velocity constant. *Note:* As E/R ranges from 5000 to 35 000 K, in practical chemical reactors only the region of $RT/E < 1$ is of interest, which is given in part (b)

The temperature dependence of the chemical reaction rate is determined by the manner in which the various velocity constants occur in the reaction rate expression. Thus, for a simple homogeneous reaction between A and B with a rate expression

$$|R| = kc_A c_B, \tag{I.17}$$

the value of $|R|_A$ will, at constant composition, be proportional to $\exp(-E/RT)$ (figure I.6, Curve 1). For an equilibrium reaction, the reaction rate is equal to the difference between a forward rate and a reverse rate, e.g.

$$|R| = k_1 c_A c_B - k_2 c_P c_Q, \tag{I.18}$$

and if E_2 is greater than E_1, $|R|$ will, at constant composition, pass through a maximum as a function of temperature; it will become zero for the temperature at which the reaction mixture is at equilibrium (figure I.6, Curve 2) and negative beyond this temperature. With solid catalysed reactions, a reaction rate expression is often encountered, e.g. of the following form:

$$|R| = \frac{kc_A c_B}{1 + K_A c_A + K_B c_B + K_{AB} c_A c_B}, \tag{I.19}$$

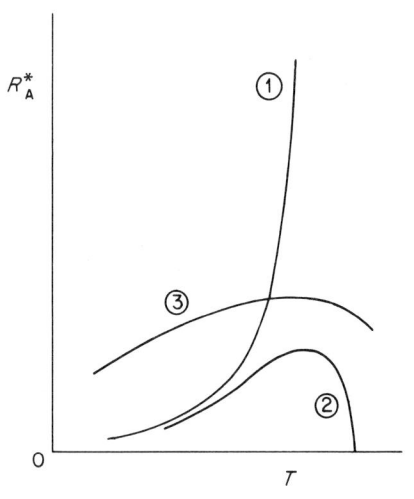

Fig. I.6 Possible temperature dependence of the chemical reaction rate at constant composition: 1—single forward reaction, 2—equilibrium reaction, 3—solid catalysed reaction, e.g. equation (I.19)

where the Ks in the denominator are constants describing the chemisorption equilibrium of A and B on the catalyst surface. Depending upon the influence of temperature on these equilibrium constants, $|R|$ may increase, remain constant, or even decrease when the temperature rises (figure I.6, Curve 3).

A further type of chemical rate equation is often encountered in homogeneous polymerization reactions; it may have the following form:

$$|R| = k_p \sqrt{\left(\frac{k_i c_B}{k_t}\right)} \times c_A \sqrt{\left(\frac{K c_A}{1 + K c_A}\right)}. \tag{I.20}$$

Here c_A is the monomer concentration and c_B the catalyst concentration; k_p, k_i and k_t are the velocity constants for the propagation, initiation and termination reactions, respectively; K is an equilibrium constant. The temperature dependence of $|R|$ is determined by the activation energies of the constants involved.

I.4.2 Influence of concentration on kinetics

Van Krevelen [8] distinguished with respect to chemical kinetics three 'scales of observation', which become ever more embracing:

A. *Elementary kinetic processes*

These comprise individual reaction steps and depend on the number of colliding molecules. The collision takes place in a more or less homogeneous reaction medium or on the reaction sites on a catalyst surface. There are only three elementary kinetic processes: mono-, bi- and trimolecular processes. However, each of these consists in an activation of the reactants, a transition state and a decomposition into reaction product(s) according to:

$$\text{Reactant(s)} \quad \left[\begin{array}{c} \xrightleftharpoons[\text{activation}]{\text{deactivation}} \cdots \text{energized molecule} \rightarrow \\ \rightarrow \text{transition state molecule} \rightarrow \end{array} \right] \rightarrow \text{Product(s)}$$
$$A + (B + C)$$

Only a few examples are known of chemical reaction rates determined by an elementary kinetic process only.

B. *Fundamental reaction mechanisms formed by a combination of a number of elementary kinetic processes*

Besides a comparatively few autokinetic reactions A → P, which are entirely spontaneous, there are a large number of reactions that require special reaction centres, which may be produced by thermal, catalytic, inductive, photochemical, electrical or radiochemical methods. After the addition of reaction centres to the system, they may be present constantly and at all places throughout the system (homogeneous catalysis) or they may be present locally (heterogeneous catalysis), or be produced by parallel reactions (entrained or induced reactions), see Table I.5. In some cases the centres are formed from the reactants by special initiating reactions (free radicals, chain reactions). If during a reaction the number of centres increases steadily, the reaction is autocatalytic, or in the case of chain reactions a branched chain reaction (most explosions are of this type). The more complex a reaction, the more involved also the kinetic relation expressing the rate of the reaction. One of the characteristics of the most complicated reaction mechanism is that these branched chain reactions nearly always require nucleation, ignition. Therefore a branched-chain reaction is also a nucleation process: as soon as they start, they are self accelerating.

C. *Gross reaction types*

These are the gross result of the fundamental mechanisms. The class of gross reaction types comprises the single and the multiple conversions (consecutive, parallel, autocatalytic, combination reactions, etc.) and are of great interest to the chemical reaction engineer.

The *single reactions* consist of three basis types:

C.I.a. *uni-directional* reactions like A + B → P + Q with chemical reaction rate equations, for example, as $|R| = kc_A c_B$

C.I.b. *reversible* reactions like A + B → P + Q with, for example, $|R_J| = k_1 c_A c_B - k_2 c_P c_Q$

Table I.5 Types of reactions promoted by reaction centres

Homogeneous catalysis	Heterogeneous catalysis	Induced reactions	Open chain reactions	Branched chain reactions
After addition of catalyst reaction centres are present constantly and everywhere in the system	Reaction centres present constantly, but localized	Reaction centres produced by foreign reaction (entrainer). Parallel reaction:	Reaction centres produced by initiation reaction; constant but limited reproduction. Initiation:	Reaction centres produced by initiation reaction (nucleation) and reproduced in constantly growing numbers. Initiation:
$A + K^* \to R_1 \ldots K^*$		$R_P + R \to E^*$	$R_i \to I_i^*$	$R_i \to I_i^*$
$A \ldots K^* + B \to K^* + P$ (etc.)		$E^* + R_P \to P_P$	Propagation:	Propagation:
		Induced reactions	$I_1^* + R_1 \to I_2^* + P_1$	$I_1^* + A \to I_2^* + P_1$
		$B + E^* \to P$	$I_2^* + B \to I_1^* + P_2$	$I_2^* + B \to I_1^* + P_2$
		Gross reaction:	(etc.)	Branching:
		$2R_P + R + B \to P_P + P$	Termination:	$I_2^* + B \to I_1^* + I_3^*$
			$I_i^* \to P$	$I_3^* + A \to I_2^* + P_3$
				Termination:
				$I_i^* \to P$

C.I.c. *autocatalytic* reactions like $A \to P$ with, for example, $R_A = -kc_A c_P$. In this case the reaction rate is zero if no product is present in the reaction mixture. The reaction rate is accelerated as soon as some product has been formed. Most microbial fermentations are autocatalytic reactions.

The composition of a reaction mixture with known starting concentrations for all simple reactions can be described with one single parameter: the degree of conversion of a key reactant.

The *multiple reactions* comprise a whole variety of possible reaction types with many different rate equations. We distinguish, however, three basic types and give some examples of rate equations:

C.II.a. *parallel* reactions like $A \xrightarrow{k_1} P$ and $A \xrightarrow{k_2} X$, for example with $R_A = -(k_1 + k_2)c_A$, $R_P = k_1 c_A$ and $R_X = k_2 c_A$.

C.II.b. *consecutive* reactions like $A \xrightarrow{k_1} P \xrightarrow{k_2} X$, for example with $R_A = -k_1 c_A$, $R_P = k_1 c_A - k_2 c_P$ and $R_X = k_2 c_P$.

C.II.c. *complex combination* reactions. Two examples in this class are $A \xrightarrow{k_1} P \xrightarrow{k_2} Q$ with also $A \xrightarrow{k_3} X$ and $P \xrightarrow{k_4} Y$ and the reaction $A + B \xrightarrow{k_1} P$ with $A + P \xrightarrow{k_2} X$. These two reactions are combinations of parallel and consecutive reactions.

The composition of a reaction mixture with known initial concentrations for multiple reactions has to be described with at least two parameters: the degree of conversion and the selectivity, which will be described in the next two paragraphs.

As far as the influence of composition is concerned, it is often possible to approximate the chemical reaction rate of a single reaction by

$$|R| = kc_A^\alpha c_B^\beta. \tag{I.21}$$

It is then said that the reaction is of the order α with respect to A and of the order β with respect to B; the total order of the reaction is $\alpha + \beta$. The orders do not have to be whole numbers. With a rate equation of the form of equation (I.19), both α and β may be <1; the order with respect to A in equation (I.20) lies between 1 and 1.5, depending on the value of K, i.e. on the temperature level.

If it is impossible to approximate the chemical reaction rate equation by equation (I.21) or a similar type for reversible reactions, often the following types of rate equations give good results:

$$|R| = \frac{kc_A}{1 + Kc_A} \tag{I.22}$$

or

$$|R| = \frac{kc_A}{(1 + Kc_A)^2} \tag{I.23}$$

Both rate equations exhibit shifting orders. Equation (I.22) results in a zero order reaction at high concentrations of reactant A with a maximum chemical reaction rate of $R_A = -k/K$ and a first order behaviour at low concentrations where $c_A \ll K^{-1}$. Equation (I.23) results in a negative reaction order of -1 at high, a zero order at intermediate and a first order behaviour at low concentration of A. For equation (I.23) the maximum rate of $0.25k/K$ is reached at a concentration $c_A = K^{-1}$.

It should be clear from equation (I.21), for example, that the reaction velocity constant has different dimensions depending on the type of reaction, whereas R always has the dimension of kmol/m^3 reaction volume s in standard units. In equation (I.21), for example, k has the dimension of

$$\frac{\text{kmoles A converted}}{\text{s} \times (\text{m}^3 \text{ reaction volume } V_R)^{1-\alpha-\beta} (\text{kmoles A in } V_R)^\alpha (\text{kmoles B in } V_R)^\beta}$$

In the case of $\alpha = 1$ and $\beta = 0$ the reaction velocity has the dimension of s^{-1}. As k is always defined in relation to a rate equation we will generally abstain from introducing special symbols distinguishing the different reaction orders, unless this causes confusion.

Till now we have based R_J on the reactor volume. However, specially in heterogeneous systems it is often more convenient to base the rate of the reaction on a different system property, which can be determined more easily. The usual properties chosen are the volume of the liquid or solid contained in the reactor, the interfacial area between two phases in the reactor, the mass of a solid or a catalyst in the reactor and the external or internal surface area of a porous catalyst. Conversion problems are simple and always of the type (for example)

$$R_{\text{massJ}} \times \begin{pmatrix} \text{mass of solids} \\ \text{in the reactor} \end{pmatrix} = R_J \times \begin{pmatrix} \text{volume of} \\ \text{the reactor} \end{pmatrix} \qquad (I.24)$$

In this equation R_{massJ} has the dimensions of kmoles J converted per unit mass of the solid per unit of time. Similar equations to (I.24) can easily be derived for the various possible system properties on which the rate equation is based.

Finally it should be mentioned that, as the conversion rate becomes more determined by physical transport phenomena in the non-reacting phase and less by chemical kinetics, the order of the reaction with respect to the reactant(s) changes, since the transport by diffusion and mass transfer is a linear function of concentration. At the same time, the temperature dependence of the conversion rate is sharply reduced; the apparent activation energy of the physical transport coefficients (e.g. the diffusivity) is one to two orders of magnitude smaller than the activation energies of the reaction velocity constants. This will be discussed extensively in this book.

In conclusion it may be said that dependable expressions for production rates or conversion rates of the species involved must be available for reactor calculations. Throughout the following chapters it will be assumed that this condition has been met; whether the information is present in the form of an analytical expression or in that of a collection of experimental results is, in principle, immaterial. Since, however, the simple analytical formulation allows for a more general quantitative treatment, it will frequently be used in the following for demonstrating general principles in reactor calculations. The need for good chemical kinetic data and for an understanding of the interfering physical phenomena for the solution of actual reactor problems cannot be sufficiently stressed. Therefore, the requirements for obtaining adequate information on chemical reaction rates are discussed separately in Chapter VII, and the influence of physical transport phenomena is treated in Chapter VII and VIII.

I.5 THE DEGREE OF CONVERSION

Since in the material balance the amounts of the chemical species involved in a reaction have to be expressed in mass units, the most logical variable for describing the composition is the *mass fraction w*. For a single reaction governed by a stoichiometric equation such as

$$v_A A + v_B B \rightarrow v_P P + v_Q Q,$$

the mass fractions of the reactants, w_A and w_B, decrease, and those of the products, w_P and w_Q, increase as the reaction proceeds from left to right. If no material is added to or withdrawn, during the reaction, from the amount of reaction mixture under consideration, we clearly have

$$w_A + w_B + w_P + w_Q = \text{constant},$$

or, in general,

$$\sum (w_J)_{\text{react}} + \sum (w_J)_{\text{prod}} = \text{constant}. \tag{I.25}$$

For such a system we can introduce the *degree of conversion*, ξ, which is a measure of the extent to which the reaction has proceeded. It is defined as the difference between the mass fraction of a component, w_J, and its initial mass fraction, w_{J_0}, in such a way that ξ_J is always positive when a reactant has been converted or a product has been formed:

$$\xi_J \equiv |w_{J_0} - w_J|. \tag{I.26}$$

So we find for a reactant A, that $\xi_A = w_{A_0} - w_A$ and for a product P, that $\xi_P = w_P - w_{P_0}$. As a consequence of this definition of the degree of conversion the sign of the chemical conversion rate $|R_J|$ in the material balance equation (I.15) is always positive when the changes in the composition of a reaction mixture are expressed in ξ_J. However, if weight fractions w_J or concentrations c_J are used in a material balance, the sign of R_J is positive if component J is produced, and negative if component J is consumed. This leads us to the convention that we will use the symbol $|R|_J$ if the material balance is expressed in a component J without specifying whether J is a reactant or a product. Consequently equation (I.1) expressed in *conversion of any component* of the reacting mixture becomes

$$\frac{dm\xi_J}{dt} = -\Delta\Phi_m \xi_J + M_J |R|_J V, \tag{I.1a}$$

in which m is the total mass contained in the considered reactor volume V. However, if we express the material balance in terms of *production*, equation (I.1) becomes

$$\frac{dmw_J}{dt} = -\Delta\Phi_m w_J + M_J R_J V \qquad (I.1b)$$

because it is still necessary to specify whether component J is being produced or consumed. However, in *specific cases* we will often start straight away with a material balance based on *production* if no confusion is possible about whether the several terms in the material balance are positive or negative.

With equation (I.25) we find for a reaction system without supply or removal of material during the reaction:

$$\sum (\xi_J)_{react} = \sum (\xi_J)_{prod}. \qquad (I.27)$$

The values of ξ of the different components taking part in a single reaction are also related to each other in consequence of the stoichiometric requirements; thus we have for the above reaction

$$\frac{\xi_A}{\nu_A M_A} = \frac{\xi_B}{\nu_B M_B} = \frac{\xi_P}{\nu_P M_P} = \frac{\xi_Q}{\nu_Q M_Q}. \qquad (I.28)$$

Accordingly, if in a single reaction the degree of conversion of one of the components is known, this is also the case for the other components. In equation (I.14), a similar expression is applied to $|R|$. This does not apply to a system of simultaneous reactions. Generally it is more convenient to use the *relative degree of conversion* of a reactant ζ_J, as a measure for the extent of the reaction. It can be quite generally defined as the fraction of the amount of a reactant fed prior to and during the reaction, which has been converted. For a reactor system without intermediate supply or removal of material, ζ_J can be expressed in terms of ξ_J:

$$[\zeta_J = \xi_J/w_{Jo} = 1 - w_J/w_{Jo}]_{react} \qquad (I.29)$$

From this equation and equation (I.28) we also find that for such a system

$$\frac{w_{Ao}\zeta_A}{\nu_A M_A} = \frac{w_{Bo}\zeta_B}{\nu_B M_B}. \qquad (I.30)$$

Now, the definitions of both ξ_J and ζ_J are based on *mass fractions*. On the other hand, the influence of the composition on the conversion rate is generally given in terms of *molar concentrations*, c_J, or of partial pressures, which can be easily converted to molar concentrations with relation $c_{Jgas} = \gamma p_J/RT$ (the fugacity coefficient $\gamma = 1$ in case the ideal gas laws apply). Since there is a single relationship between the molar concentration and the mass fraction in a mixture,

$$c_J = \rho w_J/M_J \qquad (I.31)$$

(where ρ is the specific mass or density of the mixture), concentrations can be expressed in terms of conversions or vice versa. Some of these relationships, based on equations (I.26), (I.28), (I.29) and (I.31), are shown in Table I.6.

Table I.6 Relations between concentration and conversion

Reactants	Products
$c_A = \rho\left(\dfrac{c_{Ao}}{\rho_o} - \dfrac{\zeta_A}{M_A}\right)$	$c_P = \rho\left(\dfrac{c_{Po}}{\rho_o} + \dfrac{\zeta_P}{M_P}\right)$
$c_B = \rho\left(\dfrac{c_{Bo}}{\rho_o} - \dfrac{\zeta_B}{M_B}\right)$	$c_P = \rho\left(\dfrac{c_{Po}}{\rho_o} + \dfrac{\nu_P \zeta_A}{\nu_A M_A}\right)$ †
$c_B = \rho\left(\dfrac{c_{Bo}}{\rho_o} - \nu_B \zeta_A / \nu_A M_A\right)$	$c_P = \dfrac{\rho}{\rho_o}\left(c_{Po} + c_{Ao}\dfrac{\nu_P \zeta_A}{\nu_A}\right)$ †
$c_A = \dfrac{\rho}{\rho_o} c_{Ao}(1 - \zeta_A)$	
$c_B = \dfrac{\rho}{\rho_o}\left(c_{Bo} - c_{Ao}\dfrac{\nu_B \zeta_A}{\nu_A}\right)$ †	

† Valid only for single reactions.
Note: This table indicates that ζ has a slight advantage over ξ for expressing molar concentrations in terms of conversion. The above relationships may become rather complicated for systems with a variable density, since ρ itself depends then on the degree of conversion.

For instance if the density is linearly dependent of the conversion, then $\rho = \rho_0(1 + \varepsilon\zeta_A)$ and the concentration of the reactant becomes

$$\frac{c_A}{c_{Ao}} = (1 + \varepsilon\zeta_A)(1 - \zeta_A). \tag{I.32}$$

The chemical reaction rates, we must remember, are normally dependent on the molar concentrations of the reactants and not on the weight fractions of the reactants in the reaction mixture, for they depend on the chance of collision of molecules. For the material balance we have to use weight fractions in order to prevent errors, but the conversion rate expression in the material balance will usually contain molar concentration terms. This leads us into translation troubles and either we have to translate the other material balance terms into molar concentrations or to translate the conversion rate expression in mass fractions.

I.5.1 Relation between conversion and concentration expressions

Example I.5.a
The reaction $\nu_A A + \nu_B B \to \nu_P P$ takes place in a well-mixed continuous reactor under steady state conditions ($dm/dt = 0$). The conversion rate of P is $R_P = kc_A c_B$ kmoles P/m³s. Derive

(i) the material balance equation for P based on weight fractions and on molar concentrations;
(ii) the same for $\nu_A = \nu_B$ and $c_{Bo} = c_{Ao}$;
(iii) the same as (ii) but also with $\varepsilon = 0$.

As will be shown in Chapter II the material balance for product P becomes
$\Phi_m(-w_{Pin} + w_{Pout}) = \Phi_m\zeta_P = M_P R_P V$. Introducing the equations mentioned before we find for the well-mixed reactor

$$\Phi_m\zeta_P = \Phi_m \frac{v_P M_P \zeta_A}{v_A M_A} = M_P k \rho_0^2 \left(1 + \varepsilon\frac{\zeta_A \rho_0}{c_{Ao}M_A}\right)^2 \left(\frac{c_{Ao}}{\rho_0} - \frac{\zeta_A}{M_A}\right)\left(\frac{c_{Bo}}{\rho_0} - \frac{v_B \zeta_A}{v_A M_A}\right) V.$$

Based on molar concentrations we find, if Φ_v is the volumetric flow rate,

$$\Phi_{vout}c_P - \Phi_{vin}c_{Po} = kc_{Ao}^2(1 + \varepsilon\zeta_A)^2(1 - \zeta_A)\left(\frac{c_{Bo}}{c_{Ao}} - \frac{v_B}{v_A}\zeta_A\right) V.$$

For the case $c_{Bo} = c_{Ao}$ and $v_B = v_A$, the two equations respectively reduce for this stoichiometric composition of the reactor feed to

$$\Phi_m\zeta_P = M_P k c_{Ao}^2 \left(1 + \varepsilon\frac{\zeta_A \rho_0}{c_{Ao}M_A}\right)^2 \left(1 - \frac{\rho_0 \zeta_A}{M_A c_{Ao}}\right)^2 V$$

and to

$$\Phi_{vout}c_P - \Phi_{vin}c_{Po} = kc_{Ao}^2(1 + \varepsilon\zeta_A)^2(1 - \zeta_A)^2 V.$$

If also the density of the reaction mixture is constant ($\varepsilon = 0$), we find the following expressions:

$$\Phi_m\zeta_P = M_P k c_{Ao}^2 \left(1 - \frac{\rho_0 \zeta_A}{M_A c_{Ao}}\right)^2 V$$

and

$$\Phi_v(c_{Pout} - c_{Po}) = kc_{Ao}^2(1 - \zeta_A)^2 V.$$

We see that the equations based on molar concentrations are simpler. This explains their widespread use. In Example 1.5c we will give an example of a conversion rate equation, where much care should be taken in setting up a material balance.

Example I.5.b *Conversion rate as a function of the degree of conversion, liquid reaction*

For the homogeneous equilibrium reaction in the liquid phase:

alcohol (A) + acid (B) → ester (P) + water (Q),

the rate of conversion (= rate of chemical reaction in this case) is, under certain conditions, found to be

$$|R| = k_1 c_B - k_2 c_P \, [\text{kmol/m}^3\text{s}]$$

What are the expressions for R_A, R_B and R_P in terms of the conversion of the reaction, if the density ρ remains constant and $c_{Po} = 0$?

Taking the acid B as a reference component we have, from Table I.6,

$$c_B = c_{Bo} - \frac{\rho \xi_B}{M_B} = c_{Bo}(1 - \zeta_B),$$

and

$$c_P = \rho \xi_B / M_B = c_{Bo} \zeta_B.$$

Consequently, the molar conversion rate of B is:

$$-R_B = c_{Bo}[k_1(1 - \zeta_B) - k_2 \zeta_B]$$

And because $v_A = v_B = v_P = v_Q$ we find $R_A = R_B = -R_P = -R_Q$.

Example I.5.c *Conversion rate as a function of the degree of conversion, gas reaction*

For the equilibrium reaction A → 2P in the gas phase the conversion rate is given by the formula

$$|R|_A = k_1 \left(\frac{p_A}{RT}\right) - k_2 \left(\frac{p_P}{RT}\right)^2 \quad [\text{kmol/m}^3 \text{s}]$$

The gas can be considered to be ideal, and A and P are the only components present. The total pressure p ($= p_A + p_P$) of the gas mixture is assumed to be constant, but the temperature may change with the progress of the reaction. Initially the gas consists of pure A, so that $w_{Ao} = 1$ and $\xi_A = \zeta_A$. Express $|R_A|$ in terms of ξ_A.

With a degree of conversion ξ_A, $(1 - \xi_A)$ moles of A are left in the reaction mixture per mole of A originally added, while $2\xi_A$ moles of P have been produced. Hence, the mole fractions of A and P, which in this case are equal to the relative partial pressures, become

$$\frac{p_A}{p} = \frac{1 - \xi_A}{1 + \xi_A} \quad \text{and} \quad \frac{p_P}{p} = \frac{2\xi_A}{1 + \xi_A}.$$

The rate expression then becomes

$$|R|_A = k_1 \frac{p}{RT} \frac{1 - \xi_A}{1 + \xi_A} - k_2 \left(\frac{p}{RT}\right)^2 \frac{4\xi_A^2}{(1 + \xi_A)^2},$$

where k_1 and k_2 are still functions of temperature. In this example, ξ_A, ζ_A and ξ_P go to unity on completion of the reaction. In practice, the quantity $p\xi_A / \rho_{Ao}$ is sometimes used for indicating the degree of conversion. This quantity has the disadvantage that it does not range from 0 to 1. In our example we see that for $\xi_A = 1$ and at a temperature $T_1: \rho = pM_A / RT_1 = pM_A / 2RT_1$, whereas at the initial temperature $T_0: \rho_{Ao} = pM_A / RT_0$. Hence for $\xi_A = 1: \rho \xi_A / \rho_{Ao} = T_0 / 2T_1$. Note that at complete conversion the gas consists of P only ($p = p_P$) and that from the reaction follows $M_P = M_A / 2$.

I.6 SELECTIVITY AND YIELD

In industrial chemical operations in most cases we meet with the phenomenon that from a reactant A or reactants A and B not only desired products like P and Q, but also undesired products like X and Y are formed. Therefore in chemical reactor technology the concepts of selectivity and yield are very frequently met with; they are of utmost importance for the final economic results of an entire chemical operation.

Since the reactants in a reactor are meant to be converted into desired product to the greatest possible extent, not only must a reasonably high degree of conversion be obtained, but this should be accompanied by a high selectivity of the reaction. The *selectivity* σ_P is the ratio between the amount of a desired product P obtained and the amount of a *key reactant A converted*. When these amounts are expressed in molar units and the stoichiometry of the reaction equation is taken into account, σ_P ranges between 0 and 1. The definition leads to the expression:

$$\sigma_P = \frac{\zeta_P v_A M_A}{\zeta_A v_P M_P} = \frac{\text{moles of P formed} \times v_A}{\text{moles of A consumed} \times v_P}. \tag{I.33}$$

If σ_P is based on B as key reactant a similar expression is found. For a constant density of the reaction mixture we find:

$$\sigma_P = \frac{(c_P - c_{P_o})v_A}{(c_{A_o} - c_A)v_P}. \tag{I.34}$$

Ultimately, the merits of a complex reaction operation are closely related to the amount of desired product obtained with respect to the amount of *key reactant A fed*. This ratio is called the product *yield*, η_P, and it is easily understood from the above that

$$\eta_P = \sigma_P \zeta_A. \tag{I.35}$$

The yield can likewise vary between 0 and 1; it is high when both the selectivity and the relative degree of conversion are high, but it is low when either one of them is small. They apply equally well to simpler reactor systems as to systems with feed and/or discharge distribution, provided more general definitions of ζ are used in the latter case.

If in the reactor products separation section the key reactant A can be recovered and recycled to the reactor section, the selectivity, which is based on the amount of reactant A converted in the reactor, is the key factor in the economic operation of the plant; if, however, A cannot be recovered from the reactor product stream and the non-converted A is wasted in the operation, then the yield, which is based on the amount of A fed, is the economic key factor. This will be demonstrated in the following example.

I.6.1 Selectivity and yield in a reactor section with recycle of non-converted reactant

Example I.6.a

Take a plant as given in figure I.7, where a reaction A → X and A → P takes place. The separation in the solvent recovery and in the splitters is assumed to be 100% and the fraction of the amount of A leaving the solvent recovery and recycled to the reactor section is taken as R; $R = 0$ at no recycle and $R = 1.0$ at full recycle. A material balance around the plant under steady state conditions ($\Delta\Phi_m = 0$ and $dm/dt = 0$) leads to

$$\Phi_{mAin} = \Phi_{mP} + \Phi_{mX} + \Phi_{mA}.$$

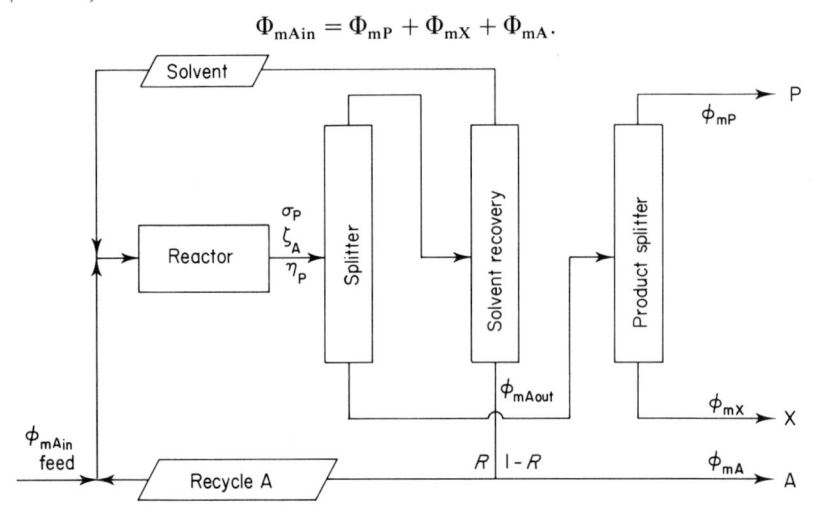

Fig. I.7 Plant design with or without recycle of non-converted reactant. The separations are 100% effective

The amount of P to be produced is constant and given by the potential sales of P on which the design is based. However, the streams Φ_{mAin}, Φ_{mA} and Φ_{mX} depend on the overall operating condition of the reactor section and whether recycle can be applied. From the definition given and from the reaction equation ($v_A = v_P$ and $M_A = M_P$) it follows that

$$\Phi_{mX} = \frac{1 - \sigma_P}{\sigma_P}\Phi_{mP}.$$

The ratio Φ_{mP}/Φ_{mX} is called the selectivity ratio. The mass flow of A at the reactor inlet must be Φ_{mP}/η_P and consists of a fresh feed stream Φ_{mAin} in the recycle stream $R\Phi_{mAout}$. We find further, for the amount of A recycled,

$$R(1 - \zeta_A)\Phi_{mP}/\eta_P$$

and for the amount of A fed as fresh reactor intake,

$$(1 - R + R\zeta_A)\Phi_{mP}/\eta_P.$$

From these equations it follows that the fresh feed intake of A equals

$$\Phi_{mAin} = (1 - R)(1 - \zeta_A)\Phi_{mP}/\eta_P + \Phi_{mP} + (1 - \sigma_P)\Phi_{mP}/\sigma_P,$$

and after elimination of ζ_A,

$$\Phi_{mAin} = \frac{\sigma_P(1-R) + \eta_P R}{\sigma_P \eta_P} \Phi_{mP}.$$

Now we take the two extremes and find at no recycle ($R = 0$)

$$\Phi_{mAin} = \Phi_{mP}/\eta_P,$$

and at full recycle ($R = 1$)

$$\Phi_{mAin} = \Phi_{mP}/\sigma_P.$$

As normally $\eta_P \leqslant \sigma_P$ (because $\eta_P = \sigma_P \zeta_A$ and $\zeta_A \leqslant 1$), we see that less reactant A is consumed in the case of full recycle. At a partial recycle the amount of fresh A intake lies between both extremes. Clearly the amount of raw material consumed plays a dominant role in the economic results of a plant. At several places in this book, the economic aspects of plant design and operation will be touched upon. An example of a simple economic consideration is given below, involving the yield as parameter.

Example I.6.b *Influence of reactor yield on plant profitability*
A plant is making ethyl acetate from acetic acid and ethyl alcohol. The average annual price of the reactants (delivered to the plant, including storage charges) were, in 1962,

acetic acid   1.172 MU/kg,
ethyl alcohol   0.678 MU/kg.

The sales price of the product, after deduction of the sales expenses is:

ethyl acetate 1.339 MU/kg.

The MU = monetary unit, is approximately equal to one Dutch guilder. It is assumed that 99% of the alcohol is converted in the installation but that the plant yield, $\bar{\eta}$ (calculated with respect to the acetic acid), is variable depending on operating conditions. It is furthermore assumed that the manufacturing costs are constant and amount to 0.100 MU/kg ethyl acetate. It is asked how the plant profitability depends on $\bar{\eta}$.

1 kg of ethyl acetate requires 0.682 kg of acetic acid and 0.523 kg of ethyl alcohol. The raw material costs per kg of ethyl acetate are:

$$\frac{0.782 \times 1.172}{\bar{\eta}} + \frac{0.523 \times 0.678}{0.99} = \frac{0.799}{\bar{\eta}} + 0.358 \text{ MU/kg ethyl acetate.}$$

The gross profit, I, of the plant per kg ethyl acetate is therefore

$$I = 1.339 - \left(0.100 + \frac{0.799}{\bar{\eta}} + 0.358\right) \text{ MU/kg ethyl acetate.}$$

This function is shown in figure I.8.

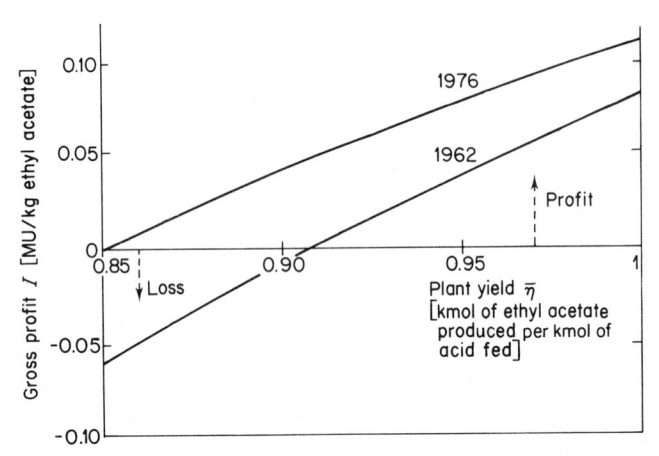

Fig. I.8 Influence of the plant yield $\bar{\eta}$ on the gross profit; Example I.6.a

It can be deduced from the figure that, most probably due to the heavy competition between ethyl acetate manufacturers, a minimum plant yield of 0.907 is needed for marginal operation and that an increase of 1% in $\bar{\eta}$ already has a marked effect on the profitability. This example also illustrates, however, the importance of the commercial aspects; if the purchasing department manages to obtain a 2% reduction in the price of the reactants, the increase in profit is the same as that achieved by a technical staff member who improves the plant yield $\bar{\eta}$ by 3%.

In 1976 the situation in the market had changed drastically, prices then were 0.92 MU/kg acetic acid, 0.54 MU/kg ethyl alcohol and 1.20 MU/kg ethyl acetate. The plant costs of the now rather obsolete unit had increased to 0.18 MU/kg ethyl acetate due to increases in labour costs and maintenance, waste disposal costs and interest. The function for the year 1976 is also shown in figure I.8. We see that the increase in plant costs and decrease in sales price is more than offset by the decrease in raw material costs. If the same plant yield can still be obtained in this old plant, the profit in 1976 will be better than it was in 1962. No allowances have been made for the revaluation of the capital costs due to inflation.

I.7 CLASSIFICATION OF CHEMICAL REACTORS

Chemical reactors exist in a wide range of forms and appearances. There is such a great variety of them that a complete systematic classification is impossible and an attempt to do so is hardly justified.

There appears to be no correlation at all between the type of reaction to be carried out (e.g. oxidation, reduction, etc.) or the complexity of the reaction (e.g. consecutive, parallel and chain reactions) and the shape and operating conditions of the equipment in which the reaction is to be carried out. Nor is the heat effect of a reaction an important shape-determining factor, except for the fact that in many cases a sufficiently large heat transfer area must be available for reactions with a great heat effect; a reactor for which these requirements are extreme will be very similar to a heat exchanger. A limited systematic survey of the reactors currently in use in the chemical industry can be made according to the two following criteria, which are related to the handling of reactants and products:

(i) One or more phases are needed for carrying out the desired reaction; *homogeneous and heterogeneous* reaction systems, respectively, are then involved.
(ii) The reaction mixture is processed in intermittent or in uninterrupted operation; if in the former case, no material is supplied or withdrawn during the reaction cycle, the reaction is carried out *batch-wise*; the other case is that of the *continuous* flow reactor.

If we apply the second criterion to *homogeneous* reaction systems, we see that batch reactions in the fluid phase (mostly the liquid phase) are carried out in vessels, tanks or autoclaves in which the reaction mixture is agitated and mixed in a suitable manner. For example, glass beakers with stirrers are the most extensively used reactors in the chemical laboratory.

Continuous flow reactors for homogeneous reaction systems already show a much greater variety. Predominant forms are the *tubular reactor* and the mixed *tank reactor*, which have essentially different characteristics. These are shown in figure I.9, together with a few intermediate forms. As was indicated by van Krevelen [8], a further distinction can be made according to whether or not the feed streams are mixed prior to entering into the reactor. This is particularly important in the case of rapid homogeneous reactions, such as the combustion of a gas. In Chapters II and III, idealized forms of the batch reactor, the stirred tank reactor and the tubular reactor will be used as the primary models for isothermal reactor calculations.

The classification of reactors for *heterogeneous* systems (in which the reaction may still proceed in a homogeneous phase) shows a great number of possibilities: either one or more phases may be processed continuously and the flow of a phase may be more or less mixed in the direction of flow (corresponding with flow in a tank reactor and in a tubular reactor as extreme cases). Furthermore, there are several ways in which two or more phases can be dispersed into each other. Wicke [33] and Van Krevelen [8] have given a survey of the combinations of these possibilities which are frequently encountered in practice. In the first instance, the flow patterns of the two continuous model reactors (tank reactor and tubular reactor) can be traced in such systems. In a burning coal stove, for example, the gas flow is similar to that occurring in a tubular reactor. The coal is slowly consumed and the reaction zone moves slowly in the direction of the gas flow. If the coal is more or less continuously supplied (and the ashes removed), the situation with respect to coal can also be handled as that prevailing in a tubular reactor. In continuous gas–solid reactions in a rotating kiln, both phases are relatively little mixed in the direction of flow, as a first approximation; when a

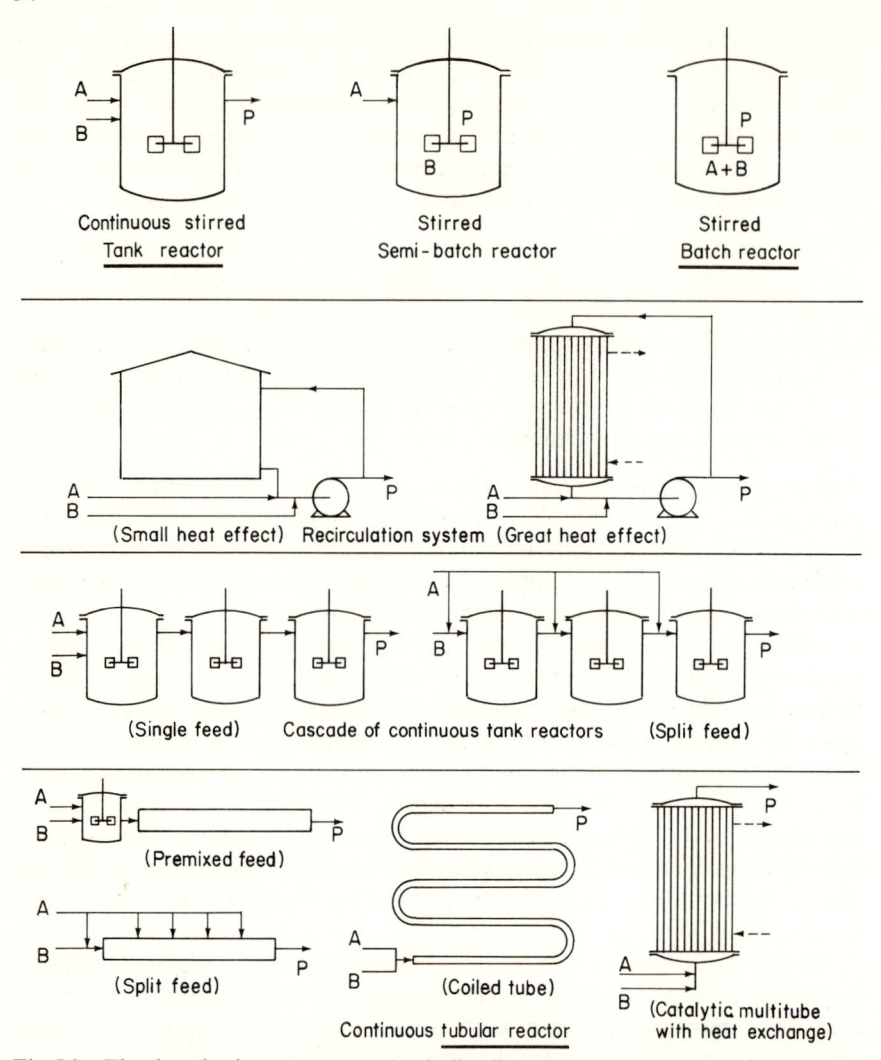

Fig. I.9 The three basic reactor types (underlined) and some related reactor systems for homogeneous or quasi-homogeneous reactions

similar reaction is carried out in a fluidized bed, the solids are well mixed (tank reactor), and the behaviour of the gas flow is intermediate between that in a tank reactor and a tubular reactor.

We see that in heterogeneous systems the contact between the different phases is a dominant factor. This leads us to a classification of reactors as a contact apparatus. Following the system used in the Ullmann encyclopedia [34] we arrive at a possible classification of reactors as in Table I.7.

Actual reactors will not be described in this book. Illustrative information of this kind is found in many articles and in books; the Ullmann encyclopedia [34] dedicates almost an entire volume to the description of industrial practice in

Table I.7 Classification of chemical reactors as contact apparatus [34]

Reactor type	Examples
I *Non catalytic homogeneous gas reactors*	
tubular reactors	
—exothermic reactions	high pressure polymerization of ethylene
—endothermic reactions	naphtha conversion to ethylene
II *Homogeneous liquid reactors*	
many types	mass polymerization of styrene; hydrolysis of ethylene oxide
III *Liquid–liquid reactors*	
columns or mixers	saponification of fats; nitration of aromatics; suspension and emulsion polymerizations
IV *Gas–liquid reactors*	
mixer	air oxidation of cyclohexame;
bubble column	absorption of isobutylene in sulphuric acid
loop reactor	oxidation of normal butane
counter current column with trays or packed	nitric acid production;
tube reactor	adipinic nitrile production
jet washer	absorption HCl, etc in $NaOH-Na_2SO_3$ solutions
liquid ring pump	acetic anhydride from ketene and acetic acid
falling film reactor	sulphonation of dodecyl benzene
jet tube reactor	oil burners
co-current column packed or with trays	desulphurization
V *Non-catalytic gas–solid reactors*	
Fixed bed reactors	
—blast furnaces	iron production
—convertors	steel production
—roasting furnaces	pyrite oxidation
—rotating kilns	cement
—coal gasifiers	synthesis gas production
Fluid bed	
—exothermic	chlorination of metals, roasting
—endothermic	oil distillation from shale oil
—multiple bed plate columns	CaO production
VI *Gas reactions catalysed in fixed bed reactors*	
adiabatic	CO shift reactor; combustion of NH_3
temperature controlled	NH_3 synthesis; SO_2 combustion
VII *Gas reactions catalysed in fluid bed reactors*	
fluid bed	phthalic anhydride production
riser reactor	catalytic cracking
VIII *Reactors for gas–liquid–solid reactions*	
fixed bed: trickle bed packed columns	hydrodesulphurization of oils
slurry reactors:	
—agitated reactors	hydrogenation of fats
—bubble columns	continuous H-oil process

IX *Special reactors*

photochemical reactors	vitamin D production
electrothermic furnaces	speciality steels (solid state reactions)
plasma reactors	acetylene production
electrochemical reactors	electrolysis
cyclone reactors	sulphonation of aromatics
solid trickle flow in packed bed reactors	gas cleaning

---

chemical reactor design. The following chapters are intended to give the principles of reactor design and operation, on the basis of which actual and more complicated reactor problems can be analysed and, to a certain extent, solved.

REFERENCES

1. Bird, R. B., Stewart, W. E., and Lightfoot, E. N.: *Transport Phenomena*, Wiley, New York, 1960.
2. Beek, W. J., and Muttzall, M. K.: *Transport Phenomena*, Wiley, New York, 1975.
3. Hougen, O. A., Watson, K. M., and Ragatz, R. H.: *Chemical Process Principles II, Thermodynamics*, 2nd edition, Wiley, New York, 1959.
4. Modell, M., and Reid, R. C., *Thermodynamics and its Applications*, Prentice-Hall, Englewood Cliffs, 1974.
5. Denbigh, K. G.: *The Principles of Chemical Equilibrium*, 3rd edition, Cambridge University Press, Cambridge, 1971.
6. Damköhler, G.: *Einfluss von Diffusion, Strömung und Wärmetransport auf die Ausbeute bei chemischtechnischen Reaktionen*, VDI, Leverkusen, 1957.
7. Frank-Kamenetski, D. A.: *Diffusion and Heat Transfer in Chemical Kinetics*, 2nd edition, Plenum New York, 1969.
8. Krevelen, D. W. van: *Chem. Eng. Sci.* **8**, 5 (1958).
9. Danckwerts, P. V.: *Gas–Liquid Reactions*, McGraw-Hill, New York, 1970.
10. Astarita, G.: *Mass Transfer with Chemical Reaction*, Elsevier, Amsterdam, 1967.
11. Hougen, O. A., Watson, K. M., and Ragatz, R. A.: *Chemical Process Principles I, Material and Energy Balances*, 2nd edition, Wiley, New York, 1962.
12. Schweyer, H. E.: *Process Engineering Economics*, McGraw-Hill, New York, 1955.
13. Peters, M. S. and Timmerhaus, K. D.: *Plant Design and Economics for Chemical Engineers*, 3rd edition, McGraw-Hill, New York, 1980.
14. Smith, J. M.: *Chemical Engineering Kinetics*, 2nd edition, McGraw-Hill, New York, 1970.
15. Reid, R. C., Prausnitz, J. M., and Sherwood, T. K.: *Properties of Gases and Liquids*, 3rd edition, McGraw-Hill, New York, 1977.
16. Hougen, O. A., and Watson, K. M.: *Chemical Process Principles II, Thermodynamics*, 1st edition, Wiley, New York, 1947.
17. Krevelen, D. W. van, and Chermin, H. A. G.: *Chem. Eng. Sci.* **1**, 66, 238 (1951–52).
18. Smith, J. M., and van Ness, H. C.: *Introduction to Chemical Engineering Thermodynamics*, 3rd edition, McGraw-Hill, New York, 1975.
19. Hougen, O. A., and Watson, K. M.: *Chemical Process Principles III, Kinetics and Catalysis*, Wiley, New York, 1947.
20. Gardiner, W. C.: *Rates and Mechanisms of Chemical Reactions*, Benjamin, New York, 1969.

21. Ashmore, P. G.: *Catalysis and Inhibitions of Chemical Reactions*, Buttersworths, London, 1963.
22. Benson, S. W.: *The Foundations of Chemical Kinetics*, McGraw-Hill, New York, 1960.
23. Boudart, M.: *Kinetics of Chemical Processes*, Prentice-Hall, Englewood Cliffs, 1968.
24. Frost, A. A., and Pearson, R. G.: *Kinetics and Mechanism*, 2nd edition, Wiley, New York, 1961.
25. Laidler, K.: *Reaction Kinetics*, Pergamon, London, 1963.
26. Walas, S. M.: *Reaction Kinetics for Chemical Engineers*, McGraw-Hill, New York, 1959.
27. Pinsent, B. R. W., Pearson, L., and Roughton, F. J. W.: *Trans. Faraday Soc.* **52**, 1512 (1956).
28. Bosscher, J. K., Ph.D. Thesis, University of Amsterdam, 1967.
29. Pinsent, B. R. W., Pearson, L., and Roughton, F. J. W.: *Trans. Faraday Soc.* **52**, 1594 (1956).
30. Manogne, W. H., and Pigford, R. L.: *A.I.Ch.E. Journal*, **6**, 494 (1960).
31. Dun, P. W., and Wood, T.: *J. Appl. Chem.*, **17**, 53 (1967).
32. Sharma, M. M., and Danckwerts, P. V.: *Chem. Eng. Sci.* **18**, 729 (1963).
33. Wicke, E.: *Z. Elektrochem.* **57**, 460 (1953).
34. *Ullmanns Encyklopaedie der technischen Chemie Band 3*, 4th edition, Verlag Chemie, Weinheim, 1973.

Chapter II

Model reactors: single reactions, isothermal single phase reactor calculations

In this chapter the material balances are applied to isothermal single phase reactions with known expressions for their conversion rates and only single reactions will be considered. The reactions are assumed to be carried out in reactor systems consisting of or derived from one of the three basic types of model reactors, namely:

(i) the well-mixed reaction vessel with uniform composition, operated batchwise: it will be referred to as a '*batch reactor*' (abbreviated BR);
(ii) the continuously operated ideal tubular reactor in which piston flow of the reacting mixture is assumed and mixing or diffusion in the direction of flow does not occur: the 'plug flow reactor' (abbreviated PFR);
(iii) the continuously operated ideally mixed tank reactor in which the composition of the reaction mixture is assumed to be uniform and equal to the composition at the outlet (abbreviated CISTR).

As a consequence of the assumptions relating to these three model reactors, the reactor calculations remain relatively simple. This makes it possible, e.g., to investigate reactions without undue complications arising from a possible departure from the ideality of the basic reactor types. The extent to which the fluid flow in actual reactors may depart from that in the models given above will be discussed in Chapters IV and V.

We will in every following section start with the material balance as discussed in Section I.2 and derive the fundamentally correct expressions for every type of model reactor. Then we will show how the equations simplify if the density of the reaction mixture (and feed and product streams) can be assumed to be constant. Finally we will give the results of the calculations for a reaction with a first order kinetic rate expression to demonstrate the fundamental behaviour of the model reactors. The following concepts will be frequently used:

reactor capacity = production rate of desired product per unit volume of reactor at a given conversion of key reactant;
throughput, feed rate = volumetric or mass flow through reactor system;
load = volumetric or mass flow per unit reactor volume or catalyst mass.

II.1 THE WELL-MIXED BATCH REACTOR

No material is supplied to or withdrawn from the reactor during the reaction so that the total mass m of the reaction mixture remains constant. The composition, which is assumed to be uniform, is only a function of time. Thus, for a component J the material balance (I.15) becomes

$$\frac{dm_J}{dt} = m\frac{d\xi_J}{dt} = M_J R_J V_r, \quad \text{(II.1)}$$

where V_r is the volume of the reaction mixture. Equation (II.1) may also be rewritten in terms of the density of the reaction mixture $\rho = m/V_r$; we thus have, for any component of the mixture,

$$\frac{d\xi_J}{dt} = \frac{M_J R_J}{\rho}. \quad \text{(II.2)}$$

Conversion of this equation to molar concentrations by means of equation (I.31) introduces an additional term since, in principle, the density of the mixture $\rho = m/V_r$ is not constant:

$$\frac{dc_J}{dt} - \frac{c_J}{\rho}\frac{d\rho}{dt} = R_J, \quad \text{(II.3)}$$

as was already shown in Example I.2. For a reactant A we have

$$\xi_A = M_A(c_{A0}/\rho_0 - c_A/\rho)$$

and for a product P

$$\xi_A = \nu_A M_A (c_P/\rho - c_{P0}/\rho_0)/\nu_P;$$

for constant density reactions we can derive from equation (II.2)

$$-\frac{dc_A}{dt} = \frac{\nu_A}{\nu_P}\frac{dc_P}{dt} = R_A. \quad \text{(II.4)}$$

It is from this expression that the current practice in physicochemical literature originates, according to which chemical reaction velocities or conversion rates are indicated by the symbol—dc_A/dt. In reactor calculations, however, this custom may cause confusion, especially when dealing with continuous operation in the steady state where concentrations are independent of time. The symbol dc_A/dt in equation (II.3) is not a reaction velocity or a conversion rate; it is the rate of concentration change in a batch reactor as a result of a chemical reaction in a constant density reaction mixture.

From equation (II.2) the reaction time t needed for a certain degree of conversion can be obtained by integration:

$$t = \int_0^{\xi_J} \frac{\rho \, d\xi_J}{M_J R_J}. \quad \text{(II.5)}$$

Other possible forms of equation (II.5) may be obtained by introducing the relative degree of conversion of one of the reactants, ζ_A:

$$t = w_{Ao} \int_0^{\zeta_A} \frac{\rho \, d\zeta_A}{M_A R_A} = \frac{c_{Ao}}{\rho_o} \int_0^{\zeta_A} \frac{\rho \, d\zeta_A}{R_A}. \tag{II.6}$$

These are greatly simplified if the density remains constant:

$$t = c_{Ao} \int_0^{\zeta_A} \frac{d\zeta_A}{R_A}. \tag{II.7}$$

If the conversion rate of reactant A is given by the expression $R_A = kc_A$ kmoles/m³ s, then we find:

$$t = -\int_0^{\zeta_A} \frac{d(1 - \zeta_A)}{k(1 - \zeta_A)} \tag{II.8}$$

because $c_A/c_{Ao} = 1 - \zeta_A$. Integrating and rearranging gives for a first order reaction in a batch reactor for the conversion

$$\zeta_A = 1 - e^{-kt}. \tag{II.9}$$

Similarly for the reactant concentration we can derive

$$\frac{c_A}{c_{Ao}} = e^{-kt}. \tag{II.10}$$

For reactions of the nth order similar expressions can be found: they will be given in Table II.1. From this table we see, that for reaction orders of $n < 1$ after a certain reaction time $t = c_{Ao}^{1-n}/k(1 - n)$ all reactant A will have disappeared, so

Table II.1 Concentration changes and half value times in a batch reactor for several reaction orders

Reaction order n	rate equation R_A	Concentration–time relation c_A/c_{Ao}	Half value time per unit of time $t_{0.5}$	dimensionless $kc_{Ao}^{n-1} t_{0.5}$
0	k	$1 - \dfrac{kt}{c_{Ao}}$	$\dfrac{c_{Ao}}{2k}$	0.5
0,5	$k\sqrt{c_A}$	$\left(1 - \dfrac{kt}{2\sqrt{c_{Ao}}}\right)^2$	$\dfrac{\sqrt{2c_{Ao}}}{k}(\sqrt{2} - 1)$	$2 - \sqrt{2}$
1	kc_A	e^{-kt}	$\dfrac{\ln 2}{k}$	$\ln 2$
2	kc_A^2	$\dfrac{1}{1 + kc_{Ao}t}$	$\dfrac{1}{kc_{Ao}}$	1
n	kc_A^n	$\left[1 + k(n-1)c_{Ao}^{n-1} t\right]^{1/(1-n)}$	$\dfrac{2^{n-1} - 1}{kc_{Ao}^{n-1}(n - 1)}$	$\dfrac{2^{n-1} - 1}{n - 1}$

that the reaction stops. In Table II.1 also the 'half value time' $t_{0.5}$ is given for reactions with the rate equation $R_A = kc_A^n$ in constant density reaction mixtures. The half value time is the time required to reduce the concentration c_A of key reactant A to half of its original value c_{Ao}, so that $c_A/c_{Ao} = 0.5$ at $t = t_{0.5}$. From isothermal single phase experiments in a BR, in which c_A as a function of the reaction time t is measured, the dependence of the conversion rate from c_A can be determined by fitting c_A or $-dc_A/dt$ as a function of time or R_{Ao} or $t_{0.5}$ as a function of c_{Ao} to one of the solutions in Table II.1 or similar possible solutions for different kinetics. An extensive treatment of these techniques is given by Levenspiel [1], among others. We will assume in this book that chemical reaction kinetics are known.

For a first order reversible reaction A → B with a conversion rate expression of the type $R_A = k_1 c_A - k_2 c_B$ and with $c_{Bo} = 0$, $v_A = v_B = 1$ and constant density we find for the conversion, if we realize that at equilibrium $R_A = 0$, so that $K \equiv c_B/c_A = k_1/k_2$:

$$+\frac{d\zeta_A}{dt} = -\frac{dc_A/c_{Ao}}{dt} = +k_1\left[1 - \zeta_A - \frac{\zeta_A}{K}\right] \tag{II.11}$$

Integration gives:

$$\zeta_A = \frac{k_1}{k_1 + k_2}(1 - e^{-(k_1 + k_2)t}) \tag{II.12}$$

At equilibrium we find the following values: $\zeta_{Aeq} = k_1/(k_1 + k_2)$ and $c_{Aeq}/c_{Ao} = k_2/(k_1 + k_2)$. From equation (II.12) it follows that equilibrium is reached after infinite time.

The general problem in design is to calculate the volume of a reactor for a certain average rate of production. The volume can be obtained from the required degree of conversion and the corresponding time (see Example II.1). This is possible since the reaction time is independent of the reactor volume; this is a consequence of the assumption that the reactor contents are well mixed and that all elements of volume behave identically. In practice, the size of a batch reactor may have some influence on the reaction time, because the conversion rate may be influenced by incompleteness of mixing intensity or temperature deviations near a heating or cooling surface. If the desired production of the product P is Φ_P and the non-productive reactor time for filling, heating, cooling, cleaning and emptying is t', we find for the reactor volume

$$V_R = \frac{v_A \Phi_P (t + t')}{v_P M_P c_{Ao} \zeta_A}$$

Example II.1 Production of ethyl acetate in a batch reactor
A daily production of 50 tonnes of ethyl acetate from alcohol and acetic acid is required. The reaction proceeds according to:

$$C_2H_5OH(A) + CH_3COOH(B) \rightarrow CH_3COOC_2H_5(P) + H_2O(Q)$$

The conversion rate in the liquid phase at 100 °C is given by Smith [2]:

$$R_A = k\left(c_A c_B - \frac{c_P c_Q}{K}\right)$$

where $k = 7.93 \times 10^{-6}$ m^3/kmol s and $K = 2.93$. The molar conversion rates of all components are equal because of the equality of the stoichiometric coefficients.

The feed solution contains 23% by weight of acid, 46% by weight of alcohol and no ester. The required relative conversion of the acid is 35%. The density may be assumed to have a constant value of 1020 kg/m^3. The plant must be operated day and night and the time for the filling, emptying and cleaning operations of a reactor is 1 hour, irrespective of its size. What is the required reaction volume if (a) one reactor vessel, (b) three reactor vessels are to be used? The concentrations in the reactor expressed in terms of ζ_B, the relative conversion of the acid, are given in Table II.2. According to equation (II.6) the reaction time t is given by

$$t = c_{Bo} \int_0^{0.35} \frac{d\zeta_B}{R_B}$$

Table II.2 Concentrations in the batch reactor of Example II.1

Component	M_J kg/mol	Composition at $t=0$, $\zeta_B = 0$ w_{J0} kg/kg	c_{J0} kmol/m^3	at conversion ζ_B c_J kmol/m^3	at $\zeta_B = 0.35$ c_J kmol/m^3
A	46	0.46	10.20	10.20 − 3.91 ζ_B	8.83
B	60	0.23	3.91	3.91 (1 − ζ_B)	2.54
P	88	0	0	3.91 ζ_B	1.37
Q	18	0.31	17.56	17.56 + 3.91 ζ_B	18.93

where R_B is expressed in terms of ζ_B. The integration can be illustrated graphically (see figure II.1). The result is $t = 7270$ s ≈ 2 hours so that 24/(2 + 1) batches can be processed per 24 hours. The daily ethyl acetate production per m^3 of reactor volume is $8 \times 1.37 \times 88 = 965$ kg/m^3 day. The total reaction volume required is $50\,000/965 \approx 52$ m^3.

It makes no difference for the total volume whether one reactor is used or three reactors in parallel. In the latter case some of the auxiliary equipment like pumps and storage vessels can be smaller, and by means of a proper time schedule the work of the operating personnel can be more evenly distributed. On the other hand, a small vessel is relatively more expensive per m^3 than a large one, and three small tanks take more space and piping than one large tank having the same volume.

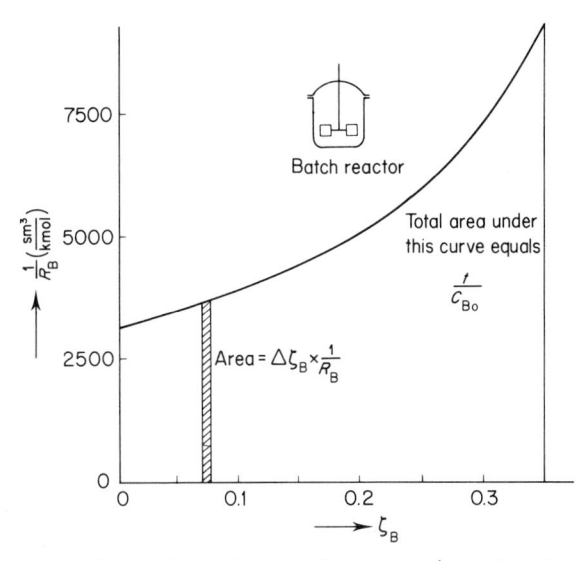

Fig. II.1 Integration of the reciprocal conversion rate R_B^{-1} as a function of the degree of conversion ζ_B, Example II.1

II.2 THE CONTINUOUSLY OPERATED IDEAL TUBULAR REACTOR

In steady state, the conditions at any point in the reactor are independent of time, and the total mass flow Φ_m through any cross section of the reactor is the same. In this reactor type the reaction mixture flows like a plug through the tube; the linear velocity u of the reacting mixture is the same at every point in a cross section S perpendicular to the direction of the flow and equal to $\Phi_m/\rho S$. Therefore we call this type of model reactor the plug flow reactor (PFR). The composition of the reaction mixture depends on the distance z from the inlet point (see figure II.2). For the calculation of the concentration distributions the material balance has to be applied to a differential section $S\,dz$, where S is the area of the cross section occupied by reacting mixture.

Under these conditions equation (I.15) becomes for a component J:

$$0 = -\Phi_m\,dw_J + M_J R_J S\,dz. \tag{II.13}$$

Introduction of the degree of conversion ξ_J yields

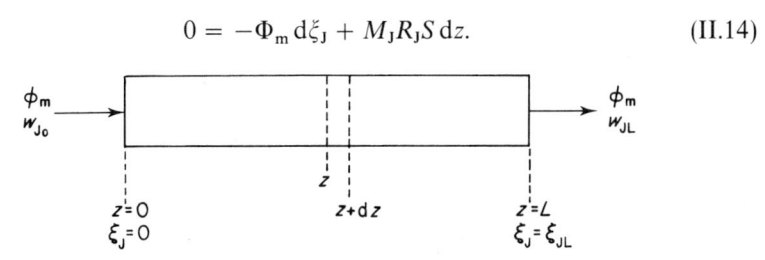

$$0 = -\Phi_m\,d\xi_J + M_J R_J S\,dz. \tag{II.14}$$

Fig. II.2 Tubular reactor with total reaction volume $V_r = SL$

The reactor volume $V_r = SL$ required for a conversion ζ_{JL} follows from integration of equation (II.14):

$$V_r = \int_0^L S\,dz = \Phi_m \int_0^{\zeta_{JL}} \frac{d\zeta_J}{M_J R_J}. \tag{II.15}$$

Equation (II.15) can be simplified in the case of a constant density of the reaction mixture; after substitution of the relations $\Phi_m = \rho\Phi_v$, $w_{Ao}\zeta_A = \zeta_A$ and $w_{Ao} = c_{Ao}M_A/\rho$, we find

$$V_r = \Phi_v c_{Ao} \int_0^{\zeta_{AL}} \frac{d\zeta_A}{R_A}, \tag{II.16}$$

or, written otherwise,

$$\tau_L = \frac{V_r}{\Phi_v} = c_{Ao} \int_0^{\zeta_{AL}} \frac{d\zeta_A}{R_A}. \tag{II.17}$$

Note that the integral on the right hand side also occurs in the calculation of the reaction time in a batch reactor (cf. equation (II.7), provided the density is constant. Apparently, in that case:

$$\left(\frac{t}{\rho}\right)_{BR} \text{ corresponds to } \left(\frac{V_r}{\Phi_m}\right)_{PFR},$$

or

$$(t)_{BR} \text{ corresponds to } \left(\frac{\rho V_r}{\Phi_m} = \tau_L\right)_{PFR}.$$

τ_L is the residence time of the reaction mixture in the tubular reactor for constant density. The correspondence between the reaction time for a batch reactor and the residence time for an ideal tubular reactor is also clear from a physical point of view; in the latter an element of volume is not supposed to mix with its surroundings. The requirement of constant density for this analogy to be valid rests in the fact that the volume of the reaction mixture in a batch reactor is in principle variable, whereas the volume of the tabular reactor has been assumed to be fixed.

For a first order kinetics reaction rate equation $R_A = kc_A$ we find for the conversion

$$\zeta_A = 1 - e^{-k\tau_L}, \tag{II.18}$$

and for the concentration of the reactant A at the reactor outlet

$$\frac{c_{AL}}{c_{Ao}} = e^{-k\tau_L}. \tag{II.19}$$

In constant density reaction mixtures for reactions of the nth order the solutions in Table I.1 can be used, provided t is replaced by τ_L.

In practice, tubular reactors are encountered which consist of one long single tube, such as in thermal cracking furnaces where the reactor may be up to 2 km long. Very often reactions are carried out in tube bundles, i.e. a number of parallel tubular reactors. These are used in cases where special heat transfer requirements have to be met (see also Chapter VI).

If the reaction in a tubular reactor proceeds under the influence of a *solid catalyst*, e.g. in the form of a packed bed of catalyst pellets, the conversion rate is often given per unit mass of solids (e.g. R_{sJ}). In that case, the total mass of solids m_{sL} required for a certain degree of conversion

$$m_{sL} = \Phi_m \int_0^{\xi_{JL}} \frac{d\xi_J}{M_J R_{sJ}}.$$

This expression is equivalent to equation (II.15) and the reactor volume may be calculated from m_{sL} and the bulk density of the catalytic material. Since a heterogeneous reaction is involved, R_{sJ} may in principle depend not only on composition and temperature, but also on the nature and size of the catalyst pellets and on the flow velocity of the mixture (see Chapter VII).

With reactors where a solid catalyst is used, the reactor load is in practice often indicated by the term *space velocity*, S.V. It is defined as the volumetric flow at the inlet of the reactor divided by the reaction volume (or by the total mass of catalyst). Thus, for a tubular reactor:

$$S.V. \equiv \Phi_m/\rho_o V_r \qquad (\text{or } \Phi_m/\rho_o m_{sL}).$$

The density ρ_o at the inlet conditions is sometimes substituted by the density at other conditions (e.g. standard temperature and pressure). This may give rise to confusion, unless the definition used for S.V. is properly specified.

The inverse value of the space velocity can have the dimension of time:

$$\frac{1}{S.V.} = \frac{\rho_o V_r}{\Phi_m} = \rho_o \int_0^{\xi_{JL}} \frac{d\xi_J}{M_J R_J}.$$

Unless ρ is constant, the value of $1/(S.V.)$ is different from the true residence time of the reaction mixture in a tubular reactor, τ_L. For the latter we have

$$\tau_L = \int_0^{V_r} \frac{\rho S \, dz}{\Phi_m} = \int_0^{\xi_{JL}} \frac{d\xi_J}{M_J R_J}.$$

Example II.2.a *Production of ethyl acetate in the liquid phase in a tubular reactor*
It is required to calculate the volume of an ideal tubular reactor for the same production and under the same conditions as given in Example II.1.

Since the density was assumed to be constant, the residence time in the tubular reactor must be equal to the time of reaction calculated in the preceding example. Thus, we have

$$\tau_L = \rho V_r/\Phi_m = 7270 \text{ s}.$$

Since the product stream contains 1.37 kmol/m³ of ethyl acetate, the volumetric flow rate for a production of 50 tonnes/day is

$$\Phi_m/\rho = \frac{50.000}{24 \times 3600 \times 1.37 \times 88} = 4.80 \times 10^{-3}\, m^3/s.$$

From the last two formulae we find:

$$V_r = 7270 \times 4.80 \times 10^{-3} = 34.8\, m^3.$$

We might have reached this result somewhat faster by realizing that we need the same reaction volume as in Example II.1. The volume calculated for the batch process was 52 m³, but about one third of the time was reserved for the operations between two batches. For a tubular reactor in which the continuous operation is not interrupted, the same production can be obtained with a volume $V_r \approx \frac{2}{3} \times 52 \approx 35\, m^3$.

Example II.2.b *Reaction at varying density*
In a tubular reactor with a length of 10 m a gas reaction A → 3B takes place at a constant temperature of 350 K. The inner diameter of the reactor tube is 0.113 m, the molecular weight of A is 90. The feed consists of pure A and the reactor load is 15.5 kg/s. The inlet pressure is 5 bars ($= 5 \times 10^5$ N/m²) and the pressure drop along the reactor length is given by $dp = 4f \times \frac{1}{2}\rho v^2 \times dz/d_t$ and $4f = 0.025$. The gas mixture behaviour is ideal and $R_A = 3.2 \frac{p_A}{RT}$. Calculate the conversion and pressure along the reactor length and give the densities and linear gas velocities at the inlet and outlet of the reactor.

We know that in this case $p = p_A + p_B$ and that one mole of A forms three moles of B, so that after a conversion of ζ_A moles of A $3\zeta_A$ moles of B have been produced. We find, for the mole fractions of A and B in the gas mixture,

$$x_A = \frac{p_A}{p} = \frac{1 - \zeta_A}{1 + 2\zeta_A} \quad \text{and} \quad x_B = \frac{p_B}{p} = \frac{3\zeta_A}{1 + 2\zeta_A}.$$

Further, we know $\rho = \bar{M}p/RT$, $\bar{M} = x_A M_A + x_B M_B$ and $M_B = M_A/3$ so that

$$\rho = \frac{M_A}{RT} \frac{p}{1 + 2\zeta_A}. \tag{a}$$

For the pressure drop we find

$$dp = \frac{4f}{d_t} \times \frac{1}{2\rho} \times \left(\frac{\Phi_m}{S}\right)^2 dz$$

and for equation (II.14):

$$\Phi_m w_{Ao}\, d\zeta_A = 3.2 \frac{M_A}{RT} p \frac{1 - \zeta_A}{1 + 2\zeta_A} S\, dz.$$

After introducing the data given we find for the pressure and conversion profiles:

$$d\zeta_A = 6.39 \times 10^{-8} \frac{p(1-\zeta_A)}{1+2\zeta_A} dz \tag{b}$$

and

$$dp = 0.858 \times 10^{10} \frac{(1+2\zeta_A)}{p} dz. \tag{c}$$

Both equation (b) and (c) have to be solved simultaneously and numerically, which should be done on a computer. The results are given in figure II.3 and we see that the $\Delta p_L = 3$ bars and $\zeta_{AL} = 0.185$.

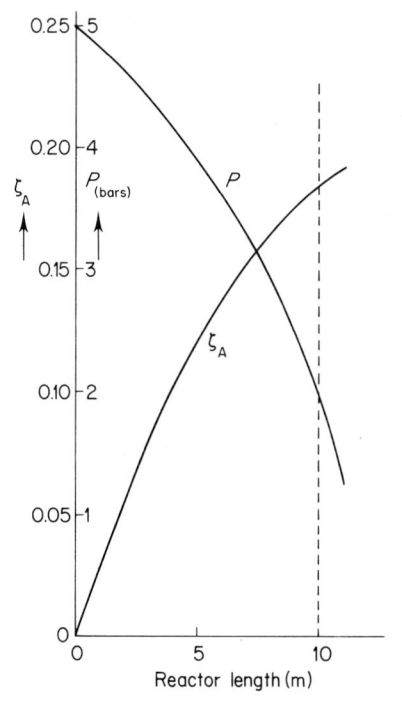

Fig. II.3 Pressure and conversion profile along te length of a PFR, Example II.2b

From equation (a) we find for the density at inlet ($\zeta_A = 0$, $p = 5.0$ bar) and outlet conditions

$$\rho_0 = 15.5 \text{ kg/m}^3 \quad \text{and} \quad \rho_{out} = 4.52 \text{ kg/m}^3.$$

The linear velocities follow from $u = \Phi_v/S = \Phi_m/\rho S$:

$$u_0 = 100 \text{ m/s} \quad \text{and} \quad u_{out} = 343 \text{ m/s}.$$

It will be evident that in this illustration the assumption of a constant density or of an average density would have led to erroneous results. Moreover, we see that gas velocities and pressure drop are unrealistically high; in practice the design engineer would install a larger diameter tube or many parallel tubes in order to save energy consumed and lost due to the friction in the reaction tubes.

Example II.2.c *Catalysed gas reaction in a tubular reactor*
The reaction

$$C_2H_5OH(A) + CH_3COOH(B) \rightarrow CH_3COOC_2H_5(P) + H_2O(Q)$$

is now carried out in the vapour phase at 277 °C and at a pressure of 1 bar. It is catalysed by silica gel having a bulk density of 700 kg/m³. According to the literature [3], the molar conversion rate at 277 °C is given by

$$R_{sB} = \frac{0.0415(0.3 + 0.9 p_Q)(p_B - p_P p_Q/9.8 p_A)}{3600(1 + 15.35 p_P)} \left[\frac{\text{kmol}}{\text{kg/s}}\right],$$

where the partial pressures are expressed in bars. What reactor volume is needed for the same production and the same feed composition as in Example II.1?

The reaction proceeds under isothermal and isobaric conditions and the flow of moles remains constant; hence the density $\rho = pM/RT$ does not change. The constant average molecular mass M is found to be 32.2 kg/kmol. The partial pressures can be expressed in terms of the relative conversion ζ_B of the acid B:

$$p_A = pMw_A/M_A, \quad p_B = pMw_B/M_B, \ldots,$$

and

$$p_A = p_{Ao} - p_{Bo}\zeta_B, \quad p_B = p_{Bo}(1 - \zeta_B), \ldots .$$

Composition data for the reaction mixture are given in Table II.3. From equation (II.15a) we derive for the total mass of catalyst

$$m_{sL} = \frac{\Phi_m}{M_B} w_{Bo} \int_0^{0.35} \frac{d\zeta_B}{R_{sB}} = \frac{\Phi_m p_{Bo}}{Mp} \int_0^{0.35} \frac{d\zeta_B}{R_{sB}}.$$

For $p = 1$ bar graphic integration yields

$$m_s L = 65.1 \times 10^3 \Phi_m/M.$$

Table II.3 Compositions in the plugflow reactor of Example II.2.c

Component	M_J kg/mol	w_{JO} kg/kg	Composition at $z = 0$, $\zeta_B = 0$ p_J bar	at conversion ζ_B p_J bar
A	46	0.46	$0.322p$	$(0.332 - 0.123\zeta_B)p$
B	60	0.23	$0.123p$	$0.123(1 - \zeta_B)p$
P	88	0	0	$0.123\zeta_B p$
Q	18	0.31	$0.55p$	$(0.555 + 0.123\zeta_B)p$
Total		1.00	p	p

For the feed rate in kmol/s we have

$$\Phi_m/M = \frac{50\,000}{88 \times 24 \times 3600 \times 0.123 \times 0.35} = 0.155\,\text{kmol/s},$$

so that $m_{sL} = 65.1 \times 10^3 \times 0.155 = 10\,100\,\text{kg}$
and $V_r = 10\,100/700 = 14.5\,\text{m}^3$.

II.3 THE CONTINUOUSLY OPERATED IDEAL TANK REACTOR

The contents of the tank reactor are assumed to be 'ideally mixed' so that the conditions throughout the tank are the same and equal to the conditions at the outlet (see figure II.4). In the steady state, the material balance (I.15) for one of the components becomes

$$0 = \Phi_m \xi_{J1} + M_J R_{J1} V_r, \qquad (\text{II.20})$$

where V_r is the constant volume of the reaction mixture. From equation (II.12) we find for any component

$$V_r = \Phi_m \frac{\xi_{J1}}{M_J R_{J1}} \qquad (\text{II.21})$$

Note that the conversion rate has to be taken at the conditions in the outlet, which are the reactor conditions. Therefore, if the degree of conversion is high, the conversion rate throughout the reactor will be relatively low. As a consequence, the continuous tank reactor requires a larger volume than a tubular reactor for the same production rate for reactions of an order $n > 0$.

In practice, for not too fast reactions, a tank reactor may be approximately considered an ideally mixed system when the mixing time is much smaller than the average residence time τ of the mixture in the reactor. This average residence time is defined as

$$\tau = V_r \rho_1 / \Phi_m \qquad (\text{II.22})$$

Voncken [4] has demonstrated that a reaction mixture in a tank reactor can be assumed to be ideally mixed if the average residence time in the tank reactor is

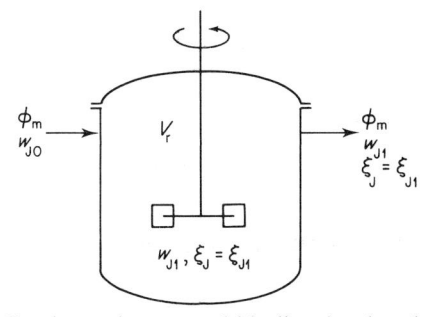

Fig. II.4 Continuously operated ideally stirred tank reactor

more than 100 times the circulation time of the reactor contents, which are pumped around by the mixer. In the turbulent region ($\text{Re}_{\text{imp}} = nd^2/v > 10^4$) Voncken proved, for turbine impellers and flat blade paddles, for the circulation time t_c, that

$$t_c \simeq \frac{\tau_c d_{\text{vessel}}^2}{nd_{\text{imp}}^2}, \tag{II.23}$$

in which $\tau_c = 1$ for turbine impellers and $\tau_c = 10$ for paddles. Moreover, Voncken showed that in the turbulent region the feed streams do not need to be premixed. On the macro scale inhomogeneities have already disappeared after four circulations; however, on the micro scale segregation may still exist after $4t_c$. This happens at low Reynolds numbers in reaction mixtures with high viscosities (see Chapter IV). So the condition for ideal mixing of low viscosity fluids in a tank reactor is

$$\tau > 100 \frac{d_{\text{vessel}}^2}{v \text{Re}_{\text{imp}}}. \tag{II.24}$$

This is not the only requirement; the "reaction time" must also fulfill certain conditions: this will be further elaborated in Chapter V. In this book we will use the abbreviation CISTR for the continuously operated ideally mixed stirred tank reactor.

If we introduce in equation (II.21) the relative degree of conversion ζ_A, we find

$$V_r = \frac{\Phi_m c_{Ao} \zeta_{A1}}{\rho_o R_A}. \tag{II.25}$$

In the case of constant density ($\rho_o = \rho_1$ and thus $\Phi_{vo} = \Phi_{v1} = \Phi_v$) equation (II.25) reduces to

$$V_r = \Phi_v \frac{c_{Ao} \zeta_A}{R_A} \tag{II.26}$$

and (II.22) to

$$\tau = V_r/\Phi_v. \tag{II.27}$$

For a first order kinetics reaction rate expression we have $R_A = kc_A$ and thus

$$V_r = \Phi_v \frac{\zeta_{A1}}{k(1 - \zeta_{A1})},$$

which leads us to the equation

$$\zeta_{A1} = \frac{k\tau}{1 + k\tau}, \tag{II.28}$$

and for the reactant concentration in the outlet,

$$\frac{c_{A1}}{c_{Ao}} = \frac{1}{1 + k\tau}. \tag{II.29}$$

For reactions of a different order some solutions are given in Table II.4. Compare this table with the results in Table II.1.

Table II.4 Outlet concentrations in a CISTR for reactions of different order

Reaction order n	Reaction rate equation R_A	Concentration residence time relation c_A/c_{Ao}
-1	kc_A^{-1}	$\frac{1}{2}\left[1 + \sqrt{\left(1 - \frac{4k\tau}{c_{Ao}^2}\right)}\right]$
0	k	$1 - \frac{k\tau}{c_{Ao}}$
$\frac{1}{2}$	$k\sqrt{c_A}$	$\sqrt{\left(1 + \frac{k^2\tau^2}{4c_{Ao}}\right)} - \frac{k\tau}{2\sqrt{c_{Ao}}}$
1	kc_A	$\frac{1}{1 + k\tau}$
2	kc_A^2	$\frac{1}{2kc_{Ao}\tau}|\sqrt{(1 + 4kc_{Ao})} - 1|$

At the end of this section we summarize the several forms of material balance equations for the three model reactors; the BR, the PFR and the CISTR. These balances for a component J can be based on degrees of conversion ξ_J, relative degrees of conversion ζ_A and on molar concentrations (or mass fractions $w_J = M_J c_J/\rho$). In Table II.5 the relevant equations are given.

Table II.5 Material Balance for model reactors

Reactor type	General form	at constant density
BR		
for component J	$m\, d\xi_J/dt = M_J R_J V_r$	—
for reactant A	—	$-d\zeta_A/dt = R_A/c_{Ao}$
for reactant A	$-d(c_A/\rho)/dt = R_A/\rho$	$-dc_A/dt = R_A$
for product P	$d(c_P/\rho)/dt = R_P/\rho$	$dc_P/dt = R_P$
PFR		
for component J	$\Phi_m\, d\xi_J = M_J R_J S\, dz$	—
for reactant A	—	$c_{Ao}\Phi_v\, d\zeta_A = R_A S\, dz$
for reactant A	$-\Phi_m d(c_A/\rho) = R_A S\, dz$	$-\Phi_v\, dc_A = R_A S\, dz$
for product P	$\Phi_m d(c_P/\rho) = R_P S\, dz$	$\Phi_v\, dc_P = R_P S\, dz$
CISTR		
for component J	$\Phi_m \xi_{J1} = M_J R_{J1} V_r$	—
for reactant A	—	$\Phi_v c_{Ao}\zeta_{A1} = R_{A1} V_r$
for reactant A	$\Phi_m(c_{Ao}/\rho_o - c_{A1}/\rho_1) = R_{A1} V_r$	$\Phi_v(c_{Ao} - c_{A1}) = R_{A1} V_r$
for product P	$\Phi_m(c_{P1}/\rho_1 - c_{Po}/\rho_o) = R_{P1} V_r$	$\Phi_v(c_{P1} - c_{Po}) = R_{P1} V_r$

Example II.3.a Check on ideal mixing
A tank reactor of $10 \, m^3$ is equipped with a six bladed turbine impeller with a diameter of $0.5 \, m$. The vessel diameter is $2.0 \, m$. The reaction mixture is an aqueous solution and the mean residence time is 2000 s. The agitator power imput into the fessel is 10 kW and the power number $P/\rho n^3 d_{imp}^5 = 5$. Check whether the reactor is ideally mixed.

From equation (II.23), it follows that for $\tau_c = 1$,

$$t_c = \frac{\tau_c d_{vessel}^2}{(Pd_{imp}/5\rho)^{1/3}},$$

provided

$$nd_{imp}^2/v = (Pd_{imp}/5\rho)^{1/3}/v > 10^4.$$

Introducing the data above, $\rho = 1000 \, kg/m^3$ and $v = 2 \times 10^{-6} \, m^2/s$, we find $Re_{imp} = 5 \times 10^5$, so that the circulation time is

$$t_c = 0.25 \, s$$

For ideal mixing the first condition is $\tau > 100 t_c$; in our case $2000 > 25 \, s$, so that this condition is fulfilled. Our tank reactor behaves as a CISTR, if also the second condition for the 'reaction time' is fulfilled, as will be explained in Chapter V.

Example II.3.b Production of ethyl acetate in a continuous tank reactor
The $52 \, m^3$ vessel of Example II.1 is to be used as a continuous tank reactor. The feed composition and the desired relative conversion of 35% of the acetic acid remain the same as in the above example. What production rate of ethyl acetate will result in this case? What volume should the tank reactor have to maintain an ester production of 50 tonnes/day?

The table in Example II.1 contains the molar concentrations at $\zeta_B = 0.35$; they can be used for calculating R_A. For the mass production rate of ester (P) at the condition $\zeta_B = 0.35$ we find

$$M_P R_{A1} = 9.5 \times 10^{-3} \, kg/m^3 s.$$

The total production rate of P is obtained with equation (II.20).

$$\Phi_m \zeta_{P1} = V_T R_{P1} = 52 \times 9.5 \times 10^{-3} = 0.493 \, kg/s,$$

which is equivalent to 42.6 tonnes/day. For a production of 50 tonnes/day the reactor volume would have to be $52 \times 50/42.6 = 61 \, m^3$, i.e. much larger than the volumes calculated for the batch reactor and the continuous tubular reactor.

Example II.3.c Start-up of a CISTR
An irreversible liquid reaction takes place in a CISTR under isothermal and constant density conditions according to:

$$A \to P$$

The conversion rate of A is $R_A = kc_A$ kmoles/m³s. Before $t = 0$ the reactor is empty, from the moment $t = 0$ the feed is pumped into the reactor at a rate of Φ_v m³/s and with a concentration of c_{Ao} of the reactant. At $t = \tau$ the liquid level has reached the overflow and after that the reaction volume V_r remains constant. The mixing is ideal after $t = 0$. Calculate the concentration c_A as a function of time (see figure II.5). For $t \leqslant \tau$ no liquid leaves the reactor and the material balance leads to

$$\frac{-dc_A V_r}{dt} = -\Phi_v c_{Ao} + kc_A V, \tag{a}$$

in which the reaction volume is $V = \Phi_v t$. This leads to the differential equation

$$\frac{d(c_A t)}{dt} = c_{Ao} - k(c_A t), \tag{b}$$

which can be solved by multiplying by the integrating factor $e^{+\int k\,dt}$. The solution is

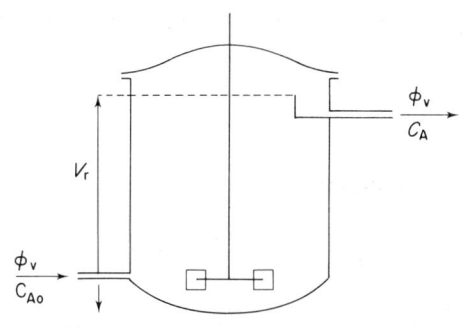

$$c_A = \frac{c_{Ao}}{kt}(1 - e^{-kt}) \quad \text{for } t \leqslant \frac{V}{\Phi_v} \tag{c}$$

Fig. II.5 A CISTR with overflow

For $t > \tau$ the reaction volume V is constant and equals V_r and the product flow out of the reactor equals the feed rate Φ_v. The material balance for reactant A now becomes

$$\frac{-d(c_A V_r)}{dt} = -V_r \frac{dc_A}{dt} = -\Phi_v(c_{Ao} - c_A) + kc_A V_r. \tag{d}$$

Simplification leads to

$$\frac{dc_A}{dt} = \frac{c_{Ao}}{\tau} - \left(\frac{1}{\tau} + k\right)c_A \tag{e}$$

and integration of

$$\int_{c_{A\tau}}^{c_A} \frac{dc_A}{c_{Ao} - (1 + k\tau)c_A} = -\int_\tau^t \frac{dt}{\tau}$$

results in

$$\frac{c_{Ao} - (1 + k\tau)c_A}{c_{Ao} - (1 + k\tau)c_{A\tau}} = e^{-(1+k\tau)(t-\tau)/\tau} \quad \text{for} \quad t > \tau \tag{f}$$

in which

$$c_{A\tau} = \frac{c_{Ao}}{k\tau}(1 - e^{-k\tau})$$

After a long time ($t \to \infty$) the right hand side of equation (f) approaches zero, so that ultimately the concentration c_A in the reactor reaches a stationary value $c_{A\infty}$. From

$$c_A - (1 + k\tau)c_{A\infty} = 0 \quad \text{for} \quad t \to \infty$$

it follows that

$$c_{A\infty}/c_{Ao} = 1/(1 + k\tau),$$

which is the normal equation for a first order reaction in a stationary CISTR.

A conservative estimate for the time required to reach steady state operation follows from equation (f) and by putting $c_{A\tau} = c_{Ao}$ (in practice of course $c_{A\tau}$ is always less than c_{Ao}). If we require that the final conversion is reached within 1% we find that

$$\frac{c_A - c_{A\infty}}{c_{Ao} - c_{A\infty}} < 0.01$$

and with $c_{A\tau} = c_{Ao}$ the criterion is

$$\left|\exp\left[-(1+k\tau)\left(\frac{t-\tau}{\tau}\right)\right]\right| < 0.01$$

or with $\ln 0.01 = -4.6$:

$$t > \tau + \frac{4.6\tau}{1 + k\tau}.$$

Thus depending on the value of $k\tau$, which is high for high conversions, steady state conditions will be reached between τ and 3τ seconds after $t = 0$. The concentration c_A in the reactor is compared with the steady state value $c_{A\infty}$ for various values of $k\tau$ in figure II.6.

II.4 THE CASCADE OF TANK REACTORS

Very often, continuous reactor operations are carried out in more than one tank reactor in cascade; this term is used to indicate that the product stream of one tank reactor is the feed stream of the next, and that the conditions in one of the reactors in the chain are not influenced by what happens in the reactors further downstream. The reactors are numbered $1, 2, \ldots, n, \ldots, N$ in the direction of flow,

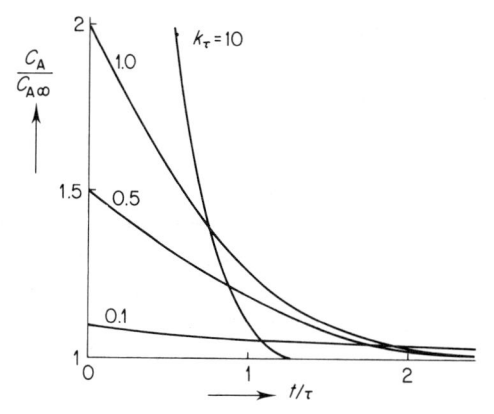

Fig. II.6 Deviation from steady state concentration as a function of t/τ, Example II.3.c

and the properties of the stream leaving the nth reactor carry the index n (see figure II.7).

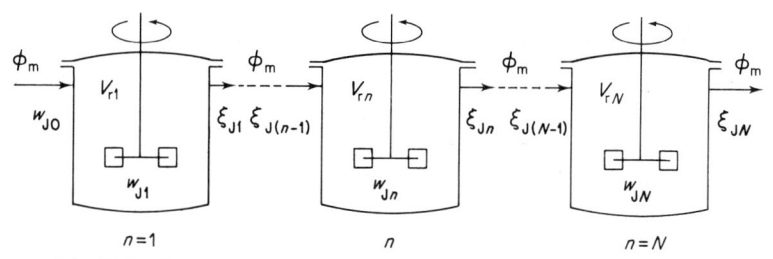

Fig. II.7 Cascade of tank reactors; N = total number of reactors

The steady state material balance (II.21) over the nth reactor gives for a component J:

$$\xi_{Jn} - \xi_{J(n-1)} = \frac{V_{rn}}{\Phi_m} M_J R_{Jn}, \qquad (II.30)$$

where $\xi_{Jn} = |w_{Jn} - w_{J0}|$. Summation over a cascade of N tank reactors yields for the degree of conversion obtained in the entire system:

$$\xi_{JN} = \frac{1}{\Phi_m} \sum_{n=1}^{N} V_{rn} M_J R_{Jn}. \qquad (II.31)$$

In deriving equation (II.31) we have assumed that no material is added to or withdrawn from the system between its inlet and its outlet.

It will be shown that as N approaches infinity (the total volume remaining constant), a cascade of tank reactors becomes identical with the ideal tubular reactor. For this purpose we put equation (II.30) into the form

$$\frac{\Delta \xi_{Jn}}{M_J R_{Jn}} = \frac{V_{rn}}{\Phi_m}.$$

By summation and taking the limit for $N \to \infty$ or $\Delta \to 0$, we get

$$\lim_{\Delta \to 0} \left[\sum_{n=0}^{N} \frac{\Delta \xi_{Jn}}{M_J R_{Jn}} \right] = \int_0^{\xi_{Jn}} \frac{d\xi_J}{M_J R_J} = \lim_{N \to \infty} \left[\frac{1}{\Phi_m} \sum_{n=0}^{N} V_{rn} \right] = \frac{V_r}{\Phi_m},$$

where V_r is the volume of the entire system. The resulting equation is the same as equation (II.15). From this result it may be concluded that the behaviour of a cascade of tank reactors, depending upon the number N of reaction vessels into which the total reaction volume is divided, may range between the behaviour of one CISTR ($N = 1$) and that of a PFR ($N \to \infty$). This is the reason that a cascade of many CISTRs is often applied in the case where a PFR would become extremely long, because the required residence time τ_L for the desired conversion is very large, or if the pressure drop in the tubular reactor would become very large (viscous reaction mixtures). We then imitate the plug flow of a tubular reactor in a cascade of many tank reactors. The behaviour of a tubular reactor with mixing in the direction of the flow is often described by a cascade of N CISTRs (see Chapter IV).

II.4.1 Calculation methods for tank reactors in cascade

Even with relatively simple expressions for the conversion rate, the set of equations describing the degree of conversion in a cascade cannot be solved analytically. One of the few exceptions is the case of a first order rate equation, as will be shown below. We take as an example a reaction $A + B \to$ products in a constant density reaction mixture, for which

$$R_A = k c_A c_B.$$

By introducing the relative degree of conversion ζ_A into equation (II.30), together with the expressions $c_A = c_{Ao}(1 - \zeta_A)$ and $c_B = c_{Bo} - c_{Ao}\zeta_A$ from Table I.6, we obtain

$$\zeta_{An} - \zeta_{A(n-1)} = \frac{V_{rn}\rho_n}{\Phi_m c_{Ao}} R_{An} = c_{Ao} k \tau_n (1 - \zeta_{An})(1 + e - \zeta_{An}), \quad (II.32)$$

where e is the relative excess of B with respect to A; $e = (c_{Bo} - c_{Ao})/c_{Ao}$.

Equation (II.32) can be solved for ζ_{An}, but the form of this solution is rather complicated. The complexity rapidly increases if one makes the next step and tries to express $\zeta_{A(n+1)}$ in terms of $\zeta_{A(n-1)}$. Therefore a numerical solution is indicated in this case, with the help of a computer.

If, however, $e \gg 1$ so that also $e \gg \zeta_A$ (since $0 < \zeta_A < 1$), the factor $(1 + e - \zeta_{An})$ in Equation (II.32) may be regarded as a constant. This means that the reaction has a pseudo first order behaviour with

$$R_A = k c_{Bo} c_A = k_1 c_A.$$

Accordingly, equation (II.32) becomes

$$\zeta_{An} - \zeta_{A(n-1)} = k_1 \tau_n (1 - \zeta_{An})$$

or

$$\frac{c_{An}}{c_{A(n-1)}} = \frac{1 - \zeta_{An}}{1 - \zeta_{A(n-1)}} = \frac{1}{1 + k_1 \tau_n}. \qquad (\text{II}.33)$$

If all reactors have the same value of $\tau_n = \tau/N$, the conversion in N tank reactors in cascade becomes

$$1 - \zeta_{AN} = \frac{c_{AN}}{c_{Ao}} = \left(1 + \frac{k_1 \tau}{N}\right)^{-N}. \qquad (\text{II}.34)$$

This relationship is shown in figure II.8, where $1 - \zeta_{AN}$ has been plotted as a function of the total average residence time τ. Note that for $N \to \infty$ we have, for a finite value of τ

$$\lim_{N \to \infty} (1 - \zeta_{AN}) = \lim_{N \to \infty} \left(1 + \frac{k_1 \tau}{N}\right)^{-N} = e^{-k_1 \tau},$$

which is the conversion in an ideal tubular reactor with a residence time $\tau_L = \tau$.

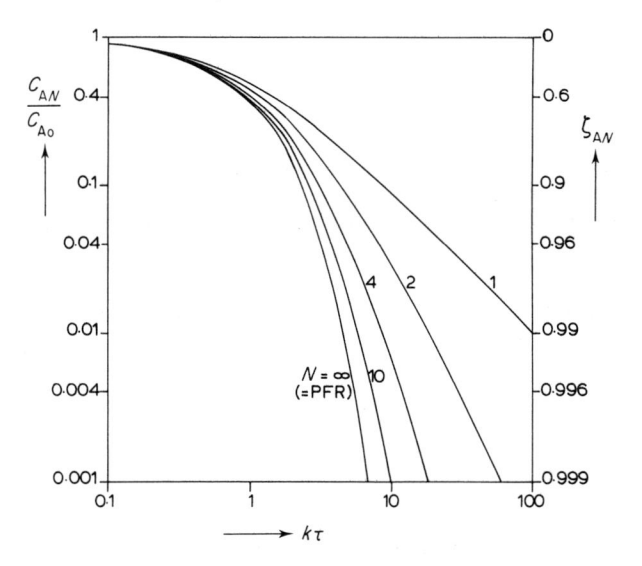

Fig. II.8 Relative degree of conversion ζ_{AN} for an isothermal first order reaction in a cascade of N equal tank reactors

Eldridge and Piret [5] have summarized a great number of design equations for various conversion rate expressions which can be solved either by an algebraic stepwise method or by a graphical method. Their paper contains a number of charts from which the conversion of second order reactions can be read. Jenney [6] has published similar data, of which figure II.9 is given as an example. It shows the relative conversion ζ_{AN} as a function of $kc_{Ao}\tau$ (with τ_n = constant) for $e = 0$ and $e = 0.3$ for the reaction system described by equation (II.32). Here τ is the total residence time in the cascade; in each CISTR the residence time is $\tau_n = \tau/N$, so that the individual tanks are equally sized. A more general graphic method for obtaining the conversion in a cascade is given by several authors, among whom Jones [7] and Weber [8]. It is based on a graphical representation of the conversion rate as a function of the degree of conversion. The conversion rate may be known empirically, e.g. from experiments in a BR or a CISTR.

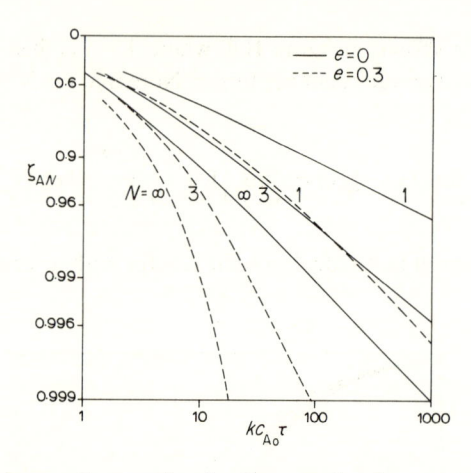

Fig. II.9 Relative degree of conversion ζ_{AN} for second order reactions in a cascade of N equal tank reactors, Jenney [6]; e is the relative excess of B [From Jenney [6]. Excerpted by special permission from *Chemical Engineering* (1955) copyright © (1955) by McGraw-Hill, Inc., New York, N.Y. 10020]

If we have determined in a CISTR the relation between R_A and c_A/c_{Ao} we can empirically plot as in figure II.10 R_A/c_{Ao} versus ζ_A. If we draw a line from ζ_{An} with a slope of

$$\tan \alpha = \frac{1}{\tau_n} = \frac{R_A}{c_{Ao}(\zeta_{A(n+1)} - \zeta_{An})}$$

and repeat this procedure, we can easily obtain the operating conditions in each tank of a cascade. This method has the advantage over the construction applied in Example II.4b, that the average residence times of the various reactors can be varied. It should be borne in mind that both procedures are limited to isothermally operating systems where the conversion rate only depends on the degree of conversion of one of the components.

Example II.4.a *Production of ethyl acetate in a cascade of three tank reactors*
In Example II.1a production of 50 tonnes/day of ester could be obtained by batchwise operation of 3 reaction vessels in parallel having a reaction volume of

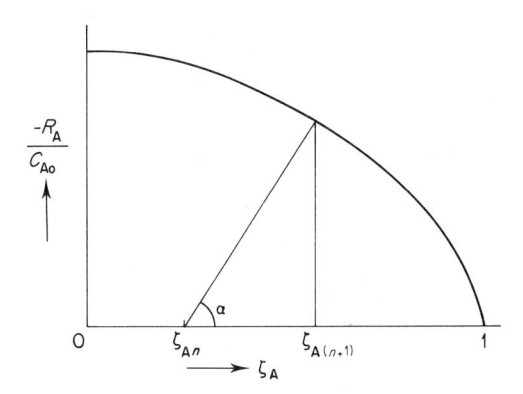

Fig. II.10 Graphical method for determining the operating conditions in a cascade

18 m³ each. What production can be obtained under the same conditions of feed composition and conversion, when the reactor operation is made continuous with the three vessels in cascade?

For the solution of this problem, a certain value of Φ_m has to be premised, whereupon ζ_{J1}, ζ_{J2} and ζ_{J3} must be calculated by means of equation (II.30). If ζ_{J3} differs from the desired value, the calculations have to be repeated with a different value of Φ_m until the desired degree of conversion has been obtained. The result of the calculations for this example is that the mass flow must be $\Phi_m = 6.23$ kg/s for the required degree of conversion; this is equivalent to the production of 63.6 tonnes/day of ethyl acetate, i.e. 27% more than with batchwise operation of these reactor vessels (where one third of the total time was not used for production). For a production of 50 tonnes/day in a cascade of three tank reactors of equal size, a volume of $18 \times 50/63.6 = 14.2$ m³ per vessel, or a total volume of 42.6 m³, would be sufficient.

Example II.4.b *Polymerization in a cascade*
For somewhat more complicated rate expressions, such as used in equation (II.32) for a second order reaction between A and B with excess B, a graphic design method was indicated by Eldridge and Piret [5]. It is, of course, based on the material balance which is now written in the form:

$$\xi_{A(n-1)} = \xi_{An} - \frac{V_{rn}}{\Phi_m} M_A R_{An}.$$

Thus, if $M_A R_{An}$ is a known function of ξ_{An}, and if V_{rn}/Φ_m has been preselected and is taken to be independent of n, the above expression gives a known relationship between $\xi_{A(n-1)}$ and ξ_{An}. This relationship is then plotted with $\xi_{A(n-1)}$ as ordinate and ξ_{An} as abscissa; it is shown as curve 1 in figure II.11. When the auxiliary straight line 2 for which $\xi_{An} = \xi_{A(n-1)}$ is also drawn, the horizontal distance between 1 and 2 is equal to $\xi_{An} - \xi_{A(n-1)}$. A stepwise construction as indicated in

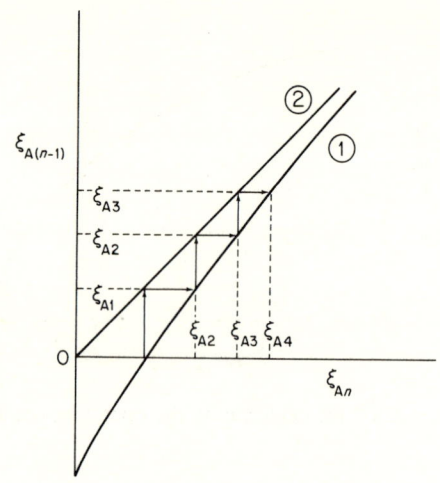

Fig. II.11 Graphic construction for obtaining the conversion in a cascade of equal size tank reactors [From Eldridge and Piret [5]. Reproduced by permission of the American Institute of Chemical Engineers]

figure II.11 can then be used in finding the number of tank reactors required for a given conversion.

In a cascade of tank reactors ($V_{rn} = 26.5\,\text{m}^3$), styrene (A) and butadiene (B) are to be copolymerized at 5 °C to a final degree of conversion $\xi_{An} = 0.0775$. The overall reaction equation is

$$A + 3.2B \rightarrow \text{polymer},$$

and the conversion rate is given as

$$R_A = kc_A c_B$$

with $k = 10^{-5}\,\text{m}^3/\text{kmol s}$. The inlet concentrations are $c_{Ao} = 0.795\,\text{kmol/m}^3$ and $c_{Bo} = 3.55\,\text{kmol/m}^3$, while $M_A = 104\,\text{kg/mol}$ and $M_B = 54\,\text{kg/mol}$. The feed rate is 19.7 tonnes/h and the density of the reaction mixture is constant at a value $\rho = 870\,\text{kg/m}^3$. Find the total number N of tank reactors required.

Since all reactors will have the same given residence time, a construction according to figure II.11 is advisable. We start from the material balance (II.30):

$$\xi_{A(n-1)} = \xi_{An} - \frac{V_{rn} M_A}{\Phi_m} R_{An}$$

From Table I.6 we have for constant density reaction mixtures:

$$c_A = c_{Ao} - \rho \xi_A / M_A \quad \text{and} \quad c_B = c_{Bo} - \rho v_B \xi_A / v_A M_A,$$

so that the material balance becomes

$$\xi_{A(n-1)} = \xi_{An} - \frac{V_{rn} M_A k c_{Ao} c_{Bo}}{\Phi_m}\left(1 - \frac{\rho \xi_{An}}{M_A c_{Ao}}\right)\left(1 - \frac{\rho v_B \xi_{An}}{v_A M_A c_{Bo}}\right)$$

$$= \xi_{An} - 0.0142(1 - 10.52\xi_{An})(1 - 7.54\xi_{An})$$

or
$$\xi_{A(n-1)} = -1.13\xi_{An}^2 + 1.26\xi_{An} - 0.0142.$$

This relation has been plotted in figure II.12 together with the auxiliary line $\xi_{A(n-1)} = \xi_{An}$. From the stepwise construction we find that for obtaining the required value of $\xi_{AN} = 0.0775$ 21 tank reactors in cascade would be needed. It is furthermore seen that in the first reactor $\xi_{A1} = 0.0117$ and in the last $\xi_{A21} - \xi_{A20} = 0.0011$, so that 15% of the total conversion occur in the first reactor and only 1.4% in the last one. Since the reaction is exothermic, this means that much more heat must be removed in the first reactor than in the last.

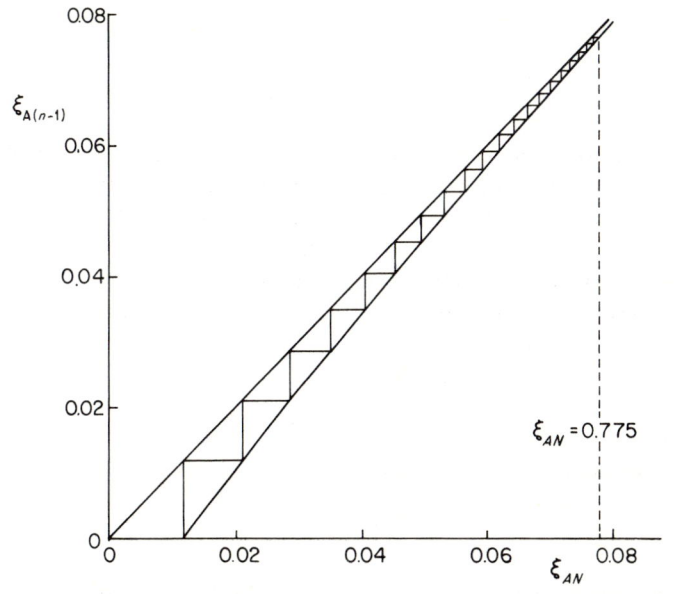

Fig. II.12 Determination of the number of tank reactors for obtaining a degree of conversion $\xi_{AN} = 0.0775$ in a polymerization reaction; Example II.4.b

The high value of N means that this system has practically the same behaviour as an ideal tubular reactor with the same total residence time. However, the required residence time is so long ($N\tau_n = \tau = 25.6$ h) that a cascade of reaction vessels is the only economical way of providing for a large reaction volume.

II.5 THE SEMI-CONTINUOUS TANK REACTOR

A tank reactor with non-steady operation can be regarded as a continuously operated reactor where the incoming and outgoing mass flows are not equal to each other; consequently, the total mass of reaction mixture is not constant. This reactor type can, however, equally well be considered a batchwise operated reactor with a feed of a reactant stream and/or a discharge of a product stream and therefore is also called a semi-batch reactor (SBR).

The method of gradual supply of reactant to a batch reactor is often adopted in practice. It is done, for example, in the case of reactions having a very high heat effect, so as to keep the temperature within certain limits. In many biological fermentation reactions nutritious matter is added at a predetermined rate in order to achieve optimum production. Reaction products are frequently removed from a batch reactor in cases of equilibrium reactions, which serves the purpose of increasing the degree of conversion. The material balance over the semi-continuous reactor for all components combined is given by equation (1.2):

$$\frac{dm}{dt} = \frac{d(\rho V_r)}{dt} = \Phi_{mo} - \Phi_{m1}, \tag{II.35}$$

while it becomes, for any component J,

$$\frac{d(\rho V_r w_{J1})}{dt} = \Phi_{mo} w_{Jo} - \Phi_{m1} w_{J1} + V_r R_{J1} M_J. \tag{II.36}$$

Subscript 1 refers to the conditions prevailing in the reactor and at the reactor outlet. The material balances for the well mixed batch reactor (II.1) and for the continuous tank reactor (II.20) are special cases of equation (II.36). Since Φ_{mo} and Φ_{m1} are as yet unspecified functions of time, a general solution of the latter equation cannot be given. However, in the case of a constant density of the reaction mixture and the feed stream, equation (II.36) reduces to

$$V_r \frac{dc_{J1}}{dt} + c_{J1} \frac{dV_r}{dt} = \Phi_{vo} c_{Jo} - \Phi_{v1} c_{J1} + V_r R_{J1}$$

For a reaction with a first order conversion rate expression $R_A = kc_A$ and for the case only reactant is added ($dV_r/dt = \Phi_{vo}$ and $\Phi_{v1} = 0$), we find

$$V_r \frac{dc_{A1}}{dt} + c_{A1} \Phi_{vo} = \Phi_{vo} c_{Ao} - V_r k c_{A1}. \tag{II.37}$$

The following examples may serve to illustrate this reactor type; equation (II.37) will be solved in Example II.5b.

Example II.5.a A semi-continuous (or semi-batch) experiment
A reaction is to be carried out in a solution which contains a catalytic component. For the determination of the conversion rate, an experimental batch reactor is used which at the start of the experiment ($t = 0$) contains only the solvent and the catalyst; the initial reaction volume is V_{ro}, and the initial density is ρ_o. The reactants are fed continuously to the reactor at a constant mass rate starting at $t = 0$; they are assumed to be mixed very fast with the reactor contents. The composition and the density of the reaction mixture is then determined as a function of time, and it is required to find an expression for the conversion rate from these data.

Since no material is removed, $\Phi_{m1} = 0$ and equation (II.35) becomes

$$\frac{d(\rho V_r)}{dt} = \Phi_{mo} \quad \text{or} \quad \rho V_r = \rho_o V_{ro} + \Phi_{mo} t.$$

The material balance for one of the reactants, A, is

$$\frac{d(\rho V_r w_{A1})}{dt} = \Phi_{mo} w_{Ao} - V_r M_A R_{A1} = \Phi_{mo} w_{A1} + \rho V_r \frac{dw_{A1}}{dt}.$$

After elimination of V_r and rearrangement, we have

$$\frac{M_A R_{A1}}{\rho} = (w_{Ao} - w_{A1})\frac{\Phi_{mo}}{\rho_o V_{ro} + \Phi_{mo} t} - \frac{dw_{A1}}{dt}$$

The first term on the right-hand side represents the rate at which the mass fraction w_{A1} would increase if the reaction were infinitely slow. Therefore R_{A1}/ρ is equal to the difference between the latter rate and the measured increase of w_{A1} per unit time. If the actual *reaction* velocity were extremely high, we would find that w_{A1} remains practically zero throughout the experiment. The amount of reactant *converted* per unit time then becomes equal to the rate at which reactant is added. Under these circumstances, the assumption of uniformity of the reaction mixture fails and the experiment is meaningless.

Example II.5.b *Slow addition of reactant to reduce heat evolution, first order reaction*

A batch reactor is filled with $2\,m^3$ solvent. The reaction rate is given by $R_A = -kc_A$, where $k = 5 \times 10^{-4}\,s^{-1}$. A solution of $1\,kmol\,A/m^3$ is added at a rate of $10^{-3}\,m^3/s$ till the reactor is filled with $4\,m^3$ reaction mixture. As the heat evolution is proportional to the reaction rate, calculate the reaction rate in the semi-batch reactor and compare it with the rate that would apply if the entire reaction mixture were put at once into the reactor, assuming that the reactors are kept isothermal. Differential equation (II.37) can be solved by multiplying all terms by the integrating factor $e^{\int P dt}$, where

$$P = \left(\frac{\Phi_{vo}}{V_r} + k\right) \quad \text{or} \quad \int P\,dt = e^{kt}(V_{ro} + \Phi_{vo} t),$$

because $V_r = V_{ro} + \Phi_{vo} t$. The solution for the concentration is

$$c_{A1} = \frac{\Phi_{vo} c_{Ao}}{k(V_{ro} + \Phi_{vo} t)}(1 - e^{-kt})$$

and the conversion rate is

$$V_r R_A = V_r k c_{A1} = \Phi_{vo} c_{Ao}(1 - e^{-kt}).$$

For a batch reactor $c_A = c_{Ao} e^{-kt}$, and the conversion rate is

$$V_r R_A = V_r k c_{A1} = V_r k c_{Ao} e^{-kt}.$$

Note that in the batch reactor $c_{Ao} = 0.5\,kmol A/m^3$. The solution is given in figure II.13. If a conversion of 99% were required, this would be reached after $\sim 8300\,s$ in the BR and after $9100\,s$ in the SBR. From the figure we see that in this case with the semi-batch mode of operation we have suppressed the maximum heat

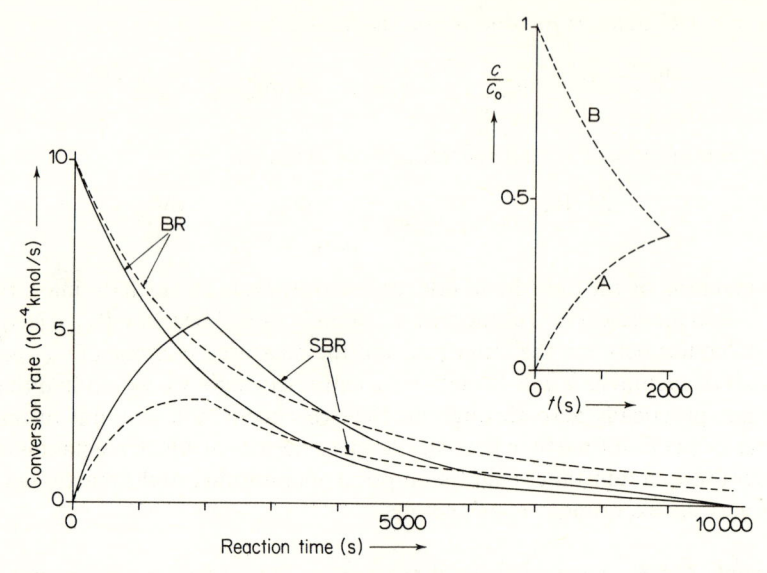

Fig. II.13   Conversion rates in a BR and a SBR; Example II.5.b (full lines) and II.5.d (dotted lines)

evolution to approximately 50 % at the expense of about 10 % additional reaction time. The cooling area in the SBR can consequently also be 50 % smaller than in the BR.

Example II.5.c   *Production of ethyl acetate in a batch reactor with removal of product*

In the system described in Example II.1, part of the product P is to be removed during the reaction in order to increase the overall conversion rate of this equilibrium reaction. The reaction is carried out at 100°C and the contents of the reactor are partly evaporated. The vapour leaving the reactor is concentrated by rectifying distillation and the top product is an azeotropic mixture with the composition $w_{A1} = 0.084$, $w_{P1} = 0.826$ and $w_{Q1} = 0.090$. The liquid hold-up of the distillation column is assumed to be very small with respect to the reaction volume. The rate of net evaporation from the reactor is chosen in such a way that the mass fraction of P in the reaction mixture does not exceed the value of $w_P = 0.02$. To what extent will the production be increased by these measures if the total reaction time is to remain the same as in Example II.1?

With a calculation as shown in Example II.1 we find that after $t = 690$ s the mass fraction of P has reached the value $w_P = 0.02$. The composition of the reaction mixture at this moment is given in Table II.6. At this moment evaporation starts, and with $\Phi_{m1} = $ the rate of removal of the product stream, the material balances are

total: $$\rho \frac{dV_r}{dt} = -\Phi_{m1},$$ (a)

alcohol (A): $\dfrac{d(V_r c_A)}{dt} = \Phi_{m1} \dfrac{w_{A1}}{M_A} - V_r R_A,$ (b)

acid (B): $\dfrac{d(V_r c_B)}{dt} = -V_r R_B,$ (c)

ester (P): $c_P \dfrac{dV_r}{dt} = -\Phi_{m1} \dfrac{w_{P1}}{M_P} + V_r R_P,$ (d)

water (Q): $\dfrac{d(V_r c_Q)}{dt} = -\Phi_{m1} \dfrac{w_{Q1}}{M_Q} + V_r R_Q.$ (e)

Table II.6 Concentration time course in the semi-batch reactor of Example II.5.c

t (s)	0	690	4000	7270	7270†
c_A (kmol/m³)	10.20	9.97	9.83	9.76	8.83
c_B (kmol/m³)	3.91	3.68	3.17	2.68	2.54
c_P (kmol/m³)	0	0.23	0.23	0.23	1.37
c_Q (kmol/m³)	17.6	17.8	19.9	21.6	18.9
V_r (m³)	52.0	52.0	47.2	44.2	52.0

† Refers to the final conditions in the batch reactor of Example II.1.

From the reaction equation we have: $R_A = R_B = R_P = R_Q$; ρ, c_P, w_{A1}, w_{P1} and w_{Q1} are constant. The sum of the four balances for the components is the total material balance. From (a) and (d) we find that

$$\dfrac{dV_r}{dt} = \dfrac{V_r R_A}{c_P - \rho w_{P1}/M_P}.$$

By combining (b), (c) and (e) with (a) and (f), respectively, we get:

$$\dfrac{dc_A}{dt} = -R_A \left[1 + \dfrac{c_A - \rho w_{A1}/M_A}{c_P - \rho w_{P1}/M_P} \right],$$

$$\dfrac{dc_B}{dt} = -R_A \left[1 + \dfrac{c_B}{c_P - \rho w_{P1}/M_P} \right],$$

and

$$\dfrac{dc_Q}{dt} = R_A \left[1 - \dfrac{c_Q - \rho w_{Q1}/M_Q}{c_P - \rho w_{P1}/M_P} \right].$$

In the last four equations, V_r, c_A, c_B and c_Q vary with time as well as R_A which is a given function of the composition (see Example II.1). The equations have to be solved by trial and error; some of the results are shown in Table II.6.

Because of the combined effect of the reaction, the reduction of the reaction volume and the partial removal of A, P and Q, the concentrations of A and B decrease more slowly and the concentration of Q increases more quickly than in the batch reactor without product removal. The amount of P produced in 7270 s in this example is:

$$\frac{\rho w_{P1}}{M_P}[(V_r)_o - V_r] + V_r c_P = 74.6 + 10.2 = 84.8 \text{ kmol}.$$

In Example II.1, $52 \times 1.37 = 71.2$ kmol of P were produced per batch in the same time, so that the capacity has been increased by 19%. The effective cycle time for one batch was 3 hours, so that in one day $8 \times 84.8 \times 0.088 = 59.7$ tonnes esters are produced. For a production of 50 tonnes ester/day in an SBR a reactor volume of $50/59.7 \times 52 = 43 \text{ m}^3$ would be required.

Example II.5.d *Slow addition of reactant to reduce heat evolution, second order reaction*

The same reaction as under Example II.5b is executed, but it now exhibits second order kinetics for a reaction

$$v_A A + v_B B \rightarrow v_P P + \cdots \text{ with } R_A = k c_A c_B \text{ and } k = 5 \times 10^{-4} \text{ m}^3/\text{kmol s}.$$

Furthermore, $v_A = v_B$. The reactor is either

(a) filled at the start with 4 m³ reaction mixture containing $c_{Ao} = c_{Bo} = 0.5$ kmol/m³; or
(b) filled with 2 m³ reaction mixture containing B with $c_{Bo} = 1$ kmol/m³ and A is added with a feed rate of 10^{-3} m³/s containing a concentration $c_{Ao} = 1$ kmol/m³. Addition stops when a stoichiometric quantity of A has been supplied.

Solution for case (a)

The reaction rate for component B is $R_B = k c_A c_B$ and because $c_{Ao}/v_A = c_{Bo}/v_B$ we have, after introducing a relative conversion ζ_B for the reaction rate, $R_B = k c_{Bo}(1 - \zeta_B)^2$. The material balance is

$$-c_{Bo}\frac{d(1-\zeta_B)}{dt} = k c_{Bo}^2 (1 - \zeta_B)^2$$

with the initial conditions $t = 0$ and $\zeta_B = 0$. The solution after a time t is

$$1 - \zeta_B = \frac{1}{1 + k c_{Bo} t}.$$

So the reaction rate at the moment t is

$$R_B = \frac{k c_{Ao} c_{Bo} V_r}{(1 + k c_{Bo} t)^2}.$$

This relation is also plotted in figure II.13.

Solution for case (b)

We first determine the volume V_r of the reaction mixture and the concentrations of A and B after a time t. The reaction mixture volume is then

$$V_r = V_{ro}\left(1 + \frac{\Phi_{vo}t}{V_{ro}}\right) \qquad (a)$$

The amount of B charged is $V_{ro}c_{Bo}$; the amount still present after a time t is

$$V_{ro}\left(1 + \frac{\Phi_{vo}t}{V_{ro}}\right)c_B. \qquad (b)$$

The amount of A present in the reaction mixture after a time t equals the amount charged minus the amount consumed by the reaction, so that

$$V_{ro}\left(1 + \frac{\Phi_{vo}t}{V_{ro}}\right)c_A = \Phi_{vo}c_{Ao}t - \frac{\nu_A}{\nu_B}V_{ro}\left[c_{Bo} - \left(1 + \frac{\Phi_{vo}t}{V_{ro}}\right)c_B\right]. \qquad (c)$$

We now can define a relative conversion of B as

$$\zeta_B = \frac{V_{ro}c_{Bo} - (V_{ro} + \Phi_{vo}t)c_B}{V_{ro}c_{Bo}} = 1 - \left(1 + \frac{\Phi_{vo}t}{V_{ro}}\right)\frac{c_B}{c_{Bo}}. \qquad (d)$$

Moreover a stoichiometric amount of A has been charged as soon as $\Phi_{vo}c_{Ao}t/\nu_A = V_{ro}c_{Bo}/\nu_B$. This leads us to introducing a dimensionless time θ, defined as:

$$\theta = \frac{\nu_B c_{Ao}\phi_{vo}}{\nu_A c_{Bo}V_{ro}}t, \qquad (e)$$

so that stoichiometric dosage is reached as soon as $\theta = 1$. The ratio of the volume added to the volume originally present in the reactor is ε, defined as

$$\varepsilon = \left(\frac{\phi_{vo}t}{V_r}\right)_{\theta=1}, \text{ which is also } = \left(\frac{\nu_A c_{Bo}}{\nu_B c_{Ao}}\right) \qquad (f)$$

after a stoichiometric quantity of A has been added. We will introduce the dimensionless quantities ζ_B, θ and ε in equations (a), (c) and (d), and after rearranging find that

$$V_r = V_{ro}(1 + \varepsilon\theta)$$
$$c_A/c_{Ao} = \varepsilon(\theta - \zeta_B)/(1 + \varepsilon\theta)$$
$$c_B/c_{Bo} = (1 - \zeta_B)/(1 + \varepsilon\theta).$$

For component B the material balance equation (II.36) becomes, for this case,

$$\frac{-d(c_B V_r)}{dt} = kc_A c_B V_r.$$

Introducing the dimensionless variables ζ_B, θ and ε, we obtain

$$\frac{d\zeta_B}{d\theta} = KT\frac{(1 - \zeta_B)(\theta - \zeta_B)}{1 + \varepsilon\theta} \qquad (g)$$

in which $KT = kc_{Bo}\varepsilon V_{ro}/\Phi_{vo}$. Hugo [9] solved this differential equation by introducing a new variable $y = (1 + \varepsilon\theta)/(1 - \zeta_B)$ which gives a linear differential equation:

$$\frac{dy}{d\theta} + \frac{KT(1-\theta) - \varepsilon}{1 + \varepsilon\theta} y = KT. \tag{h}$$

The general solution of this equation is

$$\frac{1+\varepsilon\theta}{1-\zeta_B} = y = \frac{1 + KT \int_0^\theta \left[(1+\varepsilon x)^{\frac{1+\varepsilon-\varepsilon^2/KT}{\varepsilon}} e^{-x} \right]^{KT/\varepsilon} dx}{[(1+\varepsilon\theta)^{(1+\varepsilon-\varepsilon^2/KT)\varepsilon} e^{-\theta}]^{KT/\varepsilon}}. \tag{i}$$

This equation can only be solved analytically for specific combinations of values of ε and KT, otherwise a numerical method has to be applied. In case the stream added during the reaction has a relatively high concentration of A, so that $\varepsilon \approx 0$, we can find

$$\zeta_B = 1 - \frac{e^{KT\theta(2-\theta)/2}}{1 + \sqrt{(\pi KT/2)}e^{KT/2}[\mathrm{erf}\sqrt{(KT/2)} - \mathrm{erf}(1-\theta)\sqrt{(KT/2)}]}$$

For $KT \geqslant 6$ this leads at $\theta = 1$ to

$$\zeta_B\bigg|_{\theta=1} = 1 - \sqrt{\left(\frac{2}{\pi KT}\right)}.$$

In our case

$$KT = \frac{5 \times 10^{-4} \times 1 \times 1 \times 2}{10^{-3}} = 1 \quad \text{and} \quad \varepsilon = 1.$$

The reaction rate at $t = 0$ is 0, of course. Equation (i) has been solved numerically and the reaction rate as a function of time has been plotted also in figure II.13. This reaction rate is given by

$$kc_A c_B V_r = kc_{Ao} c_{Bo} V_{ro} \frac{\varepsilon(\theta - \zeta_B)(1 - \zeta_B)}{(1 + \varepsilon\theta)}$$

After the addition of the total stoichiometric amount of A, the reaction continues, of course, as an isothermal second order reaction. At the start of this second period the concentrations of A and B are $c_A = c_B = 0.388\,\mathrm{kmol/m^3}$. So for the second period the volume is $V_{ro}(1 + \varepsilon\theta) = 2V_{ro} = 4\,\mathrm{m^3}$ and $kc_{Bo} = 0.388 \times 5 \times 10^{-4} = 194 \times 10^{-6}\,\mathrm{s^{-1}}$. Also plotted in figure II.13 are the concentrations of A and B during the addition period. We see that the maximum heat evolution by this SBR operation is reduced to 30% of that in a BR.

II.6 THE RECYCLE REACTOR

A recycle reactor is an ideal tubular reactor of which part of the product stream is recycled to the reactor inlet. This mode of operation is often applied with autocatalytic reactions, which are self accelerating, such as many fermentation reactions. Some (intermediate) reaction products are necessary to start the reaction and therefore the feed stream is mixed with a part of the reactor product (see figure II.14).

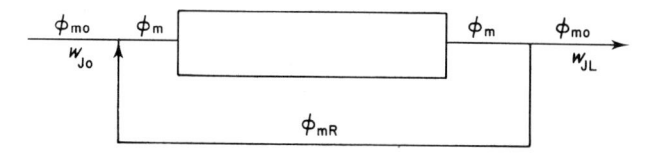

Fig. II.14 Recycle reactor

For the flow sheet in figure II.14 the flow through the reactor is

$$\Phi_m = \Phi_{mo} + \Phi_{mR} = \Phi_{mo}(1 + R), \qquad (II.38)$$

where R is the recycle ratio Φ_{mR}/Φ_{mo}. If the fresh feed Φ_{mo} contains a weight fraction w_{Jo} and the reactor product a weight fraction w_{JL} of component J, then the total reactor feed contains a weight fraction w_{Jin}, given by

$$w_{Jin} = \frac{\Phi_{mo} w_{Jo} + \Phi_{mR} w_{JL}}{\Phi_m} = \frac{w_{Jo} + R w_{J1}}{1 + R}. \qquad (II.39)$$

The degree of conversion at the reactor inlet is already $|w_{Jo} - w_{Jin}|$ due to the mixing of fresh feed with product. If the degree of conversion in the reactor section is $\xi_{JL} = |w_{Jo} - w_{JL}|$, then the degree of conversion at the reactor inlet is

$$\xi_{Jin} = |w_{Jo} - w_{Jin}| = \frac{R}{1 + R} \xi_{JL}. \qquad (II.40)$$

Introducing equations (II.38) and (II.40) into equation (II.15), we find for the reactor volume

$$V_r = \Phi_{mo}(1 + R) \int_{\frac{R}{R+1} \xi_{JL}}^{\xi_{JL}} \frac{d\xi_J}{M_J R_J}. \qquad (II.41)$$

If R approaches 0, equation (II.41) reduces to the normal equation (II.15) for the PFR; if R approaches infinity equation (II.41) reduces to $V_r = \Phi_m \xi_{J1}/M_J R_{J1}$, the material balance for the CISTR. As will be discussed in Chapter IV, the recycle PFR can be used as a concept for describing the mixing in a reactor, with R as a mixing parameter. For a first order reaction rate equation and constant density we derive from (II.41), for a reactant A,

$$V_r = \Phi_{vo}(1 + R) \int_{\frac{R}{R+1} \zeta_{AL}}^{\zeta_{AL}} \frac{d\zeta_A}{k(1 - \zeta_A)}.$$

Integration leads, for the reactor volume, to

$$V_r = \frac{\Phi_{vo}(1+R)}{k} \ln\left[\frac{R+1-R\zeta_{AL}}{(R+1)(1-\zeta_{AL})}\right], \qquad (II.42)$$

and rearranging gives, for the concentration of A in the reactor outlet,

$$\frac{c_{AL}}{c_{Ao}} = \frac{1}{(1+R)e^{kV_r/\Phi_{vo}(1+R)} - R} \qquad (II.43)$$

Rearranging equation (II.42) gives

$$e^{-kV_r/\Phi_{vo}(1+R)} = \frac{(R+1)(1-\zeta_{AL})}{1+R(1-\zeta_{AL})}.$$

For $R = 0$ this equation gives $1 - \zeta_{AL} = e^{-k\tau_L}$ as we could expect for a PFR without recycle (equation (II.18)). For almost complete recycle R approaches infinity. This extreme can be found by rewriting (II.42):

$$\frac{kV_r}{\Phi_{vo}(1+R)} = \ln\left[1 + \frac{\zeta_{AL}}{(1+R)(1-\zeta_{AL})}\right].$$

For $R \to \infty$ we can derive $\ln(1+x) \simeq x$, so that the right hand term becomes

$$\frac{\zeta_{AL}}{(1+R)(1-\zeta_{AL})} = \frac{kV_r}{(1+R)\Phi_{vo}},$$

which reduces to

$$\zeta_{AL} = \frac{k\tau}{1+k\tau}. \qquad (II.28)$$

This relation holds for a CISTR.

Example II.6 *Autocatalytic reaction in various model reactors*

The reaction $A \to P$ is autocatalytic and proceeds at constant density according to $A + P \to P + P$ with a conversion rate given by $R_A = kc_A c_P$, where $k = 10^{-3}$/kmol s. With a feed rate of $\Phi_{vo} = 0.002 \, m^3/s$, $c_{Ao} = 2 \, kmol/m^3$ and $c_{Po} = 0$, determine the reaction volumes for a conversion of 98% in:

(a) a CISTR;
(b) a cascade of two equally sized CISTRs;
(c) a PFR;
(d) a recycle reactor with $R = 1.0$;
(e) the recycle reactor with minimum volume.

The conversion rate equation, expressed in the relative degree of conversion with $c_A = c_{Ao}(1-\zeta_A)$ and $c_P = c_{Ao}\zeta_A$, is

$$R_A = -kc_{Ao}^2(1-\zeta_A)\zeta_A.$$

So we find for the various reactor types

(a) For the CISTR equation (II.21) leads to

$$V_r = \frac{-\Phi_{vo}(c_{Ao} - c_{A1})}{-kc_{A1}c_{P1}} = \frac{\Phi_{vo}}{kc_{Ao}(1 - \zeta_{A1})}.$$

We find for the volume

$$V_r = \frac{0.002}{0.001 \times 2 \times (1 - 0.98)} = 50 \, m^3.$$

(b) For a cascade of two equally sized CISTRs we can derive

$$\frac{kc_{Ao}V_r}{\Phi_{vo}} = \frac{c_{Ao} - c_{A1}}{kc_{A1}c_{P1}} = \frac{c_{A1} - c_{A2}}{kc_{A2}c_{P2}} = \frac{1}{1 - \zeta_{A1}} = \frac{\zeta_{A2} - \zeta_{A1}}{(1 - \zeta_{A2})\zeta_{A2}}.$$

From this relation the conversion in the first reactor can be determined, because $\zeta_{A2} = 0.98$ is preset. The result is $\zeta_{A1} = 0.8496$. The total volume of the cascade is

$$V_{tot} = 2V_{r1} = 2\frac{\Phi_{vo}}{kc_{Ao}(1 - \zeta_{A1})} = \frac{2 \times 0.002}{10^{-3} \times 2 \times (1 - 0.8496)} = 13.3 \, m^3.$$

(c) In a PFR at the reactor inlet $R_A = 0$ because $c_{Po} = 0$, so that the reaction cannot start. As a consequence for this reactor type $V_r = \infty$.

(d) For a recycle PFR we can derive from equation (II.41) that

$$V_r = \frac{\Phi_{vo}(1 + R)}{kc_{Ao}} \int_{\frac{R}{1+R}\zeta_{AL}}^{\zeta_{AL}} \frac{d\zeta_A}{\zeta_A(1 - \zeta_A)}.$$

We know that

$$\int \frac{d\zeta_A}{\zeta_A(1 - \zeta_A)} = \int \frac{d\zeta_A}{\zeta_A} + \int \frac{d\zeta_A}{1 - \zeta_A} = \ln\left[\frac{\zeta_A}{1 - \zeta_A}\right] + C.$$

Introducing the boundary values gives, for the recycle PFR volume,

$$V_r = \frac{\Phi_{vo}(1 + R)}{kc_{Ao}} \ln\left|\frac{1 + R(1 - \zeta_{AL})}{R(1 - \zeta_{AL})}\right|. \tag{a}$$

For $\zeta_{AL} = 0.98$ and $R = 1.0$ we find for the recycle reactor that $V_r = 7.86 \, m^3$.

(e) For the minimum reaction volume of a recycle reactor dV_r/dR must be zero; differentiation of equation (a) leads to the condition

$$\ln\left[\frac{1 + R(1 - \zeta_{AL})}{R(1 - \zeta_{AL})}\right] = \frac{R + 1}{R[1 + R(1 - \zeta_{AL})]} \quad \text{at minimum reactor volume.}$$

By trial and error we find for $\zeta_{AL} = 0.98$ a value of $R_{opt} = 0.226$. This results in $V_r = 6.62 \, \text{m}^3$. In recycle reactors in general the volumes lie in between those for the PFR and the CISTR; however, for autocatalytic reactions a recycle reactor is a must.

II.7 A COMPARISON BETWEEN THE DIFFERENT MODEL REACTORS

The reactor models discussed in the preceding sections are the well mixed batch reactor (BR), the continuously operated ideal tubular reactor (PFR), the continuously operated ideally stirred tank reactor (CISTR) and the cascade of tank reactors. The semi-continuous batch reactor (SBR) and the recycle PFR are not essentially different from the BR and the PFR, respectively. A comparison between the four basic reactor types mentioned above can be divided into two parts:

(i) a comparison between batch and continuous operation, and
(ii) a comparison between the PFR and the CISTR.

II.7.1 Batch versus continuous operation

In the design of a reactor for chemical production, the question must in principle be raised as to whether the process should be operated batchwise or continuously. Under the influence of the bulk production in the large chemical industries, there is a strong trend towards making new processes continuous. There are, however, many processes for which continuous operation would be unprofitable or even impossible. General rules for the selection of either way of operation cannot be given, but some indications are summarized below.

Batch reactors are often used for small production rates and long reaction times. They are flexible and the reaction conditions can be adjusted; they are useful in plants where they serve for the production of various different chemicals (e.g. in the pharmaceutical industry). Batch operation is often preferred for reactions where rapid fouling occurs or contamination is feared (e.g. in biological fermentations). The investment costs of a batch reactor including auxiliary equipment generally are relatively low; on the other hand, relatively extensive manual operation and supervision are required, while automation is often difficult and costly. Continuously operated reactors are capable of turning out a product of more constant quality; they require little supervision and they are well suited for the application of automatic control. Continuous reactors are especially applied at high production rates; gas phase reactions are almost exclusively carried out in a continuous process (the reciprocating internal combustion engine is one of the exceptions, but its primary purpose is to produce mechanical power and not chemicals). An important factor in the decision of whether a batch or continuous process is to be preferred is the economic flexibility of the installation. For purposes of illustration, the economic diagrams are shown schematically for a certain process in figure II.15(a and b), for batch

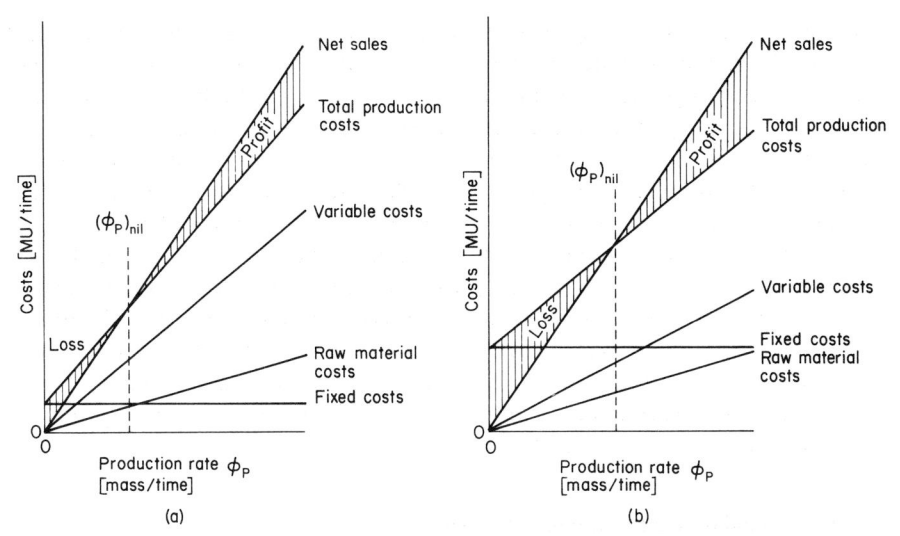

Fig. II.15 Economic diagrams typical of batch operation (a) low absolute production rate, and of continuous operation (b) high absolute production rate

and continuous operation, respectively. These diagrams give the production costs in MU (monetary units) per unit time as a function of the production rate Φ_P in an existing plant. The various lines are explained below.

(a) *The fixed costs* (depreciation, maintenance, supervision, etc.) are independent of Φ_P. They are mostly expressed as a certain percentage of these total investment costs, and they are relatively higher for continuous operation. Wages of operating personnel for the continuous plant are included in the fixed costs since it is assumed that the number of operators in this case does not depend on the production level.

(b) *The variable costs* (electric power, steam, cooling water and other utilities, laboratory service, etc.) are approximately proportional to the production rate. They are generally much higher for a batch process because of starting and stopping operations, alternate heating and cooling of the installation and the difficulty of heat recovery by heat exchange; the wages of operating personnal are included in the variable costs for the batch process.

(c) *The raw material costs* are assumed to be proportional to the production rate and equal for the two ways of operation.

(d) *The total production costs* are the sum of (a), (b) and (c).

(e) *The net sales* of the product(s) (= sales revenues − sales expenses) are assumed to be the same in both cases. Sales expenses are entirely variable if sales are effected through distributors, who work on a commission basis. Sales expenses include commissions, freight, duties, etc. If, however, a large own sales force is employed by the company, their costs should be included in the fixed costs. In the diagrams of figure II.15 the difference between (e) and (d) is equal to the loss or profit of the plant. The intersection of the lines (e) and (d) determines the break-even point $(\Phi_P)_{nil}$ below which the process is no longer profitable. Figure II.15(a and b) indicates that $(\Phi_P)_{nil}$ is smallest for batch operation, which means that it has greater economic flexibility. In other words, the economic result of a manufacturing process is less sensitive to an increase in production and raw materials costs, and to a decrease of the market value of its products, as its economic flexibility becomes greater. On the other hand, with a relatively high value of Φ_P the profit is largest with continuous operation. The economic flexibility, of course, depends very much on the market prices for raw materials and products. For large, continuously operated units for bulk chemicals $\Phi_{Pnil} \approx 0.50 - 0.85\Phi_{Pmax}$; in depressed markets with low sales prices the break even point is often even higher. In small capacity, batch operated units $\Phi_{Pnil} \approx 0.30 - 0.50\Phi_{Pmax}$. In view of the economic flexibility therefore batch processing is often preferred for small scale production, specially if more products can be made in the same plant.

II.7.2 Tubular reactor versus tank reactor

With regard to reactor capacity, the composition of the reaction mixture in a CISTR is equal to the outlet composition; it has been already demonstrated in the preceding sections that, for this reason, the capacity of a tank reactor is always smaller than that of a tubular reactor if the reaction volume and the conditions are identical in both cases, except for autocatalytic reactions and reactions of the order $n < 0$. This is also demonstrated in figure II.18 which shows the ratio between the volume of a PFR and that of a CISTR needed to obtain a certain relative degree of conversion for different order rate expressions. It is seen that the production capacity of a tank reactor is relatively small, especially at high reaction orders and high degrees of conversion, because the conversion rates in a CISTR are relatively low due to the low reactant concentrations. This can also be demonstrated in another way. For reactions of the nth order with a rate equation $R_A = -kc_A^n$ the required reactor volume for constant density reactions is as follows:

for a PFR,

$$\left(\frac{V_r}{\Phi_{vo}}\right)_{PFR} = \int_{c_{Ao}}^{c_A} \frac{dc_A}{-kc_A^n};$$

and for a CISTR,

$$\left(\frac{V_r}{\Phi_{vo}}\right)_{CISTR} = \frac{c_A - c_{Ao}}{-kc_A^n}.$$

If we plot $1/R_A$ or $-1/kc_A^n$ versus c_A as is done in figure II.16, the reactor volume of a PFR is given by the shaded area below the curve, whereas that of a CISTR is given by the area of the rectangle with the ordinate of $1/R_A$ at c_A and the abscissa $c_{Ao} - c_A$.

In figure II.16 it can be seen that in that particular case the CISTR requires a much larger reactor volume than the PFR for the same degree of conversion. In figure II.17 some representative curves for different reaction orders are given.

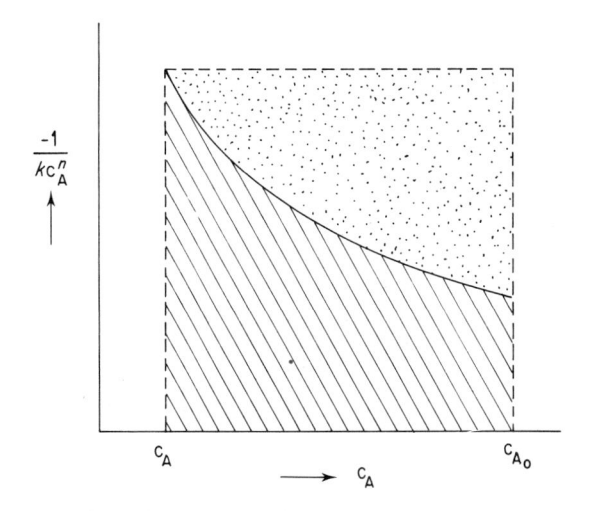

Fig. II.16 Graphical determination of the reactor volume for simple reactions

From this graph it can be concluded that for the same conversion for reactions with an order of $n > 0$ PFR's have the highest capacity, with $n = 0$ the PFR and CISTR have equal capacities and for $n < 0$ the CISTR has the highest capacity.

If a long residence time of the reaction mixture is required, a reactor consisting of a tube or tube bundle has technical and economical disadvantages. In such a case a cascade of tank reactors should be used instead. A cascade of only three reactors is in many instances considered a fair approximation of the tubular

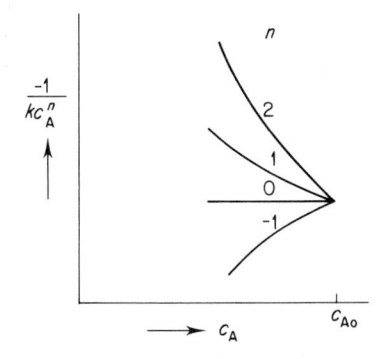

Fig. II.17 Reciprocal reaction rates versus reactant concentration for various reaction orders

reactor (see also figure II.18), although in practice cascades of 10 to 20 reactors are encountered for special purposes.

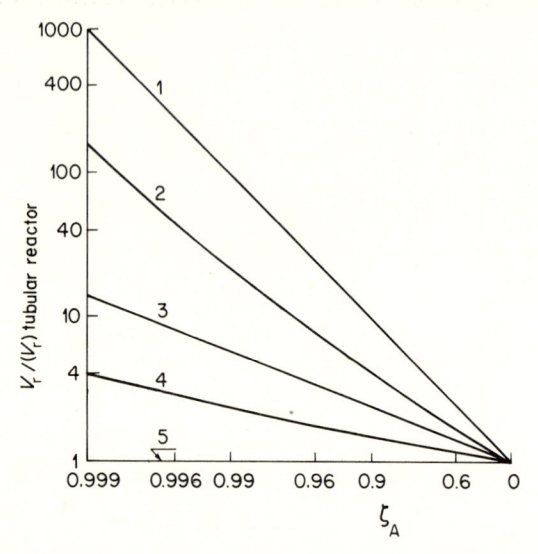

Fig. II.18 Comparison of required reactor volumes for isothermal reactions of different order as a function of ζ_A: second (1), first (2) and half (3) order reaction in one CISTR; first order reaction in cascade of these CISTR's (4); zero order reaction (5)

Moreover, we have seen that in the case of autocatalytic reactions a CISTR has a much lower capacity than a recycle reactor. For autocatalytic reactions part of the reactor product has to be recycled to the feed to start the reaction. In the case for a batch reactor part of the reactor product may be retained in the reaction vessel so that the self accelerating reaction can be started after filling the reactor with fresh reactant.

In conclusion, it should be mentioned that we have used the concepts of tubular reactor and tank reactor in an abstract sense. They do not necessarily have the shape of a tube or a tank, respectively, although in practice this is usually so. More essential, however, is the fact that a continuously operated reactor—in which mixing in the direction of main flow is negligible—has the characteristics of a PFR. If, on the other hand, mixing in the direction of flow is intensive, the properties of the CISTR are approached. In this respect, the distribution of residence times is a characteristic feature of continuously operated reactors; this subject will be treated in Chapter IV.

Example II.7 *Comparison of various model reactors*
In various sections of this chapter we have illustrated the production of ethyl acetate from ethyl alcohol and acetic acid in different types of model reactors under equal conditions of temperature and reactor feed composition. We summarize the results in Table II.7. It is seen that the PFR has the greatest, and one CISTR the smallest capacity.

Table II.7 The esterification reaction in different model reactors

Illustration	Type of operation and reactor	ester production rate (tonnes/day)	reaction volume (m^3)
II.2a	PFR	50	35
II.5c	SBR	50	42
II.4a	cascade of 3 CISTR's	50	43
II.1	BR	50	52
II.3	CISTR	50	61
II.3	CISTR	43	52
II.1	BR	50	52
II.5c	SBR	60	52
II.4a	cascade of 3 CISTR's	64	52
II.2a	PFR	75	52

II.8 SOME EXAMPLES OF THE INFLUENCE OF REACTOR DESIGN AND OPERATION ON THE ECONOMICS OF THE PROCESS

A detailed discussion of the economics of a chemical plant is beyond the scope of this book. As far as economical operation of a reactor is concerned, it will generally be sound policy to waste as little raw material and product as possible, especially when they are valuable, and to design and operate a reactor system the fixed and variable costs of which are reasonably low.

In the preceding sections we have seen that in the case of a simple reactor design the reaction conditions (such as temperature, pressure and concentration) and the reactor type already permit a certain amount of variation which will eventually influence the profitability of the installation. Apart from these there are a number of possibilities of variation which can be used either in the design stage or during operation in order to improve or adjust the plant economics. In this section mention will be made of only three of these possibilities which are frequently encountered in practice:

(i) the use of one of the reactants in excess;
(ii) the recirculation of unconverted reactant;
(iii) the policy of operation under non-steady process conditions.

II.8.1 The use of one of the reactants in excess

From the point of view of *capacity* it is always advantageous to feed the reactants in their stoichiometric ratio and at maximum concentrations. If, however, one of the reactants is relatively expensive, it is sometimes rewarding to use the other, less expensive, reactant in excess so as to obtain a higher *yield* with respect to the former reactant. This is especially true if the recovery of the more expensive reactant from the product stream is difficult.

In practice, reactants frequently must be diluted, e.g. in order to suppress undesired side reactions (see Chapter III) or temperature changes occasioned by a large heat of reaction and a limited heat exchange capacity. Supposing the more expensive reactant, A, to have a fixed feed concentration c_{Ao}, the use of the other reactant, B, in excess allows the reactor to be made smaller at the cost of extra expenses for B.

By way of illustration, let us consider the simple reaction $A + B \rightarrow P$ being carried out in a continuous tank reactor (figure II.19). We assume the density of the reaction mixture to be constant and the conversion rate to be determined by $R_A = k c_A c_B$. Using the relative excess of B with respect to A (e, as defined in Section II.4), the concentration at the inlet and at the outlet of the reactor can be expressed in terms of c_{Ao}, e and the relative degree of conversion ζ_{A1}. According to equation (II.26), the reaction volume is

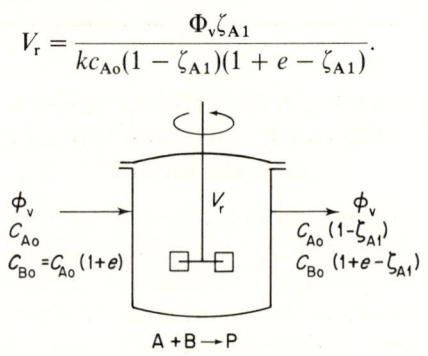

$$V_r = \frac{\Phi_v \zeta_{A1}}{k c_{Ao}(1 - \zeta_{A1})(1 + e - \zeta_{A1})}.$$

Fig. II.19 CISTR with relative excess e of reactant B

At constant values of Φ_v, k, c_{Ao} and ζ_{A1}, the reaction volume decreases as e increases. At the same time, however, the costs connected with the extra feed of B rise. Hence, there is the possibility of an optimum value of e where the total production costs pass through a minimum as a function of e.

A slightly different problem is encountered if V_r, Φ_v and c_{Ao} are given quantities and an increase in conversion is considered by using B in excess. In that case, the gain obtained by a greater degree of conversion of A is to be balanced against the expense for B.

Excess of one of the reactants is also applied in reactions where the equilibrium does not lie entirely on the product side. This likewise results in a higher degree of conversion of a possibly expensive reactant. If this reactant can be separated from the product, its recirculation to the reactor has the same effect.

II.8.2 Recirculation of unconverted reactant

As an example let us consider an equilibrium $A \rightarrow P$ carried out in a tubular reactor (cf. figure II.20). The product stream leaving the reactor is fed to a separation unit (e.g. a distillation column) where A and P are separated practically completely. Φ_m is the mass flow of fresh feed consisting only of A

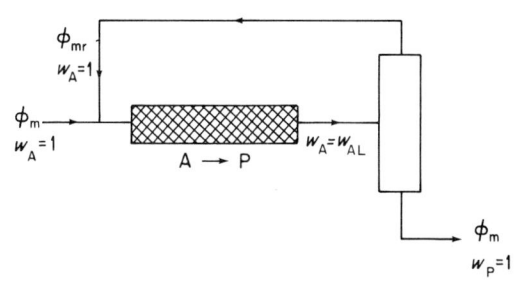

Fig. II.20 PFR with recirculation of non-converted reactant A

($w_A = 1$); the degree of conversion of A in the reactor is ξ_{AL} ($= 1 - w_{AL}$), and Φ_{mR} is the mass flow which is recirculated. From a material balance for A over the separation unit we have

$$\Phi_{mR} w_{A\,top} + \Phi_m w_{A\,bottom} = \Phi_{mR} = (\Phi_m + \Phi_{mR})(1 - \xi_{AL})$$

because we assumed $w_{A\,bottom} = 0$ and $w_{A\,top} = 1$. The reaction volume is found from equation (II.10):

$$V_r = (\Phi_m + \Phi_{mR}) \int_0^{\xi_{AL}} \frac{d\xi_A}{M_A R_A},$$

and the recycle ratio is $\Phi_{mR}/\Phi_m = (1 - \xi_{AL})/\xi_{AL}$. Suppose the conversion rate can be approximated by

$$M_A R_A = k\rho(w_A - w_P/K) = k\rho\left[1 - \frac{K+1}{K}\xi_A\right]$$

where K is the equilibrium constant of the reaction and the product $k\rho$ is constant. The last term follows from the implicit assumptions that $w_{Ao} = 1$, $v_A = v_P$ and $M_A = M_P$, moreover in this case $w_{Ao} - w_A = \xi_A$ and $w_P = \xi_A$. With the use of the last three equations we then find for the reaction volume:

$$\frac{\rho V_r k}{\Phi_m} = \frac{-K}{(K+1)\xi_{AL}} \ln\left(1 - \frac{K+1}{K}\xi_{AL}\right).$$

This quantity has been plotted in figure II.21 as a function of ξ_{AL} for $K = 1$ and for $K \gg 1$, together with the recirculation ratio Φ_{mR}/Φ_m. Since the total production costs will rise when $\rho V_r k/\Phi_m$ and Φ_{mR}/Φ_m increase, it is clear that an optimum value of ξ_{AL} exists. The position of this economic optimum can, of course, be determined only on the basis of detailed cost calculations.

II.8.3 Maximum production rate and optimum load with intermittent operation

The operation of batch reactors or of continuous catalytic processes in which the activity of the catalyst decreases as a function of time frequently poses the following two problems:

(a) What reaction or production time should be selected for obtaining maximum production with the installation?

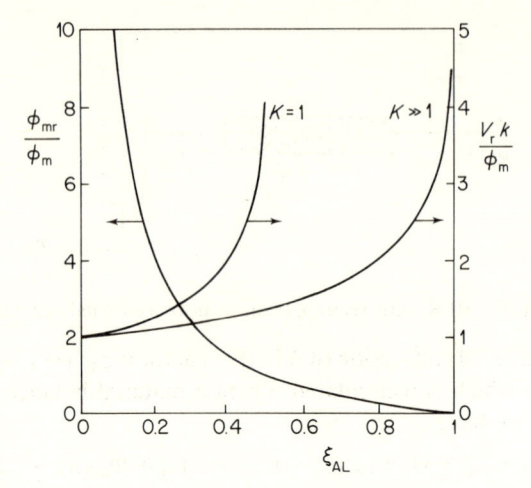

Fig. II.21 Recycle ratio and reactor volume for a first order equilibrium reaction in the installation shown in figure II.20

(b) What policy should be adopted for minimizing production costs (optimum load)?

In these installations, the cumulative mass m_P of P produced is a function of the production time, t_p, in such a way that the instantaneous production rate, $\Phi_P = dm_P/dt_p$, decreases with increasing t_p (see figure II.22). The relation between m_P and t_p may be either theoretical or empirical. If t_r is the period of time during which the reactor is out of operation for purposes of emptying, cleaning and charging, or of replacing or regenerating catalyst, the average production rate is

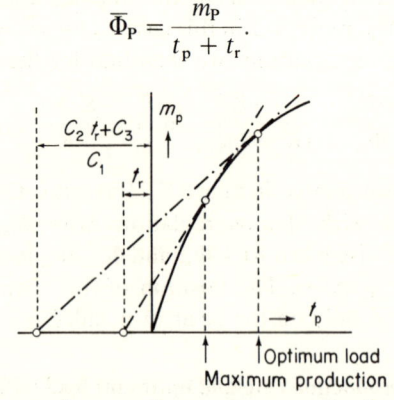

$$\overline{\Phi}_P = \frac{m_P}{t_p + t_r}.$$

Fig. II.22 Production times t_p for a maximum average production rate and for optimum load, respectively

Maximum average production rate is governed by the condition

$$d\overline{\Phi}_P/dt = \frac{1}{(t_p + t_r)} \frac{dm_P}{dt_P} - \frac{m_P}{(t_p + t_r)^2} = 0,$$

which yields

$$\frac{dm_P}{dt_P} = \frac{m_P}{t_p + t_r}.$$

It is seen from figure II.22 that the corresponding value of t_p can be obtained by drawing the tangent to the $m_P - t_p$ curve from the point $(-t_r, 0)$.

The operating costs must be available for the calculating of the *optimum load* of the installation. Suppose that
C_1 = the costs per unit time during production,
C_2 = the costs per unit time when the reactor does not produce, and
C_3 = the additional expenses per period (e.g. extra manpower, special facilities or new catalyst);
then the average total costs per unit mass of product equal $(C_1 t_p + C_2 t_r + C_3)/m_P$. For optimum load this expression must be a minimum, from which condition we find that

$$\frac{dm_P}{dt_p} = \frac{m_P}{t_p + (C_2 t_r + C_3)/C_1}.$$

The corresponding value is obtained from the point of contact of a tangent to the m_P-curve going through the point $(-(C_2 t_r + C_3)/C_1, 0)$ as shown in figure II.22. It will be clear from this that the conditions for maximum capacity and optimum load do not, in principle, coincide.

Example II.8 *Maximum production and optimum load in a batch process*
Determination of the maximum production and optimum load is required for the batch reactor treated in Example II.1. The pertinent production costs are:

During operation   27.60 MU/h (C_1)
During down-period   8.40 MU/h (C_2)
Additional per period 104.00 MU (C_3); the down period t_r is 1 hour.
MU = monetary unit, approximately equal to one Dutch guilder.

Figure II.23 shows the relative degree of conversion ζ_B of acid B as a function of

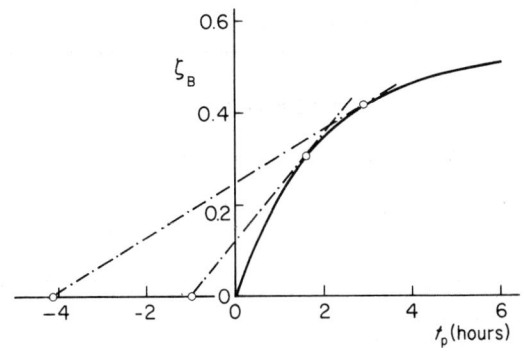

Fig. II.23 Construction of reaction times t_p for maximum production rate and optimum load, respectively, in a batch process; Example II.9

reaction time, which can be calculated from the data given in Example II.1. Since the amount of ethyl acetate produced is directly proportional to ζ_B, it is not necessary to convert the ordinate in figure II.23 to m_P.

The tangent to the ζ_B-curve through the point (-1 hour) on the abscissa gives a point of contact at $\zeta_B = 0.31$ and $t_p = 1.6$ hours. Hence, the maximum average production rate of ethyl acetate (P) is

$$\frac{24}{t_p + 1.0} \times c_{Bo}\zeta_B M_P = 985 \text{ kg/day m}^3.$$

For finding the optimum load, a tangent has to be drawn through the point on the abscissa:

$$t_p = \frac{8.40 t_r + 104.00}{27.60} = -4.1 \text{ hours},$$

which yields $\zeta_B = 0.42$ and $t_p = 2.9$ hours. A comparison of the two policies of operation is given in Table II.8. It is seen that the reaction costs per unit mass of product are 10% higher at maximum production rate than under conditions of optimum load. It also appears that the design requirement in Example II.1 ($\zeta_B = 0.35$) does not represent an economic optimum. Finally, it is to be noted that the above calculation only gives production costs for the reactor itself. The total production costs are higher by an amount which is determined by the losses of raw materials and product, and the expenses incurred in other parts of the plant (feed preparation, product separation, product storage and delivery).

Table II.8 Maximum production and optimum load of a batch reactor; Example II.8

Condition	Maximum production	Optimum load
relative degree of acid conversion ζ_B	0.31	0.42
reaction time t_p (hours)	1.6	2.9
ester production (kg/batch)	5560	7530
ester production rate (kg/m^3 day)	985	660
production costs (10^{-3} MU/kg ester)	28.1	25.5

REFERENCES

1. Levenspiel, O.: *Chemical Reaction Engineering*, second edition, Wiley, New York (1972).
2. Smith, D. F.: *J. Am. Chem. Soc.* **47**, 1862 (1925).
3. Venkatuswarlu, C., Satyanarayana, M., and Rao, M. N.: *Ind. Eng. Chem.* **50**, 973 (1958).
4. Voncken, R. M.: Ph.D. Thesis, Delft (1966).
5. Eldridge, J. M., and Piret, E. L.: *Chem. Eng. Progr.* **46**, 290 (1950).
6. Jenney, T. M.: *Chem. Eng.* **62**, 198 (1955).
7. Jones, R. W.: *Chem. Eng. Progr.* **47**, 46 (1951).
8. Weber, A. P.: *Chem. Eng. Progr.* **49**, 26 (1953).
9. Hugo, P.: *Chem. Eng. Techn.* **53**, 107 (1981).

Chapter III
Model reactors: multiple reactions, isothermal single phase reactors

In the previous chapter we discussed how in the model reactors with premixed feed the mixing influences the conversion rate per unit of reactor volume. We showed that for an equal total residence time τ in the PFR with no mixing in the direction of the flow and $\tau = \tau_L$, a considerably higher average conversion rate is obtained than in a CISTR (complete mixing and $\tau = \tau''$). This is because higher reactant concentrations for reaction orders >0 lead to higher conversion rates and because in the CISTR due to the complete mixing the reactant concentrations are lower than in the PFR at equal reactor outlet concentrations.

In this chapter we will study complex reaction systems in which more than one reaction takes place and where desired and undesired products are formed simultaneously. For two or more reactions—multiple reactions—also the relative ratio of the individual reaction rates of each reaction separately is important because this ratio eventually determines the relative amounts of desired and undesired products. In this chapter the ratio of reaction rate constants for first order reactions is defined as $\kappa = k_X/k_P$, where k_X refers to the undesired and k_P to the desired reaction. If two or more reactions occur, each of them is generally influenced by the concentrations of each component taking part in the individual reactions. The concentration levels in a reactor are influenced by the degree of mixing in the reactor. So for multiple reactions also each individual reaction rate is influenced by the degree of mixing, that is by the reactor type. So in this chapter we study the yield of desired product and the ratio of the production of desired and undesired products in various model reactors. We will still restrict ourselves in this chapter to isothermal single phase reactors.

In industrial practice not only are high conversion rates important—high conversion rates result in smaller reactors and thus in lower investment costs—but also the yield of the desired products plays a tremendous role. Higher yields result in lower raw materials cost per unit of desired product and thus in lower operating expenses for raw materials and disposal of waste products. Higher yields and selectivities indirectly also influence the investment costs for the separation of reactor products and for the disposal units. In most cases the yield and selectivity are more important for overall plant economics than a high reactor capacity.

Attempts are therefore often made, at the expense of reactor capacity, to increase relatively the production of the desired products and to suppress the formation of the undesired products. To this end, of course, catalysts and also the reaction temperature chosen play an important role, but also the choice of concentration levels and reactor type are powerful means to achieve the mentioned goals. Therefore in this chapter the cross flow reactor will also be introduced.

Of course, it is impossible to review all the possible complex reaction systems—we refer to the abundant literature to this end—but we will give a general outline of how the selectivity and the yield can be pushed in the desired direction by a correct choice of the reactor type and/or the concentration levels. For simplicity in this chapter we will consider only constant density isothermal reactions. How this has to be corrected for varying densities has already been discussed in Chapters I and II. Temperature effects on yield and selectivity will be treated also in Chapters VI and IX.

III.1 FUNDAMENTAL CONCEPTS

In Chapter I we described a general classification of complex reaction kinetics, which consist in principle of parallel reactions, consecutive reactions and combinations of these. These reactions can also be of the reversible type, as will follow from the kinetic equations. Combination reactions themselves consist of a combination of parallel and consecutive reactions. So as an example the following reaction:

$$A + B \rightarrow P + B \rightarrow X$$

is a consecutive reaction with respect to the reactant A, and a parallel reaction with respect to the reactant B:

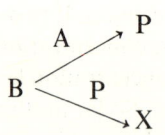

$$B \begin{cases} A \rightarrow P \\ P \rightarrow X \end{cases}$$

The different types of multiple reactions are discussed in the various sections in this chapter; a survey is given in Table III.1.

As has been outlined in Section I.6, as soon as desired and undesired products are formed not only the degree of conversion ξ_A or the relative degree of conversion of a reactant ζ_A are important for the overall reactor performance but also the yield η_P—the amount of desired product P produced per amount of key reactant A fed to the reactor, whether converted or not—and the selectivity σ_P, the amount of P produced per amount of A converted. In order to study how the parameters of model reactors can be used to monitor the selectivity and the yield of the reactor section of a plant, we have to introduce three new concepts which will be used frequently in this book: the differential selectivity, the selectivity ratio and the reaction path.

Table III.1 The structure of Chapter III

Subject	Example	Section
Fundamental concepts	$\sigma_P, \sigma'_P, \eta^P$	III.1
Parallel reactions, one reactant	$A \begin{smallmatrix} \nearrow P \\ \searrow Y \end{smallmatrix} X$	III.2
Parallel reactions, more reactants	$A + B \begin{smallmatrix} \nearrow P \\ \searrow X \end{smallmatrix}$	III.3
Consecutive reactions	$A \to P \to X$	III.4
Combination reactions	$A \to P \to X \searrow Y$	III.5
Autocatalytic reactions	$A \to P \to X$	III.6

III.1.1 Differential selectivity and selectivity ratio

For complex reactions with desired products P, Q, R, etc. and undesired products X, Y, Z, etc., we defined in Section I.6 the selectivity as the ratio of the amount of A converted into the desired product P and the total amount of A converted:

$$\sigma_P = \frac{\xi_P \nu_A M_A}{\xi_A \nu_P M_P} \tag{I.33}$$

and the yield as the amount of A converted into desired product P and the total amount of A fed into the reactor:

$$\eta_P = \sigma_p \zeta_A = \frac{\xi_P \nu_A M_A}{w_{Ao} \nu_P M_P}. \tag{I.35}$$

This relation for the yield can be derived from equation (I.33) with the defining equations (I.26) and (I.29), from which it follows that $w_{Ao}\zeta_A = \xi_A$. The definitions are entirely general: they apply equally well to simple reactor systems as to systems with feed and/or discharge distribution provided more general definitions for ξ_J and ζ_A are used in that case.

In case there is no product P present in the reactor feed ($c_{P_o} = 0$), the density ρ of the reaction mixture is constant over the entire reactor, the stoichiometric coefficients of A and P in the reaction equation are equal ($\nu_A = \nu_P$) and the molecular weights are equal ($M_A = M_P$), the defining equations above simplify to

$$\text{for the selectivity:} \quad \sigma_P = \frac{c_P}{c_{Ao} - c_A}$$

$$\text{and for the yield:} \quad \eta_P = \frac{c_P}{c_{Ao}}.$$

These simplified expressions are usually used. The selectivity as defined above is also called the integral or overall selectivity because it refers to the performance of the entire reactor as a whole. As will become clear in this chapter σ_P is dependent on the reaction conditions and on the reactor type.

However, it is often convenient to use the concept of differential selectivity or local selectivity σ'_P, which determines to what extent reactant A is being converted into P at some place or at some moment in the reactor and which is defined as:

$$\sigma'_P = -\frac{\nu_A R_P}{\nu_P R_A}. \tag{III.1}$$

Here σ'_P has been defined in such a way that $\sigma'_P = 1$ when all A is converted into P. This concept was introduced by Denbigh [1] who used the term 'instantaneous yield' for σ'_P.

For a BR or a small element of reactor volume and a constant mass flow of reaction mixture through it we can derive from eq. (III.1) and the material balance (I.15) that the following also holds:

$$\sigma'_P = \frac{M_A \nu_A \, d\xi_P}{M_P \nu_P \, d\xi_A}. \tag{III.2}$$

This differential selectivity gives a quick quantitative indication of the direction in which the desired production of species P will move when at a certain reactant concentration c_A we convert more of the reactant A. This makes the differential selectivity a useful tool for choosing local reaction conditions in any reactor.

For computational reasons it is often convenient to use a selectivity ratio instead of the selectivity. The selectivity ratio is defined as the ratio of the amount of a product P produced to the amount of a product X produced:

$$\sigma_P/\sigma_X = \frac{\nu_X M_X \xi_P}{\nu_P M_P \xi_X}. \tag{III.3}$$

In the same way as before also the differential selectivity ratio can be defined as

$$\sigma'_P/\sigma'_X = \frac{\nu_X |R|_P}{\nu_P |R|_X} = \frac{\nu_X M_X \, d\xi_P}{\nu_P M_P \, d\xi_X}. \tag{III.4}$$

In case also here $c_{P_0} = c_{X_0} = 0$, $\nu_A = \nu_P = \nu_X$ and the reaction mixture density is constant, we find the following simplifications:

—for the differential selectivity: $\qquad \sigma'_P = -\dfrac{dc_P}{dc_A}$

—for the selectivity ratio: $\qquad \sigma_P/\sigma_X = \dfrac{c_P}{c_X}$

—and for the differential selectivity ratio: $\qquad \sigma'_P/\sigma'_X = \dfrac{dc_P}{dc_X}.$

For computational purposes the (differential) selectivity ratio is sometimes easier to handle than the (differential) selectivity. Moreover, it gives a quick quantitative indication of the ratio of the productions of the desired and undesired species. This makes the (differential) selectivity ratio also a handy tool in reactor calculations. The relation between the selectivity (ratio) and the differential selectivity is often not a simple one. It depends among other things on the reaction type, the rate equations and the reactor type. From equations (III.2), (I.35) and (I.26) ($d\xi_A = w_{Ao} d\zeta_A$) we can easily derive that for a BR and a PFR we have to solve the following differential equation in order to determine the reactor yield η_P or selectivity $\sigma_P = \eta_P/\zeta_A$:

$$d\eta_P = d(\sigma_P \zeta_A) = \sigma'_P d\zeta_A. \quad (III.5)$$

In this equation often σ'_P itself is a function of the selectivity σ_P. For example for a *consecutive* reaction $A \to P \to X$ with first order kinetics and rate constants k_P and k_X equation (III.1) gives

$$\sigma'_P = + \frac{k_P c_A - k_X c_P}{k_P c_A}$$

which is a function of both σ_P and ζ_A through the concentrations of the desired product, c_P, and of the reactant, c_A. When R_P also depends on c_P, which implies that σ'_P is a function of both ζ_A and σ_P, integration of equation (III.5) can become very complicated.

Another example is a *parallel* reaction $A \to P$ and $A \to X$, which is first order to P and second order to X. Here we find, from equation (III.1),

$$\sigma'_P = \frac{k_P c_A}{k_P c_A + k_X c_A^2}$$

which is a function of ζ_A only through the reactant concentration c_A. For parallel reactions σ'_P is usually not a function of c_P, except for equilibrium and autocatalytic reactions. If the product concentration c_P does not influence the differential selectivity, σ'_P is only a function of feed conditions, kinetic constants and ζ_A, so that equation (III.5) can relatively easy be integrated. A special case are the CISTRs, where the selectivity and differential selectivity are equal. This can be easily understood by introducing the material balance equation (II.13), which reads $\xi_{J1} = V_R M_J |R|_{J1}/\Phi_m$, into equation (I.33). This leads to

$$\sigma_P = -\frac{v_A R_{P1}}{v_P R_{A1}} = \sigma'_P \quad \text{for a CISTR.} \quad (III.6)$$

For this reason a CISTR is ideally suited for selectivity determinations.

III.1.2 The reaction path

With respect to multiple reactions we shall—besides the combination reactions—mainly consider parallel and consecutive reactions which exhibit an essentially different behaviour with regard to product distribution. This is qualitatively demonstrated in Figure III.1 (a and b). For parallel reactions, figure III.1a indicates that, as soon as reactant is converted, both P and X can form in principle with a σ_P value anywhere between 0 and 1. The yield of P, of which w_P in

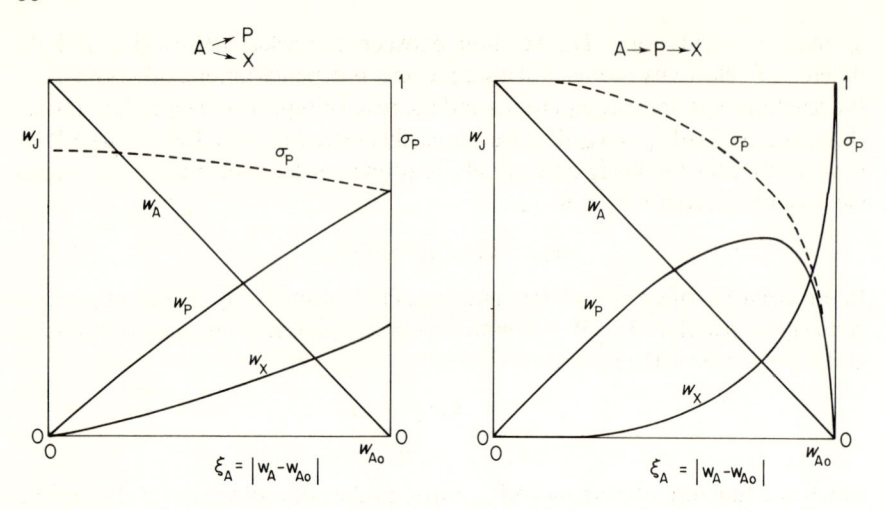

Fig. III.1 Concentrations and selectivity as a function of the degree of conversion for parallel and consecutive reactions

figure III.1a is a measure, will generally rise as the conversion proceeds. In consecutive reactions, P (in this example assumed to be the desired product) forms first, and is subsequently converted to X. Therefore, initially $\sigma_P = 1$, but in principle it decreases to 0 as the reaction nears its end. Consequently, the yield η_P passes through a maximum value as a function of the degree of conversion.

The shape of the lines in a diagram like figure III.1 depends not only on the type of complex reaction but also on the different rate constants involved (which may greatly depend on temperature) and on the type of reactor in which the reaction is carried out.

Since three kinds of substances are distinguished in multiple reactions: reactant(s), desired and undesired products, it is frequently convenient to represent the composition of the reaction mixture in a triangular composition diagram (see fig. III.2). The concentration of these substances should then be expressed in such units that their sum is constant all over the composition triangle. The line giving the course of composition as the degree of conversion increases will be called the *reaction path*. We shall adopt the convention that the left hand corner of the triangle represents 100% of reactant, the top corner 100% of desired product and the right hand corner 100% of waste product. Hence, the nearer the reaction path approaches the left hand side AP, the higher the selectivity σ_P is; furthermore, the smaller the vertical distance between a point on the reaction path and the top corner P, the higher the corresponding value of the yield η_P is.

The reaction path is primarily determined by the type of complex reaction; its shape is further influenced by the ratio of the reaction velocity constants involved (i.e., by the temperature), by the type of reactor in which the reaction is carried out, and by the operating conditions, such as feed composition, supply or removal of components during the reaction, etc. In the following we shall mainly

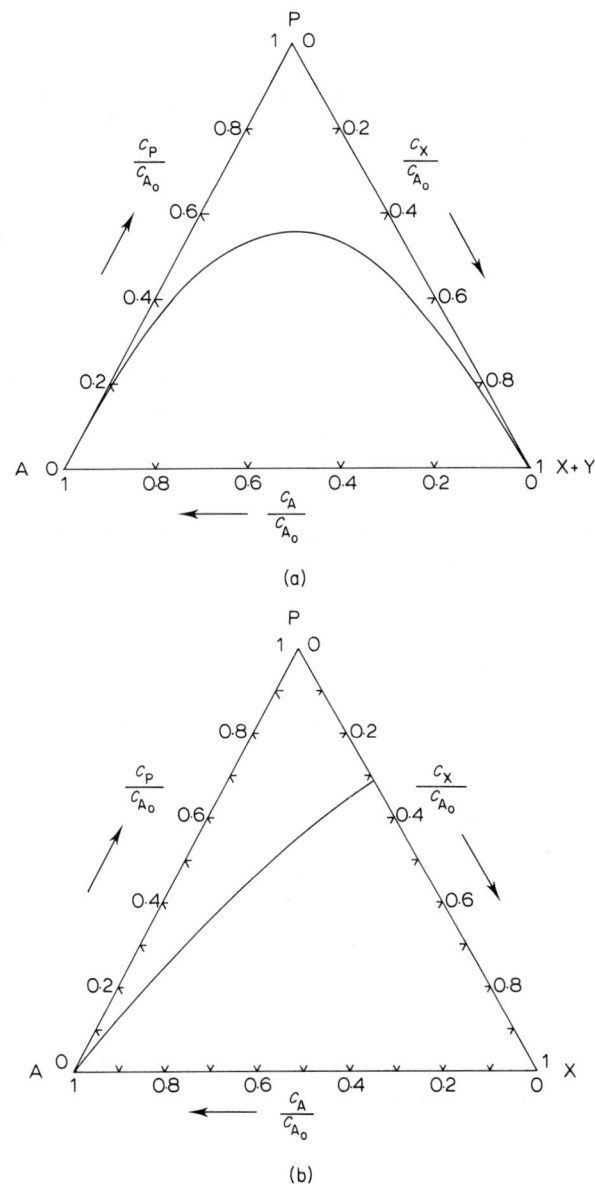

Fig. III.2   The reaction paths for the parallel and the consecutive reactions of figure III.1

discuss the influence of the reaction type and of the reactor system. For the sake of simplicity, we shall use molar concentrations, assuming that the sum of the three concentrations involved is constant. Also the reaction path is a convenient tool: it enables us to see in one glance at a single diagram what the course of a reaction is with progress of the conversion. In figure III.2 the reaction path for both reactions represented in figure III.1 are given. Notice the different shapes for parallel and consecutive reactions. The implicit variable reaction time or residence time, which determines the progress of the reaction, is not given in these diagrams, but can be derived from the value c_A/c_{Ao}, from which t, τ or τ_L can be calculated if the type of model reactor and the kinetic equations are known.

Example III.1 *Determination of selectivity and yield data in a PFR from laboratory experiments*

In an experimental laboratory reactor, which has been proven to behave like a CISTR, a series of measurements have been done at varying throughputs and at one single temperature for parallel reactions of the type A → P and A → ½X. The concentration of the reactant A in the feed has been kept constant at $c_{Ao} = 10 \text{ kmol/m}^3$ and the reactor product composition has been determined. It is known that 2 moles of A form 1 mole of X. Moreover, it was found that the reaction rates were independent of the product concentrations c_P and c_X. In a CISTR the differential selectivity can be determined as $\sigma'_P = c_{P1}/(c_{Ao} - c_{A1})$, where c_{A1} and c_{P1} refer to the reactor outlet conditions. Experimental results are given in Table III.2, where the reactor product composition is given in dimensionless form by dividing by $v_J c_{Ao}/v_A$, so that always $(c_A + c_P + 2c_X)$ $c_{Ao} = 1$.

Table III.2 Reactor product composition, experimental results; Example III.1

c_A/c_{Ao}	c_P/c_{Ao}	$2c_X/c_{Ao}$
0.950	0.026	0.024
0.900	0.053	0.047
0.800	0.111	0.089
0.700	0.176	0.124
0.600	0.250	0.150
0.500	0.333	0.167
0.400	0.429	0.171
0.300	0.538	0.162
0.200	0.667	0.133
0.100	0.818	0.082
0.050	0.905	0.045

Rate equations are unknown as is the influence of concentrations on the reaction rates. We will determine σ'_P, σ'_P/σ'_X, σ_P, η_P and reaction paths in a CISTR and a PFR from our experiments. We first calculate the differential selectivity from our experimental data. In a *CISTR* we know that $\sigma'_P = \Delta c_P/\Delta c_A$, moreover $\sigma_{P1} = \sigma'_P$. The differential selectivity ratio in our case can be found directly from the experimental results with $\sigma'_P/\sigma'_X = \Delta c_P/2\Delta c_X$, moreover in a CISTR also $\sigma_P/\sigma_X = \sigma'_P/\sigma'_X$. All the results are given in Table III.3. The yield has been calculated with $\eta_P = \sigma_P \zeta_A$.

Table III.3 Reactor performance as derived from experiments; Example III.1

c_A/c_{Ao}	ζ_A	CISTR |||| PFR					
		σ'_P	σ'_P/σ'_X	η_{P1}	η_{PL}	σ_{PL}	σ_P/σ_X	c_{PL}/c_{Ao}	c_{XL}/c_{Ao}
1.00	0.00	0.500	1.00	0.000	0.000	0.500	1.00	0.000	0.000
0.95	0.05	0.513	1.08	0.026	0.025	0.506	1.03	0.025	0.025
0.90	0.10	0.526	1.13	0.053	0.051	0.513	1.05	0.051	0.049
0.80	0.20	0.556	1.25	0.111	0.105	0.527	1.11	0.105	0.095
0.70	0.30	0.588	1.44	0.177	0.163	0.542	1.18	0.163	0.137
0.60	0.40	0.625	1.67	0.250	0.223	0.558	1.26	0.223	0.177
0.50	0.50	0.667	1.99	0.333	0.288	0.575	1.36	0.288	0.212
0.40	0.60	0.714	2.51	0.429	0.357	0.595	1.47	0.357	0.243
0.30	0.70	0.769	3.35	0.539	0.431	0.615	1.60	0.431	0.269
0.20	0.80	0.833	5.02	0.667	0.511	0.639	1.77	0.511	0.289
0.10	0.90	0.909	9.98	0.818	0.598	0.664	1.98	0.598	0.302
0.05	0.95	0.952	20.11	0.905	0.644	0.678	2.11	0.644	0.306

The results for the CISTR now have to be translated to the performance to be expected in a PFR. To this end we make use of the relation for the yield

$$\eta_P = \sigma_P \zeta_A = \int_0^{\zeta_A} \sigma'_P \, d\zeta_A.$$

This integration has to be done graphically or numerically, because we only have at our disposal an empirical set of data of σ'_P versus ζ_A as shown in figure III.3a.

By numerical integration between the boundaries 0 and varying values of ζ_A we obtain η_{PL} as a function of ζ_A. This has been indicated in figure III.3a: the area under σ'_P curve between $\zeta_A = 0$ and $\zeta_A = 0.5$ is equal to the yield in a PFR at $\zeta_A = 0.5$, for example. This procedure is repeated for sufficient different values of the total conversion ζ_A. By extrapolation we have estimated that $\sigma'_P = 0.500$ at $\zeta_A = 0$. The selectivity of the PFR is now found by dividing the values found for the yield by the relative conversion, because $\sigma_P = \eta_P/\zeta_A$. The differential selectivity (ratio) is in this case independent of the reactor type and thus equal to that of the CISTR because R_P and R_X were found to be independent of c_X and c_P.

It remains to calculate the reaction path in a PFR. According to a moles balance we have $c_{Ao} = c_A + c_P + 2c_X$, in which $c_A = c_{Ao}(1 - \zeta_A)$ and $c_P = c_{Ao}\eta_P$, because $c_{Po} = c_{Xo} = 0$. This leads to $2c_X/c_{Ao} = \zeta_A - \eta_P$, and for the selectivity ratio we find $\sigma_P/\sigma_X = \eta_P/(\zeta_A - \eta_P)$. With these data the concentrations of A, P and X can be calculated and the reaction path be drawn as has been done in figure III.3b. All results have been summarized in Table III.3.

It remains to check whether σ'_P is a function of the selectivity σ_P. Strictly speaking this can only be done in experiments where species P and X are added to the reactor feed to verify the influence of products on the selectivity itself. This was postulated not to occur at the beginning of this illustration. However, we see that σ'_P is gradually and smoothly increasing with increasing conversion. For autocatalytic reactions we would expect a very sharp increase of σ'_P as soon as some product P has been formed. For equilibrium reactions we would expect

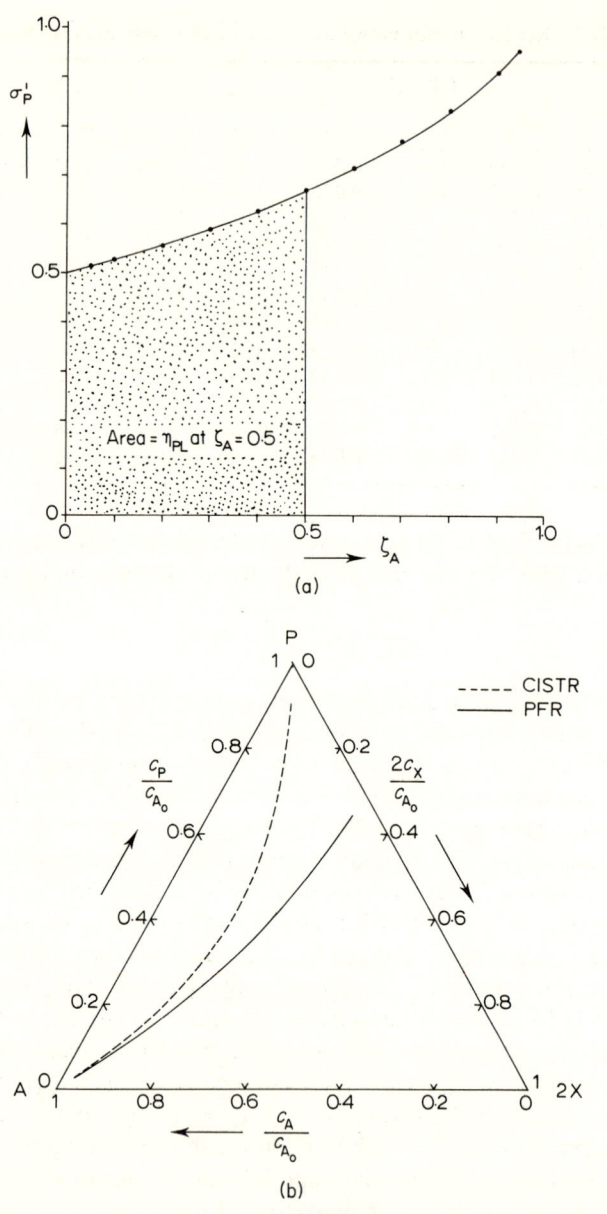

Fig. III.3 (a) Selectivity in an experimental CISTR; (b) the reaction paths in a CISTR and a PFR as derived from experiments for parallel reactions; Example III.1

decreasing values of σ'_P with increasing residence times or conversions, because on approaching the equilibrium value of c_P the production rate of P decreases and more X is formed at the expense of P. In other words, according to the first reaction, species A would always remain present if it were an equilibrium reaction, so that given sufficient reaction time the second reaction would eventually consume all reactant A. But we see that at higher conversions the selectivity improves in both reactor types.

So we can safely assume that at least in the region of our experiments σ'_P most probably is not a function of σ_P or c_X. It is, of course, dependent on how large the catalytic or inhibition effects of P and X are; at the accuracy and temperature level of our experiments they could not be observed.

III.2 PARALLEL REACTIONS

In this section we will discuss the influence of the conversion on the selectivity for parallel reactions. These reactions can be of many different types, but we will distinguish the following categories:

(i) parallel reactions with rate equations, which are a function of the concentrations of one reactant only and with equal reaction order. They are of the type

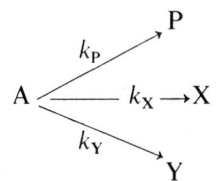

$$A \begin{array}{c} \xrightarrow{k_P} P \\ \xrightarrow{k_X} X \\ \xrightarrow{k_Y} Y \end{array}$$

with $R_A = -k_P c_A^m - k c_A^n - k_Y c_A^{n'}$, in which $m = n = n'$.

(ii) Like (i), but with different reaction orders, so that $m \neq n \neq n'$.

(iii) Like (i), but now the rate equations are a function of more than one reactant concentration, for example

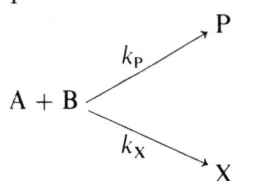

$$A + B \begin{array}{c} \xrightarrow{k_P} P \\ \xrightarrow{k_X} X \end{array}$$

with a rate equation like $R_A = -k_P c_A^m c_B^{m'} - k_X c_A^n c_B^{n'}$. This type will be evaluated in Section III.3, because in this case measures to be taken to improve the selectivity can be in conflict, so that a new reactor type—the cross flow reactor—must be introduced to cope with this particular problem.

We recall that the rate equations are representing overall results and often do not necessarily give information on the reaction mechanism itself. Moreover, we will only consider constant density reactions.

III.2.1 Parallel reactions with equal reaction order rate equations

If both reactions $A \xrightarrow{k_P} P$ and $A \xrightarrow{k_X} X$ are of equal order the differential selectivity is given by

$$\sigma'_P = \frac{k_P c_A^m}{k_P c_A^m + k_X c_A^m} = \frac{k_P}{k_P + k_X}.$$

With equations (III.5) and (III.6) this leads for both a CISTR and a PFR for the selectivity to

$$\sigma_{P1} = \sigma_{PL} = \frac{k_P}{k_P + k_X}$$

The isothermal reaction path is a straight line as shown in figure III.4, with a slope depending on the ratio of the reaction velocity constants k_P and k_X. The reaction path in this exceptional case does not depend on whether the reaction is carried out in a batch or tubular reactor, or in a continuous tank reactor. The ratio of P and X obtained upon complete conversion, i.e. the maximum yield η_P, can only be influenced by temperature. It should be borne in mind, however, that the capacity of the tank reactor is always lower than that of the tubular reactor. Capacity is always defined as the production of the desired component per unit of reactor volume.

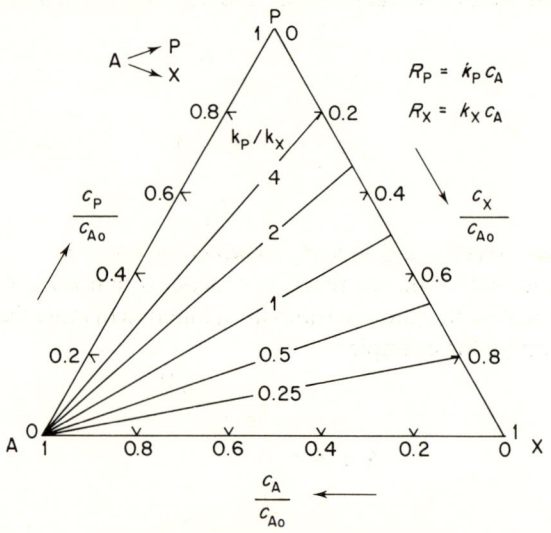

Fig. III.4 Isothermal reaction paths for parallel first order reactions

The influence of the temperature depends on the parameter κ:

$$\kappa = \frac{k_X}{k_P} = \frac{k_{X\infty}}{k_{P\infty}} \exp\left(-\frac{\Delta E}{RT}\right).$$

in which $\Delta E = E_X - E_P$. The sign of ΔE determines the choice of the temperature level. If the initially studied conditions are T_o, κ_o and σ_{Po}, we find for the selectivity σ_{PT} at a new reactor temperature T

$$\frac{\sigma_{PT}}{\sigma_{Po}} = \frac{(\kappa_o + 1)}{(\kappa_T + 1)}$$

with

$$\frac{\kappa_T}{\kappa_o} = \exp\left\{-\frac{\Delta E}{RT_o}\left(\frac{T_o}{T} - 1\right)\right\}.$$

If $\Delta E < 0$, which means that k_P is increased more by a rise in temperature than k_X, selectivity increases with increasing temperature. Vice versa, for $\Delta E > 0$ reactor temperature has to be decreased to improve the selectivity. This is shown in figure III.5 for $\sigma_{P_0} = 0.2$ and 0.5 for values of $-\Delta E/RT_0$ of -10 to $+10$. Of course if $\Delta E = 0$ there is no effect at all of a temperature change. Reactor temperatures usually can be varied only over a limited range, e.g. because of new, undesired reactions coming into play or of prohibitively large reactor volumes required.

III.2.2 Parallel reactions with differing reaction order rate equations

Parallel reaction like $A \rightarrow v_P P$ and $A \rightarrow v_X X$ of differing reaction order may have rate equations for the desired reaction of the type $R_P = k_P c_A^m$ and for the undesired reactions of the type $R = k_X c_A^n$, in which $m \neq n$. The differential selectivity is then given by

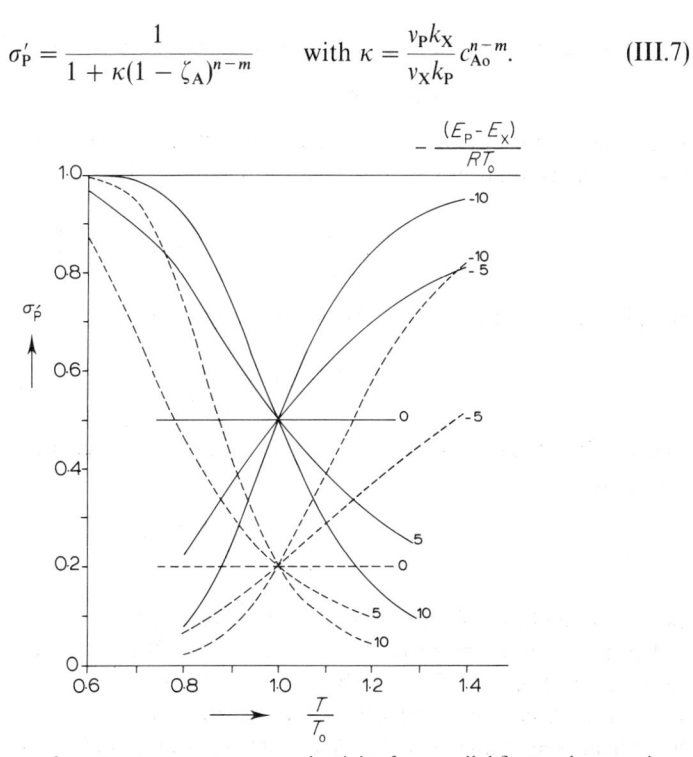

$$\sigma_P' = \frac{1}{1 + \kappa(1 - \zeta_A)^{n-m}} \quad \text{with} \quad \kappa = \frac{v_P k_X}{v_X k_P} c_{A_0}^{n-m}. \tag{III.7}$$

Fig. III.5 Influence of reactor temperature on selectivity for parallel first order reactions for $\sigma_{P_0}' = 0.2$ and 0.5. Parameter is $(E_X - E_P)/RT_0$

To study the selectivity as a function of the course of the reaction we also have to specify the inlet conditions, for example $c_A = c_{A_0}$ and $c_P = c_X = 0$. We have assumed in this chapter that the density of the reaction mixture is constant, so that, provided $v_A = v_P = v_X$, we have $c_{A_0} = c_A + c_P + c_X$ and by definition $c_P = c_{A_0} \zeta_P$ and $c_X = c_{A_0} \zeta_X$.

For a BR and a PFR we will solve equation (III.5) in this particular case by using the differential selectivity ratio. Integration of (III.5) usually requires a numerical or a graphical method, but in some cases this can be done analytically by using σ'_P/σ'_X. For the case $m - n = +1$, which means for example that the desired reaction is of second order and the undesired one of the first order, the selectivity ratio is given by

$$\sigma'_P/\sigma'_X = \frac{d\zeta_P}{d\zeta_X} = \frac{(1-\zeta_A)^2}{\kappa(1-\zeta_A)} = \frac{1-\zeta_P-\zeta_X}{\kappa}$$

in which $\kappa^{-1} = k_P c_{Ao}/k_X$. Now the following linear differential equation has to be solved:

$$\frac{d\zeta_P}{d\zeta_X} + \frac{1}{\kappa}\zeta_P = \frac{1-\zeta_X}{\kappa}.$$

With the initial conditions $c_A = c_{Ao}$ and $c_P = c_X = 0$. The solution is:

$$\zeta_P = (\kappa + 1)(1 - e^{-\zeta_X/\kappa}) - \zeta_X.$$

With the relation $\zeta_X = \zeta_A - \zeta_P$ we can rearrange this relation, so that for the reaction scheme given we find for the BR and the PFR:

$$\sigma_{Pt} = \sigma_{PL} = \frac{\zeta_P}{\zeta_A} = 1 + \frac{\kappa}{\zeta_A}\ln\left(1 - \frac{\zeta_A}{\kappa+1}\right).$$

At full conversion or $\zeta_A = 1$ this leads to the selectivity

$$\sigma_P = 1 + \kappa\ln\left(\frac{\kappa}{\kappa+1}\right).$$

For a CISTR and $m - n = +1$ it can be easily derived that in this case the selectivity is

$$\sigma_{P1} = \frac{\zeta_P}{\zeta_P + \zeta_X} = \frac{1-\zeta_A}{1+\kappa-\zeta_A}.$$

So in a CISTR at full conversion the reactor selectivity is zero.

For some values of κ the reaction paths in a PFR and a CISTR are given in figure III.6. For a PFR it can be seen that for high values of κ after a certain amount of A has been converted almost no desired product is formed and that almost all remaining reactant A is converted into undesired product X. This phenomenon is even more pronounced in a CISTR where at high conversions more and more X is produced and at full conversion no P is produced at all. This can easily be understood because the reaction rate of the desired reaction reduces with c_A^2 more quickly than of the undesired reaction with c_A as the reaction proceeds and c_A diminishes. This leads for this case to the conclusion that preferably a BR or a PFR should be chosen or if this is impossible a CISTR with low conversions and high recycle ratios.

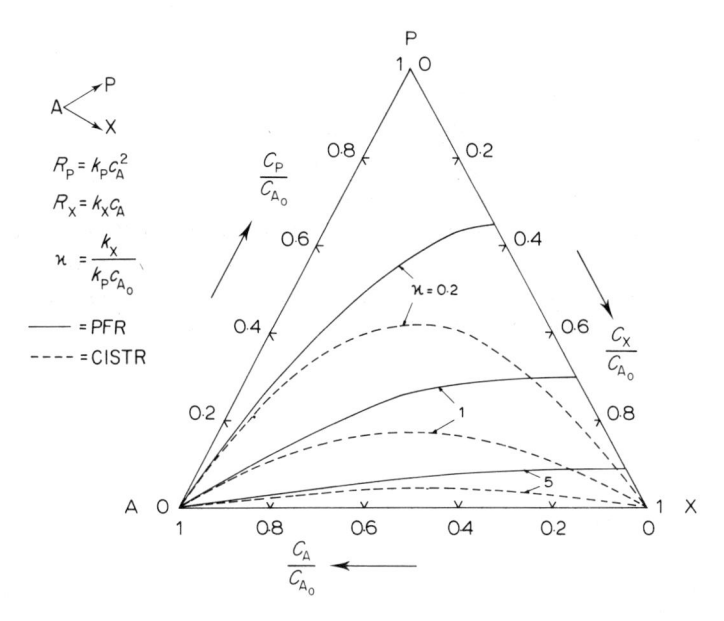

Fig. III.6 Reaction paths in a PFR and a CISTR for parallel reactions, where the desired reaction is of a higher order than the undesired one ($m - n = +1$)

In figure III.7 in a PFR and a CISTR the yields and selectivities are plotted for the same parallel reaction with $\kappa = 0.2$. We see that the differential selectivity decreases with the progress of the reaction and reaches zero at full conversion.

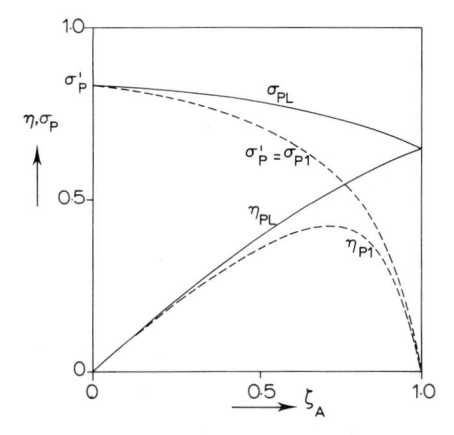

Fig. III.7 Yield and selectivity in a PFR and a CISTR for parallel reactions with $m - n = +1$ and $\kappa = 0.2$

The yield is given by $\eta_P = \sigma_P \zeta_A$. For a PFR and a BR the yield is lower than the selectivity as long as $\zeta_A < 1$. For a CISTR the yield reaches a maximum at

$$\zeta_A = (1 + \kappa) - \sqrt{[\kappa(\kappa + 1)]}$$

at which the yield equals

$$\eta_P = 1 + 2\{\kappa - \sqrt{[\kappa(\kappa + 1)]}\}.$$

This maximum can be found by differentiation of $\eta_P = \sigma_P \zeta_A$ $= \zeta_A(1 - \zeta_A)/\{(\kappa + 1) - \zeta_A\}$ towards ζ_A and setting $d\eta/d\zeta_A = 0$.

Now with the order of the reactions reversed or *the case $m - n = -1$* we find the reaction paths as shown in figure III.8. Here the CISTR gives yields considerably higher than in a PFR or BR and at full conversion only P is produced. Here the CISTR suppresses the undesired side reactions much better because of the low reactant concentrations at high conversions.

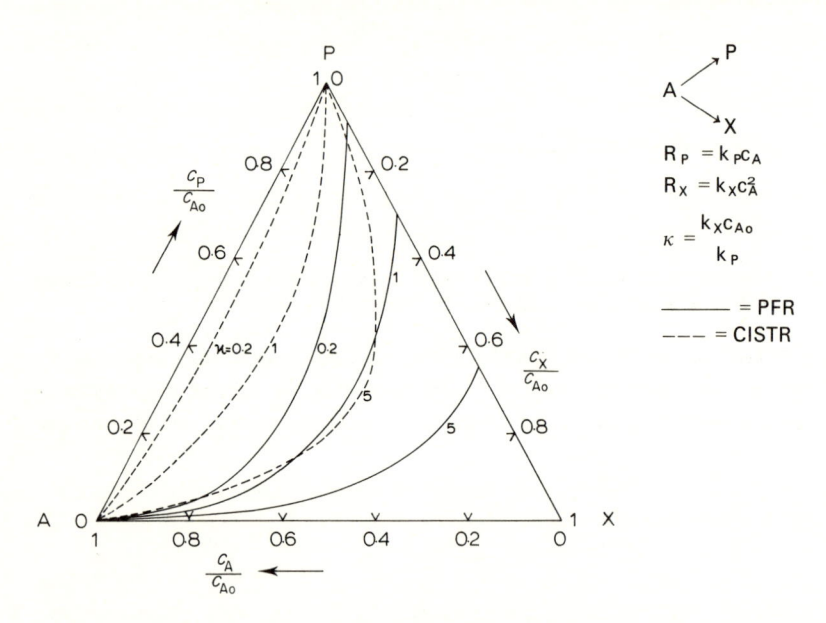

Fig. III.8 Reaction paths in a PFR and a CISTR for parallel reactions where the desired reaction is of a lower order than the desired one ($m - n = -1$)

This discussion and a study of equations (III.7):

$$\sigma'_P = \{1 + \kappa(1 - \zeta_A)^{n-m}\}^{-1}$$

leads to the following general conclusions for the choice of the reactor type for parallel reactions, depending on the orders m and n of the rate equations:

(a) If $m - n < 0$, the CISTR is the best reactor to achieve the most favourable reactor product distribution. In this case σ'_P is lower at higher reactant concentrations. Thus low reactant concentrations favour the production of P and suppress the formation of X. At the same conversion level ζ_A the CISTR has the lowest average reactant concentration level. Gas phase reactions can be executed at low partial pressures to favour the production of P and in liquid phase reactions the reactant can be diluted with an inert solvent, if only a BR or a PFR can be applied. Also an SBR can be used with a continuous addition of reactant to keep c_A low. We can understand this by realizing that conversely at increasing concentrations of A the undesired reaction increases relatively faster than the desired reaction rate.

(b) If $m - n = 0$ both the PFR and the CISTR exhibit the same yields and selectivities. The capacity of the PFR is higher, of course.

(c) If $m - n > 0$ the PFR is the preferred reactor to achieve a favourable product distribution. Reactions should be executed at high concentrations or partial pressures of the reactant. We now meet with an optimization problem because at low conversions σ'_P and σ_{PL} are the highest. If possible the reactor should be run at low conversions and after separation of the reactor products the non-converted reactant should be recirculated to the reactor inlet (see Section I.6).

III.2.3 A cascade of tank reactors

In a cascade the yield and the selectivity will be intermediate between those of a PFR and a CISTR with the same total conversion. In case a cascade has to be chosen for technical reasons e.g. because a PFR cannot be realized technically, the number of tanks has to be high in the cascade. In a cascade of N tank reactors—as can easily be shown—the yield η_{PN} is given by the relation

$$\eta_{PN} = \sum_{n=1}^{N} \sigma'_{Pn} \Delta \zeta_{An}. \tag{III.8}$$

For a given number of N tanks η_{PN} usually reaches a maximum for tanks of different sizes. The overall yield η_{PN} of a cascade with a large number of tanks has to be calculated by computer or determined graphically as shown in figure III.9.

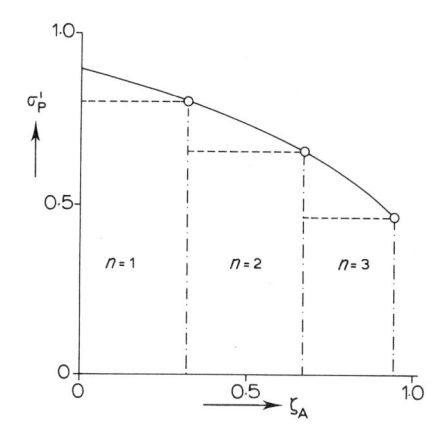

Fig. III.9 The operating points (○) of a cascade of three CISTRs in a differential selectivity–relative conversion plot. The total yield is equal to the sum of the areas for $n = 1$, $n = 2$ and $n = 3$, provided σ'_P is a unique function of ζ_A

In case no reaction rate equations are available we also can determine the relation between σ'_P and ζ_A (and other variables!) experimentally. This relation has to be determined in a CISTR as has been explained in Section III.1. We warn once more that this method works only if σ'_P is only a function of c_A. In practice the possible influence of other components should be measured in the concentration range in which these components will be present during the reaction. Once σ'_P is known as a function of all components present in the system, it is possible to calculate the conversion and the selectivity in any reactor for which the mixing characteristics are known. In the following illustration we will determine the yield and selectivity in a cascade for a reaction scheme with known kinetics.

Example III.2 *Parallel reactions in a cascade*
The parallel reactions as demonstrated in figure III.6 and with $\kappa = 0.2$ are executed in a cascade of two CISTRs. We will determine the overall selectivity and the yield in this reactor type, compare them with the PFR and one CISTR at conversions of $\zeta_A = 0.90$ and 0.95 and optimize the volume ratio for a maximum yield. We take a cascade of two equally sized tanks and after that the cascade with the highest yield. The differential selectivity is, in this case

$$\sigma'_P = \frac{\kappa(1 - \zeta_A)}{1 + \kappa(1 - \zeta_A)}.$$

The maximum yield in a cascade of two CISTRs follows from the value of ζ_{A1} in the relation

$$\frac{d\eta_{P2}}{d\zeta_{A1}} = \frac{d}{d\zeta_{A1}}\{\sigma'_{P1}\zeta_{A1} + \sigma'_{P2}(\zeta_{A2} - \zeta_{A1})\}$$

$$= \sigma'_{P1} + \zeta_{A1}\frac{d\sigma'_{P1}}{d\zeta_{A1}} - \sigma'_{P2} = 0.$$

In this relations σ'_{P1} and σ'_{P2} refer to the outlet conditions of the first and the second tank respectively. Elaboration of these equations gives

$$\frac{\kappa\{(1 + \kappa)(1 - 2\zeta_{A1}) + \kappa\zeta_{A1}^2\}}{\{1 + \kappa(1 - \zeta_{A1})\}^2} = \frac{\kappa(1 - \zeta_{A2})}{1 + \kappa(1 - \zeta_{A2})},$$

from which the unknown ζ_{A1} can be determined. Results are given in Table III.4, where also the reactor volumes are compared with that of a PFR, which is given by $k_X\tau_L = -\ln(1 - \zeta_A)$. Furthermore,

$$k_X\tau_1 = (\kappa\zeta_{A1})/[(1 - \zeta_A)^2 + \kappa(1 - \zeta_A)],$$
$$k_X\tau_2 = \kappa(\zeta_{A2} - \zeta_{A1})/[(1 - \zeta_{A2})^2 + \kappa(1 - \zeta_{A2})]$$

and

$$k_X\tau_N = k_X(\tau_1 + \tau_2).$$

Table III.4 Comparison of reactor types for Example III.2

Reactor type	ζ_{A1}	σ'_{P1}	η_{Pn}	$\dfrac{\tau_1}{\tau_2}$	$\dfrac{\tau_N}{\tau_L}$
	Final conversion $\zeta_{AN} = \zeta_{AL} = 0.90$				
PFR	—	—	0.625	—	1.00
Cascade $N = 2$: optimum	0.605	0.667	0.500	0.48	1.95
—equally sized	0.729	0.613	0.490	1.00	1.88
CISTR	0.900	0.333	0.300	—	3.91
	Final conversion $\zeta_{AN} = \zeta_{AL} = 0.95$				
PFR	—	—	0.639	—	1.00
Cascade $N = 2$: optimum	0.655	0.635	0.472	0.52	2.61
—equally sized	0.811	0.528	0.445	1.00	2.32
CISTR	0.950	0.200	0.190	—	6.34

We see that the optimal cascade gives an improved yield at the expense of a larger reactor volume. The differently sized tanks will be more expensive than two equally sized tanks.

III.3 THE CONTINUOUS CROSS FLOW REACTOR SYSTEM

The cross flow reactor is a model reactor which comes into play when the recommendations for the choice of a reactor type for parallel reactions in Section III.2 do not give one single answer. In the previous section we discussed parallel reactions with rate equations, in which appeared the concentration of one reactant only. Often the concentrations of two reactants play a role, as for example in the following reaction scheme:

$$A + B \begin{cases} \nearrow P \text{ with } R_P = k_P c_A^m c_B^{m'} \\ \searrow X \text{ with } R_X = k_X c_A^n c_B^{n'} \end{cases}$$

The conclusions reached in Section III.2 with respect to the desired reactor type and the difference in reaction order are also valid here, if we take into account the difference in reaction order for each reactant concentration separately. So if both $m - n$ and $m' - n'$ are >0, then c_A and c_B should be as high as possible and we should apply a PFR or BR. Conversely if $m - n$ and $m' - n'$ are <0, then c_A and c_B should be as low as possible and a CISTR should be chosen.

More ambiguous is the case where $m - n < 0$ and $m' - n' > 0$ or vice versa. If the concentration of one reactant should be kept low and the other kept high a different approach leads to very useful results.

In the foregoing discussions on continuously operated reactors we assumed that the total feed stream was supplied to the reactor system at one point only, and that the product stream left the system at one outlet point. A distributed feed and/or a distributed take-off along the length of the reactor system (as shown in figure III.10) also belongs to the possibilities. We shall call such a system a cross-flow reactor. There are, of course, many possible variations in the location of the various inlet and outlet points, and also in the distribution of the flow rates over these points.

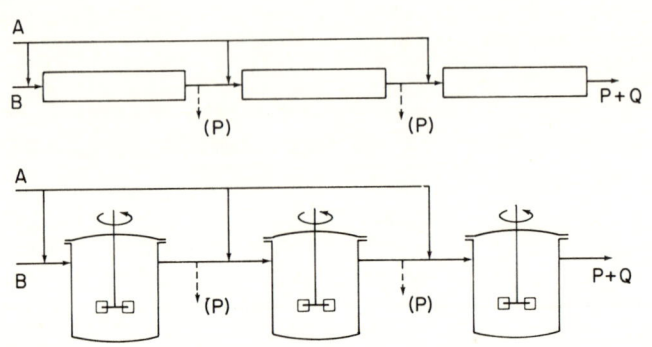

Fig. III.10 Possible cross flow reactor systems consisting of tubular reactors and tank reactors

Such applications may be profitable in particular cases, e.g. for the purpose of suppressing undesired side reactions. It will be clear that similar results can be obtained with a programmed feed into or discharge from a semi-continuous batch reactor. For purposes of calculation, material balances have to be applied to appropriate sections of the reactor system. For a tubular reactor, a section between two inlet and/or outlet points can be treated with the theory of Section III.2; for a cross-flow system consisting of tank reactors, the material balance over each reactor is essentially given by equation (II.20), provided the possibility of more than one feed or discharge stream is taken into account. Solutions for specific problems of this kind can be built up accordingly.

A system which is slightly more accessible to a generalized treatment is the ideal tubular reactor in which the number of side feed points (or discharge points, respectively) is infinite. We call this the idealized cross flow reactor. Such a gradual distribution of a part of the feed (or the discharge) over the entire reactor length cannot, of course, be realized in practice, but as an extreme model it may serve to investigate problems of reactor optimization. This has been shown by van de Vusse and Voetter [2], and will be elucidated in the following illustration.

Example III.3 *Parallel reactions in an idealized cross flow reactor*
The following two reactions occur simultaneously in a process:

$$A + B \to P, \text{ with } R_P = k_P c_A c_B$$
$$A + A \to X, \text{ with } R_X = \tfrac{1}{2} k_X c_A^2$$
$$\left. \begin{array}{l} R_A = -R_P - 2R_X \\ R_B = -R_P. \end{array} \right.$$

P is the desired product and X a waste product, while A is an expensive reactant which cannot be easily removed from the product stream. It is therefore required to obtain a high relative degree of conversion of A (say $\zeta_{AL} = 0.98$) together with a relatively high production, i.e. a high yield of P. The rate equations indicate that the latter may be achieved by keeping the concentration of A in the reactor system relatively low. Accordingly, an idealized cross flow tubular reactor will be considered where A is fed along the length of the reactor and B is fed at the inlet end only. We do not yet wish to optimize the system, but will arbitrarily require an injection distribution of A causing constant concentration $c_A = c_{AL}$ over the entire reactor length. If the total molar feed rates of A and B are equal and $k_P = k_X$, what is the relative degree of conversion of B to P in this system? How does this result compare with the performance of a batch reactor and a continuous tank reactor under the same conditions?

The diagram of the reactor, together with the symbols to be used, is shown in figure III.11. If we assume the density of all streams to be identical, the following set of equations has to be solved:

material balance of A for a section dz:

$$0 = -d(\Phi_v c_A) + c_{Aw} d\Phi_v + R_A S\, dz; \qquad (a)$$

material balance of B:

$$0 = -d(\Phi_v c_B) + R_B S\, dz. \qquad (b)$$

Additional requirements are the constancy of c_A:

$$c_{Ao} = c_A = c_{AL} \qquad (c)$$

and the equality of the feed rates of A and B:

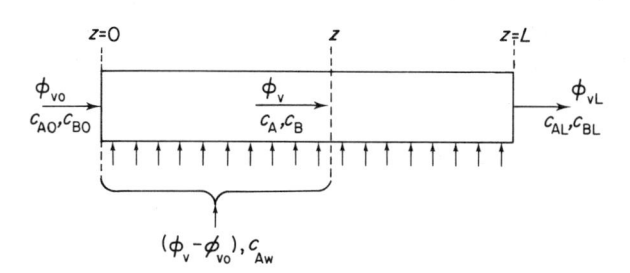

$$\Phi_{vo} c_{Bo} = \Phi_{vo} c_{Ao} + (\Phi_{vL} - \Phi_{vo}) c_{Aw}. \qquad (d)$$

Fig. III.11 The idealized cross flow tubular reactor

If the size of the reactor has to be calculated together with the required distribution of the side injection of A, equations (a) and (b) have to be integrated separately. In this example we shall, however, only work out the relationship between the conversions of A and B for the conditions specified above. To this end, we eliminate $S\,dz$ from equations (a) and (b) and substitute (c) and the conversion rate expressions. The resulting differential equation is

$$(c_{AL} - c_{Ao}) d\Phi_v = \left(1 + \frac{k_X c_{AL}}{k_P c_B}\right) d(\Phi_v c_B) \simeq \left\{1 + \Phi_{vL} \frac{k_X c_{AL}}{k_P(\Phi_v c_B)}\right\} d(\Phi_v c_B). \quad (e)$$

The latter approximation is valid only if the variation in Φ_v is small. This would mean that $\Phi_{vo} \simeq \Phi_v \simeq \Phi_{vL}$ or $\Phi_{vL} - \Phi_{vo} \ll \Phi_{vo}$; i.e., the total flow of the side injection is relatively small, but its content in A relatively high so that also $c_{Bo} \ll c_{Aw}$. On this assumption equation (e) can be integrated between $z = 0$ and $z = L$, which leads to

$$(c_{AL} - c_{Aw})(\Phi_{vL} - \Phi_{vo}) = (\Phi_{vL} c_{BL} - \Phi_{vo} c_{Bo}) + \frac{\Phi_{vL} c_{AL} k_X}{k_P} \ln \frac{\Phi_{vL} c_{BL}}{\Phi_{vo} c_{Bo}}.$$

Combining the result with equation (d), we finally obtain

$$\ln \frac{\Phi_{vL} c_{BL}}{\Phi_{vo} c_{Bo}} = \frac{k_P}{k_X} \left\{1 - \frac{c_{BL}}{c_{AL}}\right\}. \quad (f)$$

We now introduce the relative degrees of conversion of A and B at the reactor outlet. Since the molar feeds of A and B are equal and the total amounts of A and of B are $\Phi_{vo} c_{Bo}$, we have

$$\zeta_{AL} = 1 - \Phi_{vL} c_{AL}/\Phi_{vo} c_{Bo}$$

and

$$\zeta_{BL} = 1 - \Phi_{vL} c_{BL}/\Phi_{vo} c_{Bo}.$$

Substitution of these expressions into equation (f) leads to the final result:

$$\ln(1 - \zeta_{BL}) = \frac{k_P}{k_X} \frac{\zeta_{BL} - \zeta_{AL}}{1 - \zeta_{AL}}. \quad (g)$$

It is seen that, for this reactor and on the assumptions (c) and (d), a single relationship exists between ζ_{AL} and ζ_{BL}, with the value of k_P/k_X as a parameter. For the dimensionless reactor volume of this reactor we find with

$$\Phi_{vo} \simeq \Phi_v \simeq \Phi_{vL} \quad \text{and} \quad c_A = c_{AL} \simeq c_{Bo}(1 - \zeta_{AL})$$

and by substituting in and integrating of equation (b):

$$k_P c_{Bo} \frac{V}{\Phi_{vo}} = k_P c_{Bo} \tau_L = -\frac{\ln(1 - \zeta_{BL})}{(1 - \zeta_{AL})}.$$

If the same reaction is carried out in a *batch reactor*, we may write, with a constant density (see equation (II.3))

$$\frac{dc_A}{dt} = -k_P c_A c_B - k_X c_A^2$$

and

$$\frac{dc_B}{dt} = -k_P c_A c_B.$$

In order to obtain a relationship between c_A and c_B, the time is eliminated from these equations:

$$\frac{dc_A}{dc_B} = 1 + \frac{k_X c_A}{k_P c_B}.$$

After introducing the new variable c_A/c_B and integrating with $c_A = c_{Ao}$ when $c_B = c_{Bo}$, the solution is

$$\left(\frac{k_X}{k_P} - 1\right) \ln \frac{c_B}{c_{Bo}} = \ln \left[\frac{1 + \left(\frac{k_X}{k_P} - 1\right)\frac{c_A}{c_B}}{1 + \left(\frac{k_X}{k_P} - 1\right)\frac{c_{Ao}}{c_{Bo}}}\right]$$

and, in terms of relative conversions, when $c_{Ao} = c_{Bo}$,

$$\left(\frac{k_X}{k_P} - 1\right) \ln (1 - \zeta_B) = \ln \left[1 + \left(\frac{k_X}{k_P} - 1\right)\frac{1 - \zeta_A}{1 - \zeta_B}\right] - \ln \frac{k_X}{k_P}. \tag{h}$$

For $k_P/k_X = 1$ this expression is reduced to

$$1 - \zeta_A = (1 - \zeta_B)\{1 + \ln (1 - \zeta_B)\}.$$

For a continuous *tank reactor* the material balances for A and B become

$$c_{Ao} - c_{A1} = k_P \tau c_{A1} c_{B1} + k_X \tau c_{A1}^2$$

and

$$c_{Bo} - c_{B1} = k_P \tau c_{A1} c_{B1}.$$

Division of the first equation by the second yields

$$\frac{c_{Ao} - c_{A1}}{c_{Bo} - c_{B1}} = 1 + \frac{k_X c_{A1}}{k_P c_{B1}}$$

and with $c_{Ao} = c_{Bo}$

$$\frac{\zeta_{A1}}{\zeta_{B1}} = 1 + \frac{k_X}{k_P} \frac{1 - \zeta_{A1}}{1 - \zeta_{B1}}. \tag{k}$$

We can compare the solutions (h) and (k) with the result (g) for the idealized cross flow reactor with uniform concentration of A. With $k_P/k_X = 1$, stoichiometric feed quantities of A and B, and $\zeta_A = 0.98$, we find:

batch (or tubular) reactor: $\zeta_B = 0.61$
continuous tank reactor: $\zeta_{B1} = 0.86$
idealized cross-flow reactor: $\zeta_{B1L} = 0.93$.

It is seen that the continuous tank reactor gives a smaller conversion of B, and consequently a lower yield of B than the cross flow reactor. This results from the fact that the ratio c_B/c_A, averaged over the whole reactor content, is highest in the latter case. The higher this ratio, the more the desired reaction is favoured. Since this ratio is lowest in a batch or normal tubular reactor, a relatively large portion of A is here converted to the undesired product.

This is also demonstrated in figure III.12, where the yields and selectivities are shown for the three different reactor types. The reaction paths are shown in figure III.13 for the given reaction scheme. From this figure the same conclusions can be drawn.

(*Example III.3 ended*)

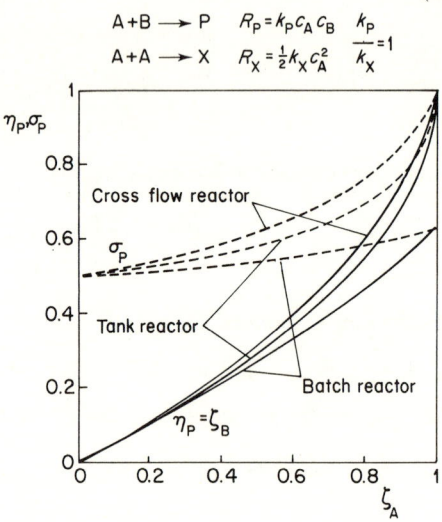

Fig. III.12 Selectivity and yield with parallel reactions; Example III.3

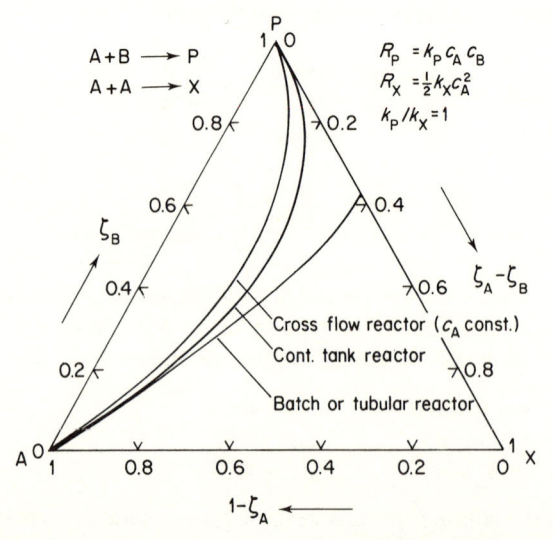

Fig. III.13 Isothermal reaction paths for parallel reactions of different order with respect to one of the reactants; Example III.3. Stoichiometric quantities of A and B fed

For the reaction scheme $A + B \to P$ and $A + B \to X$ with $R_P = k_P c_A c_B$ and $R_X = \frac{1}{2} k_X c_A^2$ and with isothermal operation in a batch reactor or a continuous tubular reactor, it is shown in figure III.13 that the rate of the undesired reaction decreases as the conversion increases in consequence of the decrease of c_A. With the use of a continuous tank reactor or a properly operated cross flow reactor, the concentration of A is intentionally kept relatively low so that the desired reaction, which is only first order in A, is favoured. Here, with an extremely low value of c_A, practically pure P can be produced; this would, however, require an extremely large reaction volume. Examples of these reactions are the synthesis of methanol ($CO + 2H_2 \to CH_3OH$) with the side reaction $CO + 3H_2 \to CH_4 + H_2O$ and the isomerization of xylenes (first order) with second order side reactions leading to the decomposition of the xylenes.

The simultaneous or parallel reactions in Example III.3 were of the type

$$A + B \to P, \quad R_P = k_P c_A c_B$$
$$A + A \to X, \quad R_X = \tfrac{1}{2} k_X c_A^2.$$

A high yield can be obtained with this system if the ratio c_B/c_A is as high as possible throughout the reactor. It was shown (also see figure III.12) that a tank reactor gives a higher yield than a tubular reactor, but that even better results can be obtained if a distributed feed of reactant A is properly applied as an additional degree of freedom. This was put forward in a paper by van de Vusse and Voetter [2] who for the above reactions studied the strategy of multiple injection of component A for obtaining a high yield of P. When the feed distribution to a tubular reactor was made in such a way that the ratio c_B/c_A remained constant over the reactor length, the yield of a tank reactor with the same value of c_B/c_A could be matched, but the capacity was much higher. This can be readily understood, since the value of σ'_P then has the same value everywhere in both reactors, while the absolute reactant concentrations are, on the average, higher in the cross flow tubular reactor.

If, on the other hand, c_A is made constant over the length of the tubular reactor, the ratio c_B/c_A averaged over the length is greater than for a tank reactor operating with the same c_A value. Accordingly, the yield is higher than that of a tank reactor with the same final degree of conversion of A. Its capacity is also higher. This indicates that the cross flow reactor has possibilities for obtaining reasonable capacities and high yields with this type of reaction. This is of importance for some catalysed gas reactions since the stirred tank is not a suitable type of reactor for these.

For practical reasons, the side injection of A should be concentrated at a limited number of feed points. Messikommer [9] investigated how reactant A should be distributed over a given number of feed points in order to obtain the maximum reactor yield η_P. He found that its value was rather insensitive to the feed distribution and that a feed of equal fractions gave almost the same result as the optimum feed distribution.

The performance of a number of reactor systems and injection policies for the above reaction has been compared in figure III.14. Figure III.14a shows the idealized cross flow tubular reactor in which c_A is constant over the reactor length. Figures III.14b and III.14c give the results of Messikommer's calculation for five injection points; there is hardly any difference between the optimum and the equal feed distributions. The systems a, b, c and d all show better yields and capacities than the single tank reactor in figure III.14e. The ordinary tubular reactor (f) has a much greater capacity than the tank reactor, but its yield is considerably lower. All strategies of figure III.14, except d and e, can also be carried out in a semi-batch reactor.

	$\dfrac{k_1 c_{Bo} V_r}{\phi_v}$	η_P
(a) Idealized cross flow reactor, c_A = constant	39	0.855
(b) Optimum feed distribution of A 0.38 0.24 0.17 0.12 0.09	25	0.830
(c) Feed of A distributed in equal parts 0.20 0.20 0.20 0.20 0.20	25	0.827
(d) Feed of A distributed in equal parts 0.20 0.20 0.20 0.20 0.20	34	0.818
(e) Tank reactor	70	0.780
(f) Tubular reactor	4.1	0.595

Dimensionless volume Yield

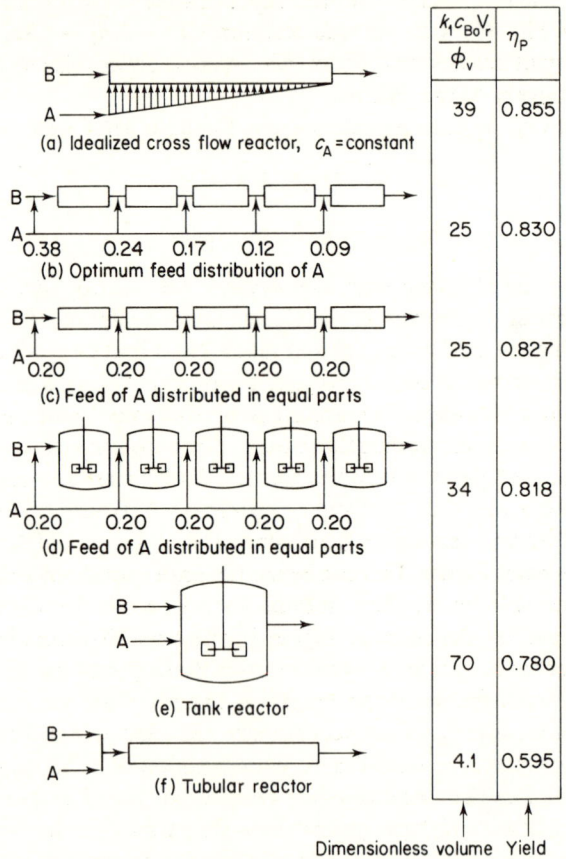

Fig. III.14 Dimensionless reactor volume and yield for various cross flow reactor systems; reaction of Example III.3 with $k_P = k_X$ and equal molar flow rates of A and B, final degree of conversion $\zeta_A = 0.95$

III.4 CONSECUTIVE REACTIONS

Pure consecutive reactions occur very often in industrial practice. They are of the general type

$$A \rightarrow P \rightarrow X$$

with rate equations like

$$R_A = -k_P c_A^m, \quad R_P = k_P c_A^m - k_X c_P^n \quad \text{and} \quad R_X = k_X c_P^n.$$

From those rate equations we find, for the differential selectivity,

$$\sigma_P' = \frac{k_P c_A^m - k_X c_P^n}{k_P c_A^m}. \tag{III.9}$$

From this equation it follows that whatever the value of m and n may be, the differential selectivity always decreases with increasing conversion (lower c_A) and increasing concentrations c_P of P. For a good selectivity, therefore, it is important that P once it is formed leaves the reactor as quick as possible. Any form of backmixing, therefore, is undesirable. This leads to the conclusion that the highest selectivity at equal conversions is obtained in a PFR or BR, because in these reactors at equal final conversions the concentration of the desired product, c_P, averaged over the reactor volume or reaction time, and the averaged concentration of the reactant, c_A are respectively lower and higher than in any other reactor type, provided P cannot be removed during the reaction. We discussed in Section I.6 that cutting off the reaction quickly, separating the reactor products and recirculation of unconverted reactant leads to the best results.

With consecutive reactions, the reaction path always ends in the final product(s) of the reaction chain. In many cases one of the intermediate products is the one desired. The maximum yield depends on the rates of formation and consumption of P, respectively, and on the way in which the reaction is carried out.

III.4.1 First order consecutive reactions in a plug flow reactor

For a PFR and similarly for a BR the relation between the yield and the conversion is given by

$$\sigma_P' = -\frac{dc_P}{dc_A} = \frac{d\eta_P}{d\zeta_A}. \tag{III.5}$$

Introduction of equation (III.9) for $m = n = 1$ into equation (III.5) and with $c_A = c_{Ao}(1 - \zeta_A)$ and $c_P = c_{Ao}\eta_P$ we have to solve the following differential equation:

$$\frac{d\eta_P}{d\zeta_A} = 1 - \kappa \frac{\eta_P}{1 - \zeta_A} \tag{III.10}$$

in which $\kappa = k_X/k_P$, if the index P refers to the desired reaction and X to the undesired one. The differential equation can also be solved by first determining the concentration profiles along the reactor length. The relation between the concentrations and location in the reactor are governed by the following differential equations:

for A $\quad : -dc_A = k_P c_A \, d\tau$

and for P: $\quad dc_P = (k_P c_A - k_X c_P) \, d\tau$

in which $\tau = zS/\Phi_v$. Integration with the boundary conditions $c_A = c_{Ao}$ and $c_P = 0$ for $z = 0$ leads to

for A $\quad : c_A/c_{Ao} = e^{-k_P \tau_L} = (1 - \zeta_{AL})$

and for P: $c_P/c_{Ao} = \eta_{PL} = \dfrac{1}{\kappa - 1} [e^{-k_P \tau_L} - e^{-k_X \tau_L}]$

$$= \dfrac{1 - \zeta_{AL}}{\kappa - 1} [1 - (1 - \zeta_{AL})^{\kappa - 1}] \quad \text{(III.10a)}$$

if

$$\kappa \neq 1 \quad \text{and} \quad c_P/c_{Ao} = (1 - \zeta_{AL}) \ln \dfrac{1}{1 - \zeta_{AL}} \quad \text{for} \quad \kappa = 1.$$

The reactor selectivity can be found with $\eta_{PL} = \sigma_{PL} \zeta_{AL}$, so that

$$\sigma_{PL} = \dfrac{1 - \zeta_{AL}}{(\kappa - 1) \zeta_{AL}} [1 - (1 - \zeta_{AL})^{\kappa - 1}]$$

and the differential selectivity by differentiation of η_P. This gives

$$\sigma'_P = \dfrac{1}{\kappa - 1} [\kappa (1 - \zeta_{AL})^{\kappa - 1} - 1]. \quad \text{(III.10b)}$$

For $\kappa = 0.2$ concentration profiles and a σ'_P profile are given as a function of the location in a PFR in figure III.15. We see that the yield reaches a maximum, at which also $\sigma'_P = 0$, which can be readily understood from the discussion above. As shown after a certain reactor length (or reaction time in a BR) all A has been converted and only P is still being converted into undesired product X.

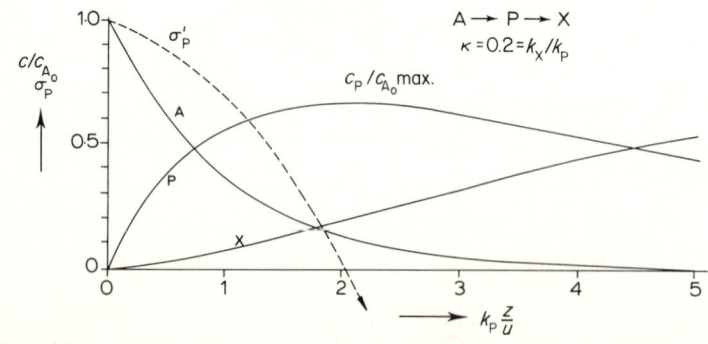

Fig. III.15 Concentration and differential selectivity profiles in a PFR for first order consecutive reactions

As discussed in Section I.6, the maximum yield is most important for plant economics. This maximum can be determined by differentiation of c_P/c_{Ao} and setting the derivative equal to zero. We then find:

for the optimum residence time:

$$k_P \tau_{L\,opt} = \frac{1}{\kappa - 1} \ln \kappa$$

and for the maximum yield:

$$\eta_{P\,max} = \left(\frac{c_P}{c_{Ao}}\right)_{max} = \kappa^{\kappa/(1-\kappa)}$$

and the reactant conversion at the optimum is

$$1 - \zeta_{AL} = e^{-k_P \tau_{L\,opt}} = \kappa^{1/(1-\kappa)}.$$

In figures III.16 and III.17 the relevant parameters for the optimum PFR are

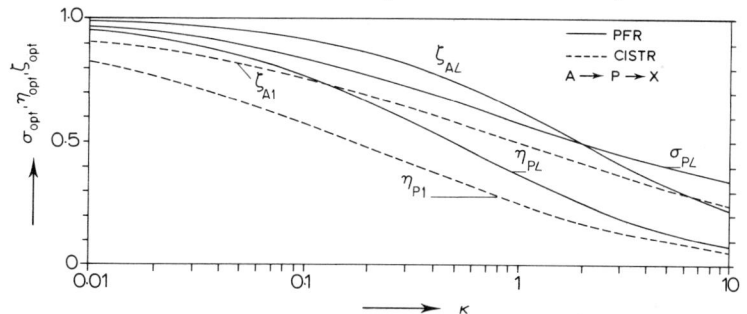

Fig. III.16 Optimal conversions, yields and selectivities in two model reactors for first order consecutive reactions as a function of the ratio $\kappa = k_X/k_P$

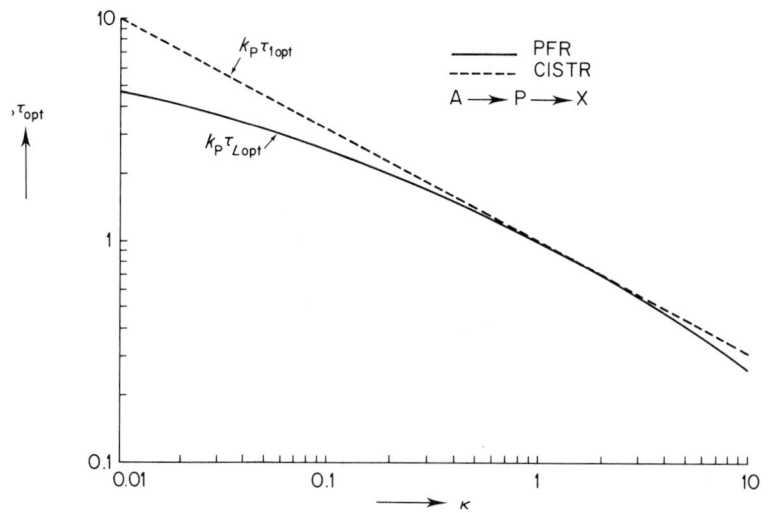

Fig. III.17 Optimal residence times in a PFR and a CISTR for first order consecutive reactions as a function of the ratio $\kappa = k_X/k_P$.

plotted as a function of κ. We see that in order to achieve high yields κ must be low. For yields above 75% the desired reaction must be at least ten times faster than the undesired reaction ($\kappa < 0.1$). The reactor selectivity is always higher than the yield, so that in case nonconverted reactant can be recovered and recycled to the reactor, higher values of κ are permitted in order to achieve equal overall plant outputs of desired product per unit of reactant feed intake, e.g. for $\sigma_{PL} = 0.75$ the value of κ should be lower than 0.3 (see Section I.6). We further see that at higher values of κ the conversion ζ_{AL} is considerably below 1, so in the optimum reactor large amounts of reactant remain unconverted after passing the reactor. Further it is shown that at lower values of κ the reactor becomes relatively larger. In order to change the value of κ into the desired direction either temperature has to be changed or catalysts have to be sought.

Example III.4.a *Relation between reactor selectivity and relative conversion*
The selectivity of a reactor σ_P can be easily derived from the local or differential selectivity only if σ'_P is not a function of σ_P, that is only if it is not influenced by products concentrations. This is in general the case only with irreversible parallel reactions. With consecutive reactions σ'_P generally is a function of both ζ_A and σ_P, so that the selectivity has to be found by solving the differential equation

$$\frac{d\sigma_P \zeta_A}{d\zeta_A} = \sigma'_P = f(\zeta_A, \sigma_P). \tag{III.5}$$

We will derive the selectivity for a PFR for a reaction $A \to P \to X$ with the rate equations

$$R_A = -k_P c_A$$
$$R_P = k_P c_A - k_X c_P + k_{-X} c_X$$
$$R_X = k_X c_P - k_{-X} c_X.$$

At the inlet of the PFR $c_P = c_X = 0$, the density of the reaction mixture is constant, $v_A = v_P = v_X = 1$ and $M_A = M_P = M_X$. In this reaction scheme the differential selectivity at a certain place in the reactor, where a relative conversion of ζ_A and selectivity of σ_P have been reached, is

$$\sigma'_P = \frac{k_P c_A - k_X c_P + k_{-X} c_X}{-k_P c_A}$$

where

$$c_A = c_{Ao}(1 - \zeta_A),$$
$$c_P = c_{Ao} \sigma_P \zeta_A$$

and

$$c_X = c_{Ao} - c_A - c_P = c_{Ao} \zeta_A (1 - \sigma_P).$$

Elimination of c_P and c_X and rearranging leads to the following differential equation:

$$\frac{d(\sigma_P \zeta_A)}{d\zeta_A} = \sigma'_P = -\frac{k_X + k_{-X}}{k_P(1-\zeta_A)}(\sigma_P \zeta_A) + 1 + \frac{k_{-X}}{k_P}\frac{\zeta_A}{1-\zeta_A}.$$

This linear first order differential equation can be solved analytically only for discrete values of $(k_X + k_{-X})/k_P$, otherwise it has to be integrated numerically by computer.

In case $k_P = k_X = k_{-X}$ the solution is, for the PFR,

$$\sigma_{PL} = \frac{1}{2\zeta_A}[1-(1-\zeta_A)^2].$$

For a PFR with a very short residue time ($\zeta_A \simeq 0$) we find $\sigma_{PL} = 1.0$, and for very long residence times, where ζ_A is approximately 1.0 and the reactant is almost completely converted, we find that an equilibrium between the products P and X has been established. Because $k_X = k_{-X}$ in this case the selectivity is $\sigma_{PL} = 0.5$ and $c_P = c_X$. The reaction path is shown in figure III.18.

$A \to P \to X$
$R_A = -k_P c_A$
$R_P = k_P c_A - k_X c_P + k_{-X} c_P$

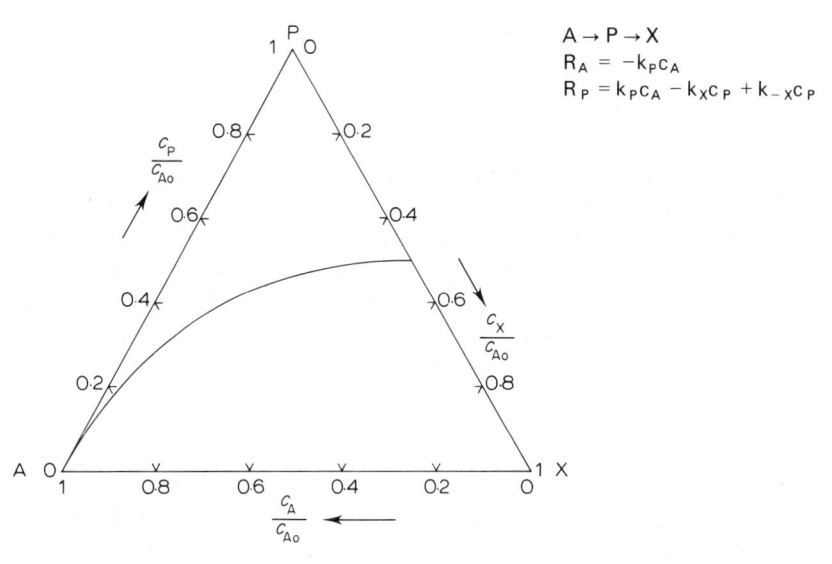

Fig. III.18 The reaction path for the reaction $A \to P \rightleftarrows X$ in a PFR. Here $k_P = k_X = k_{-X}$

III.4.2 First order consecutive reactions in a tank reactor

The relation equation (III.9) reduces for first order consecutive reactions in a CISTR to

$$\sigma_{P1} = \frac{1-\zeta_{A1}}{(\kappa-1)\zeta_{A1}+1} \qquad (III.10c)$$

because for a CISTR $\sigma'_P = \sigma_P$. From this we immediately see that the yield in a CISTR is given by

$$\eta_{P1} = \frac{\zeta_{A1}(1 - \zeta_{A1})}{(\kappa - 1)\zeta_{A1} + 1}$$

and for $\kappa = 1$ we find

$$\sigma_{P1} = 1 - \zeta_{A1}.$$

Also a CISTR demonstrates a maximum yield as a function of the average residence time τ_1. We can easily derive this by setting $d\eta_{P1}/d\zeta_{A1} = 0$, so that for the optimum CISTR the maximum yield is given by

$$\eta_{P1\,max} = \frac{1}{(1 + \sqrt{\kappa})^2}$$

and the relative conversion at the optimum by

$$\zeta_{A1\,opt} = \frac{1}{1 + \sqrt{\kappa}}.$$

The residence time in the optimal reactor is

$$k_P\tau_{1\,opt} = \frac{1}{\sqrt{\kappa}} \quad \text{or} \quad \tau_{1\,opt} = \frac{1}{\sqrt{(k_P k_X)}}.$$

The selectivity in a CISTR at optimum conditions cannot be equal to zero as in a PFR. From the relations above we see $\sigma'_P = \sigma_{P1} = \zeta_{A1}$ at the optimum. In figures III.16 and III.17 we see that for consecutive reactions the CISTR gives lower yields at lower relative conversions at optimum conditions than a PFR, whereas it requires a larger reactor volume, provided both reactors work at the same value of κ. The same can be seen from the reaction path, as shown in figure III.19.

III.4.3 General discussion

From the previous discussion we have seen that for consecutive reactions we always have to apply a PFR in order to achieve high yields and selectivities or—if a PFR cannot be realized technically—we have to apply a cascade with as many CISTRs as economically justified. In the latter case raw material and energy costs have to be balanced against the investment related costs of the reactor section.

For consecutive reactions of different reaction orders calculations are straightforward. Generally, a computer will be required. Chermin and van Krevelen [4] have extensively discussed non-reversible different order consecutive reactions in an isothermal PFR and Jungers et al. [5] have done so for reversible consecutive reactions. However, in this case, we introduce for κ:

$$\kappa = \frac{k_X}{k_P} c_{Ao}^{n-m}.$$

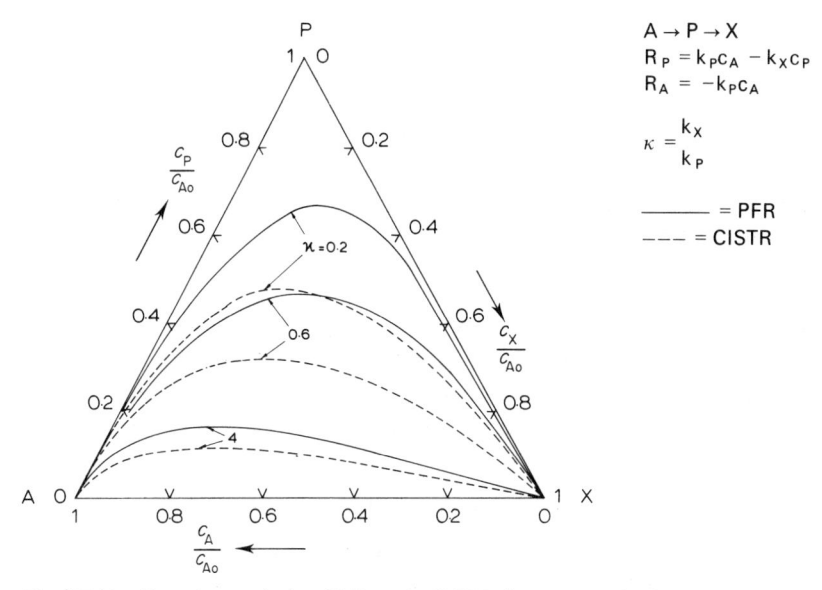

Fig. III.19 Reaction paths in a PFR and a CISTR for consecutive first order reactions and for varying values of κ

This κ gives us an additional tool to influence the isothermal reactor performance into the direction of the desired reaction apart from using catalysts: now by increasing the reactant feed concentration or reactor pressure if $n - m < 0$ and by decreasing if $n - m > 0$, we lower the value of κ and consequently yields and selectivities are improved. Because of non-linearity of the differential equations we will generally have to use computers. For a CISTR and a cascade of CISTRs equations are algebraic and can be solved more easily. Graphical or numerical solutions are recommended if the calculations become too involved in case of a cascade.

In the following three illustrations we will elaborate more on two important aspects of consecutive reactions executed in an isothermal PFR.

Example III.4.b *Instructions on reactor operation*
Suppose you have designed a PFR for the manufacture of a costly product, which decomposes. Research and development provided the kinetics: reactions are of the first order consecutive type with $\kappa = 2$ at design conditions. Yield is very poor, but acceptable because the product is an expensive fine chemical. You designed the reactor for optimum conditions and have to instruct the operator on the allowable flow rates to the reactor, which is of the PFR type. The first thing is to point out that the reactor gives only the design yield at the design feed flow rate. However, if a loss in yield of 5% of the maximum is permissible, the flow rate limits can be calculated from equation (III.10a):

$$\eta_{PL}/\eta_{PL\,max} = \frac{1}{\kappa - 1}\left[e^{-k_P\tau_L} - e^{-k_X\tau_L}\right]\kappa^{\kappa/(\kappa - 1)} \geq 0.95.$$

This leads to the requirement

$$0.4913 < k_P\tau_L < 0.9462,$$

whereas at optimum conditions $k_P\tau_L = 0.6931$. So the feed rate should not be lower than $0.4913/0.6931 = 71\%$ or higher than 137% of the design value, if yields are to be kept within 5% of the maximum value. Because $\kappa = 2$ the roots of the equation could be found easily. For other values of κ a computer has to be used.

Example III.4.c *The chlorination of benzene in various reactor systems*

The desired product monochlorobenzene (P) is formed in the chlorination of benzene, but further chlorination resulting in dichloro- and trichloro-benzene (undesired products X and Y, respectively) takes place as well. The equations of the three consecutive reactions are:

$$C_6H_6(A) + Cl_2(B) \rightarrow C_6H_5Cl(P) + HCL$$
$$C_6H_5Cl(P) + Cl_2(B) \rightarrow C_6H_4Cl_2(X) + HCl$$
$$C_6H_4Cl_2(X) + Cl_2(B) \rightarrow C_6H_3Cl_3(Y) + HCl.$$

The reaction is carried out in large stirred reaction vessels in which gaseous Cl_2 is dispersed. Hydrochloric acid, some chlorine and a small amount of volatile by-products leave the reactor as gases. In the most favourable temperature range (around 50°C) the reaction is so slow that the conversion rate is determined entirely by the chemical reaction velocity, and that the liquid may be assumed to be saturated with Cl_2 at the prevailing temperature and pressure. Under these circumstances the various net production rates can be written in the following form:

$$R_A = -k_P c_A c_B, \qquad R_P = k_P c_A c_B - k_X c_P c_B$$
$$R_X = k_X c_P c_B - k_Y c_X c_B, \qquad R_Y = k_Y c_X c_B$$

where k_P, k_X and k_Y are functions of temperature. McMullin [3] gives for 55°C:

$$\frac{k_P}{k_X} = 8.0 \quad \text{and} \quad \frac{k_X}{k_Y} = 30.$$

As the reaction proceeds, the density of the reaction mixture increases. For the sake of convenience we shall postulate the approximation that the total number of moles per unit volume (disregarding the dissolved chlorine) remains constant:

$$c_A + c_P + c_X + c_Y = c_{Ao}.$$

The above data are not sufficient for the calculation of the reaction volume needed for a certain production rate of P. We can, however, calculate the product distribution for various reactor types as a function of the extent of the reaction; for this we may take either the relative degree of conversion of benzene,

$$\zeta_A = 1 - c_A/c_{Ao},$$

or the number of moles of Cl_2 used per mole of benzene,
$$\gamma_B = (c_P + 2c_X + 3c_Y)/c_{Ao};$$
the latter quantity may exceed unity. The reactor types to be discussed are the batch reactor, the continuous tank reactor and a cascade of three tank reactors.

Strictly speaking, the *batch reactor* is here a semi-continuous one, since Cl_2 is continuously supplied and HCl is removed; as a result the total mass of the reaction mixture increases. The above assumption that the sum of molar concentrations remains constant, however, involves that the total reaction volume V_r remains constant. Application of the material balance (II.1) for benzene (A), which is neither added to nor discharged from the system during the reaction, then yields

$$\frac{d(\rho V_r w_A)}{dt} = M_A R_A V_r.$$

With $w_A = M_A c_A/\rho$ and V_r = constant, we get

$$\frac{d(V_r c_A)}{dt} = R_A V_r, \quad \text{or} \quad \frac{dc_A}{dt} = R_A.$$

Similarly

$$\frac{dc_P}{dt} = R_P, \quad \frac{dc_X}{dt} = R_X \quad \text{and} \quad \frac{dc_Y}{dt} = R_Y.$$

This shows that the system may be treated as a normal batch reactor. After substitution of the rate expressions we form the ratio

$$\frac{dc_P}{dt} \Big/ \frac{dc_A}{dt} = \frac{dc_P}{dc_A} = \frac{k_P c_A - k_X c_P}{-k_P c_A}.$$

This differential equation must be solved with the boundary condition: $c_P = 0$ for $c_A = c_{Ao}$. Following introduction of c_P/c_A as a new dependent variable, the equation can be integrated:

$$\frac{c_P}{c_{Ao}} = \frac{k_P}{k_P - k_X} \frac{c_A}{c_{Ao}} \left[\left(\frac{c_A}{c_{Ao}}\right)^{(k_X - k_P)/k_P} - 1 \right]. \quad (a)$$

Similarly, an expression for c_X can be found. Here, however, we shall consider X and Y together as the undesired products, with concentration $c_X + c_Y = c_{Ao} - c_A - c_P$. Since k_X/k_Y is large and an excessive degree of chlorination will not be considered ($c_Y \ll c_X$) we have approximately

$$\gamma_B \approx (c_P + 2c_X + 2c_Y)/c_{Ao} = (2c_{Ao} - 2c_A - c_P)/c_{Ao}.$$

In accordance with the definition (II.23), the selectivity is given by

$$\sigma_P = c_P/(c_{Ao} - c_A) = c_P/c_{Ao}\zeta_A,$$

and the yield of monochlorobenzene becomes

$$\eta_P = c_P/c_{Ao} \approx 2\zeta_A - \gamma_B.$$

Some values of the product distribution calculated from equation (a) with $k_P/k_X = 8.0$ are shown in Table III.5. The product distributions, σ_P and η_P, have been plotted in figure III.20 as a function of the degree of chlorination γ_B.

Table III.5 Chlorination in a batch reactor; Example III.4.c

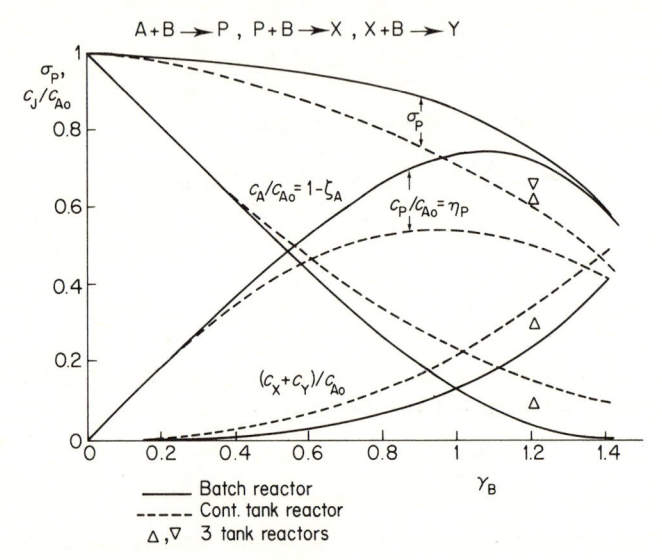

$c_A/c_{Ao} = 1 - \zeta_A$	1.0	0.8	0.6	0.4	0.2	0.1	0.05	0.01
$c_P/c_{Ao} = \eta_P$	0	0.195	0.384	0.562	0.706	0.745	0.732	0.632
$(c_X + c_Y)/c_{Ao}$	0	0.005	0.016	0.038	0.094	0.155	0.218	0.358
γ_B	0	0.21	0.42	0.64	0.89	1.06	1.17	1.35
$c_P/c_{Ao}\zeta_A = \sigma_P$	1.0	0.98	0.96	0.94	0.88	0.83	0.77	0.64

Fig. III.20 Concentrations and selectivity for the chlorination of benzene, $k_P/k_X = 8.0$; Example III.4.c

If the reaction is carried out in a single *tank reactor*, the following material balances apply to the liquid phase:

$$0 = \Phi_v(c_{Ao} - c_{A1}) - k_P c_{A1} c_{B1} V_r$$
$$0 = -\Phi_v c_{P1} + (k_P c_{A1} c_{B1} - k_X c_{P1} c_{B1}) V_r$$
$$0 = -\Phi_v c_{X1} + (k_X c_{P1} c_{B1} - k_Y c_{X1} c_{B1}) V_r$$
$$0 = -\Phi_v c_{Y1} + k_Y c_{X1} c_{B1} V_r.$$

The ratio of the first two expressions yields the relation between c_{P1} and c_{A1}:

$$\frac{c_{P1}}{c_{Ao} - c_{A1}} = 1 - \frac{k_X}{k_P} \frac{c_{P1}}{c_{A1}}. \tag{b}$$

Formulae for c_{X1} and c_{Y1} can similarly be found, but we only use their sum, as before. Some values of the composition calculated from equation (b) are given in

Table III.6. The composition according to equation (b) has also been plotted in figure III.20. It is seen that in this case the maximum yield of monochlorobenzene (P) is much smaller than in the case of the batch reactor and that the selectivity rapidly diminishes as a function of increasing conversion.

Table III.6 Chlorination in a tank reactor; Example III.4.c

$c_{A1}/c_{Ao} = 1 - \zeta_{A1}$	1.0	0.8	0.6	0.4	0.2	0.1	0.05
$c_{P1}/c_{Ao} = \eta_{P1}$	0	0.195	0.369	0.505	0.534	0.425	0.281
$(c_{X1} + c_{Y1})/c_{Ao}$	0	0.005	0.031	0.095	0.266	0.475	0.669
γ_{B1}	0	0.21	0.43	0.70	1.07	1.38	1.62
$c_{P1}/c_{Ao}\zeta_{A1} = \sigma_{P1}$	1.0	0.98	0.92	0.84	0.67	0.47	0.30

The maximum yield of monochlorobenzene (P) can be increased with respect to the above process by carrying out the reaction in a cascade of tank reactors instead of in one tank reactor. We select as an example a cascade of *three tank reactors*. From the many possible distributions of the total reaction volume and the chlorine feed over the three tanks we arbitrarily select a system where the chlorine consumption in each tank reactor has the same value. The material balances for A and P over the nth tank reactor give the relationship

$$\frac{c_{Pn} - c_{P(n-1)}}{c_{A(n-1)} - c_{An}} = 1 - \frac{k_X\, c_{Pn}}{k_P\, c_{An}}.$$

In view of the requirement of equal chlorine consumption we eliminate c_{Pn} and $c_{P(n-1)}$ by introducing the relative degrees of chlorination:

$$\gamma_{Bn} \approx (2c_{Ao} - 2c_{An} - c_{Pn})/c_{Ao},$$

and

$$\gamma_{B(n-1)} \approx (2c_{Ao} - 2c_{A(n-1)} - c_{P(n-1)})/c_{Ao},$$

after which the above formula becomes

$$\frac{\gamma_{Bn} - \gamma_{B(n-1)}}{c_{A(n-1)} - c_{An}} c_{Ao} = 1 - \frac{k_X}{k_P}\left[2 - (2 - \gamma_{Bn})\frac{c_{Ao}}{c_{An}}\right]. \tag{c}$$

It has to be applied to the three successive reactors; Table III.7 gives the results for $\quad k_P/k_X = 8.0 \quad$ and $\quad \gamma_{B1} = \gamma_{B2} - \gamma_{B1} = \gamma_{B3} - \gamma_{B2} = 0.4.$

Table III.7 Chlorination in a cascade of three tank reactors

Reactor number:	$n = 0$	$n = 1$	$n = 2$	$n = 3$
γ_{Bn}	0	0.4	0.8	1.2
$c_{An}/c_{Ao} = 1 - \zeta_{An}$	1	0.627	0.305	0.090
$c_{Pn}/c_{Ao} = \eta_{Pn}$	0	0.346	0.590	0.620
$(c_{Xn} + c_{Yn})/c_{Ao}$	0	0.027	0.105	0.290
$c_{Pn}/c_{Ao}\zeta_{An} = \sigma_{Pn}$	1.0	0.93	0.85	0.68

The result for $N = 3$ is represented by one point in figure III.21. It is seen that the performance of the cascade lies between that of the single tank reactor and the batch reactor.

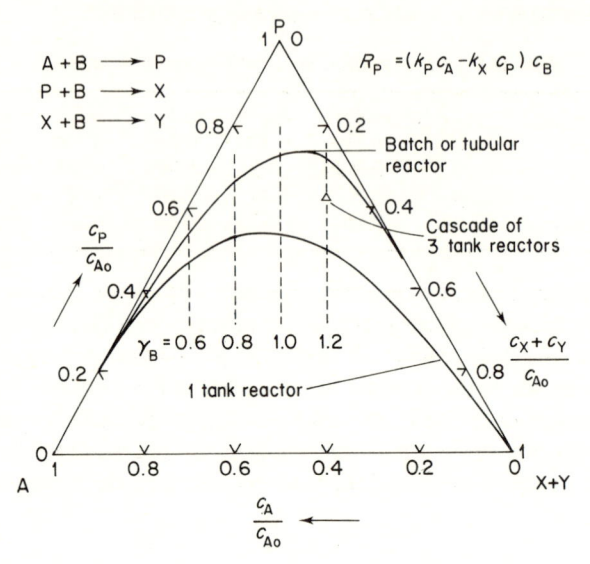

$$A + B \longrightarrow P$$
$$P + B \longrightarrow X$$
$$X + B \longrightarrow Y$$

$$R_P = (k_P c_A - k_X c_P) c_B$$

Fig. III.21 Isothermal reaction paths for the chlorination of benzene; Example III.4.c

It is a general feature of consecutive reactions that—if an intermediate product is the one desired—a batch or tubular reactor, or a cascade approaching its behaviour, gives the highest conversion to the desired product. It is clear from figure III.21 that the reaction should be stopped when the concentration of the desired product has reached its maximum value. For continuous operation this means that the maximum yield of P is obtained only at one particular value of the throughput. Any departure from this value involves a decrease in the yield.

The fact that the selectivity is still large at a small degree of conversion can be used in principle if the non-converted reactant can be separated in an auxiliary unit from the products and recirculated to the reactor. In this way, both the selectivity and the degree of conversion (calculated over the combination of reactor and separating unit) can be kept at a high value. The proper selection of the conditions to be used is governed entirely by economic considerations.

Example III.4.d *A chain of consecutive reactions, the chlorination of methane* In Example III.4c we studied the chlorination of benzene, where the monochlorobenzene was the desired product. That problem could be solved straightforwardly, but the solution becomes more complicated if a whole chain of consecutive reactions occurs and intermediate products cannot be described as desired or undesired ones. This, for example, is the case in the chlorination of methane (0), where methylchloride CH_3Cl (1), methylenechloride CH_2Cl_2 (2), chloroform $CHCl_3$ (3) and carbontetrachloride CCl_4 (4) are formed successively according to:

$$CH_4 \xrightarrow[1]{+Cl_2} CH_3Cl \xrightarrow[2]{+Cl_2} CH_2Cl_2 \xrightarrow[3]{+Cl_2} CHCl_3 \xrightarrow[4]{+Cl_2} CCl_4.$$

Market demand is relatively high for CH_2Cl_2 and CCl_4 and low for CH_3Cl and $CHCl_3$. So in general we want to suppress the formation of 1 and 3 and increase the production of 2 and 4. This problem was first studied by Küchler and Langbein [6] and later extended by Kobelt and Troltenier [7]. The reactions are carried out in a tubular furnace in the gas phase; they proceed above 400°C at reasonable rates, and reaction temperatures should be kept below 500°C to prevent soot formation. By applying excess CH_4 and recirculation of the CH_4 the relative heat capacity of the reaction mixture is increased and temperature changes are kept reasonably low, so that in the first instance the reactor conditions can be approximated by an isothermal PFR. With the assumption of a constant density we have to solve the following set of differential equations:

$$dc_0 = -k_1 c_0 c_{Cl_2} \, d\tau$$
$$dc_1 = (k_1 c_0 - k_2 c_1) c_{Cl_2} \, d\tau$$
$$dc_2 = (k_2 c_1 - k_3 c_2) c_{Cl_2} \, d\tau$$
$$dc_3 = (k_3 c_2 - k_4 c_3) c_{Cl_2} \, d\tau$$
$$dc_4 = k_4 c_3 c_{Cl_2} \, d\tau$$

and for the chlorine consumption:

$$c_{Cl_2}(0) - c_{Cl_2} = \Sigma v[c_v - c_v(0)] = \Delta c_1 + 2\Delta c_2 + 3\Delta c_3 + 4\Delta c_4.$$

(o) refers to inlet conditions.

Further $\tau = z/u$ and inlet conditions have to be specified. The overall reactions are first order in chlorine and (chloro)hydrocarbon concentrations as has been proved experimentally [8]. Further, Küchler and Langbein determined that $k_2/k_1 : k_3/k_1 : k_4/k_1 = 3 : 1.5 : 0.375$ under actual reactor conditions. The equations above have to be solved by computer. Results for a PFR fed with methane are given in figure III.22, where product concentrations are shown as a function of the specific chlorine consumption γ_B. The specific chlorine consumption is defined as the number of moles of Cl_2 consumed per mole of (chloro)hydrocarbons fed to the reactor.

In figure III.20 the feed consists solely of CH_4. We see that a fixed reactor product composition is obtained as a function of γ_B. In order to reach a high yield of CH_2Cl_2, γ_B should be around 1.2. At $\gamma_B = 1.2$ the reactor product composition is:

$$c_0/c_0(o) = 0.410$$
$$c_1/c_0(o) = 0.172$$
$$c_2/c_0(o) = 0.248$$
$$c_3/c_0(o) = 0.153$$
$$c_4/c_0(o) = 0.017.$$

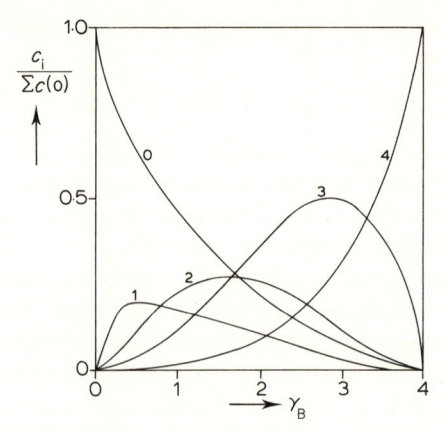

Fig. III.22 Concentration profiles for the chlorination of methane; Example III.4.d. γ_B is the chlorine consumption in moles per mole of feed [6] [From Küchler and Langbein [6]. Reproduced by permission of Pergamon Press Ltd]

It is clear that normally this product distribution will not conform to market demand and that excess amounts of CH_3Cl and $CHCl_3$ are produced. Now part of the reactor products after separation may be recirculated to the reactor section as shown in figure III.23. If CH_3Cl is recycled, for example, a possible concentration profile is given in figure III.24. At the γ_B value indicated in this figure no net CH_3Cl is produced at all in the plant; at lower values of γ_B net production is positive; at higher values of γ_B negative. If the fractions R_1 of CH_3Cl and R_3 of $CHCl_3$ produced are recycled to the reactor section, the inlet concentrations will be $R_1 c_{1L}$ and $R_3 c_{3L}$ respectively. The total amount of (chloro)hydrocarbons fed to the reactor now equals $c_0(o) + R_1 c_{1L} + R_3 c_{3L}$ and is kept constant in the following. This means that less CH_4 is fed and more chlorine is consumed to chlorinate CH_3Cl and $CHCl_3$. In figure III.25 the fraction of CH_4 converted into consecutive products is shown for varying values

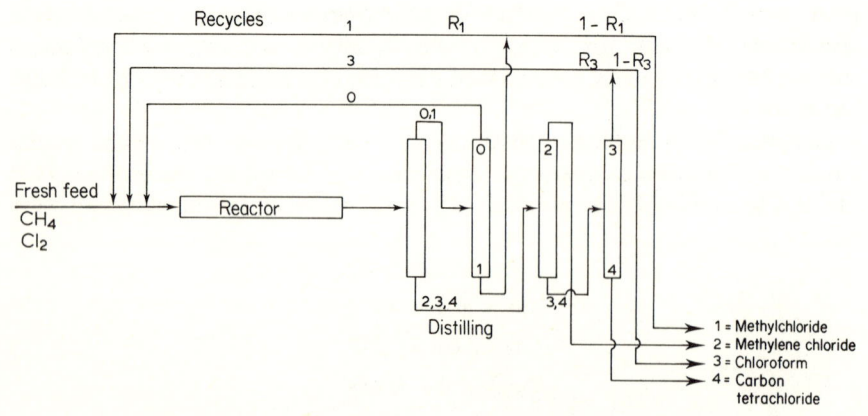

Fig. III.23 Possible line-up of a methane chlorination plant

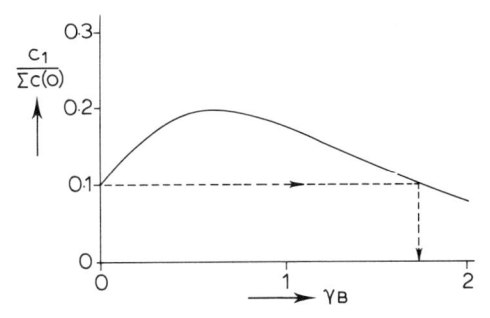

Fig. III.24 Concentration profile of component 1, if it is (partly) recycled; Example III.4.d. γ_B is the specific chlorine consumption [6] [From Küchler and Langbein [6]. Reproduced by permission of Pergamon Press Ltd]

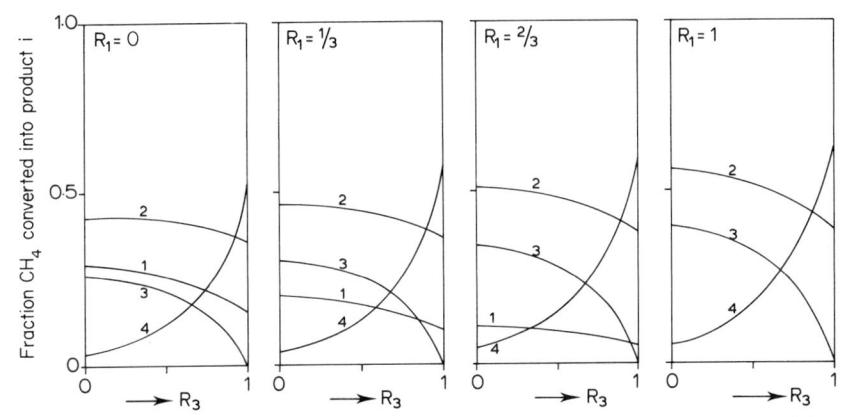

Fig. III.25 Yields in a methane chlorination plant, if intermediate products 1 and 3 are recycled. See also figure III.23 for symbols used

of R_1 and R_3 at constant $\gamma_B = 1.2$. The range of product distributions, which can be covered at this set value of γ_B by recycling of CH_3Cl and/or $CHCl_3$ is shown in Table III.8.

By also recycling CH_2Cl_2 many possible combinations of market demands can be covered. Of course other processes are available if only one or two products out of the whole range have to be made.

Table III.8 Influence of recycle on product distribution; Example III.4.d

Recycle ratios		Fraction of CH_4 converted into			
R_1	R_3	CH_3Cl	CH_2Cl_2	$CHCl_3$	CCl_4
0	0	0.29	0.42	0.26	0.03
1	0	0	0.56	0.39	0.05
0	1	0.15	0.36	0	0.49
1	1	0	0.38	0	0.62

III.5 COMBINATION REACTIONS

With combination reactions are meant combinations of parallel and consecutive reactions, which can be of either the reversible or the irreversible type. As an example the reaction scheme

$$2A \to P \xrightarrow{+A} X$$

has a consecutive character with respect to $A \to P \to X$ and a parallel character with respect to species $A: A + A \to P$ and $A + P \to X$. So this is a combination of reaction types. We find the same in

$$B \rightleftarrows A \to P \to X$$
$$\searrow Y$$

Here we have also a consecutive character with $A \to P \to X$ and with $A \to B \to Y$ and a parallel behaviour with regard to

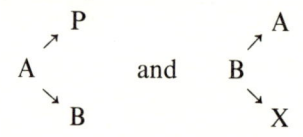

$$A \begin{matrix} \nearrow P \\ \searrow B \end{matrix} \quad \text{and} \quad B \begin{matrix} \nearrow A \\ \searrow X \end{matrix}$$

A great number of combination reaction schemes, often complex, occur in practice.

General

Each individual part of the reaction scheme consisting only either of parallel or of consecutive reactions, of course is subject to the rules given in Sections III.2–4, but just the combination of them prevents us from giving general rules for controlling the yield or selectivity. Solving of reactor problems for combination reactions is quite complicated and a general solution can seldom be found. By giving illustrations we will demonstrate how the relevant problems have to be approached. In general we have to try to acquire as much information as possible on the reaction scheme and the rate equations and then try to find more generally valid conclusions for our particular case by intelligently manipulating the available information. In this section we restrict ourselves to graphical and algebraic methods in order to illustrate how to approach yield and selectivity questions for combination reactions.

Graphical methods are the only resource if rate equations or rate constants are lacking, but they can help us to understand some underlying principles and to determine the effect on reactor performance of certain measures taken. If as a minimum all the reaction rate constants at one temperature are known, we can apply direct mathematical methods, although equations often get so complicated that a computer has to be applied. Nevertheless, we can obtain much information when the rate constants are available. We will give some illustrations and further leave it to the reader to develop his equations along the lines of this chapter and find his own solutions for his particular problems at hand.

Also for combination reactions, the *reaction path* is a useful tool. As an example, the reaction path is shown in figure III.26 for a reaction where the primary product P reacts further with reactant. If it is carried out in a batch reactor, only the first reaction occurs initially since only a small amount of P is present. At the end of the reaction, more P is present than A so that dc_P/dt is negative; however, depending on the value of k_P/k_X, more or less of P may be left after A has been completely converted, so that the reaction path does not necessarily end in the right-hand corner of the diagram. In a continuous tank reactor the composition is equal to that at the outlet; even at a relatively small degree of conversion of A, the composition of the reaction mixture favours the conversion of A and P to X.

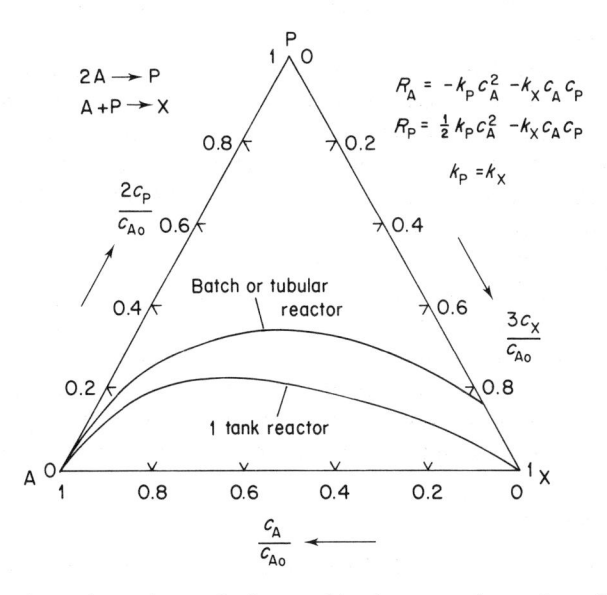

Fig. III.26 Isothermal reaction paths for combined consecutive and parallel reactions

If the reaction scheme is known and the expressions for the conversion rates are available, reaction paths for batch (or tubular) and tank reactors can, in principle, be predicted; for the former reactor types this entails the solution of differential equations and for the latter algebraic equations only. If only the reaction path in a BR were available, a graphical method would sometimes provide information on the behaviour of continuously operated reactors like the PFR or the CISTR, as will be shown later.

With respect to the *rate equations* we have assumed in this book that this information has already been put at the disposal of the chemical reaction engineer. If possible, we always recommend experimental work to obtain information on reaction kinetics. The manipulation of the rate equations is often complicated. Many complex reaction schemes are analysed in the book of

Rodiguin and Rodiguina [16]. The kinetics of coupled first order reactions, e.g. of the scheme

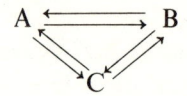

$$A \rightleftarrows B$$
$$\searrow C \nearrow$$

can be handled by linear algebra, as has been shown in a classical work of Wei and Prater [10]; provided the rate coefficients are constant. This provision excludes non isothermal operation or reactions with catalyst deactivation, where kinetic coefficients are a function of time. The decay of catalyst activity can also often be described with a slow undesired side reaction. Mah and Aris [11] used a consecutive and Froment and Bischoff [12] a parallel reaction to describe the catalyst deactivation. Wei and Kuo [13] considered both reaction types for catalyst fouling.

Often many reactions occur at the same time. Luss and Hutchinson [14] and also Wei and Kuo [13] have demonstrated how a mixture of many parallel reactions—as often occurs in petroleum or biological chemistry—can be lumped together. When the decay rate of the lump is empirically fitted to an nth order rate expression, this order is usually higher than that of the individual reacting species. For example, Wolfbauer *et al.* [15] showed that a group of zero order reactions under certain mixing conditions can be lumped to a first order rate expression.

For actual performance of a chemical plant the reactor yield and the reactor selectivity are of decisive importance. We will demonstrate that it is not always the PFR or the CISTR that are the best reactors to obtain maximum yields or selectivities, but that also other parameters like feed programming in an SBR, recycle of reactor product to the reactor inlet, concentration of the reactant in the feed and other measures can have beneficial effects on yield and selectivity.

III.5.1 Graphical methods

If kinetic data are absent, graphical methods can be helpful to give guidance as to reactor type and required residence times for optimum performance. Data will have to be procured in a model reactor—either an isothermal BR for slow reactions and a PFR for fast reactions, or a CISTR. If only *conversion as a function of batch or residence time is known*, we can determine the reaction rate from the conversion–'time' plot. For constant density reactions the derivative of the concentration–reaction time plot equals the reaction rate, because then $dc_J/dt = |R|_J$ (equation II.3). From a reaction rate versus conversion plot we can determine graphically the operating conditions: for a PFR the area under the curve is a measure for the required residence time, for a CISTR the area of $(1/R_1)\zeta_1$, and for a cascade see Section II.4. This is shown in figure III.27. If the reaction path in a PFR or a BR is known as the result of a number of laboratory experiments and no corresponding rate expressions are available, it is possible to find the reaction path for a tank reactor by a graphic construction. The method can only be used for reactions where at least one of the products, P or X, does not take part in the reaction and is a true final product.

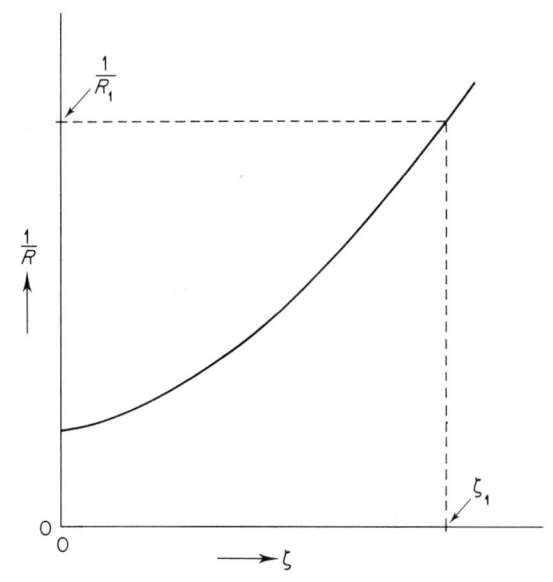

Fig. III.27 Reciprocal reaction rate versus conversion

If we consider a point 1 on an empirical batch reaction path, then we have (in molar units)

$$\left(\frac{dc_A}{dt}\right)_1 = R_{A1} \quad \text{and} \quad \left(\frac{dc_P}{dt}\right)_1 = R_{P1}.$$

We do not need to consider c_X since $c_A + c_P + c_X$ is assumed to be constant.
After elimination of dt, we find

$$\left(\frac{dc_A}{dc_P}\right)_1 = \left(\frac{R_A}{R_P}\right)_1. \tag{III.11a}$$

If the same point in the triangular diagram represents the composition at which a continuous tank reactor operates, we have (with $\rho = $ constant):

$$\Phi_v(c_{A1} - c_{Ao}) = R_{A1} V_r \quad \text{and} \quad \Phi_v(c_{P1} - c_{Po}) = R_{P1} V_r,$$

so that

$$\frac{c_{A1} - c_{Ao}}{c_{P1} - c_{Po}} = \left(\frac{R_A}{R_P}\right)_1. \tag{III.11b}$$

A comparison of equations (III.11a) and (III.11b) shows that the feed composition (c_{Ao}, c_{Po}) of a tank reactor operating in point 1 must lie on the tangent to the batch reaction path at point 1 (see figure III.28). Conversely, one point of the reaction path for a tank reactor is found by drawing the tangent from the point (c_{Ao}, c_{Po}) to a batch reaction path. Evidently, in order to construct a number of such points, more than one batch reactor path must be available. They can be found by considering that, if the initial reaction mixture of the batch experiment already contained some of the final product (not taking part in the reaction), the batch reaction path would be entirely similar. In figure III.28, for example, there is complete similarity between the reaction path starting from point A' (50% A and 50% X) in the secondary triangle A'P'X and the one starting with pure A. Point 2 in this diagram is one of the possible working points of a tank reactor with a feed containing component A only because the tangent in point 2 passes through the left hand corner. Thus, after construction of a number of 'secondary' batch reaction paths it is possible to construct the reaction path for a tank reactor by the method of tangents described above. Figure III.29 shows another illustration of this procedure, which naturally can be applied equally well when the composition is given in mass units. In principle, the method can be extended to a cascade of tank reactors.

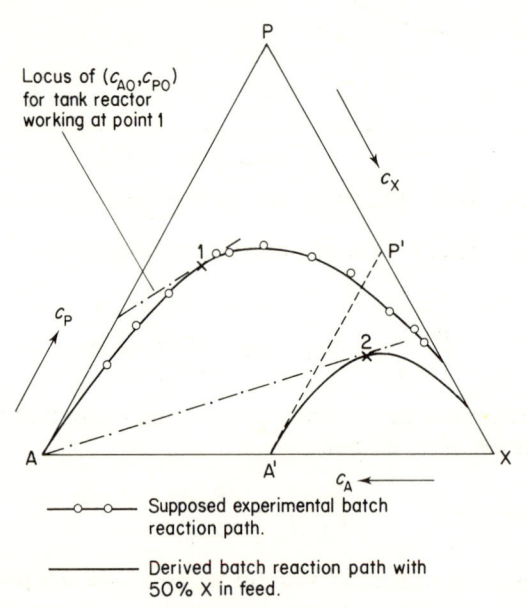

Fig. III.28 Principle of the method for finding the reaction path in a CISTR from an empirical batch reaction path

For processes where the reaction mixture consists of different components and a whole range of products is formed (such as in the petroleum industry), a triangular diagram in mass fractions may be useful, provided suitable discrimination between desired and undesired products is possible. By carrying out such reactions to different degrees of conversion and by variation of temperature, catalyst, feed composition, etc., one may readily see from the reaction paths which possibilities of increasing the yield of the desired product(s) exist. For a proper interpretation of such experiments it is essential, however, that the reaction paths obtained are isothermal.

This technique has been extensively used in a book by Waterman *et al.* [19]. These authors frequently approximate the reaction path by means of a hyperbola

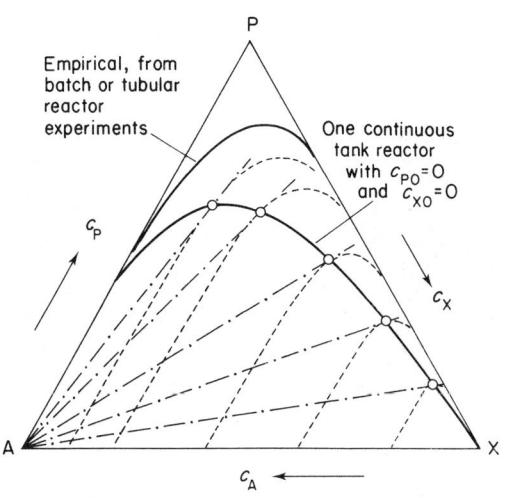

Fig. III.29 Construction of isothermal reaction path for one tank reactor from an empirical batch reaction path

which passes through points A and X. For consecutive reactions, where X stands for the final product(s), such a mathematical description is always possible by proper selection of the constants involved. In principle, such a description does not apply to reactions where the end of the reaction path lies anywhere between the corners P and X. With this reaction path analysis we can quickly determine whether a PFR or a CISTR gives the highest yields. We must, however, be aware of the fact that under special conditions a cross flow reactor or SBR or a recycle reactor can give even higher yields than one of the model reactors mentioned.

With the occurrence of wasteful side reactions or consecutive reactions, the solution to a yield optimization problem greatly depends on the relative rates of the desired and undesired reactions. In this respect in a complex reaction system with a principal reaction proceeding according to

$$v_A A + v_B B \to v_P P,$$

undesired species X, Y etc. are produced as well. The value of $-v_A R_P/v_P R_A \equiv \sigma'_P$ determines at any place in a reactor to what extent A is being converted to P.

In case *experimental data in a tank reactor* (CISTR) are available and results are accurate enough to determine the differential selectivity σ'_P as a function of the conversion, we have another tool for the choice of the reactor type and to determine the yield. Since the value of σ'_P depends on the composition of the reaction mixture, analysis of this dependence generally indicates the circumstances under which a high yield may be expected. In particular, when the composition is entirely characterized by the degree of conversion of one of the reactants, σ'_P can be plotted as a function of, say, ζ_A for a given feed composition and temperature. A relation of this kind can be obtained from experiments performed in a continuous tank reactor. Such a graph is shown in figure III.30a for a reaction system in which the differential selectivity decreases with increasing

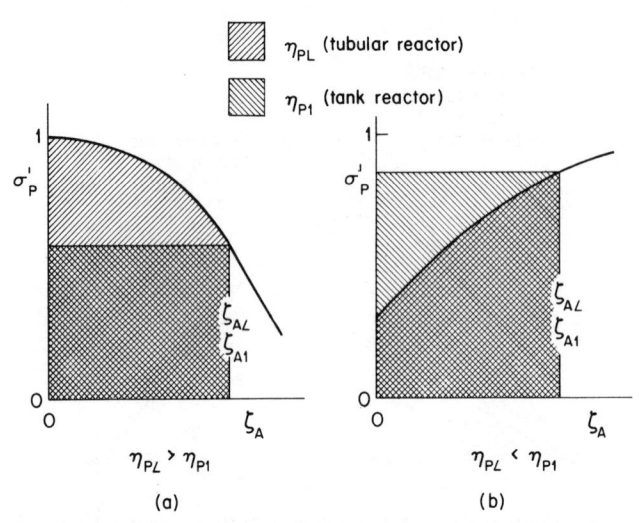

Fig. III.30 The course of the differential selectivity with ζ_A determines what type of continuous reactors gives the highest yield

conversion. The yield obtainable with this system in a tubular reactor and in a tank reactor can be easily derived from the graph, as shown below. For a tubular reactor, we have the material balances (with constant density):

$$\Phi_v \, dc_A = -R_A S \, dz,$$

and

$$\Phi_v \, dc_P = R_P S \, dz.$$

By taking the ratio of these equations, and with the definitions of the relative degree of conversion ζ_A and of σ'_P, we find

$$\frac{v_A \, dc_P}{v_P \, c_{Ao}} = -\sigma'_P \frac{dc_A}{c_{Ao}} = \sigma'_P \, d\zeta_A,$$

and hence for the yield obtained with the tubular reactor:

$$\eta_{PL} = \frac{v_A c_P}{v_P c_{Ao}} = \int_0^{\zeta_{AL}} \sigma'_P \, d\zeta_A. \tag{III.5}$$

It is seen that η_{PL} is equal to the area below the σ'_P curve in figure III.30a enclosed between $\zeta_A = 0$ and $\zeta_A = \zeta_{AL}$. Similarly, it can be shown that, for a single tank reactor,

$$\eta_{P1} = (\sigma'_P \zeta_A)_1; \tag{III.5}$$

the index 1 refers to the outlet conditions. The latter equation also follows from the definition of the yield in Section I.6. Clearly, the yield obtainable with a single tank reactor is equal to the area of the rectangle with sides σ'_{P1} and ζ_{A1}.

In the example of figure III.30a a tubular reactor gives a higher yield than a tank reactor for the same final degree of conversion; the latter reactor operates at the outlet conditions involving a low selectivity. If, on the other hand, as shown in figure III.30b, the curve rises with increasing degree of conversion, a tank reactor will give a higher yield than a tubular reactor. Thus, the sign of the slope of the σ'_P–ζ_A curve presents a standard for the selection of the basic reactor type giving the highest yield. This standard is equivalent to that given earlier by Denbigh [17] and Trambouze and Piret [18], who considered the sign of the quantity $d^2\xi_P/d\xi_A^2$ of a reaction system. When it is negative, the formation of desired product is favoured at a low degree of conversion (tubular reactor preferable) and when it is positive most of the desired product can be made at a high degree of conversion (tank reactor preferable). In the latter case, the capacity of the reactor is necessarily small so that a larger reaction volume is always required in comparison with a tubular reactor. Whether and to what extent some of the yield has to be sacrificed to an increase in reactor capacity, e.g. by using a cascade of tank reactors, can only be judged on economic grounds.

III.5.2 Optimum yield in a cascade of tank reactors

The σ'_P–ζ_A graph can likewise be used for finding the yield obtainable with a cascade of tank reactors. It can easily be shown that, for the nth reactor in cascade, equation (III.8) applies in the form

$$\eta_{Pn} - \eta_{P(n-1)} = \sigma'_{Pn}[\zeta_{An} - \zeta_{A(n-1)}]$$

and hence for a cascade of N reactors:

$$\eta_{PN} = \sum_{n=1}^{N} \sigma'_{Pn}[\zeta_{An} - \zeta_{A(n-1)}]. \quad (III.8)$$

η_{PN} can thus be represented as the sum of the areas of a number of rectangles. It will be clear that the value of η_{PN} will now depend on the selection of the values of ζ_{An}; they can be chosen in such a way that the yield has a maximum value for a given number of reactors in a cascade, i.e. the sum of the areas of a given number of rectangles under the curve is a maximum.

The condition for an optimal set of N reactors in a cascade implies that the following N relations between the conversion or reactant concentration in each of the consecutive stages must exist:

$$(\zeta_n - \zeta_{n-1})\left(\frac{d\sigma'_P}{d\zeta}\right)_n + \sigma'_{Pn} - \sigma'_{P(n-1)} = 0 \quad \text{for} \quad 1 < n < N. \quad (III.12)$$

This relation is obtained by differentiating equation (III.8). This equation can be solved graphically as shown in figure III.31. So by choosing a value for ζ_1 and drawing the tangent to the σ'_P–ζ_A curve at the point (σ'_{P1}, ζ_1) we can determine graphically the conditions in the following tanks in an optimal cascade. Whatever the choice of ζ_1, a fixed value of the overall yield η_{PN} and of the exit conversion ζ_N

step 1: $a = \xi_n - \xi_{n-1}$ gives point P

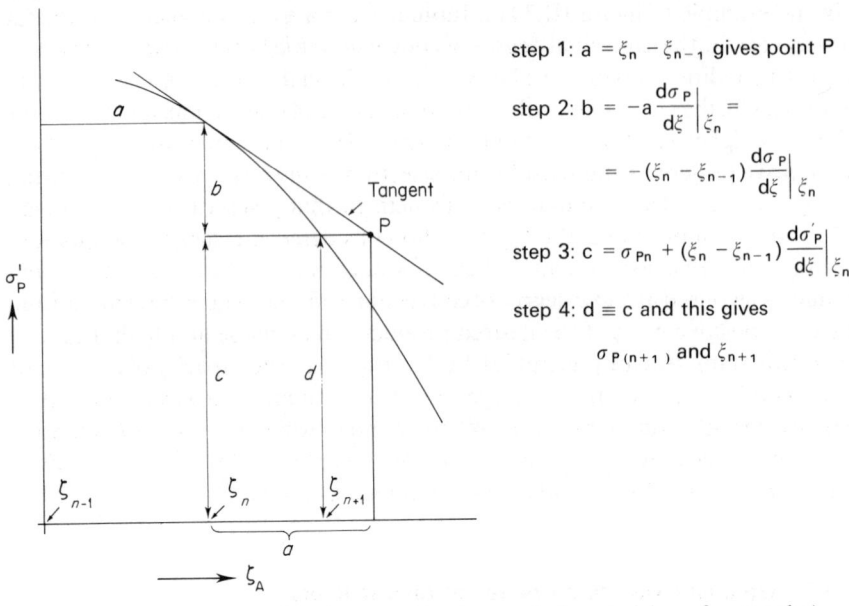

step 2: $b = -a \dfrac{d\sigma_P}{d\xi}\bigg|_{\xi_n} = -(\xi_n - \xi_{n-1})\dfrac{d\sigma_P}{d\xi}\bigg|_{\xi_n}$

step 3: $c = \sigma_{Pn} + (\xi_n - \xi_{n-1})\dfrac{d\sigma_P'}{d\xi}\bigg|_{\xi_n}$

step 4: $d \equiv c$ and this gives
$\sigma_{P(n+1)}$ and ξ_{n+1}

Fig. III.31 Construction for an optimal cascade (maximizing of rectangles)

accompany it. This provides a graph as figure III.32 of η_{PN} as a function of ζ_N with N as a parameter. From the same graph a curve can be derived of the maximum yield obtainable in a cascade as a function of the number of tanks in the cascade. As a result of this method we obtain for a cascade of N tanks the value of ζ_{AN} for which the yield reaches a maximum value and all the intermediate values of ζ_{AN}, at which each nth tank should operate. Besides this, we need a ζ_A–τ plot in order to determine the residence time τ_n in every tank of the optimal cascade with help of the relation

$$\zeta_{An} - \zeta_{A(n-1)} = k_1 \tau_n (1 - \zeta_{An}) \tag{II.33}$$

Most probably the various tanks in the cascade will have different volumes.

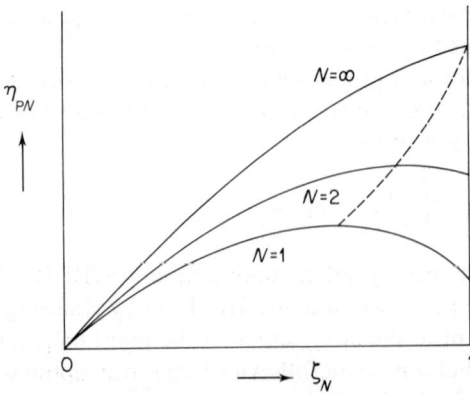

Fig. III.32 Optimal reactor yield in a cascade

This way of putting the problem of yield optimization was illustrated by Denbigh [1] with regard to the production of Cyclonite (P) from Hexamine and nitric acid at constant temperature. It was found experimentally that the differential selectivity of the reaction has a maximum as a function of the acid strength of which the degree of conversion ζ ($=$ mass of Hexamine reacted/mass of acid fed) is a measure; the final degree of conversion was set at $\zeta = 0.111$. As can be seen from figure III.33, one tank reactor would give a yield $\eta_{P1} = 0.58$. With two tank reactors in cascade, a maximum value of $\eta_{P2} = 0.73$ would be obtainable with $\zeta_1 = 0.082$; since the reactors were large enough to ensure practically complete conversion of the Hexamine supplied, optimum conditions for two tank reactors could be achieved by feeding 74% ($=0.082/0.111$) of the Hexamine into the first reactor and the rest into the second one.

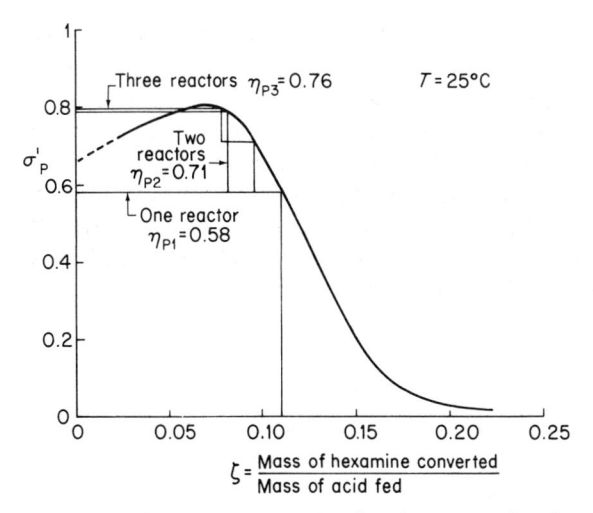

Fig. III.33 Optimization of yield in a cascade of tank reactors for the production of cyclonite [1]

If three reactors in cascade were used, it would be possible to increase the maximum yield still further to $\eta_{P3} = 0.76$; 71, 16 and 13% of the Hexamine feed should then be injected into the first, second and third reactor, respectively.

Some of the methods which can be used for maximizing the total area of the rectangles have been indicated in Denbigh's paper [1] and in the ensuing discussion.

In this example it should be noted that a distributed feed of one of the reactants had to be applied for proper adjustment of the composition of the reaction mixture in the different reactors. A similar case is treated in the following example.

Example III.5.a Alkylation of isobutane with propylene
From a study by Oden [20] on the alkylation of isobutane with propylene to light alkylate at 5°C, a relation between the differential selectivity and the excess

isobutane can be derived (see figure III.34). It is required to calculate the yield of light alkylate η_{P1} in a single tank reactor, and also η_{P3} for a cascade of three tank reactors where 50, 25 and 25 % of the olefins are fed to the first, second and third reactor, respectively. The residence times are so long that the olefins are converted practically completely in each tank; the ratio of the mass feed rate of isobutane and the total feed rate of olefins is to be 5.

It can be directly concluded from figure III.34 that

$$\eta_{P1} = \frac{0.86 \times 0.20}{0.20} = 0.86,$$

and

$$\eta_{P3} = 0.5 \times 0.96 + 0.25 \times 0.915 + 0.25 \times 0.86 = 0.924.$$

To illustrate the economic implications of a change from the single tank reactor to the cascade, we will assume the difference between the manufacturing costs in the two installations to be negligibly small. If the net sales value of light alkylate ($M = 100$ kg/kmol) is 340 MU/tonne and that of the by-products ($M = 100$ kg/kmol) 280 MU/tonne, a sales value of 39 410 MU/day becomes available with a daily feed of 50 tonnes of propylene ($M = 42$ kg/mol) in a single tank reactor; in the second case this figure will be 39 870 MU/day. Accordingly, an extra gross profit of 158 000 MU/year can be made with the cascade of three tank reactors, on the basis of 345 stream-days per year.

(*Example III.5.a ended*)

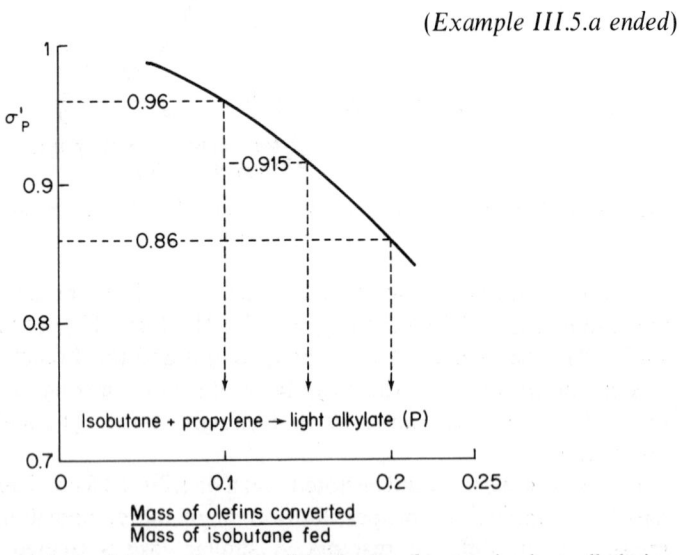

Fig. III.34 Differential selectivity as a function of the degree of conversion in an alkylation process [20]; Example III.5.a

In many cases the differential selectivity of an isothermal reaction system depends on more than one composition variable; the method described above for calculating and optimizing yields is then no longer applicable. Suppose that σ'_P depends on both the reactant concentrations c_A and c_B which are not uniquely related to each other but can be varied at will, and that σ_P' is large when one of these concentrations is relatively small. It is then of interest to examine how the proper reactant concentrations can be ensured in a reactor so as to obtain a high yield of the desired product. For several basic types of kinetics this has been shown in Section III.3.

III.5.3 Algebraic methods

In case rate expressions for all reactions in a combination reaction scheme are available we can refer to mathematical methods to determine yields and selectivities in the various model reactors. Because there are numerous combinations of reactions, we will restrict ourselves to the reaction scheme with the so-called van de Vusse kinetics and to a reaction scheme where the reactant concentration in the feed is an important optimization parameter. Van de Vusse [21] considered the following reaction scheme with a consecutive and a parallel reaction:

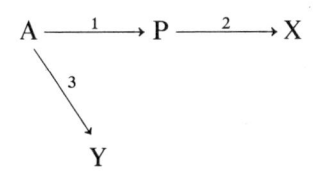

$$A \xrightarrow{1} P \xrightarrow{2} X$$
$$A \xrightarrow{3} Y$$

where reaction 3 is of second order and reactions 1 and 2 of first order. In this case a PFR with short residence times is preferred for a high yield of P and a CISTR with long residence times in order to suppress the undesired formation of Y. An example of this reaction scheme is the gas-phase chlorination of propylene with chlorine to produce alkyl chloride, where X = 1.3 dichloropropylene and Y = 1.2 dichloropropane [22]. Both requirements for an optimum yield are conflicting so that no conclusions can be drawn as yet with respect to the choice of the reactor type.

The rates of production are given by

$$R_A = -k_Y c_A^2 - k_P c_A$$
$$R_P = k_P c_A - k_X c_P.$$

We then find, for the differential selectivity σ_P',

$$\sigma_P' = -\frac{R_P}{R_A} = \frac{k_P c_A - k_X c_P}{k_P c_A + k_Y c_A^2} = \frac{1}{1 + \kappa_2 \dfrac{c_A}{c_{Ao}}} - \frac{\kappa_1 c_P/c_{Ao}}{\dfrac{c_A}{c_{Ao}} \left(1 + \kappa_2 \dfrac{c_A}{2 c_{Ao}}\right)}, \quad (III.13)$$

in which $\kappa_1 = k_X/k_P$ and $\kappa_2 = k_Y c_{Ao}/k_P$. The reactor inlet conditions are $c_A/c_{Ao} = 1$ and $c_P/c_{Ao} = 0$. After introducing $c_A/c_{Ao} = 1 - \zeta_A$ we find, for the differential selectivity σ_P',

$$\sigma_P' = \frac{1}{1 + \kappa_2(1 - \zeta_A)} - \frac{\kappa_1 c_P/c_{Ao}}{(1 - \zeta_A)[1 + \kappa_2(1 - \zeta_A)]}. \quad (III.14)$$

For a PFR $\sigma'_P = -dc_P/dc_A$ and for a CISTR

$$\sigma_{P1} = \frac{c_P}{c_{Ao} - c_A} = \eta_{P1}/\zeta_{A1},$$

so that for a CISTR we find

$$\frac{c_P}{c_{Ao}} = \eta_{P1} = \frac{\zeta_A(1 - \zeta_A)}{(1 - \zeta_A) + \kappa_2(1 - \zeta_A)^2 + \kappa_1\zeta_A} \quad \text{(III.15)}$$

and, for the selectivity σ_{P1},

$$\sigma_{P1} = \frac{(1 - \zeta_A)}{(1 - \zeta_A) + \kappa_2(1 - \zeta_A)^2 + \kappa_1\zeta_A}. \quad \text{(III.16)}$$

By differentiating and setting $d\eta_{P1}/d\zeta_A = 0$, we find the conversion ζ_{A1}, at which the *yield* reaches a maximum in a CISTR. This is

$$\zeta_{A1\,\text{max}} = \frac{1}{1 + \sqrt{[\kappa_1/(1 + \kappa_2)]}}.$$

By substituting this value in equation (III.15) we obtain the maximum yield in a CISTR:

$$\eta_{P1\,\text{max}} = \frac{1}{1 + \kappa_1 + 2\sqrt{[\kappa_1(1 + \kappa_2)]}}.$$

If, however, recirculation of nonconverted reactant A is possible, the selectivity $\sigma_{P1} = \eta_{P1}/\zeta_{A1}$ is the relevant parameter, which determines plant performance (see Section I.6). The maximum *selectivity* in a CISTR is reached at

$$\zeta_{A1\,\text{max}} = 1 - \sqrt{\left(\frac{\kappa_1}{\kappa_2}\right)}.$$

If $\kappa_1 > \kappa_2$, the maximum selectivity is reached at $\zeta_{A1} = 0$, and the maximum selectivity in a CISTR is found, after substitution into equation (III.16) to be

$$\sigma_{P1\,\text{max}} = \frac{1}{1 - \kappa_1 + 2\sqrt{(\kappa_1\kappa_2)}}$$

If $\kappa_1 > \kappa_2$, the selectivity is $\sigma_{P1\,\text{max}} = (1 + \kappa_2)^{-1}$ at $\zeta_{A1} = 0$.

In a PFR we have to integrate $-dc_P/dc_A = \sigma'_P$ in order to find the reaction path. This is a linear differential equation, which gives as a general solution

$$\eta_{PL} = \frac{c_P}{c_{Ao}} = -\left[\frac{c_A/c_{Ao}}{1 + \kappa_2 c_A/c_{Ao}}\right]^{\kappa_1} \int_1^{c_A/c_{Ao}} \frac{(1 + \kappa_2 c_A/c_{Ao})^{\kappa_1 - 1}}{(c_A/c_{Ao})^{\kappa_1}} d(c_A/c_{Ao}). \quad \text{(III.17)}$$

This equation can only be integrated analytically for some specific values of κ_1, such as $\kappa_1 = \frac{1}{6}, \frac{1}{4}, \frac{1}{3}, \frac{1}{2}, 1, 2, 3, 4, 6$, etc. [21].

For $\kappa_1 = 1$ and for $c_P = c_{P_0}$ in the reactor feed we find, for example,

$$\frac{c_P}{c_{Ao}} = \frac{c_A/c_{Ao}}{1 + \kappa_2 c_A/c_{Ao}} \left[\frac{c_{P_0}}{c_{Ao}}(1 + \kappa_2) - \ln \frac{c_A}{c_{Ao}} \right]$$

By differentiation of equation (III.17) we find that between the maximum yield and the corresponding conversion at the maximum the following relation holds:

$$\eta_{PL} = \frac{1 - \zeta_{AL}}{\kappa_1}.$$

This leads to the conclusion that in a PFR the yield never will be higher than κ_1^{-1}. The maximum yield can be found by numerically integrating equation (III.17) from 0 to ζ_{AL}; the maximum depends on the value of κ_2. Results for selected values are given in Table III.9.

Table III.9 Maximum yield in a PFR for van de Vusse kinetics

κ_1	κ_2	$\eta_{PL\,max}$	ζ_{AL} at $\eta_{PL\,max}$
0	0.5	0.81	0
	1.0	0.69	0
	2.0	0.55	0
0.5	0.5	0.42	0.21
	1.0	0.37	0.19
	2.0	0.30	0.15
1.0	0.5	0.31	0.31
	1.0	0.27	0.27
	2.0	0.24	0.24
2.0	0.5	0.22	0.44
	1.0	0.20	0.40
	2.0	0.17	0.34

By substituting $\eta_{PL} = \sigma_{PL}\zeta_{AL}$ into equation (III.17), dividing by ζ_{AL} and then differentiating we find the following relation between the maximum selectivity in a PFR and the value of the outlet conversion, where the maximum selectivity is reached:

$$\sigma_{PL\,max} = \frac{1 - \zeta_{AL}}{\kappa_2(1 - \zeta_{AL})^2 + (1 - \kappa_1)(1 - \zeta_{AL}) + \kappa_1}.$$

The selectivity of a PFR can be determined with the following relation:

$$\sigma_{PL} = -\frac{1}{\zeta_{AL}} \left[\frac{1 - \zeta_{AL}}{1 + \kappa_2(1 - \zeta_{AL})} \right]^{\kappa_1} \int_0^{\zeta_{AL}} \frac{[1 + \kappa_2(1 - \zeta_A)]^{\kappa_1 - 1}}{(1 - \zeta_A)^{\kappa_1}} \, d\zeta_A.$$

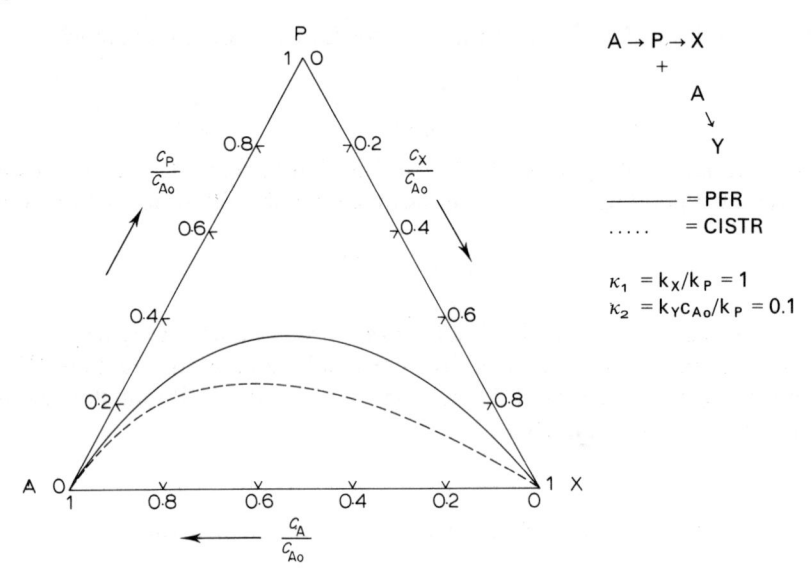

Fig. III.35 The reaction paths in a PFR and a CISTR for a reaction with van de Vusse kinetics

In figure III.36 a selectivity versus reactor outlet conversion plot for some values of κ_1 and κ_2 are shown.

Solutions for $0.01 < \kappa_1 < 2$ were obtained with a computer by van de Vusse [21]. As an example the reaction path for $\kappa_1 = 1$ and $\kappa_2 = 0.1$ is given in figure III.35. In figure III.36 the selectivity and the yield are plotted versus the conversion for different values of κ_1 and κ_2.

When $\kappa_2 \leq \kappa_1$ (or $\kappa_Y c_{Ao} \leq \kappa_X$) the selectivity was highest at $\zeta_{AL} = 0$. However, if $\kappa_2 > \kappa_1$, then the selectivity passes through a maximum for $\zeta_A > 0$. From the computer curves van de Vusse found with respect to the yield that $\eta_{PL\,max} \geq \eta_{P1\,max}$ if $\kappa_1 \geq 0.04\kappa_2^{1.4}$. A further analysis by van de Vusse showed that both reactor types give equal maximum selectivity if $\kappa_2 \leq \kappa_1$. However, if $\kappa_2 > \kappa_1$, the maximum selectivity is always higher in the CISTR. This now gives three domains for the choice of the reactor type. From figure III.37 we can read that (provided non-converted reactant can be recycled) for combinations of κ_1 and κ_2 in area III a PFR or a CISTR should be chosen and in areas II and I a CISTR. In case non-converted reactant recycle is impossible, yield is the relevant parameter and then in areas III and II a PFR and in area I a CISTR should be chosen in order to obtain the highest output per unit of reactant in the plant feed.

Reactor residence time ratio can be derived to be

$$\frac{\tau_1}{\tau_L} = \frac{\zeta_{A1}}{(1 - \zeta_{A1})[1 + \kappa_2(1 - \zeta_{A1})]\ln\left[\dfrac{(1 + \kappa_2)(1 - \zeta_{AL})}{1 + \kappa_2(1 - \zeta_{AL})}\right]}.$$

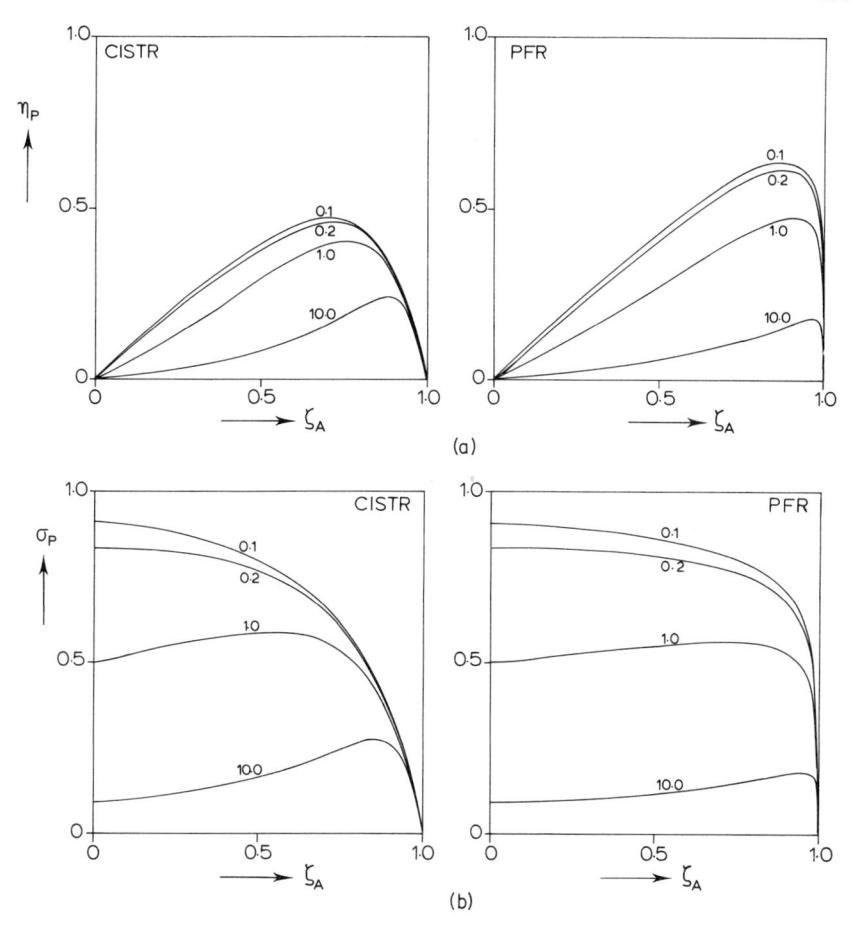

Fig. III.36 Yields and selectivities in a CISTR and a PFR for van de Vusse kinetics. $\kappa_1 = k_X/k_P = 0.2$, $\kappa_2 = k_Y c_{A0}/k_P$ is the parameter

More reactor volume is required for the CISTR. Relative reactor volumes are given in figure III.38.

The van de Vusse kinetics have been studied in a recycle PFR by Gillespie and Carberry [23]; they showed that for $\kappa_2 \gg \kappa_1$ even higher yields can be obtained in a recycle reactor. The reactor is given in figure III.39. In this reactor $(\Phi_R + \Phi_v)c_{in} = \Phi_R c + \Phi_v c_o$, and if $\Phi_R/\Phi_v = R$, the recycle ratio, the concentration at the reactor inlet is given by

$$c_{in} = \frac{Rc + c_o}{R + 1}.$$

For $R = 0$ we have the PFR and for $R = \infty$ a CISTR, in practice the CISTR is indicated for values of R of, say, above 20. The recycle reactor has the advantage of lower inlet concentrations of the reactant, which is advantageous in our case. We

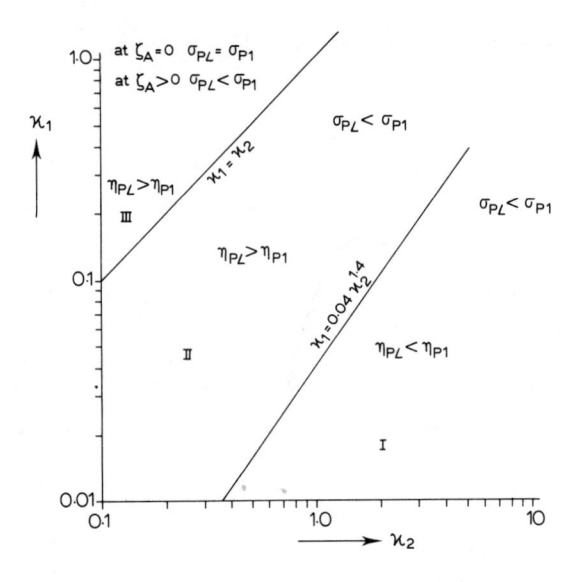

Fig. III.37 Reactor choice for van de Vusse kinetics

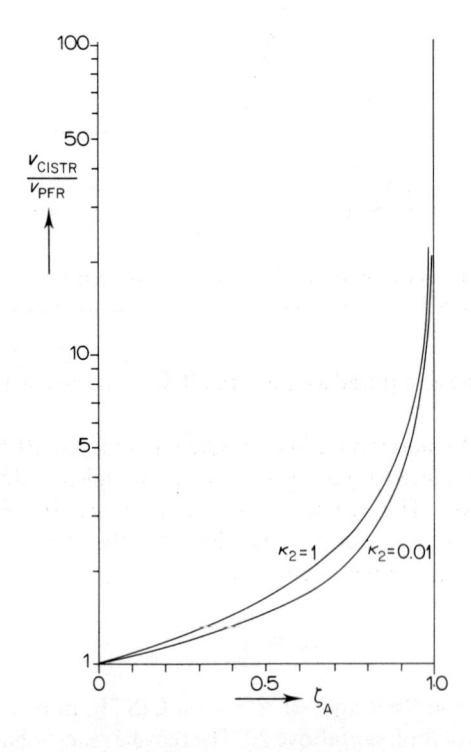

Fig. III.38 Ratio of the reactor volumes of a CISTR and a PFR for van de Vusse kinetics

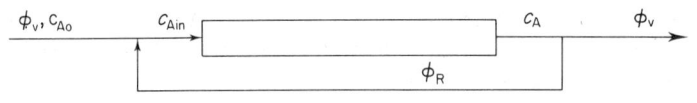

Fig. III.39 Recycle reactor

have to realize that the reactor residence time increases. Now the concentrations at the reactor inlet are, if we denote the converted fraction of reactant A in the feed by

$$(\Phi_v c_{Ao} - \Phi_v c_A)/\Phi_v c_{Ao} = \zeta_A$$

or $c_A/c_{Ao} = 1 - \zeta_A$,

$$\frac{c_{A\,in}}{c_{Ao}} = \frac{R(1-\zeta_A)+1}{R+1} \quad \text{and} \quad \frac{c_{P\,in}}{c_{Ao}} = \frac{Rc_P/c_{Ao}}{R+1} \quad \text{at} \quad \tau = 0.$$

By using these inlet conditions to determine the integration constant after integrating $-dc_P/dc_A = \sigma'_P$, we find the yield in a recycle PFR. For example, if $\kappa_1 = 1$ the result is

$$\frac{c_P}{c_{Ao}} = \frac{\dfrac{c_A/c_{Ao}}{1+\kappa_2 c_A/c_{Ao}} \ln \dfrac{c_{A\,in}/c_{Ao}}{c_A/c_{Ao}}}{1 - \dfrac{c_A/c_{Ao}}{1+\kappa_2 c_A/c_{Ao}} \dfrac{R}{R+1} \dfrac{1+\kappa_2 c_{A\,in}/c_{Ao}}{c_{A\,in}/c_{Ao}}}.$$

Gillespie and Carberry have shown that at relatively high values of κ_2, where the desired reaction 1 is much slower than the undesired reaction 3 and relatively high values of κ_2/κ_1, where the undesired reaction 3 is much faster than the undesired reaction 2, it is advantageous to dilute the fresh reactor feed with recycled reactor product, so that the undesired reaction 3 is more suppressed than the desired reaction 1. In this region at low recycle ratios of e.g. $R = 1.5$ a recycle reactor shows a better yield than a PFR or a CISTR. However, in the region with high values of κ_2 and of κ_2/κ_1, yields are very low so that better a selective catalyst or another synthesis route can be searched for.

Ridlehoover and Seagrave [24] and Lund and Seagrave [25] have shown that in an SBR with a special programmed feeding and filling of the reactor also higher yields can be obtained than in either a PFR or a CISTR in the region, where $0.3 < \kappa_2 < 3$. Pasquali and Lelli [26] have shown that in the region of high yields and selectivities, which are industrially important, a cascade with backmixing between the stages as shown in figure III.40 may give higher yields than either a PFR or a CISTR. From their work it can be concluded that high yields

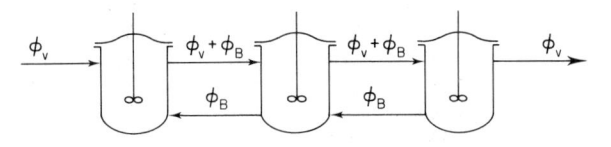

Fig. III.40 Cascade with backmixing between vessels

($\eta_P > 0.50$) can be obtained only if $\kappa_1 \kappa_2 < 0.2$. Yield in a cascade with backmixing can be slightly higher (5–10%) than in the PFR or CISTR at high backmixing ratios (Φ_B/Φ_v).

In the following illustration we will discuss an example taken from the industrial practice of one of the authors, where the concentration of the reactant in the feed is an important parameter for yield optimization.

Example III.5.b *Yield improvement with increase of reactant concentration in the feed*

A fine chemical P is made from raw material A in a cascade of three equally sized CISTRs at a practically constant temperature of 30 °C. The feed concentration of the reactant is 0.2 kmol/m³ and the total residence time in the cascade is 14400 s (4 hours). The reactant cannot be recovered from the reactor product. The development chemical engineer is asked to study the reactor performance and to make recommendations for yield improvement, if possible.

Laboratory experiments have revealed that species P decomposes according to zero order kinetics into an undesired product Y. The reaction rate constant has been determined and is $k_Y = 1.75 \times 10^{-7}$ kmol/m³s. Moreover, it was found that the reactant A is in equilibrium with an isomer B. The equilibrium is established infinitely fast and $K = c_A/c_B = 1.41$. Species B decomposes into an undesired product X according to a second order reaction. This leads to the following reaction scheme:

$$B \leftrightarrows A \xrightarrow{1} P \xrightarrow{2} Y$$
$$\downarrow 3$$
$$X$$

Experimentally the following values for the other kinetic coefficients were the best fit: $k_P = 2.81 \times 10^{-4}$/s and $k_X = 5.21 \times 10^{-3}$ m³/kmol s. In Table III.10 plant tests in a cascade of three tanks are compared with the values calculated with the above given coefficients.

We see that experiments and calculations fit sufficiently well to have confidence in the postulated reaction scheme and the derived kinetic data. Our chemical

Table III.10   Calculated and experimental concentrations in the cascade of Example III.5.b

Residence time (s) dimensionless concentrations	Tank 1 4800 Meas.	Calc.	Tank 2 4800 Meas.	Calc.	Tank 3 4800 Meas.	Calc.
A	0.265	0.267	0.135	0.135	0.066	0.071
B	0.191	0.190	0.094	0.096	0.047	0.050
P	0.358	0.357	0.537	0.535	0.665	0.627
X	0.180	0.182	0.226	0.226	0.237	0.239
Y	0.004	0.004	0.008	0.008	0.012	0.013

engineer now calculates the performance of a BR (for the long reaction times required a PFR is not adequate) and of a CISTR. Also the performance of a cascade of 2 and 3 tanks, all of equal volume, was studied. For a *batch reactor* the following equations—the concentrations have been made dimensionless by dividing by the total concentration c_o in the feed of A + B, so that $c_o = c_{Ao} + c_{Bo}$—have to be solved:

$$c_A/c_o = Kc_B/c_o$$

$$-\frac{dc_B/c_o}{dt} = \frac{1}{K+1}[k_PKc_B/c_o + k_Xc_o(c_B/c_o)^2]$$

$$\frac{dc_P/c_o}{dt} = k_PKc_B/c_o - \frac{k_Y}{c_o}$$

$$\frac{dc_X/c_o}{dt} = k_Xc_o(c_B/c_o)^2$$

$$\frac{dc_Y/c_o}{dt} = \frac{k_Y}{c_o}$$

$$-\frac{dc_A/c_o}{dt} = \frac{K}{K+1}[k_PKc_B/c_o + k_Xc_o(c_B/c_o)^2].$$

The starting conditions are: $t = 0$; $c_P/c_o = c_X/c_o = c_Y/c_o = 0$, $c_B/c_o = 1/(1+K)$ and $c_A/c_o = K/(1+K)$. Notice further that $c_A + c_B = (K+1)c_B$.

The solution of these equations is

$$c_B/c_o = \frac{A_1 \exp(-Bt)}{1 - A_2 \exp(-Bt)}$$

$$c_A/c_o = Kc_B/c_o$$

$$c_P/c_o = \frac{1-A_2}{A_2}\ln\left[\frac{1-A_2\exp(-Bt)}{1-A_2}\right] - \frac{k_Y}{c_o}t$$

$$c_X/c_o = \frac{1-A_2}{A_2}\left\{\frac{1}{1-A_2} - \frac{1}{1-A_2\exp(-Bt)} - \ln\left[\frac{1-A_2\exp(-Bt)}{1-A_2}\right]\right\}$$

$$c_Y/c_o = \frac{k_Y}{c_o}t,$$

in which

$$A_1 = \frac{k_PK}{k_Xc_o + k_PK(1+K)}, \quad A_2 = \frac{k_Xc_o}{k_Xc_o + k_PK(1+K)}$$

and $B = k_PK/(K+1)$.

The results, given as the yield as a function of the reaction time, are given in figure III.41 for three different reactant feed concentrations c_o. In the solvent applied the highest solubility of reactants is $c_o = 0.4 \text{ kmol/m}^3$. We see that the

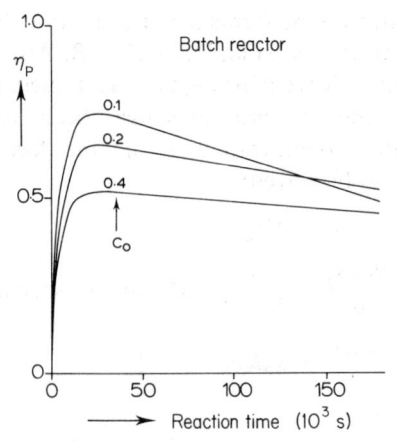

Fig. III.41 Yield of desired component in a BR as a function of the feed concentration for the reaction scheme of Example III.5.b

highest yield is obtained at the lowest reactant concentration in the inlet. This is expected because by lowering c_o the undesired second order reaction 3 is more suppressed than the desired first order reaction 1.

For a **CISTR** and a *cascade* the following equations have to be solved:

$$c_A/c_o = K c_B/c_o$$

$$c_{Bn}/c_o - c_{B(n-1)}/c_o = -\frac{\tau_n}{K+1}\left[k_P \frac{c_{An}}{c_o} + k_X c_o \left(\frac{c_{Bn}}{c_o}\right)^2\right]$$

$$c_{Pn}/c_o - c_{P(n-1)}/c_o = \left[k_P\left(\frac{c_{An}}{c_o}\right) - \frac{k_Y}{c_o}\right]\tau_n$$

$$c_{Xn}/c_o - c_{X(n-1)}/c_o = k_X c_o \left(\frac{c_{Bn}}{c_o}\right)^2 \tau_n$$

$$c_{Yn}/c_o - c_{Y(n-1)}/c_o = \frac{k_Y}{c_o}\tau_n.$$

Inlet conditions are equal as in the BR. Further $\tau_n = \tau/N$. For reactant B the solution is after elimination of c_{An}/c_o:

$$\frac{c_{Bn}}{c_o} = \frac{\sqrt{[(k_P K \tau_n + 1 + K)^2 + 4 k_X c_o \tau_n (1 + K) c_{B(n-1)}/c_o]} - (k_P K \tau_n + 1 + K)}{2 k_X c_o \tau_n}.$$

The solution of these equations is straightforward. Results are given in figure III.42 and are summarized in Table III.11 and compared with the yields in the BR.

Comparing all results we see that in the plant the total residence time is too short for the optimum yield. The plant operators had been increasing the production of the plant by removing bottlenecks without being aware of the disastrous effect of a considerable throughput increase on the yield in the reactor section. Further we see that at shorter residence times reaction 2 is not yet

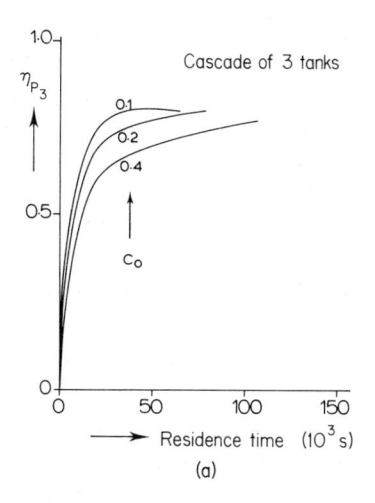

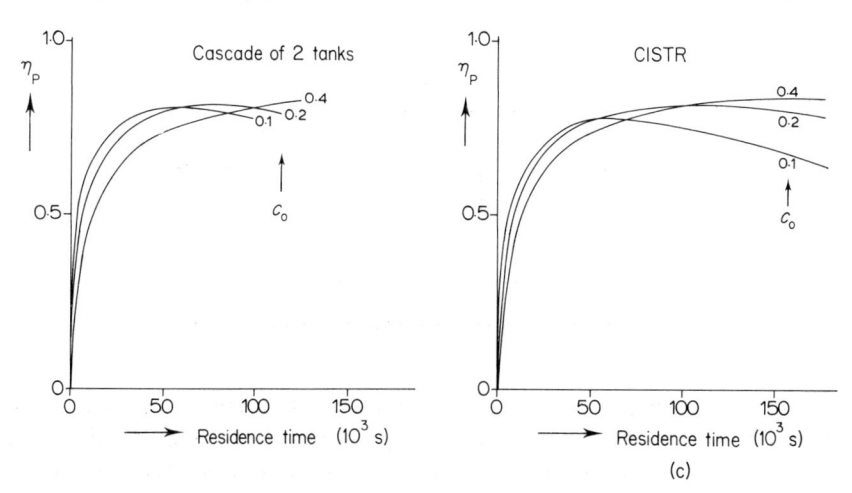

Fig. III.42 Yield in a cascade or a CISTR as a function of feed concentration for the reaction scheme of Example III.5.b

Table III.11 Maximum yields ($\eta_P = c_{P\max}/c_o$) and corresponding reaction or residence times t_{opt} and τ_{opt} (in seconds × 10^3) for Example III.5.b

	BR		Cascade				CISTR	
			3 vessels		2 vessels			
C_o (kmol/m³)	$\eta_{P\max}$	(t_{opt})	$\eta_{P\max}$	(t_{opt})	$\eta_{P\max}$	(t_{opt})	$\eta_{P\max}$	(t_{opt})
0.10	0.744	(19.8)	0.801	(47)	0.802	(54)	0.765	(61)
0.20	0.647	(27.0)	0.794	(83)	0.814	(94)	0.805	(97)
0.40	0.515	(30.6)	0.764	(104)	0.815	(158)	0.830	(180)

important, but reaction 3 is. The influence of the feed concentration is similar to that in a BR for a three-vessel cascade, but is reversed in a CISTR. In a CISTR due to the low reactant concentrations reaction 3 is effectively suppressed; meanwhile at higher c_o the reaction rate of reaction 2 is reduced with respect to the desired reaction 1 with a factor $1/c_o$. We see that increasing c_o has a less unfavourable effect in a cascade of three tanks than in a BR. Doubling c_o gives the opportunity to about halve the feed rate in order to maintain the same output of product P, which equals $\Phi_v c_o \eta_P$.

Based on the considerations above, the development engineer gave the following recommendations:

—increase the feed concentration to $c_o = 0.4 \, \text{kmol/m}^3$ and reduce the feed rate to a half;
—remove at the next plant shut-down the separating walls between the compartments (see figure III.43) to change the reactor into a CISTR;
—study in the laboratory the influence of the temperature on the kinetics.

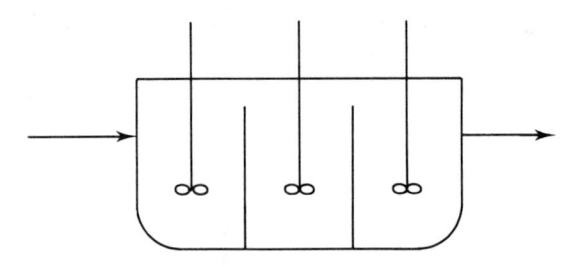

Fig. III.43 The compartmented cascade vessel

Eventually the laboratory experiments indicated that temperature had to be lowered drastically to obtain a noticeable yield improvement. So finally a very large CISTR was purchased with a design residence time of two days; this caused problems in reaching steady state conditions during start-up in the whole plant, but this could be solved by special measures. The actual yield increases were successively:

after step 1 ($c_o = 0.4 \, \text{kmol/m}^3$ and halving Φ_v) to 0.665
after step 2 (removal of separating walls)   to 0.73
after step 3 (installation of large CISTR)   to 0.80.

Example III.5.c *Lumping of kinetics for the desulphurization of gas oil* [35]
For the desulphurization of one single sulphur containing component in a gas oil it has been found experimentally [35], that the reaction rate is first order with $R_J = -k_J c_J$. Here k_J is expressed in kmoles of sulphur component J converted per kg gas oil per second. The gas oil, however, contains a whole variety of sulphur components of different types such as thiols, sulphides, disulphides, thiophenes, benzotiophenes, etc. Each individual component J will have its own rate constant

k_j. The reaction rate distribution at the time t is now defined as the frequency distribution $f(k, t)$, for which

$$f(k, t) \Delta k = \frac{c_j(t)}{c(t)}.$$

Here $c_j(t)$ is the sulphur concentration of the components which react with a rate constant between k and $k + \Delta k$, $c(t)$ is the total sulphur concentration at time t. At $t = 0$, before the reaction starts, we have

$$\frac{c}{c_o} = \int_0^\infty f(k, 0) \, dk = 1.$$

In a BR (or a PFR) the average concentration c after a reaction time t (or a residence time τ_L) follows from

$$\frac{c}{c_o} = \int_0^\infty e^{-kt} f(k, 0) \, dk,$$

whereas the average reaction rate constant at time t becomes

$$\bar{k}(t) = \frac{\int_0^\infty k f(k, t) \, dk}{\int_0^\infty f(k, t) \, dk}$$

The distribution of the rate constants remains constant during the reaction and equals $f(k, 0)$; however, the average rate constant diminishes with the course of time, because the fast reacting components will have disappeared soon and only the slower reacting species will have remained. Koetsier et al. [35] proved that a modified Poisson distribution gives a sufficiently accurate description of the gas oil desulphurization kinetics:

$$f(k, 0) = \frac{n^n}{\bar{k} \Gamma(n)} \left(\frac{k}{\bar{k}}\right)^{n-1} e^{-nk/\bar{k}},$$

in which $\bar{k}$ is the average k value at $t = 0$. Moreover, they found in their experiments that $n = 1$ fitted their results well, so that

$$f(k, 0) = \frac{1}{\bar{k}} e^{-k/\bar{k}}.$$

This leads to

$$\frac{c}{c_o} = \int_0^\infty e^{-kt} \frac{1}{\bar{k}} e^{-k/\bar{k}} \, dk.$$

The solution of this integral can be found easily:

$$\frac{c}{c_o} = \frac{1}{1 + \bar{k}t}.$$

Now if $\bar{k}$ is taken equal to k_2c_o we obtain the conversion equation for an isothermal second order reaction in a BR. So by lumping a large number of first order reactions we obtain an overall second order behaviour. For a CISTR with $c_J/c_{Jo} = (1 + k_J\tau)^{-1}$ we have

$$\frac{c}{c_o} = \int_0^\infty f(k,0)\frac{1}{1+k\tau}dk.$$

Substitution of the modified Poisson distribution with $n = 1$ and integration gives as a solution

$$\frac{c}{c_o} = \frac{\exp\left(\frac{1}{\bar{k}t}\right)}{\bar{k}t}\operatorname{Ei}\left(\frac{1}{\bar{k}t}\right)$$

In figure III.44 some experimental results are given for a BR and a CISTR.

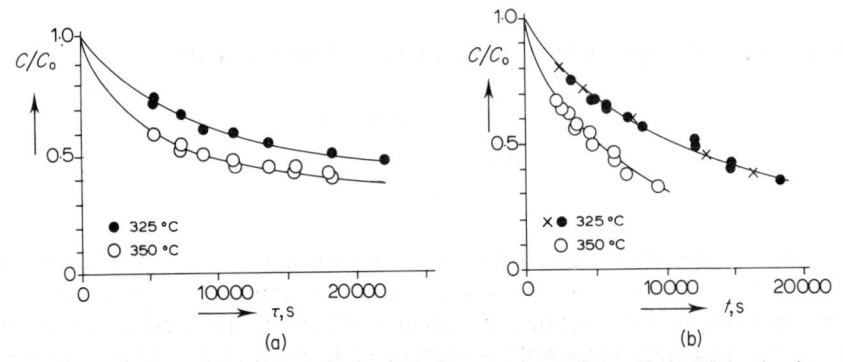

Fig. III.44 Conversions observed in (a) a CISTR and (b) a BR for the desulphurization of gas oil [35]. Drawn lines are calculated from reaction rate distribution function, points are experimental results [From Koetsier et al. [35]. Reproduced by permission of W. P. M. van Swaaij]

III.6 AUTOCATALYTIC REACTIONS

Autocatalytic reactions of a multiple character occur frequently in biotechnology. They are, for example, of the type

$$v_A A \rightarrow v_P P \begin{array}{c} \nearrow v_X X \\ \searrow v_Q Q \end{array}$$

In this scheme the reaction $A \rightarrow P$ is autocatalytic with a rate equation in its simplest form of $R_P = k_P c_A c_P$. In biotechnology A is the substrate and P are the enzymes or living cells. The substrate is consumed, the cells grow and a desired product Q is formed. However, the enzymes or cells also decompose or die according to the reaction P → X. In waste water treatment A represents the water contaminants, in ethanol production A represents the sugar compounds, P the yeast cells and Q the ethanol. The autocatalytic reaction A → P exhibits special kinetics, which have a peculiar influence on reactor performance, so that in this section we will elaborate first on the special features of this reaction before discussing multiple autocatalytic reactions.

In biochemical reactions a living micro-organism P—a type of bacteria or a cell—is fed with a substrate A. As a consequence, the cell grows and the number of cells increases. The possible reaction mechanism was first postulated by Michaelis and Menten [27] and based on their mechanism a kinetic equation was first proposed by Monod [28] as follows:

$$R_P = k_P \frac{c_A c_P}{K_A + c_A} \tag{III.18}$$

We notice a certain resemblance with the so-called Langmuir kinetics. The growth rate of cells R_P equals a kinetic coefficient k_P:

—times the cell concentration c_P at high values of c_A, the substrate concentration for which $K_A \ll c_A$;
—times the product of the cell concentration c_P and of c_A at low values of c_A, for which $K_A \gg c_A$.

In equation (III.18) K_A is often called the 'saturation' constant: this is the concentration at which R_P reaches half of its maximum value for a given value of c_P. The reaction rate of cell growth varies between zero and first order in the substrate concentration and depends on the amount of feed available for the cells or the ratio c_A/K_A. So we can distinguish two extreme kinetic regimes:

—the uninhibited growth phase, where more than sufficient feed is available for the cells, so that they will reproduce in accordance with the number of cells present or $R_P = k_P c_P$;
—the feed limited growth phase with $R_P = (k_P/K_A)c_A c_P$, where insufficient feed is available and the cells will grow in accordance with the feed and the number of cells available.

Concentrations and K_A are usually expressed in weight per unit volume (kg/m³) because of the difficulty in determining molar concentrations in biotechnology. The dimension of k_P is per unit time (s⁻¹). The substrate consumption rate according to Monod [28] equals

$$R_A = -\frac{k_P}{v'} \frac{c_A c_P}{K_A + c_A}. \tag{III.19}$$

In this equation v' is a kind of a stoichiometric coefficient and is often called the yield factor; it is expressed as kg cells formed per kg substrate consumed. The value of k_P is determined in experiments with a large excess of substrate, and that of K_A by determining the value of c_A at which $R_P = \frac{1}{2}k_P c_P$. With the two rate expressions (III.18) and (III.19) in principle the performance of a biochemical reactor can be calculated.

Many extensions of the Monod kinetics have been proposed in the past. A survey is given by Humphrey [29]. One simple extension is the supposition that part of the micro-organisms die equal to the rate $k_X c_P$, so that now

$$R_P = k_P \frac{c_A c_P}{K_A + c_A} - k_X c_P. \tag{III.20}$$

The rate expressions are very sensitive to temperature and to the pH of the medium as shown in figure III.45. We see that k_P reaches an optimum temperature where the micro-organisms grow most comfortably; above that temperature degradation sets in. Many micro-organisms feel best at pH values of $6.5 \pm 1 = \text{pH}_{\text{opt}}$. Typically at a value of $\text{pH} = \text{pH}_{\text{opt}} \pm 1.0$ to 1.5 activity is already reduced to half its value at pH_{opt}.

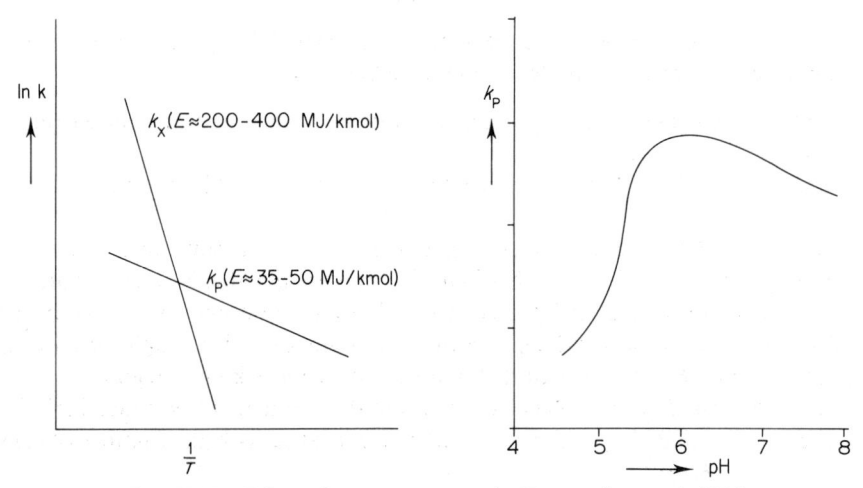

Fig. III.45   Effect of temperature and pH on cell growth [29]

III.6.1   Single biochemical reactions

In this part we assume that in equation (III.20) $k_X = 0$ and moreover that the residence time distribution of the cells is equal to that of the liquid in the reactor. The reason for the last assumption will become clear in Chapter IV. For single biochemical reactors we will illustrate the phenomena of a minimum residence time in a reactor section and of multiplicity in a CISTR.

Under the assumption that rate expressions (III.18) and (III.19) are valid and that v' is constant, Bischoff [36] has shown that a special configuration is required in order to obtain a *minimum residence time in a biochemical reactor section*. If v' is constant, then

$$-\frac{dc_P}{dc_A} = v'.$$

Integration gives $c_P = -v'c_A + \text{const}$. At the reactor inlet or at the start of the reaction $c_P = c_{Po}$ and $c_A = c_{Ao}$, so that the integration constant equals $c_o = c_{Po} + v'c_{Ao}$, so that $c_P = v'(c_{Ao} - c_A) + c_{Po}$ or

$$c_A = \frac{c_o - c_P}{v'}. \tag{III.21}$$

Substitution of (III.21) in (III.18) gives, for the production rate of the micro-organisms,

$$R_P = k_P c_P \frac{c_o - c_P}{v'K_A + c_o - c_P}.$$

We see that the reaction rate is zero for $c_P = 0$ and for $c_P = c_o = c_{Po} + v'c_A$. For a CISTR this leads to a residence time of

$$\tau_1 = \frac{c_{P1} - c_{Po}}{k_P c_{P1}} \frac{v'K_A + c_o - c_{P1}}{c_o - c_{P1}} \tag{III.22}$$

and for a PFR after integration to

$$\tau_L = \int_{c_{Po}}^{c_{PL}} \frac{dc_P}{R_P} = \frac{v'K_A + c_o}{k_P c_o} \ln \frac{c_{PL}}{c_{Po}} + \frac{v'K_A}{k_P c_o} \ln \left[\frac{v'c_{Ao}}{c_o - c_{PL}} \right].$$

In figure III.46 the reciprocal micro-organism production rate is plotted versus the micro-organism concentration. Intuitively it can be felt that a combination of first a CISTR and then a PFR will give the minimum residence time in the total reactor system. A CISTR has to be chosen for a conversion from c_{Po} to $c_{P\min}$, where R_P^{-1} reaches a minimum, and further a PFR for the conversion from $c_{P\min}$ to c_{PL}, the desired outlet concentration. By differentiation it can be proved that at the minimum for R_P^{-1} the cell concentration is

$$(c_P)_{\text{at min}} = (v'K_A + c_o)\left[1 - \sqrt{\left(\frac{v'K_A}{c_o + v'K_A}\right)}\right].$$

Further, we have to realize that for the situation in figure III.46, where $c_{Po} = 0$, the reaction would not start in a PFR even with an infinite reaction volume because of the absence of micro-organisms in the reactor feed to the PFR. So we always need some backmixing in a continuously operated biochemical reactor. The reaction would not start in any reactor if there were no inoculation with micro-organisms. Hence every reactor needs inoculation or at least some backmixing.

Due to the peculiar form of the rate equation, the CISTR as a biochemical reactor will demonstrate the special behaviour of having two stable operating points at one and the same reactor temperature. We can demonstrate this by considering the consumption of the substrate A. The substrate concentration in the reactor product is formed by solving for c_A of the equation $\Phi_v(c_{Ao} - c_A) = -R_A V_R$, which, if $c_{Po} = 0$ so that $c_o = v'c_{Ao}$, can be written as

$$c_{Ao} - c_A = \frac{k_P \tau_1 c_A (c_{Ao} - c_A)}{k_A + c_A}. \tag{III.23}$$

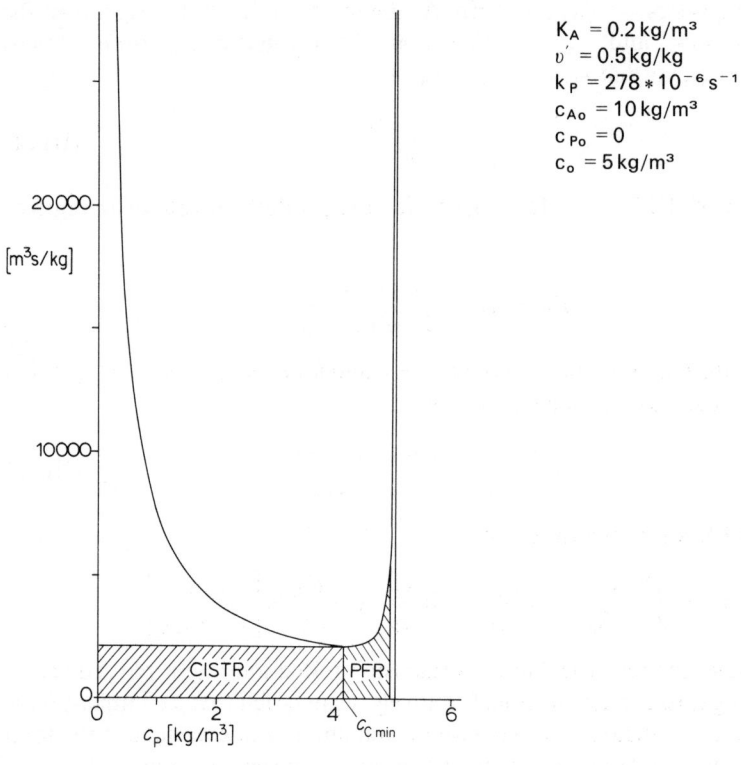

$K_A = 0.2 \text{ kg/m}^3$
$v' = 0.5 \text{ kg/kg}$
$k_P = 278 * 10^{-6} \text{ s}^{-1}$
$c_{Ao} = 10 \text{ kg/m}^3$
$c_{Po} = 0$
$c_o = 5 \text{ kg/m}^3$

Fig. III.46 Reciprocal cell production rate vs. cell concentration

This equation has two solutions: the first is $c_A = c_{Ao}$, so that no conversion takes place and no micro-organisms are produced. The second solution is

$$c_A = \frac{K_A}{k_P \tau_1 - 1}, \qquad \text{(III.24)}$$

which is only a real solution for $k_P \tau_1 > 1$. This is shown in figure III.47. The straight line through the points $(c_{Ao}, 0)$ and $(0, c_{Ao})$ represents the amount of substrate disappeared in the reactor and the curved line the amount of substrate consumed in the reactor. The intersections of both lines represent the conditions where both are equal and therefore the possible operating points. For the curve $k_P \tau_1 = 2$ there are two operating points 1 and 2 in figure III.47. According to equation (III.23) there will be two operating points as long as $\tau_1 > k_P^{-1}$. We call this phenomenon, when there is more than one possible operating point, *multiplicity*. We can easily understand that the point 2 is an unstable operating point as long as $\tau_1 > k_P^{-1}$ because if after a disturbance in a reactor operating in point 2, a small amount of substrate is consumed by the reaction so that $c_{A1} < c_{Ao}$, then $\Phi_v(c_{Ao} - c_A)$, the net supply of A to the reactor is smaller than $R_A V_R$, the amount of A consumed in the reactor by the reaction, so that the reactor outlet concentration must necessarily decrease further. The reactant

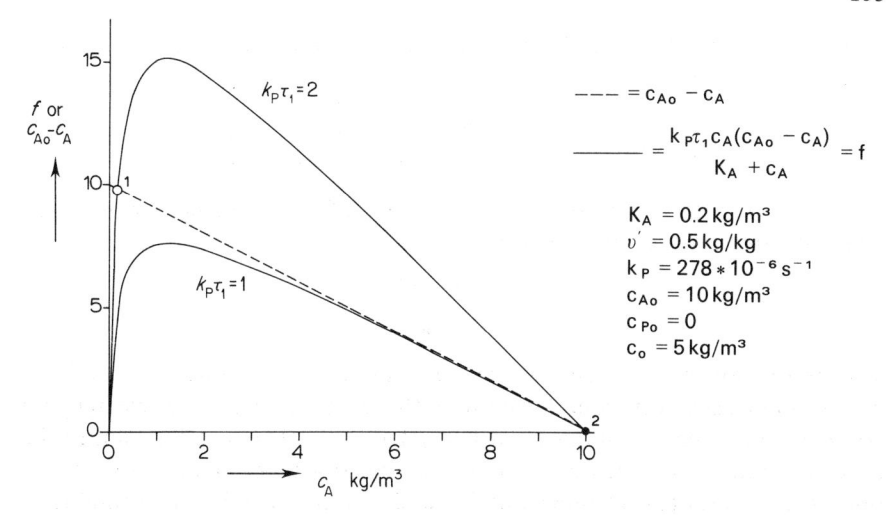

Fig. III.47 Operating points and multiplicity in a biochemical reactor

concentration c_A will continue to decrease towards and up to point 1, where both net supply and consumption are in balance again. That point 1 is the only stable operating point as long as $k_P \tau_1 > 1$ has been shown by Humphrey et al. [29–31], see also Tsuchija et al. [32].

However, if the feed flow rate is increased and consequently the residence time is reduced, the operating point 1 moves into the direction of point 2 and as soon as $k_P \tau_1$ equals or becomes lower than 1, the point 1 disappears completely and point 2 becomes the only stable operating point. When τ_1 becomes equal to k_P^{-1} *wash out* takes place; that is, the micro-organisms are washed out of the reactor.

From the above it can be derived, using equation (III.21), that the concentrations of the micro-organisms in the reactor for operating point 1 in figure III.47, provided $c_{P_0} = 0$, equals

$$c_P = v'\left(c_{Ao} - \frac{K_A}{k_P \tau_1 - 1}\right). \qquad (III.25)$$

and the production rate of micro-organisms per unit of reactor volume is $\Phi_v c_P / V_R = c_P / \tau_1$. For a reactor with a volume V_R the production reaches a maximum at a certain value of Φ_v. The production is

$$\Phi_v c_P = \Phi_v v'\left(c_{Ao} - \frac{K_A \Phi_v}{k_P V_R - \Phi_v}\right)$$

and by differentiation and the condition $d\Phi_v c_P / d\Phi_v = 0$ we find for the *optimum residence time*

$$(k_P \tau_1)_{opt} = \left[1 - \sqrt{\left(\frac{K_A}{K_A + c_{Ao}}\right)}\right]$$

and also we can derive for the maximum production of P per unit of reactor volume—provided $c_{Po} = 0$—that

$$\left(\frac{c_P}{\tau_1}\right)_{max} = v'c_{Ao}\left\{1 + \frac{K_A}{c_{Ao}} - \sqrt{\left[\frac{K_A}{c_{Ao}}\left(\frac{K_A}{c_{Ao}} + 1\right)\right]}\right\}$$

From equation (III.25), by setting $c_P = 0$ we find, for the *critical residence time*, where the micro-organisms are washed out,

$$\tau_{1\,crit} = \frac{K_A + c_{Ao}}{k_P c_{Ao}}.$$

The production rate of the cells per unit of reactor volume c_P/τ_1 and the corresponding substrate concentrations in the reactor outlet are shown in figure III.48 for the same conditions as in Fig. III.47 and III.46 as a function of the residence time τ_1. We see that for this case the critical and optimum residence time are quite near to each other and that under optimal conditions around 13 % of the substrate is not converted for the reaction chosen.

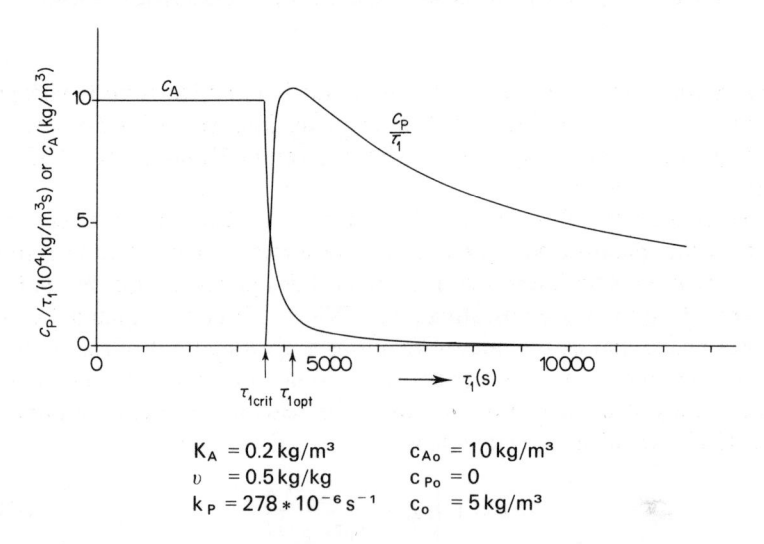

Fig. III.48 Substrate concentration and reactor capacity for a biochemical CISTR

We conclude that, due to the peculiar reaction rate equations in biochemical reactions, a biochemical reactor demonstrates the following special features:

—a multiplicity of operating points;
—at too short residence times the reaction stops, the active cells are 'washed out';
—for one CISTR the reactor production of biomass reaches a maximum at a specific residence time.

After this introduction on single autocatalytic reactions with Monod kinetics we will continue with multiple biochemical reactions.

III.6.2 Multiple autocatalytic reactions

From the above it follows that the CISTR as a fermentation reactor is very sensitive to the reactor residence time. In a PFR or a BR the reaction would not start if no product were present at the reactor entrance: the reactor feed has to be inoculated with micro-organisms ($c_{P_0} \neq 0$) or reactor product. This is due to the autocatalytic character of biochemical reactions. As has been outlined in Section II.4 in this case recycle of reactor product to the reactor inlet has a favourable influence on the reaction and the required residence time. We will study the influence of recycle in a waste water treatment plant, where biologically oxidizable materials are consumed by an active sludge, which is supplied with oxygen from air. The bacteriological mass of the active sludge is also auto-oxidized, which causes degradation. The auto-oxidation is proportional to the amount of bacteria present. Now the rate equation for the active sludge (P) formation is given by

$$R_P = \frac{k_P c_P c_A}{K_A + c_A} - k_X c_P.$$

The last term represents the degradation, A represents the water contaminants. In a heterogeneous population of contaminants K_A, k_P and k_X refer to the mean values for the population. A simplified flow scheme is given in figure III.49.

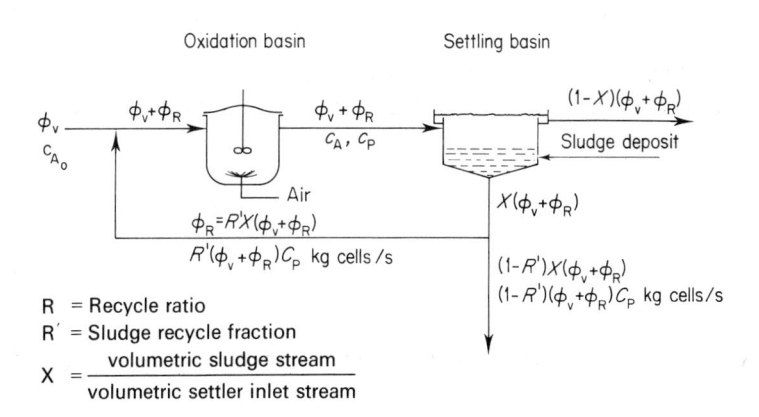

Fig. III.49 Simplified flow scheme of a waste water treatment plant

The biochemical reactor is considered to behave as a CISTR. In the settling tank all the sludge settles on the bottom and a volume fraction X of the incoming stream is carried away in the bottom stream as a sludge. The sludge contains all P, the overflow of the settling tank is clear. A volume fraction R' of the settling tank bottom stream is recycled to the reactor inlet, which contains

$$(\Phi_v + \Phi_R)c_P/x(\Phi_v + \Phi_R) = c_P/x \text{ kg/m}^3$$

sludge. So the recycle stream is $\Phi_R = R'x(\Phi_v + \Phi_R)$ and the recycled active sludge stream is $R'(\Phi_v + \Phi_R)c_P$ kg/s. The active sludge stream $(1 - R')(\Phi_v + \Phi_R)c_P$ kg/s is removed from the plant. We have made implicitly the following simplifying assumptions:
— there are no physical transport limitations for oxygen and substrate, either in the active sludge cells or in the liquid;
— degradation of substrate and cells in the settling tank is negligible. This is reasonable because of the oxygen deficit in the settler;
— complete separation of the total sludge in the settler. Total sludge is active sludge P + dead cells + water;
— the oxygen concentration in the aeration tank reactor is so high that the reaction rates are independent of the oxygen concentration.

For the inlet concentrations to the reactor we find, if $R = \Phi_R/\Phi_v$,

$$c_{A\,in} = c_{Ao}/(R + 1) \quad \text{and} \quad c_{P\,in} = R'c_P$$

For a CISTR we now find, for the substrate,

$$(R + 1)\Phi_v(c_A - c_{A\,in}) = -\frac{k_P}{v'}\frac{c_A c_P}{K_A + c_A}V_R \tag{III.26}$$

and for the active sludge,

$$(R + 1)\Phi_v(c_P - c_{P\,in}) = \left[\frac{k_P c_P c_A}{K_A + c_A} - k_X c_P\right]V_R. \tag{III.27}$$

After introducing $c_{P\,in} = R'c_P$ into equation (III.27), c_P disappears out of the equation, so that we find for the concentration of the substrate in the reactor outlet, if $\tau_1 = V_R/\Phi_v$:

$$c_A = \frac{K_A}{\dfrac{k_P\tau_1}{(1 + R)(1 - R') + k_X\tau_1} - 1}. \tag{III.28}$$

Similarly we find for the active sludge concentration from Eq. III.26 the following relation:

$$c_P = \frac{v'[c_{Ao} - c_A(R + 1)]}{(1 + R)(1 - R') + k_X\tau_1} \tag{III.29}$$

From equation (III.28) some interesting conclusions can be drawn as was done by, among others, Ottengraf and Rietema [33, 34]:

1. for full recycle ($R' = 1$) or in an extremely large reactor $\bar{\tau}_1 \simeq \infty$) the substrate conversion reaches a minimum of

$$c_{A\,min} = \frac{K_A}{(k_P/k_X) - 1}$$

so that all substrate can never be converted. This is a consequence of the undesired consecutive reaction: the unconverted amount of substrate is higher, the larger k_X with respect to k_P or the more important the undesired degradation reaction is.

2. The substrate conversion is independent of the substrate inlet concentration, provided the simplifying assumptions are valid, especially that of ample oxygen supply.
3. For a certain value of $\tau_1 = V_R/\Phi_v$ there is a minimum value of the sludge recycle ratio R, below which the substrate consumption or $c_{Ao} - c_A$ becomes zero, which means that the reactor is washed out by a too large recycle stream. From $\Phi_R = R'x(\Phi_R + \Phi_v)$ it follows that $1 + R = (1 - R'x)^{-1}$, so that by putting $c_A = c_{A\,in}$ in equation (III.28) we can find the minimum value R'_{min} of the sludge recycle ratio by trial and error from

$$c_{Ao}(1 - R'x)\left[\frac{k_P\tau_1}{\frac{(1 - R')}{(1 - R'x)} + k_X\tau_1} - 1\right] = K_A$$

or

$$\frac{1 - R'_{min}}{1 - xR'_{min}} = k_P\tau\left[\frac{(c_{Ao}(1 - xR'_{min}))}{K_A + c_{Ao}(1 - xR'_{min})} - \frac{k_X}{k_P}\right].$$

In figure III.50 the concentration of substrate A and living micro-organisms P are given for a certain waste water treatment plant. The minimum recycle ratio is $R = 0.34$ or $R' = 0.42$. We see that below $R = 0.34$ the reactor is washed out and that above $R = 0.34$ the active sludge concentration c_P steadily increases. Further data are given in the figure; remember that $v' = 0.25$. The total sludge removed consists of P, active and X, non-active sludge.

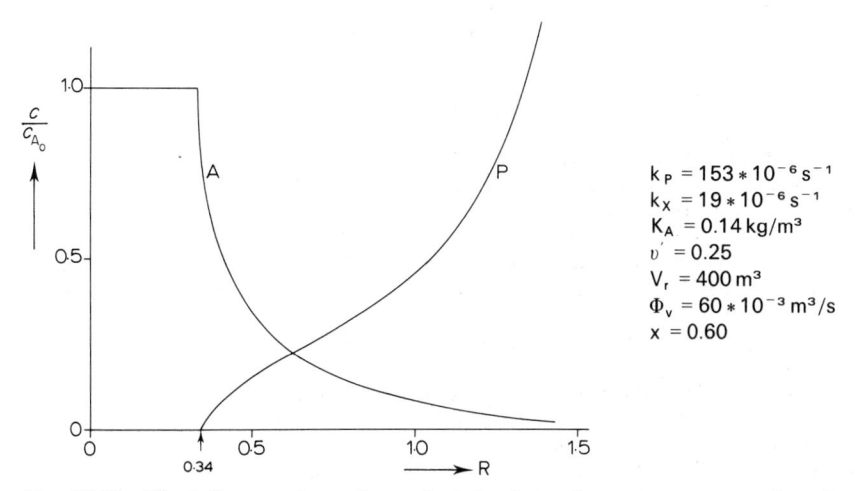

$k_P = 153 * 10^{-6}\,s^{-1}$
$k_X = 19 * 10^{-6}\,s^{-1}$
$K_A = 0.14\,kg/m^3$
$v' = 0.25$
$V_r = 400\,m^3$
$\Phi_v = 60 * 10^{-3}\,m^3/s$
$x = 0.60$

Fig. III.50 The influence of recycle on the behaviour of a waste water treating plant

If the waste water treater behaves as a cascade, a series of algebraic equations as above has to be solved. In the extreme case a PFR is approached. No analytical expressions can be obtained for a PFR and numerical solutions are required. Ottengraf and Rietema [33] have shown that in industrial practice, where sludge recycle ratios are high, a constant sludge concentration along the reactor volume can be reasonably assumed. By taking $c_P = c_{PL}$ constant an analytical expression is obtained for a PFR, from which the substrate conversion can be derived by trial and error:

$$\frac{c_{Ain} - c_{AL}}{1 + R}\left[1 - \frac{k_P \tau_L}{(1 + R)(1 - R') + k_X \tau_L}\right] + K_A \ln\left[\frac{c_{Ain} + R c_{AL}}{(1 + R) c_{AL}}\right] = 0$$

and

$$c_{PL} = \frac{v'(c_{Ain} - c_{AL})}{(1 + R)(1 - R') + k_X \tau_L}.$$

Results according to this consideration are quite dependent on the model reactor which approaches the actual reaction conditions best. This seems to be contradictory to industrial practice, where waste water treating plants of widely varying designs nevertheless give similar results. This was explained by Wolfbauer et al. [15] as being caused by zero order kinetics for single substrates. In practice we have many substrates, but for each individual substrate both a PFR and a CISTR require the same residence time $\tau_1 = \tau_L = (c_{Ao} - c_A)/k$ for a certain conversion in case the reaction is of zero order. In our case at high conversions ($c_A \ll K_A$) the conversion rate for the substrate is zero order in the substrate concentration with $R_A = -k_P c_P/v' K_A$. As was discussed in Section III.5, lumping of many parallel reactions for a mixture of substrates would result in an overall reaction rate of a higher order than that of the individual reactions. This would explain the fact that in practice usually the same volume for an aeration basin is regarded as necessary to achieve a specified conversion, independently of the concentration of the contaminants.

REFERENCES

1. Denbigh, K. G.: *Chem. Eng. Sci.* **14**, 25 (1961).
2. Vusse, J. G., van de and Voetter, H.; *Chem. Eng. Sci.* **14**, 90 (1961).
3. McMullin, R. B.; *Chem. Eng. Progr.* **44**, 183 (1948).
4. Chermin, H. A. G. and van Krevelen, D. W., *Chem. Eng. Sci.* **14**, 58 (1961).
5. Jungers, J. C. with Balaceanu, J. C., Coussemant, F., Eschard, F., Giraud, A., Hellin, M., Leprince, P., and Limido, G. E.: *Cinetique chimique appliquée*, Société des Editions Technip, Paris (1958).
6. Küchler, L., and Langbein, D., *Proc. Chem. React. Eng. Congress*, 267, Pergamon (Oxford) (1965).
7. Kobelt, D., and Troltenier, U.: *Chem. Ing. Techn.* **38**, 134 (1966).
8. McBee, E. T., Hass, H. B., Neher, G. M., and Strickland, H., *Ind. Eng. Chem.* **34**, 296 (1942).
9. Messikommer, B. H.: *Proc. Int. Seminar on 'Analogue Computation Applied to the Study of Chemical Processes'*, Brussels (Nov. 1960).

10. Wei, J., and Prater, C. D.: *Adv. Catal.* **13,** 203 (1962).
11. Mah, R. S. H., and Aris, R.: *Chem. Eng. Sci.* **19,** 541 (1964).
12. Froment, G. F., and Bischoff, K. B., *Chem. Eng. Sci.* **17,** 105 (1962).
13. Wei, J., and Kuo, J. C. W., *Ind. Eng. Chem. Fund* **8,** 114, 128 (1968).
14. Luss, D., and Hutchinson, P., *Chem. Eng. Journal* **1,** 129 (1970); **2,** 172 (1971).
15. Wolfbauer, O., Klettner, H., and Moser, F.: *Chem. Eng. Sci.* **33,** 953 (1978).
16. Rodiguin, N. M., and Rodiguina, E. N., *Consecutive Chemical Reactions*, Van Nostrand, Princeton, N.J. (1964).
17. Denbigh, K. G.: *Trans. Far. Soc.* **40,** 352 (1944).
18. Trambouze, P. J. and Piret, E. L., *A.I.Ch.E. Journal* **5,** 384 (1959).
19. Waterman, H. I., Boelhouwer, C., and Huibers, D. T. A., *Process Characterization*, Elsevier, Amsterdam (1960).
20. Oden, E. C.: *Petr. Refiner* **29**(4), 103 (1950).
21. Van de Vusse, J. G.: *Chem. Eng. Sci.* **19,** 994 (1964).
22. Smith, J. M.: *Chemical Engineering Kinetics*, second ed., McGraw-Hill, New York (1970).
23. Gillespie, B. M., and Carberry, J. J.: *Chem. Eng. Sci.* **21,** 472 (1966).
24. Ridlehoover, G. A., and Seagrave, R. C.: *Ind. Eng. Chem. Fundam.* **12,** 444 (1973).
25. Lund, M., and Seagrave, R. C.: *Ind. Eng. Chem. Fundam.* **10,** 494 (1971).
26. Pasquali, G., and Lelli, U.: *Chem. Eng. Sci.* **29,** 1291 (1974).
27. Michaelis, L., and Menten, M. M. L.: *Biochem. Z.* **49,** 333 (1913).
28. Monod, J.: *Ann. Rev. Microbiol.* **3,** 371 (1949).
29. Humphrey, A. E.: *Chemical Reaction Engineering Reviews*—Houston, A.C.S. Symp. Series **72,** 262 (1978).
30. Koga, S., Burg, C., and Humphrey, A. E.: *Appl. Microbiol.* **15,** 493 (1967).
31. Humphrey, A. E., and Yang, R. D.: *Biotechnol. Bioengr.* **9,** 1976 (1975).
32. Tsuchija, H. M., Frederikson, A. G., and Aris, R.: *Adv. Chem. Eng.* **6,** 125 (1966).
33. Ottengraf, S. P. P., and Rietema, K.: *J. Water Poll. Contr. Fed.* **41,** R282 (1969).
34. Ottengraf, S. P. P.: *De Ingenieur* **83,** Ch. 39 (1971).
35. Koetsier, W. T., Pniak, B. J., Boegborn, J. F., and van Swaaij, W. P. M.: *Proc. CHISA congress*, paper H3.22 (1981).
36. Bischoff, K. B.: *Can. J. Chem. Eng.* **44,** 281 (1966).

Chapter IV

Residence time distribution and mixing in continuous flow reactors

The purpose of this chapter and Chapter V is to discuss to what extent continuous reactor systems used in practice are different from the models treated in the previous chapters. As a starting point we take the residence time distribution (RTD) which is largely characteristic of the reactor type and on the basis of which a comparison with model reactors is possible. Next the influence of the residence time distribution on the chemical reaction is discussed for the case of simple chemical reactions. Unless otherwise stated we will assume a micro fluid, which means that the concentration is assumed to be uniform at the scale of a small volume element with a dimension down to almost molecular scale. If this is not the case and the fluid shows concentration differences on a small scale the fluid is called a macro fluid or a (partially) segregated fluid. To visualize a segregated fluid and a micro mixed fluid one could think of a viscous fluid partially mixed with a coloured fluid with the same properties. On a macro scale the fluid may seem of a uniform colour but viewed through a microscope one might see streaks and isolated domains of more intensive colour. This is illustrated in figure IV.1. If these inhomogenous domains exist the fluid is called (partially) segregated fluid or macro fluid. Only if the mixture is uniform on a scale down to say 100 times the dimension of the molecules can the liquid be considered as a microfluid. On a larger scale there can be average concentration gradients in both segregated and micro fluids. Partially segregated fluids are difficult to deal with, especially if fast and other than first order chemical reactions occur. These last cases will be considered in Chapter V.

IV.1 THE RESIDENCE TIME DISTRIBUTION (RTD)

IV.1.1 The *E* and the *F* diagram

Because in many cases it is not possible to predict or even to describe the hydrodynamic behaviour of the fluid flow through a real reactor there is often a need for a simplified flow model in combination with experimental data on the fluid flow. A very powerful concept is the residence time distribution, which is essentially a statistical approach for the fluid flow. A typical element of the fluid is considered while it enters the reactor. The probability to leave the reactor after a certain residence time is expressed as a residence time distribution function (RTD). The

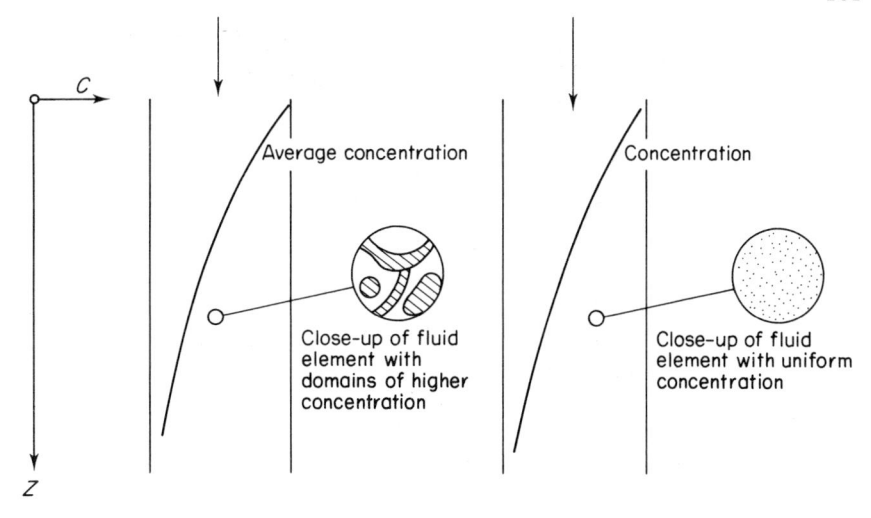

Fig. IV.1 Macro and micro fluids in chemical reactors

idea has been first clearly explained by Danckwerts [1]. To define this RTD function more exactly we will consider a reactor operating under steady state conditions. We first have to look at the boundaries of the reactor as illustrated in figure IV.2.

We will first assume plug flow in both the inlet tube and the outlet tube. Such boundary conditions are called closed–closed. This refers to the fact that a fluid element can enter and leave the reactor only once. This is not evident for all boundaries. It will be clear that for a reactor with mixing also in the inlet tube a fraction of the elements that entered the reactor may be mixed back to the inlet tube and later enter the reactor again. Similar arguments hold for the outlet of the reactor. Figure IV.2 gives the different possibilities. Conditions other than closed–closed are important for certain experimental techniques to assess the mixing parameters inside the reactor but we will concentrate on the closed–closed conditions.

The elements of a fluid entering the reactor at a certain moment $t = 0$ will generally not stay in the reactor for exactly the same period; there will be a distribution of residence times. This can be represented in different ways. Figure IV.3 represents the frequency function often used in statistics and called the E diagram in RTD descriptions. The shaded area $E \, dt$ represents the fraction of fluid elements that entered the reactor at $t = 0$ which leaves the reactor between t and $t + dt$. Or, otherwise stated, it represents the chance for a fluid element to have a residence time in the reactor between t and $t + dt$. The E diagram thus defined has an area of unity below the curve because all fluid elements will have a residence time between 0 and ∞. Therefore we may write

$$MO_0 = \int_0^\infty E(t) \, dt = 1.$$

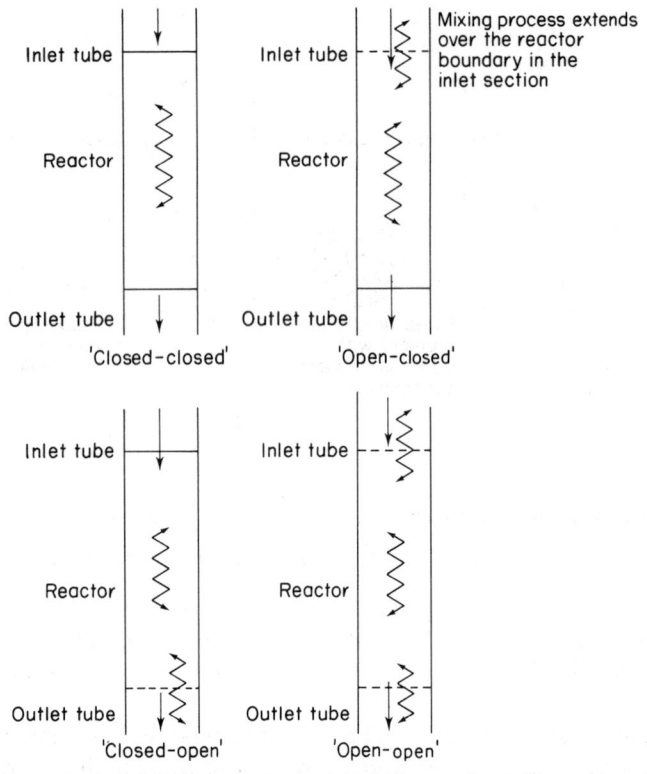

Fig. IV.2 Boundary conditions of reactors in relation to the residence time distribution

MO_0 is the area under the distribution curve or the zero-th moment. Moments are used to characterize the residence time distribution function in terms of statistical parameters such as the average residence time, spread in residence time, skewness of distribution, etc. The moments are defined as

$$MO_k \equiv \int_0^\infty t^k E(t)\, dt, \qquad (IV.1)$$

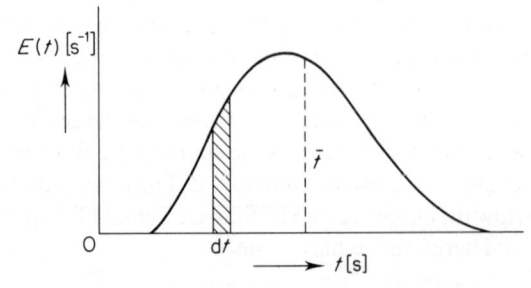

Fig. IV.3 The RTD represented in an $E(t)$ diagram. The shaded area $E(t)\,dt$ represents the chance for a fluid element to have a residence time between t and $t + dt$

in which MO_k is the kth moment around the origin. The mean residence time is equal to the first moment:

$$\bar{t} = MO_1 = \int_0^\infty tE(t)\,dt. \tag{IV.2}$$

The spread in residence time is characterized by the standard deviation σ_t or the variance σ_t^2:

$$\sigma_t^2 = \int_0^\infty (t - \bar{t})^2 E(t)\,dt = MO_2 - MO_1^2, \tag{IV.3}$$

using

$$MO_2 = \int_0^\infty t^2 E(t)\,dt.$$

Higher moments, which are used in quantities like skewness, peakedness etc. of the RTD function are less often used. The E diagram is often also presented, using a reduced time, as $E(\theta)$ versus θ, where $\theta = t/\tau$, in which τ is the holding time. For a fluid of constant density we may write:

$$\tau = \frac{V}{\Phi_v}$$

in which V is the volume accessible for the fluid and Φ_v the volumetric flow rate. Sometimes the time is reduced by the average residence time $\bar{t}$ resulting in an E diagram: $E(\theta_t)$ versus θ_t. The average residence time $\bar{t}$ can differ from the holding time τ (which is also often called average residence time) for example in case of adsorption on a catalyst or on the reactor walls, etc. In the case of adsorption the average residence time $\bar{t}$ is larger than the holding time τ. Also for other reactor boundaries than closed–closed there can be a difference between $\bar{t}$ and τ as will be demonstrated later on (Example IV.4.c).

The different E diagrams are related by

$$E(t) = \frac{1}{\bar{t}} E(\theta_t) = \frac{1}{\tau} E(\theta). \tag{IV.4}$$

Similar to equation (IV.1) the moments for the $E(\theta)$ diagrams are found from

$$MR_k = \int_0^\infty \theta^k E(\theta)\,d\theta$$

the average is

$$MR_1 = \int_0^\infty \theta E(\theta)\,d\theta = \bar{\theta}$$

and the variance is

$$\sigma_\theta^2 = MR_2 - MR_1^2 = \int_0^\infty (\theta - \bar{\theta})^2 E(\theta)\,d\theta.$$

If two closed vessels are put in series the resulting $E(t)$ diagram of the cascade can be calculated from the E diagram of the first and the second vessel. This is illustrated in figure IV.4.

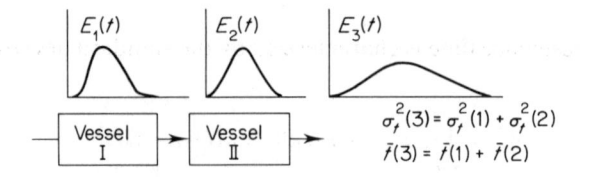

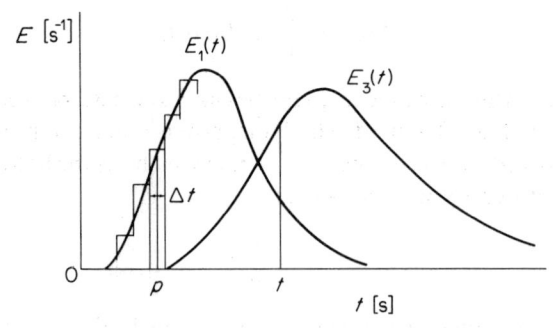

Fig. IV.4 The convolution of two residence time distribution functions when two reactors are put in series. $E_3(t)$ is the RTD of the two reactor cascade with $E_1(t)$ and $E_2(t)$ as RTD functions respectively

The different fractions of a fluid element entering the first vessel at $t = 0$ will come out of this vessel in fractions as indicated by $E_1(t)$. Therefore, $E_1(p)\,\Delta p$ will be the fraction of the original element which will leave vessel I and enter vessel II around $t = p$. This fraction can again be considered as a fluid element entering vessel II and subject to the next residence time distribution E_2. The fraction of this element that will leave the second vessel at time t will be $E_2(t - p)\,\Delta t$. Counted as a fraction of the original element that entered vessel I this is:

$$E_1(p)\,\Delta p\, E_2(t - p)\,\Delta t.$$

Not only the element entering vessel II at $t = p$ will contribute to the exit stream of vessel II but in principle all other fractions as given by E_1 from $p = 0$ to $p = t$. Therefore the value of $E_3(t)$ is given by

$$E_3(t)\,\Delta t = \sum_{p=0}^{t} E_1(p)\,\Delta p\, E_2(t - p)\,\Delta t$$

or, written as an integral,

$$E_3(t) = \int_0^t E_1(p) E_2(t - p)\,\mathrm{d}p. \qquad (IV.5)$$

Equation (IV.5) is called the convolution integral, sometimes written as $E_1 * E_2 = E_3$ and is useful if separate reactors or flow regions are put in series. The sequence of reactors or regions does not influence the RTD result and we may also write, for $E_3(t)$

$$E_3(t) = E_2 * E_1 = \int_0^t E_2(p) E_1(t-p)\,dp.$$

Generally for N reactors in series we may write

$$E_{cascade} = E_1 * E_2 * E_3 * \cdots E_N. \qquad (IV.6)$$

For the mean residence time and the variances of a cascade the following simple equations can be obtained:

$$\bar{t}_{cascade} = \bar{t}_1 + \bar{t}_2 + \bar{t}_3 + \cdots + \bar{t}_N \qquad (IV.7)$$

$$\sigma^2_{t\,cascade} = \sigma^2_t(1) + \sigma^2_t(2) + \sigma^2_t(3) + \cdots \sigma^2_t(N). \qquad (IV.8)$$

These equations for reactor cascades can be very useful though applicability is limited to statistically independent flow regions only. Generally, for closed–closed systems these expressions can be applied without danger but, especially for open boundaries, this has to be further verified.

A complementary presentation of the residence time distribution can be made by a cumulative function called the $F(t)$ diagram. Figure IV.5 gives an example.

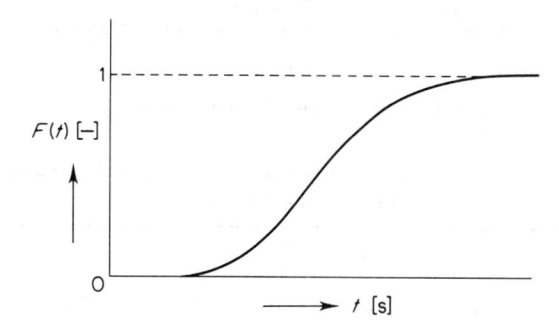

Fig. IV.5 Cumulative RTD function or F diagram. $F(t)$ represents the fraction of fluid elements in the vessel exit stream with a residence time shorter than t seconds

Here the fraction $F(t)$ is plotted of elements with a residence time shorter than t seconds. The E and F diagrams are related by

$$F(t) = \int_0^t E(t)\,dt. \qquad (IV.9)$$

The average residence time can be calculated from

$$\bar{t} = \int_0^\infty t\,dF(t) \qquad (IV.10)$$

and the variance from:

$$\sigma_t^2 = 2\int_0^\infty t\{1 - F(t)\}\,dt - \bar{t}^2.$$

The functions F and E are equivalent but some properties are more easily demonstrated with an E diagram, so we use them more frequently in this book.

Example IV.1.a *Some numerical examples of E and F diagrams and properties* Three examples of residence time distributions will be discussed. The E diagrams are given in figure IV.6. For the distribution E_1 numerical values are given in Table IV.1 and we will show in detail how the different characteristic parameters are calculated.

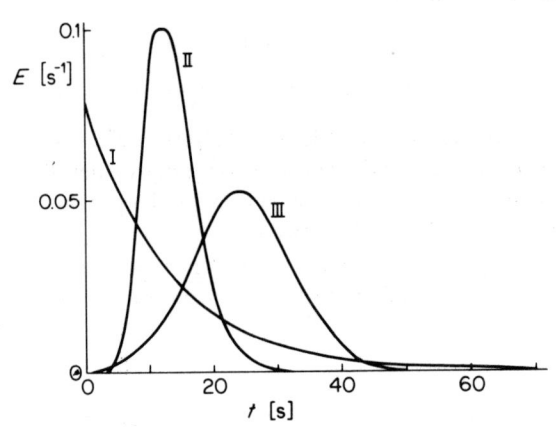

Fig. IV.6 Examples of $E(t)$ diagrams

Table IV.1 Numerical values for the residence time distribution E_1; Example IV.1.a

No.	t [s]	E [s^{-1}]	No.	t [s]	E [s^{-1}]
1	0.00	0.0769	16	34.8	0.0053
2	2.32	0.0643	17	37.1	0.0044
3	4.64	0.0538	18	39.5	0.0037
4	6.96	0.0450	19	41.7	0.0031
5	9.29	0.0377	20	44.1	0.0026
6	11.6	0.0315	21	46.4	0.0022
7	13.9	0.0263	22	48.8	0.0018
8	16.3	0.0220	23	51.1	0.0015
9	18.6	0.0184	24	53.4	0.0013
10	20.9	0.0154	25	55.7	0.0011
11	23.2	0.0129	26	58.0	0.0009
12	25.5	0.0108	27	60.4	0.0007
13	27.9	0.0090	28	62.7	0.0003
14	30.2	0.0076	29	65.0	0.0000
15	32.5	0.0063			

Firstly, the area below the curve is calculated, e.g., by the Simpson rule or simply by

$$MO_0 \simeq \sum_0^{28} \tfrac{1}{2}\{E(i) + E(i+1)\}\Delta t_i =$$

$$\tfrac{1}{2}\{0.0769 + 0.0643\}2.32 + \tfrac{1}{2}\{0.0643 + 0.0538\}2.32$$
$$+ \cdots + \tfrac{1}{2}\{0.0003 + 0.0000\}2.32 = 0.995 \simeq 1.00$$

The area is almost unity which means that the curve has already been normalized. Otherwise E values should be divided by the area under the curve MO_0. The average residence time $\bar{E} = MO_1$ can be approximated by

$$\bar{E} \simeq \sum_0^{28} \tfrac{1}{2}\{t(i)E(i) + t(i+1)E(i+1)\}\Delta t_i$$
$$= \tfrac{1}{2}\{0.0 \times 0.0769 + 2.32 \times 0.0643\}2.32$$
$$+ \tfrac{1}{2}\{2.32 \times 0.0643 + 4.64 \times 0.0538\}2.32$$
$$+ \cdots + \tfrac{1}{2}\{62.7 \times 0.0003 + 65.0 \times 0.0\}2.32 = 12.4 \text{ s}.$$

The standard deviation can be found from the second moment around the origin and MO_1, or alternatively directly from the equation for the second moment around the average (equation (IV.3)).

$$MO_2 \simeq \sum_0^{28} \tfrac{1}{2}\{t^2(i)E(i) + t^2(i+1)E(i+1)\}\Delta t_i = 291.9 \text{ s}^2,$$

from which it follows that the variance is equal to

$$\sigma_t^2 = MO_2 - MO_1^2 = 291.9 - (12.4)^2 = 138.1 \text{ s}^2$$

and the standard deviation

$$\sigma_t = 11.8 \text{ s}.$$

Also the reduced distribution function and its parameters can be easily found

$$E(\theta_t) = E_t MO_1 = E_t \bar{t}$$
$$\theta_t = t/MO_1 = t/\bar{t}$$

and if $\bar{t} = V/\Phi_v = \tau$ (constant density, no adsorption, etc.) then we may write

$$E_\theta = E_t \bar{t}$$
$$\theta = t/\bar{t}.$$

This reduced function is given as curve I in figure IV.7.

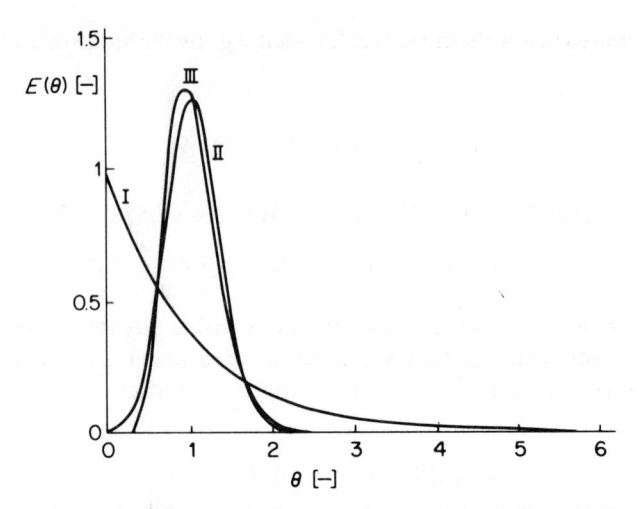

Fig. IV.7 Reduced E diagrams for examples I, II and III

The variance and the standard deviation from this reduced distribution can be found directly from the previous results via the reduced moments:

$$MR_k = \int_0^\infty \theta^k E(\theta)\, d\theta$$

$$\bar{\theta} = MR_1 = MO_1/MO_1 = 1$$

$$\sigma_\theta^2 = MR_2 - MR_1^2 = MO_2/MO_1^2 - \frac{MO_1^2}{MO_1^2}$$

$$= 291.9/(12.4)^2 - 1 = 0.8984$$

$$\sigma_\theta = 0.95.$$

Of course $\bar{\theta} = 1$ only holds for the condition $\tau = \bar{t} = MO_1$. If $\tau \neq \bar{t} = MO_1$ (e.g. in the case of adsorption on walls), MR_1 may be much larger than 1 for the $E(\theta)$ diagram. For the $E(\theta_t)$ diagram $MR_1 = 1$ always holds.

In the same way as presented here the parameters of distributions II and III have been calculated and are given in Table IV.2.1. As can be seen from Table IV.2.1, all E diagrams are normalized ($MO_0 = 1$), distributions I and II have nearly the same

Table IV.2.1 Parameters of RTD functions I, II, and III; Example IV.1.a

Distribution	MO_0	MO_1 or $\bar{t}$ [s]	MO_2 [s^2]	σ_t [s]	σ_θ [—]
I	1.00	12.4	291.9	11.8	0.952
II	1.00	12.9	183.1	4.09	0.317
III	1.00	23	582	7.27	0.316

average residence time and distribution III nearly twice as long. The relative spread in residence time σ_θ is almost the same for distributions II and III. Distribution I has a much larger spread, $\sigma_\theta \simeq 1$, and as will be shown later it is almost identical to the residence time distribution of an ideal mixer. Distribution II is somewhat less symmetrical than III and this type of asymmetry is normally encountered in practice. It is possible to characterize these functions further by the higher moments MO_3 and MO_4, etc. However, it is often not possible to obtain accurate figures for these parameters from experimental curves. Due to the factors t^3 and t^4 respectively, heavy weight is put on the end fraction of the curves which is often not sufficiently accurate for experimentally obtained distributions. Figure IV.8 shows the F diagrams for distributions I, II and III obtained from

$$F(t) = \int_0^t E(t)\,dt$$

with numerical integration. If the distributions are nearly symmetrical the properties of the normal distribution curves can be applied for estimation of average and variance:

If $E(t) \simeq E$ (normal dist.) $= \dfrac{1}{\sqrt{(2\pi\sigma_t^2)}} e^{-(t-\bar{t})^2/2\sigma t^2}$ the value $2\sigma_t$ can be found from the width of the $E(t)$ curve at a position of $0.61\,E_{\max}$. For the F diagram the slope at $t = \bar{t}$ can be related to σ_t via

$$\sigma_t = \frac{1}{\sqrt{(2\pi)} \left(\dfrac{dF}{dt}\right)_{t=\bar{t}}}.$$

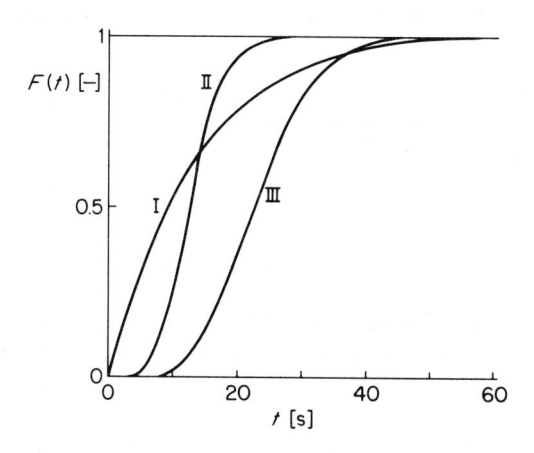

Fig. IV.8 $F(t)$ diagrams for distributions I, II and III

The two short-cut methods, taking the width of the E diagram or the slope from the F diagram to obtain σ_t for symmetrical curves approaching a normal distribution, are illustrated in figure IV.9 for curve III. It should be noted that RTD functions cannot be strictly symmetrical, specially not at large spread in residence time, as this would imply the occurrence of negative residence times. However, for a small spread in residence time the normal distribution function can be a good approximation.

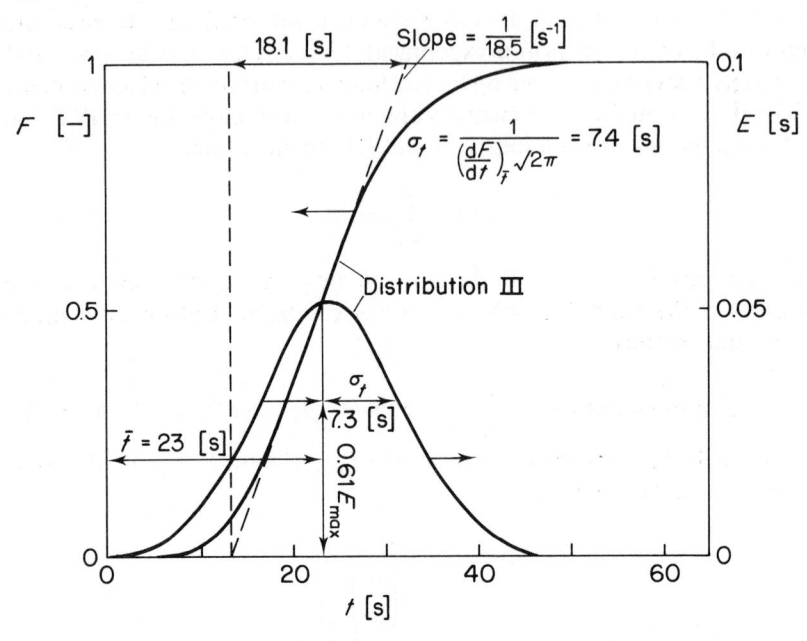

Fig. IV.9 Short-cuts for σ_t evaluation from E or F diagrams approaching normal distributions. Distribution III is used as an example

Example IV.1.b *Applications of the convolution integral to flow regions in series*
A reactor (I) with an RTD as given in figure IV.10, curve I, is put in series with a second reactor (II) with RTD curve II. The resulting RTD curve III has been calculated by

$$E_{III}(t) = \int_0^t E_i(t-p)E_{II}(p)\,dp.$$

This integral is evaluated numerically for a large number of points as follows: The time axis is divided into N discrete intervals from $t = 0$ to $t = t_N$. Then $E_{III}(t_N)$ can be found from

$$E_{III}(t_N) \simeq \sum_{n=0}^{N-1} \tfrac{1}{2}\{E'_{III}(t_n) + E'_{III}(t_{n+1})\}(t_{n+1} - t_n)$$

in which E'_{III} is given by

$$E'_{III}(t_n) = E_1(t_N - t_n)E_2(t_n).$$

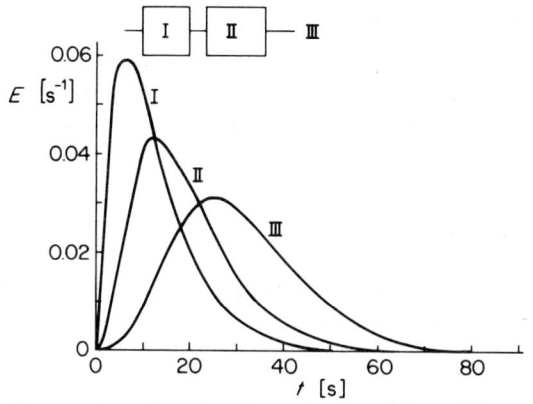

Fig. IV.10 Flow regions in series. Convolution of E_I and E_{II} producing E_{III}

To demonstrate the addition rules the moments, calculated as explained in Example IV.1.a, are given in Table IV.2.2

Table IV.2.2

Curve	MO_0	$MO_1 = \bar{t}$	σ_t	$\sigma_\theta = \dfrac{\sigma_t}{\bar{t}}$
	[—]	[s]	[s]	[—]
I	1.0	12.6	8.4	0.667
II	1.0	18.9	10.8	0.571
III	1.0	31.5	14.09	0.447

According to the addition rules of average residence times and variances for cascades we may write

$$\bar{t}(III) = \bar{t}(II) + \bar{t}(I)$$
$$31.5 = 18.9 + 12.6$$
$$\sigma_t^2(III) = \sigma_t^2(II) + \sigma_t^2(I)$$
$$198.5 \simeq 116.6 + 70.6 - 187.2.$$

Due to numerical errors there is a slight deviation of $\sigma_t(III)$ as calculated from the left hand and right hand side of the equality.

$$\frac{\Delta\sigma_t^2(III)}{\sigma_t^2(III)} \simeq 6\% \quad \text{or} \quad \frac{\Delta\sigma_t(III)}{\sigma_t(III)} \simeq 3\%.$$

Often it is attractive to 'subtract' one flow region from two or more in series, e.g. to eliminate end effects by deconvolution. This is generally a very complicated problem, especially if accurate results are desired; assistance of a mathematical expert will be required. Only if the flow region to be subtracted is ideally mixed is the solution simple, and it has been given by Hanhart et al. [2] for the F diagram and by de Groot [3] for the E diagram.

IV.1.2 The application of the RTD in practice

Before we go into detail about tracer techniques used for experimental determination of RTDs and discuss the different models that are applied for a description of RTD we will briefly summarize the type of problems for which the RTD concept is of interest in practice.

Identification of the mixing in the reactor

Experimentally observed RTDs can be compared with those predicted for plug flow, ideal mixing or with those predicted for intermediate mixing models to be treated later. Experimentally observed RTD functions can be used for identification of malfunctioning of a reactor. Two cases are illustrated in figure IV.11. Short cutting will be visible in the front part of the E diagram as a relatively large fraction of the fluid with a very short residence time. If dead zones are present a relatively large fraction of the fluid will have a very long residence time which is indicated by an excessively long tail at the E curve. In extreme cases the tail may not be detectable because of very slow renewal of the dead zone by the active part of the fluid. In that case the average residence time observed from the E diagram is much smaller than $\tau = V/\Phi_v$.

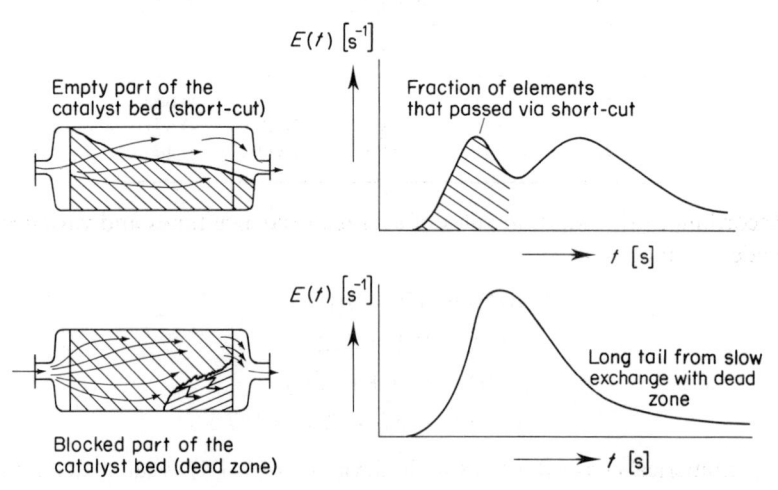

Fig. IV.11 Examples of short-cutting and dead zones and their appearances in the $E(t)$ diagram

Conversion prediction for first order homogeneous reactions

In the calculation method the fluid elements entering the reactor are considered as separate batch reactors keeping their identity during their stay in the reactor and having a residence time probability as given by the RTD. After taking the average composition of all fluid elements leaving the reactor the steady state conversion is obtained. Of course, this picture is not consistent with the general assumption of a microfluid used in this chapter but, as will be shown in Chapter V for the special case of a first order reaction, the microfluid and the macro fluid give the same conversion result. Consider the fluid elements leaving the reactor after a residence time of t seconds. We will assume constant density and temperature. If the reactant concentration of these elements at the inlet was c_o and a first order chemical reaction takes place with a rate constant k (1/s) the concentration at the outlet, if we consider the element as a small batch reactor, would be:

$$c = c_o e^{-kt}.$$

The average outlet concentration for all fluid elements is

$$\bar{c} = c_o \int_0^\infty e^{-kt} E(t) \, dt. \tag{IV.11}$$

Calculations of parameters of complex flow models

For many complex flow models the residence time distribution can be calculated and compared with experimentally observed RTD curves or related properties such as variance and mean residence time to evaluate parameters such as exchange rates between different zones. Examples can be found in the literature, e.g. those given by van Deemter [4] and van Swaaij [5].

Calibrating flow rate measurements and reactor volume and/or tests for occurrence of adsorption

For a noncompressible fluid and a closed vessel the relation

$$\bar{t} = \tau = \frac{V}{\Phi_v}$$

should hold. Therefore the average residence time $\bar{t}$ experimentally derived from tracer measurements can be used to check flow rates or to measure the volume V accessible for the fluid. If these quantities are known, tracer test results can be used to check for adsorption of the tracer component. In case of adsorption the average residence time is larger than the holding time:

$$\bar{t}(\text{experimental}) > \frac{V}{\Phi_v}.$$

IV.2 EXPERIMENTAL DETERMINATION OF THE RESIDENCE TIME DISTRIBUTION

IV.2.1 Input functions

The general procedure for RTD determination is that fluid elements flowing into the reactor or the measuring section are marked and detected at the outlet. This marking is done by changing some property with time. Mostly the concentration of some tracer substance is varied. The requirements for such a substance are that it should be easily detectable and measurable at low concentrations, e.g. by electrical conductivity, light absorption or nuclear radiation. Furthermore, as far as possible, the tracer solution should have the same properties, such as density, as the fluid, to avoid flow changes due to viscosity and molecular diffusion of the tracer in the solution, which is specially important in laminar flow. Both in theoretical and experimental studies four types of incoming tracer signals are frequently used:

(i) A sharp pulse, where at $t = 0$ a certain amount of tracer is injected into the inlet in the shortest possible time. Mathematically this is represented by $n_{inj}\delta(t)$, where n_{inj} is the number of moles injected and $\delta(t)$ is the Dirac delta function. This function is explained in figure IV.12, where the delta function is depicted as an extreme case of a block function. While keeping the surface area unity and reducing the width of the block, the limit at an infinitely small block is obtained. The delta function is a very practical tool because a mathematically well defined function is obtained which can be incorporated in mass balances, integrated, Laplace transformed and so on. An advantage of the δ-pulse injection is that in case of a reactor with closed–closed boundaries after normalization to $MO_0 = 1$ the $E(t)$ diagram is directly obtained as the response. However, it is difficult to realize a sufficiently close approximation to a δ-pulse. At least the injection time should be negligible with respect to the average residence time.

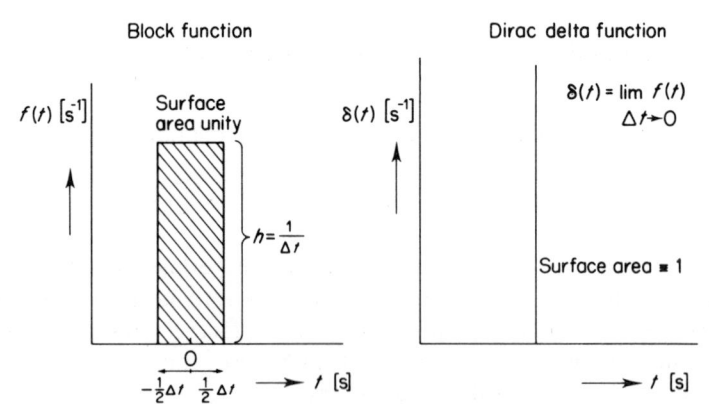

Fig. IV.12   The idealized pulse injection function δ, the Dirac delta function

(ii) A step function where at $t = 0$ the incoming concentration changes from one steady value, usually zero, to another one. Mathematically this is represented by $\Phi_v c_{inj} U(t)$. This injection function is depicted in figure IV.13. The unit step function can also be handled as a mathematically well defined function. It is often more simple to approach in practice than a δ-pulse. For a closed–closed vessel the direct response is the F diagram.

(iii) A steady sinusoidal variation of the incoming signal. Here the response is likewise sinusoidal with the same frequency. By plotting the amplitude attenuation and the phase shift of the outgoing signal with respect to the incoming one as a

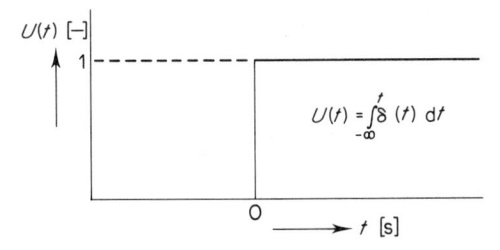

Fig. IV.13 The idealized step injection function $U(t)$, the Heaviside step function

function of the applied frequency ω the so-called frequency response diagram is obtained. This way of presentation is customary in the theory of automatic control and servomechanisms for characterizing the dynamic behaviour of the various elements in a control circuit. The sinusoidal input has experimental advantages for systems with very short average residence times where other ideal injection functions are difficult to realize.

(iv) Non ideal signals such as an arbitrary pulse of tracer can be used provided the incoming signal is measured separately at the inlet. With various methods such as the convolution integral (equation (IV.5)) the response can be interpreted. See also Example IV.6.b.

Figure IV.14 shows as an example the response of two CISTRs in series on the four types of input signals mentioned above. The frequency response diagrams essentially contain the same information as the others. The translation of the frequency function response diagram into a step function or pulse response and vice versa is complicated and will not be treated here. Kramers and Alberda [6] introduced this technique. In practice it is often difficult to produce signals which approach the mathematical descriptions mentioned under (i) and (ii). Therefore, they only give direct results in experiments with systems having a relatively long average residence time and rather large variances. In other cases the application of (iv) should be considered. The frequency response measurements are time consuming and require special equipment for the production of sinusoidal signals in a wide frequency range. Examples of applications of these techniques are given by Kramers and Alberda [6] and Fontaine and Harriot [7]. Sometimes data are obtained from experiments in which tracer is injected and/or detected at some position within the reactor. This is done in an attempt to eliminate the 'end' effects of inlet or distribution sections outside the reactor section proper. The 'closed–closed' vessel conditions are then not applicable because tracer can move upstream out and into the test section. Such a situation requires a careful consideration of the boundary conditions which are full of mathematical subtleties. An example is given in Example IV.6.b. Mathematical treatment has been provided by Aris [8] and Wen and Fan [9].

Another complication is the difficulty of marking the fluid elements properly in a representative manner. Also the detection within a test section can be a complicated problem and improper sampling can lead to large errors. This is elucidated in the following example.

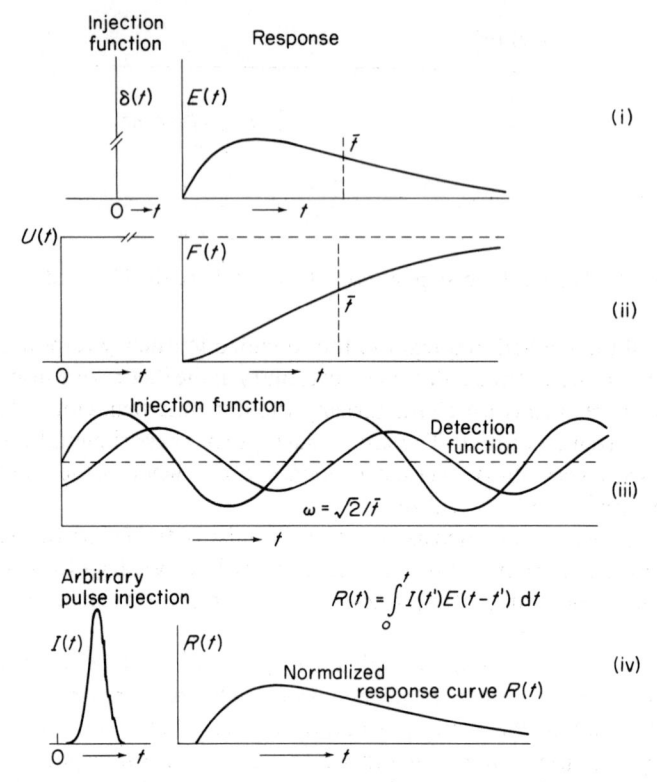

Fig. IV.14 Different injection functions and corresponding responses. In the example given here the reactor consists of two ideal mixers in series

Example IV.2 *Injection and sampling of tracer within a flow section*
A typical experimental set-up for injection and measurement of a tracer within the test section is given in figure IV.15. The distribution of the tracer at the injection plane is difficult, especially in view of the velocity profile. To obtain a representative marking of fluid elements the tracer injection rate at each point should be

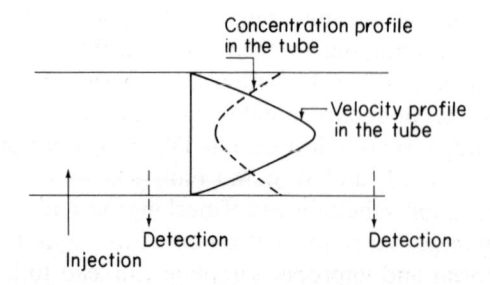

Fig. IV.15 Tracer injection and detection within a test section

proportional to the local velocity. The correct way of sampling on the other hand would be to find the flow averaged or mixed cup concentration:

$$\bar{c} = \frac{\iint_s cv\,ds}{\iint_s v\,ds}$$

in which s is the cross sectional area of the fluid. In practice only an average of the concentration over the cross section is measured or even an average over a line in case of a 'through the wall' measurement. If averaged over the surface area we obtain

$$\bar{c} = \frac{\iint_s c\,ds}{\iint_s ds}$$

instead of the former equation. This may lead to large deviations. Levenspiel and Turner [10] have calculated the different 'through the wall' solutions and the mixed cup solution for the RTD of a laminary flow through an empty tube. For this case the differences can be vary large indeed. A correct method of tracer injection and detection is shown in figure IV.16. Ideal mixers with negligible residence time are added for proper marking and detection. In practice this can be approached by converting the end sections outside the reactor zone proper into mixers and correcting the overall result for the RTD of the end sections. Examples of these techniques can be found in the literature. See de Groot [3] and Hanhart et al. [2].

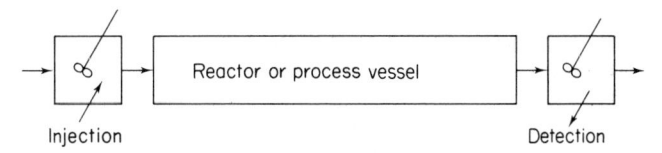

Fig. IV.16 Example of a correct way of injection and detection of tracer. Ideal mixers with negligible volumes are placed in the inlet and outlet streams

IV.3 RESIDENCE TIME DISTRIBUTION IN A CONTINUOUS PLUG FLOW REACTOR AND IN A CONTINUOUS IDEALLY STIRRED TANK REACTOR

A plug flow reactor has no spread in residence time because the fluid is supposed to flow like a plug through the reactor. This means that the input signal can be found back in the outlet with a time delay equal to the residence time. This is shown for the

$E(\theta)$ and $E(t)$ diagram in figure IV.17. It can be easily understood that the variance has to be zero in a PFR:

$$\sigma_t^2 = \int_0^\infty (t - \bar{t})^2 E(t)\,dt = 0$$

The E diagram for the continuous stirred tank reactor can be derived from a mass balance for the mixer after an ideal pulse has been injected at the inlet. The pulse is immediately distributed over the reactor; moreover the concentration remains everywhere the same in the reactor and equals the outlet concentration. The mass balance reads

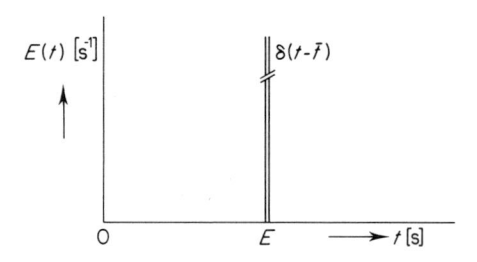

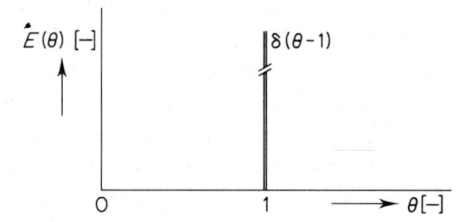

$$\frac{d(\rho V_r w_1)}{dt} = \Phi_{mo} w_o - \Phi_{m1} w_1. \tag{IV.12}$$

Fig. IV.17 E diagrams for a plug flow reactor

If the total mass accumulation is zero, then $\Phi_{mo} = \Phi_{m1}$ holds. Further, for constant density and taking $w_o = 0$ for $t > 0$ we may write

$$V_r \frac{dc_1}{dt} = -\Phi_v c_1 \tag{IV.13}$$

while at $t \to 0$

$$c_1 = c_o = \frac{n_{inj}}{V_r},$$

with n_{inj} = total number of moles injected in the pulse at $t = 0$. After integration of equation (IV.13), this leads to

$$\frac{c_1}{c_o} = e^{-t\Phi_v/V_r} = e^{-t/\tau} = e^{-\theta}.$$

Because c_1/c_o represents the reduced response at the outlet to a δ-injection at $\theta = 0$ in the inlet stream, c_1/c_o is equal to the $E(\theta)$ diagram. Therefore, for a continuous stirred tank we have:

$$E(\theta) = e^{-\theta}. \tag{IV.14}$$

This result is plotted in figure IV.18. The relative moments of the distribution are

$$MR_0 = \int_0^\infty e^{-\theta}\,d\theta = 1 \quad \text{(normalization condition)}$$

$$MR_1 = \bar{\theta} = \int_0^\infty \theta e^{-\theta}\,d\theta = 1$$

and the variance

$$\sigma_\theta^2 = MR_2 - MR_1^2 = 1.$$

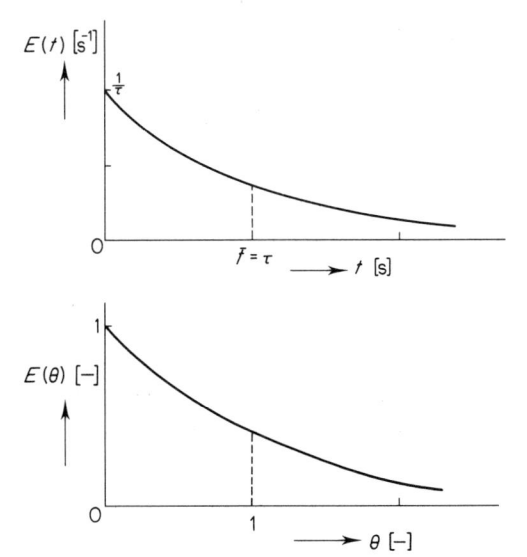

Fig. IV.18 E diagrams for a single continuous stirred ideal tank reactor

The $E(t)$ diagram reads, according to equation (IV.4),

$$E(t) = \frac{1}{\tau}e^{-t/\tau} \tag{IV.15}$$

and the moments for the $E(t)$ diagram are

$$MO_0 = \int_0^\infty \frac{1}{\tau}e^{-t/\tau}\,dt = 1$$

$$MO_1 = \bar{t} = \int_0^\infty \frac{t}{\tau}e^{-t/\tau}\,dt = \tau$$

and the variance

$$\sigma_t^2 = MO_2 - MO_1^2 = \tau^2$$

Example IV.3.a *E-diagram and moments derived from Laplace transformation*
Especially with more complex models the differential equations describing the mass balance are conveniently handled by Laplace transformation (see, for example, Spiegel [11]). The differential equations can then be solved in the Laplace domain and the solution transformed back to the time domain. The latter step is often complicated but the moments can be solved without back transformation via the Laplace transform property:

$$MR_k = \int_0^\infty \theta^k E(\theta)\,d\theta = \lim_{s \to 0} (-1)^k \frac{d^k \underline{E}}{ds^k} \quad (IV.16)$$

$\underline{E}$ = Laplace transform of $E(\theta)$, defined by

$$\underline{E} = \int_0^\infty e^{-s\theta} E(\theta)\,d\theta. \quad (IV.17)$$

For the ideal mixer the differential equation

$$\frac{dC_1}{d\theta} = -C_1 \quad \text{with} \quad C_1 = \frac{c_1}{c_o} \quad \text{and} \quad c_o = \frac{n_{inj}}{V_r}$$

becomes after Laplace transformation

$$s\underline{C}_1 - C_1(0) = -\underline{C}_1.$$

$C_1(0)$ = relative concentration at $t = 0$ and therefore $C_1(0) = 1$. So

$$\underline{C}_1 = \frac{1}{1+s}.$$

$\underline{C}_1$ represents the Laplace transform of the reduced response at the outlet to a δ-injection at $\theta = 0$ at the inlet and therefore it is equal to the Laplace transform $\underline{E}$ of the $E(\theta)$ function. So

$$\underline{E} = \frac{1}{1+s}.$$

From back transformation one finds:

$$E(\theta) = e^{-\theta}$$

in accordance with equation (IV.14). The moments are found taking the limits for $s \to 0$:

$$MR_1 = \lim_{s \to 0} (-1)^1 \frac{d\underline{E}}{ds} = \lim_{s \to 0} \frac{1}{(1+s)^2} = 1$$

$$MR_2 = \lim_{s \to 0} (-1)^2 \frac{d^2\underline{E}}{ds^2} = \frac{2}{(1+s)^3} = 2$$

$$\text{var} = \sigma_\theta^2 = MR_2 - MR_1^2 = 1.$$

Equation (IV.16) is identical to equation (IV.11) if one substitutes $k\tau$ by s and therefore the Laplace transform of the E diagram can also be interpreted as the relative reactant slip (c/c_o) function for a first order homogeneous reaction carried out in that reactor.

Example IV.3.b *Modelling a circulation reactor* (Brothman et al. [12]).
We will consider an agitated vessel with a well defined internal circulation flow (see figure IV.19a) or an external loop in which fluid recirculates (figure IV.19b) as an approximation of a well mixed tank reactor. The volumetric circulation flow rate Φ_{vc} is assumed to be much greater than the steady flow Φ_v in and out of the system.

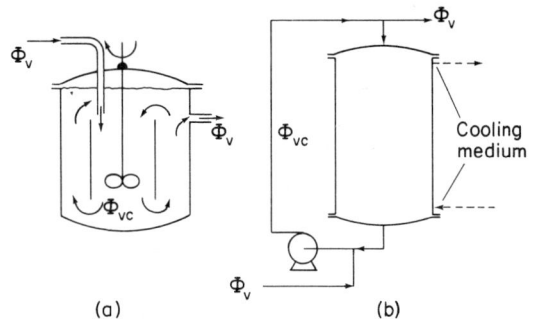

Fig. IV.19 Circulation systems used to approximate a continuous ideal stirred tank reactor

The probability p of a volume element leaving the system after one cycle is assumed to be equal to $\Phi_v/(\Phi_{vc} + \Phi_v)$. With $\tau = V_r/\Phi_v$ and with the time for one circulation

$$t_{circ} = \frac{V_r}{\Phi_{vc} + \Phi_v},$$

we get

$$p = \frac{t_{circ}}{\tau}.$$

The probability that a fluid element is still present after m cycles is given by

$$(1-p)^m = \int_{mt_{circ}}^{\infty} E(t)\,dt = 1 - \int_0^{mt_{circ}} E(t)\,dt$$

For very small values of p and with $t = mt_{circ}$, we may write

$$\int_0^t E(t)\,dt = \lim_{p \to 0}\{1 - (1-p)^m\} = 1 - e^{-t/\tau}$$

or

$$E(t) = \frac{1}{\tau}e^{-t/\tau}.$$

This result is identical to equation (IV.15), which means that a recycle system as described above approximates an ideal tank reactor if the recycle ratio is sufficiently high. For practical applications a value of $p < 1/10$ has been suggested for a good approximation of ideal mixing (see Chapter III).

If a chemical reaction takes place the reaction time scale should also be compared with the circulation time as a second criterion: $t(\text{reaction})/t_{\text{circ}} \gg 1$. An important application is the laboratory reactor for kinetic measurements where the mass transfer between a liquid and catalyst particles and also, if necessary, a gas phase, can be set at different values by varying the recycle ratio. The conversion per pass should be kept low, whereas the overall fluid mixing can be close to that of an ideal mixer. This is shown in figure IV.20.

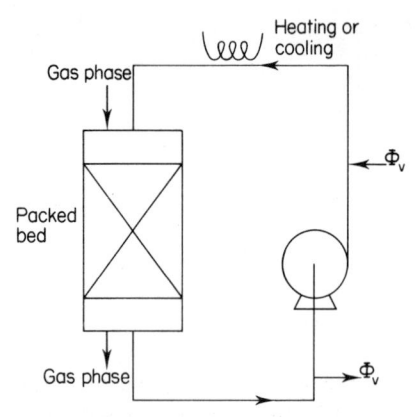

Fig. IV.20 Experimental set-up simulating an ideal continuous stirred tank reactor in a three phase system, where the mass transfer rates can be varied

IV.4 MODELS FOR INTERMEDIATE MIXING

In many cases the RTD of an actual reactor is neither that of a CISTR nor that of a PFR, but shows an intermediate behaviour. In reactors with a high L/d_t ratio longitudinal mixing may occur which can often be traced back to the following causes:

(i) Vortices and turbulent eddies may produce convective mixing in the direction of the flow.
(ii) The velocity distribution over a cross section is generally not uniform and consequently the residence times of fluid particles travelling along different stream lines is different.
(iii) In principle molecular diffusion always takes place. In the majority of practical cases, the effects of (i) and (ii) are much greater than the effect of (iii). Furthermore there is an essential difference between (i) and (ii) in the sense that phenomenon (i) can give rise to 'back-mixing', which is upstream transport of material under the influence of a concentration gradient. In case (ii), on the other hand, back-mixing is possible only if the direction of the flow is opposite at different positions in the cross section of the reactor. If the spread in residence time is relatively small it appears that the $E(\theta)$ curve is

somewhat similar to a normal error or Gaussian distribution curve. There are two simple, one parameter models in use that predict E curves like these: the tanks in series model and the plug flow with superimposed axial dispersion model.

IV.4.1 Model of a cascade of N equal ideally mixed tanks

This model can be used for a real cascade of mixers but it has also been applied to other reactors which in that case are considered to be equivalent to a cascade of an appropriate number of CISTRs in series. The basis for comparison is usually the same average residence time and the same spread in residence time. An advantage of this model is that it can be easily applied for complex reactions by simple calculations from mixer to mixer. The $E(\theta)$ diagram can be derived from the mass balance which reads, for the nth mixer,

$$\frac{d\rho V_{r,n} w_n}{dt} = \Phi_{m,n-1} w_{n-1} - \Phi_{m,n} w_n, \qquad (\text{IV.18})$$

in which V_r = volume of one mixer
w_n = weight fraction of tracer substances in the nth mixer.

If the total mass accumulation in every CISTR is zero, the density is constant, the tank volume is constant and equal for all vessels, the mass balance reduces to

$$V_{r,n} \frac{dc_n}{dt} = \Phi_v (c_{n-1} - c_n).$$

With a pulse injection of n_{inj} moles in the feed stream to the first vessel the initial condition for the inlet becomes

$$c_{\text{inj}} = \frac{n_{\text{inj}} \delta(t)}{\Phi_v}$$

or, after introduction of dimensionless units,

$$\frac{1}{N} \frac{dC_n}{d\theta} = C_{n-1} - C_n \quad \text{with} \quad \theta \leq 0 \rightarrow C_n = 0 \quad \text{and} \quad C_{\text{inj}} = \delta(\theta). \qquad (\text{IV.19})$$

In these equations

$$C_n = c_n/c_o, \quad c_o = \frac{n_{\text{inj}}}{N V_{r,n}},$$

in which

$$V_{r,n} = V_r/N \quad \text{and} \quad \theta = t/\tau,$$

where $\tau = V_r/\Phi_v$,

The solution of equation (IV.19) for the Nth mixer is given by

$$C_N(\theta) = E(\theta) = \frac{N^N \theta^{N-1}}{(N-1)!} e^{-N\theta}. \qquad (\text{IV.20})$$

This result can be obtained from the E diagram of one single mixer by successive convolution, but is readily obtained via Laplace transformation. This will be demonstrated in Example IV.4.a. Some results for $E(\theta)$ are given in figure IV.21. In figure IV.22 $E(t)$ is given for cascades of 1, 2, 3 and 5 mixers respectively, each with an average residence time of one second per mixer. The moments for the reduced diagram $E(\theta)$ are

$$MR_1 = \bar{\theta} = 1$$

and the variance

$$\sigma_\theta^2 = \frac{1}{N}. \tag{IV.21}$$

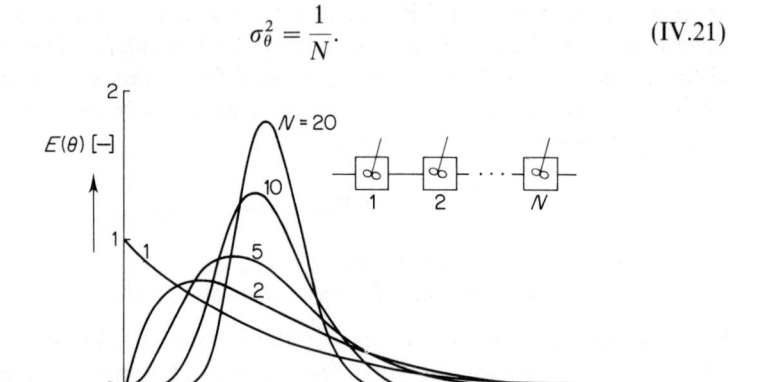

Fig. IV.21 $E(\theta)$ diagram for a cascade of N CISTRs

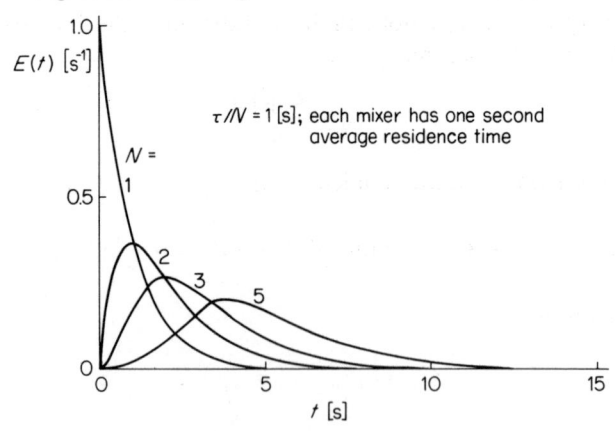

Fig. IV.22 $E(t)$ diagram for a cascade of N CISTRs

For the $E(t)$ diagram these values are

$$MR_1 = \tau \quad \text{with} \quad \tau = NV_{r,n}/\Phi_v$$

and the variance is

$$\sigma_t^2 = \frac{\tau^2}{N}. \tag{IV.22}$$

These latter equations can be derived from the addition rules for flow regions in series. See equations (IV.7) and (IV.8). The average residence time for one mixer of the cascade is $\tau_n = \tau/N$. The variance of one CISTR is $\sigma_{t,n}^2 = (\tau/N)^2$. The variance of the cascade is therefore $\sigma_t^2 = N(\tau/N)^2 = \tau^2/N$.

As can be seen from equations (IV.21) and (IV.22), when putting more mixers in series the spread as measured in σ_t increases, but because the average residence time increases more $\sigma_\theta = \sigma_t/\tau$ becomes smaller. This can also be seen from figures IV.21 and IV.22. Therefore the larger the number of mixers in series that are equivalent to a reactor of a given volume, the better plug flow is approached in that reactor. Plug-flow is in fact reached for $N \to \infty$ and $\sigma_\theta \to 0$. In practice in many cases as few as 5 mixers in series give a reasonable approach to plug flow.

Example IV.4.a *E-diagram and the moments of a cascade of mixers from Laplace transformation*

The mass balance equation (IV.19) becomes, after Laplace transformation,

$$\frac{-C_n(0)}{N} + s\frac{\underline{C}_n}{N} = \underline{C}_{n-1} - \underline{C}_n, \qquad C_n(0) = 0$$

or

$$\frac{\underline{C}_n}{\underline{C}_{n-1}} = \frac{1}{1 + s/N};$$

therefore

$$\frac{\underline{C}_N}{\underline{C}_{inj}} = \left(\frac{1}{1 + s/N}\right)^N \text{ with } \underline{C}_{inj} = 1.$$

By back transformation [11] equation (IV.20) is found. The moments can also be found from the Laplace transform via

$$MR_k = (-1)^k \lim_{s \to 0} \frac{d^k \underline{C}_N}{ds^k}$$

which leads to

$$MR_1 = \bar{\theta} = 1$$

$$MR_2 = 1 + \frac{1}{N} \quad \text{or} \quad \sigma_\theta^2 = \frac{1}{N}.$$

IV.4.2 The axially dispersed plug flow model

Just like the model of ideal mixers in series the axially dispersed plug flow model can be used to describe intermediate cases between ideal mixing and plug flow. In many cases the deviation from plug flow is not too large and the residence time distribution observed is highly symmetrical and approaches the normal (also called the error or Gaussian) distribution curve. This suggests that the residence time distribution of such a reactor may be considered as being approximately the result of piston flow with a superposition of longitudinal dispersion. The latter is taken into account by means of a constant effective longitudinal dispersion coefficient, D_1, which has the same dimension as the molecular diffusivity. It may be—and usually is—much larger than the molecular diffusivity because it also incorporates all other effects that cause deviation from plug flow like velocity differences, eddies and vortices. The mass flux caused by this dispersion is described by Fick's law which leads in the one-dimensional case to

$$J = -D_1 \frac{\partial c}{\partial z}. \qquad (IV.23)$$

The residence time distribution as predicted by the axially dispersed plug flow model can be derived from a mass balance over a volume element $S\,\Delta z$ as indicated in figure IV.23. We will assume constant density and constant D_1, then the mass balance reduces to

$$\frac{\partial c}{\partial t} S\,\Delta z = \langle v \rangle (c_z - c_{z+\Delta z})S + \left(-D_1 \frac{\partial c}{\partial z}\bigg|_z + D_1 \frac{\partial c}{\partial z}\bigg|_{z+\Delta z}\right)S$$

or with $\Delta z \to 0$

$$\frac{\partial c}{\partial t} = -\langle v \rangle \frac{\partial c}{\partial z} + D_1 \frac{\partial^2 c}{\partial z^2}. \qquad (IV.24)$$

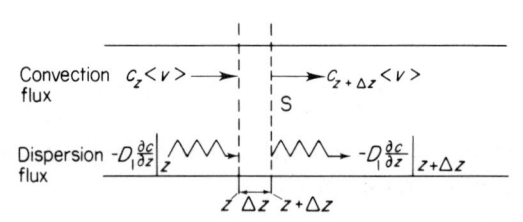

Fig. IV.23 Mass balance for the axially dispersed plug flow model. S is the cross sectional area of the fluid perpendicular to the fluid flow

Now we can introduce the following dimensionless variables:

$$\theta = \frac{t\langle v \rangle}{L}, \quad Pe = \frac{\langle v \rangle L}{D_1}, \quad Z = \frac{z}{L}, \quad C = \frac{c}{c_0}$$

$$c_0 = \frac{n_{\text{inj}}}{LS}.$$

In dimensionless form equation (IV.24) now reads

$$\frac{\partial C}{\partial \theta} = -\frac{\partial C}{\partial Z} + \frac{1}{Pe}\frac{\partial^2 C}{\partial Z^2} \qquad (IV.25)$$

The Peclet number Pe can be considered as the ratio between the transport rate by convection and the transport rate by dispersion. The physical meaning can be understood from the two extreme cases:

— In case Pe is infinite the dispersion rate is negligible compared to the convection rate. This is plug flow.
— In case Pe approaches zero the convection rate is much slower than the dispersion rate so that the flow region is completely mixed.

The boundaries are taken here as closed for dispersion at inlet and outlet (See figure IV.24). A pulse injection takes place at the inlet plane of n_{inj} kmoles. At the reactor side of the injection plane or $0 \leftarrow z$, two parallel transport mechanisms play a role, viz. dispersion and convection, so that

$$n_{inj}\delta(t) = \langle v \rangle cS - D_1 \frac{\partial c}{\partial z} S \quad \text{at} \quad 0 \leftarrow z$$

or in dimensionless units this boundary condition is

$$\delta(\theta) = C - \frac{1}{Pe}\frac{\partial C}{\partial Z} \quad \text{at} \quad 0 \leftarrow Z \tag{IV.26}$$

For the outlet boundary we have

$$Z \to 1 \quad \frac{\partial C}{\partial Z} = 0 \quad \text{and} \quad C(\text{out}) = C(Z = 1). \tag{IV.27}$$

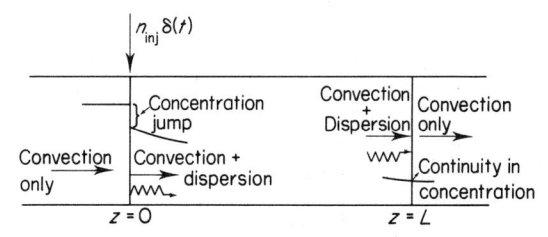

Fig. IV.24 Boundary conditions 'closed–closed vessel' for the axially dispersed plug flow model

This boundary condition can be understood by realizing that no dispersion flux is assumed to occur through the plane $Z = 1$ and continuity in concentration is to be expected on both sides of the plane at $Z = 1$. Furthermore we have the initial condition

$$\theta \leqslant 0, \quad C = 0. \tag{IV.28}$$

A solution of equation (IV.25) with its boundary conditions as given here was presented by Otake and Kunigita [13]:

$$E(\theta) = \exp\left[\frac{Pe}{2}\left(1 - \frac{\theta}{2}\right)\right] \sum_{n=1}^{\infty} \frac{\delta_n[Pe \sin \delta_n + 2\delta_n \cos \delta_n]}{\left[\delta_n^2 + \left(\frac{Pe}{2}\right)^2 + Pe\right]} \times \exp\left[-\frac{\delta_n^2 \theta}{Pe}\right], \tag{IV.29}$$

Where δ_n is given by the nth root of the transcendental equation

$$\cot \delta_n = \frac{\delta_n}{Pe} - \frac{Pe}{4\delta_n}.$$

The equation is plotted in figure IV.25. The curves are skewed especially for low values of Pe. This is mainly due to the fact that tracer arriving at the outlet has been subjected for a longer period of time to the dispersion process than tracer arriving earlier. The skewness is also influenced by the boundary conditions. If there were no boundaries (open–open) the concentration of tracer would by symmetrically distributed with regard to position. The concentration curve measured at a fixed position as a function of time would still be skewed. For $Pe = 0$ the E diagram for an ideal mixer is found. Equation (IV.29) is rather difficult to manipulate. Simpler equations have been given in the literature for other boundary conditions. Some are presented in Example IV.4.c. For medium and high Peclet numbers these solutions are useful also for closed–closed vessels. However, for low Pe numbers they deviate considerably from the exact solution and for Pe approaching zero they do not tend to the limit of the E diagram for one ideal mixer.

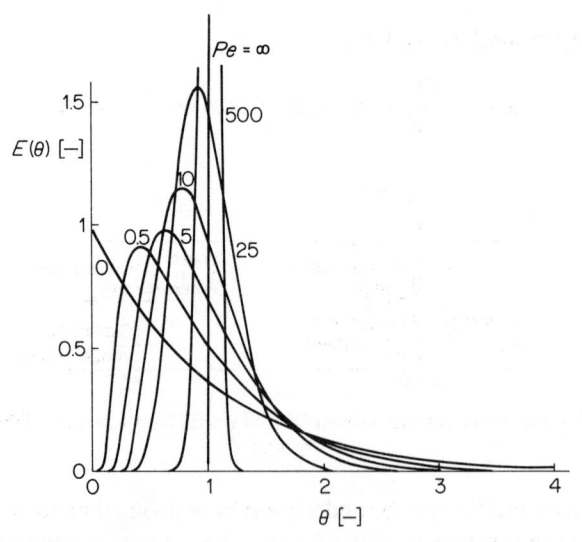

Fig. IV.25 $E(\theta)$ diagram for the axially dispersed plug flow model with closed–closed boundary conditions

The average reduced residence time is given by

$$\bar{\theta} = MR_1 = 1$$

and the reduced variance

$$\sigma_\theta^2 = \frac{2Pe - 2 + 2e^{-Pe}}{Pe^2}. \tag{IV.30}$$

By comparison of figures IV.25 and IV.21 it is clear that the $E(\theta)$ diagram for the mixer cascade and the axially dispersed plug flow model are similar. Comparison of both models can be made by matching the variances as given by equations (IV.30) and (IV.21), which for small deviations from plug flow, or large values of Pe, leads to

$$\frac{Pe}{2} = N.$$

For the whole range of the Pe the expression $\frac{Pe}{2} = N - 1$ could be used instead because it leads to $Pe = 0$ for $N = 1$. However, for the range of Pe from 0 to 10 no unique comparison is possible because the distribution curves of both models have somewhat different shapes. For the closed–closed boundary conditions the F diagram of the axially dispersed plug-flow model has been given by Yagi [14] but could of course also be derived from equation (IV.29). Some examples are given in figure IV.26.

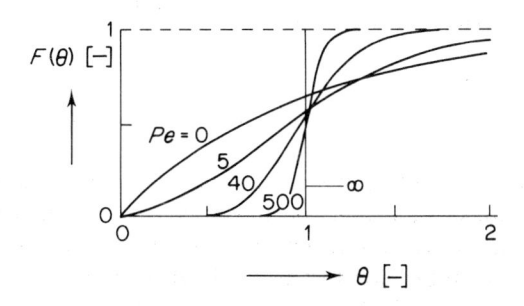

Fig. IV.26 $F(\theta)$ diagram for the axially dispersed plug flow model

Example IV.4.b *Moments of the RTD of the axially dispersed plug flow model by Laplace transformation*

Equation (IV.25) with boundary conditions equations (IV.26) and (IV.27) and initial condition equation (IV.28) are Laplace transformed as follows:

$$\frac{\partial C}{\partial \theta} = -\frac{\partial C}{\partial Z} + \frac{1}{Pe}\frac{\partial^2 C}{\partial Z^2} \rightarrow -C(0) + s\underline{C} = -\frac{d\underline{C}}{dZ} + \frac{1}{Pe}\frac{d^2\underline{C}}{dZ^2}$$

at $Z \rightarrow 0$: $\delta(\theta) = C - \frac{1}{Pe}\frac{\partial C}{\partial Z} \rightarrow 1 = \underline{C} - \frac{1}{Pe}\frac{d\underline{C}}{dZ} \qquad 0 \leftarrow Z$

at $Z \rightarrow 1$: $\frac{\partial C}{\partial Z} = 0 \rightarrow \frac{d\underline{C}}{dZ} = 0.$

Further $\theta \leqslant 0$, $C = 0$ means $C(0) = 0$.

The Laplace transform is a simple second order differential equation which leads, after substitution of an exponential function $\underline{C} = A e^{-BZ}$ and introduction of the boundary conditions, to the following solution for $Z = 1$

$$\underline{E} = \underline{C} = \frac{4q \exp\dfrac{Pe}{2}}{(1+q)^2 \exp\left\{\dfrac{Pe}{2}q\right\} - (1-q)^2 \exp\left\{-\dfrac{Pe}{2}q\right\}}$$

with

$$q = \sqrt{\left(1 + \frac{4s}{Pe}\right)}. \qquad (\text{IV.31})$$

From this equation the moments can be derived via

$$MR_k = \lim_{s \to 0} (-1)^k \frac{d^k E}{ds^k},$$

which leads after some tedious calculations to

$$MR_1 = \bar{\theta} = 1 \qquad \sigma_\theta^2 = \frac{2Pe - 2 + 2e^{-Pe}}{Pe^2}.$$

Example IV.4.c *Different boundary conditions for the axially dispersed plug flow model*

Many solutions for different boundary conditions have been put forward in the literature. If conditions are different from the closed–closed boundaries, we should realize that the result is not a residence time distribution function $E(t)$ or $E(\theta)$. There are conceptual problems as to whether the elements that enter and leave the reactor several times should be counted several times or not at the open boundaries. Besides, the concentration in a non-closed boundary plane is not necessarily related to the flux because the flux is the sum of the convective and dispersive flux. The latter depends on the concentration gradient and not on the concentration as such. This area of boundary conditions is full of pitfalls and mathematical subtleties. Examples of discussions on this subject can be found in the literature: see, e.g., Gibilaro [15] and Nauman [16].

The solutions for different boundary conditions are sometimes useful if the experimental set up for a tracer pulse test is representative for the solution adopted. Furthermore they can be used as a simplified approach to equation (IV.29) for high Peclet numbers. To distinguish these solutions from the real RTD functions we will indicate them by $C(\theta)$ instead of $E(\theta)$. Two cases are given in Table IV.3. Especially at low values of Pe, the solutions for different boundary conditions differ; for high values they are almost identical. This is demonstrated in figure IV.27. The first moments of the $C(\theta)$ curves do deviate from $\bar{\theta} = 1$. This can be understood from the fact that fractions of the injected tracer can move outside the test section and then return. If the detection of tracer takes place within the test section at two different positions simultaneously, it may be useful to apply the so-called transfer function. This function simply relates the concentration at one position to that at another.

Table IV.3 Distributions and moments of the axially dispersed plug flow model for different boundary conditions: Example IV.4.c

Boundary conditions	Boundary condition equations	Reduced concentration at $Z = 1$	Relative moments
closed/open	$Z = 0,\ \delta(\theta) = C - \dfrac{1}{Pe}\dfrac{dC}{d\theta}$ $Z \to \infty,\ C \to 0$ $\theta \leq 0,\ C = 0$	$C(\theta) = \left(\sqrt{\dfrac{Pe}{\pi\theta}}\right)\exp\left[\dfrac{-Pe(1-\theta)^2}{4\theta}\right]$ $- Pe\exp(Pe)\,\mathrm{erfc}\,A$ $A = \left(\sqrt{\dfrac{Pe}{\theta}}\right)\left(\dfrac{1+\theta}{2}\right)$ Villermaux and van Swaaij [17]	$MR_1 = 1 + \dfrac{1}{Pe} = \bar{\theta}$ $\sigma_\theta^2 = \dfrac{2}{Pe} + \dfrac{3}{Pe^2}$
open/open	$Z = -\infty,\ \dfrac{dC}{dZ} = 0$ $Z = +\infty,\ \dfrac{dC}{dZ} = 0$	$C(\theta) = \tfrac{1}{2}\sqrt{\dfrac{Pe}{\pi\theta}}\exp\left[-\dfrac{Pe}{4\theta}(1-\theta)^2\right]$ Levenspiel and Smith [18]	$MR_1 = 1 + \dfrac{2}{Pe} = \bar{\theta}$ $\sigma_\theta^2 = \dfrac{2}{Pe} + \dfrac{8}{Pe^2}$

Except for the closed–closed conditions this transfer function may be somewhat different from the pulse responses discussed here; see Aris [8], Gibilaro [15] and Example IV.6.b.

Example IV.4.d *Dead zones, two region models and combined models*
In some cases of intermediate mixing the shape of the RTD curves cannot be approximated accurately by the axially dispersed plug flow nor by the mixer in

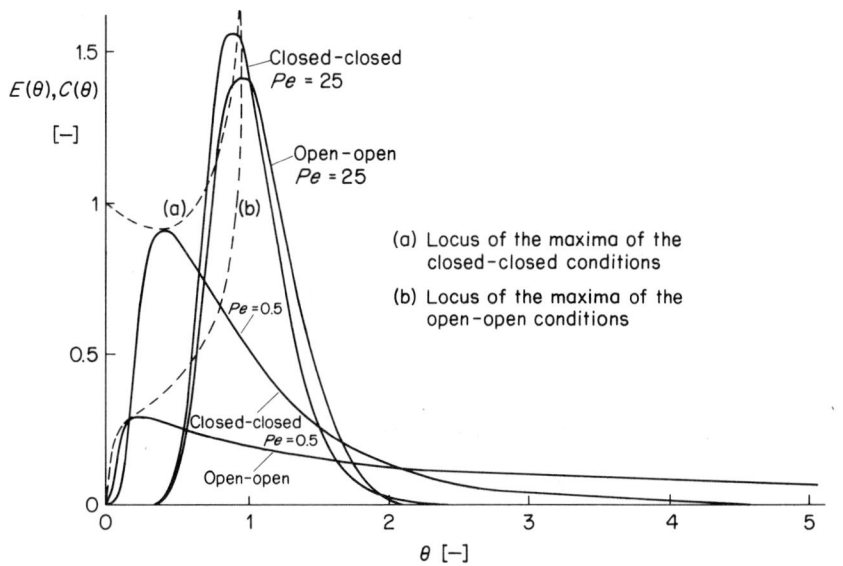

Fig. IV.27 Axially dispersed plug flow model for different boundary conditions

series model. Especially if the reactor contains certain fluid areas with a very long residence time, the models mentioned are inadequate for a correct description of the RTD. Figure IV.28 gives a typical example of such a RTD with clearly visible tailing. Such phenomena have been observed in several types of chemical reactors and special models have been set up for their description. We will discuss two cases. The first one was set up for trickle flow through packed columns, where stagnant zones have been observed in the liquid phase by Hoogendoorn and Lips [19] and van Swaaij et al. [20].

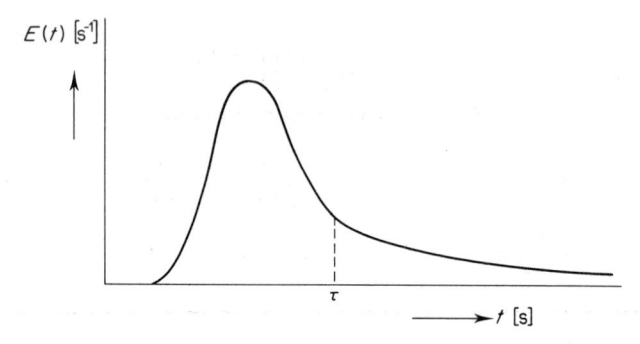

Fig. IV.28 The tailing phenomenon

The liquid in the stagnant areas, which has sometimes been identified with the capillary hold-up at the contact points of the packing elements, is only slowly refreshed by the freely flowing liquid which can be considered to be in axially dispersed plug flow. The model is indicated in figure IV.29. The liquid hold-up, $\beta_{tot} = $ m^{3}liquid/m^{3}column, is divided into a dynamic hold-up β_{dyn} and a static or stagnant hold-up β_{stat}. The local rate of exchange between the static and dynamic

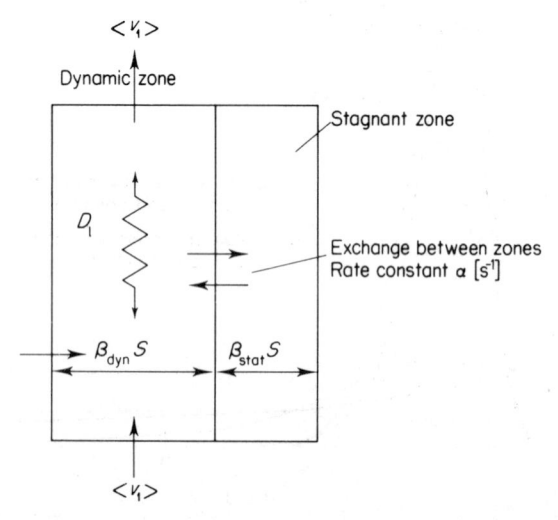

Fig. IV.29 Plug flow with axial dispersion and exchange with dead zones [20], [17]

hold-up is supposed to be proportional to the concentration difference in the dynamic and static phases and can be characterized by an exchange coefficient $\alpha_m\,[1/s]$, defined by

$$Ja = \frac{\text{number of moles exchanged}}{\text{m}^3\text{ liquid . second}} = \alpha_m(c_1 - c_2),$$

where c_1 and c_2 are the concentrations in the dynamic and static phase respectively, whereas α_m can be considered as the product of a mass transfer coefficient and the specific interfacial area between the flowing and the stagnant zones, $k_L a$. The model equations can be found from a mass balance for the dynamic (1) and static (2) phases over a thin slice perpendicular to the direction of the flow, whereas we assume the density and other variables to be constant over the length coordinates:

$$\beta_{\text{dyn}} S \frac{\partial c_1}{\partial t} = \beta_{\text{dyn}} S D_1 \frac{\partial^2 c_1}{\partial z^2} - \beta_{\text{dyn}} S \langle v_1 \rangle \frac{\partial c_1}{\partial z} - \beta_{\text{tot}} S \alpha_m (c_1 - c_2)$$

$$\beta_{\text{stat}} S \frac{\partial c_2}{\partial t} = -\beta_{\text{tot}} S \alpha_m (c_2 - c_1)$$

S = cross sectional area of the column (m²)
$\langle v_1 \rangle$ = velocity in dynamic phase = $\Phi_v / S \beta_{\text{dyn}}$ (m/s)
D_1 = dispersion coefficient in dynamic phase (m²/s)

We will introduce the following dimensionless units:

$$Z = \frac{z'}{L}, \quad \Phi_\beta = \beta_{\text{dyn}} / \beta_{\text{tot}},$$

$$N_\alpha = \frac{\alpha_m L}{\Phi_\beta \langle v_1 \rangle}, \quad Pe_1 = \frac{\langle v_1 \rangle L}{D_1},$$

$$\theta = t/\tau \quad \text{with} \quad \tau = \frac{L}{\Phi_\beta \langle v_1 \rangle}, \quad C_1 = \frac{c_1}{c_0}, \quad C_2 = \frac{c_2}{c_0}.$$

N_α is the number of mass transfer units for the mass exchange between the dynamic and the static phase. Pe_1 is the Peclet number for the dynamic phase. L is the length of the column packing. The mass balances can now be rewritten as

$$\Phi_\beta \frac{\partial C_1}{\partial \theta} = \frac{1}{Pe_1} \frac{\partial^2 C_1}{\partial Z^2} - \frac{\partial C_1}{\partial Z} - N_\alpha (C_1 - C_2),$$

$$(1 - \Phi_\beta) \frac{\partial C_2}{\partial \theta} = -N_\alpha (C_2 - C_1).$$

The boundary conditions at the inlet can be found in a similar way as for equation (IV.26):

$$Z = 0, \quad \delta(\theta) = C_1 - \frac{1}{Pe_1} \frac{\partial C_1}{\partial Z} \quad \text{for} \quad \theta = 0, \quad C_1 = C_2 = 0.$$

For the simplicity we take open boundary conditions at the outlet:

$$Z \to \infty, \quad C_1 = C_2 = 0$$

The solutions can be written as

$$C_1(\theta) = f_1(\theta, \Phi_\beta, N_\alpha, Pe_1)$$

$$\bar{\theta} = 1 + \frac{1}{Pe_1},$$

$$\sigma_\theta^2 = f_2(\Phi_\beta, N_\alpha, Pe_1).$$

f_1 and f_2 are complicated functions for which analytical expressions have been given by Villermaux and van Swaaij [17]. A few interesting cases are given in figures IV.30 and IV.31. In figure IV.30 it can be seen that the large spread in

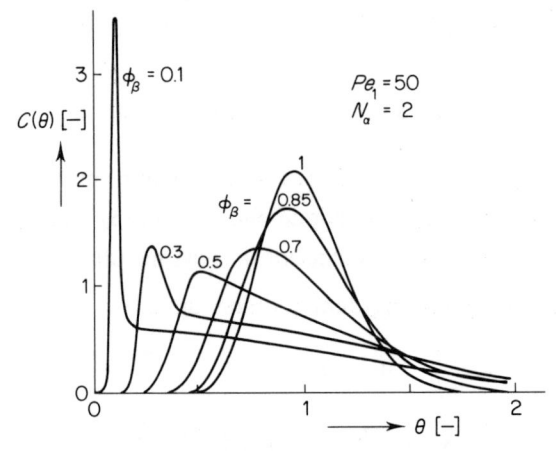

Fig. IV.30 Typical residence time distribution curves for the plug flow with axial dispersion and exchange with stagnant zones model. Influence of Φ_β. [From Villermaux and van Swaaij [17] Reproduced by permission of Pergamon Press Ltd]

residence times that occurs for low dynamic hold-ups, which is rather a short-cut stream, is reduced if the relative fraction of the dynamic hold-up increases. Figure IV.31 shows the influence of the exchange rate between dynamic and stagnant hold-up. If there is no exchange or $N_\alpha = 0$, the curve is shifted towards shorter times θ, and the average time will only indicate a dynamic phase. Because in figure IV.31 $\Phi_\beta = 0.5$ the average reduced residence time for $N_\alpha = 0$ will then only be $\bar{\theta}(N_\alpha = 0)$ $= 0.5 \times \bar{\theta}(N_\alpha \neq 0)$. This can be understood by realizing that τ in $\theta = t/\tau$ is based on the total hold-up and only half of it, the dynamic part, is 'seen' by a tracer. As soon as there is some exchange, at least theoretically, the average residence time immediately becomes $\bar{\theta} = 1 + 1/Pe_1 \simeq 1$ (for $Pe_1 = 50$ and $N_\alpha \neq 0$). In practice for low values of N_α it is difficult to obtain accurate values for $\bar{\theta}$ because of a long tail where $C(\theta) \simeq 0$. For high values of N_α the stagnant area is effectively absent and the solution for the axially dispersed plug flow for closed–open boundary conditions will be obtained.

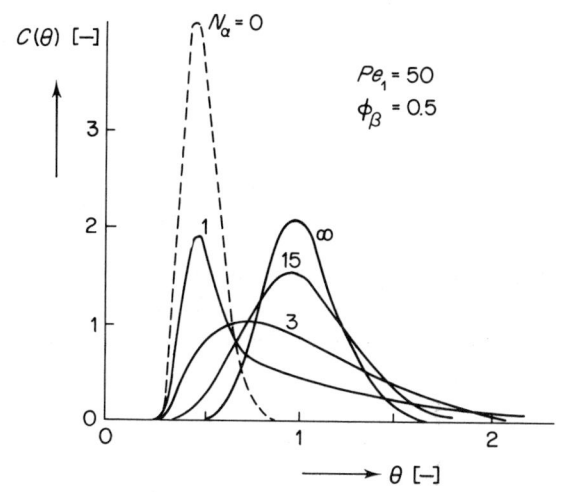

Fig. IV.31 Typical residence time distribution curves for the model of a plug flow with axial dispersion and exchange with stagnant zones. [From Villermaux and van Swaaij [17] Reproduced by permission of Pergamon Press Ltd]

A second example of a model with interchange between a flowing and a relative stagnant area has been published by May [21] and van Deemter [4] to describe the mixing in the gas flow in a gas–solids fluid bed; see also de Groot [3] and van Swaaij and Zuiderweg [22].

The dynamic phase is the bubble stream which is considered to be in plug flow. The static phase is the interstitial gas and the gas inside the porous particles in the dense phase. In this dense phase axial mixing by dispersion occurs. Furthermore there is a (slow) gas through flow which will be neglected in this illustration. The model is schematically shown in figure IV.32.

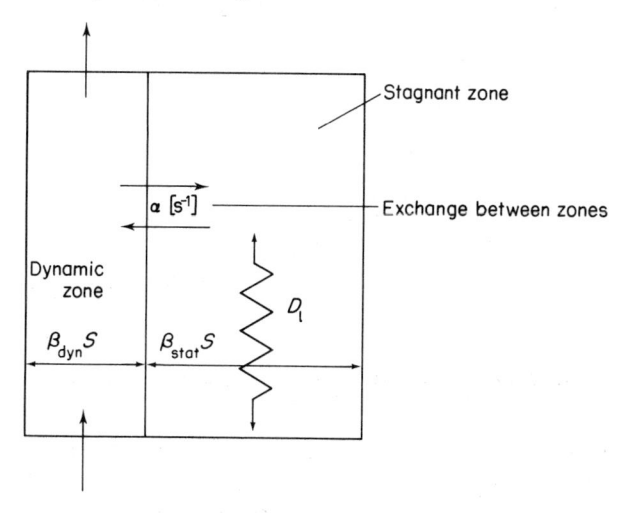

Fig. IV.32 Simplified version of the two zones model of van Deemter [22]

The differential equations for the van Deemter model can be derived in a similar way as in the previous model:

$$\Phi_\beta \frac{\partial C_1}{\partial \theta} = -\frac{\partial C_1}{\partial Z} - N_\alpha(C_1 - C_2)$$

$$(1 - \Phi_\beta)\frac{\partial C_2}{\partial \theta} = \frac{1}{Pe_2}\frac{\partial^2 C_2}{\partial Z^2} - N_\alpha(C_2 - C_1),$$

where the dimensionless groups are the same as in the former model. Moreover, here $Pe_2 = \langle v_2 \rangle L/D_1$, in which $\langle v_2 \rangle = \Phi_v/\beta_{stat}S$ and β_{stat} is the stagnant gas hold-up, being equal to the reactor volume fraction occupied by the gas in the dense phase. The boundary conditions are

$$Z = 0, \quad C_1 = \delta(\theta) - \frac{1}{Pe_2}\frac{\partial^2 C_2}{\partial Z^2} = 0$$

$$Z = 1, \quad \frac{1}{Pe_2}\frac{\partial^2 C_2}{\partial Z^2} = 0$$

and the initial conditions: $\theta \leq 0$, $C_1 = C_2 = 0$.

No analytical solution for this model is yet available, but the moments have been calculated from the Laplace transforms of the model equations by van Deemter [4]. Because of the closed–closed boundary conditions, $\bar{\theta} = 1$ and the reduced concentration at the outlet would give the $E(\theta)$ diagram directly. Bohle [23] found some approximate numerical solutions for the outlet concentrations and some E diagrams are given in figure IV.33.

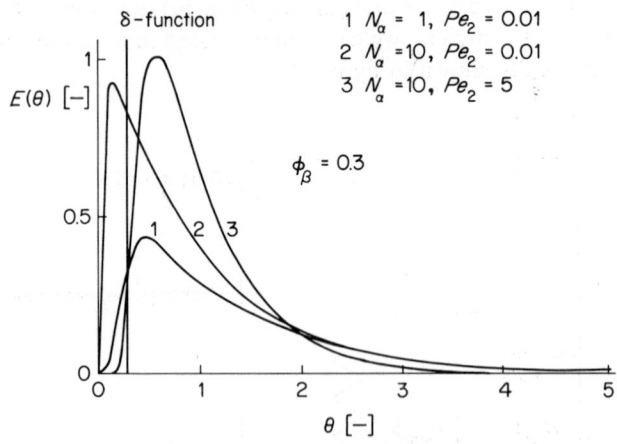

Fig. IV.33 E diagram for the two region model of van Deemter [4]. The numerical solutions are from Bohle [23] [Reproduced by permission of W. Bohle]

For slow exchange rates ($N_\alpha = 1$) and high mixing rates ($Pe_2 = 0.01$) it can be seen that the remaining concentration δ-pulse arrives at $\theta = 0.3$. Because $\Phi_\beta = 0.3$ the passing time of the dynamic phase is 0.3 times the average residence time. A part of the fluid elements has a shorter residence time, however, because it travelled by means of the rapid mixing process through the stagnant hold-up. This is a very special situation, which is still more pronounced if $N_\alpha \to \infty$, when an ideal mixer is almost approached. At high Peclet numbers there is practically no such rapid breakthrough in the E diagram. For the particular case of $Pe_1 = Pe_2 = \infty$ the two models discussed here become identical, because in that case we have plug flow with exchange with dead zones. The $E(\theta)$ diagram then becomes as shown by Villermaux and van Swaaij [17]:

$$E(\theta) = \exp[-N_\alpha] \times \delta(\theta - \Phi_\beta) + \exp[-N_\alpha(1 + \theta - 2\Phi_\beta)/(1 - \Phi_\beta)]$$
$$\times [N_\alpha(1 - \Phi_\beta)(\theta - \Phi_\beta)]^{-1/2} \times I_1(2N_\alpha[(\theta - \Phi_\beta)/(1 - \Phi_\beta)]^{1/2})$$
$$\times U(\theta - \Phi_\beta)$$

in which the modified Bessel function is

$$I_1(y) = \frac{y}{\pi} \int_{-1}^{+1} (1 - p^2)^{1/2} \exp(py) \, dy$$

and U is the unit step function. The moments are

$$\theta = 1 \quad \text{and} \quad \sigma_\theta^2 = \frac{2(1 - \Phi_\beta)^2}{N_\alpha}.$$

Some E diagrams for this model are given in figure IV.34.

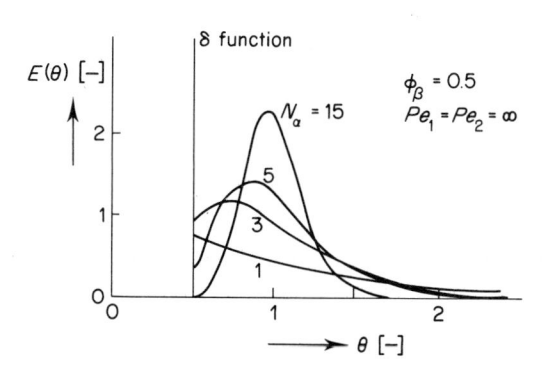

Fig. IV.34 E diagrams for the model of plug flow with exchange with dead zones. The non-transferred fraction of the fluid appears at $\theta = \Phi_\beta$ at the outlet [From Villermaux and van Swaaij [17]. Reproduced by permission of Pergamon Press Ltd]

For high values of the exchange rate (high N_α) the solutions become very similar to the simple axially dispersed plug flow model and for $N_\alpha \to \infty$ even plug flow is obtained. As in the van Deemter model, at low values of N_α a fraction of the δ-pulse passes without transfer to the stagnant zone. This is represented by a δ-function in figure IV.34 appearing at $\theta = \Phi_\beta$ as explained above. For the trickle flow model such sharp pulses at the outlet cannot be expected because the pulses are broadened by the dispersion process. For $N_\alpha = 1$ in figure IV.31 such a peak can still be observed, however.

Similar models to those discussed here can be derived with the mixers in series concept; quite a few have been published: see e.g. van Swaaij *et al.* [20]. We will call such models combined models, because they consist of a combination of the following elementary elements: plug flow zones, ideally mixed zones, bypass flow elements and dead zones. Many of such models have been assembled by Levenspiel and Bischoff [24], by Levenspiel [25] and Wen and Fan [9]. Combination models may have to be taken into consideration if a great accuracy is required in the modelling of the state of mixing. The large number of parameters involved can become a considerable problem. An early example is the model for a single mixer of Cholette and Cloutier [26] given in figure IV.35. They considered an ideally mixed zone A, a dead zone C and a short cut stream B and correlated the different parameters with, among others, the stirring speed. In reality the situation is still more complicated because often there is both a time lag between the inlet and outlet streams and an exchange between the dead zone and the ideally mixed section. In practice it will be difficult, however, to obtain accurate values for the different relevant parameters. Slightly different models have been devised based on the analysis of the physical behaviour of a real mixed vessel with its circulation loops and exchanges. Examples are given by van der Vusse [27] and Wen and Chung [28]. More complex cascades have also been introduced. Figure IV.36 shows the model of Adler and Hovorka [29]. Such models are to a certain extent equivalent to the dispersion–exchange models, but cascade models in principle do not show

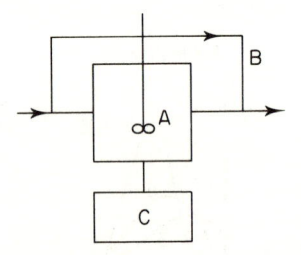

Fig. IV.35 Mixer model of Cholette and Cloutier [26]. A = ideally mixed zone, B = short-cut stream, C = dead space

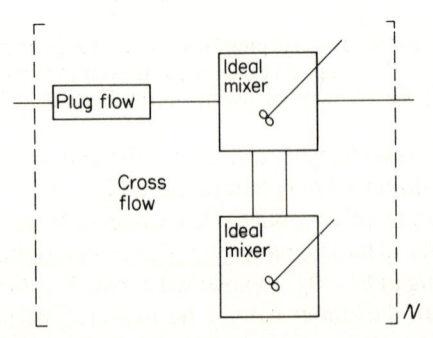

Fig. IV.36 Model of Adler and Hovorka [29] for the description of the state of mixing in a system

real back-mixing. Real back-mixing is a property of the dispersion models. It has also been tried to introduce back-mixing in a cascade of ideal mixers by insertion of back-flow. Figure IV.37 shows an example presented by Klinkenberg [30]. Specially in countercurrent exchange processes, accompanied by chemical reactions and carried out in plate or compartmented columns, e.g. a rotating disc contactor, these models with back-flow can be useful.

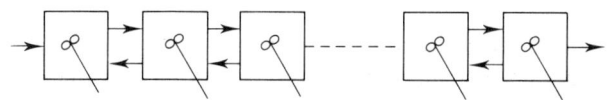

Fig. IV.37 Cascade of mixers with backmixing as proposed by Klinkenberg [30]

IV.5 CONVERSION IN REACTORS WITH INTERMEDIATE MIXING

Both the axially dispersed plug flow model and the mixers in series models can be used for the conversion prediction of an actual reactor to cover the mixing states between ideal plug flow and ideal mixing. We will treat in this section only the isothermal steady state and constant density cases. The calculations for the cascade of CISTRs are simple because they can be based on tank to tank calculations, as outlined in Chapters II and III. Especially with complex kinetics, this model has advantages over the axially dispersed plug flow model. It has already been demonstrated in Chapter II that in case of a first order reaction the conversion can be found from

$$1 - \zeta = \left[\frac{1}{1 + k\tau/N}\right]^N \quad \text{(equation II.34)}.$$

Here τ is the average residence time of the cascade, k the first order rate constant and N the number of mixers of the cascade. If N approaches infinity the solution is $1 - \zeta = e^{-k\tau}$, which is the result for a plug flow reactor. The calculations with the axially dispersed plug flow model are more complicated. Nevertheless, this model is often used because it is in many cases more realistic, especially with regard to back-mixing and because it is more suitable for non-isothermal cases with axial and radial temperature gradients. These problems will be treated in Chapter IX. The disadvantage of the complexity of calculations is less important nowadays, thanks to modern high-speed computers. In case of isothermal reactions, a constant dispersion coefficient, a constant density and a steady state, the mass balance of a component A over a thin slice of reaction phase $S\,\Delta z$ reads, for $\Delta z \to 0$,

$$D_1 \frac{d^2 c_A}{dz^2} - \langle v \rangle \frac{dc_A}{dz} + R_A = 0. \tag{IV.32}$$

We will again make the assumption of no dispersion in the inlet and outlet sections. At $z = 0$ the continuity of fluxes law must hold. Therefore, the boundary condition at the inlet is

$$\langle v \rangle c_{A_o} = \langle v \rangle c_A - D_1 \frac{dc_A}{dz} \quad \text{at} \quad z = 0. \tag{IV.33}$$

For a non-zero value of D_1 this condition implies a discontinuous decrease in c_A at $z = 0$, which can be compared with the concentration jump between the inlet and the bulk of an ideal mixer as the extreme case. A similar relationship should apply for $z = L$:

$$\langle v \rangle c_A - D_1 \frac{dc_A}{dz} = \langle v \rangle c_{AL} \quad \text{at} \quad z = L.$$

However, because D_1 is finite and dc_A/dz is negative due to the chemical reaction, this outlet condition would mean that c_{AL} is higher than the value in the reactor at $z = L$. This is in conflict with physical intuition and therefore in general the following is taken as boundary condition at the outlet:

$$\frac{dc_A}{dz} = 0 \quad \text{at} \quad z = L. \tag{IV.34}$$

Wehner and Wilhelm [31] proved that this equation is correct if no dispersion occurs for $z < 0$ and $z > L$ (closed–closed boundaries). For a first order reaction ($R_A = -kc_A$) and after introduction of the dimensionless variables,

$$Z = \frac{z}{L}, \quad \tau = \frac{L}{\langle v \rangle} \quad \text{and} \quad Pe = \frac{\langle v \rangle L}{D_1},$$

the system to be solved becomes

$$\frac{1}{Pe} \frac{d^2 c_A}{dZ^2} - \frac{dc_A}{dZ} - k\tau c_A = 0 \tag{IV.35}$$

$$c_A - c_{Ao} + \frac{1}{Pe} \frac{dc_A}{dZ} = 0 \quad \text{at} \quad Z = 0,$$

$$\frac{dc_A}{dZ} = 0 \quad \text{at} \quad Z = 1.$$

The following solution was given by Danckwerts [1] and by Wehner and Wilhelm [31]:

$$\frac{c_{AL}}{c_{Ao}} = \frac{4q}{(1+q)^2 \exp\left\{\frac{Pe}{2}(1-q)\right\} - (1-q)^2 \exp\left\{-\frac{Pe}{2}(1+q)\right\}}$$

$$\text{with} \quad q = \sqrt{\left(1 + \frac{4k\tau}{Pe}\right)} \tag{IV.36}$$

Equation (IV.36) is valid for the extreme values of Pe. For the plug flow limit or $Pe \to \infty$ this can be seen as follows:

if $Pe \gg 1$ and $Pe \gg k\tau$, then

$$q \simeq 1 + \tfrac{1}{2}\left(\frac{4k\tau}{Pe}\right) - \frac{1}{8}\left(\frac{4k\tau}{Pe}\right)^2 + \cdots$$

and equation (IV.35) becomes

$$\frac{c_{AL}}{c_{Ao}} = \left[1 + \frac{(k\tau)^2}{Pe}\right] e^{-k\tau},$$

which for $Pe \to \infty$ leads to $c_{AL}/c_{Ao} = e^{-k\tau}$.

This is the result for an isothermal plug flow reactor. If Pe approaches 0, Pe and $k\tau Pe$ are smaller than one and the exponential forms in the denominator of equation (IV.36) can be expanded in series, with the result

$$\frac{c_{AL}}{c_{Ao}} \simeq \left[1 + k\tau + \tfrac{1}{6} Pe \cdot k^2\tau^2 \left(1 - \frac{Pe}{4}\right) + \cdots\right]^{-1}.$$

For $Pe = 0$, we find $c_{AL}/c_{Ao} = 1/(1 + k\tau)$, which is the result for a first order reaction in a CISTR. Results for intermediate values of Pe are given in figure IV.38.

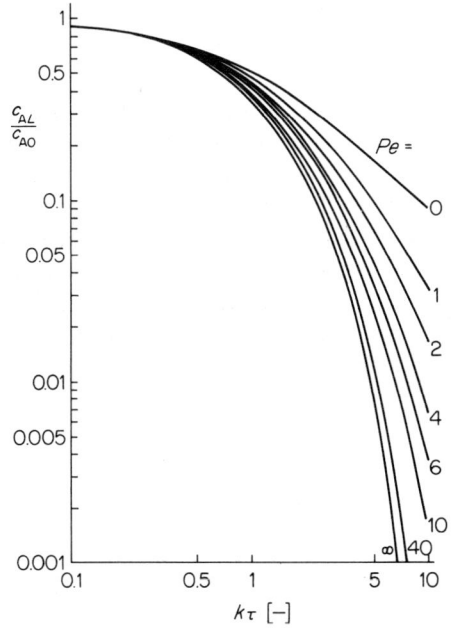

Fig. IV.38 Conversion for a first order reaction for the axially dispersed plug flow model

With more complex reactions the solutions are more difficult to obtain and numerical procedures have to be applied. Figure IV.39 shows some results of the influence of the Peclet number for a second order reaction and Figure IV.40 gives the volume ratio of the residence time required in a reactor with dispersion compared to that in a plug flow reactor. It can be seen that for a higher order reaction the conversion is more affected by the Peclet number representing the spread in residence time. This is because with a larger dispersion, or lower Peclet number at the same outlet concentration, the average concentration in the reactor

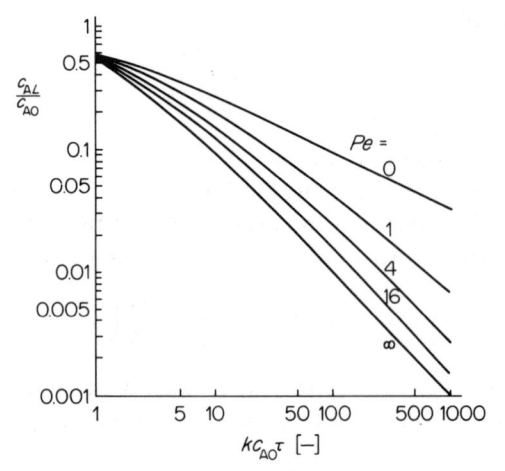

Fig. IV.39 Conversion for a second order reaction for the axially dispersed plug flow model

is lower, whereas the rate of the higher order reaction is reduced more on lowering the concentration than for a lower reaction order. Of course with complex reactions the conversion is not the only factor; selectivity is often more important. Such a case will be discussed in Example IV.5.b. With complex reactions there is no unique relation between the RTD and the conversion or selectivity, these will depend to some extent on the details of the mixing model applied, of which the axially dispersed plug flow model and mixers in cascade model are only specific cases. An exception is the first order reaction, for which the E diagram gives all the information required for the prediction of the conversion of a homogeneous isothermal reaction. An example will be given in the next Example.

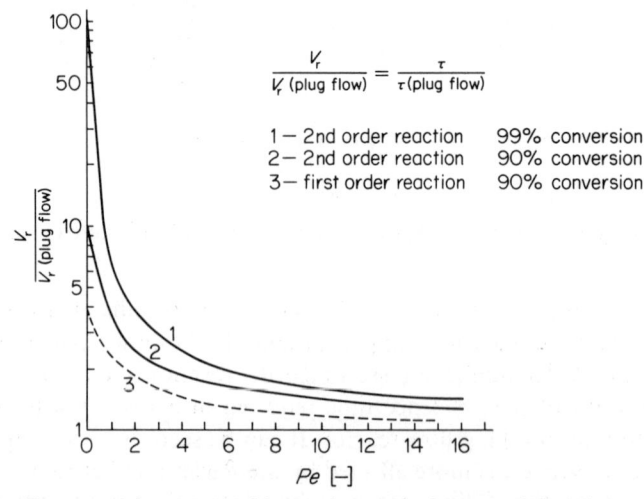

Fig. IV.40 Reactor volume or residence time ratio for an axially dispersed plug flow reactor as compared to a plug flow reactor

Example IV.5.a *Prediction of the conversion for a first order chemical reaction*
If experimental data on the residence time distribution of a reactor are available, the prediction of the conversion of a first order reaction in that reactor can be based on two different approaches:

A. With the help of a mixing model. The model parameter N or Pe can be derived from the E diagram and with known N or Pe value the conversion can be calculated via

$$\frac{c_{AL}}{c_{Ao}} = \left[\frac{1}{1 + k\tau/N}\right]^N$$

for the mixer cascade, or via equation (IV.36) for the axially dispersed plug flow model.

B. Directly from the $E(t)$ diagram via equation IV.11:

$$\frac{c_{AL}}{c_{Ao}} = \int_0^\infty e^{-kt} E(t)\, dt.$$

Figure IV.41 shows an experimental residence time distribution curve obtained from a pulse response test, represented by curve 1. The moments of the $E(t)$ diagram have been calculated as in Example IV.1, which leads to

$$MO_1 = \tau = 6.57\,\text{s}$$
$$\sigma_t = 2.79\,\text{s}$$
$$\sigma_\theta = 0.4247.$$

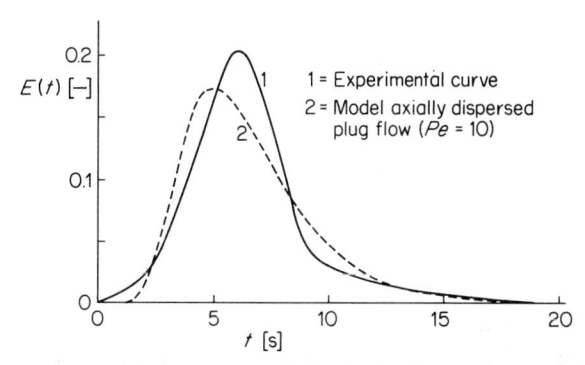

Fig. IV.41 Experimental curve (1) and axially dispersed plug flow model curve with the same variance and average residence time, curve 2

For the cascade model we found, for the variance, with equation (IV.21), $\sigma_\theta^2 = 1/N$ which leads for $\sigma_\theta = 0.4247$ to $N = 5.5$. The axially dispersed plug flow model gives for σ_θ^2 with equation IV.30:

$$\sigma_\theta^2 = \frac{2Pe - 2 - 2e^{-Pe}}{Pe^2},$$

which results in $Pe = 10$. In figure IV.41 also the model function $E(t)$ for the axially dispersed plug flow with $Pe = 10$ is indicated. As can be seen, the fit is not very good. To illustrate route B we will use figure IV.42. In figure IV.42 first the experimental $E(t)$ diagram is given again, in this case $k = 0$. We will consider a typical element entering the continuously operated reactor at $t = 0$. Its fragments will leave the reactor partially converted. If we consider the fragments as individual isothermal constant density batch reactors with a reaction time equal to their residence times their reactant conversion is given by

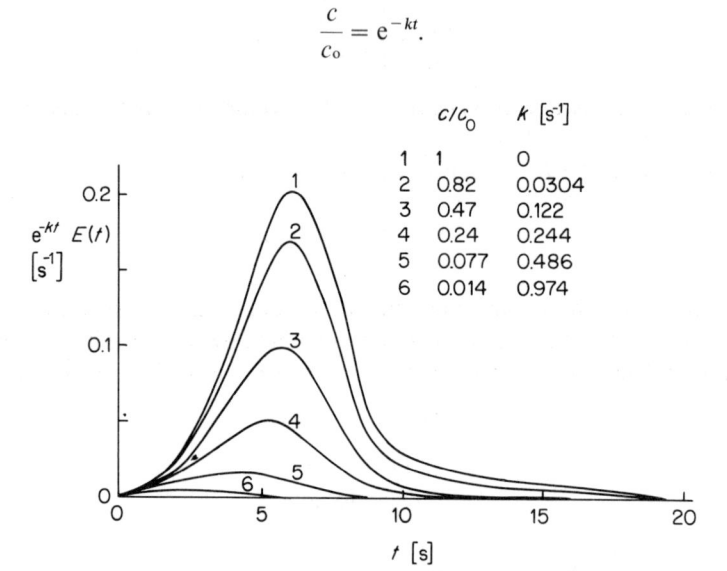

$$\frac{c}{c_0} = e^{-kt}.$$

	c/c_0	k [s^{-1}]
1	1	0
2	0.82	0.0304
3	0.47	0.122
4	0.24	0.244
5	0.077	0.486
6	0.014	0.974

Fig. IV.42 Exit distribution of unconverted reactant for different first order rate constants

From the original element entering at $t = 0$ the fraction of fragments arriving at the exit between $t = t$ and $t + \Delta t$ is equal to $E(t)$ and therefore the amount of unconverted reactant is given by

$$\frac{c(t)}{c_0} = e^{-kt} E(t) \Delta t$$

in which c_0 is the initial reactant concentration in the original fluid element. The function $e^{-kt} E(t)$ is plotted in figure IV.42 for different reaction rate constants. As can be seen, fragments that arrive relatively early contain more unconverted reactant. To obtain the conversion in the steady state we need to calculate the average conversion for the typical fluid element and therefore the average for all its fragments:

$$\frac{\bar{c}}{c_0} = \int_0^\infty e^{-kt} E(t) \, dt.$$

This integral is plotted as a function of $k\tau$ in figure IV.43 together with the results of the calculation via the models. As can be seen, the conversion predicted by the

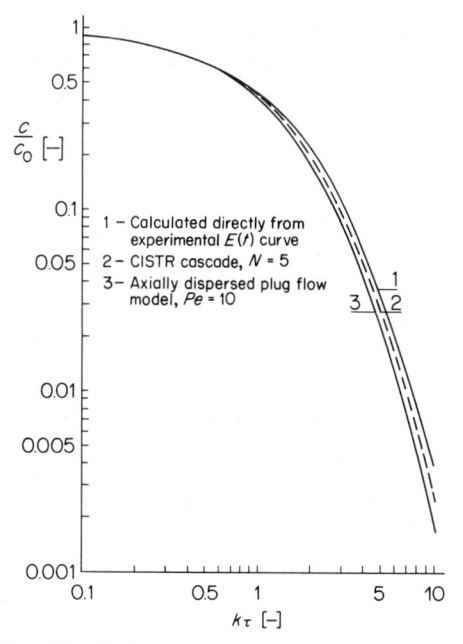

Fig. IV.43 Conversion directly from the $E(t)$ diagram and from model prediction. First order isothermal reaction

1 – Calculated directly from experimental $E(t)$ curve
2 – CISTR cascade, $N = 5$
3 – Axially dispersed plug flow model, $Pe = 10$

axially dispersed plug flow model and the mixer cascade model is somewhat higher than that predicted directly with the E diagram. This is due to the fact that for the same spread in residence time the models predict lower values for $E(t)$ at short residence times than actually observed from the experimental curve. The front part of the $E(t)$ diagram contributes more to the unconverted reactant concentration than the tail. The tail on the other hand contributes more to the variance which is used for the calculation of the parameters N or Pe. These two effects do not compensate each other entirely.

For other reaction kinetics in fact the same conversion calculation could be carried out directly with the $E(t)$ diagram. As will be discussed in Chapter V, the intermixing of the fragments has a strong influence for other kinetics. It will be shown that the calculation given here would then correspond to one of the extremes: the complete segregation. Except in case of first order reactions, intermixing influences the overall conversion.

Example IV.5.b *Consecutive reactions in a reactor with axially dispersed plug flow*

We consider the reactions

$$A \to P \to X$$

with the rate equations

$$R_A = -k_P c_A \quad \text{and} \quad R_P = k_P c_A - k_X c_P.$$

The reaction is carried out isothermally in a tubular reactor with longitudinal dispersion. The influence of the dispersion on the maximum yield of P is to be studied. At constant density the concentration distribution along the reactor length z is given by the set of equations (IV.35) (where k has to be replaced by k_P) and a similar set for species P:

$$\frac{d^2c_P}{dZ^2} - Pe\frac{dc_P}{dZ} + Pe(k_P\tau c_A - k_X\tau c_P) = 0$$

$$Pe \cdot c_P - \frac{dc_P}{dZ} = 0 \quad \text{at} \quad z = 0, \qquad \frac{dc_P}{dZ} = 0 \quad \text{at} \quad Z = 1.$$

Solution of these equations for the yield at the end of the reactor ($Z = 1$) results in

$$\eta_P = \frac{c_P(Z=1)}{c_{A0}} = \frac{4k_P/k_X}{k_P/k_X - 1}\left[\frac{-q}{(1-q)^2 e^{-Pe(1-q)/2} - (1-q)^2 e^{-Pe(1+q)/2}}\right.$$

$$\left. + \frac{r}{(1+r)^2 e^{-Pe(1-r)/2} - (1-r)^2 e^{-Pe(1+r)/2}}\right],$$

where

$$q = \sqrt{\left(1 + \frac{4k_P\tau}{Pe}\right)} \quad \text{and} \quad r = \sqrt{\left(1 + \frac{4k_X\tau}{Pe}\right)}.$$

If the second reaction does not occur η_P reduces to (IV.34). If the second reaction is much faster than the first ($k_P/k_X \to 0$) η tends towards zero as expected. When ideal plug flow is approached or Pe approaches infinity we have

$$\lim_{Pe \to \infty} \eta_P = \frac{k_P/k_X}{1 - k_P/k_X}(e^{-k_P\tau} - e^{-k_X\tau}).$$

On the other hand, in the extreme case of a CISTR or $Pe = 0$ the yield becomes

$$\lim_{Pe \to 0} \eta_P = \frac{k_P\tau}{(1 + k_P\tau)(1 + k_X\tau)}.$$

These results can be found in Chapter III. The yield η_P has a maximum value $(\eta_P)_{max}$ as a function of $k_P\tau$ which can be evaluated as a function of Pe and k_P/k_X. Some results are given in figure IV.44. For a constant value of k_P/k_X the ideal plug flow reactor ($Pe = \infty$) gives the highest maximum yield. The maximum yield is decreased by the longitudinal dispersion and $k_P\tau_{max}$ increases at the same time, which means that the reactor load has to be reduced. This effect of the spread in residence time can only be counteracted by increasing k_P/k_X, e.g. by a change in reaction temperature, since k_P and k_X will generally be different functions of temperature because of different activation energies.

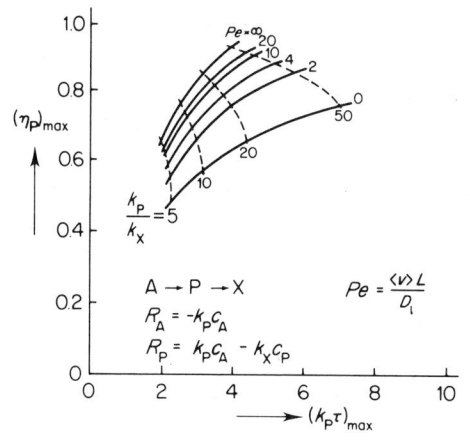

Fig. IV.44 Influence of axial dispersion and the selectivity parameter k_P/k_X on the maximum yield and corresponding residence time for consecutive first order reactions

IV.6 SOME DATA ON THE LONGITUDINAL DISPERSION IN CONTINUOUS FLOW SYSTEMS

Many data on residence time distribution have been published in the literature. Reviews can be found in Levenspiel and Bischoff [24] and Wen and Fan [9]. We will only discuss briefly those cases that have been interpreted by the axially dispersed plug flow model, in order to allow a quick estimate of the effect of dispersion in different types of reactors.

If the cascade model is preferred, it is easy to calculate the corresponding parameter N of this model, the number of mixers in series. Strictly speaking, no unique comparison is possible, but because the majority of data on dispersion have been obtained from residence time distribution measurements by tracer methods, it seems logical to base the comparison on the same spread in residence time, which leads to the approximate equation

$$\frac{Pe}{2} = N - 1.$$

Often the axial dispersion data are correlated on the basis of a special Peclet number which has a characteristic length d, the Bodenstein number, $Bo = \langle v \rangle d/D_l$. This d is not the length of the experimental reactor L but a dimension significant in view of the mechanism causing the dispersion, e.g. a tube diameter or the particle diameter of the packing.

IV.6.1 Flow through empty tubes

From both a theoretical and a practical point of view it is important to distinguish between laminar flow and the turbulent regime. We will treat *laminar flow* first. Due to the velocity profile a considerable spread in residence time can be expected.

An important role is played by molecular diffusion which takes place in both axial and radial direction. The diffusion in radial direction tends to counteract the spreading effect of the velocity profile, while in the axial direction the molecular diffusion process increases the dispersion rate. The equation to be solved follows from the mass balance of a component, e.g. a tracer, over a concentric ring in the tube. For isothermal, incompressible flow the differential equation reads:

$$\frac{\partial c}{\partial t} = D\frac{\partial^2 c}{\partial x^2} + D\left(\frac{\partial^2 c}{\partial r^2} + \frac{1}{r}\frac{\partial c}{\partial r}\right) - v(r)\frac{\partial c}{\partial x}. \tag{IV.37}$$

We will discuss here only some limiting cases for Newtonian fluids. Non-Newtonian fluids, which may be very important, e.g. in polymerization reactors, are treated in the literature, e.g. by Wen and Fan [9].

a. Negligible molecular diffusion in both axial and radial direction

This situation may be approached in short tubes in which a highly viscous liquid flows at low Reynolds numbers. The calculated residence time distribution is here strongly dependent on the injection and detection functions which should be 'mixed cup' or flow averaged functions, as shown by Levenspiel and Turner [10]. The correct solution is

$$E(\theta) = \frac{1}{2\theta^3} \quad \text{for } \theta > 0.5 \tag{IV.38}$$

$$E(\theta) = 0 \quad \text{for } \theta < 0.5.$$

This is shown in figure IV.45. It is not possible to apply the axially dispersed plug flow model for this case because of the different shapes of the E diagram. Conversion prediction for a first order reaction, calculated by means of the equation (IV.11),

$$\frac{c_{AL}}{c_{Ao}} = \int_0^\infty e^{-kt} E(t) \, dt$$

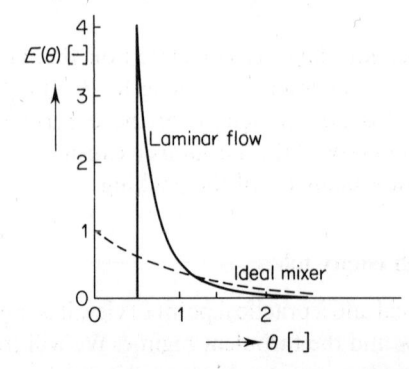

Fig. IV.45 Residence time distribution for laminar flow without molecular diffusion

is given in figure IV.46 which shows a behaviour somewhat different from the axially dispersed plug flow model.

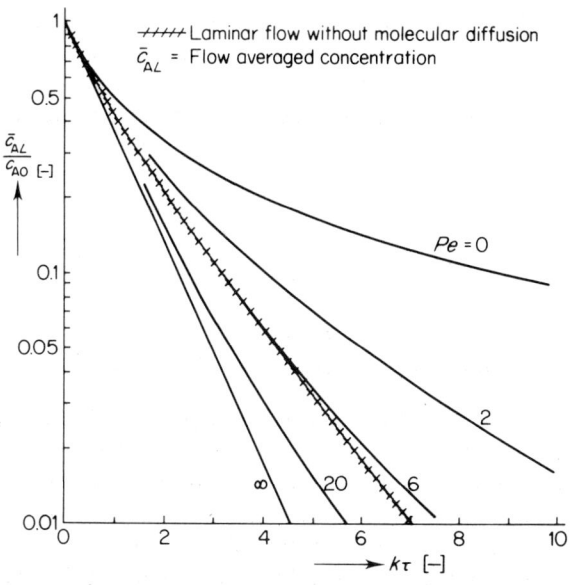

Fig. IV.46 First order reaction in a laminar flow reactor without molecular diffusion. To compare, some predictions for the axially dispersed plug flow model are also presented

b. Relatively fast molecular diffusion in radial direction is relatively fast

This means that the time constant for radial diffusion or the Fourier time should be much shorter than the average residence time:

$$\frac{L}{\langle v \rangle} \gg \frac{d_t^2}{D} \quad \text{or} \quad \frac{L}{d_t} \gg \frac{\langle v \rangle d_t}{D} \quad \text{(IV.39)}$$

Taylor [32] and van Deemter et al. [33] found for this case that the axially dispersed plug flow model could be applied and the corresponding dispersion coefficient can be derived from

$$Bo = \frac{\langle v \rangle d_t}{D_1} = \frac{192 D}{d_t \langle v \rangle} = 192 (ReSc)^{-1} \quad \text{(IV.40)}$$

with the restriction of equation (IV.39) which can be modified with equation (IV.40) to

$$Pe = Bo \times \frac{L}{d_t} > 200.$$

More refined criteria for different regimes are given in the literature by Wen and Fan [9].

It follows from the *Bo* values in the laminary region that condition (IV.40) can be obtained in practice for gases, but leads to excessively high values of L/d_t for liquids where $Sc \simeq 1000$. Note that D_l will decrease as the molecular diffusivity increases because the exchange of material between streamlines with different velocity counteracts the effect of the spreading in residence times caused by the velocity profile. The effect of molecular diffusion in the axial direction can usually be neglected but, if necessary, it can be added to the dispersion coefficient as shown by Aris [34]:

$$D_l = D + \frac{d_t^2 \langle v \rangle}{192D}, \qquad (IV.41)$$

which after some rearranging leads to

$$\frac{1}{Bo} = \frac{1}{ReSc} + \frac{ReSc}{192} \quad \text{for} \quad Re < 2000. \qquad (IV.42)$$

If the viscous flow of a reaction mixture in a straight tube would produce too great a dispersion, it can be considerably reduced by spiralling the tube. Without secondary flow one would expect an increase in the spread of residence time. It appears, however, that due to the secondary flow in a curved pipe the spread in residence time for the now spiralling streamlines is $\frac{1}{2}$ to $\frac{1}{3}$ of that of a straight tube with Poiseuille flow. Moreover, the influence of radial diffusion is enhanced by the secondary flow. As a consequence the apparent longitudinal dispersion coefficient for laminar flow in curved tubes is relatively small.

For *turbulent flow* in tubes the *Bo* number for longitudinal dispersion is much higher than for laminar flow. Owing to the turbulence the transverse transport of both momentum and matter is facilitated so that the velocity profile is flattened out and radial mixing is improved. Taylor [35] made an estimate of D_l in the region

$$10^4 < Re = \frac{\langle v \rangle d_t}{v} < 10^6$$

and found:

$$Bo = \frac{\langle v \rangle d_t}{D_l} \simeq 0.28 f^{-1/2}, \qquad (IV.43)$$

where f is the Fanning friction factor. A more precise analysis along the same lines was published by Tichacek et al. [36]. The results of their correlations are given in figure IV.47 together with a recent correlation given by Wen and Fan [9]:

$$\frac{1}{Bo} = \frac{3 \times 10^7}{Re^{2.1}} + \frac{1.35}{Re^{1/8}}. \qquad (IV.44)$$

In general the *Bo* number for turbulent flow in pipes at $Re > 10^4$ is so large that even with an L/d_t ratio of e.g. 50 the value of *Pe* becomes 100 or higher and the ideal tubular reactor is approached. Some information is available on the longitudinal dispersion of single phase flow in tubular systems with mechanical agitation; a

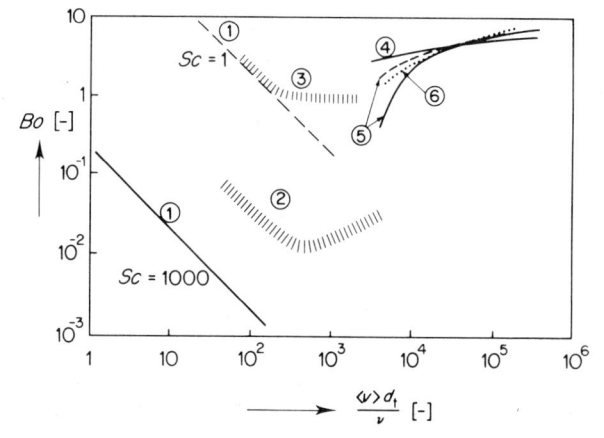

Fig. IV.47 Bodenstein number $Bo = \dfrac{\langle v \rangle d_t}{D_l}$ for single phase flow through tubes.

Laminar flow
1. Taylor [32]
2. Experimental, curved tubes, liquid
3. Experimental, curved tubes, gas

Turbulent flow
4. Taylor [35]
5. Tichacek et al. [36] --- gas–liquid
6. Wen and Fan [9]

special feature of these systems is that the longitudinal dispersion can be greatly influenced by the angular velocity ω of the agitator.

For the flow through the annulus between a rotating cylinder (radius r_1) and a stationary outer cylinder (radius r_2) Croockewit et al. [37] established that

$$Bo = \frac{\langle v \rangle (r_2 - r_1)}{D_l} \simeq \frac{38 \langle v \rangle}{\omega r_1} \qquad (IV.45)$$

in the range of speeds where the flow exhibits laminar toroidal vortices. The vortices break down into general turbulence at higher angular speeds and Bo becomes fairly independent of ω. For a single phase 'rotating disc contactor', a tube divided into compartments by stator rings with a central rotating disc in each compartment, early work by Stemerding et al. [38], Westerterp and Landsman [39] and Westerterp and Meyberg [40] leads to the correlation

$$Bo = \frac{2}{1 + 0.002 \omega r / \langle v \rangle}, \qquad (IV.46)$$

where $Bo = \langle v \rangle L / D_l$ and L = height of compartment.

Example IV.6.a Blending in pipeline transport

In order to illustrate the application of the F curve and the correlations for mixing in empty tubes, we shall discuss the following problem. Two different liquids A and B are transported in succession over a distance of 1 km through a pipeline with an internal diameter d_t of 0.10 m at an average velocity of 1 m/s. The two liquids have approximately the same density and viscosity $v = 5 \times 10^{-6}$ m^2/s. If the transition from A to B is sharp at the inlet of the pipeline, during what period should a mixture of A and B be discharged separately at the end in order not to let the contamination of A by B exceed 4% and that of B by A 1%?

D_1 in this case is given by the relationship equation (IV.43):

$$\frac{\langle v \rangle d_t}{D_1} = \frac{0.283}{\sqrt{f}},$$

where f is the Fanning friction factor. If we denote the mass fraction of B by w, w_0 changes from 0 to 1 at $t = 0$ at the inlet of the tube; the problem to be solved is to find the times at which w_L has the values 0.04 and 0.99 respectively at the end of the tube. This is easily solved by application of the F diagram of the plug flow with axial dispersion model. From the above data we find $Re = 2 \times 10^4$ and the corresponding value of f for smooth pipes is 0.0062.

Hence

$$Pe = \frac{\langle v \rangle L}{D_1} = \frac{L}{d_t} \frac{\langle v \rangle d_t}{D_1} = 10^4 \times \frac{0.283}{\sqrt{0.0062}} = 3.74 \times 10^4,$$

which indicates that the dispersion is relatively small. The F diagram is therefore not sensitive for the boundary conditions and we can select the simple double open tube, for which the C diagram was given in Table IV.3. The F diagram is given by

$$F = \tfrac{1}{2}\left\{1 - \operatorname{erf}\left(\tfrac{1}{2}\sqrt{Pe} \times \frac{(1-\theta)}{\sqrt{\theta}}\right)\right\},$$

in which the error function erf is defined as

$$\operatorname{erf} y = \int_0^y e^{-a^2}\, da.$$

This function is tabulated in many mathematical handbooks. Some values are erf $(\pm\infty) = \pm 1$ and erf$(0) = 0$. The average residence time: $\tau = L/\langle v \rangle = 1000$ s.

From the F diagram we can calculate the data given in Table IV.4. It is seen that the discharge of undesired mixture should start at about 13 seconds before $t = \tau$, and continue until 17 seconds after this moment. The amount discharged is the fraction $(13 + 17)/1000$ of the pipe volume, i.e. 0.24 m^3, corresponding to a length of 30 m pipe.

IV.6.2 Packed beds

When a fluid is flowing through a packed bed of solid non-porous particles, the variations in the local velocity cause a dispersion in the direction of flow. In not too short beds ($L/d_p > 10$) this dispersion can be described by means of a longitudinal dispersion coefficient, although in reality no backmixing takes place. If one

Table IV.4 Numerical values for the blending in pipeline transport problem: Example IV.6.a

Property	Value for mass fraction of B at the outlet	
	$w_L = 0.04$	$w_L = 0.99$
$\mathrm{erf}\left(\frac{1}{2}(\sqrt{Pe})\frac{1-\theta}{\sqrt{\theta}}\right)$	0.92	-0.98
$\frac{1}{2}(\sqrt{Pe})\frac{1-\theta}{\sqrt{\theta}}$	1.24	-1.65
$\frac{1-\theta}{\sqrt{\theta}} \simeq 1-\theta$	-12.8×10^{-3}	17×10^{-3}
$t - \tau$	-12.8	17.0

considers the empty spaces of a packed bed as ideal mixers and the number of voids roughly equal to

$$N \simeq \frac{L}{d_p}$$

and uses the relation $N = Pe/2 = vL/(2D_1)$, the following expression for the axial dispersion in packed beds is found:

$$Bo = \frac{\langle v \rangle d_p}{D_1} = 2. \quad (IV.47)$$

It can be seen from figure IV.48 that for gases and for sufficiently high Reynolds numbers this is indeed the case. For lower Re values the axial mixing due to molecular diffusion becomes important and Bo becomes proportional to $ReSc$. At lower Reynolds numbers apparently also some kind of channelling occurs as explained by Schlünder [41]. This results in surprisingly low Bo numbers. Examples are given by Moulijn and van Swaaij [42].

For liquids $Bo = 0.5$ for the lower Re numbers, which corresponds to four particle layers for one ideal mixer; only at very high Reynolds numbers is $Bo = 2$ approached. Apparently the lack of mixing on a micro scale by molecular diffusion, which is much slower for liquids, prevents complete mixing within the voids.

Although the mechanisms discussed here are simplifications, they nevertheless are convenient for estimating Pe numbers in a packed bed. Reviews of experimental findings are given by Wen and Fan [9] and Levenspiel and Bischoff [24]. It is seen from the order of magnitude of Bo that the longitudinal dispersion in packed beds is so low that the ideal tubular reactor is closely approached for $L/d_p > 100$. In shallow beds, containing, say, not more than 10 layers of particles as used for extremely rapid catalysed gas reactions or solid-gas reactions, the spread in residence time may be considerable. Also when there is a large wall effect, say

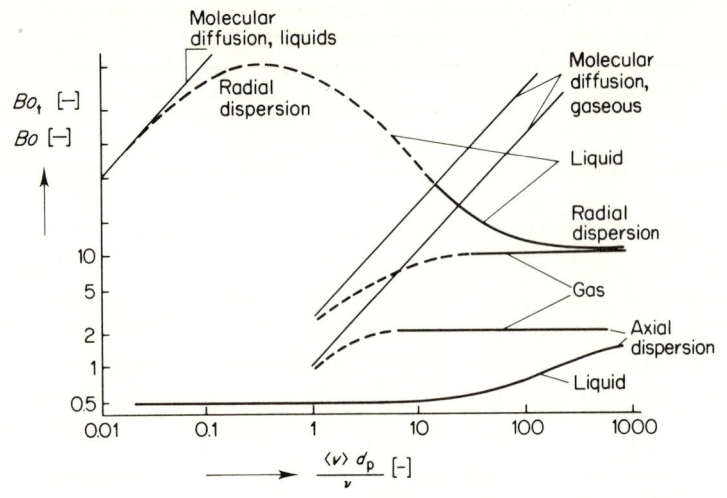

Fig. IV.48 Axial and radial dispersion for single phase flow through a packed bed of non-porous spheres.

$$Bo = \frac{\langle v \rangle d_p}{D_l}, Bo_t = \frac{\langle v \rangle d_p}{D_t}$$

$d_p/d_t > 0.1$, the axial mixing may be much higher due to the increased wall flow. For porous particles the axial mixing is higher than the value indicated in figure IV.48, depending on particle size and effective internal diffusivity, as was shown by Moulijn [43]. Correspondingly, Bo will be decreased. However, in such cases it is often of little value to lump these effects in the axial dispersion coefficient; two region models are more appropriate. Also indicated in figure IV.48 is the radial or transversal dispersion for which $Bo_t \simeq \langle v \rangle d_p/D_t$ approaches 10–12 at $Re > 100$.

So the coefficient of transverse dispersion D_t is about 6 times smaller than D_l. This also follows from theoretical analyses, see e.g. Levenspiel and Bischoff [24]. The radial transport of heat and matter is very important in non-isothermal packed bed reactors; these problems will be treated in Chapter IX.

IV.6.3 Fluidized beds

With a gas flowing through fluidized beds, the model of piston flow with superimposed longitudinal dispersion is merely a very rough description of what is actually happening. The excess of fluidizing gas is rapidly segregated from the rather dense particle phase and ascends quickly through the bed in the form of bubbles, which may become very large. As a consequence, there is an intense top-to-bottom mixing of the solids together with the interstitial and possibly adsorbed gas. If a residence time distribution experiment is performed or if a chemical reaction occurs in the gas phase, the compositions of the bubbles and of the interstitial gas will be different and material will be exchanged between the two phases. The bypassing effect of large bubbles, the longitudinal mixing in the dense phase and the mass transfer between these phases are very sensitive to the density of the particles, their average size and particularly their size distribution. It will be clear that with rapid reactions the conditions in the dense particle phase will be near equilibrium, resulting in a relatively small capacity of the bed as a whole; with slow reactions the short circuiting effect of the large bubbles will cause a much lower conversion than in the case of a uniform solid–gas distribution. Van Deemter [4] mathematically formulated the above model as was shown in Example IV.4.d. His theory allows us to estimate the degree of conversion for a gas reaction with a known reaction rate on the basis of an experimental residence time distribution function. Examples of its application are given by De Groot [3] and van Swaaij and Zuiderweg [22, 44]. In liquid fluidized beds where in most cases fluidization is homogeneous, the axially dispersed plug flow model has been applied successfully. It appears that at incipient fluidization the spread in residence time corresponds to that of the fixed bed. It increases with increasing fluidization velocity but then at higher porosities it will tend towards the value for empty tubes. As a consequence, at higher porosities the Bo number cannot be described on the basis of the fluid and particle properties alone but will also depend on the vessel dimensions. This is reflected in the Bo number as demonstrated in figure IV.49.

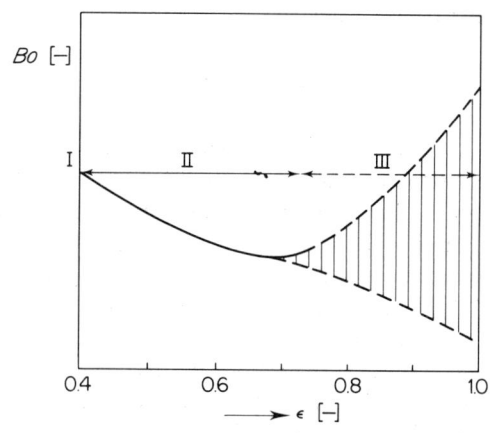

Fig. IV.49 Qualitative picture of $Bo = \langle v \rangle d_p / D_1$ in a liquid–solid fluid bed
Region I is identical to the packed bed, II is the fluidization region in which particle properties have a dominant influence and III is the region in which the Reynolds number based on the vessel diameter is the dominant parameter

Chung and Wen [45] have correlated the axial dispersion coefficients from many literature data both for fixed and fluidized beds for $Re_{mf} \leq Re$ by

$$\frac{\varepsilon Bo}{(Re)_{mf}/Re} = 0.20 + 0.011 \, Re^{0.48}, \tag{IV.48}$$

in which $Bo = \langle v \rangle d_P / D_1$, $Re = u d_P / v$, ε is the porosity and $(Re)_{mf}$ is the Reynolds number at incipient fluidization. Their correlation is assumed to be valid for $\varepsilon = 0.4$–0.8, Re from 10^{-3} to 10^3 and particle densities up to $8000 \, kg/m^3$. At

$(Re)_{mf}/Re \simeq 1$ approximately the same results as presented in figure IV.48 are found for liquid flow through packed beds.

IV.6.4 Mixing in gas–liquid reactors

The axial mixing in both phases of gas-liquid reactors may be important criteria for the selection of a specific type of reactor. Other criteria are interfacial area, mass transfer rates, etc. A recent excellent review on this subject has been given by Shah et al. [46].

In vertically sparged bubble columns Baird and Rice [47] derived theoretically by applying the isotropic turbulence model of Hinze [48] that, for the liquid phase,

$$D_l \propto d_t^{1.33} u_G^{0.33}$$

whereas Deckwer [49] found that experimental results reported in literature could be correlated by

$$D_l = 0.678 \, d_t^{1.4} u_G^{0.3}, \qquad (IV.49)$$

which is an excellent agreement. Experiments and correlations show no influence of the liquid flow rate. For the gas phase in a bubble column few data are available. Diboun and Schügerl [50] suggested

$$Bo_G = \frac{u_r d_t}{D_l} = 0.2, \qquad (IV.50)$$

in which u_r is the relative velocity between the gas and the liquid. In gas–liquid–solid systems the situation is complicated by the large number of variables. In three-phase fluid beds data on liquid back-mixing have been reported by Kim et al. [51] and by Kato et al. [52]. Schügerl [53] and Michelsen and Østergaard [54] gave data on the axial mixing in the gas phase and Kato et al. [52] and Imafuku et al. [55] on the axial mixing of the particles. No generally valid correlations are available yet.

In trickle flow reactors even with non-porous particles the axially dispersed plug flow can only be considered as a rough approximation. Van Swaaij et al. [20] demonstrated that for the liquid phase the more complicated model of Example IV.4.d gives a better representation than the axially dispersed plug flow model. If the axially dispersed plug flow model is to be applied, the static liquid hold-up which is mainly located on the contact points of the particles has to be taken into account. If this static hold-up is less than, say, 10% of the total hold-up, the correlation given in figure IV.48 for liquids can be applied. For higher static hold-up values the *Bo* number decreases as indicated in figure IV.50. The static hold-up can be estimated from the Eötvos number as shown by van Swaaij et al. [20]. Many more results from experimental work on trickle flow have been summarized by Shah et al. [46]. Nearly all data refer to water–air systems and data on other systems are almost absent. Especially at higher gas velocities, the gas phase has some influence on the liquid phase *Bo* number, as shown by van Swaaij [5]. The data on mixing within the gas phase for countercurrent and co-current trickle flow

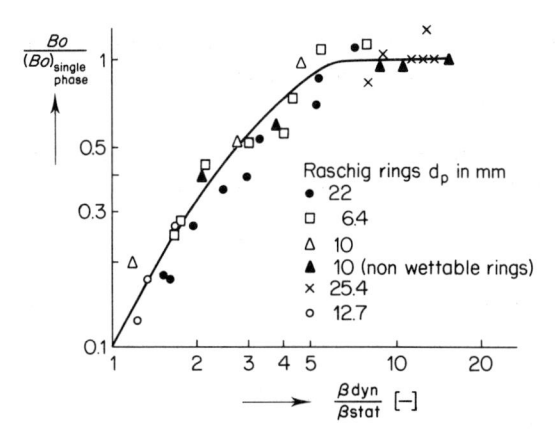

Fig. IV.50 Axial dispersion in packed columns at trickle flow as a function of the ratio of dynamic and static hold-up. $Bo = \langle v \rangle d_p/D_1$ (Bo) single phase can be found from figure IV.48
[From Van Swaaij et al. [20]. Reproduced by permission of Pergamon Press Ltd]

are still scattered. The Bo number for the gas phase is influenced both by the liquid and the gas flow rate. Recent experiments leading to correlations are given by Hochman and Effron [56], Woodburn [57], van Swaaij [5], Midoux [58, 59], de Maria and White [60], Sater and Levenspiel [61] and Britton and Woodburn [62] for countercurrent flow. For packed columns with gas–liquid flowing co-currently in the upward direction Stiegel and Shah [63] recently proposed for the liquid phase

$$Bo = 0.128\, Re_L^{0.245} Re_G^{-0.16} (a_p d_p)^{0.53}$$

in which

$$Bo = \frac{u_L d_p}{D_1}, \quad Re_L = \frac{\rho u_L d_p}{\eta_L} \quad \text{and} \quad Re_G = \frac{\rho u_G d_p}{\eta_G}.$$

For agitated gas–liquid contactors the residence time distribution curve for both gas and liquid approaches that of an ideal mixer. For the gas phase this has been shown by Hanhart et al. [2]. This does not mean, of course, that the gas phase is ideally mixed, because the gas bubbles are separated entities with at best a finite coalescence/redispersion rate.

Chen et al. [64] gave data on back-mixing in countercurrent bubble columns packed with screen cylinders. Many data on staged gas–liquid contactors such as multistaged columns, sieve plate columns, rotating disc contactors and impellor segmented contactors have been reviewed by Shah et al. [46]. For liquid–liquid contacting equipment a recent review on the axial mixing is given by Marr and Moser [65].

Example IV.6.b *Determination of parameters from pulse response tests*
Parameters of RTD models are usually obtained with tracer tests with input functions resembling a δ-pulse. There are several ways to derive the model parameters from these types of tests. In this illustration we will give an example for

the axially dispersed plug flow model. We will consider a non ideal pulse injection with detection in one or two positions. Two distinct experimental set-ups are shown in figure IV.51:

(i) Both injection and detection take place in piston flow regions as shown in figure IV.51a, which results in a closed–closed test section.

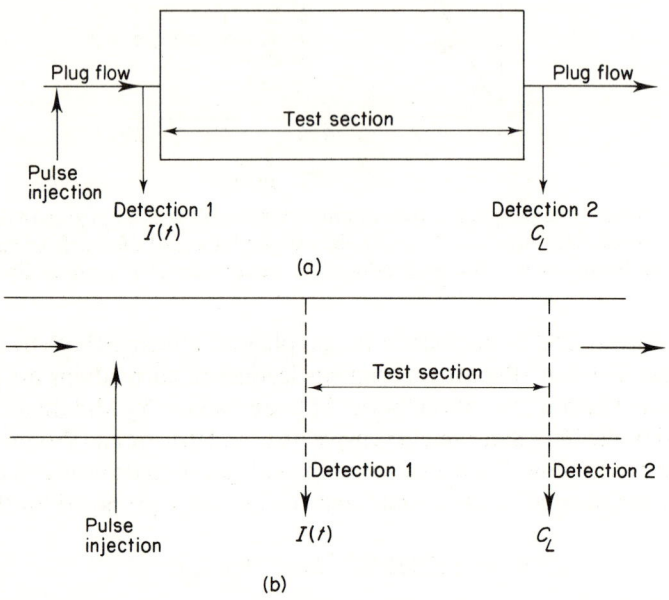

Fig. IV.51 Experimental set-up for RTD parameter determination with the imperfect pulse technique: (a) closed–closed test section; (b) open–open test section

(ii) Both detections take place within the flow region to be investigated, as shown in figure IV.51b, which results in an open–open test section.

In case of a closed–closed test section the situation is simple, especially if the pulse can be considered to be practically ideal, which means that the variance of the pulse is much smaller than the variance of the response. The moments of the response can be directly compared with those predicted by the model. Alternatively, the whole response curve could be matched with the model E-curve. Because the model curve will probably deviate from the experimental response, a best fit criterion is required. As an example the following goal function could be minimized:

$$\Delta = \int_0^\infty \{E(t) - C_L(t)\}^2 \, dt.$$

In this equation $C_L(t)$ is the normalized response at the exit of the reactor and $E(t)$ is the model E diagram. The Peclet number for which Δ is minimal is then taken as a best fit value. Sometimes both Pe and τ are optimized, especially if τ is not known beforehand. If the input pulse is not close enough to a δ-input, the moments method can still be applied; the moments of the response curve have to be corrected for those of the input curve via the addition rules for the moments as follows from equation (IV.7) and (IV.8) and Example IV.1.b. For a direct comparison with the model the output curve could be corrected for the input curve by deconvolution. However, this is a cumbersome procedure. Generally it is, therefore, better to find the convolution of the input pulse $I(t)$ and the model E curve to produce $E^*(t)$ by

$$E^*(t) = \int_0^t E(p)I(t-p)\,dp.$$

This $E^*(t)$ can be inserted in the goal function together with the experimental response:

$$\Delta = \int_0^\infty \{E^*(t) - C_L(t)\}^2\,dt$$

and optimal values of Pe and τ can be found.

For complex models often the $E(t)$ function is extremely difficult to find and even the $E(t)$ diagram for the axially dispersed plug flow with closed–closed boundary conditions (Eq. IV.29) is difficult to handle. Therefore, often the Laplace transform $\underline{E}$ of the $E(t)$ diagram of the model is compared with the Laplace transform of the pulse response evaluated numerically:

$$\underline{C_L} = \int_0^\infty e^{-st} C_L\,dt,$$

$\underline{C_L}$ now being a function of s. The goal function then becomes

$$\Delta = \int_0^\infty (\underline{E} - \underline{C_L})^2\,ds.$$

For the dispersion model the Laplace transformed $E(t)$ diagram is for closed–closed boundaries, as was shown in Example IV.4.b:

$$\underline{E} = \frac{4q\exp(Pe/2)}{(1+q)^2 \exp\left\{\dfrac{Pe}{2}q\right\} - (1-q)^2 \exp\left\{-\dfrac{Pe}{2}q\right\}}$$

with $q = \sqrt{\left(1 + \dfrac{4s}{Pe}\right)}$.

If the pulse is not ideal the response can now be easily corrected for the non-ideal pulse. For a cascade of independent flow regions the convolution integral corresponds to simple multiplication in the Laplace domain [11] or

$$\underline{E}_{cascade} = \underline{E}_1 * \underline{E}_2 * \underline{E}_3 * \cdots * \underline{E}_N.$$

Now the corrected response can be obtained in the Laplace domain

$$\underline{C_L}(\text{test section}) = \frac{\underline{C_L}(\text{total response})}{\underline{I}},$$

in which $\underline{I}$ is the Laplace transform of the normalized non-ideal pulse. For a perfect pulse $\delta(t)$, $\underline{\delta} = 1$. Then again the parameters Pe and τ can be found from a best fit criterion. Although this method and related methods have often been used, generally it is uncertain whether the best fit in the Laplace form corresponds to the best fit in the normal time domain. Therefore, sometimes Fourier transforms are used instead, for which the Parseval's theorem [11] states that the best fit in the Fourier domain corresponds to the best fit in the time domain. Model Fourier transforms can be obtained by replacing s by $j\omega$ in the Laplace transforms, but further manipulation is somewhat more complicated, see Clements [66]. The methods for parameter evaluation discussed so far are summarized in Table IV.5.

Table IV.5 Different methods for the parameter determination of residence time distribution functions. For open test sections transfer functions should be used instead of $E(t)$ diagrams: Example IV.6.b

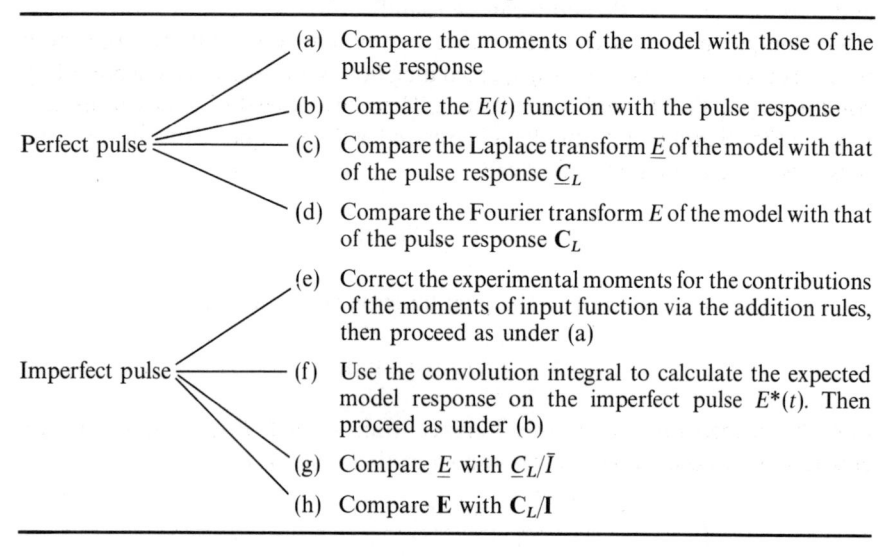

Perfect pulse	(a)	Compare the moments of the model with those of the pulse response
	(b)	Compare the $E(t)$ function with the pulse response
	(c)	Compare the Laplace transform $\underline{E}$ of the model with that of the pulse response $\underline{C}_L$
	(d)	Compare the Fourier transform E of the model with that of the pulse response $\mathbf{C}_L$
Imperfect pulse	(e)	Correct the experimental moments for the contributions of the moments of input function via the addition rules, then proceed as under (a)
	(f)	Use the convolution integral to calculate the expected model response on the imperfect pulse $E^*(t)$. Then proceed as under (b)
	(g)	Compare $\underline{E}$ with $\underline{C}_L/\underline{I}$
	(h)	Compare $\mathbf{E}$ with $\mathbf{C}_L/\mathbf{I}$

If the measurements are taken from positions with *open boundaries* as in figure IV.51b, which might be attractive to eliminate end effects, the situation is somewhat more complicated. The regions to be subtracted are not separated by plug flow but do interfere. For example, a δ-pulse injected at position 'detection 1' in figure IV.51b would travel partially outside of the test zone before entering this section. This situation was examined more closely by Gibilaro [15] and Nauman [16]. To handle the responses obtained at 'detection 1' and 'detection 2' in a similar way as for the closed–closed case we need the transfer function $TR(t)$ which simply translates the concentrations in point 1 to those in point 2. For the closed–closed case the transfer function is identical with the $E(t)$ diagram, but for open systems it generally differs from the response to an ideal pulse function for reasons described above. For the axially dispersed plug flow model with open–open boundary conditions the transfer function is, as shown by Aris [8] and Gibilaro [15],

$$TR(\theta) = \sqrt{\left(\frac{Pe}{4\pi\theta^3}\right)} \exp\left\{\frac{-Pe(1-\theta)^2}{4\theta}\right\},$$

while the average $\bar{\theta} = MR_1 = 1$ and variance $\sigma_\theta^2 = 2/Pe$. This equation is different from the δ-pulse response in an open–open system given in Example IV.4.c. The transfer function in time units is

$$TR(t) = \sqrt{\left(\frac{Pe\tau}{4\pi t^3}\right)} \exp\left\{\frac{-Pe}{4t/\tau}\left(1-\frac{t}{\tau}\right)^2\right\}.$$

With these equations we can use the same techniques as described for the closed–closed test section of figure IV.51. We will illustrate the different evaluation techniques with an example taken from Roes [67] where axial mixing is measured in a gas–solids trickle flow reactor as published by Roes and van Swaaij [68]. The experimental set-up used by Roes was identical to that of figure IV.51b. The distance between the detection points in the packed column reactor was 0.9 m and both detection points were far enough away from the end sections of the column to consider the open–open boundary conditions appropriate. The results obtained at detection points 1 and 2 are plotted in figure IV.52. The various methods lead to the following results.

(i) *Pe and τ from the moments*

From the moments of the I and C_L curves evaluated numerically it was derived that at detection point 1 $MO_1 = \bar{t} = 0.95$ s and $\sigma_t = 0.528$ s, and at detection point 2 $MO_1 = \bar{t} = 5.87$ s and $\sigma_t = 1.179$ s. The average residence time in the test section then follows from the moments of the transfer function $TR(t)$:

$$\tau = \frac{V(\text{test section})}{\Phi_v} = \bar{t}_2 - \bar{t}_1 = 5.87 - 0.95 = 4.92 \text{ s}$$

The Peclet number follows from

$$\frac{2}{Pe} = \{\sigma_t^2(2) - \sigma_t^2(1)\}/(\bar{t}_2 - \bar{t}_1)^2$$

$$= \frac{(1.179)^2 - (0.528)^2}{(4.92)^2} = 0.0459$$

which leads to $Pe = 44$.

(ii) *Pe and τ from the convolution integral*

The convolution integral was applied with the transfer function $TR(t)$ and the function $I(t)$, equation (IV.5). This was done for different values of Pe and τ. The best fit result is also plotted in figure IV.52. The best fit parameters thus obtained were:

$$\tau = 4.88\,\text{s} \quad \text{and} \quad Pe = 50.3$$

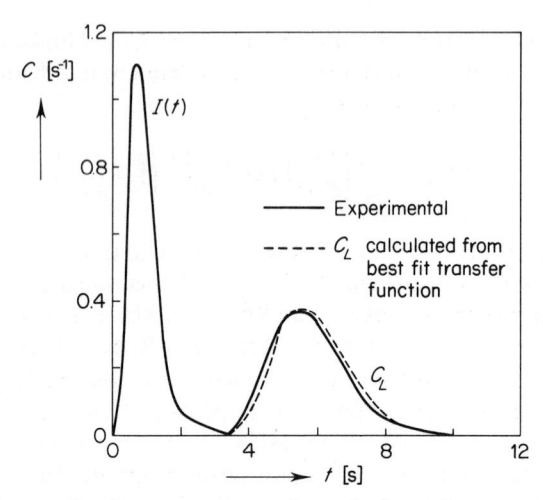

Fig. IV.52 Set of normalized response curves from the imperfect tracer pulse response technique [From Roes [67]. Reproduced by permission of A. W. M. Roes]

(iii) Pe and τ from the Laplace transforms

The Laplace transform of the transfer function of the axially dispersed plug flow model with open–open boundaries has been given by Michelson and Østergaard [69]:

$$\underline{TR} = \exp\left\{\frac{Pe}{2}\left[1 - \left(1 + \frac{4s\tau}{Pe}\right)^{1/2}\right]\right\}.$$

The expected Laplace transformed response at position 2 can now be found from

$$\underline{C_L}(\text{expected}) = \underline{TR} * \underline{I}$$

$\underline{C_L}(\text{exp})$ and $\underline{I}$ are found from

$$\underline{C} = \int_0^\infty e^{-st} C \, dt.$$

Figure IV.53 shows the result for the best fit. The parameters obtained this way were $\underline{\tau = 5.15}$ and $\underline{Pe = 44}$.

(iv) Pe and τ from the Fourier transforms

In principle the procedure here is the same as for the Laplace transforms. Because Fourier transforms have real and imaginary parts, the goal function has to be split

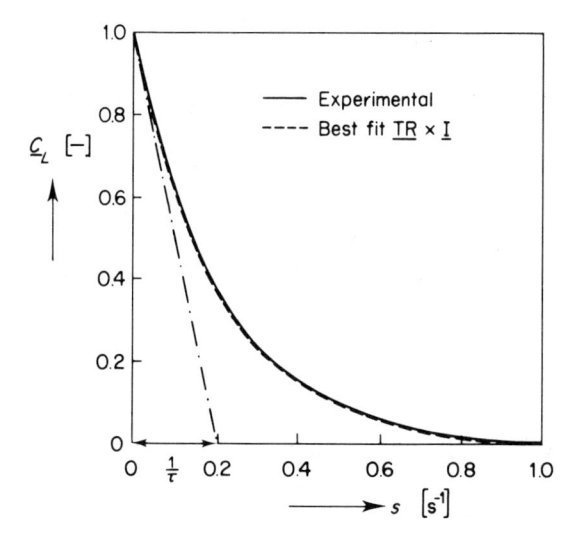

Fig. IV.53 Parameters from Laplace domain fit [From Roes [67]. Reproduced by permission of A. W. M. Roes]

up in two parts to allow separate minimization of the real and imaginary parts. Clements [66] provided for the transfer function of the axially dispersed plug flow model, the real and imaginary parts of the Fourier transform being obtained separately.

$$\mathbf{TR}(j\omega) = \mathrm{RE}_{\mathbf{TR}} + j = \sqrt{-1}\mathrm{IM}_{\mathbf{TR}}$$

in which

$$\mathrm{RE}_{\mathbf{TR}} = \exp\left\{\frac{Pe}{2}[1 - (A^2 + 1)^{1/4}\cos B]\right\}$$
$$\times \cos\left\{\frac{Pe}{2}[1 - (A^2 + 1)^{1/4}\sin B]\right\}$$

and

$$\mathrm{IM}_{\mathbf{TR}} = \exp\left\{\frac{Pe}{2}[1 - (A^2 + 1)^{1/4}\cos B]\right\}$$
$$\times \sin\left\{-\frac{Pe}{2}(A^2 + 1)^{1/4}\sin B\right\}$$

with

$$A = \frac{4\omega\tau}{Pe} \quad \text{and} \quad B = \tfrac{1}{2}\arctan A.$$

With fast Fourier transform techniques [11] the Fourier transforms $\mathbf{I}$ and $\mathbf{C}_L$ can be found and furthermore:

$$\mathbf{C}_L(\text{expected}) = \mathbf{TR} \times \mathbf{I}$$

The goal function to be minimized is now:

$$\int_0^{\omega_{max}} \{\text{RE}_{\text{TR}}C_L(\text{observed}) - \text{RE}_{\text{TR}}(\text{expected})\}^2$$
$$+ \{\text{IM}_{\text{TR}}C_L(\text{observed}) - \text{IM}_{\text{TR}}C_L(\text{expected})\}^2 \, d\omega.$$

With the specific data of Roes ω_{max} was limited to the value where the **TR** C_L (experiment) began to oscillate.

Figure IV.54 gives the polar plot of the best fit. The best fit parameters were:

$$\tau = 4.89 \quad \text{and} \quad Pe = 50.3$$

The results of the different methods are summarized below:

	Moments	Transfer function	Laplace	Fourier
τ [s]	4.92	4.88	5.15	4.89
Pe [—]	44	50	44	50

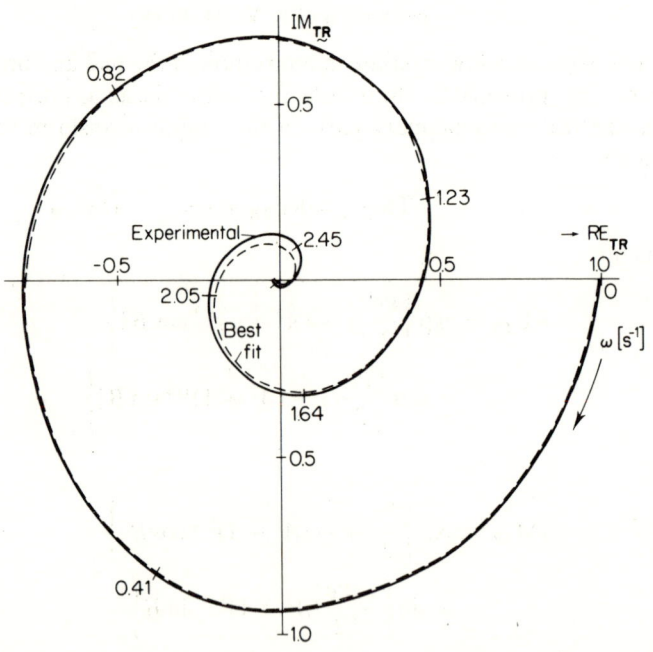

Fig. IV.54 Polar plot of the experimental and best fit Fourier transformed transfer functions for the test section [From Roes [67]. Reproduced by permission of A. W. M. Roes]

As can be expected from Parseval's theorem the transfer function and the Fourier transformed transfer function gave almost the same results. Both the moments method and the Laplace transform technique, if fitted for low values of s, put relatively heavy weight on the tail fraction of the curve and happen to give nearly the same Peclet numbers in our particular example. If the model is a perfect representation of the experimental curve, all methods should give the same values of the parameters. If this is not the case, the preferred method will depend on the envisaged application of the results, e.g. a conversion prediction, and the effort required for obtaining the parameters. The Fourier transform technique is much more complicated than the moments technique and is often not worth the effort.

If the time domain functions $E(t)$ or $TR(t)$ can be made available, method (b) or (f) of Table IV.5 should be preferred.

Other methods derived from the techniques described above have been proposed such as the method of the weighted moments, proposed by Michelson and Østergaard [69], Gunn [70] and Abbi and Gumm [71].

REFERENCES

1. Danckwerts, P. V.: *Chem. Eng. Sci.* **2**, 1 (1953)
2. Hanhart, J., Kramers, H., and Westerterp, K. R.: *Chem. Eng. Sci.* **18**, 503 (1963)
3. de Groot, J. H.: *Proc. Int. Symp. on Fluidization Eindhoven*, 1967, Neth. Univ. Press, Amsterdam 348 (1967)
4. van Deemter, J. J.: *Chem. Eng. Sci.* **13**, 143 (1961)
5. van Swaaij, W. P. M.: Thesis, Eindhoven University of Technology, Eindhoven (1967)
6. Kramers, H., and Alberda, G.: *Chem. Eng. Sci.* **2**, 173 (1953)
7. Fontaine, R. W., and Harriot, P.: *Chem. Eng. Sci.* **27**, 2189 (1972)
8. Aris, R.: *Chem. Eng. Sci.* **9**, 266 (1959)
9. Wen, C. Y., and Fan, L. T.: *Models for Flow Systems and Chemical Reactors*, Marcel Dekker, New York (1975)
10. Levenspiel, O., and Turner, J. C. R.; *Chem. Eng. Sci.* **25**, 1605 (1970)
11. Spiegel, M. R.: *Theory and Problems of Laplace Transforms*, Schaum's Outline Series, McGraw-Hill, New York (1965)
12. Brothman, A., Weber, A. P., and Barish, E. Z.: *Chem. Met. Eng.* **7**, 111 (1943)
13. Otake, T., and Kunigita, E.: *Kagaku Kogaku* **22**, 144 (1958)
14. Yagi, T.: *Chem. Engng. Japan* **17**, 382 (1953)
15. Gibilaro, L. G.: *Chem. Eng. Sci.* **33**, 487 (1978)
16. Nauman, E. B.: *Chem. Eng. Sci.* **36**, 957 (1981)
17. Villermaux, J., and van Swaaij, W. P. M.: *Chem. Eng. Sci.* **24**, 1097 (1969)
18. Levenspiel, O., and Smith, W. K.: *Chem. Eng. Sci.* **6**, 227 (1957)
19. Hoogendoorn, C. J., and Lips, J.: *Can. J. of Chem. Eng.* **43**, 125 (1965)
20. van Swaaij, W. P. M., Charpentier, J. C., and Villermaux, J.: *Chem. Eng. Sci.* **24**, 1083 (1969)
21. May, W. G.: *Chem. Eng. Prog.* **55**, 49 (1959)
22. van Swaaij, W. P. M., and Zuiderweg, F. J.: *Proc. 5th Europ. Symp. Reaction Engng*, Elsevier, Amsterdam (1972) B9
23. Bohle, W.: Private communication (1981)
24. Levenspiel, O., and Bischoff, K. B.: *Patterns of Flow in Chemical Process Vessels. Advances in Chemical Engineering* IV, Academic Press, London (1963)
25. Levenspiel, O.: *Chemical Reaction Engineering*, second ed., Wiley, New York (1972)
26. Cholette, A., and Cloutier, L.: *Can. J. Chem. Eng.* **37**, 105 (1959)
27. van der Vusse, J. G.: *Chem. Eng. Sci.* **17**, 507 (1962)
28. Wen, C. Y., and Chung, S. F.: *Can. J. of Chem. Eng.* **43**, 101 (1965)
29. Adler, R. J., and Hovorka, R. B.: *Second Joint Automatic Control Conference*, Denver, Colorado, June 1961
30. Klinkenberg, A.: *Chem. Eng. Sci.* **2**, 172 (1968)

31. Wehner, J. F., and Wilhelm, R. H.: *Chem. Eng. Sci.* **6**, 89 (1956)
32. Taylor, G. I.: *Proc. Roy. Soc.* **A219**, 186 (1953)
33. van Deemter, J. J., Broeder, J. J., and Lauwerier, H. A.: *Appl. Sci. Res.* **A5**, 374 (1956)
34. Aris, R.: *Proc. Roy. Soc.* **235A**, 67 (1956)
35. Taylor, G. I.: *Proc. Roy. Soc.* **A223** 446 (1954)
36. Tichacek, L. J., Barkelew, C. H., and Baron, T.: *A.I.Ch.E.J.* **3**, 439 (1957)
37. Croockewit, P., Honig, C. C., and Kramers, H.: *Chem. Eng. Sci.* **4**, 11 (1965)
38. Stemerding, S. F., Zuiderweg, F. J., van der Vusse, J. G., Thegze, V. B., Wall, R. J., Tram, K. E., and Olney, R. B.: *Erdöl Zeitschrift* **77**, 401 (1961)
39. Westerterp, K. R., and Landsman, P.: *Chem. Eng. Sci.* **17**, 363 (1962)
40. Westerterp, K. R., and Meyberg, W. H.: *Chem. Eng. Sci.* **17**, 373 (1962)
41. Schlünder, E. U., *Chem. Eng. Sci.* **32**, 845 (1977)
42. Moulijn, J. A. and van Swaaij, W. P. M.: *Chem. Eng. Sci.* **31**, 845 (1976)
43. Moulijn, J. A.: Thesis, University of Amsterdam, Amsterdam (1975)
44. van Swaaij, W. P. M., and Zuiderweg, F. J.: *Proc. Int. Conf. on Fluidization and its Applications*, 454, Cepadues, Toulouse, 1973
45. Chung, S. F., and Wen, C. Y.: *A.I.Ch.E. J.* **14**, 857 (1968)
46. Shah, Y. T., Stiegel, G. J., and Sharma, M. M. *A.I.Ch.E. J.* **24**, 369 (1978)
47. Baird, M. H. I., and Rice, R. G.: *Chem. Eng. J.* (Lausanne) **9**, 17 (1975)
48. Hinze, J. O., *Turbulence*, McGraw-Hill, New York (1958)
49. Deckwer, W. D., Bruckhan, R., and Zoll, G.: *Chem. Eng. Sci.* **29**, 2177 (1974)
50. Diboun, M., and Schügerl, K.: *Chem. Eng. Sci.* **22**, 147 (1967)
51. Kim, S. D., Baker, C. G. J., and Bergougnou, M. A.: *Can. J. Chem. Eng.* **50**, 695 (1972)
52. Kato, Y., Fukuda, T., and Tanaka, S.: *J. Chem. Eng. Jpn.* **5**, 112 (1972)
53. Schügerl, K.: *Proc. Int. Symp. on Fluidization*, Eindhoven, 1967, Neth. Univ. Press, Amsterdam 782 (1967)
54. Michelson, M. L., and Østergaard, K.: *Chem. Eng. J.* (Lausanne) **1**, 37 (1970)
55. Imafaku, K., Wang, T. Y., Koide, K., and Kubota, H.: *J. Chem. Eng. Jpn.* **1**, 153 (1968)
56. Hochman, J. M., and Effron, E.: *Ind. Eng. Chem. Fund.* **8**, 63 (1969)
57. Woodburn, E. T.: *A.I.Ch.E. J.* **20**, 1003 (1974)
58. Midoux, N., and Charpentier, J. C.: *Chem. Eng. J.* (Lausanne) **4**, 287 (1972)
59. Midoux, N.: Thesis, Université de Nancy, Nancy (1971)
60. de Maria, K., and White, R. R.: *A.I.Ch.E. J.* **6**, 473 (1960)
61. Sater, V. E., and Levenspiel, O.: *Ind. Eng. Chem. Fund.* **5**, 86 (1966)
62. Britton, M. J., and Woodburn, E. T.: *A.I.Ch.E. J.* **12**, 541 (1966)
63. Stiegel, G. J., and Shah, Y. T.: *Can. J. Chem. Eng.* **55**, 3 (1977)
64. Chen, B. H., Manna, B. B., and Hines, J. W.: *Ind. Chem. Eng. Proc. Design and Dev.* **10**, 341 (1971)
65. Marr, R., and Moser, F.: *Chem. Ing. Tech.* **50**, 90 (1978)
66. Clements, W. C.: *Chem. Eng. Sci.* **39**, 957 (1969)
67. Roes, A. W. M.: Thesis, Twente University of Technology, Enschede (1978)
68. Roes, A. W. M., and van Swaaij, W. P. M.: *Chem. Eng. J.* (Lausanne) **18**, 29 (1979)
69. Michelson, M. L., and Østergaard, K.: *Chem. Eng. Sci.* **25**, 583 (1970)
70. Gunn, D. J.: *Chem. Eng. Sci.* **29**, 957 (1969)
71. Abbi, Y. P., and Gunn, D. J.: *Trans. Inst. Chem. Engrs.* **54**, 225 (1976)

Chapter V
Influence of micromixing on chemical reactions

The flow of a reacting fluid through a reactor is a very complex process. The interaction of a complex flow pattern with the chemical reaction, even in the simple case of a single isothermal reaction, cannot be analysed rigorously. Therefore simplified and/or idealized models are necessary.

Apart from the ideal flow reactors, plug flow reactor and continuous ideally stirred tank reactor, in Chapter IV models which can describe intermediate mixing have been discussed. When applying these models to real reactor systems, the question remains whether these models do reflect the real mixing behaviour sufficiently closely to predict accurately the reactor performance. Especially in the case of non-first order reactions or reactions with separate reactant inlets, the models should describe sufficiently accurately not only the gross overall mixing behaviour but also the detailed intermixing of the different fluid elements. To avoid problems related to micromixing, in Chapters I–IV the assumption of an ideal microfluid was made. This concept is explained in the introduction of Chapter IV and means that each fluid element is homogeneous down to almost molecular scale. This is not necessarily true in practice.

For a continuously operated mixed vessel which shows a residence time distribution of an ideal mixer:

$$E = \frac{1}{\tau} e^{-t/\tau}$$

the different states of micromixing are illustrated in figure V.1:

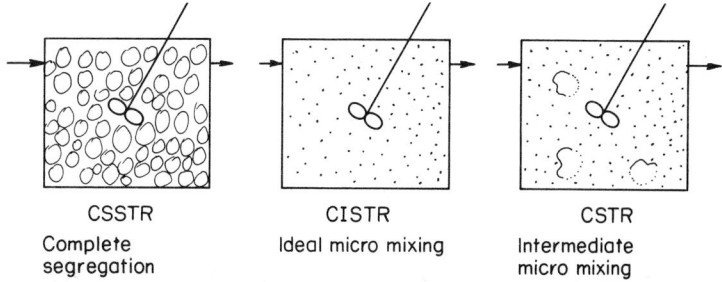

CSSTR	CISTR	CSTR
Complete segregation	Ideal micro mixing	Intermediate micro mixing

Fig. V.1 Possible states of micromixing in a stirred vessel

— CISTR, the ideal mixer processing a micro fluid;
— CSSTR, the completely segregated mixer, where the fluid consists of lumps

which travel through the reactor without exchange of matter with their neighbours;
— CSTR, the partially segregated mixer.

The behaviour of the CSTR falls between that of the CISTR and the CSSTR. However, as will be explained in this chapter, a unique description of the micromixing with one parameter is generally not possible.

For non-first order reactions, apart from the reaction rate not only the intensity of exchange of mass between the fluid elements in the reactor is important, but also the timing: will the mixing be late or early? In other words will entering fluid elements immediately mix with old ones or first stay together for some time. As will be explained, for a reactor with a given RTD these factors can vary between certain boundaries. This makes micromixing a difficult subject. Fortunately the maximum influence of micromixing effects can be calculated relatively simply and often these effects can be neglected. Typical practical problems in which micromixing may have to be taken into account are:

— reactors processing viscous fluids in which rapid non-linear reactions take place such as second order reactions, autocatalytic reactions, consecutive reactions, etc.;
— polymerization reactions;
— reactors in which rapid reactions take place between reactants fed through separate feed inlet points;
— non linear reactions in emulsions with finite coalescence/redispersion rates;
— precipitation reactions;
— non-isothermal reactors such as flames, etc.

For these systems an exact description is not possible and will probably not be practical for a long time to come. In many cases the boundaries between which conversion and selectivity could be influenced by micromixing effects can be calculated, as will be demonstrated in this chapter. With special models more accurate predictions are possible. We will consider only isothermal constant density systems in this chapter.

V.1. NATURE OF THE MICRO MIXING PHENOMENA

Generally the mixing in a chemical reactor can be described by three different factors which are not completely independent. These factors are the macromixing as represented by the RTD, the state of aggregation and the earliness of mixing. They will be discussed separately.

V.1.1 Macro or gross overall mixing as characterized by the residence time distribution

Extreme cases are plug flow and ideal mixing at least on a macro scale: the CSTR. The residence time distribution describes the residence times of the molecules while traveling through the reactor, but not the environment with which they have to react. If the detailed flow pattern of a reactor is given, the RTD function can be calculated in principle. However, the reverse statement is not true as different flow patterns may lead to the same RTD function. As will be shown later, only in the case of a simple first order reaction is the RTD information sufficient to give a unique prediction of the chemical conversion in the reactor considered. Generally the aspect of gross overall mixing as characterized by the RTD is important for all systems in which a high relative conversion is aimed for.

V.1.2 The state of aggregation of the reacting fluid

It has to be known to what extent the fluid travels as lumps of molecules together. In a dispersed system such as an emulsion of oil in water, the oil is segregated into clearly distinguishable domains; the oil drops. The effect of this segregation can be counteracted by coalescence/redispersion. Also in single fluids, especially in viscous ones, such aggregates of molecules can be distinguished. They may have a variable life span and they can influence the reactor performance if local concentration gradients are present.

The extreme form of the state of aggregation is called a *macro* fluid, the dispersed phase of an emulsion with no coalescence is an example. If no aggregates are present or effective, the fluid is said to behave like a *micro* fluid. The effective state of aggregation generally can only be judged in relation to reaction rate and residence time because the same fluid at the same flow conditions may behave like a micro fluid or a macro fluid. This can be seen as follows.

If the reaction rate is low and thus the reaction time is long in comparison to the residence time or to the life span of an aggregate, conversion can be calculated assuming the fluid to behave like a micro fluid. However, if we have an extremely short reaction time or a fast reaction and a short residence time, the life span of the agglomerate may be long in comparison and conversion can be found assuming that the fluid behaves effectively as a macro fluid. The size of the agglomerates in a single fluid depends on hydrodynamical factors such as energy input and on fluid properties such as viscosity and diffusivity. Based on the work of Kolmogoroff [1], Rosensweig [2], Batchelor [3] and Corrsin [4] it is possible to make an estimate on the relevant dimension of the agglomerates in the fluid.

V.1.3 The earliness of the mixing

The earliness of mixing is a separate factor which can be distinguished, but it cannot always be separated from the two previous factors. In a continuous flow system molecules leaving together generally did not enter together. They may have become mixed earlier or later during their residence period in the reactor, without influence on the RTD function. The possibilities of early or late mixing are influenced by the gross overall mixing pattern and by the state of segregation of the fluid or by a combination of both. With premixed feedstock and a plug flow reactor there is of course no effect of early or late mixing, but with a broad spectrum of residence times such as in a CSTR there can be an important effect. Here the processing of a *micro* fluid would result in the earliest possible mixing, whereas for the complete *macro* fluid the latest possible mixing is obtained, e.g. only at the outlet point. However, the state of aggregation is not the only determining factor in late or early mixing as can easily be demonstrated. In figures V.2 and V.3 the influence of the early or late mixing on a second order chemical reaction is demonstrated. The early or late mixing is here only affected by the arrangement of the overall mixing pattern while the fluid is assumed to be an ideal microfluid. The RTD of a cascade of a mixer and a plug flow reactor is not influenced by the sequence of the apparatus; only the earliness of mixing is changed, as demonstrated in figure V.2.

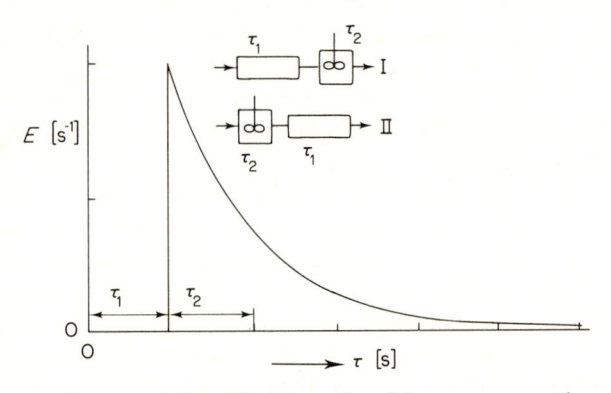

Fig. V.2 E diagram for cascade I and II. The order of the arrangement has no influence on the RTD

As can be seen from figure V.3, the earliness of mixing has a pronounced influence on the conversion of a second order reaction. The conversion equations leading to figure V.3, derived for the ideal reactors, are

$$\zeta_A = 1 + \frac{1}{2kc_{A_0}\tau_2}\left[1 - \sqrt{\left(1 + 4\frac{kc_{A_0}\tau_2}{1 + kc_{A_0}\tau_1}\right)}\right] \quad \text{for arrangement I,}$$

$$\zeta_A = 1 + \frac{1 - \sqrt{(1 + 4kc_{A_0}\tau_2)}}{2kc_{A_0}\tau_2 - kc_{A_0}\tau_1[1 - \sqrt{(1 + 4kc_{A_0}\tau_2)}]} \quad \text{for arrangement II.}$$

By a simple example it can be demonstrated how the earliness of mixing influences the conversion rate of a chemical reaction and why there is no such influence with a first order reaction. Figure V.4 gives the reaction rate as a function of the concentration for an elementary nth order chemical reaction with $|R_A| = kc_A^n$. We consider two equal volume elements 1 and 2 from a continuous reactor and we suppose that they will leave the reactor at the same moment. The concentration of the remaining reactant is not the same because at the moment of observation they have already different residence times in the reactor. The

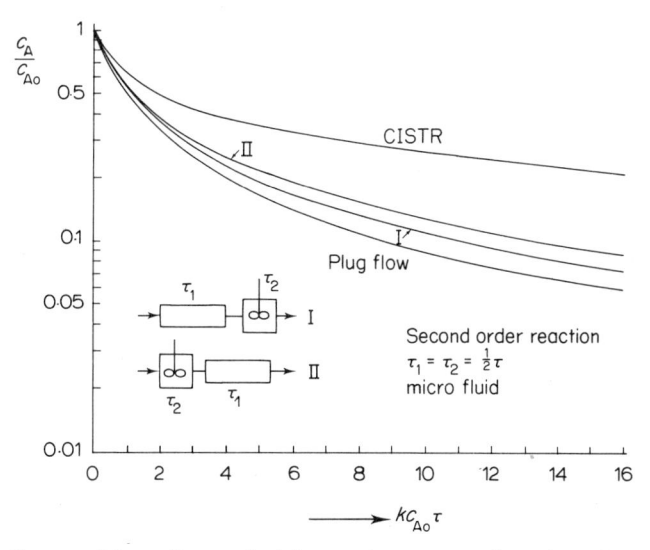

Fig. V.3 Influence of the earliness of mixing on the reactant slip c_A/c_{A0} of a second order reaction. Cascade I and II have the same RTD but differ in earliness of mixing

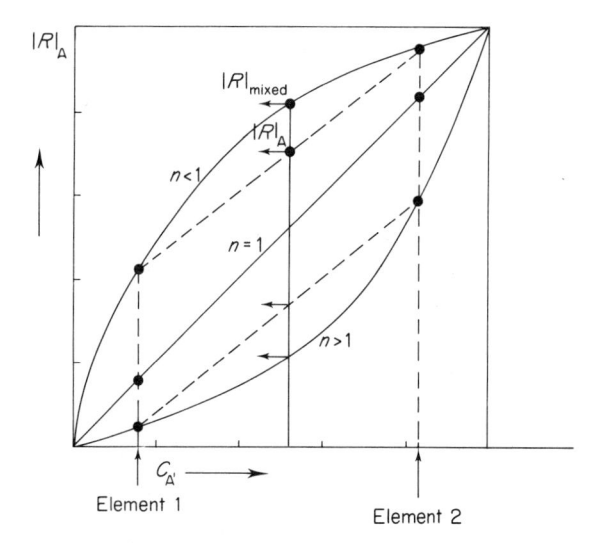

Fig. V.4 Influence of mixing on the average reaction rate of two equal volume fluid elements 1 and 2 that will leave the reactor at the same moment in the future

average instantaneous conversion rate for the two elements $|\bar{R}|_A$ is different from the conversion rate obtained if the reactants were actually mixed:

$$|\bar{R}|_A = \frac{kc_{A1}^n + kc_{A2}^n}{2} \quad \text{and} \quad |R|_{\text{mixed}} = k\left(\frac{c_{A1} + c_{A2}}{2}\right)^n.$$

This leads to the conclusion that:

$$|\bar{R}|_A < |R|_{mixed} \quad \text{for } n < 1$$
$$|\bar{R}|_A = |R|_{mixed} \quad \text{for } n = 1$$
$$|\bar{R}|_A > |R|_{mixed} \quad \text{for } n > 1.$$

So only for a first order reaction does the earliness of mixing have no influence on the reaction, as can also be seen in figure V.4. The intermixing of elements 1 and 2, which will leave the reactor at the same moment, is not detectable by measuring the RTD function, so that for non-first order reactions the RTD provides insufficient data for a unique prediction of the conversion.

Nevertheless, it is possible to calculate boundaries for the conversion and the selectivity of a chemical reaction with premixed reactants carried out in a chemical reactor with a given RTD, provided the kinetics of the reaction are known. These boundaries take into account the combined effect of the state of agglomeration and the earliness of mixing which are together called the micromixing. Fortunately often the boundaries are rather close together, so that the effect of micromixing is of only minor importance. The boundaries become wider apart if the spread in residence time increases. They coincide for plug flow and are at their maximum distance for a mixed vessel.

V.2 BOUNDARIES TO MICROMIXING PHENOMENA

The method of calculating the boundaries for arbitrary RTD functions was first given by Zwietering [5] and will be explained in Section V.2.2. First we will consider the ideal plug flow reactor and the CSTR with *premixed* reactants. For reactors with *separate* feed inlet points for different reactants the situation is completely different and the micromixing effects may be as important for plug flow reactors as for CSTRs. In the extreme case of agglomeration of the fluids hardly any reaction can take place because the different reactants remain in different aggregates and cannot come into contact with each other.

V.2.1 The model tubular and tank reactors

We will first consider the ideal reactors. For a plug flow reactor with premixed reactants the concept of micromixing is irrelevant because all elements that pass through the reactor can only mix with elements that have exactly the same age. Otherwise the plug flow assumption is violated.

In a continuous stirred tank reactor we can consider two extreme states of aggregation: the completely segregated fluid and the micro fluid. These two limits are equivalent to the latest possible mixing for complete segregation and to the earliest possible mixing for the microfluid, as was shown by Zwietering [5].

For the mixed tank reactor the calculations for conversion and selectivity in the two extremes of micromixing are simple. For the microfluid or maximum mixedness, assuming constant density, the conversion is identical to that of the CISTR which we described in Chapter II. For the macrofluid the situation is different. The average conversion of the aggregates leaving the reactor should be evaluated. Each aggregate in the outlet fluid can be considered as a batch reactor with a reaction time equal to its residence time.

The fraction of aggregates having a residence time between t and Δt is $E\,\Delta t$. This follows directly from the definition of the E diagram given in Chapter IV. If the remaining reactant concentration after conversion in a batch reactor with a residence time of t seconds is $c_A^{Batch}(t)$, the average concentration in the outlet of the segregated continuous flow reactor CSSTR is

$$\frac{\bar{c}_A}{c_{Ao}} = \frac{1}{c_{Ao}}\int_0^\infty c_A^{Batch}(t)\,E\,dt. \tag{V.1}$$

The residence time distribution function for the CSTR is equal to

$$E = \frac{1}{\tau}e^{-t/\tau}.$$

Therefore the conversion in a CSSTR is given by

$$1 - \zeta_A = \frac{\bar{c}_A}{c_{Ao}} = \frac{1}{c_{Ao}}\int_0^\infty c_A^{Batch}(t)\frac{1}{\tau}e^{-t/\tau}\,dt. \tag{V.2}$$

This equation can be used in the complete segregation case in a continuous tank reactor, both for conversion and selectivity predictions and for any type of reaction.

A special case is a simple first order chemical reaction for which

$$\frac{c_A^{Batch}(t)}{c_{Ao}} = e^{-kt}.$$

Inserting this equation into equation (V.2) leads to

$$\frac{\bar{c}_A}{c_{Ao}} = \frac{1}{\tau}\int_0^\infty e^{-kt}e^{-t/\tau}\,dt = \frac{1}{1 + k\tau}. \tag{V.3}$$

This is the same result as for the ideally mixed reactor, the CISTR processing a micro fluid, and is given in Chapter II. Equation (V.3) demonstrates again that for a first order reaction the micromixing effects are absent or that the boundaries for the conversion coincide. Some typical results for the two extremes for other reaction orders are given in Table V.1. Figure V.5 and figure V.6 show the boundaries for first order, second order, half order and zero order reactions. It can be seen that the effect for reaction orders smaller than one is reversed relative to those larger than one. This can be understood from figure V.4 for higher order reactions: mixing the elements first before reaction occurs results in a lower average reaction rate. For reaction orders smaller than one the reverse occurs. A special case is the zero order reaction. In a micro fluid in a tank the conversion can reach 100% for a finite average residence time $\tau = c_{Ao}/k$ as follows also from Table V.1. For a segregated tank reactor conversion in principle will never reach 100%. Zero order reactions are therefore very sensitive to micromixing at high

Table V.1 Conversion equations for stirred vessels with microfluids, or CISTRs, and with completely segregated fluids, or CSSTRs

Kinetics	Micro fluid	Macro fluid with complete segregation
$R_A = -kc_A$	$\dfrac{c_A}{c_{Ao}} = \dfrac{1}{1+N_r}$	$\dfrac{\bar{c}_A}{c_{Ao}} = \dfrac{1}{1+N_r}$
$N_r = k\tau$		
$R_A = -k$	$\dfrac{c_A}{c_{Ao}} = 1 - N_r,\ N_r \leqslant 1;\ \dfrac{c_A}{c_{Ao}} = 0,\ N_r \geqslant 1$	$\dfrac{\bar{c}_A}{c_{Ao}} = 1 - N_r + N_r e^{-1/N_r}$
$N_r = \dfrac{k\tau}{c_{Ao}}$		
$R_A = -kc_A^{0.5}$	$\dfrac{c_A}{c_{Ao}} = 1 - \dfrac{N_r^2}{2}[-1 + \sqrt{(1 + 4/N_r^2)}]$	$\dfrac{c_A}{c_{Ao}} = 1 - N_r\left[1 - \dfrac{N_r}{2}(1 - e^{-2/N_r})\right]$
$N_r = \dfrac{k\tau}{\sqrt{c_{Ao}}}$		
$R_A = -kc_A^2$	$\dfrac{c_A}{c_{Ao}} = \dfrac{-1 + \sqrt{(1 + 4N_r)}}{2N_r}$	$\dfrac{\bar{c}_A}{c_{Ao}} = \dfrac{-1}{N_r} e^{1/N_r} Ei(-1/N_r)$*
$N_r = k\tau c_{Ao}$		
$R_A = -kc_A^n$	$\left(\dfrac{c_A}{c_{Ao}}\right)^n N_r + \dfrac{c_A}{c_{Ao}} - 1 = 0$	$\dfrac{\bar{c}_A}{c_{Ao}} = \dfrac{1}{\tau}\displaystyle\int_0^\infty [1 + (n-1)c_{Ao}^{n-1} kt]^{1/1-n} e^{-t/\tau}\,dt$
$N_r = c_{Ao}^{n-1} k\tau$		

* $Ei(-x) = \displaystyle\int_x^\infty \dfrac{e^{-y}}{y}\,dy$

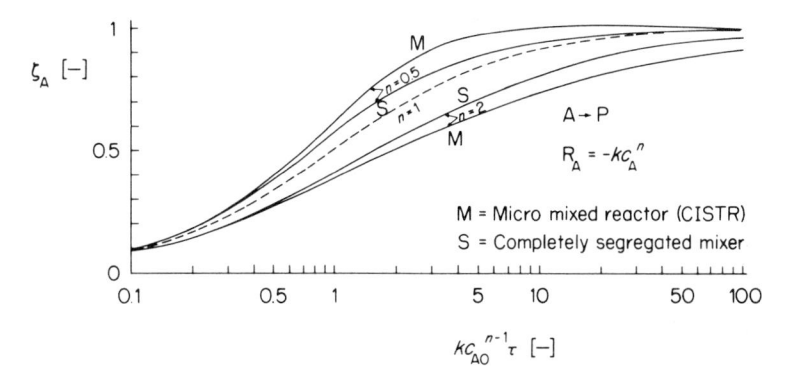

Fig. V.5 Influence of micromixing on the conversion of a nth order chemical reaction in a stirred vessel

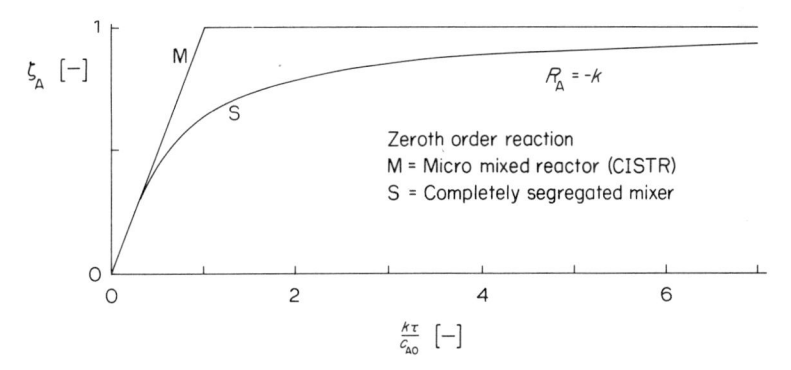

Fig. V.6 Influence of micromixing on the conversion of a zero order chemical reaction in a stirred vessel

conversions, so they are attractive as model reactions for assessing micromixing. Regretfully many reactions which at high reactant concentrations appear to be zero order, increase in apparent reaction order at high conversions. This is the case, for example, if the reaction can be described by Langmuir or Michaelis–Menten kinetics.

With complex reactions in a stirred tank the yield and selectivity is also influenced by the micromixing. As an example, figure V.7 shows the reaction paths for a set of parallel reactions which are of different order in reactant A. In case of ideal mixing, a CISTR, and a high degree of conversion, the production of P is favoured because the concentration of A is low in the entire reactor volume; even pure P can be produced at $\tau = \infty$ although at an infinitely slow rate. In case of segregation, each volume element (or each droplet) is essentially a small batch reactor in which the concentration of A diminishes with time. In the initial stage the undesired product X is formed as well as P, so that the product stream consisting of drops of various ages always contains both species. Figure V.8 gives

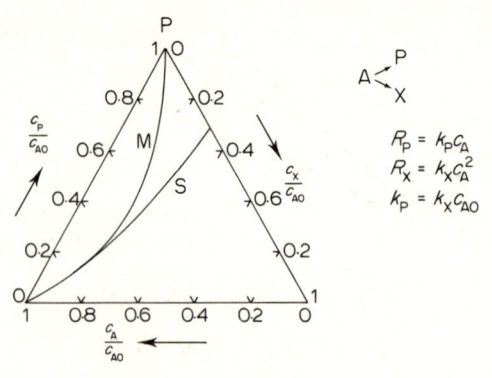

Fig. V.7 Reaction path for parallel reactions of different order in a segregated tank, a CSSTR (curve S) and an ideally mixed micro fluid, a CISTR (curve M)

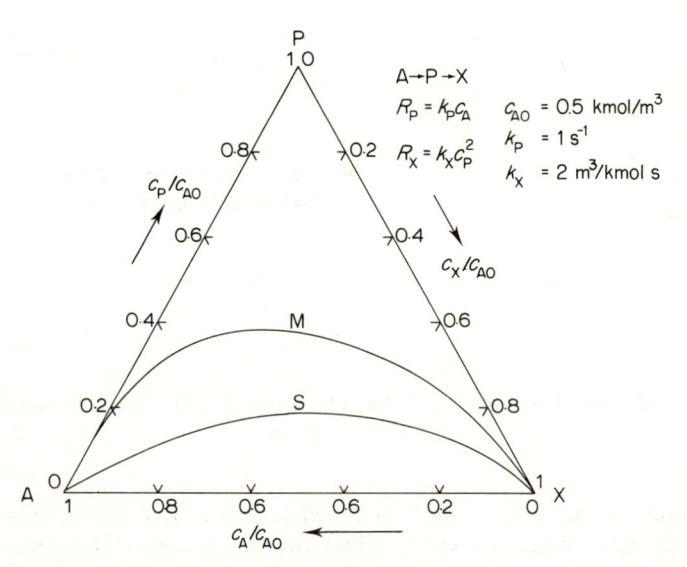

Fig. V.8 Reaction path for a consecutive reaction in a tank reactor with segregated flow, a CSSTR (curve S) and an ideally mixed micro fluid, a CISTR (curve M)

the result for a consecutive reaction scheme. The by-product formation reaction is second order in P, i.e. $R_X = k_X c_P^2$. Therefore the concentration of P should be kept as low as possible. This can be better realized in the CISTR than in the CSSTR. Therefore, at the same average conversion level less by-product is formed in the CISTR.

For a first order consecutive reaction scheme A → P → X, there is no difference in performance between a CISTR and CSSTR. The rate of production of P in a batch reactor would be

$$R_P = \frac{dc_P}{dt} = k_P c_{A_0} e^{-k_P t} - k_X c_P$$

which results in
$$\frac{c_P}{c_{Ao}} = \frac{k_P}{k_P - k_X}[e^{-k_P t} - e^{-k_X t}]$$
substitution of this equation into equation (V.2) leads after solution to
$$\frac{\bar{c}_P}{c_{Ao}} = \frac{k_P \tau}{(1 + k_P \tau)(1 + k_X \tau)},$$
which is the same result as given for the CISTR in Chapter III.

For *intermediate* degrees of micromixing no unique solution can be given because both the degree of segregation and the earliness of mixing may interfere as more or less independent factors. Therefore one has to rely on models for intermediate micromixing which have to be confirmed experimentally. These models will be discussed in Section V.3.

Example V.1 *Emulsion polymerization in a continuous tank reactor*

A reactant A ($M_A = 104$ kg/kmol) is to be polymerized in the dispersed phase in a stirred tank reactor. The results of batch experiments are available for the reaction conditions to be used. The following wholly empirical conversion rate formula has been derived from these results:

$$|R| = kc_A^{3/2}, \text{with } k = 2.5 \times 10^{-4} \text{s}^{-1} \left(\frac{\text{kmol}}{\text{m}^3}\right)^{-1/2}.$$

These experiments furthermore yielded the information that the spread in the degree of polymerization, molecules of monomer per molecule of polymer, is rather small and that the average degree of polymerization can be approximated by:

$$p(t) = 14.1\sqrt{t},$$

where t is the reaction time in the batch reactor in seconds.

It is required to calculate the size of the tank reactor, the average degree of polymerization and the distribution of p for a production of 10 tonnes of polymer per day with a relative degree of conversion $\zeta_A = 0.9$, under the assumption that coalescence between the drops does not occur. The feed of the dispersed phase consists of pure A, and the density of this phase during the reaction remains constant at a value of 832 kg/m³. The volume fraction of the dispersed phase in the reactor is 0.165.

The material balance for a batch or an ideal tubular reactor together with the given rate equation yields for the relative degree of conversion $\zeta_A(t)$, after a reaction time t,

$$\zeta_A(t) = 1 - \left(\frac{2}{2 + kt\sqrt{c_{Ao}}}\right)^2$$

On the basis of equation (V.2) the average conversion ζ_{A1} in the stream leaving the reactor, in this case a CSSTR, becomes

$$\zeta_A = \int_0^\infty \left\{ 1 - \left(\frac{2}{2 + k\tau\sqrt{c_{Ao}}}\right)^2 \right\} e^{-t/\tau} \, d(t/\tau).$$

From this equation it is found for $\zeta_A = 0.9$ that $k\tau\sqrt{c_{Ao}} = 15.4$, from which it follows that $\tau = 21\,600\,\text{s} = 6$ hours. The required production of polymer is $0.116\,\text{kg/s}$ and is equal to $0.9 \times \rho_A V_r/\tau$; it follows that the volume of the dispersed phase $V_r = 3.34\,\text{m}^3$, and that the total liquid volume in the reactor has to be $3.34/0.165 = 20.2\,\text{m}^3$.

A drop leaving the reactor after a residence time t has an average degree of polymerization $p(t) = 14.1\sqrt{t} = 2070\sqrt{(t/\tau)}$. The average degree of polymerization $\langle p \rangle$ can be calculated by equation (V.1):

$$\langle p \rangle = \int_0^\infty 2070\sqrt{(t/\tau)}\, e^{-t/\tau}\, d\frac{t}{\tau} = 2070\frac{\sqrt{\pi}}{2} = 1830.$$

If we write this equation in terms of p only, we obtain

$$\langle p \rangle = \int_0^\infty \frac{2p^2}{2070^2} e^{-(p/2070)^2}\, dp.$$

Because we may also write $\langle p \rangle = \int_0^\infty p f(p)\, dp$, in which $f(p)$ is the frequency function of the degree of polymerization, we find for

$$f(p) = \frac{2p}{(2070)^2} e^{-(p/2070)^2}.$$

This function has been plotted in figure V.9 together with the cummulative degree of polymerization function

$$F(p) = 1 - e^{-(p/2070)^2}.$$

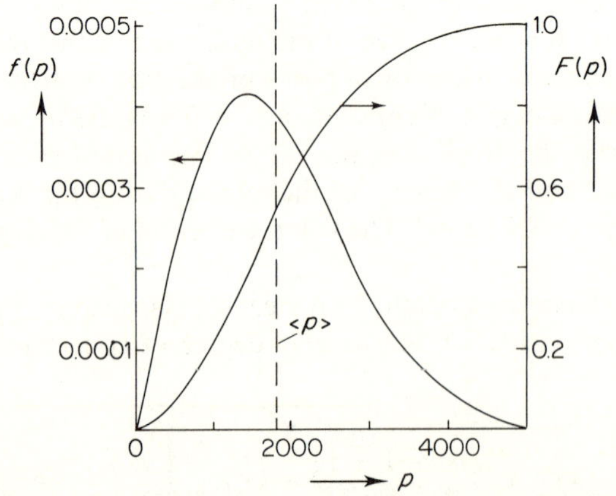

Fig. V.9 Distribution function $f(p)$ and cummulative distribution function $F(p)$ of the degree of polymerization or chainlength p. Example V.1

These chain length distributions are only approximate, since the degree of polymerization in the batch reactor may have had a relatively wide spread itself. This effect, however, is to some extent obliterated by the very large spread in residence time of the tank reactor.

It is also questionable whether the assumption of no coalescence between the different emulsion drops is valid, particularly in view of the long average residence time in this system. Regretfully it is not possible to make a parallel calculation for a tank reactor in which the reaction mixture is an ideally mixed microfluid. The data required for this calculation cannot be derived from simple batch experiments because the conversion rate and polymer growth do not depend only on the reactant concentration but also on the product distribution itself. Since the latter is entirely different in a completely mixed tank reactor a profound knowledge of the kinetics of all reactions involved should be available. Such models should involve, for example, the initiation, propagation and termination reactions, the influence of viscosity on kinetic parameters such as the gel or Tromsdorff effect, etc.

Only few reliable experimental investigations have been published on the influence of micromixing on polymerization reactions, despite its importance for industrial polymerization. Harada *et al.* [6] studied the polymerization of styrene in a CSTR and Nagasubramanian and Greassly [7] that of vinyl acetate. Reviews on these subjects were given by Gerrens [8] and Villermaux [9].

V.2.2 Boundaries for micromixing for reactors with arbitrary RTDs

The limit conditions for micromixing for a reactor with an arbitrary RTD can still be calculated although the factors RTD, segregation and earliness of mixing interact. This was demonstrated by Zwietering [5]; his method at least in principle allows the determination of the boundaries of the conversion and the selectivity for a reactor with an arbitrary RTD and reactions with known kinetics. Zwietering defined these boundaries as the states of maximum and minimum mixedness. Figure V.10 demonstrates the concept of *minimum mixedness*. A reactor with an arbitrary RTD can be simulated by a plug flow reactor with exit streams of negligible volume located along the reactor so that the RTD exactly matches the RTD of the reactor to be simulated. Because of the plug flow nature of the main flow and the premixing of the feed we may consider the split streams as already separated at the reactor inlet. Therefore no elements of different residence times (or 'age') are mixed up except when they have left the reactor. Therefore this model contains both complete segregation and the latest possible mixing, namely only at the outlet of the reactor. Therefore the conversion is described by equation (V.1):

$$1 - \zeta_A = \frac{\bar{c}_A}{c_{Ao}} = \frac{1}{c_{Ao}} \int_0^\infty c_A^{\text{Batch}}(t) E \, dt.$$

With similar reasoning the other extreme of *maximum mixedness* can be derived.

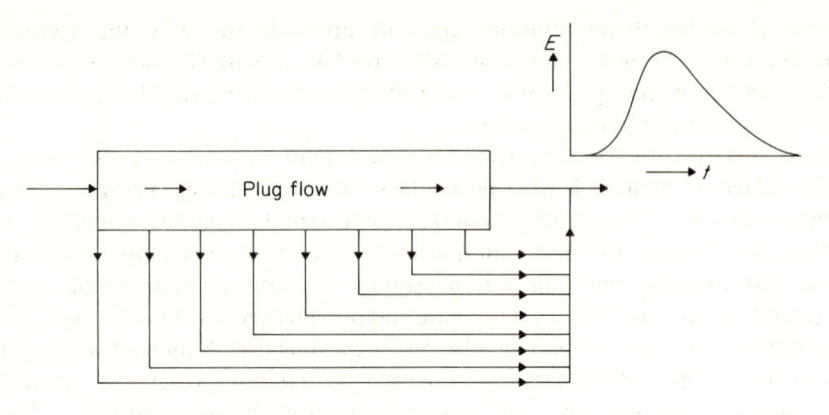

Fig. V.10 Minimum mixedness according to Zwietering [5]. Side streams are situated along the plug flow reactor to match a given RTD

This is illustrated in figure V.11. Now the feed inlets are distributed along the plug flow reactor, the volumetric flow rates are chosen to match the RTD curve.

The side inlet streams are assumed to mix ideally in the radial direction. In this arrangement the inlet elements mix as early as possible with the elements with which they jointly leave the reactor. We denote by t_e the period that a fluid element will still be in the reactor: t_e ranges from zero at the outlet to its maximum at the first inlet point. To derive a mass balance for a reactant A over the reactor element corresponding to the values of t_e and $t_e + dt$, see figure V.12, we first will formulate the different volumetric flow rates. For the side entry flow rates matching the E diagram we find

$$\Phi_v E(t_e)\, dt_e = \Phi_v E(t)\, dt.$$

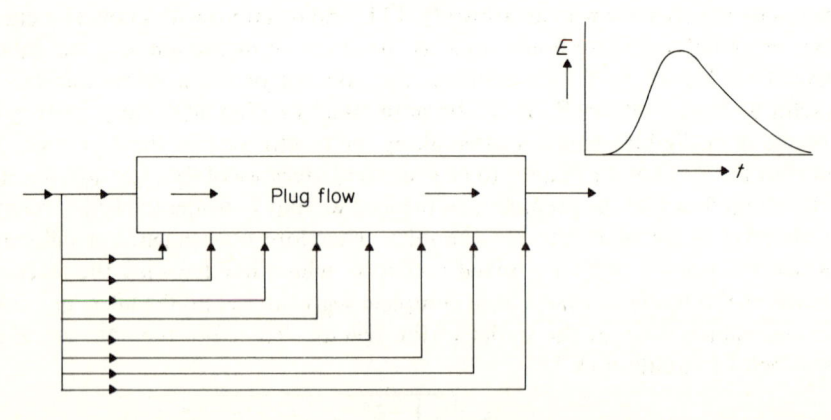

Fig. V.11 Maximum mixedness according to Zwietering [5]. Side streams are situated along the plugflow reactor to match a given RTD. Fluid elements entering the reactor are immediately mixed with the elements that will leave the reactor at the same moment in the future

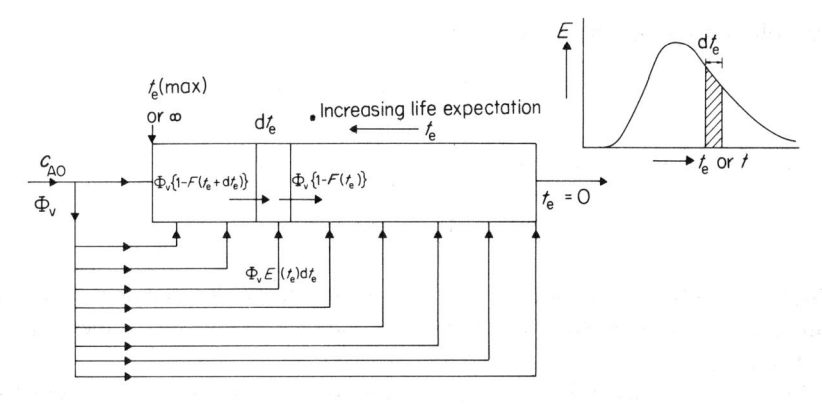

Fig. V.12 The mass balance for the maximum mixedness

The accumulated flow rate in the plug flow region is

$$\Phi_v \int_\infty^{t_e} E(t_e)\,dt_e = \Phi_v(1 - F(t_e)) = \Phi_v(1 - F(t)).$$

In these equations E and F are the E and F functions of the residence time distribution of the reactor. The elements of the mass balance can now be formulated:

inflow at $t_e + dt_e$: $\Phi_v[1 - F(t_e + dt_e)]c_A(t_e + dt_e)$
outflow at t_e : $\Phi_v[1 - F(t_e)]c_A(t_e)$
side entry by inflow   : $\Phi_v E(t_e)c_{Ao}\,dt_e$
disappearance by reaction: $-R_A(c_A(t_e))[1 - F(t_e)]\Phi_v\,dt_e$,

in which E and F are the E and F diagram of the RTD respectively. The mass balance then reads

$$\frac{dc_A(t_e)}{dt_e} = -R_A(c_A(t_e)) - \frac{E(t_e)}{1 - F(t_e)}[c_{Ao} - c_A(t_e)]. \tag{V.4}$$

The boundary condition is

$$\frac{dc_A(t_e)}{dt_e} = 0 \quad \text{for} \quad t_e \to \infty.$$

For very large life expectations, $t_e \to \infty$, experimental data are often difficult to obtain in practice, but very small fractions with extremely long residence times have little effect on the averaged total conversion. Usually elements with $t_e > 4\tau$ can be neglected and $t_e = 4\tau$ can be taken as the boundary for the integration of equation (V.4). For a first order reaction equation (V.4) leads to

$$\frac{dc_A(t_e)}{dt_e} = kc_A(t_e) + \frac{E(t_e)}{1 - F(t_e)}(c_A(t_e) - c_{Ao}).$$

The solution was given by Zwietering [5]:

$$c_A(t_e) = \frac{c_{Ao} e^{kt_e}}{1 - F(t_e)} \int_{t_e}^{\infty} E(t) e^{-kt} \, dt$$

so we have at the reactor exit, where $t_e = 0$,

$$\frac{c_A}{c_{Ao}} = \int_0^{\infty} E(t) e^{-kt} \, dt.$$

This result is identical to the case of minimum mixedness for a first order reaction and shows once more that first order reactions are insensitive to the state of micromixing.

Zwietering [5] gave a few examples for a residence time distribution predicted for three mixed tanks in series, which have been included in figure V.13. In this figure we take as a base case a reactor with a residence time distribution identical to that of three tanks in series. Curves 1 to 4 give the conversion prediction for a second order reaction for different states of micromixing. For comparison also the conversion predictions for an ideal plug flow reactor, a CISTR and a CSSTR have been added. Curve 1 gives the conversion for the maximum mixedness case, curve 2 for three CISTRs in series, which is in fact the conversion prediction as calculated for the tanks in series model, curve 3 is the conversion prediction for three CSSTRs with mixing in the transfer lines between the tanks and curve 4 represents the prediction for the limit of complete segregation. It should be noted that the model of three CISTRs in series does not give the conversion boundary of maximum mixedness. With the given RTD of three tanks in series earlier mixing

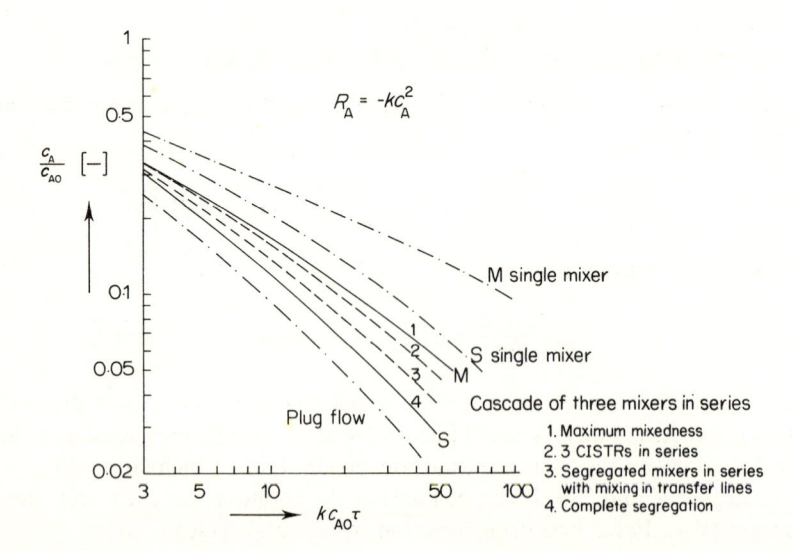

Fig. V.13 Influence of micromixing on the conversion of a second order chemical reaction in a cascade of three mixed tank reactors. For comparison results for the plug flow reactor and single tank are also given

is possible than realized in a cascade of three CISTRs. The larger the spread in residence time, the wider the range of possible conversions between the boundaries for maximum and minimum mixedness.

This is illustrated in figure V.14 for a second order and a half order reaction. It is assumed that the residence time distribution can be represented by that of the tanks in series model and therefore the number of tanks in series is used as a parameter for the spread in residence time. Generally the effect of the RTD is more important than the exact state of micromixing and uncertainties due to an unknown state of micromixing can be tolerated taking into account the usually limited accuracy of kinetics.

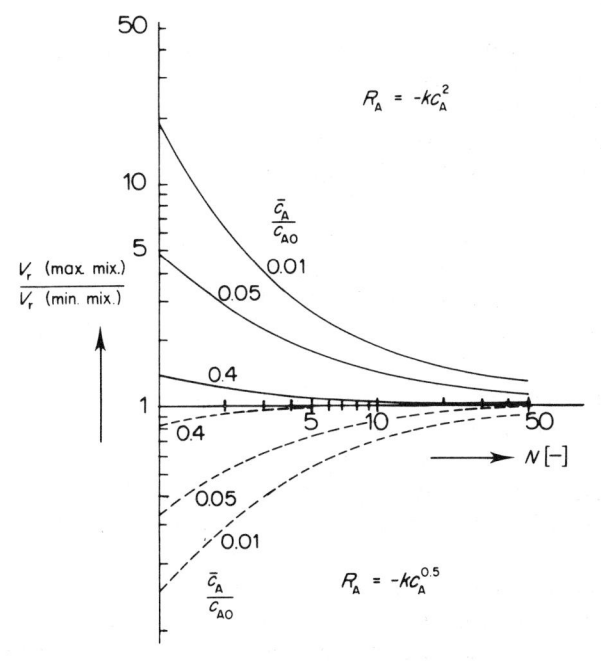

Fig. V.14 Ratio of reactor volume required for maximum and minimum mixedness extremes as a function of the macromixing parameter. The RTD of the reactor is assumed to be identical to that of N tanks in series and therefore N is the macromixing (or RTD) parameter

With premixed feedstocks micromixing effects are most important in single mixed tanks, so that henceforth we will concentrate on the behaviour of CSTRs both for premixed and non-premixed systems.

V.3 INTERMEDIATE DEGREE OF MICROMIXING IN CONTINUOUS STIRRED TANK REACTORS

Already the case of intermediate degree of micromixing in a CSTR is very complex. Different situations should be distinguished as shown in Table V.2. Reactants can be premixed or not, one or more phases can be present and the

Table V.2 Classes of problems in a stirred tank reactor with one or more fluid phases

	Premixed reactants	Non-premixed reactants
Single phase	A	B
Two or more phases	C reaction in one phase only	D reaction in one phase only
	E reactions in two or more phases	F reaction in two phases

reactions can take place in one or more phases. For two phase systems with premixed feedstocks and only one reaction phase without transfer of reactants to the other phase, the single phase description can be used. Such systems are seldom used in practice. An inert phase is introduced sometimes to absorb the reaction heat, e.g. in exothermic suspension polymerization reactions. The same holds for system D. Such systems, however, may be very convenient for measuring the degree of micromixing within one phase of a multiphase system. For example, an oil feed stream containing only reactant A is fed, separately from another oil feed stream containing only reactant B, to a water–oil suspension in a stirred vessel. By measuring the conversion rate the state of micromixing within the oil phase can be followed.

In this case the micromixing is usually expressed in a coalescence–redispersion rate. In systems E and F the exchange rate from one phase to the other can play an important role and via this exchange the degree of micromixing of both phases may be mutually influenced. Other effects also may play a role in these systems such as enhancement of the mass transfer between the phases due to rapid chemical reactions. This will be explained in Chapter VII. Such enhancements may also occur in system B (single phase systems) with separated reactants feed inlets. Due to aggregate formation and fast reaction, mass exchange between the aggregates and their surroundings may be accompanied by reactions in parallel; this may enhance the transfer rate by increasing the concentration gradients. This is illustrated in figure V.15. Often these effects are neglected in micromixing models. The models that have been set up to describe intermediate mixing can be divided into formal models and agglomeration models. We will discuss a few examples from both categories.

V.3.1 Formal models

Formal micromixing models usually do not give a clear physical picture of the actual phenomena but only try to cover the span between maximum and minimum mixedness by one or two adjustable parameters. These parameters are determined experimentally and depend on the geometry of the experimental installation. A very simple approach is to express the degree of micromixing by the position of the conversion with respect to the prediction of the boundaries for

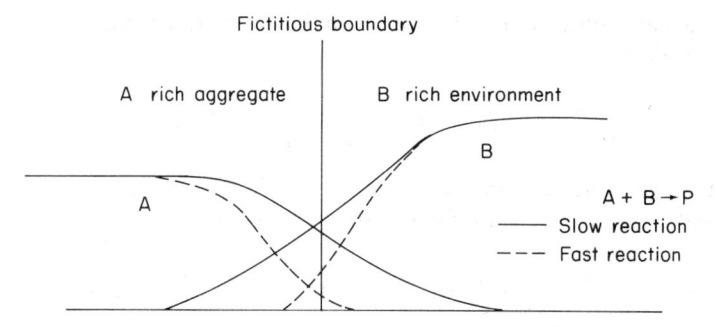

Fig. V.15 The influence of a fast chemical reaction on the mixing process. Due to the fast reaction the concentration profiles become steeper and the molecular diffusion faster. Selectivity may also be influenced

maximum and minimum mixedness, after measuring the conversion of a chemical reaction with non-first order kinetics. A disadvantage is that the resulting micromixing parameter is just a fitting parameter which can have different values, e.g. for different conversions or reaction orders.

A somewhat more refined concept for a mixed vessel and its feed inlet has been proposed by Weinstein and Adler [10] and corrected and improved by Villermaux and Zoulalian [11]. Lumps of feed are entering the mixed vessel in the entering environment which is supposed to be in a state of complete segregation. The material from the entering zone is transferred to the exit zone which is supposed to be in a state of maximum mixedness, and partly directly to the exit. The model is indicated in figure V.16. The maximum residence time in the segregated zone t_s forms the principal model parameter which can be varied between zero and any value up to $t_s = \infty$ to allow for the maximum and minimum mixedness respectively. The conversion can be calculated from a mass balance over the two environments if the kinetic relation for the chemical reaction is given. If we call $c_A^{\text{Batch}}(t)$ the concentration of the reactant in a batch reaction with a reaction time t, the concentration at the end of the segregated section is $c_A^{\text{Batch}}(t_s)$. The average side stream concentration for this section is

$$c_A^{\text{Seg}} = \int_0^{t_s} c_A^{\text{Batch}}(t) E(t) \, dt \bigg/ \int_0^{t_s} E(t) \, dt$$

in which $E(t)$ is the E diagram of the RTD. The concentration at the exit of the

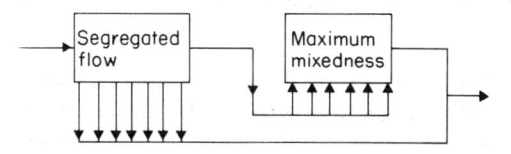

Fig. V.16 Consecutive one parameter model for intermediate micromixing according to Weinstein and Adler [10]

maximum mixedness section follows from the integration from $t_e = 0$ to $t_e = t_s$ of equation (V.4):

$$\frac{dc_A^{mm}(t_e)}{dt_e} = -R(c_A^{mm}(t_e)) - \frac{E(t_e)}{1 - F(t_e)}[c_A^{Batch}(t_s) - c_A^{mm}(t_e)].$$

The average outlet concentration of the whole system then follows from the weighted addition of the outflow of the two sections:

$$\frac{\bar{c}_A}{c_{Ao}} = \frac{1}{c_{Ao}}\left[\int_0^{t_s} c_A^{Batch}(t)E(t)\,dt + c_A^{mm}(t_s)\int_{t_s}^{\infty} E(t)\,dt.\right] \quad (V.5)$$

This equation is valid for any RTD. For a stirred vessel the $E(t)$ diagram is given by

$$E(t) = \frac{1}{\tau}e^{-t/\tau}$$

and the F diagram by

$$F(t) = 1 - e^{-t/\tau}.$$

These equations should be inserted in equation (V.5) in the case of a CSTR. Figure V.17 shows a few results for a second order reaction.

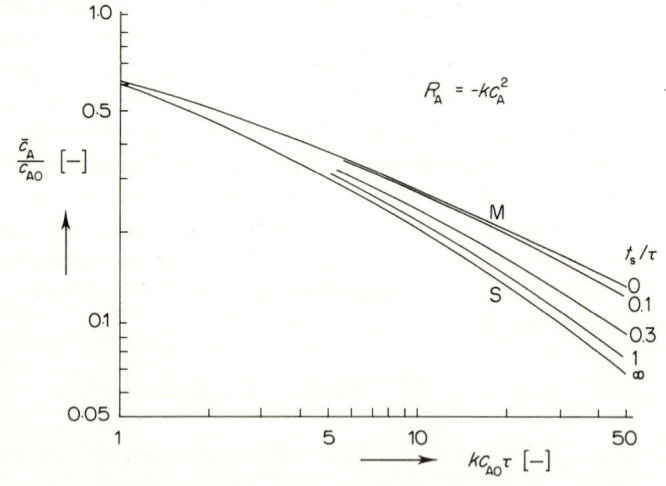

Fig. V.17 The conversion prediction for a second order reaction in a tank reactor with the consecutive micromixing reactor model of Weinstein and Adler [10]

In some polymerization reactors producing a viscous fluid it may be better to assume the leaving environment in a segregated state—the reverse of the previous model. Such models have also been published by Chen and Fan [12]. Many other formal models have been published such as the statistical fluid mixing network model of Asbjørnsen [13] and models based on the degree of segregation as introduced by Danckwerts in a pioneering article on micromixing [14]. We will not discuss these formal models any further because they do not present a clear physical picture of the micromixing process.

V.3.2 Agglomeration models

The agglomeration models describe the degree of micromixing in terms of the exchange with the environment of material from agglomerates with a fixed identity and a variable life-span. Although again we will consider only mixed vessels, agglomeration models also have been used to describe the mixing in other systems. Some of the agglomeration models can be used for both single and more phase systems. Especially with single flow systems, the nature of the agglomerates should be carefully taken into account because here the agglomerates, their size and life span are related to the hydrodynamic conditions during the process of mixing. This mixing process as described by Beek and Miller [15], takes place in three stages:

1. distribution of one fluid through the other to achieve an uniform concentration on a macro scale without decrease of local concentration gradients;
2. reduction of the size of the regions with uniform concentration;
3. mixing by molecular diffusion.

The rates of steps 1 and 2 depend strongly on the hydrodynamics and are of different natures in laminar and turbulent regions. In laminar flow the fluid elements will be stretched whereas in turbulent flow large eddies will be broken down to smaller ones. In simple agglomeration models only the third step is considered; the small agglomerates are created instantaneously and the time required for steps 1 and 2 is regarded to be zero. Sometimes only step 2 is considered as in the model of eroding aggregates of Plasari et al. [16]. Here we will concentrate on models taking into account only one single agglomerate type. With these models mixing states between maximum mixedness (or ideal mixing) and the minimum mixedness (or complete segregation) can be described in a stirred vessel by means of an exchange rate of the agglomerates. The faster this exchange rate the closer the maximum mixedness will be approached. This exchange rate must be compared to both the rates of the chemical reaction and the rate of the flow through the vessel. This is shown in Table V.3, as was originally suggested by David and Villermaux [17]. In this table the three characteristic times are compared:

$$\tau = \frac{V_r}{\Phi_v}, \quad \text{the average residence time}$$

$$t_r = \frac{1}{kc_{Ao}^{n-1}}, \quad \text{the reaction time for a } n\text{th order reaction}$$

$$t_m = \frac{l}{12D}, \quad \text{the micromixing time.}$$

Table V.3 Influence of three characteristic time constants on the state of mixing in a tank reactor [From David and Villermaux [17]. Reproduced by permission of *Journal de Chimie physique*]

	$t_m \ll t_r$	$t_m \approx t_r$	$t_m \gg t_r$
$t_m \ll \tau$	MM	PS	S
$t_m \approx \tau$	PS	PS	S
$t_m \gg \tau$	S	S	S

t_r = reaction time constant, τ = average residence time
t_m = micromixing time constant
MM = ideal micromixing, PS = partial segregation
S = complete segregation

t_m is a Fourier time for the agglomerates as introduced by Corrsin [4] and adapted by Nauman [18]. l is the characteristic size of the agglomerates. In the expression for t_m it is assumed that pure molecular diffusion is the transport mechanism for exchange of mass of an agglomerate and its surroundings. If a different mechanism occurs the expressions should be modified. If the micromixing time t_m is small in comparison to both τ and t_r the stirred vessel can be considered an ideal mixer, a CISTR. If t_m is not small in comparison to τ and/or t_r the system will behave as segregated or partially segregated and will be in a state of intermediate to minimum mixedness.

A major problem is the formulation of the rate of exchange of the aggregates with their surroundings, because it is difficult to define the surroundings exactly. This is even more complicated in a reactor with an arbitrary RTD. We will shortly discuss four approaches that have been proposed for stirred vessels: (1) random coalescence, (2) erosion of aggregates, (3) mass exchange with the average tank content (IEM model), and (4) enhanced mass exchange with the tank contents.

Random coalescence

In this approach the whole tank is considered to contain only aggregates. These aggregates coalesce with each other and after each coalescence, the two aggregates immediately split again into two new aggregates having the average composition of the two original aggregates. The whole process can be described by the coalescence frequency ω, the number of coalescences per unit of time as used e.g. by Curl [19]. This parameter has a clear physical meaning in an emulsion reactor but is more or less a fitting parameter if applied to describe micromixing in a single phase system. A statistical approach can be used to keep record of the different coalescence events, the so-called Monte Carlo technique by which a random coalescence can be simulated for a large number of agglomerates. Figure V.18 gives typical results for a second order chemical reaction as was calculated by Spielman and Levenspiel [20].

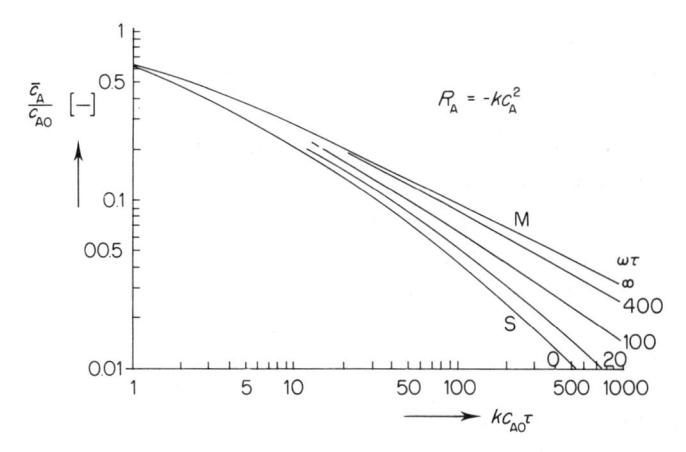

Fig. V.18 Influence of intermediate mixing on the conversion of a second order chemical reaction in a tank reactor as predicted by the random coalescence model calculated by Spielman and Levenspiel [20]. $\omega\tau$ = average number of coalescences experienced by the individual drops or agglomerates [From Spielman and Levenspiel [20]. Reproduced by permission of Pergamon Press Ltd]

Erosion of aggregates

In this approach the agglomerates are considered to be large lumps of fluid of dimension L_c, originating from the inlet streams. L_c is small in comparison with the vessel diameter but large in comparison with the smallest scales of turbulence where molecular diffusion gives a rapid elimination of concentration gradients. The exchange now does not occur between the agglomerates, but the agglomerates are eroded away to produce a well mixed microfluid. Such a model has been put forward by Plasari et al. [16].

Mass exchange with the average tank content (IEM model)

In this model a single typical agglomerate is followed from its birth at the feed inlet. This agglomerate exchanges mass with the average tank content. The average tank content is derived from the residence time distribution for the CSTR and from the concentration in typical aggregates with different residence times, similar to equation (V.2). This requires an iteration procedure.

Enhanced mass exchange with the tank contents

In this approach the chemical enhancement during diffusion of a reactant in an agglomerate is taken into account. This enhancement is specially important for vessels with separate reactant inlet points and has already been visualized in figure V.15. The agglomerate structure can be considered as an intermediate between a homogeneous and a heterogeneous system. This enhancement more often occurs in two phase systems as will be explained in Chapter VII. Examples of models including enhancement in micromixing have been given by Bourne, Kozicki and Rys [25].

V.3.3 Model for micromixing via exchange of mass between agglomerates and their 'average' environment, the IEM model

Especially in stirred tank reactors the simplification of considering only the exchange of a typical agglomerate with the average field of all the other agglomerates is a very useful one, because the conversion can be calculated without Monte-Carlo techniques or other complicated techniques such as those given by Curl [19].

The IEM model can be used for single phase systems, but also for dispersions if an appropriate mass exchange relation is introduced. We define the age t_α of an agglomerate as the time this agglomerate is already in the reactor immediately after its birth at the feed inlet. We now can write a balance for component A within this agglomerate:

$$\frac{dc_A}{dt_\alpha} = \frac{\bar{c}_A - c_A}{t_m} + R_A \tag{V.6}$$

in which R_A is the production rate of A inside the agglomerate, t_m the time constant for mass exchange between agglomerate and the rest of the reactor per unit driving force, the micromixing parameter, and $\bar{c}_A$ the average concentration of A in the reactor. For the CSTR we have

$$\bar{c}_A = \frac{1}{\tau} \int_0^\infty c_A(t_\alpha) e^{-t_\alpha/\tau} dt_\alpha \tag{V.7}$$

in which $c_A(t_\alpha)$ is the concentration inside an agglomerate with an age t_α. Equation (V.7) is similar to equation (V.1) but here $c_A(t_\alpha)$ is not determined uniquely by the reaction rate and the residence time, but also by the exchange parameter t_m. Furthermore, in equation V.7 the age distribution function should be inserted. For a stirred vessel this function is identical to the RTD function, because in the exit stream the same fluid is present as anywhere in the mixed vessel. From equations (V.6) and (V.7) $\bar{c}_A$, which is also the concentration of A at the exit, can be found by iteration techniques. The state of micromixing in the stirred vessel model can now be modified by varying t_m. If $t_m \to 0$ the microfluid and the maximum mixedness limit is reached, while for $t_m \to \infty$ the system is completely segregated and in the state of minimum mixedness. In the intermediate cases, to determine the effect of the micromixing state on the course of the chemical reaction, one should also consider the two other factors t_r and τ.

The micromixing parameter t_m can be related to hydrodynamical quantities. In homogeneous systems:

$$t_m = \frac{l^2}{12D} \tag{V.8}$$

in which l is the dimension of the agglomerates. By experimental techniques or from turbulence theories as presented by Kolmogoroff [1], Rosensweig [2], Batchelor [3] and Corrsin [4] the dimension l of the characteristic eddies, for which concentration differences can be eliminated only or mainly by molecular diffusion, can be estimated.

On a formal basis the IEM model can be compared to the random coalescence model as was shown by Ritchie and Tobgy [26] and by Villermaux [27]. By matching the relative amount of matter lost from a single non-reacting aggregate the relation is

$$t_m \approx \frac{4}{\omega}.$$

The same value for t_m is found if the two models are used to predict the conversion of an infinitely fast reaction of two reactants A and B, fed in stoichiometric amounts to a CSTR in separate streams. This will be demonstrated in the following example.

Example V.3 *Application of the IEM model to the conversion calculation for a second order reaction (after Villermaux [27]).*
We will consider the reaction $A + B \to P$, with $R_A = -kc_A c_B$. Two separate feed streams with the same volumetric flow rate and containing equal concentrations of either A or B are fed into a mixed tank reactor. The mixed cup concentrations of these combined streams would be c_{Ao} and c_{Bo} respectively. These are fictitious concentrations excluding reaction. We will consider two cases with and without actually premixing of reactant streams prior to feeding them to the reactor.

Premixed feeding of A and B
The mass balance for A in this case reads

$$\frac{dc_A}{dt} = \frac{1}{t_m}(\bar{c}_A - c_A) - kc_A c_B$$

and at $t = 0$ we have $c_A = c_{Ao}$ and $c_B = c_{Bo}$; moreover, the equation $c_A = c_B$ always holds. The average concentration of A is

$$\bar{c}_A = \int_0^\infty c_A \frac{1}{\tau} e^{-t/\tau} dt$$

or in dimensionless form, with $\theta = t/\tau$

$$\frac{dc_A/c_{Ao}}{d\theta} = -N_r(c_A/c_{Ao})^2 + \frac{\tau}{t_m}(\bar{c}_A/c_{Ao} - c_A/c_{Ao})$$

$$\bar{c}_A/c_{Ao} = \int_0^\infty (c_A/c_{Ao})e^{-\theta}d\theta.$$

With the boundary condition $c_A/c_{Ao} = (1 - \zeta_A) = 1$ at $\theta = 0$, and further the equations $c_A/c_{Ao} = c_B/c_{Bo}$ and $N_r = kc_{Ao}\tau$ both hold for this case of a second order reaction with a stoichiometric feed ratio of A and B.

The solution of these equations can be found by iteration. Some results are given in the shaded area of figure V.19. The boundaries of this area are given by the maximum mixedness and the minimum mixedness or complete segregation state respectively assuming premixed feed.

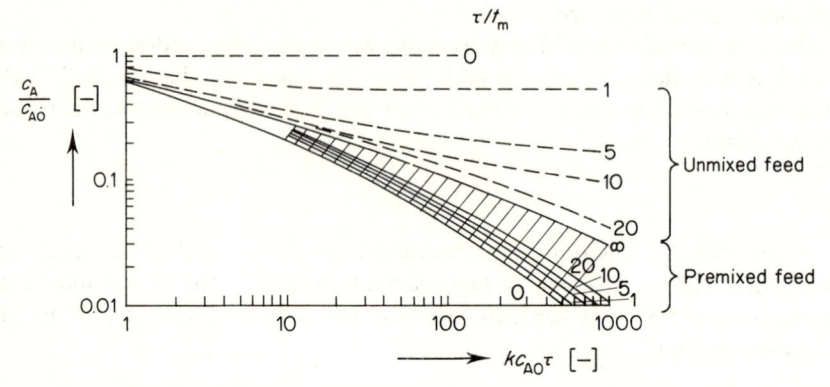

Fig. V.19 Influence of the micromixing parameter τ/t_m on the conversion in a tank reactor for the second order reaction $A + B \to P$ as predicted by the IEM model. [Adapted from Villermaux [27] by permission of Martinus Nijhoff, Publishers, The Hague, The Netherlands]

Separate feed streams of A and B

For this case we have to follow two agglomerates, one initially only containing A and one initially only containing B. Again we consider only the case of equal amounts of A and B in separate and equal feed streams. For the agglomerates containing initially only A we can derive

$$\frac{dc_A^A}{dt} = \frac{1}{t_m}(\bar{c}_A - c_A^A) - kc_A^A c_B^A$$

$$\frac{dc_B^A}{dt} = \frac{1}{t_m}(\bar{c}_B - c_B^A) - kc_A^A c_B^A.$$

with the boundary conditions $c_{Ao}^A = 2c_{Ao}$ and $c_{Bo}^A = 0$ at $t = 0$. Here c_{Ao}^A is the concentration of A in feed stream A and c_{Ao} is the fictitious mixed cup concentration of A in the combined feed streams. Similarly for the agglomerates containing initially only B we can write

$$\frac{dc_A^B}{dt} = \frac{1}{t_m}(\bar{c}_A - c_A^B) - kc_A^B c_B^B$$

$$\frac{dc_B^B}{dt} = \frac{1}{t_m}(\bar{c}_B - c_B^B) - kc_A^B c_B^B,$$

and at $t = 0$, $c_{Ao}^B = 0$ and $c_{Bo}^B = 2c_{Bo}$. For the average exit concentration of A we find

$$\frac{c_A}{c_{Ao}} = \frac{0.5}{\tau}\int_0^\infty \left(\frac{c_A}{c_{Ao}}\right)_A e^{-t/\tau}\,dt + \frac{0.5}{\tau}\int_0^\infty \left(\frac{c_A}{c_{Ao}}\right)_B e^{-t/\tau}\,dt.$$

Results have been given by Villermaux [27] and are included in figure V.19. The premixed feedstock and the non-premixed feedstocks share the same boundaries of maximum mixedness. However, the results for partial or complete segregation are very different; moreover the micromixing parameter has a much larger influence in the case of a separate reactant inlet.

For extremely high reaction rates the mass balance for the agglomerates can be simplified further. The mass balances for an agglomerate containing initially only A become

$$\frac{dc_A^A}{dt} = \frac{1}{t_m}(\bar{c}_A - c_A^A) + R_A$$

$$\frac{dc_B^A}{dt} = \frac{1}{t_m}(\bar{c}_B - c_B^A) + R_B$$

with $R_A = R_B$ and $\bar{c}_A = \bar{c}_B$. After subtraction of these balances the following expression is obtained:

$$\frac{d(c_A^A - c_B^A)}{dt} = -\frac{1}{t_m}(c_A^A + c_B^A).$$

If the reaction rate is very large, $c_B^A = 0$ and we can write

$$\frac{dc_A^A}{dt} = -\frac{1}{t_m}c_A^A \quad \text{or} \quad c_A^A/c_{Ao}^A = e^{-t/t_m}.$$

Averaging over the RTD of the reactor leads to

$$\bar{c}_A^A = \frac{c_{Ao}^A}{\tau_A}\int_0^\infty e^{-t/t_m} e^{-t/\tau_A} dt \quad \text{or} \quad \frac{\bar{c}_A}{c_{Ao}} = \frac{1}{1 + \tau/t_m}.$$

It is interesting to compare this result with that of the random coalescence model, as is done in figure V.20, adapted from Villermaux [27]. The results are almost

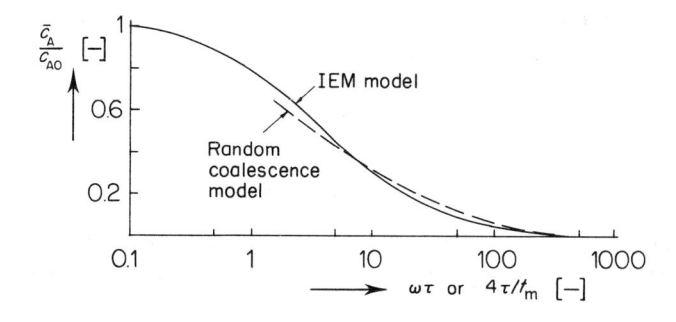

Fig. V.20 Equivalence between the IEM model and the random coalescence model for an infinitely fast reaction $A + B \to P$, $c_{Ao} = c_{Bo}$. Reactants are fed separately to the system. [Adapted from Villermaux [27] by permission of Martinus Nijhoff, Publishers, The Hague, The Netherlands]

the same except for low coalescence frequencies. It should be realized that for very low values of $\omega\tau$ the model becomes questionable because of the statistical requirement of a large number of coalescences for meaningful results. The IEM model is much simpler to deal with than the random coalescence model. Therefore it has been applied to many reaction systems. Villermaux [27] has given results for the reaction $A + B \to P$ for different feed rates and stoichiometry and for separated and premixed feeds. Also results for consecutive reactions schemes have been given [27]. An important assumption in these models is that the micromixing parameter t_m is not influenced by the chemical reaction rate, so that chemical enhancement factors (see Chapter VII) are not taken into account. Especially with complex reaction schemes and extremely fast reactions, these complications may have to be taken into account. An interesting example has been given by Bourne et al. [25].

V.4 EXPERIMENTAL RESULTS ON MICROMIXING IN STIRRED VESSELS

In many investigations it has been attempted to relate the micromixing parameter t_m to the hydrodynamical and geometrical variables of the stirred vessel, such as stirrer and vessel size, stirring speed and fluid properties. As a theoretical background the theory of isotropic turbulence introduced by Kolmogoroff [1] is often used. According to this theory the kinetic energy introduced e.g. by the stirrer or by the fluid inlet jets is transferred from the larger fluid eddies to smaller ones until a size is reached where the kinetic energy is rapidly dissipated to heat. The final stage of energy transfer/dissipation takes place in a universal way, being a function only of the energy input rate per mass of fluid ε, and the kinematic viscosity v. The final stage can be characterized by the size of the smallest eddies or the so called Kolmogoroff scale l_K, which is given by

$$l_K = (v^3/\varepsilon)^{1/4}.$$

In an analogous way the unmixedness or concentration differences are eliminated by molecular diffusion in the final stage of mixing. Also here scales have been introduced e.g. by Batchelor [3]:

$$l_B = (vD^2/\varepsilon)^{1/4}$$

and by Corrsin [4]:

$$l_C = (D^3/\varepsilon)^{1/4},$$

in which D is the molecular diffusion coefficient. There is no general agreement on which of these scales, if any, should be used in the micromixing parameter:

$$t_m = \frac{l^2}{12D}.$$

Even in a standard mixing vessel equipped with a standard Rushton turbine, it is still difficult to apply these relations because the actual values of ε vary strongly with the position in the vessel. Moreover the average size of the agglomerates will be higher because the scales are referring to the final stage of the mixing process only.

A complicating factor is that the effective agglomerate size may depend on the residence time in the vessel, especially in case of short residence times. This can be understood as follows. At the feed inlet points relatively large regions of unmixed fluid elements are present. If the average residence time is extremely short, many of the large agglomerates can leave the vessel practically unmixed. At very long average residence times the large agglomerates of the first mixing stages will occupy only a very small volume fraction of the mixed vessel and almost none will escape unmixed from the vessel. Klein et al. [28] analysed this situation on the basis of experiments with premixed and non-premixed reactants in a stirred vessel. They concluded that the mixing process can be split up into two successive stages:

1. formation of big aggregates at the inlet and subsequent erosion of these aggregates;
2. formation by erosion products of new micro scale agglomerates with premixed reactants.

Plasari et al. [16] gave a description of the process of erosion of the big aggregates. The diameter of the aggregates was found to decrease linearly with time; therefore the remaining large agglomerate fraction ε_a in the mixer can be found similarly to equation (V.1):

$$\varepsilon_a = \frac{1}{\tau} \int_0^{t_{er}} \left(1 - \frac{t}{t_{er}}\right)^3 e^{-t/\tau} \, dt.$$

In this equation $(1/\tau)e^{-t/\tau}$ is the RTD of the mixer and

$$\left(1 - \frac{t}{t_{er}}\right)^3 = \frac{V_a}{V_{a0}},$$

in which V_a is the agglomerate volume remaining after t seconds of residence time in the vessel, and t_{er} is the time necessary for complete erosion of the aggregate. According to Plasari et al. [16], t_{er} can be found from a mass transfer relation of Calderbank and Moo Young [29]. The relation for t_{er} becomes

$$t_{er} = \frac{l_0}{0.26 D (\varepsilon/v^3)^{1/4} Sc^{1/3}}.$$

The initial agglomerate dimension l_0 has to be estimated and some recommendations for its value were given by Plasari et al [16].

The micromixing parameter for the second stage of mixing referring to the erosion product agglomerates was taken from Corrsin [4] in the work of Klein et al. [28]:

$$t_m = 18 \left(\frac{5}{\pi}\right)^{2/3} \frac{L_c^{2/3}}{\varepsilon^{1/3}}. \tag{V.9}$$

L_c is the Corrsin macro scale of mixing [4] which can be taken to be roughly $(0.3-1.0)d_{imp}$ in which d_{imp} is the stirrer blade width (see Rao and Brodkey [30]). The value of t_m given by equation (V.9) can be inserted in the IEM model to describe the micromixing in the erosion products fluid fraction. Of course in limiting cases the two stage model can be simplified to a single stage model with one type of agglomerate only. In these cases the simple IEM models can be used. Examples of applications were given by Zoulalian and Villermaux [31] and Truong and Methot [32] for the second stage only, and by Plasari et al. [16] for the erosion of large aggregates. A somewhat similar two step approach has been proposed by Pohorecki and Baldyga [33]. The segregated part of the fluid is derived from the macromixing time constant measured by classical methods, while the micromixing time constant was derived from turbulence spectrum calculations. Such complex models contain a large number of parameters, some of which have to be adjusted, so that extrapolation to other systems may remain difficult. One of the adjustable parameters is the efficiency of the energy input by the stirrer and inflow jets [28], which may depend on the general geometry of the mixed vessel.

Also in the simple case of a single agglomeration model, such as the IEM model, the prediction of the agglomerate size is not yet possible. It has been suggested by Bourne et al. [25], Belevi et al. [34] and Bourne et al. [35] that the product distribution for complex reaction schemes can be a sensitive parameter for the study of micromixing in a stirred vessel. In their work they also took the influence of the chemical reaction on the mass transfer to the agglomerates into account. They derived an agglomerate size of 0.2–0.4 times the Kolmogoroff scale based an average power input to the mixer.

For *liquid–liquid* emulsion systems the agglomerate size is well defined and is equal to the drop size for the discontinuous phase. At the present state of the art it is however not yet possible to predict accurately either the drop size or the interaction parameter. Excellent reviews on micromixing in these emulsion systems have been given by Rietema [36] and Villermaux [27].

Example V.4   Macro and micromixing in a stirred tank reactor
It is required to check whether a 1 m^3 reactor can be considered as a CISTR. The data are the following: inner tank diameter $d_T = 0.95$ m, liquid height 1.43 m, the agitator is a Rushton six blade flat turbine with a diameter $d_a = 0.32$ m and the vessel is equipped with standard baffles. The power input rate per unit mass of liquid is given by

$$\varepsilon = \frac{P}{\rho V} = P_o n_s^3 d_a^5 / V,$$

in which n_s is the stirring speed [s^{-1}] and $P_o = 6$ for $Re = \rho n_s d_a^2 / \mu > 10^4$. We will consider two liquids to be processed by this reactor: Liquid I with water type properties and Liquid II with a higher viscosity. Three agitator power input rates are chosen: 10, 1 and 0.1 W/kg. The residence time in the vessel $\tau = 2000$ s and the reaction time constant is

$$t_r = 1/kc_{Ao}^{n-1} = 700 \text{ s}.$$

Bourne et al. [35] suggested that the Kolmogoroff scale of turbulence based on the average power input would be a measure of the micromixing agglomerate size:

$$l_K = \left(\frac{v^3}{\varepsilon}\right)^{1/4}.$$

The results of the calculations are given in Table V.4.

Table V.4 Macro-and micromixing time constants in a stirred vessel (Example V.4)

Liquid I: $\rho = 1000 \text{ kg/m}^3$, $v = 10^{-6} \text{ m}^2/\text{s}$, $D = 10^{-9} \text{ m}^2/\text{s}$

ε (W/kg)	n_s (s^{-1})	Re	l_K (m)	t_m (s)	t_{er} (s)	$4t_C$ (s)
10	7.92	8.1×10^5	1.8×10^{-5}	0.027	0.6	4.4
1	3.68	3.7×10^5	3.2×10^{-5}	0.085	1.2	9.6
0.1	1.71	1.8×10^5	5.6×10^{-5}	0.26	2.2	20.6

Liquid II: $\rho = 1000 \text{ kg/m}^3$, $v = 10^{-5} \text{m}^2/\text{s}$, $D = 10^{-10} \text{m}^2/\text{s}$

ε (W/kg)	n_s (s^{-1})	Re	l_K (m)	t_m (s)	t_{er} (s)	$4t_C$ (s)
10	7.92	8.1×10^4	1×10^{-4}	8.3	38	4.4
1	3.68	3.7×10^4	1.8×10^{-4}	27	68	9.6
0.1	1.71	1.8×10^4	3.2×10^{-4}	85	121	20.6

If we insert these values in an agglomeration model taking into account only molecular diffusion, we find for the micromixing time constant (equation (V.8))

$$t_m = \frac{l_K^2}{12D}.$$

Especially for liquid I, the micromixing times thus obtained are very short in comparison with the average residence time and the reaction time constant. With the exception of liquid II at a low power input, the reactor can be considered as a CISTR. For short residence times and/or fast non-linear reactions, especially with liquid II, micromixing effects may be important, however. The calculation of the micromixing time constant can at best only be considered as an approximation because as reported by Bourne et al. [35] the actual power input varies strongly with the position inside the stirred vessel. Actually these authors suggested for the agglomerate size 0.2–0.4 l_K as an effective value for the whole vessel. According to Plasari et al. [16] and to Pohorecki and Baldyga [33] for rapid reactions and short residence times also the initial stage of the mixing process becomes important. This first stage is the breaking down of lumps of the inlet fluid. The characteristic erosion time of these fluid lumps has been calculated from the equation recommended by Plasari et al. [16]:

$$t_{er} = \frac{l_0}{0.26D(\varepsilon/v^3)^{1/4}Sc^{1/3}}.$$

The initial lump size l_0 was taken as 10^{-4} m and was based on the average power input rate. The values of t_{er} are given in Table V.4 and are larger than the micromixing times. Another estimate of this macromixing time scale was made from the circulation time constant of the vessel t_c; see Example II.3.a. $4t_c$ should be a measure for the macromixing time. The values of $4t_c$ are also indicated in Table V.4. In contrast with the t_{er} values, $4t_c$ values do not depend on the viscosity of the liquid.

For a rapid reaction with *separate* feed inlet points for the different reactants the initial mixing of the inlet fluid lumps can be simply accounted for as an inert fraction of the reactor volume. Here no reaction can take place because the reactants are still separated in this stage of the mixing process. For the hydrodynamical conditions chosen here this inert fraction of the reactor is small. For *premixed* reactants it forms a segregated volume fraction of the reactor.

For the hydrodynamical and the reaction conditions considered in this Example, the deviation from the CISTR is small, but in principle it can be accounted for easily with the model of Weinstein and Adler [10] or with the model of Klein et al. [28].

The estimates given here can be considered as indicative only, because the time scales of micro- and macromixing differ much from the average residence time and the reaction time constant. Generally experiments are necessary if micromixing can be critical.

V.5 CONCLUDING REMARKS ON MICROMIXING

It can be concluded that with non-first order reactions micromixing may influence both the conversion and the selectivity in chemical reactors. Boundaries for the influence of micromixing can be calculated if the residence time distribution of the reactor and the kinetics of the chemical reactions are known. These boundaries may be far apart in case of fast reactions, reactions with separate feed inlet points of the reactants and in case of a large spread in residence times such as in a CSTR. Models for intermediate micromixing can be simply fitting models or agglomeration models. The latter type is to be preferred, although at the present state of the art it is not yet possible to predict the agglomerate size for given conditions. By comparing the different characteristic times for reaction and micromixing with the average residence time in the tank reactor the state of micromixing can be estimated.

With low viscosity fluids and slow and moderately fast reactions, in most cases a stirred tank can be considered to be a CISTR. With polymerization reactions the micromixing in a CSTR reactor can have an important influence on the polymer properties. An accurate knowledge of the polymerization kinetics and of the influence of the product composition on the fluid properties is required to predict the impact of micromixing. Much investigation is still required in this area.

REFERENCES

1. Kolmogoroff, A. N.: *Compt. rend. Acad. Sci. URSS.* **30,** 301 (1941); **32,** 16 (1941)
2. Rosensweig, R. E.: *A.I.Ch.E. Journal,* **10,** 92 (1964)
3. Batchelor, G. K. J.: *Fluid Mech.* **5,** 113 (1959)
4. Corrsin, S., *A.I.Ch.E. Journal* **10,** 87 (1964)
5. Zwietering, T. N.: *Chem. Eng. Sci.* **11,** 1 (1959)
6. Harada, M., Yamadam, T., Tanaka, K., Eguchi, W., and Nagata, S.: *Kagaku Kogaku* **2,** 227 (1967)
7. Nagasubramanian, K., and Greassly, W. W.: *Chem. Eng. Sci.* **25,** 1549 (1970)
8. Gerrens, H.: *Proc. 4th Int. Symp. Chem. React. Eng. Heidelberg (1976),* Dechema Frankfurt, p. 421 (1976)
9. Villermaux, J.: *Proc. 8th Int. Symp. Chem. Reaction Eng. Boston (1982)*
10. Weinstein, H. and Adler, R. J.: *Chem. Eng. Sci.* **22,** 65 (1967)
11. Villermaux, J. and Zoulalian, A.: *Chem. Eng. Sci.* **24,** 1513 (1969)
12. Chen, M. S. K. and Fan, L. T.: *Can. J. Chem. Eng.* **49,** 704 (1971)
13. Asbjørnsen, O. A.: Paper presented at the A.I.Ch.E.–I.Ch.E. Joint meeting, 39, London (June 1965)
14. Danckwerts, P. V.: *Chem. Eng. Sci.* **8,** 93 (1958)
15. Beek, J. and Miller, R. S.: *Chem. Eng. Prog. Symp. Ser.* **55,** 33 (1959)
16. Plasari, E., David, R. and Villermaux, J.: *Proc. 5th Int. Symp. Chem. React. Eng. Houston ACS Symp. Ser.* **65,** 11 (1978)
17. David, R. and Villermaux, J.: *J. de chimie Physique,* **75,** 656 (1978)
18. Nauman, E. B.: *Chem. Eng. Sci.* **30,** 1135 (1975)
19. Curl, R. L.: *A.I.Ch.E. Journal* **9,** 175 (1963)
20. Spielman, L. A. and Levenspiel, O.: *Chem. Eng. Sci.* **20,** 247 (1965)
21. Harada, M., Arima, K., Eguchi, W. and Harada, S.: *The memoirs of the Faculty of Eng., Kyoto Univ.* **24,** 431 (1962)
22. Villermaux, J. and Devillon, J. C.: *Proc. 2nd Int. Symp. Chem. React. Eng. B1-13 (1972),* Amsterdam. Elsevier (1972)
23. Aubry, C. and Villermaux, J.: *Chem. Eng. Sci.* **30,** 457 (1975)
24. David, R. and Villermaux, J.: *Chem. Eng. Sci.* **30,** 1309 (1975)
25. Bourne, J. R., Kozicki, F. and Rys, R.: *Chem. Eng. Sci.* **36,** 1643 (1981)
26. Ritchie, B. W. and Tobgy, A. H.: *Proc. Int. Symp. Chem. React. Eng. Evanston. ACS Series* **133,** 376 (1974)
27. Villermaux, J.: In *Multiphase Chemical Reactors,* Rodrigues, Sweet and Calo (eds.) Noordhof (1981)
28. Klein, J. P., David, R. and Villermaux, J.: *Ind. Eng. Chem. Fundam.* **19,** 373 (1980)
29. Calderbank, P. H. and Moo Young, M. B.: *Chem. Eng. Sci.* **16,** 39 (1961)
30. Rao, M. A. and Brodkey, R. S.: *Chem. Eng. Sci.* **27,** 137 (1972)
31. Zoulalian, A. and Villermaux, J.: *Proc. 3rd Int. Symp. Chem. React. Eng. Evanston. ACS Series* **133,** 348 (1974)
32. Truong, K. T. and Methot, J. C.: *Can. J. Chem. Eng.* **54,** 572 (1976)
33. Pohorecki, R. and Baldyga, J.: *7th Int. Congress CHISA'81,* Praha, paper H 1.7 (1981)
34. Belevi, H., Bourne, J. R. and Rys, P.: *Chem. Eng. Sci.* **36,** 1649 (1981)
35. Bourne, J. R., Kozicki, F., Moergeli, U. and Rys, P.: *Chem. Eng. Sci.* **36,** 1655 (1981)
36. Rietema, K. *Adv. Chem. Eng. V,* 237 (1964)

Chapter VI

The role of the heat effect in model reactors

All chemical reactions are in principle accompanied by evolution or absorption of heat. In practice, therefore, a reaction hardly ever proceeds entirely under isothermal conditions, as was assumed in Chapters II and III. A certain temperature level and a certain temperature range at which the reaction must be carried out is generally specified. Consequently, the heat effects must be taken into account when removing or supplying heat during the reaction.

The choice of the desired temperature level and temperature range is determined by one or more of the following factors:

— the chemical and physical properties of the reaction mixture, such as the position of the chemical equilibrium, the conversion rate of the main reaction, the occurrence of undesired reactions, the dewpoint of a gaseous reaction mixture, the boiling point and the solidification point of a liquid reaction mixture, etc;
— the properties of any catalyst used, such as sintering of solid catalyst, decomposition, aging, etc;
— the cost of bringing the reactants to the reaction temperature and of keeping the reaction mixture at the desired temperature level.

Several means are available for keeping a reaction mixture between prescribed temperature limits:

— adjustment of the initial or entry temperature of the reaction mixture, e.g. in case of adiabatic operation;
— addition of an inert compound for reducing temperature changes;
— division of a reactor into adiabatic sections with intermediate cooling or heating of the mixture;
— supply or withdrawal of heat during the reaction, e.g. by heat transfer from or to another medium or by evaporation of a volatile compound of the reaction mixture;
— circulation through an external heat exchanger of a separate heat carrier, e.g. a solid catalyst;
— dilution of solid catalyst particles with inert particles, e.g. with a good heat conductivity, to reduce the (local) conversion rate in a catalyst in a cooled reactor.

In the study of these many possible alternatives it is essential that the energy balance be properly applied in conjunction with the material balance. This is the main theme of this chapter, which deals with applications to the more current types of model reactors.

VI.1 THE ENERGY BALANCE AND THE HEAT OF REACTION

For a non-flow process, which is intermittent in character and in which no continuous streams of material enter or leave the system during the course of the operation as in a batch reactor, the energy balance is given for a certain period of time by

$$\Delta U = Q - W.$$

For flow processes, equation (I.3) conveniently formulates the principle of conservation of energy per unit of time for a finite mass m with volume V_r of a reaction mixture:

$$\frac{d(\langle u \rangle m)}{dt} = -\Delta(h\Phi_m) + \dot{Q} - \dot{W}, \qquad (VI.1)$$

where u and h represent the internal energy and the enthalpy, respectively, per unit mass of the reaction mixture. $\langle u \rangle$ is the averaged internal energy over the reactor space. They are interrelated according to

$$h = u + p/\rho. \qquad (VI.2)$$

If we denote the internal energy and the enthalpy per unit mass of species J by u_J and h_J, we can write for u and h, respectively,

$$u = \sum_J u_J w_J \quad \text{and} \quad h = \sum_J h_J w_J.$$

For additional use of equation (VI.1) it is important to relate a change in u or h to a change in the physical conditions and composition of the reaction mixture. We shall do this below for the enthalpy h, which is a function of pressure, temperature and composition:

$$dh(p, T, w_J) = \sum_J w_J \frac{\partial h_J}{\partial p} dp + \sum_J w_J \frac{\partial h_J}{\partial T} dT + \sum_J d(h_J w_J)_{p,T}. \qquad (VI.3)$$

For the first two sums we may write

$$\sum_J w_J \frac{\partial h_J}{\partial p} dp = \frac{1}{\rho} dp \quad \text{and} \quad \sum_J w_J \frac{\partial h_J}{\partial T} dT = c_p \, dT,$$

where c_p is the specific heat per unit of mass at constant pressure of the reaction mixture. The last term of equation (VI.3) represents the enthalpy change at constant temperature and pressure due to a change in the degree of conversion. In order to relate this term to the heat of reaction, we first consider, as an example, the reaction equation:

$$v_A A + v_B B \to v_P P + v_Q Q,$$

with which a heat effect ΔH_r is associated when the reaction moves completely from the left side to the right side at constant temperature and pressure (see Section I.3). If we realize that $H_J = M_J h_J$, then we have for ΔH_r in this case, according to equation (I.4)

$$\Delta H_r = v_P H_P + v_Q H_Q - v_A H_A - v_B H_B.$$

A small change in composition due to conversion can be characterized by a change in one of the mass fractions, say w_A. The changes in the other mass fractions are related to dw_A (see equation (I.28)) by

$$\frac{dw_A}{v_A M_A} = \frac{dw_B}{v_B M_B} = -\frac{dw_P}{v_P M_P} = -\frac{dw_Q}{v_Q M_Q}.$$

If we assume the values of h_J to be independent of composition, this means that the heats of mixing are disregarded in this derivation, the last term of equation VI.3 can now be written as

$$\sum_J d(h_J w_J)_{p,T} = \sum_J (h_J dw_J)_{p,T}$$

$$= (v_A M_A h_A + v_B M_B h_B - v_P M_P h_P - v_Q M_Q h_Q) \frac{dw_A}{v_A M_A}$$

$$= (v_A H_A + v_B H_B - v_P H_P - v_Q H_Q) \frac{dw_A}{v_A M_A}$$

$$= -\Delta H_r \frac{-d\xi_J}{v_J M_J} \equiv \frac{(\Delta H_r)_J}{M_J} d\xi_J \equiv (\Delta h_r)_J d\xi_J. \tag{VI.4}$$

Therefore, while ΔH_r is the heat of reaction associated with the stoichiometric formula, $(\Delta H_r)_J$ is the heat of reaction per mole of J converted, and $(\Delta h_r)_J$ the heat of reaction per unit mass of J converted. The sign of ΔH_r, $(\Delta H_r)_J$ and $(\Delta h_r)_J = (\Delta H_r)_J/M_J$ is always the same; it is negative for exothermic reactions and positive for endothermic reactions, because in the first law of thermodynamics the heat flow is defined as the heat flow *into* the system. As a result of the above considerations we find for single reactions the following relation for the enthalpy per unit of mass:

$$dh = \frac{1}{\rho} dp + c_p dT + \frac{(\Delta H_r)_J}{M_J} d\xi_J. \tag{VI.5}$$

For n simultaneous reactions we find similarly

$$dh = \frac{1}{\rho} dp + c_p dT + \sum_n \frac{(\Delta H_r)_J}{M_J} d\xi_J. \tag{VI.5a}$$

If we realize that $U = f(\rho, T, w_J)$, then we find in the same way, for the internal energy per unit of mass,

$$du = -p\, d\left(\frac{1}{\rho}\right) + c_v\, dT + \frac{(\Delta U_r)_J}{M_J}\, d\xi_J. \tag{VI.6}$$

$(\Delta U_r)_J/M_J$ is the heat of reaction at constant temperature and density per unit of mass of J converted: it is related to the integral heat effect ΔU_r according to

$$\frac{(\Delta U_r)_J}{M_J} = \frac{\Delta U_r}{\nu_J M_J}.$$

The last two terms in equation (VI.1) are still to be discussed. For the *rate of heat supply* to the reaction mixture, $\dot{Q}$, several expressions can be used, depending on the method by which heat is supplied or removed. For example, if heat is withdrawn through a heat exchange surface with an area A by means of a coolant having a temperature T_c, we can write, for the heat flow into the system

$$\dot{Q} = -UA(T - T_c). \tag{VI.7}$$

U is the overall heat transfer coefficient and T the temperature of the reaction mixture. In case the reaction mixture is heated with a heating medium having the temperature T_h, we can write

$$\dot{Q} = UA(T_h - T). \tag{VI.7.a}$$

In adiabatic reactors $\dot{Q} = 0$, of course.

The *work done* by the system per unit of time, $\dot{W}$, includes all possible forms of work [1], such as shaft and mechanical, electrical or magnetic work. The flow work to overcome forces flowing into and out of the system is already accounted for in the enthalpy terms, so flow work is excluded in the term $\dot{W}$. Usually we are concerned only with shaft or mechanical work. In reversible batch processes the shaft work during a certain period of time is $W_s = \int_1^2 p\, dV$—e.g. in combustion engines—and in reversible continuous processes in the absence of changes in potential surface and external kinetic energies the shaft work is $W_s = -\int_1^2 V\, dp$, e.g. in expansion turbines. In a continuous process the shaft work per unit of time is

$$\dot{W}_s = -\int_1^2 \Phi_m \frac{dp}{\rho}.$$

Very seldom, however, is a combustion engine or expansion turbine included in the reactor system, so in general $W_s = 0$. In this book we will exclude shaft work in our discussions.

In batch processes with constant volume reactions the mechanical work is zero and with constant pressure reactions the mechanical work per unit of time is

$$\dot{W} = p\frac{dV_r}{dt} \tag{VI.8}$$

if the reaction volume V_r changes during the reaction.

VI.2 THE WELL-MIXED BATCH REACTOR

Since the total mass of the reaction remains constant and the conditions are uniform over the entire contents of the reactor, equation (VI.1) becomes

$$m\frac{du}{dt} + p\frac{dV_r}{dt} = \dot{Q}. \tag{VI.9}$$

For reactor calculations this equation must be considered in conjunction with the material balance for the batch reactor (see equation (II.1)):

$$m\frac{d\xi_J}{dt} = M_J R_J V_r. \tag{VI.10}$$

In accordance with normal operating procedures, a distinction can be made between operation at constant pressure (e.g. in an open vessel) and at constant volume and density (e.g. in an autoclave).

At *constant pressure* we can combine the left-hand side of equation (VI.9) with equation (VI.2) and realizing that $\rho = m/V_r$ or $V_r = m/\rho$, we find that

$$m\frac{du}{dt} + p\frac{dV_r}{dt} = m\frac{du}{dt} + mp\frac{d}{dt}\left(\frac{1}{\rho}\right) = m\frac{d}{dt}\left(u + \frac{p}{\rho}\right) = m\frac{dh}{dt}.$$

After introducing equation (VI.5), in which at constant pressure $dp = 0$, we find, for the energy balance in a well-mixed batch reactor operating at constant pressure,

$$mc_p\frac{dT}{dt} + m\frac{(\Delta H_r)_J}{M_J}\frac{d\xi_J}{dt} = \dot{Q}. \tag{VI.11}$$

It is seen from this equation that, if the reaction proceeds at constant temperature and pressure, the amount of heat to be supplied per unit mass of species J converted, $\dot{Q}\, dt/m\, d\xi_J$, is equal to $(\Delta h_r)_J$; this is in agreement with the definition VI.4. Furthermore, it should be borne in mind that the enthalpy is a point property so that the heat supply per unit mass of mixture necessary for a change in temperature from T_a to T_b and in conversion from ξ_{Ja} to ξ_{Jb},

$$\int_{T_a,\xi_{Ja}}^{T_b,\xi_{Jb}} [c_p\, dT + (\Delta h_r)_J\, d\xi_J] = \int_a^b \frac{\dot{Q}}{m}\, dt,$$

is independent of the path of integration. This is illustrated in figure VI.1, which shows two particular paths of integration. In Path 1 the reaction mixture is assumed to be heated from T_a to T_b at a degree of conversion ξ_{Ja}, whereupon it reacts at T_b with a change of conversion from ξ_{Ja} to ξ_{Jb}; according to Path 2 the conversion takes place entirely at T_a and it is followed by the temperature rise of a mixture having the final composition characterized by ξ_{Jb}.

With a *constant volume* of the reaction mixture, its density remains constant so that $d(1/\rho) = 0$. With the use of equation (VI.6) now the heat balance (equation

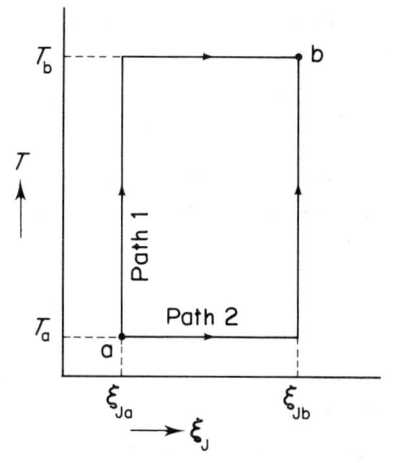

Fig. VI.1 Two paths for integrating equation (VI.11)

(VI.9)) for a well-mixed batch reactor operating at constant volume, becomes

$$m\frac{du}{dt} = mc_v\frac{dT}{dt} + m(\Delta u_r)_J\frac{d\xi_J}{dt} = \dot{Q}. \qquad (VI.12)$$

The heat supply per unit mass of J converted, $\dot{Q}\,dt/m\,d\xi_J$, is equal to the heat of reaction $(\Delta u_r)_J$ if temperature and density remain constant. For the integration of equation (VI.12) we can take advantage of the fact that the internal energy—at constant density only a function of temperature and composition—is a point property.

In general, when dealing with a batch reactor, it will be required to calculate the reaction temperature and the degree of conversion as a function of time under prescribed conditions of heat supply or removal, or to calculate the heating or cooling requirements for obtaining a desired temperature range and reactor capacity. To this end, the material balance (VI.10) has to be used together with the energy balance (VI.9) or one of its derived forms such as (VI.11) or (VI.12). A straightforward simultaneous solution is generally not feasible since the conversion rate depends both on temperature and on the degree of conversion itself. Accordingly, numerical methods of solution have to be used.

Equation (VI.1) for constant pressure reactions and equation (VI.12) for constant volume reactions are generally valid. In practice molar concentrations are frequently used. From Table I.3 we find, for a reactant A,

$$\xi_A = M_A\left(\frac{c_{Ao}}{\rho_o} - \frac{c_A}{\rho}\right)$$

and for a product P

$$\xi_P = M_P\left(\frac{c_P}{\rho} - \frac{c_{Po}}{\rho_o}\right).$$

For a reactant A in a constant pressure reaction we find for equation (VI.11), based on molar concentrations,

$$mc_P \frac{dT}{dt} - m(\Delta H_r)_A \frac{d(c_A/\rho)}{dt} = \dot{Q}.$$

If the density remains constant during the reaction, as is approximately the case with liquid-phase reactions, and after introducing the relative conversion $\zeta_A = (c_{A_0} - c_A)/c_{A_0}$, we find

$$\rho c_P \frac{dT}{dt} + (\Delta H_r)_A c_{A_0} \frac{d\zeta_A}{dt} = \frac{\dot{Q}}{V_r}$$

and after dividing by ρc_P, the heat capacity per unit of volume of the reaction mixture, and rearranging,

$$\frac{dT}{dt} = -\frac{(\Delta H_r)_A c_{A_0}}{\rho c_P} \frac{d\zeta_A}{dt} + \frac{\dot{Q}}{\rho c_P V_r}$$

For a cooled batch reactor we introduce equation (VI.7) and then finally find

$$\frac{dT}{dt} = -\frac{(\Delta H_r)_A c_{A_0}}{\rho c_P} \frac{d\zeta_A}{dt} - \frac{UA(T - T_c)}{\rho c_P V_r}. \tag{VI.13}$$

For an exothermic reaction ($\Delta H_r < 0$) this equation states that the rate of temperature increase is linearly dependent on the rate of conversion and on the rate of heat removal by the cooling medium. Similar equations for other conditions have been summarized in Table VI.1.

Table VI.1 Balances for a batch reactor

Conditions	Material balance	Energy balance
Constant pressure reactions for reactant A	$\dfrac{d(c_A/\rho)}{dt} = -\dfrac{R_A}{\rho}$	$c_P \dfrac{dT}{dt} - (\Delta H_r)_A \dfrac{d(c_A/\rho)}{dt} = \dfrac{\dot{Q}}{m}$
and constant density	$\dfrac{dc_A}{dt} = -R_A$	$\rho c_P \dfrac{dT}{dt} - (\Delta H_r)_A c_{A_0} \dfrac{d\zeta_A}{dt} = \dfrac{\dot{Q}}{V_r}$
for product P	$\dfrac{d(c_P/\rho)}{dt} = \dfrac{R_P}{\rho}$	$c_P \dfrac{dT}{dt} + (\Delta H_r)_P \dfrac{d(c_P/\rho)}{dt} = \dfrac{\dot{Q}}{m}$
and constant density	$\dfrac{dc_P}{dt} = R_P$	$\rho c_P \dfrac{dT}{dt} + (\Delta H_r)_P \dfrac{dc_P}{dt} = \dfrac{\dot{Q}}{V_r}$
Constant volume reactions for reactant A	$\dfrac{dc_A}{dt} = -R_A$	$\rho c_v \dfrac{dT}{dt} - (\Delta U_r)_A \dfrac{dc_A}{dt} = \dfrac{\dot{Q}}{V_r}$
for product P	$\dfrac{dc_P}{dt} = R_P$	$\rho c_v \dfrac{dT}{dt} + (\Delta U_r)_P \dfrac{dc_P}{dt} = \dfrac{\dot{Q}}{V_r}$

Note: As follows from these formulae only one reaction takes place.

In case the reactor operates adiabatically ($\dot{Q} = 0$) we find, after integrating equation (VI.13) with the initial conditions $T = T_o$ and $\zeta_A = 0$ at $t = 0$,

$$T - T_o = -\frac{(\Delta H_r)_A c_{A_o}}{\rho c_P} \zeta_A.$$

At full conversion $\zeta_A = 1$ we have the so-called *adiabatic temperature rise* ΔT_{ad} of the reaction mixture:

$$\Delta T_{ad} = \frac{-(\Delta H_r)_A c_{A_o}}{\rho c_P}, \tag{VI.14}$$

which indicates how much the temperature of the reaction mixture would rise if the reaction were to proceed adiabatically to completion. So the relation between the temperature and the conversion in an adiabatic reactor is

$$T = T_o + \Delta T_{ad} \zeta_A. \tag{VI.15}$$

We see that, by dilution, which means reducing c_{A_o}, we can reduce the adiabatic temperature rise and so keep reactor temperature within desired limits.

The adiabatic temperature rise ΔT_{ad} can have different meanings. In a strict sense the true ΔT_{ad} of a reaction mixture should be calculated with the values of ΔH_r, ρ and $\bar{c}_P$ averaged over the whole interval from the initial to the final temperature with

$$-\int_{T_o}^{T_o + \Delta T_{ad}} \frac{\rho(T) \bar{c}_P(T) \, dT}{(\Delta H_r)_A(T)} = c_{A_o}.$$

However, as in most cases we want to prevent excessive temperature increases in chemical reactors, we will normally try to maintain the reactor temperature in the neighbourhood of T_o. Therefore it is often more appropriate to base the value of ΔT_{ad} on the initial conditions. In this case the initial ΔT_{ad} is defined by

$$\Delta T_{ado} = -\frac{(\Delta H_r)_{Ao} c_{Ao}}{\rho_o \bar{c}_{Po}},$$

where the subscript o refers to the initial conditions. We will not distinguish between the true or the initial ΔT_{ad}: it will be clear from the case under consideration whether the true or the initial value has to be used. The ΔT_{ad} can be very strongly dependent on the initial conditions, especially for gas reactions near to or above the critical point of the reaction mixture. In figure VI.2 the initial ΔT_{ad} for the NH_3 synthesis from a stoichiometric mixture of N_2 and H_2 is shown as a function of the pressure and initial temperature.

The adiabatic temperature rise is an important property of a reaction mixture; it will be used many times in this chapter. Especially to make equations for the heat production dimensionless, ΔT_{ad} is very useful, as will be further explained in Section VI.4, where four different expressions for the heat production will be presented.

Because of the similarity between the BR and PFR we will discuss adiabatic reactors more extensively in the following section. In well-agitated tanks, either

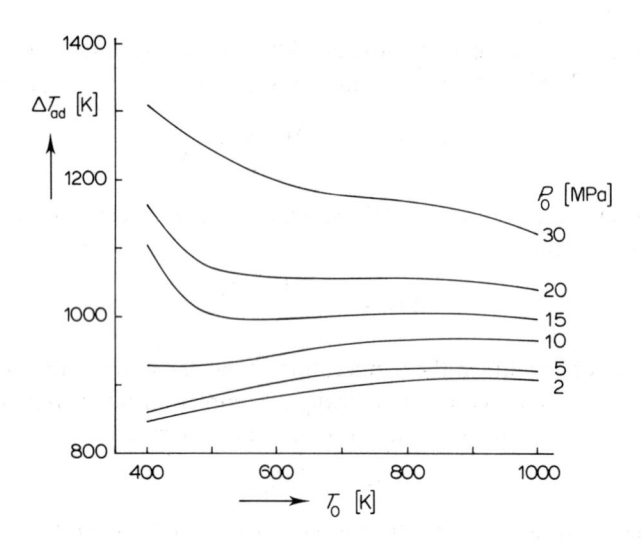

Fig. VI.2 The initial adiabatic temperature rise for the NH$_3$ formation from a stoichiometric mixture of N$_2$ and H$_2$. Parameters are the initial temperature T_0 and the pressure

batch wise or continuously operated, the temperature T of the reaction mixture is uniform all over the tank. The temperature of the cooling or heating medium, however, varies. To prevent cumbersome calculations with varying cooling or heating medium temperatures we will use the inlet temperatures of the medium and assume that the overall heat transfer coefficient is a corrected U, according to

$$U = \frac{U_{\text{true}}}{1 + \dfrac{U_{\text{true}}A}{2\Phi_{\text{mc}}c_{\text{pc}}}}. \tag{VI.16}$$

This correction can be derived by eliminating the outlet temperature T_{cout} of the cooling medium in the heat balance for the cooling medium:

$$U_{\text{true}}A(T - \bar{T}_c) = U_{\text{true}}A[T - \tfrac{1}{2}(T_{\text{cout}} + T_{\text{cin}})] = \Phi_{\text{mc}}c_{\text{pc}}(T_{\text{cout}} - T_{\text{cin}}).$$

This is only an approximation, because the logarithmic mean temperature difference should have been used, of course. We then find that

$$UA(T - T_c) = UA(T - T_{\text{cin}}) = U_{\text{true}}A(T - \bar{T}_c) = \frac{U_{\text{true}}A(T - T_{\text{cin}})}{1 + \dfrac{U_{\text{true}}A}{2\Phi_{\text{mc}}c_{\text{pc}}}},$$

where Φ_{mc} and c_{pc} refer to the cooling medium and T_{cin} to the inlet temperature of the cooling medium.

We shall give below two illustrative examples from the many possible problems in this field.

Example VI.2.a *Influence of the policy of operation on the reaction time*
In a batch reactor, having a volume $V_r = 5 \, m^3$, a constant density exothermic reaction A → P is carried out in the liquid phase. For the conversion rate we have

$$R_A = -kc_A \, [kmol/m^3 s],$$

with

$$k = 4 \times 10^6 \exp(-7900/T) \, s^{-1}.$$

Further data are:

$$(\Delta h_r)_A = -1.67 \times 10^6 \, J/kg,$$
$$\rho c_p = 4.2 \times 10^6 \, J/m^3 {}^\circ C$$
$$M_A = 100 \, kg/kmol \text{ and}$$
$$c_{Ao} = 1 \, kmol/m^3.$$

The initial temperature T_o of the reaction mixture is 20 °C and the maximum allowable reaction temperature is 95 °C. The reactor contains a spiral for heat exchange purposes; its surface area A is 3.3 m² and it can be operated with steam ($T_s = 120$ °C, $U = 1360 \, W/m^2 \, °C$) and with cooling water ($T_c = 15$ °C, $U = 1180 \, W/m^2 \, °C$). The times required for filling and emptying the reactor are 600 and 900 s, respectively.

The problem to be solved is to calculate the duration of one reaction cycle and the steam consumption for a relative conversion $\zeta_A \geq 0.90$ and for the following policies of operation:

I. Preheat to 55 °C, let the reaction proceed adiabatically, start cooling when 95 °C or $\zeta_A = 0.90$ is reached, cool down to 45 °C.
II. Heat up to 95 °C, let the reaction proceed isothermally until $\zeta_A = 0.90$, cool down to 45 °C.

We shall use the energy balance in the form (VI.11) for this problem. It appears that for the calculation of the *heating up period* the chemical conversion and its heat effect can be neglected. For this period, therefore, equation (VI.11) becomes

$$\rho c_p V_r \frac{dT}{dt} = UA(T_s - T).$$

After integration between $T = 20$ °C and $T = 55$ °C it is found that the latter temperature is reached after 2030 seconds.

For the *adiabatic* reaction operation (Case I), the material and heat balances become, in terms of the relative degree of conversion $\zeta_A = 1 - c_A/c_{Ao} = \rho \xi_A/c_{Ao} M_A$,

$$\frac{d\zeta_A}{dt} = k(1 - \zeta_A)$$

and

$$\rho c_p \frac{dT}{dt} + (\Delta H_r)_A c_{Ao} \frac{d\zeta_A}{dt} = 0.$$

The latter equation constitutes a linear relation between T and ζ_A. In this example $\zeta_A = 0$ at $T = 55\,°C$, so that integration yields

$$T = 55 + \frac{(\Delta H_r)_A c_{Ao}}{\rho c_p} \zeta_A \; °C.$$

In this case $\Delta T_{ad} = 39.8\,°C$; hence it follows that at $\zeta_A = 0.9$ the temperature is 90.8 °C, i.e., below the maximum value of 95 °C. Moreover, the reaction velocity constant k can now be expressed as a function of c_A only. Hence, the material balance can be integrated, preferably by means of a numerical procedure. The results are shown in figure VI.3. Both the ζ_A–t and the T–t curves have an inflection point. The slope of the curves [proportional to $k(1 - \zeta_A)$] is relatively small in the initial stage of the reaction because of the low temperature (k small) and in the final stage because of the high degree of conversion.

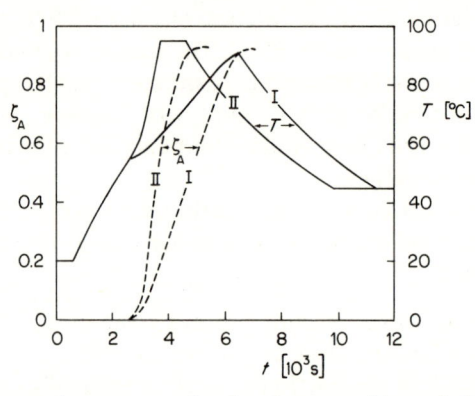

Fig. VI.3 Conversion and temperature in a batch reactor; Example VI.2.a, Cases I and II

For Case II, *simultaneous heating and reaction*, the equations to be solved are

$$\frac{d\zeta_A}{dt} = k(1 - \zeta_A) \quad \text{and} \quad \rho c_p \frac{dT}{dt} + (\Delta H_r)_A c_{Ao} M_A \frac{d\zeta_A}{dt} = \frac{UA}{V_r}(T_s - T).$$

In this case the solution has to be arrived at with the aid of a step-by-step method on a computer. Starting from the initial concentration and temperature, a small temperature interval ΔT is selected. The equations are then solved for Δt with the average values of T and k and the initial value of ζ_A for that interval. The value of Δt thus estimated is then used to calculate $\Delta \zeta_A$ from the first equation, so that a better approximation of the average concentration in the interval becomes available. A second approximation for Δt is then found and the procedure is repeated until the solution is consistent within the desired limits. In the present example, this calculation was carried out with temperature intervals of 5 °C until the maximum temperature of 95 °C was reached; the results are shown in figure VI.3. For the subsequent completion of the reaction under isothermal conditions, only the material balance is needed with the value of k at 95 °C. The time required for cooling down the reaction mixture to 45 °C can be calculated in a similar manner as the heating-up period, if the effects of the small amount of additional conversion are neglected.

The period of a complete reaction cycle, as obtained from the above calculations, is given in Table VI.2.

Table VI.2

Operation	Time (s) Case I	Case II
filling	600	600
preheating	2030	2030
reaction	3590	1930
cooling	5100	5300
emptying	900	900
Total	12 220 s	10 760 s
or	3.4 h	3.0 h

It can be calculated that the steam consumption per cycle is 330 and 510 kg for Cases I and II, respectively. Consequently, an extra steam consumption of 180 kg involves a gain in time of 24 minutes. Apart from this, it is evident that the production capacity is greatly limited by the long cooling period in this installation. If, for example, the reaction product were discharged through a separate cooler immediately after the desired conversion was reached, the capacity could be almost doubled. The production per cycle is $5 \times 100 \times 0.9 = 450$ kg P and for operation in three shifts the production per day is $450 \times 24/3.4 = 3182$ kg P per day. If the BR were originally designed and operated according to method I, we can conclude that the original capacity can be increased successively:

1. by method II by 13.6% to 3613 kg P/day;
2. if moreover the BR is emptied at 95°C and the product cooled with a cooler installed in the discharge line by 72% to 5461 kg P/day;
3. if the reactor is filled with a reaction mixture preheated to 50 °C by another 20%;
4. by taking measures (2) and (3) simultaneously, by 140% to 7639 kg P/day.

Capacity can therefore be increased by higher energy consumption and additional capital expenditure or both. Depending on individual costs an economic optimum exists between additional expenditures and capacity increase. We also see that a vessel well suited as a reactor is a bad heat exchanger, because of the small heat transfer area that can be installed in it.

Example VI.2.b *An explosion*
A closed spherical tank with a volume of 250 m³ contains air with 4% by volume of n-butane vapour at $T_o = 18\,°C$ and a pressure p_o of 1 bar. The mixture is

ignited due to the presence of pyrophoric iron sulphide. The combustion of butane (B) follows the equation:

$$C_4H_{10} + 6\tfrac{1}{2}O_2 \rightarrow 4CO_2 + 5H_2O.$$

The standard heat of combustion (18 °C, 1 bar) of butane according to this reaction equation is equal to $\Delta H_r = -2.88 \times 10^6$ kJ. What will be the maximum pressure in the tank after the explosion, if the gas mixture is assumed to follow the ideal gas law?

Since the reaction proceeds adiabatically and at constant volume, according to equation (VI.9) the total internal energy of the reaction mixture does not change upon conversion. The temperature rise due to the explosion can best be calculated from equation (VI.12) with $\dot{Q} = 0$:

$$\int_{T_0}^{T} c_v \, dT + \int_{\xi_{B0}}^{\xi_B} (\Delta u_r)_B \, d\xi_B = 0. \tag{a}$$

Since the internal energy is a point property, the second integral can be calculated under standard conditions at 18 °C and 1 bar, and consequently the specific heat for the composition of the final reaction mixture has to be taken for c_v in the first integral.

In calculating $(\Delta u_r)_B$, we recall that the difference between ΔH_r for the above reaction (p and T constant) and ΔU_r (V and T constant) is the work needed for expansion of the gas mixture at standard temperature and pressure when the reaction goes to completion. For an ideal gas $PV = nRT$ so that from $H = U + pV$ we can derive at constant temperature

$$\Delta H_r = \Delta U_r + RT \Delta n.$$

Since the number of moles increases by $\tfrac{3}{2}$ in the above reaction equation and the ideal gas law has been assumed to apply, we have:

$$RT \Delta n = \tfrac{3}{2} RT = 3.63 \times 10^3 \text{ kJ}.$$

Hence, the correction to be applied is only of the order of 0.1%, so that we may write, with $M_B = 58$ kg/kmol B,

$$\frac{(\Delta U_r)_B}{M_B} \approx \frac{(\Delta H_r)_B}{M_B} = \frac{\Delta H_r}{M_B} = -49.7 \times 10^3 \text{ kJ/kg B}.$$

The initial and final gas compositions can be easily calculated; they are listed in Table VI.3, where air is taken as 21 vol% O_2 and 79 vol% N_2. Since $\xi_B = 0.077 - 0.017 = 0.06$, the second term of equation (a) is now calculated at 18 °C as $-49.7 \times 10^3 \times 0.06 = -2980$ kJ/kg mixture. The specific heat at constant volume of the product mixture must be calculated as a function of temperature. After some trials it appears that the average value of c_v over the temperature range to be considered is 1.06 kJ/kg °C, so that equation (a) indicates that

$$2980 = 1.06 (T - T_0) \quad \text{or} \quad T = 2830 \,°C.$$

Table VI.3 Gas compositions before and after explosion (Example VI.2.b)

	M_J (kg/kmol)	before explosion mole fraction	w_{Jo}	after explosion kmol/kmol of original mixture	w_J
$C_4H_{10}(B)$	58	0.040	0.077	0.0089	0.017
N_2	28	0.758	0.708	0.758	0.708
O_2	32	0.202	0.215	—	—
CO_2	44	—	—	0.124	0.182
H_2O	18	—	—	0.155	0.093
Total		1.000	1.000	1.046	1.000

Assuming the ideal gas law to apply, the explosion pressure can be calculated as a result of the temperature rise and the relative increase of the number of moles in the mixture:

$$p = p_o \times \frac{291 + 2810}{291} \times 1.046 = 11.1 \text{ bar.}$$

The spherical tank even if designed to withstand this pressure under static conditions, may still rupture due to the sudden shock.

VI.2.1 Batch versus semi-batch operation

Reaction vessels for batch operation are usually purchased in standard sizes, provided with standard cooling jackets and/or internal cooling coils. For strongly exothermic reactions the heat transfer area available may be too small to keep the reaction mixture at the required reaction temperature. Required reaction temperatures have usually been determined in isothermally conducted laboratory experiments, in which the optimum temperature range and reaction times have been established to obtain the highest yield of the desired product in the desired quality. In large industrial reactors, the cooling area per unit of reaction volume is much lower than in laboratory reactors. If reaction rates are so high that the reaction heat in an industrial reactor cannot be removed at the required rate, an SBR can be chosen in which the heat production rate can be reduced by slowly adding (one of) the reactant(s).

Hugo [2] has determined some rules for conditions under which a BR still can be applied or an SBR has to be chosen. In figure VI.4 the usual mode of operation of a BR is depicted for highly exothermic reactions. The reactor is filled and reaction starts adiabatically at the temperature T_o. When the reaction mixture has heated itself to the temperature T_e cooling is started at maximum flow of the cooling medium. After a certain period the maximum (allowable) temperature T_m is reached. Now the reaction gradually slows down and when T starts to drop the cooling flow rate is gradually reduced to maintain T at the desired level. It can be

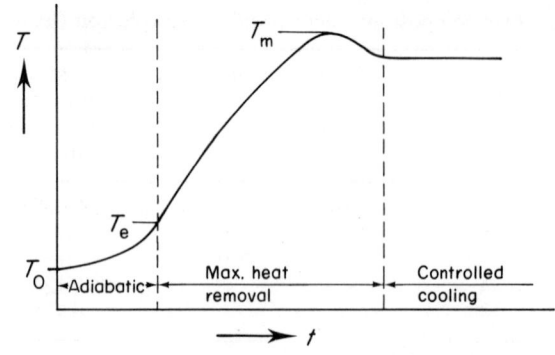

Fig. VI.4 Temperature history in a batch reactor for a highly exothermic reaction [From Hugo [2]. Reproduced by permission of Verlag Chemie GmbH]

intuitively felt that the moment to start the cooling is critical. If we start too soon, reaction time will be long and capacity reduced; with a too late start the maximum allowable temperature will be overrun.

In figure VI.5 the corresponding temperature-conversion trajectory for the reaction is shown. In general, at T_m for the reaction,

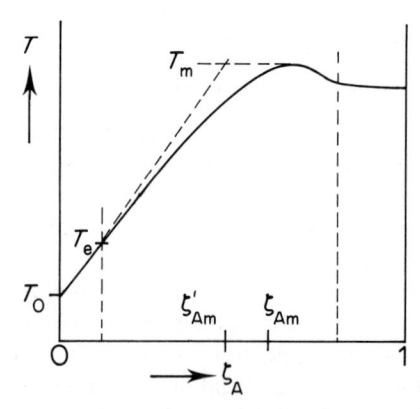

$$UA(T_m - T_c) = -(\Delta H_r)_A V_r R_{Am}. \tag{VI.17}$$

Fig. VI.5 Temperature-conversion trajectory in a batch reactor for a highly exothermic reaction

In a batch reactor the cooling area A is constant for a certain reactor volume, but, of course, can depend on the degree of filling of the reaction vessel. In a semi-batch reactor A generally is a function of time. The difficulty now is that the conversion ζ_{Am} is unknown at the point where the maximum temperature has been reached. A too conservative value ζ'_{Am} found by extrapolating the adiabatic line to T_m, so that $\zeta'_{Am} = (T_m - T_o)/\Delta T_{ad}$. The parameter ζ'_{Am} represents the fraction of the chemical heat production required to heat the reaction mixture from T_o to T_m and that consequently does not need to be removed by cooling. If $\zeta'_{Am} \geq 1$, the reactor can work adiabatically or even has to be heated. Hugo has established, by computation, that if $\zeta'_{Am} < 0.25$ for second order reactions, it is hardly possible to execute the reaction batchwise.

If batch reaction is feasible, then another necessary condition is that the maintaining of the maximum temperature T_m is warranted by the cooling. Then the sensitivity of T_m towards the cooling medium inlet temperature T_c, dT_m/dT_c, should be lower than 1 in order to have a dampening effect.

We will consider a second order reaction with

$$R_A = -kc_{Bo}c_{Ao}(1-\zeta_A)\left(1 - \frac{c_{Ao}}{c_{Bo}}\zeta_A\right).$$

In the adiabatic period after the start up, thus for $T < T_e$ and no cooling, the temperature increase can be calculated from

$$\frac{dT}{dt} = -\Delta T_{ad}\frac{R_A}{c_{Ao}} \quad \text{and} \quad \zeta_A = (T - T_o)/\Delta T_{ad},$$

with the initial conditions $T = T_o$, $\zeta_A = 0$, at $t = 0$. After $t = t_e$ the values $T = T_e$ and $\zeta_A = \zeta_{Ae}$ have been reached and cooling is started at full flow. Then the temperature increase is governed by

$$\frac{dT}{dt} = \frac{UA}{\rho c_p V_r}(T - T_c) + \Delta T_{ad}\frac{R_A}{c_{Ao}} \tag{VI.18}$$

and the conversion increase by

$$\frac{d\zeta_A}{dt} = \frac{R_A}{c_{Ao}}. \tag{VI.19}$$

The temperature conversion trajectory can be found by dividing equation (VI.18) by equation (VI.19):

$$\frac{dT}{d\zeta_A} = +\frac{UAc_{Ao}(T - T_c)}{\rho c_p V_r R_A} + \Delta T_{ad}. \tag{VI.20}$$

From equation (VI.17) we can derive (because at T_m, the highest reaction temperature, $dT/dt = 0$ and $d\zeta_A/dt \neq 0$) that

$$\frac{UA}{\rho c_p V_r} = \frac{\Delta T_{ad} R_{Am}}{c_{Ao}(T_m - T_c)}$$

so that for a second order reaction with stoichiometric feed ($c_{Bo} = c_{Ao}$), we find

$$\frac{dT}{d\zeta_A} = \Delta T_{ad}\left[1 - \frac{R_{Am}}{R_A}\frac{T - T_c}{T_m - T_c}\right]$$

$$= \Delta T_{ad}\left[1 - \frac{T - T_c}{T_m - T_c}\left(\frac{1 - \zeta_{Am}}{1 - \zeta_A}\right)^2 \exp\frac{E(T_m - T)}{RTT_m}\right]. \tag{VI.21}$$

Hugo integrated this equation for many combinations of $(T_m - T_c)$ and $(1 - \zeta_{Am})$ and different initial conditions ζ_{Ao} and T_o and also for values of $c_{Bo}/c_{Ao} \neq 1$, and he determined the sensitivity dT_m/dT_c. For a sensitivity >1 he established that the BR is critical, which means that the maximum temperature is difficult to maintain; for a sensitivity <1 the BR is non-critical. His results are given in figure VI.6, where the value $B_{crit} = E \Delta T_{ad}/RT_m^2$ is plotted versus $\zeta'_{Am} = (T_m - T_o)/\Delta T_{ad}$. In the non-critical region, by carefully choosing the cooling temperature T_c and the temperature T_e at which to start cooling the reaction mixture, the reaction temperature will not surpass the prescribed value of T_m and operation is safe. On the boundary line B_{crit} versus ζ'_{Am}, the values of T_e and T_o coincide. By introducing the value of ζ'_{Am} into equation (VI.17), we can calculate the required cooling area A. Summarizing, if for a certain value of ζ'_{Am} the value of $B = E \Delta T_{ad}/RT_m^2$ is below B_{crit}, the reaction can be executed in a BR. If $B > B_{crit}$ then a SBR has to be chosen. In case a BR can be chosen; detailed calculations for T_e and t_e are required or the approximations of Hugo can be used [2, 3].

Example VI.2.c *Synthesis of a maleic acid mono-ester in a BR*
Hugo [2] studied the synthesis of hexyl monoester of maleic acid according to

No solvent is used. Maleic anhydride (A) melts at 53 °C; a slight excess of hexanol (B) is used. Both reactants are dumped into the reaction kettle and heated up till all A has melted. In

$$R_A = -kc_{Ao}c_{Bo}(1 - \zeta_A)(1 - c_{Ao}\zeta_A/c_{Bo})$$

the rate coefficient is

$$k = 1.37 \times 10^{12} \exp(-12\,628/T)\,\text{m}^3/\text{kmol s}.$$

The heat of reaction is $\Delta H_r = -33.5$ MJ/kmol, the heat capacity of the reaction mixture $\rho c_p = 1980$ kJ/m³ K. After melting under agitation $c_{Ao} = 4.55$ and $c_{Bo} = 5.34$ kmol/m³. The temperature of the reaction mixture should not exceed $T_m = 100$ °C. Can this reaction be executed in a BR?

We first calculate the adiabatic temperature rise:

$$\Delta T_{ad} = \frac{33.5 \times 10^6 \times 4.55}{1980 \times 10^3} = 77\,°C$$

The reaction starts after A has melted and has dissolved into B, say at 55 °C. Under adiabatic conditions the final temperature would be $77 + 55 = 133$ °C, so that we have to cool. In this case

$$B = \frac{12\,628 \times 77}{373^2} = 6.99.$$

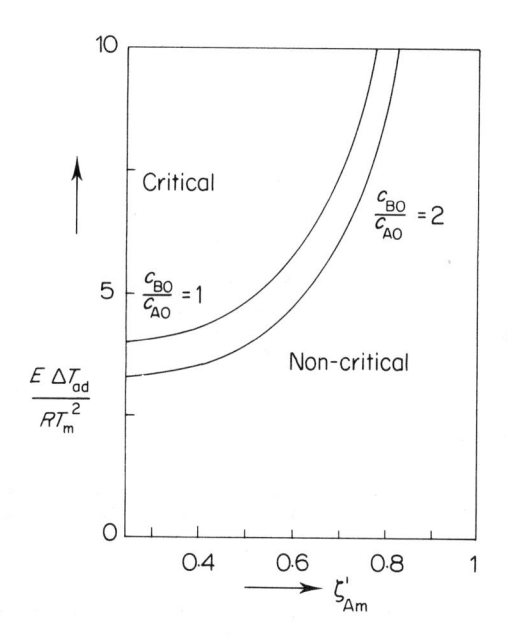

Fig. VI.6 Regions where it is critical whether a second order reaction can be executed in a batch reactor. $\zeta'_{Am} = (T_m - T_o)/\Delta T_{ad}$ and $B_{crit} = E\Delta T_{ad}/RT_m^2$ [From Hugo [2]. Reproduced by permission of Verlag Chemie GmbH]

From figure VI.6 we read that ζ'_{Am} should be at least 0.70. With $\zeta'_{Am} = (T_m - T_o)/\Delta T_{ad}$ we then find that $T_m - T_o = 0.70 \times 77 = 54\,°C$, hence for $T_m < 100\,°C$ the reaction should start at or below $100 - 54 = 46\,°C$. At that temperature A is not yet melted, which leads us to the constraint that the temperature of the cooling medium for safe operation should be 55 °C or higher in order to prevent the crystallization of A on the cooling surface with the consequent reduction in heat transfer coefficient. With the data given above and with $U = 250\,W/m^3\,K$, we find at $\zeta'_{Am} = 0.70$,

$$\frac{A}{V} = \frac{\rho c_P \Delta T_{ad} k_{cBo}(1 - \zeta'_{Am})(1 - c_{Ao}\zeta'_{Am}/c_B)}{U(T_m - T_c)}$$

$$= \frac{1980 \times 10^3 \times 77 \times 1.37 \times 10^{12} \times \exp(-33.86) \times 5.34 \times 0.30 \times 0.404}{250 \times (100 - 55)}$$

$$= 24\,m^2/m^3.$$

A cooling area of $24\,m^2/m^3$ is far too high for normal standard reaction kettles available on the industrial market. Depending on the kettle size $2-5\,m^2/m^3$ can be installed in reaction vessels with only jacket cooling and $4-8\,m^2/m^3$ if the vessel is also provided with an internal cooling coil.

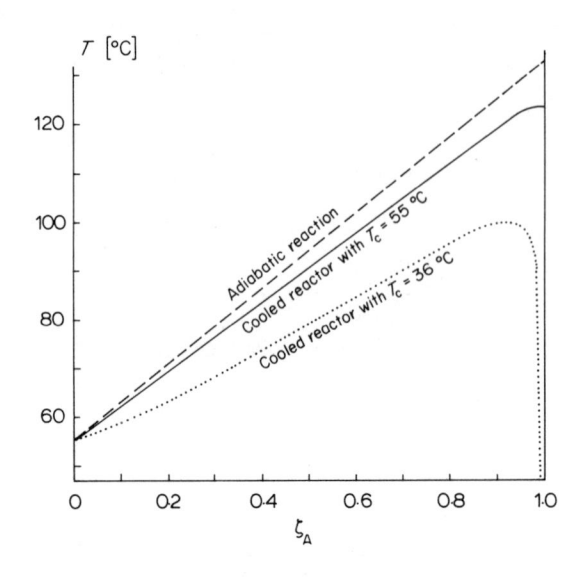

Fig. VI.7 Temperature–conversion trajectories in a batch reactor for the reaction in Example VI.2.c

For a batch size of $5\,\text{m}^3$ and a cooling area of $10\,\text{m}^2$, the temperature conversion trajectory is given in figure VI.7. The reaction mixture reaches a maximum temperature of 123 °C, if the temperature of the cooling medium $T_c = 55\,°C$. If $T_c = 36\,°C$ then under the conditions given, the maximum temperature is not higher than 100 °C. This T_c, however, cannot be chosen, because the cooling surface will get covered with crystals! In an SBR we will first melt A and maintain the reaction temperature by a controlled addition of B at full flow of the cooling medium. This will be shown in Example VI.2.d. Of course, as alternatives to an SBR, the use of a solvent and a larger batch reactor, the recirculation of the reaction mixture over an external heat exchanger with a sufficiently high cooling area, etc. can also be chosen. A final choice can only be made on economic grounds.

Example VI.2.d   *The synthesis of a maleic acid mono-ester in an SBR*
The reaction of the previous illustration is now carried out in an SBR. The reactor is filled with A and heated up till all A has melted. The volume V_{ro} is now $2.20\,\text{m}^3$. When the reactor temperature has reached 55 °C a constant flow $\Phi_B\,\text{m}^3/\text{s}$ of B is fed into the reactor and maintained constant until 1.20 kmol of B have been fed per kmol of A originally present. At 55 °C $c_{A0} = 10.1\,\text{kmol/m}^3$, the density of the reaction mixture remains constant. The feed flow of B is introduced at the temperature T_B and contains $c_{BL} = 9.7\,\text{kmoles B/m}^3$. Further conditions are the same; also the same vessel is used as in the previous example. We will calculate the value of Φ_B at which the maximum reaction temperature is $T_m = 100\,°C$.

The material balances for A and B are respectively

$$\frac{d(c_A V_r)}{dt} = c_A \frac{dV_r}{dt} + V_r \frac{dc_A}{dt} = -kc_A c_B V_r$$

$$\frac{d(c_B V_r)}{dt} = c_B \frac{dV_r}{dt} + V_r \frac{dc_B}{dt} = \Phi_B c_{BL} - kc_A c_B V_r,$$

and further the reaction volume is $V_r = V_{ro} + \Phi_B t$. The heat balance is

$$\rho c_p V_r \frac{dT}{dt} = (-\Delta H_r) kc_A c_B V_r - \rho c_p \Phi_B (T - T_B) - U\frac{A}{V} V_r (T - T_c).$$

Note that the cooling area increases with the filling of this reactor. After some rearrangements we find

$$\frac{dc_A}{dt} = -\frac{c_A}{A+t} - kc_A c_B$$

$$\frac{dc_B}{dt} = \frac{c_{BL} - c_B}{A+t} - kc_A c_B$$

$$\frac{dT}{dt} = Bkc_A c_B - \frac{T - T_B}{A+t} - C(T - 328).$$

In our case the values of the constants in these equations are:

$A = V_{ro}/\Phi_B$, the time required to double the originally present reaction volume. The higher the value of A, the slower reactant B is added.

$$B = -\Delta H_r/\rho c_p = \frac{33.5 \times 10^6}{1.98 \times 10^6} = 16.92 \, \text{m}^3 \, \text{K/kmol}$$

$$C = \frac{U}{\rho c_p} \times \frac{A}{V} = \frac{250 \times 2}{1.98 \times 10^6} = 0.253 \times 10^{-3} \, \text{s}^{-1}$$

The initial conditions at $t = 0$ are: $c_{Ao} = 10.1 \, \text{kmol/m}^3$, $c_{Bo} = 0 \, \text{kmol/m}^3$ and $T_o = 328 \, \text{K}$. As in the previous illustration we stop the feeding of B when $\Phi_B c_{BL} t = 1.20 \times V_{ro} c_{Ao}$. In that case $t = 1.20 \times \frac{10.1}{9.7} \times A = 1.2A$. The three simultaneous differential equations have been solved numerically on a computer for different values of A. Results for full flow of the cooling medium are given in figure VI.8, where reactor temperature–reaction time profiles are given for different values of A. We have to realize that at the start of the reaction the available cooling area is $(A/V_r) \times V_{ro} = 4.5 \, \text{m}^2$ and at the end of the addition period $A = 10 \, \text{m}^2$. From figure VI.8 we see that the maximum reactor temperature $T_{max} = 373 \, \text{K}$ is reached for $A = 6400$ seconds. At this value of A the feed rate is $\Phi_B = V_{ro}/A = 0.355 \times 10^{-3} \, \text{m}^3/\text{s}$. At higher feed rates the maximum allowable temperature is overshot. At lower feed rates cooling at full flow is too strong, reaction temperatures are too low and the total reaction time too long. In the

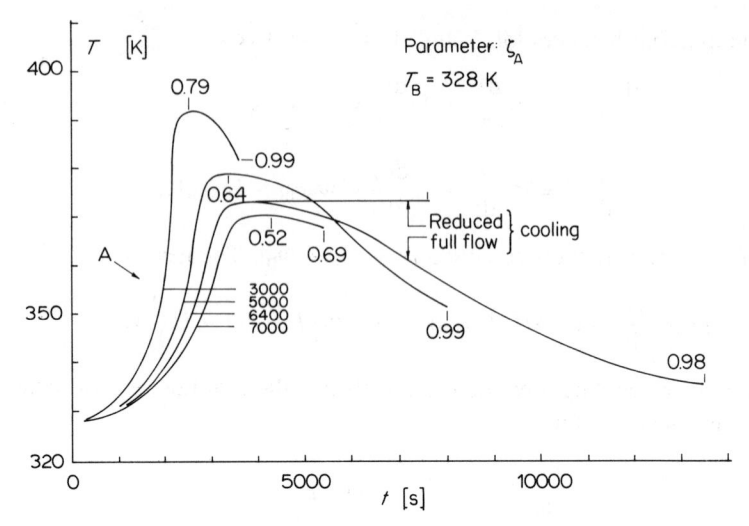

Fig. VI.8 Temperature–time histories in a semi batch reactor; Example VI.2.d. A is the time required to double the reaction volume—a measure for the addition rate of reactant B—and T_B the temperature of the stream added. The reactor is cooled at maximum cooling rate during the addition period

optimum case, with $T_{max} = 373$ K, reaction time is the shortest possible and reactor capacity highest. In that case after reaching $T_{max} = 375$ K the cooling flow, of course, is reduced to maintain the reactor temperature at 373 K in order to shorten the reaction time.

In figure VI.9 the concentrations of A and B and the conversion are given as a function of time in the optimal case. Here the conversion is defined as $\zeta_A = 1 - (V_{rt}c_{At})/V_{ro}c_{Ao}$, in which $V_{rt} = V_{ro} + \Phi_B t$, so that also

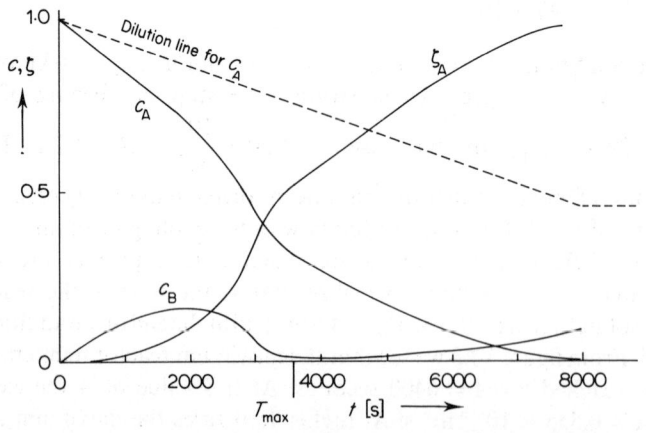

Fig. VI.9 Concentration–time and conversion–time profiles for the maximum reaction temperature of 373 K in the semi-batch reactor of Example VI.2.d

$\zeta_A = 1 - c_{At}(A + t)/Ac_{Ao}$ and c_{At} is the concentration of A at the moment t. Also a dilution line for c_A is drawn, which indicates the effect of dilution only. The value of c_A would follow this line due to the dilution by B, if no reaction took place. We see that the conversion rate increases sharply in the beginning due to the temperature increase. After that the conversion rate is almost constant, due to the increase of the concentration of B in the final part of the run. At the end of the addition period of 8000 seconds a conversion of $\zeta_A = 0.99$ is already reached.

VI.3 THE TUBULAR REACTOR WITH EXTERNAL HEAT EXCHANGE

In this section we shall assume that the temperature and the composition of the reaction mixture are uniform over the cross section of the reactor tube and depend only on the distance z from the feed point. The complications arising from a radial temperature gradient and a radial change of composition—frequently encountered in fixed bed catalytic reactors—will not be considered here but in Chapter IX.

According to equation (VI.1), the steady state energy balance over a differential volume of the reactor becomes, if no work is done by the reaction mixture:

$$0 = -\Phi_m \, dh + d\dot{Q}. \tag{VI.1}$$

If the pressure changes relatively little over the reactor, the enthalpy change dh is made up of two contributions: one due to a temperature change dT, and one due to a conversion characterized by the change in degree of conversion of one of the components. We thus have, according to equation (VI.5),

$$dh = c_p \, dT + (\Delta h_r)_J \, d\xi_J.$$

The heat supply (which may be positive or negative) is assumed to be provided by heat transfer from a medium flowing outside the tube with a temperature T_c. The overall heat transfer coefficient U is assumed to be localized at the tube wall. By introducing the hydraulic radius of the tube, r_h, which is the ratio between the cross sectional area S and the circumference, we obtain for $d\dot{Q}$ ($r_h = d_t/4$ for tubes):

$$d\dot{Q} = \frac{U(T_c - T)}{r_h} S \, dz.$$

Following the above substitutions, equation (VI.1) becomes

$$0 = \Phi_m c_p \, dT + \Phi_m (\Delta h_r)_J \, d\xi_J + \frac{U(T - T_c)}{r_h} S \, dz. \tag{VI.22}$$

This relationship must be used in conjunction with the material balance (II.14):

$$0 = -\Phi_m \, d\xi_J + M_J |R_J| S \, dz,$$

for the calculation of the temperature and the degree of conversion along the reactor. In some cases the flow through a tubular reactor is associated with a considerable pressure drop due to frictional losses and to the acceleration of the reaction mixture (e.g. in a cracking furnace); under such circumstances the equations of motion or the momentum balance have to be taken into account also, as well as the influence of pressure on the enthalpy of the mixture.

If we consider a reactant A in a PFR with a heated or cooled tube wall, the equations can be simplified. Generally for a reactant $d\xi_A = -M_A d(c_A/\rho)$, and if the density of the reaction mixture is constant $d\xi_A = +(M_A c_{Ao} d\zeta_A)/\rho$, so we find, after dividing equation (VI.22) by $c_p \Phi_m$, for a reaction in a tube,

$$dT = \frac{-(\Delta H_r)_A c_{Ao}}{\rho c_p} d\zeta_A - \frac{U(T-T_c)S}{\rho c_p \Phi_v r_h} dz = \Delta T_{ad} d\zeta_A - \frac{4U}{\rho c_p d_t}(T-T_c) d(z/u)$$

(VI.23)

and for the material balance we have

$$d\zeta_A = \frac{R_A}{c_{Ao}} d(z/u).$$

(VI.24)

A straightforward solution of the last two equations is only possible in a few special cases. For instance, if the reactor runs under adiabatic conditions, the last term in equation (VI.23) vanishes and a single relation is left between the temperature and the degree of conversion. The conversion rate can hence be expressed as a function of the degree of conversion only, so that the material balance can be directly integrated. Another case where the reactor calculation does not present special difficulties is the rather trivial one where the heat effect is small and the heat transfer with the surrounding medium is extremely good. Under such conditions the reactor works isothermally at $T \approx T_c$ and equation (VI.23) can further be disregarded.

In general, however, the calculation of a non-isothermal cooled or heated tubular reactor requires the simultaneous solution of (VI.23) and (II.14), e.g. by means of the numerical method indicated in Example VI.2a, which is also employed in one of the following Examples where an endothermic catalytic reaction is dealt with.

If we wish to determine the temperature profile along the reactor length in a PFR, we have to solve the following differential equation in conjunction with equation (VI.24) simultaneously:

$$\frac{dT}{d\tau} = -\Delta T_{ad} \frac{R_A}{c_{Ao}} - \frac{4U}{\rho c_p d_t} \Delta T.$$

Here $\Delta T = T - T_c$ and $\tau = z/u$. In an isothermal PFR $dT/d\tau = 0$, so that in this case the time constant for the required heat removal capacity is given by

$$\frac{\rho c_p d_t \Delta T_{ad}}{4U \Delta T} = -\frac{c_{Ao}}{R_A}.$$

As R_A varies strongly along the reactor length this requires a continuous drastic change of $d_t/\Delta T$ along the reactor (or $\Delta T \approx 0$ or $U \approx \infty$, of course), which in practice cannot be realized. For a first order reaction $R_A = - kc_A$. For k we can write

$$k = k_c(k/k_c) = k_c \exp\left(\frac{E\Delta T}{RTT_c}\right) \approx k_c \exp\frac{E\Delta T}{RT_c^2}.$$

In this relation k_c, the reaction rate constant at the temperature T_c of the cooling medium, is given by

$$k_c = k_\infty \exp\left(-\frac{E}{RT_c}\right),$$

so that

$$\frac{1}{d_t} = \frac{\rho c_p}{4U}(1 - \zeta_A)\exp\left(\frac{E\Delta T}{RT_c^2}\right) \times \frac{\Delta T_{ad}}{\Delta T} k_\infty \exp\left(-\frac{E}{RT_c}\right).$$

In this relation U is still a function of the tube diameter d_t. We see that it is impossible to shape the tube diameter in such a way that ΔT remains constant, especially if the reactor also is required to operate satisfactorily at varying loads.

Example VI.3.a *Tube diameters required in an isothermal laboratory plug flow reactor*

A first order reaction is carried out in a tube reactor in a thermostatic bath in the laboratory. The reactor can be assumed to be a PFR. Data on the reaction are $k = 3.94 \times 10^{12} \exp(-11\,400/T)$ s and $\Delta T_{ad} = 146\,°C$. The overall heat transfer coefficient, assumed to be independent of d_t, is $U = 400\,W/m^2\,°C$ and for the reaction mixture $\rho c_p = 1760\,kJ/m^3\,°C$. The feed to the reactor is preheated, to the reaction temperature. We will determine the required tube diameters, if the reaction temperature is not allowed to surpass the temperature of the thermostatic bath by more than $3\,°C$. The heat evolution rate is highest at the reactor inlet, where $\zeta_A = 0$. At the inlet the diameter is determined by

$$\frac{1}{d_t} = \frac{1.76 \times 10^6}{4 \times 400} \times 1 \times \exp\left(\frac{11\,400 \times 3}{T_c^2}\right) \times \frac{146}{3} \times 3.94 \times 10^{12} \exp\left(-\frac{11\,400}{T_c}\right).$$

After substituting the bath temperatures T_c we find the following results:

T_c (K)	320	330	340	350
d_t (mm)	10.1	3.5	1.3	0.5

So we can conclude that if, for example, a tube diameter of 3 mm is used, the reaction conditions cannot be kept sufficiently isothermal above 330 K.

Example VI.3.b *A catalytic dehydrogenation reaction in a tubular reactor*
It is required to calculate a tubular reactor for the production of methyl-ethyl ketone (P) from 2-butanol (A) according to the endothermic reaction

$$CH_3CHOHC_2H_5 \rightarrow CH_3COC_2H_5 + H_2.$$
$$\quad\quad\text{(A)} \quad\quad\quad\quad \text{(P)} \quad\quad \text{(Q)}$$

The reaction is to be carried out in 25 parallel tubes (internal diameter 0.10 m) filled with brass spheres (having a diameter of 3.43 mm) as a catalyst. Perona and Thodos [4] studied the conversion rate for this catalytic material between 350 and 400 °C at atmospheric pressure. They arrived at the expression

$$-R_A'' = \frac{k''(p_A - p_P p_Q/K)}{p_P(1 + K_A p_A + K_{AP} p_A/p_P)},$$

where the reaction rate is given on the basis of the external surface of the catalytic spheres; k'', K_A and K_{AP}, and K represent the reaction velocity constant, the adsorption equilibrium constants, and the reaction equilibrium constant, respectively; these quantities are given as a function of temperature. The feed is 100 tonnes/day of butanol, which has to attain a degree of conversion ζ_{AL} of 0.95. Further data are:

$$(\Delta H_r)_A = +59 \times 10^6 \text{ J/kmol A}$$

or

$$(\Delta h_r)_A = +0.80 \times 10^6 \text{ J/kg A},$$
$$C_p = 178 \times 10^3 \text{ kJ/kmol mixture, assumed constant,}$$
$$p = \text{average pressure in reactor} = 5 \text{ bars,}$$
$$T_o = 450 \,°C,$$
$$T_h = 800 \,°C \quad \text{(heating medium)}$$

and

$$U = 200 \text{ W/m}^2 \,°C.$$

It is assumed that temperature and composition are uniform over the cross section of the tube and that the kinetic data referred to above may be extrapolated to higher and lower temperatures and pressures outside the range of 350–400 °C. This reactor, of course, is a heterogeneous reactor, but—as the heat conductivity of brass spheres is high—we assume we can treat this reactor as an ideal PFR. This means that concentrations and temperatures of the gas and the catalyst particles are uniform over the cross section. The rate equation is first rewritten in terms of the degree of conversion of A, $\zeta_A = 1 - w_A$, by means of the procedure indicated in Section 1.5. Then, since the amount of catalyst per tube is to be calculated, the conversion rate is based on the unit mass of catalyst; this amounts to a division of R_A' by the ratio mass/surface area of the catalyst, which can be calculated from the sphere diameter (3.43 mm) and the density of brass $\rho_s = 8850 \text{ kg/m}^3$. The mass per particle is $\pi d_p^3 \rho_s/6$ and the surface area is πd_p^2, so that the ratio is $d_p \rho_s/6 = 4.89 \text{ kg/m}^2$. Finally, the conversion rate is expressed in mass units of A by multiplying by the molecular weight $M_A = 74 \text{ kg A/kmol A}$, so that

$$M_A R_{SA} = \frac{74}{4.89} \times R_A'' = 15.1 R_A'' \text{ kg A converted/s kg catalyst}.$$

If the partial pressures are given in terms of the total pressure and ξ_A, we can plot at a pressure of 5 bars $M_A R_{SA}$ as a function of ξ_A and T. This is shown in figure VI.10.

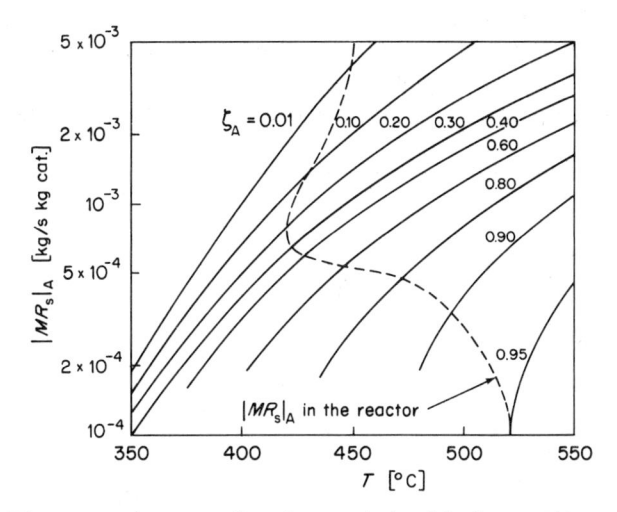

Fig. VI.10 The conversion rate for the catalytic dehydrogenation of 2-butanol (component A) at 5 bars, calculated and extrapolated from data of Perona and Thodos [4]; Example VI.3.b, an endothermic equilibrium reaction

Instead of the distance z from the reactor inlet, we now take as a variable the mass of catalyst m, between the inlet and the point under consideration. The equations to be solved then become (equation (II.11))

$$0 = d\xi_A - R_{SA} M_A \, d(m_s/\Phi_m), \tag{a}$$

and from equation (VI.22) with $m_s = \rho_s(1 - \varepsilon)Sz$,

$$0 = c_p \, dT + (\Delta h_r)_A \, d\xi - \frac{U(T_h - T)}{\rho_s(1 - \varepsilon)r_h} d\left(\frac{m_s}{\Phi_m}\right). \tag{b}$$

The boundary conditions for $m_s = 0$ are

$$\xi_A \equiv 0 \quad \text{and} \quad T = T_o = 450 \,°\text{C}.$$

Because of the assumed constancy of molar specific heat we may write for c_p

$$c_p = C_p(1 + \xi_A)/M_A = 2.40(1 + \xi_A) \, 10^3 \text{ J/kg °C}.$$

If the porosity of the catalytic bed, ε, is assumed to be 0.4, we find that the quantity $U/\rho_s(1 - \varepsilon)r_h = 1.58$ W/kg °C. A step-by-step method should again be followed for the simultaneous solution of (a) and (b):

— consider a 'slice' Δm_s at the feed point;
— find $-R_{SA}M_A$ from figure IV.4 at $\xi_A = 0$ and $T = T_o$;
— calculate $\Delta \xi_A$ from (a) and ΔT from (b);
— calculate the average values of ξ_A and T over the interval m_s, and with these recalculate $\Delta \xi_A$ and ΔT;
— repeat this until (a) and (b) have been consistently solved, and then proceed to the next interval.

The results of these calculations are shown in figure VI.11, where ξ_A and T have been plotted against m_s/Φ_m. In this case of an endothermic reaction, the reaction temperature T decreases rapidly at first and then passes through a minimum. In figure VI.10 the path of $-R_{SA}M_A$ has been indicated as a dotted line.

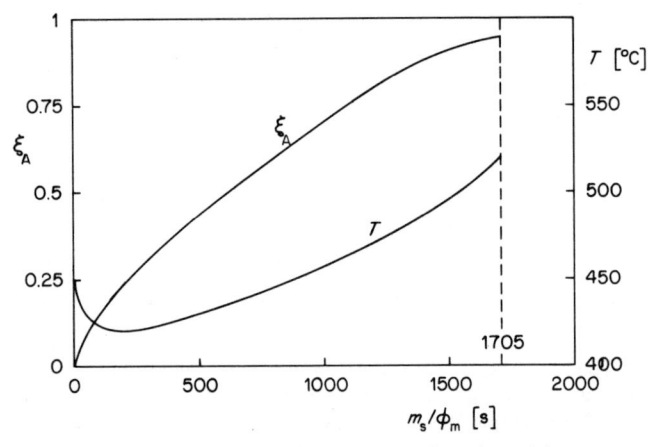

Fig. VI.11 Degree of conversion and temperature as a function of the amount of catalyst in a tubular reactor with an endothermic reaction; Example VI.3.b

From figure VI.11 it appears that for $\xi_{AL} = 0.95$ the value of m_{sL}/Φ_m must be 1705 kg catalyst per kg/s of feed. Since the mass flow Φ_m per tube is

$$\Phi_m = \frac{100.000}{25 \times 24 \times 3600} = 46.3 \times 10^{-3} \text{ kg/s},$$

the amount of catalyst required per tube is

$$m_{sL} = 1705 \times 46.3 \times 10^{-3} = 79 \text{ kg}.$$

A tube length of 2 m is needed to accommodate this amount of catalyst:

$$L = \frac{m_{sL} \times 4}{\rho_s(1 - \varepsilon)\pi d_t^2} = \frac{79 \times 4}{8550 \times (1 - 0.4) \times \pi \times (0.1)^2} = 1.96 \text{ m}.$$

The pressure drop over this tube can be calculated to be approximately $9 \times 10^4 \, \text{N/m}^2 \approx 0.9$ bar.

The outlet temperature of the reaction mixture is about 520 °C. This may well be too high a value in view of thermal decomposition of the organic material involved. This can be avoided by lowering the temperature T_h of the heating medium and/or by diminishing the heat consumption per unit volume (e.g. by diluting the feed or the catalyst). These changes would result in a considerable increase in reaction volume and pressure drop.

Could this dehydrogenation also be carried out in an adiabatic PFR? In that case the last term in equation (b) would disappear, so that $c_p \, dT = -(\Delta H_r)_A \, d\zeta_A/M_A$, which leads to

$$T - T_o = -\int_0^{0.95} \frac{(\Delta H_r)_A}{c_p(1 + \zeta_A)} \, d\zeta_A$$

or

$$T = T_o - \frac{590 \times 10^6}{178 \times 10^3} \ln 1.95 = T_o - 221.$$

This temperature drop is far too high for the reaction to proceed at reasonable rates (see figure VI.10). The feed could be diluted with a heat carrier in order to reduce the adiabatic temperature drop. The dilution would reduce the reaction rate because of the lower partial pressures, but far less than the decrease in temperature would do. The adiabatic reactor would have a completely different geometric configuration: we would have one large diameter packed bed instead of a series of parallel packed tubes.

Example VI.3.c *First order reaction in an adiabatic PFR*

The first order reaction of Example VI.3a is carried out in a PFR adiabatically. Which reactor length is required if the inlet temperature is $T_o = 330 \, \text{K}$? In this case equation (VI.15) gives the relation between the reactor temperature and the conversion:

$$T = T_o + \Delta T_{ad} \zeta_A.$$

Substitution in the material balance gives

$$d\tau = \frac{c_{Ao} \, d\zeta_A}{k c_{Ao}(1 - \zeta_A)},$$

in which

$$k = k_\infty \exp\left(-\frac{E}{RT}\right).$$

For the temperature profile we have

$$\frac{dT}{d\tau} = \Delta T_{ad} k(1 - \zeta_A) = \frac{\Delta T_{ad}(\Delta T_{ad} - T + T_o)}{\Delta T_{ad}} k_\infty \exp\left(-\frac{E}{RT}\right)$$

$$= (T_{ad} - T) k_\infty \exp\left(-\frac{E}{RT}\right).$$

With the initial conditions $\tau = 0$ and $T = T_o$ we find

$$k_\infty \tau = \int_{T_o}^{T} \frac{\exp(E/RT) \, dT}{T_{ad} - T}.$$

Here $T_{ad} = T_o + \Delta T_{ad}$. If we substitute $y = E/RT$, then $T = E/Ry$ and $dT = -E \, dy/Ry^2$ and with $y_{ad} = E/RT_{ad}$ this becomes

$$k_\infty \tau = -\int_{y_o}^{y} \frac{e^y \, dy}{y\left(\dfrac{y}{y_{ad}} - 1\right)}.$$

By splitting in fractions we find

$$k_\infty \tau = \int_{y_o}^{y} \frac{e^y}{y} \, dy - \int_{y_o}^{y} \frac{e^y \, dy}{y - y_{ad}}.$$

On substitution of $x = y - y_{ad}$ the second integral becomes

$$\int_{y_o}^{y} \frac{e^y \, dy}{y - y_{ad}} = e^{y_{ad}} \int_{x_o}^{x} \frac{e^x}{x} \, dx.$$

The exponential integral is defined by

$$\int_{-\infty}^{a} \frac{e^y}{y} \, dy = \text{Ei}(a),$$

so that the solution now is

$$k_\infty \tau = \exp\left(\frac{E}{RT_{ad}}\right)\left[\text{Ei}\left(\frac{E}{RT_o} - \frac{E}{RT_{ad}}\right) - \text{Ei}\left(\frac{E}{RT} - \frac{E}{RT_{ad}}\right)\right]$$
$$- \text{Ei}\left(\frac{E}{RT_o}\right) + \text{Ei}\left(\frac{E}{RT}\right).$$

Herewith the temperature profile $T - \tau$ can be calculated for first order reactions in an adiabatic PFR. Values of Ei can either be found e.g. in Jahnke et al. [5] or be calculated on a computer. For solutions for adiabatic reactors with zero and second order kinetic equations executed at constant pressure or with constant density in a PFR, see e.g. Almasy [6] and Douglas and Eagleton [7].

For the reaction type in this Example the results are given in figure VI.12.

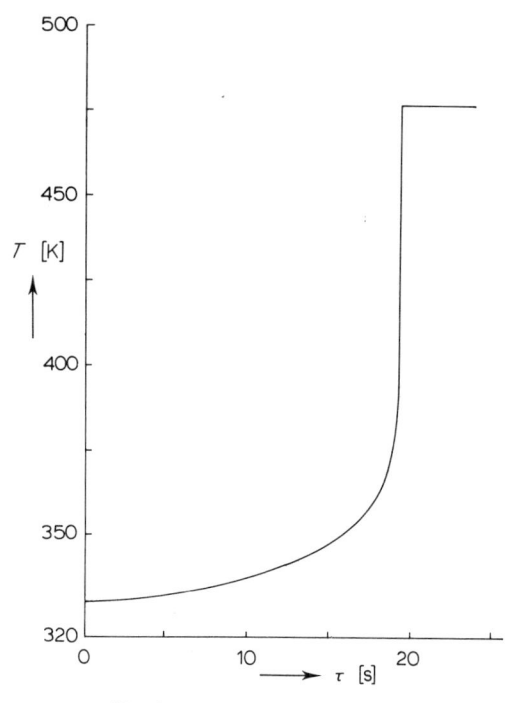

Fig. VI.12 Temperature profile along an adiabatic plug flow reactor for a first order reaction; Example VI.3.c

VI.3.1 Maximum temperature with exothermic reactions; parametric sensitivity

In the endothermic reaction of Example VI.3.b it was seen that the reaction temperature passes through a minimum. Similarly, the temperature in a cooled tubular reactor may pass through a maximum value in the case of exothermic reactions. This is qualitatively shown in figure VI.13, which demonstrates that the

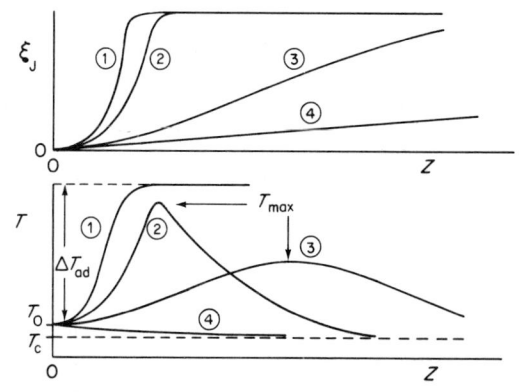

Fig. VI.13 Qualitative conversion and temperature profiles for an exothermic reaction in a cooled tubular reactor; 1–adiabatic operation; 2, 3 and 4—low, rather high and extremely high capacity for heat removal, respectively

value of this maximum temperature T_{max} lies between T_o and $T_o + \Delta T_{ad}$; its actual value depending on the heat effect and the possibilities for heat exchange. As in the batch reactor, so in the tubular reactor in many cases it is required that T_{max} (often called the 'hot spot temperature') does not exceed a prescribed value, because undesired side reactions might otherwise occur, such as thermal decomposition of the product. Since this highest permissible temperature is often much lower than the temperature obtainable under adiabatic conditions, it is of great importance to know the influence on the value of T_{max} of the properties of the reaction mixture, of the operating variables and of the capacity for heat exchange.

Not only dT/dz but also $dT/d\zeta_A$ is zero at the 'hot spot'. Therefore, after eliminating the variable $d(z/u)$ from equations (VI.23) and (VI.24) by means of the result

$$\frac{dT}{d\zeta_A} = \Delta T_{ad} - \frac{4U(T - T_c)c_{Ao}}{d_t \rho c_p |R_A|}, \quad (VI.25)$$

we may, if $T = T_{max}$, put $dT = 0$, so that

$$T_{max} - T_c = \frac{|R_A|\rho c_p d_t \Delta T_{ad}}{4U c_{Ao}} = \frac{(-\Delta H_r)_A |R_A| d_t}{4U}.$$

In this equation, the reaction rate $-R_A$ must be taken at the temperature and the degree of conversion in the hot spot; since these quantities are determined by the entire history of the reaction mixture between the feed point and the hot spot, the latter equation does not permit direct calculation of T_{max}. It can be seen, however, that, with a large capacity for heat exchange, U/d_t and a low coolant temperature T_c, T_{max} may in principle be kept below a certain value. It is not practical to choose too low a value of T_c because this will appreciably reduce the rate of conversion in the final section of the reactor; this would involve a large reactor volume. For this reason, high temperature coolants (e.g. molten salts) are often used for cooling tubular (catalytic) reactors. The most powerful variable at our disposal for keeping the reaction temperature between desired limits is r_h, the ratio of the reactor volume and the heat transfer surface. The smaller its value, i.e. the smaller the tube diameter, the better the reaction temperature can be controlled. Sometimes the reactor is also divided into sections with different tube diameters. If tube diameters become too small with respect to the pressure drop over the reactor, other measures have to be taken, e.g. dilution of the feed, in order to reduce the maximum reaction temperature.

In a certain range the temperature and conversion profiles in a PFR with external heat exchange are very sensitive to a small change in the values of their parameters. This can be determined by the simultaneous solution of the differential equations (VI.23) and (VI.24). It is often more convenient to determine first the temperature-conversion trajectories by using equation (VI.25) and then solving equations (VI.23) and (VI.24). For first order reactions with

$R_A = -kc_{Ao}(1-\zeta_A)$ we have in a cooled PFR:

$$\frac{dT}{d\zeta_A} = \Delta T_{ad} - \frac{4U(T-T_c)}{d_t \rho c_p k(1-\zeta_A)} \tag{VI.26}$$

with the initial conditions $\zeta_A = 0$ and $T = T_o$. The controlling parameters are ΔT_{ad}, T_o, T_c, $4U/d_t\rho c_p$ and, of course, also the activation energy of the kinetic rate constant. In ΔT_{ad} are included the inlet concentration of the reactant c_{Ao}, the heat of reaction $(-\Delta H_r)_A$ and the volumetric heat capacity of the reaction mixture ρc_p. By solving equation (VI.26), we find that for small changes in one of the parameters under certain conditions suddenly the temperature increases very rapidly along the reactor length, the temperature 'runs away'. In the reactor a 'hot spot' develops and, at very insufficient cooling, the temperature continues increasing until all reactant has been consumed.

Example VI.3.d *Parametric sensitivity*
We will calculate for the first order exothermic reaction, with

$$k = 3.94 \times 10^{12} \exp\left(-\frac{11\,400}{T}\right) s^{-1},$$

the temperature and conversion profiles for cooling temperatures of 339 and 337 K. These temperatures have been chosen in the region where the reactor is sensitive towards the cooling medium temperature. The reactor inlet temperature is $T_o = 340$ K, further $\Delta T_{ad} = 146$ K and $4U/d_t\rho c_p = 0.20\,s^{-1}$. After that we will also study the sensitivity towards the reactor inlet temperature at a constant cooling temperature of $T_c = 337$ K.

Introducing the data given into equation (VI.25):

$$\frac{dT}{d\zeta_A} = 146 - \frac{5.076 \times 10^{-14}(T-337)}{(1-\zeta_A)\exp(-11\,400/T)}.$$

With the T–ζ_A relation obtained we can calculate the T–τ and ζ_A–τ profiles by introducing it into the equations

$$\frac{d\zeta_A}{d\tau} = k(1-\zeta_A) = 3.94 \times 10^{12} \exp\left(-\frac{11\,400}{T}\right) \times (1-\zeta_A)$$

and

$$\frac{dT}{d\tau} = \Delta T_{ad} k(1-\zeta_A) - \frac{4U}{d_t \rho c_p}(T-T_c)$$

$$= 146 \times 3.94 \times 10^{12} \times \exp\left(-\frac{11\,400}{T}\right) - 0.20\,(T-337).$$

Calculations have been done by computer. Results are given in figures VI.14.a, b and c. We see that the reactor is extremely sensitive towards the cooling temperature in the range of 337 to 339 K. At $T_c = 337$ K the reactor exhibits a hot spot of $T_{max} = 353$ K; beyond this point the temperature further decreases. At this cooling temperature for a conversion of $\zeta_A = 0.95$ the reactor requires a residence time of $\tau_L = 235$ s. At a cooling temperature of $T_c = 339$ K the reactor temperature completely runs away to above 420 K. The same conversion of $\zeta_A = 0.95$ is now reached after $\tau_L \approx 19$ s. Still part of the reaction heat is removed by the cooling medium as can be seen from the adiabatic line in figure VI.14.a. The constancy of the cooling medium temperature is very critical in this case: its control can best be accomplished by applying a boiling liquid as cooling medium and carefully controlling the boiling pressure.

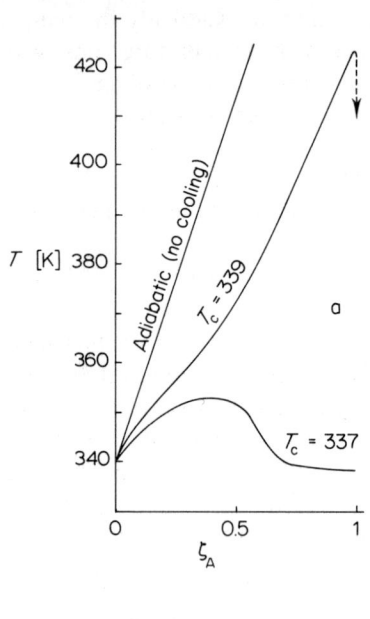

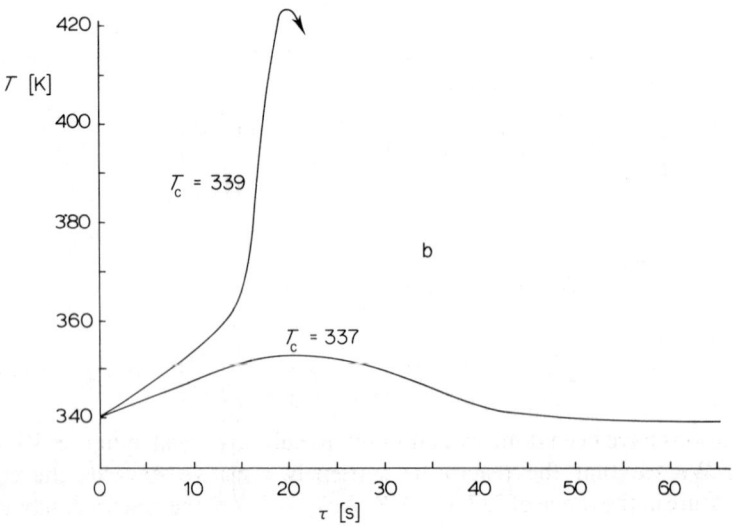

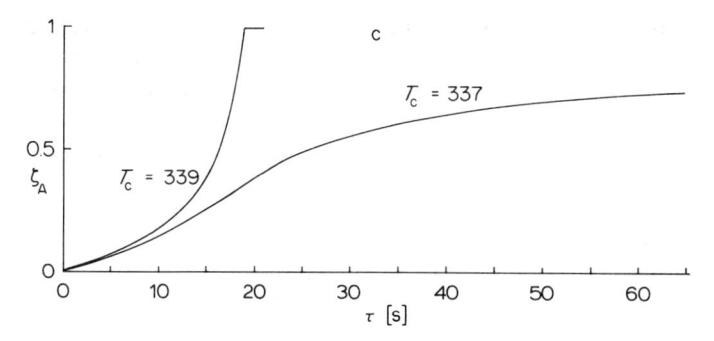

Fig. VI.14 Parametric sensitivity towards the cooling temperature T_c in a cooled plug flow reactor; Example VI.3.d. $k = 3.94 \times 10^{12} \exp(-11\,400/T)\,\text{s}^{-1}$, $T_o = 340\,\text{K}$, $\Delta T_{ad} = 146\,\text{K}$ and $4U/d_t\rho c_p = 0.20\,\text{s}^{-1}$: (a) T–ζ_A trajectories; (b) T–τ profiles; (c) ζ_A–τ profiles

The cooling temperature $T_c = 337\,\text{K}$ prevents a run away of the reactor under the conditions given. We will now vary the reactor inlet temperature T_o at a constant cooling temperature $T_c = 337\,\text{K}$. Results have been plotted in figure VI.15.a and 15.b. We see that at an inlet temperature above $T_o = 342\,\text{K}$ the reaction is already that fast at the reactor inlet, that with the cooling capacity provided we can no longer keep the reaction temperatures within reasonable limits and the hot spot occurs practically at full conversion. Between $T_o = 342$ and 343 K the temperature difference between the hot spot and the inlet suddenly becomes 3.5 times higher.

The fact that, in a certain range of parameters, the hot spot temperature T_{max} is very sensitive to a change in these parameters ('parametric sensitivity') has also been very clearly demonstrated by Bilous and Amundson [12]. These authors

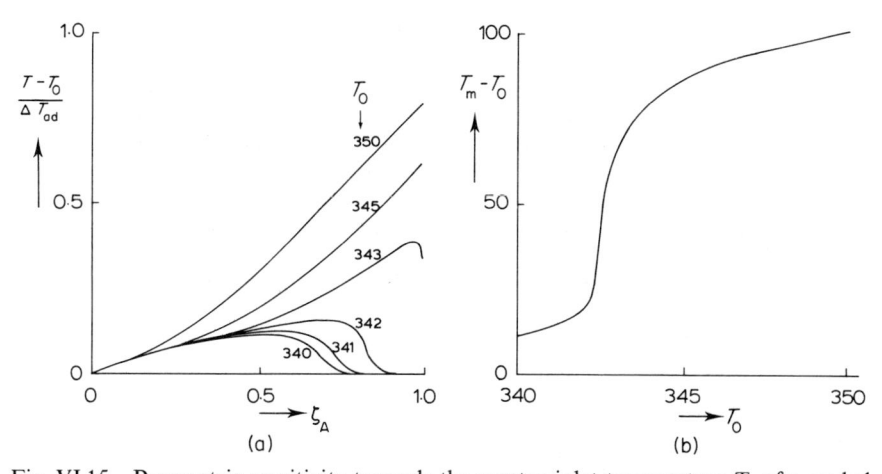

Fig. VI.15 Parametric sensitivity towards the reactor inlet temperature T_o of a cooled tubular reactor; Example VI.3.d. $T_c = 337\,\text{K}$: (a) T–ζ_A trajectories with T_o as parameter; (b) maximum temperature increase in the reactor as a function of the inlet temperature T_o

examined the concentration and temperature profiles for a first order exothermic reaction A → P under various circumstances by means of an analog computer. Some of their results have been collected in figures VI.16 and 17. In the first figure

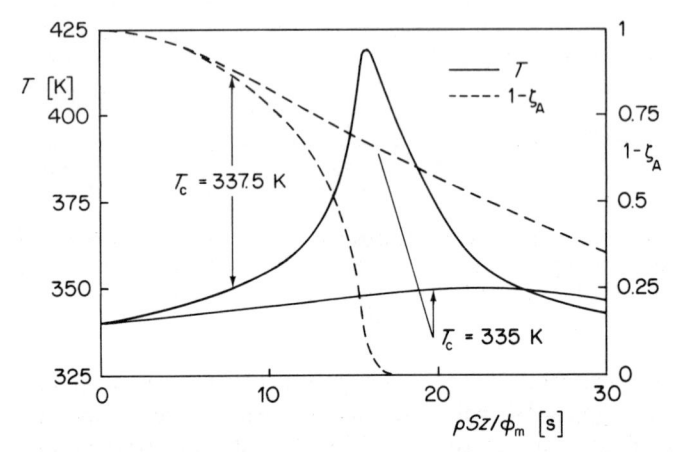

Fig. VI.16 Influence of a twofold change in cooling capacity on the temperature and concentration distribution in a plug flow reactor, Bilous and Amundson [12], Amundson [13]; exothermic first order reaction with $k = 3.94 \times 10^{12} \exp(-11\,400/T)\,\text{s}^{-1}$, $\Delta T_{ad} = 146\,°C$, $T_o = 340\,K$ and $4U/d_t\rho c_p = 0.20\,\text{s}^{-1}$ [From Bilous and Amundson [12]. Reproduced by permission of the American Institute of Chemical Engineers]

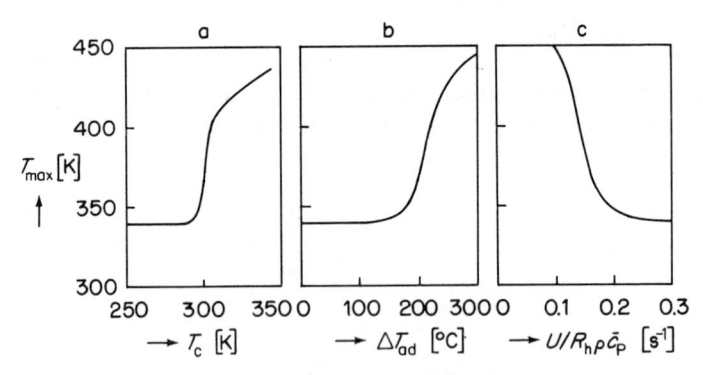

Fig. VI.17 Influence of T_c, ΔT_{ad} and the heat removal capacity on T_{max} in a plug flow reactor [12, 13]; conditions:

	a	b	c	
k	15.9	15.9	$7.89 \times 10^3 \exp\left(-\dfrac{5700}{T}\right)$	s^{-1}
ΔT_{ad}	200	var.	200	°C
T_o	340	340	340	K
T_c	var.	300	320	K
$4U/d_t\rho c_p$	0.20	0.20	var.	s^{-1}

it is seen that the very high maximum in T practically disappears when the rate of heat removal at the inlet $[\sim(T_0 - T_c)]$ is doubled; figure VI.17 shows that in a certain range a relatively small increase in T_c, ΔT_{ad} and r_h/U may cause T_{max} to change from T_0 to a value near $T_0 + \Delta T_{ad}$. This phenomenon is closely related with the course of the $\Delta\vartheta$–ζ_A curve relative to the locus of $\Delta\vartheta_{max}$, as will be demonstrated in figure VI.18. The parametric sensitivity towards T_0 has already been demonstrated in Example VI.3.d.

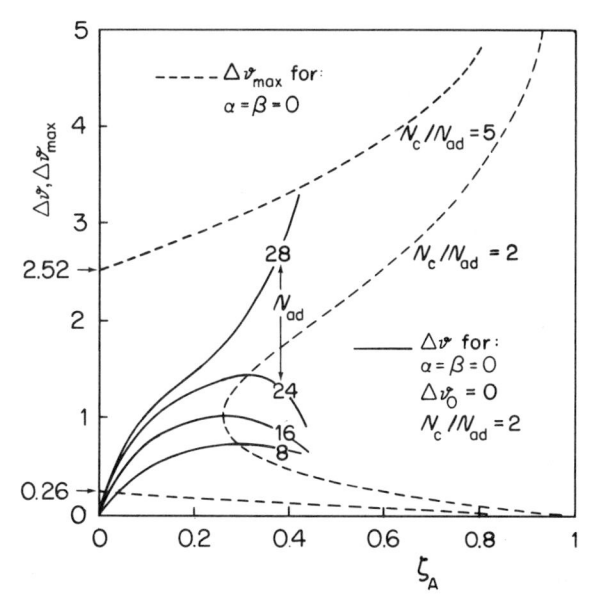

Fig. VI.18 Temperature increase and locus of maximum temperatures for first order exothermic reactions in cooled tubular reactors (equations (VI.27) and (VI.28)) [After Barkelew [8]. Reproduced by permission of the American Institute of Chemical Engineers]

Because of the analogy between the ideal tubular reactor and the mixed batch reactor, the above considerations apply equally well to the latter reactor type. A practical difference between the two types is that a predetermined temperature and conversion course with time can be enforced in a batch reactor by properly programming the rate of heat supply and/or removal; this was demonstrated in Example VI.2.a. In a continuously operated tubular reactor similar variable heat exchange requirements can only be realized if the reactor is divided into different sections, or if it is approximated by a cascade of tank reactors operated under different thermal conditions. The choice of the proper temperature sequence in such systems offers great possibilities for obtaining high product yields with complicated reactions. This aspect is connected with reactor optimization and will be discussed in Chapter X.

Although no general expression can be given for the quantitative evaluation of T_{max}, it is possible to approach it to some extent by solving (VI.23) after the introduction of a few approximations. We shall here follow the procedure developed by Barkelew [8]. The reaction temperature will be taken with reference to the cooling temperature T_c, which is assumed to be constant. A dimensionless temperature difference $\Delta\vartheta$ is defined as

$$\Delta\vartheta \equiv \frac{E}{RT_c^2}(T - T_c).$$

The following rate equation will furthermore be used:

$$R_A = -kc_{A0}(1 - \zeta_A)g(\zeta_A).$$

The function $g(\zeta_A)$ serves to introduce deviations from first order kinetics:

$$g(\zeta_A) = \frac{1 + \alpha\zeta_A}{1 + \beta\zeta_A}.$$

When $\alpha = \beta = 0$ the reaction is of the first order; with $\alpha = -1$ and $\beta = 0$ it is of second order in A, and with $\beta > 1$ a product inhibited reaction can be simulated. For the reaction velocity constant k we write

$$k = k_\infty \exp(-E/RT)$$
$$= k_\infty \exp(-E/RT_c)\exp\left[\frac{E}{RTT_c}(T - T_c)\right]$$
$$\approx k_c \exp\left[\frac{E}{RT_c^2}(T - T_c)\right] = k_c e^{\Delta\vartheta}.$$

The reaction rate constant $k_c = k_\infty e^{-E/RT_c}$ is based on the cooling medium temperature. The approximation used here is that the ratio of the absolute temperatures T and T_c has been made equal to unity.

If a maximum error of 100A % is allowed, this approximation can be used—as can be easily derived—for values of

$$\frac{T}{T_c} \leq 1 - \frac{A}{2}\frac{RT_c}{E} + \sqrt{\left(A\frac{RT_c}{E}\right)}.$$

By introducing ΔT_{ad} and the above substitutions in equation (VI.25), we obtain the following result:

$$\frac{d(\Delta\vartheta)}{d\zeta_A} = N_{ad} - N_c \frac{\Delta\vartheta e^{-\Delta\vartheta}}{(1 - \zeta_A)g(\zeta_A)}, \tag{VI.27}$$

where

$$N_{ad} = \frac{E\,\Delta T_{ad}}{RT_c^2} \quad \text{and} \quad N_c = \frac{4U}{k_c \rho c_p d_t}.$$

The dimensionless number N_{ad} accounts for the adiabatic temperature rise; N_c is a measure of the cooling capacity by means of which the temperature rise can be reduced. N_c is very sensitive to the coolant temperature (since it contains k_c, which is exponentially dependent on T_c) whereas N_{ad} is not.

Equation (IV.27) can be solved by numerical methods for a given set of values of N_{ad}, N_c, α and β, and for the boundary condition

$$\Delta\vartheta = \Delta\vartheta_o \quad \text{for} \quad \zeta_A = 0.$$

Such solutions are shown in figure VI.18 for $\alpha = \beta = 0$ (first order reactions), $\Delta\vartheta_o = 0$, $N_c/N_{ad} = 2$ and various values of N_{ad}. This figure also contains the locus of the maxima in the $\Delta\vartheta - \zeta_A$ curves for $N_c/N_{ad} = 2$. This locus can be calculated from equation (VI.27) with $d(\Delta\vartheta)/d\zeta_A = 0$; on its left $d(\Delta\vartheta)/d\zeta_A$ is positive and on its right $\Delta\vartheta$ decreases with increasing conversion. It is seen from figure VI.18 that as N_{ad} increases, $\Delta\vartheta_{max}$ becomes higher. Particularly, above a certain value of N_{ad} (in figure VI.18 between 24 and 28) the $\Delta\vartheta-\zeta_A$ curve completely misses the lower part of the locus of $\Delta\vartheta_{max}$ with the result that $\Delta\vartheta_{max}$ will become very high and close to the value for adiabatic operation. We shall return later to this great sensitivity of T_{max} to a variation in parameters.

It follows from equation (VI.27) that the curve of $\Delta\vartheta_{max}$ as a function of ζ_A is given by the equation

$$1 - \zeta_A = \frac{N_c}{N_{ad}} \Delta\vartheta_{max} \, e^{-\Delta\vartheta_{max}} \tag{VI.28}$$

for first order reactions. For one value of ζ_A either two solutions (one with $\Delta\vartheta_{max} > 1$ and one $\Delta\vartheta_{max} < 1$) or none is found. The curve reaches a maximum for a lowest value of ζ_A, which can be determined by setting $d\zeta_A/d\Delta\vartheta = 0$ or

$$\frac{d\zeta_A}{d\Delta\vartheta} = -\frac{N_c}{N_{ad}} [e^{-\Delta\vartheta_{max}} - \Delta\vartheta_{max} \, e^{-\Delta\vartheta_{max}}] = 0.$$

The solution is $\Delta\vartheta_{max} = 1$. Above this maximum, that is at higher values of N_{ad}, very rapidly run away conditions are reached. Van Welsenaere and Froment [9] therefore set as a criterion for run away conditions

$$\Delta\vartheta_{max} = 1. \tag{VI.29.a}$$

From equation (VI.28) we see that in this maximum $\zeta_{Amax} = 1 - N_c/eN_{ad}$. For $\Delta\vartheta_{max} \leq 1$ the reactor temperatures will not run away, above $\Delta\vartheta_{max} = 1$ they will run away or stay very close to run away conditions.

The locus of $d\Delta\vartheta/d\zeta_A = 0$ (equation (VI.28)), for $N_c/N_{ad} = e$ and 5 is shown in figure VI.19. For $N_c/N_{ad} = e$ the value of the minimum of ζ_A is $\zeta_A = 0$ as can easily be derived from equation (VI.28). For values of $N_c/N_{ad} > e$ the locus in the $\Delta\vartheta-\zeta_A$ plane is split into two branches. Below the upper branch and above the lower branch the value of $d\Delta\vartheta/d\zeta_A$ is negative.

The curves of $\Delta\vartheta_{max}$ as a function of ζ_A as shown in figure VI.19 can be very helpful in estimating the requirements for keeping $\Delta\vartheta_{max}$ below a desired value. When the cooling capacity is relatively small (say $N_c/N_{ad} < 2$), the locus lies very much to the right and the chance that, at constant N_{ad}, $\Delta\vartheta_{max}$ will exceed a certain value is increased. On the other hand, if N_c/N_{ad} is large, a large region exists where $d(\Delta\vartheta)/d\zeta_A$ must be negative. Thus, for $N_c/N_{ad} = 5$ (see figure VI.19) and with a

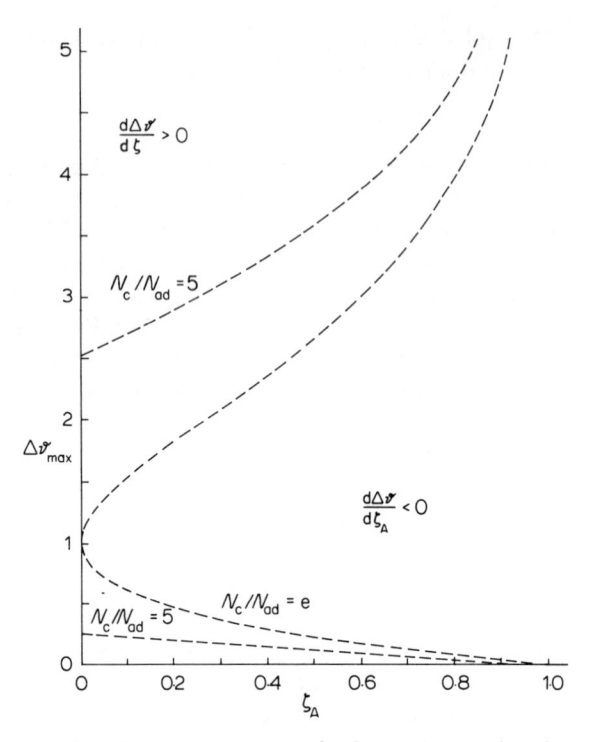

Fig. VI.19 Locus of maximum temperatures for first order reactions in a cooled tubular reactor, the influence of N_c/N_{ad} in equation (VI.28)

value of $\Delta\vartheta_o$ between 0.26 and 2.52 no maximum in $\Delta\vartheta$ will occur; accordingly $\Delta\vartheta$ will be consistently lower than $\Delta\vartheta_o$. In case $\Delta\vartheta_o$ lies below 0.26, the reaction temperature will initially rise with increasing conversion, but $\Delta\vartheta_{max}$ will remain below the value of 0.26. Finally, if the feed temperature corresponds to $\Delta\vartheta_o > 2.52$, the reaction temperature will rise to a high value.

Since the locus of $\Delta\vartheta_{max}$ does not intersect the $\zeta_A = 0$ axis for $N_c/N_{ad} < e$, particular care must be taken in this range not to let the reaction temperature run away. Starting from equation (VI.27), Barkelew [8] had a large number of $\Delta\vartheta$–ζ_A curves computed for different values of α, β, N_c/N_{ad}, N_{ad} and $\Delta\vartheta_o$. The results of these calculations were used by him for the derivation of a criterion for the boundary between a region where $\Delta\vartheta_{max}$ remains limited (the 'stable' region) and a region where $\Delta\vartheta_{max}$ may easily become excessively high (the 'unstable' region). Figure VI.20 shows three such boundary lines for a few values of α, β and $\Delta\vartheta_o$; lines for other conditions may be found in Barkelew's original paper. As to the significance of his criterion, it can be seen from the figure that for the conditions of figure VI.18 ($N_c/N_{ad} = 2$) N_{ad} should be smaller than about 20, while the latter figure indicates that $\Delta\vartheta_{max}$ will rapidly increase when $N_{ad} > 24$.

Many other authors [9–11] have been challenged to predict safe operating conditions to prevent run away of the reactor temperature. Van Welsenaere and

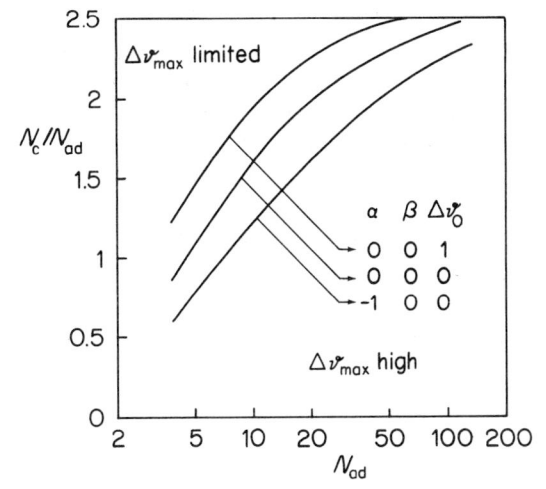

Fig. VI.20 Lines above which $\Delta\vartheta_{max}$ remains small and below which $\Delta\vartheta_{max}$ may become large [After Barkelew [8]. Reproduced by permission of the American Institute of Chemical Engineers]

Froment [9] studied exothermic first order reactions under the condition $T_o = T_c$, the reactor inlet temperature equals the cooling medium temperature. In principle their method is the same as that of Barkelew, the only difference being that they used the maximum allowable temperature T_m as the reference temperature instead of the cooling medium temperature T_c. So the condition $\Delta\vartheta_{max} \leq 1$ to prevent run away according to these authors is

$$\frac{E}{RT_m^2}(T_m - T_c) \leq 1. \tag{VI.29.b}$$

They called this their first criterion for no run away. It gives higher results for the allowable $T_m - T_c$, because $T_m > T_c$. More explicitly equation (VI.29.b) leads to the criterion

$$\frac{RT_m}{E} \leq 0.5\left[1 - \sqrt{\left(1 - \frac{4RT_c}{E}\right)}\right]. \tag{VI.29.c}$$

By carefully studying the $T-\zeta_A$ trajectories which also develop inflection points under run away conditions, van Welsenaere and Froment derived empirically by computation a second criterion for no run away:

$$\frac{N_{ad}}{\Delta\vartheta_{max}} = \frac{\Delta T_{ad}}{T_m - T_c} \leq 1 + \sqrt{\left(\frac{N_c}{e}\right)} + \frac{N_c}{e} \tag{VI.30}$$

if $T_o = T_c$.

Here $N_c/e = 4U/\rho c_p d_t k_m$, because $k_m = k_c e^{\Delta\vartheta_{max}} = k_c e$ is the reaction rate constant at the maximum allowable temperature T_m. This condition is very helpful in determining maximum allowable reactant concentrations in the reactor feed, because $c_{Ao} = \rho c_p \Delta T_{ad}/(-\Delta H_r)_A$. The applicability of the method of van Welsenaere and Froment is somewhat limited because especially their second criterion is derived for the condition $T_o = T_c$. In practice for very exothermic reactions at higher temperature levels the reactor feed is preheated to the reactor inlet temperature by the coolant so that $T_o \approx T_c$.

Example VI.3.e   *Maximum reaction temperature*
Find, for the reaction system specified in figure VI.16,
(i) below which value of T_c no hot spot will occur;
(ii) for what value of T_c the reactor will still have a reasonably low value of T_{max} according to Barkelew's criterion.

From the data in figure VI.19 we have

$$E/R = 11\,400 \text{ K}, \quad k_c = 3.94 \times 10^{12} \exp\left(-\frac{11\,400}{T_c}\right) \text{s}^{-1},$$

$$\Delta\vartheta_o = \frac{11\,400}{T_c^2}(340 - T_c), \quad N_{ad} = \frac{11\,400}{T_c^2} \times 146,$$

$$N_c = 0.20/k_c.$$

With these data Table VI.4 can be drawn up. Since a first order reaction is involved, $\alpha = \beta = 0$, and figures VI.18 and VI.20 may be used. As to question (i), for no hot spot to develop, the reactor inlet temperature $\Delta\vartheta_o$ must lie on or below the upper branch of the $\Delta\vartheta_{max}$–ζ_A locus (see figure VI.17). At inlet conditions $\zeta_A = 0$, so that $\Delta\vartheta_o$ must be equal to or lower than $\Delta\vartheta_{max}$ at $\zeta_A = 0$, in which case $\Delta\vartheta_{max}$ is still a function of N_c/N_{ad}. Table VI.4 gives $\Delta\vartheta_o$ as a function of N_c/N_{ad}; $\Delta\vartheta_{max}$ for $\zeta_A = 0$ as a function of N_c/N_{ad} can be calculated from equation (VI.28).

Table VI.4   Data for Example VI.3.e

T_c K	E/RT_c^2 K^{-1}	k_c 10^3 s^{-1}	$\Delta\vartheta_o$ —	N_{ad} —	N_c —	N_c/N_{ad} —
345	0.096	19.6	−0.48	13.9	10.2	0.73
340	0.098	12.1	0	14.4	16.5	1.15
338	0.100	9.9	0.20	14.5	20.2	1.39
336	0.101	8.1	0.40	14.7	25	1.68
334	0.102	6.6	0.61	14.9	31	2.0
330	0.104	4.4	1.04	15.2	46	3.0
320	0.111	1.5	2.2	16.2	133	8.2

For $N_c/N_{ad} = 2.72$ we find at $\zeta_A = 0$ that $\Delta\vartheta_o = \Delta\vartheta_{max} = 1.00$. We find that the highest cooling temperature at which no maximum is developed in the temperature profile is 330.4 K, corresponding to $\Delta\vartheta_o = 1.00$ at $N_c/N_{ad} = e$. For higher values of N_c/N_{ad}, $\Delta\vartheta_o$ will fall below the upper branch of the locus.

Question (ii) can be answered by interpolation between the two upper curves of figure VI.20, from which it appears that a value of $N_c/N_{ad} = 2.0$ with $T_c = 334$ K would just be safe according to Barkelew's criterion. It can be deduced from figure VI.16 that T_{max} would then remain somewhat below 350°C. Since this hypothetical reaction exhibits a high conversion rate at the inlet, a change of only a few degrees in the cooling temperature profoundly influences the temperature profile and the average rate of conversion. It can thus be estimated from the k_c values listed in Table VI.4 that for $T_c = 331$ K the reactor would have to be about 3 times longer than with $T_c = 334$ K.

We must keep in mind that we have used an approximation method. By means of the relation

$$T = T_c \left[1 - \frac{ART_c}{2E} + \sqrt{\left(\frac{ART_c}{E}\right)} \right]$$

we can calculate that in answering question (i) the error is $A = 0.03$, so that the error is 3% in the calculated value of $T - T_c$.

Example VI.3.f *Preliminary reactor design to prevent runaway*
For an exothermic first order reaction carried out in a cooled PFR, the reaction velocity constant is given by $k = 7.4 \times 10^8 \exp(-13\,600/T)\,\text{s}^{-1}$. Further data are $U = 100\,\text{W/m}^2\,\text{K}$, $\rho c_p = 1300\,\text{J/m}^3\,\text{K}$ and $(\Delta H_r)_A = -1300\,\text{MJ/kmol A}$. We will assume that $T_o = T_c$ and use the method of van Welsenaere and Froment to answer some preliminary design questions:

1. If the reactor diameter is $d_t = 25 \times 10^{-3}$ m and $T_o = T_c = 635$ K, what is the maximum allowable reactor temperature and the maximum feed inlet concentration?
   T_m can be calculated from equation (VI.29.c):

$$T_m = 13\,600 \times 0.5 \left[1 - \sqrt{\left(1 - \frac{4 \times 635}{13\,600}\right)} \right] = 667.8 \text{ K}.$$

In this case $k_c = 7.4 \times 10^8 \exp(-13\,600/635) = 0.37\,\text{s}^{-1}$. This leads to

$$N_c = \frac{4U}{d_t \rho c_p k_c} = \frac{4 \times 100}{1300 \times 25 \times 10^{-3} \times 0.37} = 33.3.$$

From their second criterion we find

$$\Delta T_{ad} = (667.8 - 635)[1 + \sqrt{(33.3/e)} + 33.3/e] = 549 \text{ K}.$$

Thus

$$c_{Ao} = \frac{\rho c_p \Delta T_{ad}}{(-\Delta H_r)_A} = \frac{549 \times 1300}{1300 \times 10^6} = 0.55 \times 10^{-3} \text{ kmol/m}^3.$$

So the reactant concentration in the feed should be kept at or below 0.55×10^{-3} kmol/m³.

2. The reactant concentration in the feed is 1×10^{-3} kmol/m³ and $T_o = T_c = 635$ K, what is the maximum allowable reactor diameter?
   From (1) we know that $T_m = 667.8$ K. Further we calculate $\Delta T_{ad} = 1 \times 10^{-3} \times 1.300 \times 10^6/1300 = 1000$ K. We now can solve the condition N_c/e

$+ \sqrt{(N_c/e)} + 1 - 1000/32.8 = 0$ for N_c. This leads to $N_c = 66.7$. Therefore
$$d_t = 4U/\rho c_p k_c N_c = 4 \times 100/1300 \times 66.7 \times 0.37 = 12.5 \times 10^{-3}\,\text{m}.$$
So the tube diameter should be equal to or smaller than 12.5 mm.

3. The same reaction is carried out with $c_{A_0} = 0.5 \times 10^{-3}\,\text{kmol/m}^3$ in a reactor with a diameter $d_t = 25$ mm. The maximum allowable temperature is $T_m = 675$ K. What is the maximum value of T_c, the cooling temperature?

In this case $T_m - T_c = RT_m^2/E = 33.5\,K$ or $T_c = 675 - 33.5 = 641.5\,K$. So the cooling temperature should be at or below 641.5 K.

VI.4 THE CONTINUOUS TANK REACTOR WITH HEAT EXCHANGE

For the steady operation of a continuous tank reactor provided with a heat exchange area A, the energy balance (VI.1) becomes

$$0 = -\Phi_m(h_1 - h_o) + UA(T_c - T_1), \tag{VI.31}$$

where the subscript 1 refers to the conditions at the outlet and in the reactor itself and T_c to the temperature of the cooling or heating medium.

The difference in enthalpy between the outgoing and the incoming stream can be obtained by integration of equation (VI.5). Assuming the pressure difference between inlet and outlet to be negligible we arrive at

$$h_1 - h_o = \int_{T_0,0}^{T_1,\xi_{J1}} [c_p\,dT + (\Delta h_r)_J\,d\xi_J].$$

The value of $h_1 - h_o$ is independent of the path of integration. We may therefore choose a path where the reactor feed is heated from T_o to T_1 and the heat of reaction is absorbed or liberated at the outlet conditions ($T = T_1$); this integration path is the same as Path 1 in figure VI.1. Accordingly, we find for equation (VI.13),

$$0 = \int_{T_0}^{T_1} [c_p]_{\xi_J=0}\,dT - [(\Delta h_r)_J]_{T=T_1}\xi_{J1} + \frac{UA}{\Phi_m}(T_c - T_1). \tag{VI.32}$$

If path 2 in figure VI.1 had been used, then we would have found

$$0 = \int_{T_0}^{T_1} [c_p]_{\xi=\xi_1}\,dT - [(\Delta h_r)_J]_{T=T_0}\xi_{J1} + \frac{UA}{\Phi_m}(T_c - T_1). \tag{VI.32.a}$$

This implies that, if we use average values of c_p and ΔH_r, we have to take either c_p averaged over the feed composition in the temperature range T_o to T_1 and ΔH_r at T_1, or c_p averaged over the reactor product composition in the same temperature range and ΔH_r at T_o. The relation (VI.32) can be used in conjunction with the isothermal material balance II.12 for calculating the reaction temperature and the degree of conversion in tank reactors where the heat effect and the external supply or removal of heat are taken into account. Some aspects of the stability of such a reactor with exothermic reactions will be discussed in Section VI.7. If now the conditions in the reactor are such that we can consider the density of the reaction mixture to be constant—equal densities of the feed and product stream of the reactor—and can use an averaged value of c_p, we find after introducing molar concentrations for the material balance VI.32, for a reactant A,

$$0 = c_p \rho (T_1 - T_o) - (\Delta H_r)_A (c_{Ao} - c_{A1}) + \frac{UA}{\Phi_v}(T_c - T_1), \qquad (VI.33)$$

and for a product P,

$$0 = c_p \rho (T_1 - T_o) - (\Delta H_r)_P (c_{P1} - c_{Po}) + \frac{UA}{\Phi_v}(T_c - T_1). \qquad (VI.33.a)$$

Extension of the theory to a cascade of tank reactors does not involve essential difficulties. A simple illustration is given below.

Example VI.4 *Heat transfer requirements in a cascade of tank reactors*
The reaction of Example VI.2.a is to be carried out continuously in a cascade of three tank reactors of equal size at 95°C; the feed conditions and the required relative conversion ζ_{A3} are the same as in Example VI.2.a, and the production of the installation should be that of Case 1, i.e. $0.9 \times 5/12\,220 = 0.375 \times 10^{-3}$ kmol/s of A converted. Calculate the reactor volume and the heat exchange requirements.

With constant density, temperature and reaction volumes we have

$$1 - \zeta_{A1} = \left(1 + \frac{k\tau}{3}\right)^{-1}$$

$$1 - \zeta_{A2} = (1 - \zeta_{A1})\left(1 + \frac{k\tau}{3}\right)^{-1} = \left(1 + \frac{k\tau}{3}\right)^{-2}$$

$$1 - \zeta_{A3} = (1 - \zeta_{A2})\left(1 + \frac{k\tau}{3}\right)^{-1} = \left(1 + \frac{k\tau}{3}\right)^{-3}.$$

With $\zeta_{A3} = 0.9$ we find $\zeta_{A1} = 0.538$, $\zeta_{A2} = 0.785$ and $k\tau/3 = 1.15$. The value of k at 95°C is 1.92×10^{-3} s^{-1}, so that the average residence time τ in the whole system must be 1800 s. The volumetric rate of flow through the system is determined by the required degree of conversion and the production rate; it is equal to $0.375 \times 10^{-3}/0.9 = 0.416 \times 10^{-3}$ m³/s. Hence, the total reaction volume is $0.416 \times 10^{-3} \times 1800 = 0.75$ m³, distributed over three reactors of equal size.

With regard to the heat exchange requirements, it is decided to operate the first reactor adiabatically and to provide cooling coils for the two other tanks. For calculating the required inlet temperature of the first reactor, we use equation (VI.32) without the last term:

$$0 = c_p(T_1 - T_o) - (\Delta h_r)_A \frac{c_{Ao} M_A}{\rho} \zeta_{A1}$$

or

$$T_o = 95 + \frac{-1.67 \times 10^6 \times 1 \times 100 \times 0.538}{4.2 \times 10^6} = 73.6\,°C.$$

For the second reactor, we calculate the required heat transfer area A_2 from equation (VI.33), omitting the first term ($T_1 = T_2$):

$$0 = -(\Delta h_r)_A \frac{c_{Ao} M_A}{\rho}(\zeta_{A2} - \zeta_{A1}) + \frac{UA_2}{\Phi_m}(T_o - T_2).$$

Values of $T_c = 20\,°C$ and $U = 1180\,W/m^2\,°C$ result in $A_2 = 0.19\,m^2$, and after a similar calculation for the third tank, $A_3 = 0.09\,m^3$. These cooling surfaces can be easily installed.

Figure VI.21 shows a possible flow sheet for this continuous process. It has several advantages over the batch process treated in Example VI.2.a:

— No steam is consumed and the heat of reaction is completely utilized in the first reactor.
— The heat transfer requirements are not excessive.
— The total reaction volume is only $0.75\,m^3$, compared with $5\,m^3$ for batch operation.
— The reaction temperature can be kept within the desired limits by simple automatic control.

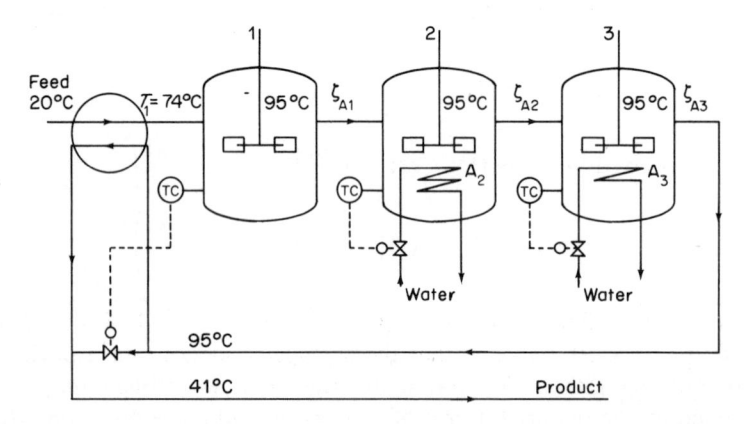

Fig. VI.21 Cascade of three tank reactors with heat exchanges. Example VI.4

The stability of these reactors will be discussed in Example VI.5.b.

(*Example VI.4 ended*)

As has been said, equation (VI.33) has to be solved in conjunction with the material balance, which for a first order irreversible reaction results for a CISTR in

$$c_{Ao}\left(1 - \frac{c_{A1}}{c_{Ao}}\right) = c_{Ao} - c_{A1} = c_{Ao}\frac{k\tau}{1 + k\tau}.$$

We can now rewrite equation (VI.33) for the first order reaction as

$$-(\Delta H_r)_A c_{Ao} \frac{k\tau}{1 + k\tau} = \rho c_p (T_1 - T_o) + \frac{UA}{\Phi_v}(T_1 - T_c). \quad \text{(VI.34)}$$

This equation has to be solved by trial and error, either numerically or graphically, because k is a function of temperature. An illustration of the graphical solution is given in figure VI.22. Evidently, the reaction temperature T_1 in the steady state must have a value which satisfies this equation. Here $k = k \exp\left(-\frac{E}{RT_1}\right)$.

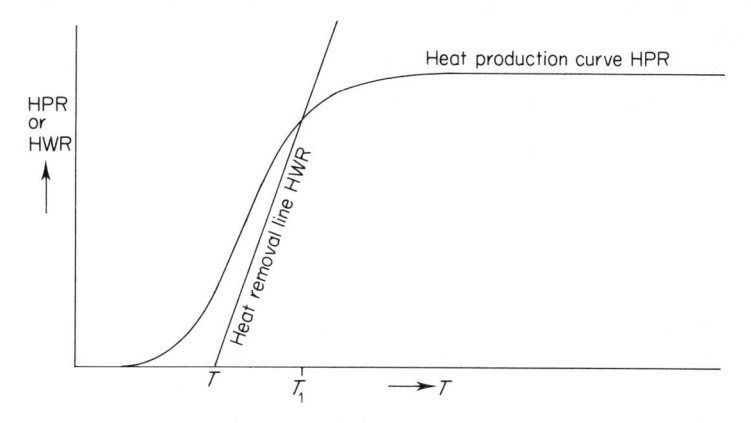

Fig. VI.22 Graphical solution of the heat balance for a tank reactor. The solution is found at the intersection of the heat production curve and the heat removal line, plotted as a function of the reactor temperature

The left-hand side of equation (VI.34) represents the heat produced per unit volume of reaction mixture. This specific production of heat is proportional to the heat of reaction and the degree of conversion, which in turn is determined by the residence time τ and the conversion rate. The two right-hand terms of equation (VI.34) represent the heat removed per unit volume of the reaction mixture as a result of the heat absorbed by the cold feed and of the heat transferred to the cooling medium, respectively. For reasons of convenience we will use several different expressions for 'heat production' and 'heat removal', which are interchangeable. For the heat production with constant density reactions we have:

1. the heat production per unit of time: $(-\Delta H_r)_A \Phi_v c_{Ao} \zeta_A$;
2. the heat production per unit of volume of the reaction mixture: $(-\Delta H_r)_A c_{Ao} \zeta_A$;
3. the dimensionless heat production, defined as the actual heat production divided by the maximum heat production at full conversion: ζ_A;
4. the heat production per unit of volume of the reaction mixture divided by

ρc_p, the sensible heat per unit of volume of the reaction mixture: $(-\Delta H_r)_A c_{A_o} \zeta_A / \rho c_p = \Delta T_{ad} \zeta_A$. This represents the fraction of the maximum possible temperature increase, which is actually reached in the reactor.

For the heat removal similar expressions can be defined. As long as the expressions for both the production and removal of heat are consistent, all will lead to the same conclusions with respect to the reactor design and operation.

If we divide equation (VI.34) by $c_p \rho$, we obtain

$$\Delta T_{ad} \frac{k\tau}{1 + k\tau} = (T_1 - T_o) + \frac{UA\tau}{\rho c_p V_r}(T_1 - T_c). \qquad (VI.35)$$

We will now study this equation more in detail by discussing both sides separately.

The *right hand side* can—if we call $UA\tau/\rho c_p V_r = St'$, a modified Stanton number for a CISTR—be rewritten as

$$\text{heat removal} = (1 + St')(T_1 - T_R), \qquad (VI.36)$$

where

$$T_R = \frac{T_o + St' T_c}{1 + St'}. \qquad (VI.36.a)$$

This has been shown in figure VI.23.a and b. In figure VI.23.a the lines for the heat removal by the reaction mixture and by the cooling medium in a cooled reactor are drawn. The total heat removal is the sum of the two individual parts, of course. The heat removal by the cooling medium depends on the residence time in the reactor, because τ appears in the modified Stanton number St'. The total heat removal line as shown in figure VI.23.b becomes steeper and moves to the left if

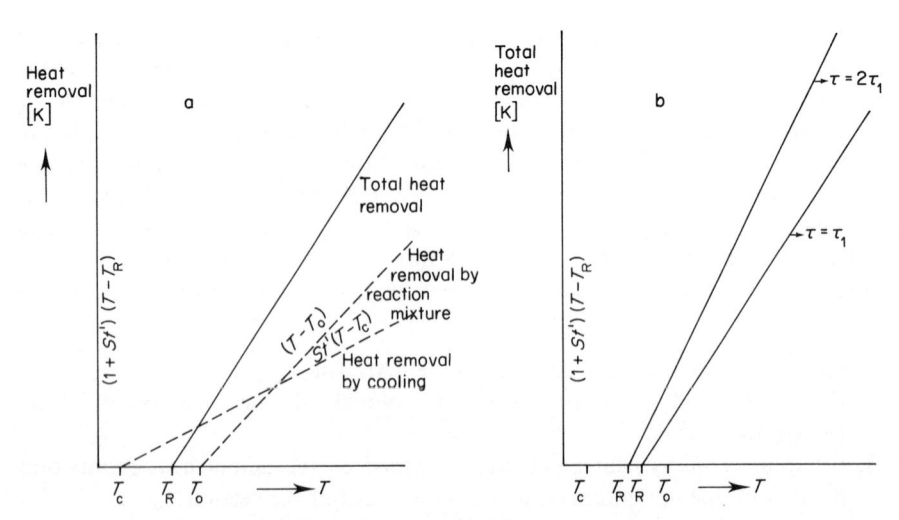

Fig. VI.23 Heat removal in a tank reactor

the residence time is increased. In an adiabatically operated CISTR $St' = 0$ and heat is removed only by the reaction mixture itself. For endothermic reactions in a heated CISTR a similar graph for the heat supply lines can be derived.

For an exothermic first order irreversible reaction with $R_A = -kc_A$ the left-hand side of equation (VI.34) in a CISTR is given by

$$\text{heat production} = \Delta T_{ad} \frac{k\tau}{1 + k\tau} \qquad (VI.37)$$

with $\Delta T_{ad} = -(\Delta H_r)_A c_{Ao}/\rho c_p$ and $k = k_\infty \exp(-E/RT)$. The heat production is a function of the properties of the reaction mixture $((\Delta H_r)_A, \rho, c_p$ and $k)$, the concentration of the reactant in the feed c_{Ao}, the residence time in the reactor τ and the reactor temperature, which directly influences the value of k. For a particular reaction the most important design variables are c_{Ao} and τ. This is shown in figure VI.24. For the same reaction at the same reactor temperature T_1 the heat production increases with feed concentration and residence time. For one single CISTR generally we are only interested in the upper part of the S-shaped curve, where conversions are high. For a CISTR in a cascade also the lower parts in the heat production curve are of interest.

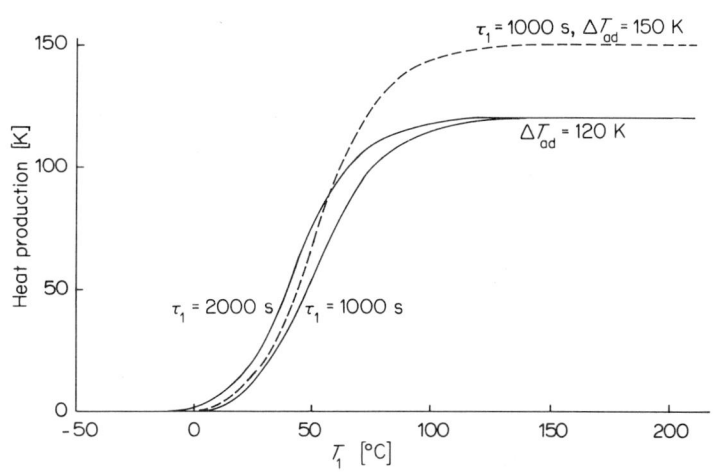

Fig. VI.24 Heat production in a tank reactor for a first order reaction with $k = 4.71 \times 10^7 \exp(-8000/T) \, \text{s}^{-1}$

The shape of the heat production curve is also dependent on the kinetics, that is on the activation energy E and on the temperature level where the reaction starts, which means the value of k_∞ in $k = k_\infty \exp[-E/(RT_1)]$. In figure VI.25 heat production curves are shown for different temperature levels and different activation energies. For first order reversible reactions of the type $A \rightleftharpoons B$ with $-R_A = +(k_A c_A - k_{-A} c_B)$ the material balance gives $\Phi_v (c_{Ao} - c_A) = -R_A V_r$. After substitution of $c_A = c_{Ao}(1 - \zeta_A)$ and $c_B = c_{Ao}\zeta_A$ (no species B in the feed) we find, for the conversion,

$$\zeta_A = \frac{k_A \tau}{1 + (k_A + k_{-A})\tau}.$$

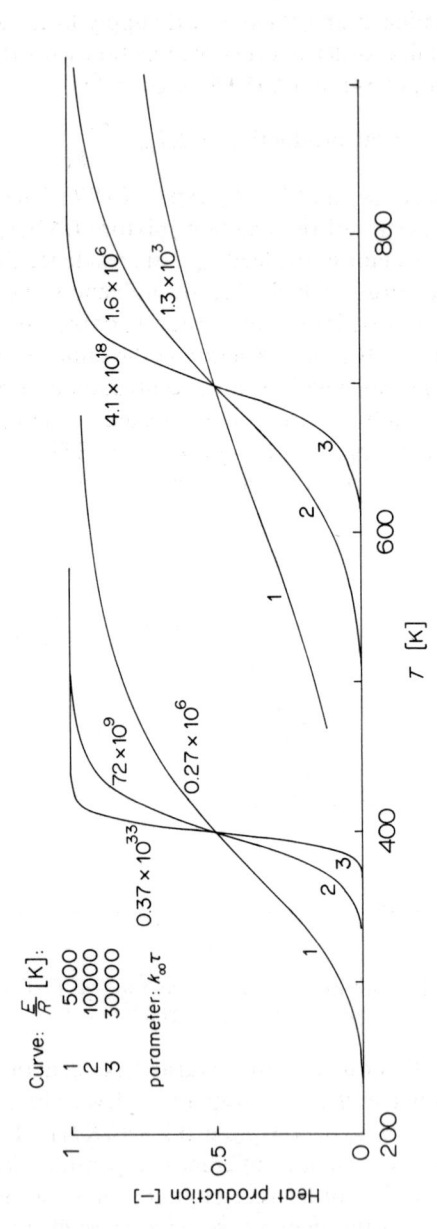

Fig. VI.25 Heat production curves for first order reactions with different kinetic constants and activation energies. The parameter is $k_\infty \tau$

In this case it is convenient to take as a reference temperature T_R, at which $k_A = k_{-A} = k_R$, so that $k_A = \kappa k_R$ and $k_{-A} = \kappa^p k_R$, in which $p = E_{-A}/E_A$. We then find

$$\zeta_A = \frac{\kappa k_R \tau}{1 + (\kappa + \kappa^p)k_R\tau} = \left[\frac{1}{\kappa k_R \tau} + 1 + \kappa^{p-1}\right]^{-1}.$$

By differentiating we find that ζ_A reaches a maximum at $\kappa = [k_R(p-1)]^{-1/p}$ and for the maximum we then have

$$\zeta_{A\max} = [(p-1)^{1/p}(k_R\tau)^{(1-p)/p} + [k_R\tau(p-1)]^{(p-1)/p}]^{-1}.$$

The relation between the reactor temperature and κ is

$$T_1 = \frac{T_R}{1 - (E_A/RT_R)\ln\kappa}.$$

Expressed in the same units as above the heat production in K is now given by $\Delta T_{ad}\zeta_A$. So a ζ_A–T plot is indicative for the heat production (see figure VI.26, where we observe bell-shaped curves). The heat production is low at lower and higher temperatures: at lower values of T_1 the reaction rate is still too low and at higher temperature levels the equilibrium conversion is low. The maximum in the bell-shaped curve depends on the value of $E_{-A}/E_A = p$ and the residence time in the reactor; at low temperature levels and long residence times almost complete conversion can be obtained. For other reaction orders except for zero order reactions, the heat production curves have similar shapes; also for endothermic reactions they can easily be derived. For first order endothermic reactions their general shape is shown in figure VI.27. We will continue our discussion of the CISTR in the next section.

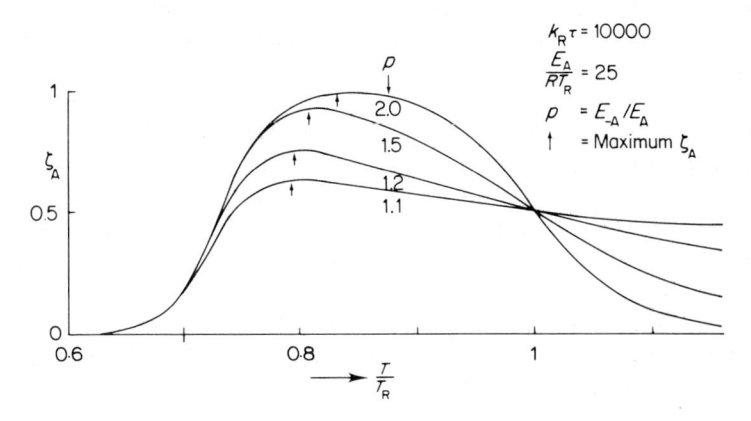

Fig. VI.26 Conversion versus reactor temperature in a CISTR for a reversible first order reaction

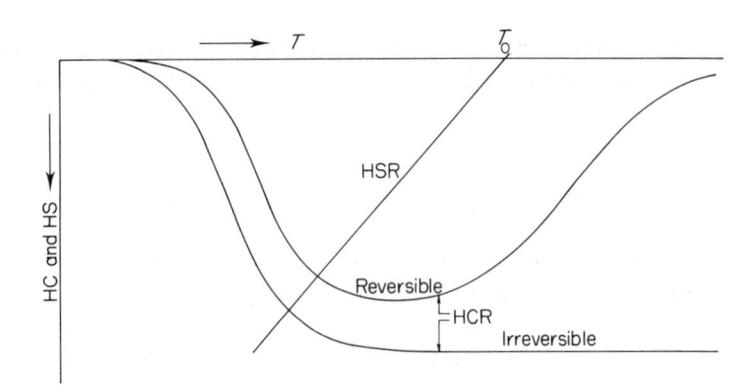

Fig. VI.27 Heat consumption curves (HCC) for endothermic first order reaction and the heat supply line (HSL)

VI.5 AUTOTHERMAL REACTOR OPERATION

When an exothermic reaction is carried out, it is generally desirable to employ the liberated heat of reaction somewhere in the plant in a useful manner. The heat of reaction is often used for preheating the feed to the reactor, particularly when the reaction takes place at a high temperature level and the reactants are available at a much lower temperature. A reactor system in which such a feed-back of reaction heat to the incoming reactant stream is applied is said to be operated under autothermal conditions. Generally the reactants are fed to the reactor at such a low temperature that the reaction rate at this inlet temperature is practically zero, whereas the high temperature at which the reaction eventually takes place is achieved in the autothermal reactor exclusively by the heat evolved due to the reaction. This distinguishes the autothermal reactor from the continuously operated adiabatic reactor without feed back of reaction heat, because in such a reactor the inlet temperature necessarily must be high enough for the reaction to start. Essential for the autothermal reactor is the feed back of reaction heat to the cold reactor feed.

Example VI.5.a *Comparison of a non-autothermal and an autothermal reactor*
A liquid reaction mixture with $c_{Ao} = 2.0\,\text{kmol/m}^3$ and $\rho c_p = 4.19\,\text{MJ/m}^3\,\text{K}$ is available at 12 °C. The data on the first order irreversible reaction are $\Delta H_{rA} = -251\,\text{MJ/kmolA}$ and $k = 12.8 \times 10^{30} \exp(-30\,000/T)\,\text{s}$, and we wish to reach a conversion of 90 %. We will study whether the reaction can be executed adiabatically.

In a *tank reactor* the reactor operates at the final temperature and the cold feed is immediately dispersed over the hot reactor contents and so immediately heated up to the final temperature. The final temperature is

$$T = T_o + \Delta T_{ad}\zeta_A = 12 + \frac{251 \times 2.0}{4.19} \times 0.90 = 120\,°\text{C}.$$

At this temperature $k = 9.01 \times 10^{-3}$ s and the reactor residence time can be found by solving

$$(-\Delta H_r)_A c_{A_o} k\tau/(1 + k\tau) = \rho c_p (T - T_o),$$

which leads to

$$\tau = 1000 \text{ s}.$$

In a *tubular reactor* the cold reaction mixture enters the reactor at 12 °C and is not heated up by the heat evolved upstream in the reactor. The reaction rate constant at the reactor inlet is $k = 2.47 \times 10^{-15}$ s, that is a factor 274×10^{-15} lower than in the CISTR. The outlet temperature in the adiabatic PFR also will be 120 °C. We now can calculate the required residence time in the adiabatic PFR by means of the equation derived in Example VI.3.c:

$$k_\infty \tau_L = \exp\left(\frac{E}{RT_{ad}}\right)\left[\text{Ei}\left(\frac{E}{RT_o} - \frac{E}{RT_{ad}}\right) - \text{Ei}\left(\frac{E}{RT} - \frac{E}{RT_{ad}}\right)\right]$$
$$- \text{Ei}\left(\frac{E}{RT_o}\right) + \text{Ei}\left(\frac{E}{RT}\right).$$

By substituting $T_o = 285$ K, $T = 393$ K and $T_{ad} = 405$ K and $E/R = 30\,000$ K into this equation we find:

$$\tau_L = 9.54 \times 10^{12} \text{ s} \simeq 302\,000 \text{ years}.$$

We see that due to the extremely low reaction rate at the inlet of the PFR the reaction does not start. On the contrary in the CISTR, which behaves autothermally due to the complete back-mixing in an ideal mixer, the reaction rate is sufficiently high to permit a reasonable residence time for a liquid-phase reactor.

(Example VI.5.a ended)

The term 'autothermal' stems from the fact that such a system is to a great extent self-supporting in its thermal energy requirements, so that it is possible to operate at a high temperature level without preheating the feed by external means. As a consequence of this feed-back, an autothermally working reactor has the property that in many cases the reactor has to be 'ignited' in order to attain steady state operation. In this respect, the ordinary flame and, in general, all rapid combustion reactions are autothermal as well, since the reactants are likewise preheated to the reaction temperature by thermal conduction and radiation (see also Chapter IX).

The theory of steady state behaviour of reactors under autothermal conditions was first proposed by Wagner [14] and was later treated extensively by van Heerden [15, 16]. Of the different possibilities for obtaining the required feedback, we shall discuss in this section:

(i) the tank reactor in which the cold feed is heated by mixing with the reactor contents;
(ii) an adiabatic tubular reactor with heat exchange between the incoming stream and the reaction mixture;
(iii) a multi-tube reactor with internal heat exchange between the reaction mixture and the feed.

VI.5.1 The tank reactor

For the purpose of this discussion we shall simplify equation (VI.32) by assuming that the specific heat of the reaction mixture does not depend on the temperature and the composition and that, consequently, the heat of reaction is independent of temperature. The heat balance can then be written in the form:

$$-\xi_{J1}(\Delta h_r)_J = c_p(T_1 - T_o) + \frac{UA}{\Phi_m}(T_1 - T_c). \tag{VI.38}$$

Evidently, the reaction temperature T_1 in the steady state must have a value which satisfies this equation. The reaction temperature T_1 is obtained, for example, graphically by combining the two diagrams VI.23 and VI.24 into one graph, as shown in figure VI.28 (the heat production and heat removal functions can be plotted in various ways; each has its own advantages and several methods will be applied in the following). At an intersection of a heat production curve and a heat removal line, equation (VI.38) is obeyed, so that steady operation would be possible at the corresponding reaction temperature. It is seen from figure VI.28 that several solutions may exist: this is called the *multiplicity* of steady states.

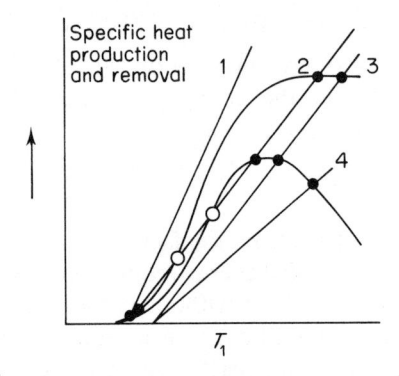

Fig. VI.28 Heat production and heat removal in an autothermally operating tank reactor and different possibilities of operation; ●—stable solutions, ○—unstable solutions

If the heat removal is represented by line 1 in figure VI.28, the reaction mixture is so effectively cooled that steady operation is possible only at a low temperature and a very low degree of conversion. If the cooling capacity is reduced, line 2 is obtained which has three points of intersection with the heat production curves; those at the lowest and highest reaction temperature represent stable conditions, and especially the latter point is of practical interest since it represents a high degree of conversion. The stability of these two operating conditions is related to the fact that the slope of the heat removal line is greater than that of the heat production line. Thus, on a positive deviation from the intersection temperature more heat is removed than produced, so that the reaction temperature will return to the steady state value; similarly, a negative deviation will cause more heat to be produced than removed so that the temperature will rise. In the intermediate intersection point, on the other hand, the situation is reversed. Since the slope of the heat production curve is here greater than that of the heat removal line, any positive temperature deviation will be amplified until the reactor works in the upper stable operating point and any negative temperature deviation will cause the lower stable operating point to be attained finally; in the latter case we may say that the reaction is 'extinguished'. The intermediate intersection point is unstable because of this behaviour. If it is desired to operate the reactor at the upper stable point, the lower stable condition has to be by-passed, e.g. by temporarily preheating the feed or by operating under a smaller load; such a procedure is called the 'ignition' of the reaction.

If the feed temperature T_o and the cooling capacity are such that the heat removal line 3 in figure VI.28 prevails, only the upper stable working point is a possible solution to equation (VI.38) and no special measures need be taken to start the reaction. Finally, line 4 shows that a relatively small slope of the heat removal line (which may mean that the cooling capacity is too low) may give rise to high reaction temperatures and, in the case of an equilibrium reaction, to a reduction in the degree of conversion.

It is of interest to study the sensitivity of the autothermal tank reactor to a change in operating variables, such as the feed temperature and concentration, the throughput and the cooling capacity. To this end, we further simplify equation (VI.38) by assuming constant density of the reaction mixture and by introducing the relative degree of conversion ζ_A of a reactant A and the adiabatic temperature rise ΔT_{ad}. With these substitutions (VI.38) can be written as

$$\zeta_{A1} = \frac{T_1 - T_o}{\Delta T_{ad}} + \frac{UA(T_1 - T_c)}{\rho c_p V_r \Delta T_{ad}} \tau. \tag{VI.39}$$

If a reaction is of the first order in A with a reaction rate constant k, we have from the material balance (e.g. equation (II.28))

$$\zeta_{A1} = \frac{k\tau}{1 + k\tau} \tag{VI.39.a}$$

in which k is still a function of T_1. As an example, figure VI.29 shows the functions (VI.39) and (VI.39.a) as a function of T_1 for $\tau = 0.5$, 1 and 2 seconds under otherwise specified conditions in order to demonstrate the influence of a fourfold variation in reactor load. At $\tau = 1$ s the reactor has three possible operating points (of which the intermediate one is unstable). When the reactor throughput is halved ($\tau = 2$ s) the reaction is extinguished; when it is doubled ($\tau = 0.5$ s), stable operation results with a high degree of conversion. Under the latter conditions the reactor can be started without any special measures.

Figure VI.29 then shows that—if a throughput corresponding to $\tau = 0.5$ s is the normal operating condition—underloading of the reactor results in a *decrease* of conversion because the cooling becomes relatively too effective. This situation could be improved by lowering the feed temperature and at the same time diminishing the rate of heat removal by heat transfer, $UA(T_1 - T_c)$.

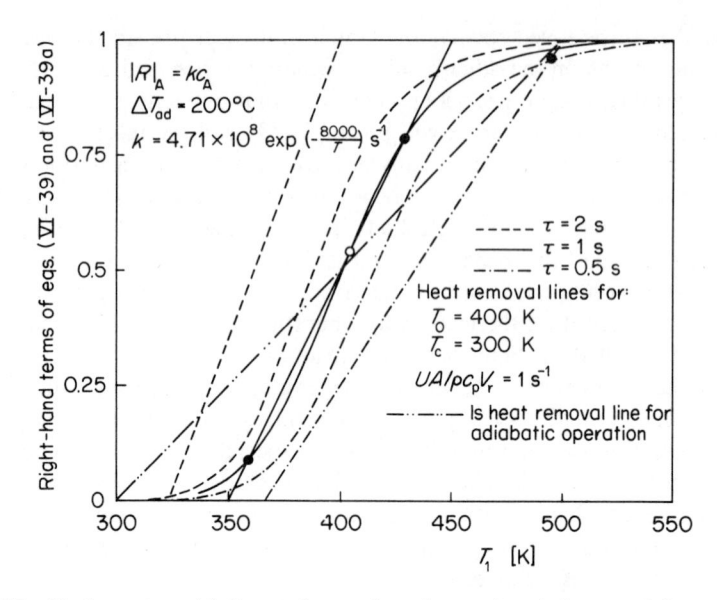

Fig. VI.29 Tank reactor with first order exothermic reaction; influence of the residence time τ on stable autothermic operation

Figure VI.29 shows that if the reactor is operated without heat exchange with a cooling medium (i.e., adiabatically) and if $T_o = 300\,\text{K}$, then the reaction temperature and the degree of conversion slightly *increase* when the throughput is diminished; in this case, however, the reactor would have to be ignited. In view of the above considerations, a proper combination of inlet temperature and cooling facilities can bring about a situation of minimum sensitivity to load variations.

More properties of the heat production–heat removal diagram are of importance. In figure VI.30 two heat removal lines a and b are drawn touching the heat production line. According to equation (VI.35) the slope of these lines is $(1 + St')/\Delta T_{ad}$, with $St' = UA\tau/\rho c_p V_r$, and they start in the point $(0, T_R)$ on the

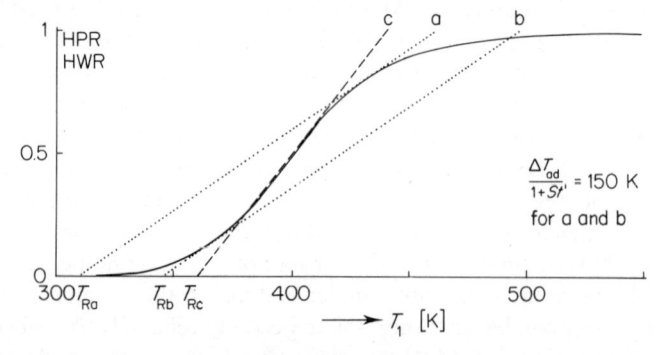

Fig. VI.30 Demarcation of the multiplicity of steady states. Same data as in figure VI.29 for $\tau = 1\,\text{s}$

horizontal axis; both lines have the same slope. For heat removal lines with the slope $(1 + St')/\Delta T_{ad}$, we now can *distinguish three regions:*

(i) $T_R < T_{Ra}$: in this region there is only one stable operating point with a low conversion. For $T_R = T_{Ra}$ we have a border line a with two operating points.
(ii) $T_{Ra} < T_R < T_{Rb}$: in this region there will be three operating points; one stable at low conversion, one unstable and one stable at high conversion. This region passes into the third one with borderline b with two operating points.
(iii) $T_{Rb} < T_R$: in this region there is again only one stable operating point with a high conversion.

We now can make a cross plot figure VI.31, where we plot either T_R or T_1 as a function of $(1 + St')/\Delta T_{ad}$, which is a measure of the heat removal capacity. $\Delta T_{ad}/(1 + St')$ is the maximum temperature increase in the reactor at full conversion $\zeta_A = 1$ and depends on the residence time in the reactor $(St' = UA\tau/\rho c_p V_r)$. In the areas enveloped by the curves multiplicity will occur. Outside of these areas only one (stable) operating point is possible with high conversions above the areas and low conversions below the areas. The point with the lowest value of $\Delta T_{ad}/(1 + St')$ is often called the *trifurcation point*, the curves departing from it the *bifurcation lines*.

In the trifurcation point the heat production curve reaches its highest slope. This slope must be equal to the slope of the heat removal line—this is line c in figure VI.30—so that here

$$\frac{d}{dT}\left(\frac{k\tau}{1+k\tau}\right)_{max} = \frac{k\tau}{(1+k\tau)^2}\frac{E}{RT^2}\bigg|_{T=T_{max}} = \frac{1+St'}{\Delta T_{ad}}.$$

The maximum slope of the heat production curve is found by setting

$$\frac{d^2}{dT^2}[k\tau/(1+k\tau)] = 0.$$

The temperature T_1 at this maximum is found by solving the condition

$$\frac{RT_{max}}{E} = \frac{1-k\tau}{2(1+k\tau)} \quad \text{with} \quad k = k_\infty \exp\left(-\frac{E}{RT_{max}}\right) \tag{VI.40}$$

by trial and error with a computer. Now for all values of

$$\frac{1+St'}{\Delta T_{ad}} > \frac{d}{dT}\left(\frac{k\tau}{1+k\tau}\right)_{T_1=T_{max}}$$

only one stable operating point will be obtained. So for the data in figure VI.31 we obtain for $\Delta T_{ad}/(1 + St') < 79.5$ for $\tau = 1$ s and < 85 for $\tau = 0.5$ s only one stable solution. The corresponding values of $T_R = (T_o + St'T_c)/(1 + St')$ are 362 and 372 K respectively and for $T_1 = T_{max}$ they are 396 K and 412 K respectively.

Other interesting conclusions can be drawn from figure VI.31. For example we can read from the graph that, for a value of $\Delta T_{ad}/(1 + St') = 150$ K and $\tau = 1.0$ s,

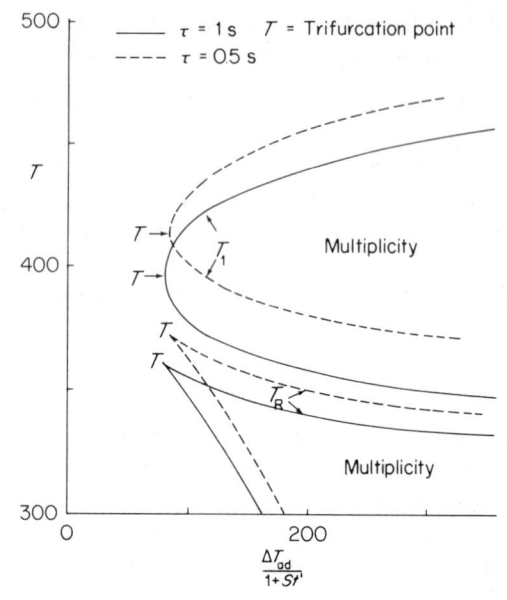

Fig. VI.31 Regions of multiplicity for a CISTR for reactor temperatures T_1 and pseudo reactor inlet temperatures $T_R = (T_o + St'T_c)/(1 + St')$. First order reaction with $k = 3.94 \times 10^{12} \exp(-11\,400/T)\,\text{s}^{-1}$, $T =$ trifurcation point

the reactor will exhibit (follow a vertical line through $\Delta T_{ad}/(1 + St') = 150\,\text{K}$) multiplicity as long as $310 < T_R < 346\,\text{K}$ and that the corresponding range of reactor temperatures is $365 < T_1 < 431\,\text{K}$. Also, if we want to run the reactor at a residence time of $\tau = 0.5\,\text{s}$, at a reactor temperature $T_1 = 450\,\text{K}$ and outside of the multiplicity region, then the design or operating parameters should be $\Delta T_{ad}/(1 + St') < 170\,\text{K}$ and $T_R > 354\,\text{K}$. If in this case $St' = 0.30$ and $T_c = 320\,\text{K}$, then the conditions lead to $\Delta T_{ad} < 221\,\text{K}$ and $T_o > 364\,\text{K}$. Operating in the region of multiplicity is not dangerous as such, as long as we make sure that we operate in the upper stable operating point, where the conversion is high. So the conclusion that T_o should be higher than $364\,\text{K}$ can better be reduced to a statement that the reactor should be provided with a start-up feed preheater with sufficient capacity to bring the feed to a temperature $T_o > 364\,\text{K}$ during the start-up, in order to ensure that the upper stable operating point is reached. Once the reactor operates at a high conversion the feed temperature can be reduced again, as will be explained next.

Another aspect of autothermal operation is the occurrence of *hysteresis*. If the feed temperature T_o is increased at a constant feed rate, the reaction will be ignited at a certain feed temperature $(T_o)_{ign}$ and the reactor will work at the upper stable operating point. If T_o is then lowered again, the reactor continues operating at a high conversion level until extinction finally takes place at a feed temperature $(T_o)_{ext}$. Figure VI.32 shows such a hysteresis diagram for adiabatic operation; it was constructed on the basis of the data of figure VI.29. In this particular case the hysteresis interval is $(T_o)_{ign} - (T_o)_{ext} = 67\,°\text{C}$; consequently, the ignited adiabatic

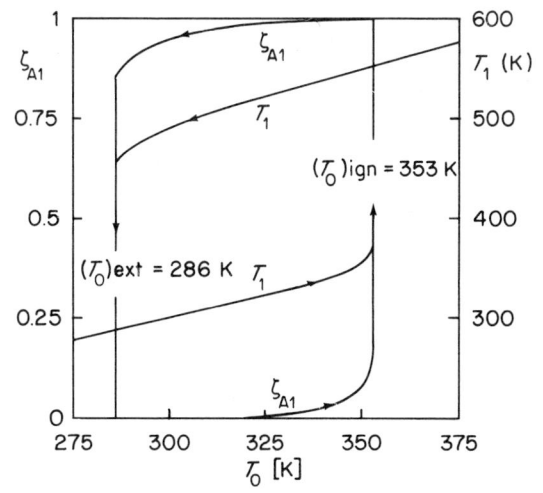

Fig. VI.32 Hysteresis with autothermal operation of a tank reactor; constructed from figure VI.29 for adiabatic operation and $\tau = 0.5$ s

tank reactor is rather insensitive to disturbances in T_o. When additional cooling is applied, the hysteresis interval is generally narrowed, and the sensitivity to feed temperature changes increased, as can be seen from figure VI.29.

It is, of course, equally possible to examine the influence of a change in other operating variables such as the feed concentration. This will not be discussed further, but it should be noted that the method of analysis used above is only applicable to changes which are so slow that the corresponding change in reactor operation can be regarded as a succession of pseudo-stationary states (static stability). The dynamic behaviour and stability of an autothermally operating tank reactor under the influence of relatively rapid fluctuations of the various parameters are of great interest, especially for the purpose of automatic control; they will be treated separately in Section VI.7.

Example VI.5.b *Static stability of tank reactors in a cascade*
It is required to check whether the three tank reactors of Example VI.4 will perform under statically stable conditions.
  After introducing

$$\Delta T_{ad} = -\frac{c_{Ao} M_A}{\rho c_p} (\Delta h_r)_A \quad (=39.9\,°C)$$

the heat balances become (equation (VI.39))

Reactor 1: $\zeta_{A1} \Delta T_{ad} = T_1 - T_o + \dfrac{UA_1}{\Phi_m c_p}(T_1 - T_c)$

Reactor 2: $(\zeta_{A2} - \zeta_{A1})\Delta T_{ad} = T_2 - T_1 + \dfrac{UA_2}{\Phi_m c_p}(T_2 - T_c)$

Reactor 3: $(\zeta_{A3} - \zeta_{A2})\Delta T_{ad} = T_3 - T_1 + \dfrac{UA_3}{\Phi_m c_p}(T_3 - T_c).$

The left-hand and the right-hand side of each equation have to be examined as a function of the corresponding reaction temperature (T_1, T_2 and T_3, respectively). It is found from Example VI.4 that

$$\zeta_{A1} = k\tau/(1 + k\tau)$$
$$\zeta_{A2} - \zeta_{A1} = (1 - \zeta_{A1})k\tau/(1 + k\tau)$$
$$\zeta_{A3} - \zeta_{A2} = (1 - \zeta_{A2})k\tau/(1 + k\tau).$$

With these substitutions, the heat balances can be written in the form

$$\frac{k\tau}{1 + k\tau} = \frac{1}{\Delta T_{ad}}\left[T_1 - T_0 + \frac{UA_1}{\Phi_m c_p}(T_1 - T_c)\right]$$

$$\frac{k\tau}{1 + k\tau} = \frac{1}{\Delta T_{ad}(1 - \zeta_{A1})}\left[T_2 - T_1 + \frac{UA_2}{\Phi_m c_p}(T_2 - T_c)\right]$$

$$\frac{k\tau}{1 + k\tau} = \frac{1}{\Delta T_{ad}(1 - \zeta_{A2})}\left[T_3 - T_2 + \frac{UA_3}{\Phi_m c_p}(T_3 - T_c)\right].$$

These equations can be solved by plotting the left-hand and the right-hand sides as a function of the operating temperatures T_1, T_2 and T_3, respectively. The left-hand terms are equal; they are shown in figure VI.33 for $\tau = 600\,\mathrm{s}$ and

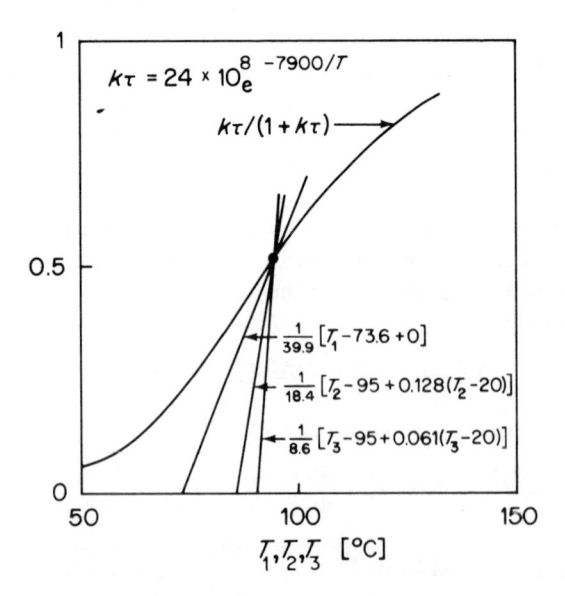

Fig. VI.33 Investigation into the static stability of three tank reactors in cascade; Example VI.5.b

$k = 4 \times 10^6 \exp(-7900/T)\,\text{s}^{-1}$. With the data of Example VI.4 the right-hand terms become, respectively:

Reactor 1: $\dfrac{1}{39.9}[T_1 - 73.6 + 0]$

Reactor 2: $\dfrac{1}{18.4}[T_2 - 95 + 0.128(T_2 - 20)]$

Reactor 3: $\dfrac{1}{8.6}[T_3 - 95 + 0.061(T_3 - 20)]$.

The three corresponding heat removal lines have been drawn in figure VI.33, from which it is seen that all three tank reactors in the cascade show static stability.

Example VI.5.c Design of a CISTR
It is required to design a CISTR for a reaction with $R_A = -kc_A$ and $k = 4.71 \times 10^7 \times \exp(-8000/T)\,\text{s}^{-1}$; further $\overline{\rho c_p} = 3.5\,\text{MJ/m}^3\,\text{K}$ and $(\Delta H_r)_A = -168\,\text{MJ/kmol A}$ converted. Reactor feed is available at 10°C and contains $c_{Ao} = 3\,\text{kmol A/m}^3$. The construction material considered has a maximum allowable temperature of 130 °C. A residence time $\tau = 1000$ s is considered a good choice. The conversion should be $\zeta_A \geq 0.98$ for economic reasons.

We first consider a reactor without cooling and draw a heat production curve for $\tau = 1000$ s as is done in figure VI.34. The adiabatic temperature increase is $\Delta T_{ad} = 3 \times 168/3.5 = 144$ °C. The heat removal line a in figure VI.34 represents the case without cooling and we find three operating points:

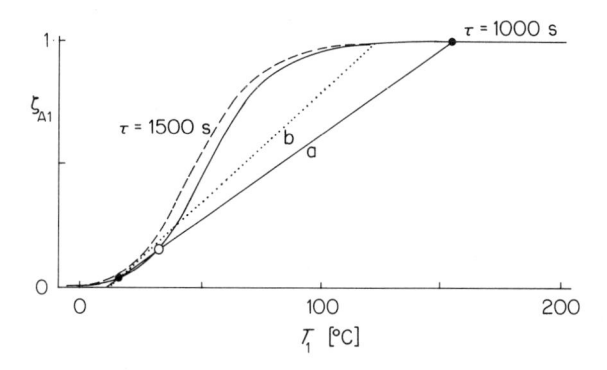

$$15\,°C \quad \text{and} \quad \zeta_A = 0.04, \quad \text{stable}$$
$$35\,°C \quad \text{and} \quad \zeta_A = 0.16, \quad \text{unstable}$$
$$154\,°C \quad \text{and} \quad \zeta_A = 1.00, \quad \text{stable}.$$

Fig. VI.34 Heat production–heat removal diagram for the reactor design of Example VI.5.c

For operation without cooling at the required conversion level the reactor surpasses the maximum allowable temperature. From the diagram we read that at $\tau = 1000$ s a conversion of $\zeta_A \geq 0.98$ requires a reactor temperature of $T_1 \geq 115\,°C$. Several measures can be taken to operate at, say, $T_1 = 120\,°C$:

(i) The reactor feed is diluted, so that $\Delta T_{ad} = 120 - 10 = 110\,°C$. Then c_{Ao} becomes $110 \times 3.0/144 = 2.29$ kmol A/m³. This reduces the reactor production $\Phi_v c_{Ao} \zeta_{A1}$, so that the reactor volume must be increased by a factor $144/110 = 1.31$. At the same time solvent recovery expenses increase.

(ii) Cooling facilities are built in into the reactor. This requires $1 + St' = 144/110 = 1.31$ or $St' = 0.31$. For an overall heat transfer coefficient of $500\,W/m^2\,K$ we then find $A/V_r = St' \rho c_p / U\tau = 0.31 \times 3.5 \times 10^6/500 \times 1000 = 2.17\,m^2/m^3$. Probably only a cooling jacket is sufficient dependent on the reactor vessel size. With cooling water available at 20°C we find for $T_R = (10 + 0.31 \times 20)/1.31 = 12.4\,°C$. The corresponding heat removal line b in figure VI.29 also exhibits these operating points.

(iii) The reactor is operated at a higher residence time. Calculations reveal that $\tau = 1500$ s suffices, provided the vessel is equipped with a cooling jacket. The cooling area required (see under (ii)) is $A/V_r = 0.31 \times 35 \times 10^6/500 \times 1500 = 1.45\,m^2/m^3$. This operation has the advantage that we will always work stable at high conversions with $T_o = 10\,°C$ and $T_c = 20\,°C$.

The most probable solution (iii) will be the cheapest one despite its larger reactor volume, because in case (ii) facilities are required for preheating the feed during start-up in order to reach the upper stable operating point.

VI.5.2 An adiabatic tubular reactor with heat exchange between reactants and products

Figure VI.35 shows a system in which the heat of reaction is used to preheat the feed by countercurrent heat exchange between the feed and product streams. It is

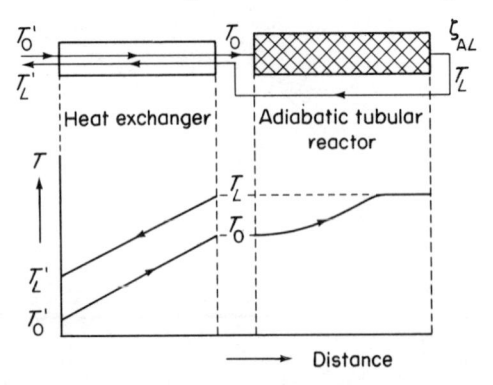

Fig. VI.35 Adiabatic tubular reactor with external heat exchange between feed and reactor product

seen that the temperature rise in the reactor, $T_L - T_o$, is equal to the temperature difference available for heat transfer. Thus, the feed will not be preheated if no conversion takes place. On the other hand, when the reactor has been ignited a temperature rise $T_L - T_o$ results which makes it possible to preheat the reactor feed to a temperature level which may be sufficiently high to keep the reactor going with a high degree of conversion.

After introduction of the adiabatic temperature rise of the reaction mixture ΔT_{ad} we obtain from the heat balance over the reactor:

$$T_L - T_o = \zeta_{AL}\Delta T_{ad}. \tag{VI.41}$$

Here, ζ_{AL} has to be determined from the conversion rate expression and the operating conditions in the manner indicated in Section VI.3 for adiabatic tubular reactors. With a given feed composition and conversion rate formula, ζ_{AL} depends on T_o and the residence time τ_L in the reactor. Furthermore, we have the heat balance over the heat exchanger:

$$\Phi_m c_p (T_o - T_o') = UA(T_L - T_o),$$

and after elimination of $(T_L - T_o)$ by means of equation (VI.41), we obtain:

$$T_o - T_o' = \zeta_{AL}\Delta T_{ad} \frac{UA}{\Phi_m c_p}. \tag{VI.42}$$

It is seen that the temperature rise of the feed stream in the heat exchanger is equal to the temperature rise in the reactor multiplied by the factor $UA/\Phi_m c_p$; in practice, this factor is often greater than 1.

The left-hand term of equation (VI.42) is proportional to the heat taken up in the heat exchanger per unit mass of feed, and the right-hand term is proportional to the heat production by chemical conversion. Both terms can be plotted separately as a function of the reactor inlet temperature T_o, which in this case is the primary variable. The points of intersection between the two functions represent possible points of operation of the system. This has been done in figure VI.36 for a first order reaction of which the relevant parameters have been indicated in the graph. Various values of τ_L were selected in order to demonstrate the influence of the throughput. It was assumed that the volume of the reactor was fixed, so that τ_L is inversely proportional to Φ_m; moreover, the overall heat transfer coefficient U was assumed to be proportional to $\Phi_m^{3/4}$. The S-shaped heat production curves reach a horizontal level at high temperatures where the conversion is complete; this level depends on the residence time τ_L in the reactor, because the multiplying factor $UA/c_p\Phi_m$ is higher at higher values of τ_L. ζ_{AL} as a function of the reactor inlet temperature T_o at a constant value of τ_L is calculated as was demonstrated in Example VI.3.c.

It is seen from figure VI.36 that stable operation with a high degree of conversion is possible above a certain value of τ_L. In this region, the temperature level of the reactor is rather sensitive to the throughput. When the mass flow through the system is increased, the reaction is extinguished at a certain value of

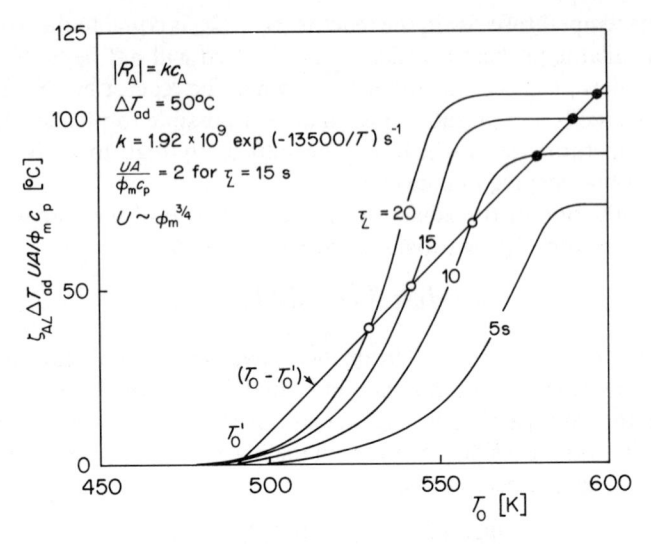

Fig. VI.36 Operating points of system shown in figure VI.35 for first order reaction at different residence times in the reactor

τ_L. This occurs in this example when τ_L is about 8.6 s, as shown in figure VI.37. The reaction can be started by temporarily preheating the feed (for example, with an electric heater) at a relatively low value of Φ_m. Shah [17] made a simulation of an NH_3 reactor with two adiabatic beds, with cold shot cooling and with feed–product heat exchange as shown in figure VI.38. f is the fraction of the feed flowing through the heat exchanger; the cold fraction $(1 - f)$ of the feed is injected at the entrance of the second adiabatic bed. Shah demonstrated that if $(1 - f)$ is too large, no stable operation is possible and the reaction is extinguished or 'blown' out; at sufficiently low values of $(1 - f)$ two stable

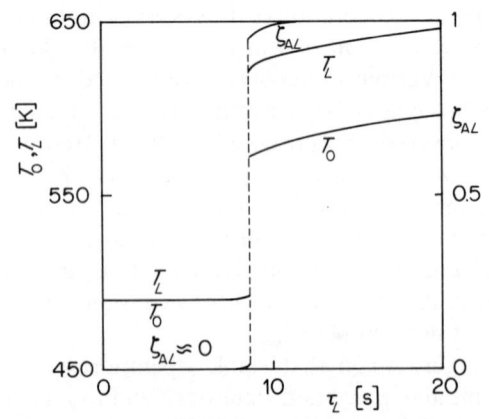

Fig. VI.37 Relative conversion, and inlet and outlet temperatures as a function of residence time; from figure VI.36

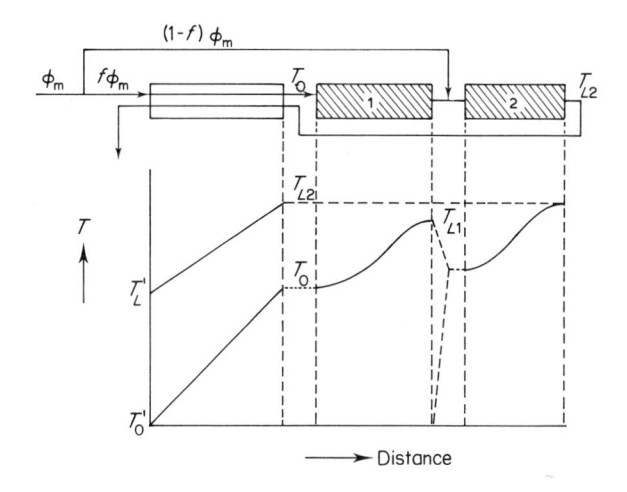

Fig. VI.38 Adiabatic two bed reactor with cold shot cooling

operating points exist, one at lower—still reasonable—conversions and exit temperatures of the second adiabatic bed and one with higher values of T_{L2} and ζ_{AL2}. He showed the same to hold for the inlet temperature, the reactor load and the reactor pressure.

VI.5.3 A multi-tube reactor with internal heat exchange between the reaction mixture and the feed

The autothermal behaviour of a catalytic tubular reactor in which the gas feed is taking up heat from the reacting mixture was also discussed by van Heerden [15]. Figure VI.34 shows a schematic diagram of such a reactor and its temperature distribution, which is typical of a converter for NH_3 synthesis from H_2 and N_2 at high pressure and temperature. Since this is an exothermic equilibrium reaction, a high capacity and a high degree of conversion necessitate a high reaction temperature when the degree of conversion is still low, as well as a drop in temperature as the conversion proceeds. This is achieved in a reactor of the type shown in figure VI.39.

For the calculation of the conversion and the temperature distribution, a material balance (e.g., equation (II.14)) has to be combined with differential energy balances, one for the gas which is preheated:

$$0 = f\Phi_m c_p \, dT' + U(T - T')\frac{A}{L}dz \qquad (VI.43)$$

and one of the reaction mixture:

$$0 = \Phi_m c_p \, dT + \Phi_m(\Delta h_r)_A \, d\xi_A + U(T - T')\frac{A}{L}dz. \qquad (VI.44)$$

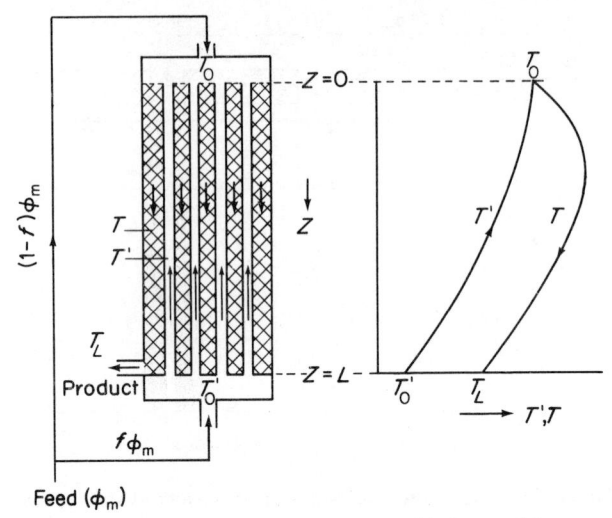

Fig. VI.39 Autothermal multi-tube reactor with internal heat exchange

Here T' refers to the temperature inside the tubes and T to the temperature in the catalyst bed outside the tubes. In these equations, c_p has been assumed to be constant, and the heat transfer resistance $1/U$ between the two streams has been supposed to be localized at the dividing wall; A is the total heat exchange area, f is the fraction of feed flowing through the heat exchanger, and the fraction $(1-f)$ is injected directly at the feed end of the catalyst bed. The boundary conditions are

$$z = L: \quad T' = T'_L$$
$$z = 0: \quad T = T_o = (1-f)T'_L + fT'_o$$

and

$$\zeta_A \equiv 0.$$

With a known rate expression, the above system of equations can be solved with a computer, e.g. for various values of the main parameters ΔT_{ad}, $UA/c_p\Phi_m$ and f.

For a qualitative discussion of the autothermal character of this reactor type, the problem is simplified by taking $f = 1$ (no cold feed injection at the entrance of the catalyst bed) and then combining equations (VI.43) and (VI.44) to yield

$$\Phi_m c_p \, d(T - T') = -\Phi_m (\Delta h_r)_A \, d\xi_A = -\Phi_m (\Delta h_r)_A w_{Ao} \, d\zeta_A.$$

In view of the last two boundary conditions and of $f = 1$, we find at any cross section in the reactor:

$$T - T' = -\frac{(\Delta h_r)_A w_{Ao}}{c_p} \zeta_A = \Delta T_{ad} \zeta_A.$$

In this relation ζ_A is a function of z and has to be found by means of the relation

$$\zeta_A(z) = \int_0^z \frac{M_A R_A(T)}{\Phi_m w_{Ao}} S \, dz$$

By substitution of $(T - T')$ into equation (VI.43) we obtain

$$\frac{c_p \Phi_m}{UA} dT' = -\frac{\Delta T_{ad}}{L} \zeta_A dz,$$

and after integration between $z = 0$ and $z = L$,

$$\frac{(T_o - T'_L)}{UA/c_p \Phi_m} = \frac{\Delta T_{ad}}{L} \int_0^L \zeta_A \, dz = \Delta T_{ad} \langle \zeta_A \rangle, \tag{VI.45}$$

where $\langle \zeta_A \rangle$ is the degree of conversion averaged over the reactor length, and $\Delta T_{ad} \langle \zeta_A \rangle$ the average driving force for the heat exchange (compare equation (VI.42), where the temperature difference between both streams in the heat exchanger was constant and equal to $\Delta T_{ad} \zeta_A$). For a given system $\langle \zeta_A \rangle$ depends on T_o and on $UA/c_p \Phi_m$, and for an equilibrium reaction it passes through a maximum as a function of T_o.

In a similar manner as in equation (VI.42), the first term in equation (VI.45) is proportional to the heat taken up by the feed before it enters into the catalyst bed $[c_p \Phi_m (T_o - T'_L)]$; the total temperature increase of the feed is equal to the average driving force $\Delta T_{ad} \langle \zeta_A \rangle$, multiplied by a factor $UA/c_p \Phi_m$. Both terms of equation VI.45 have been plotted in figure VI.40 as a function of the inlet temperature of the catalyst bed T_o for several values of $UA/c_p \Phi_m$ (after van Heerden [15]). It is seen that not only the heat removal lines but also the heat production curves depend on the value of $UA/c_p \Phi_m$. With too small a heat exchange capacity, the bed entry temperature becomes too low and no conversion results. In the situation 1 in figure VI.40, the reactor is on the verge of stable operation. If too much heat exchange capacity has been built in (e.g., situation 3 in figure VI.40),

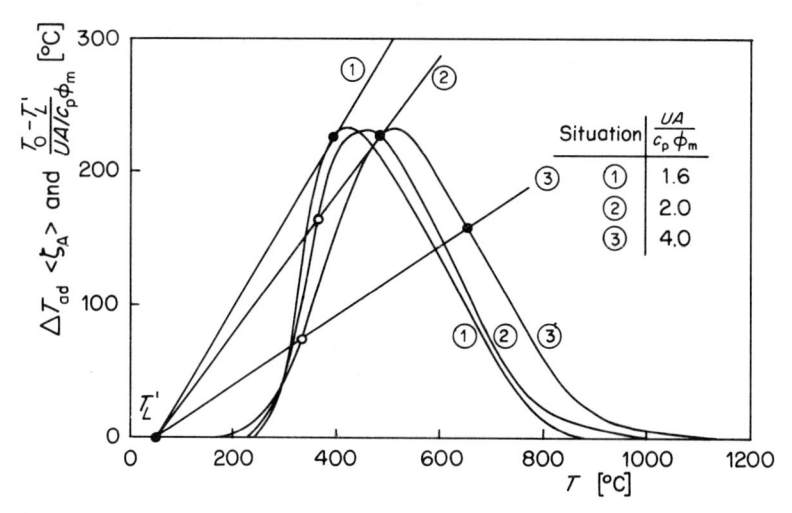

Fig. VI.40 Possible operating points of the reactor of figure VI.39 for an equilibrium reaction with $f = 1$; NH_3 converter [Reprinted with permission from van Heerden [15]. Copyright 1982 American Chemical Society]

the reaction temperature is too high and the equilibrium is adversely affected. Accordingly, it is desirable not to make the heat exchange capacity much greater than necessary, as e.g. in situation 2 in figure VI.40.

In actual practice, once the reactor has been designed, the parameter $UA/c_p\Phi_m$ is rather insensitive to load variations, since the overall heat transfer coefficient U is approximately proportional to $\Phi_m^{3/4}$. A reactor will, therefore, be designed with too large a heat exchange area A, and part of the cold feed will be by-passed ($f < 1$) and injected directly at the entrance of the catalyst bed. As the heat exchange area becomes fouled or the catalyst becomes less active during operation, f is steadily increased in order to maintain high conversion levels. The simplified qualitative discussion for $f = 1$ is no longer valid under such circumstances.

Baddour et al. [18] have made a simulation study of the ammonia converter of figure VI.39. They used a simplified model which compared well with actual plant performance and presented sensitivity data of optimal reactors for the sensitivity of the reactor production, the blow-off temperature and the hot spot towards the reactor load, mole fractions of NH_3 and inert material in the reactor feed, the catalyst activity and the heat exchange capacity UA. They presented the relation between the reactor inlet temperature T'_L of the gas and the entry temperature T_o of the catalyst bed; no feed was bypassed or $f = 1$. Their results are shown in figure VI.41. The branch to the left of the minimum of T'_L corresponds to the unstable operating points in figure VI.40 and the right-hand side to the stable points. We see that for an inlet temperature T'_L below the blow-off temperature no stable operation is possible. The optimum condition for maximum production in figure VI.41 lies very near to the minimum—the catalyst bed entry temperature is about 15 °C higher than the minimum of T_o, where blow-off occurs—as is also indicated in figure VI.40, in which the maximum reactor production coincides with the maximum of the $\Delta T_{ad}\langle\zeta_A\rangle$ curve at the given residence time in the reactor.

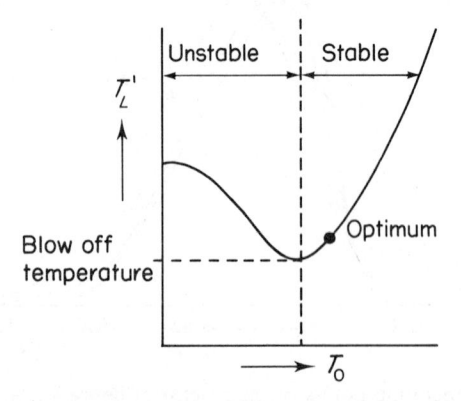

Fig. VI.41 Relation between gas inlet temperature T'_L and catalyst bed entry temperature T_o for the reactor of figure VI.39 at constant reactor load

VI.5.4 Determination of safe operating conditions

In this section we have considered separately the heat production and heat withdrawal in autothermal reactors; with graphical methods determining points of intersection we could find the operating points where the heat production rate is balanced by the heat withdrawal rate. Once the reactor has been designed and constructed, only the residence time, the temperature and the reactant concentration of the feed and the cooling temperature as operating variables can be corrected to prevent the reactor to go to undesired operating conditions. We will therefore discuss the method originally derived by Slinko and Muler [37], which enables us to determine necessary conditions to avoid multiplicity. We will restrict ourselves to autothermal tank reactors.

For a first order reaction in a CISTR the basic equations are

$$\text{heat production } HPR: \Delta T_{ad} \frac{k\tau}{1 + k\tau}$$

$$\text{heat withdrawal } HWR: (T - T_o) + \frac{UA}{\rho c_p \Phi_v}(T - T_c)$$

In the operating points $HWR = HPR$ or $HWR/HPR = 1$. So it is of interest to investigate the function HWR/HPR. We can also use the new function $\Delta T_{ad} HWR/HPR$, which is very helpful to determine the range of feed concentrations or equivalently the values of ΔT_{ad}, where multiplicity occurs. We first write the function in dimensionless form by introducing a dimensionless temperature $\vartheta = \frac{E}{RT_o}(T - T_o)$, so that $T = \frac{RT_o^2}{E}\vartheta + T_o$ and $\frac{T}{T_o} = 1 + \frac{RT_o}{E}\vartheta$.

We will call $RT_o/E = \beta$ and also have

$$k = k_o \exp\left(\frac{\vartheta}{1 + \beta\vartheta}\right),$$

in which $k_o = k_\infty \exp\left(-\frac{1}{\beta}\right)$.

Introducing the new variables we obtain, after some rewriting,

$$\frac{E}{RT_o}\Delta T_{ad}\frac{HWR}{HPR} = \frac{\frac{E}{RT_o}\left[(T - T_o) + \frac{UA}{\rho c_p \Phi_v}(T - T_c)\right](1 + k\tau)}{k\tau} =$$

$$\vartheta_{ad} = \left[1 + \left\{k_o\tau \exp\left(\frac{\vartheta}{1 + \beta\vartheta}\right)\right\}^{-1}\right](\vartheta - Q)(1 + U^*k_o\tau) = F(\vartheta) \quad (VI.46)$$

in which

$$Q = \frac{U^*k_o\tau\vartheta_c}{1 + U^*k_o\tau}, \quad \vartheta_c = \frac{E}{RT_o^2}(T_c - T_o), \quad U^* = \frac{UA}{k_o\rho c_p \Phi_v}$$

and

$$\vartheta_{ad} = \frac{E}{RT_o^2} \frac{(-\Delta H_r)_A c_{Ao}}{\rho c_p}.$$

This function $F(\vartheta)$ is represented in figure VI.42 for different values of the parameter $k_o\tau$. For a low value of $k_o\tau$, either three or only one solution will exist, depending on whether the value of ϑ_{ad} is above or below the local maximum or local minimum respectively in $F(\vartheta)$. For the data given in figure VI.41, which relate to Example VI.5.c, the value of $\vartheta_{ad} = 14.4$. We see that only the F curves for $\tau = 500$ and $\tau = 1000$ seconds have three points of intersection with $\vartheta_{ad} = 14.4$, so that multiplicity occurs. For the other values of τ only one point of intersection of the F curve with $\vartheta_{ad} = 14.4$ is found.

For increasing values of $k_o\tau$ the local maximum and local minimum in $F(\vartheta)$ get less pronounced and finally for high values of $k_o\tau$ they will disappear. Moreover, for high values of ϑ the $F(\vartheta)$ will approach the line: $(1 + U^*k_o\tau)\vartheta - U^*k_o\tau\vartheta_c$ which passes through the point $F(\vartheta) = 0$, $\vartheta = Q$ with a slope $(1 + U^*k_o\tau)$. The function becomes monotonic when the local maximum, the local minimum and the inflection point coincide. This implies that multiplicity is no longer possible. The first and second derivatives are

$$\frac{dF}{d\vartheta} = \exp\left(-\frac{\vartheta}{1 + \beta\vartheta}\right)\left[1 + k_o\tau \exp\frac{\vartheta}{1 + \beta\vartheta} + \frac{Q - \vartheta}{(1 + \beta\vartheta)^2}\right]\frac{(1 + U^*k_o\tau)}{k_o\tau} \quad \text{(VI.47)}$$

$$\frac{d^2F}{d\vartheta^2} = \frac{-\exp\left(-\frac{\vartheta}{1 + \beta\vartheta}\right)}{(1 + \beta\vartheta)^4}[2(1 + \beta\vartheta)^2 + (Q - \vartheta)\{1 + 2\beta(1 + \beta\vartheta)\}]\frac{1 + U^*k_o\tau}{k_o\tau}.$$

(VI.48)

Only the terms between the large brackets can become zero. If we now assume that $\beta \ll 1$ and $\beta\vartheta \simeq 0$, then we find from the conditions $dF/d\vartheta = 0$ and $d^2F/d\vartheta^2 = 0$, that

$$1 + k_o\tau \exp \vartheta + Q - \vartheta = 0 \quad \text{(VI.49)}$$

$$2 + Q - \vartheta = 0 \quad \text{(VI.50)}$$

After subtraction we find

$$k_o\tau e\vartheta = 1, \quad \text{(VI.51)}$$

so that in this point

$$\vartheta = -\ln k_o\tau. \quad \text{(VI.52)}$$

Substitution of this value of ϑ into equation (VI.49) gives us the boundary dividing the region where multiplicity is possible from that where it is not. This leads to the following *condition for multiplicity never to be possible*:

$$k_o\tau > \exp - (2 + Q). \quad \text{(VI.53)}$$

This is also called a 'uniqueness' criterion. However, Q itself is also dependent on τ and T_o, so we have to solve the implicit equation

$$k_o\tau - \exp\left[-2 + \frac{UA\tau}{\rho c_p V_r + UA\tau}\frac{E}{RT_o^2}(T_c - T_o)\right] = 0. \qquad (VI.54)$$

So multiplicity in a single phase CISTR can never occur if

— for fixed inlet temperature T_o and cooling temperature T_c the residence time in the reactor is higher than the root τ of equation (VI.54),
— for a fixed residence time τ_1 and cooling temperature T_c the feed inlet temperature T_o is higher than the root T_o of equation (VI.54).
— for fixed T_o and τ the cooling temperature T_c is higher than the root of equation (VI.54).

In practice too conservative values will be found for the minimum τ or T_o because of the assumption that $\beta = 0$. For the reactor data of figure VI.42, we find that $\tau > 3200\,\text{s}$ for $T_o = 10\,°\text{C}$ or $T_o > 22\,°\text{C}$ for $\tau = 1000\,\text{s}$, both for a cooling temperature of $T_c = 20\,°\text{C}$, in order to avoid multiplicity under all circumstances.

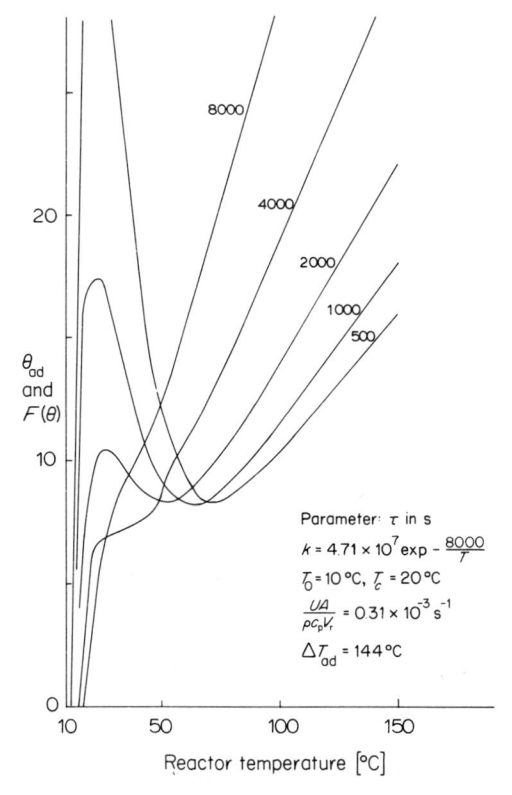

Fig. VI.42 F curves for different residence times in an autothermal tank reactor. Same data as in Example VI.5.c

Another question is what reactant concentration in the feed should be chosen to avoid multiplicity. In this case we should choose ϑ_{ad} at such a value that we are below the local minimum or above the local maximum in the $F(\vartheta)$ curve, because then only one operating point will be found as shown in figure VI.43. *Below the local minimum* we will find a *lower* stable operating point which is not of interest in a single CISTR because of the low conversion. Remember, however, that in a cascade of tank reactors we often operate in the lower stable operating point! *Above the local maximum* in the $F(\vartheta)$ curve we will find only *upper* stable operating points with high conversions.

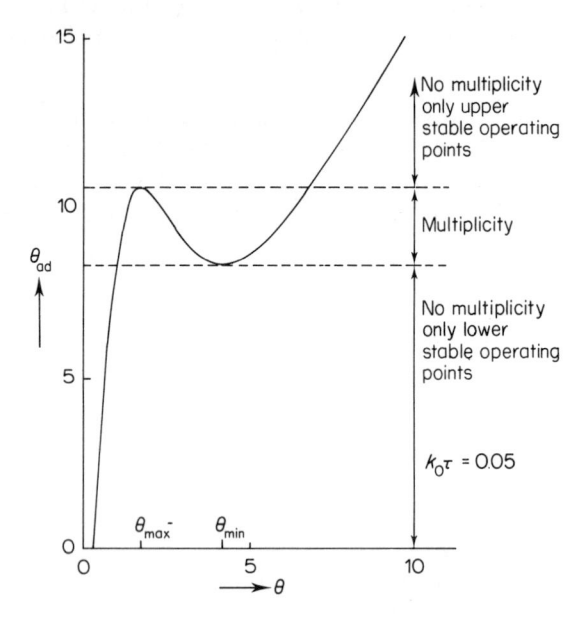

Fig. VI.43 Determination of safe reactant concentrations in the reactor feed to avoid multiplicity. Same data as in figure VI.42

We can determine the values of ϑ in the local maximum and local minimum of the $F(\vartheta)$ curve from the condition

$$k_o\tau \exp\frac{\vartheta}{1+\beta\vartheta} - \frac{\vartheta - Q}{(1+\beta\vartheta)^2} + 1 = 0. \qquad (VI.55)$$

This equation has two roots, of which the smaller root corresponds with the local maximum. Substituting this root ϑ_{max} into equation (VI.46) gives us the value of ϑ_{ad}, above which multiplicity no more can occur and at the same time a high conversion will take place. In this case the exact value of ϑ_{ad} is found, because here we did not assume $\beta = 0$. In Example VI.5.c this corresponds to $\Delta T_{ad} = 176\,°C$. This would give a far too high a reactor temperature in view of the limit of $130\,°C$

set by the construction material. In figure VI.43 the F curve is given for the same Example, but with $\tau = 2000$ seconds. We see that multiplicity can occur for $8.3 < \vartheta_{ad} < 10.6$. This corresponds to reactant concentrations in the feed between 1.48 and 1.90 kmol/m^3.

VI.6 MAXIMUM PERMISSIBLE REACTION TEMPERATURES

Although high reaction temperatures may have the advantage of producing high rates of conversion, there may be several reasons for keeping the reaction temperature below, and sometimes above, a certain value. Once temperature limits have been set, it is possible to predict the measures to be taken in order to comply with these requirements; the methods outlined in the preceding sections can be used for this purpose, provided enough information is available on the reaction and on the reactor to be used.

A very important argument for keeping the reaction temperature within a limited region originates from the possible occurrence of undesired side reactions. If, for example, the selectivity of the desired reaction in a complex reaction decreases with increasing temperature, an upper temperature limit must be adhered to if both the yield and the capacity of the installation are to have acceptable values. In most cases, however, a statement that the reaction temperature in a given system should not exceed a certain fixed value would be entirely inadequate.

The yield of a desired product in complex reactions is determined not only by the temperature (range) at or in which the reaction is carried out, but also by other variables, such as the reaction or residence time and the reactor type used. Especially if the side reactions are accompanied by important heat effects, the manner of heat supply or removal may likewise have a great effect on the result of the reaction. Since the above circumstances are generally entirely different in laboratory experiments and under actual plant conditions, temperature specifications derived from laboratory data may not be valid for practical use. A knowledge of the reaction scheme and of the rate expressions for the reactions involved as a function of temperature is a much more dependable basis for finding the appropriate reaction conditions. It is realized that such information is hard to obtain in many practical cases; a few indications of a more or less qualitative nature may nevertheless be helpful in this respect.

We shall demonstrate the points raised above by considering a set of parallel reactions and a set of consecutive reactions with heat effects carried out in a continuous *tank reactor*. These examples were treated in a paper by Westerterp [19], to which the reader is referred for more details.

Suppose that a key reactant A can be converted to a desired product P and an undesired product X. Molar concentrations will be used and it will be assumed that ρ and c_p remain constant, that the feed contains neither P nor X ($c_{Po} = c_{Xo} = 0$) and that the sum of the concentrations of A, P and X is constant

($c_{Ao} = c_A + c_P + c_X$). The material balances for the species P and X become, for steady state operation in a tank reactor,

$$0 = -c_{p1} + R_P \tau$$
$$0 = -c_{X1} + R_X \tau$$

in which R_P and R_X have to be taken at the reaction conditions. Assuming first order kinetics, we have

$$\left. \begin{array}{l} R_P = k_P c_A \\ R_X = k_X c_A \end{array} \right\} \text{ for parallel reactions,}$$

and

$$\left. \begin{array}{l} R_P = k_P c_A - k_X c_P \\ R_X = k_X c_P \end{array} \right\} \text{ for consecutive reactions}$$

In examining the influence of temperature it is convenient to introduce the reference temperature T_R at which k_P and k_X are equal and have the value k_R. The values of T_R and k_R follow from

$$k_R = k_{P|T_R} = k_{P\infty} e^{-E_P/RT_R} = k_{X|T_R} = k_{X\infty} e^{-E_X/RT_R}$$

The composition of the reaction mixture can now be calculated as a function of

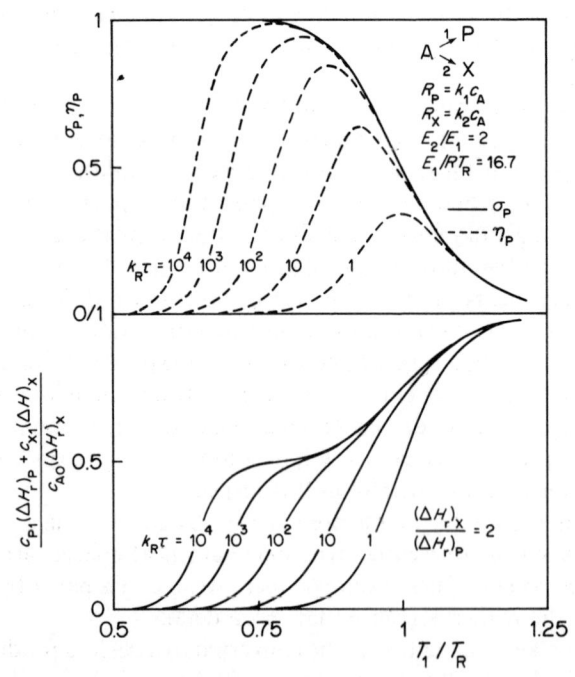

Fig. VI.44 Selectivity, yield and relative heat production for exothermic parallel reactions in a tank reactor [From Westerterp [19]. Reproduced by permission of Pergamon Press Ltd]

the dimensionless absolute reactor temperature T_1/T_R and the dimensionless average time of residence $k_R\tau$, while the reaction system is characterized by k_R, T_R and the ratio of the activation energies E_X/E_P which influences the selectivity. As a result of such calculations the upper parts of figures (VI.44) and (VI.45) show, for a parallel and a consecutive reaction system respectively, the selectivity

$$\sigma_P = c_{P1}/(c_{A0} - c_A)$$

and the yield

$$\eta_P = c_{P1}/c_{A0}$$

as a function of the dimensionless temperature T_1/T_R and for $E_X/E_P = 2$. The relative degree of conversion of A, ζ_A, equals the ratio of σ_P and η_P; the reactant A is completely converted, therefore, if the latter two magnitudes coincide. It is seen that for both reaction systems the yields have a maximum value which is higher as the reaction temperature is further removed from T_R. An increase in maximum yield, in this case, can only be obtained at the expense of a great increase in residence time and a corresponding reduction in reactor capacity. It is also observed from these figures that the same value of σ_P can be obtained in a whole range of operating temperatures. Thus, an *a priori* specification of a maximum temperature at which a side reaction would be sufficiently suppressed has no significance.

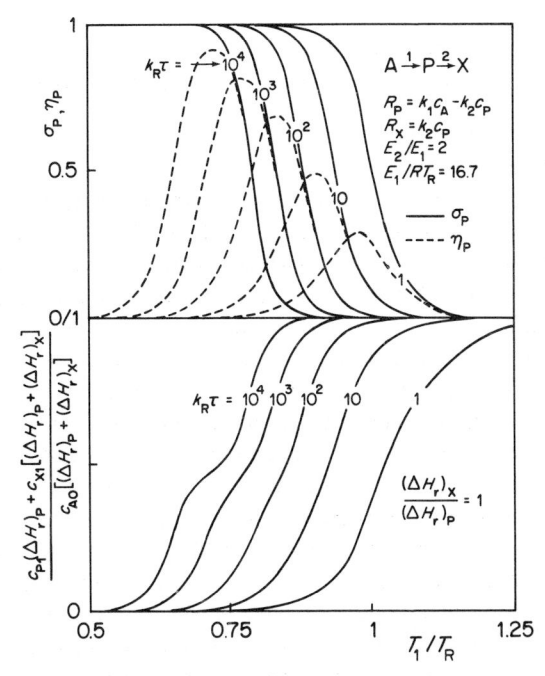

Fig. VI.45 Selectivity, yield and relative heat production for exothermic consecutive reactions in a tank reactor [From Westerterp [19]. Reproduced by permission of Pergamon Press Ltd]

Another point is that, if both the main reaction and the side reaction have a heat effect, these effects have to be taken into account, so that the reactor may be operated in a proper (i.e., stable) manner. In writing the heat balance over a tank reactor in a form similar to equation (VI.38), the heat production of all reactions has to be considered. We thus obtain, for the above parallel reactions,

$$-c_{P1}(\Delta H_r)_P - c_{X1}(\Delta H_r)_X = \rho c_p(T_1 - T_o) + \frac{\rho UA}{\phi_m}(T_1 - T_c), \quad \text{(VI.56.a)}$$

and for the consecutive reactions

$$-c_{P1}(\Delta H_r)_P - c_{X1}\{(\Delta H_r)_P + (\Delta H_r)_X\} = \rho c_p(T_1 - T_o) + \frac{\rho UA}{\Phi_m}(T_1 - T_c).$$
(VI.56.b)

$(\Delta H_r)_X$ is the heat of reaction associated with the formation of one molar unit of X. Division of the heat production terms on the left-hand side of these equations by the highest heat production yields

$$\frac{-c_{P1}(\Delta H_r)_P - c_{X1}(\Delta H_r)_X}{-c_{Ao}(\Delta H_r)_X} = \eta_P\left[\frac{(\Delta H_r)_P}{(\Delta H_r)_X} + \frac{1-\sigma_P}{\sigma_P}\right] \quad \text{(VI.57.a)}$$

for the relative heat production with parallel reactions [assuming $(\Delta H_r)_X > (\Delta H_r)_P$], and

$$\frac{-c_{P1}(\Delta H_r)_P - c_{X1}[(\Delta H_r)_P + (\Delta H_r)_X]}{-c_{Ao}[(\Delta H_r)_P + (\Delta H_r)_X]} = \eta_P\left[\frac{(\Delta H_r)_P}{(\Delta H_r)_P + (\Delta H_r)_X} + \frac{1-\sigma_P}{\sigma_P}\right]$$
(VI.57.b)

with consecutive reactions.

If η_P and σ_P are known as a function of $k_R\tau$ and of the dimensionless reaction temperature T_1/T_R, these relative heat production functions can be plotted for fixed values of $(\Delta H_r)_X/(\Delta H_r)_P$. This has been done in figures (VI.44) and (VI.45) for the reactions considered, on the assumption that they are exothermic. Under conditions where a high yield η_P is obtained, the relative heat production curves show a tendency to form a plateau. As a consequence, 5 solutions of the heat balance can in principle be found by comparing such a curve with the heat removal line when the autothermal behaviour of the reactor is studied (compare Section VI.5). Two solutions are unstable; of the 3 stable ones, the lowest corresponds to practically no conversion and the highest to an almost complete conversion of reactant to the product involving the highest heat of reaction. If this product is not wanted, the reactor must work at the intermediate stable operating point. It will be clear that, especially with reactions producing large heat effects, a proper combination of feed concentration and temperature, heat transfer area and cooling temperature have to be chosen to keep the reactor operating under the desired conditions. An example will be given in the illustration at the end of this section.

It should be noted that this section deals with very simple kinds of complex reactions carried out in a continuous tank reactor. Much more complicated reactions may occur in practice (such as shown by Hoftijzer and Zwietering [21]), and a similar treatment for, say, tubular reactors with heat exchange will also become extremely involved. However, a qualitative analysis of the influence of side reactions with a heat effect on the autothermal behaviour of tank reactors and on the parametric sensitivity of tubular reactors can lead to valuable results, even if simplified rate equations are introduced in order to make the problem more tractable (van Heerden [16]).

Example VI.6 The oxidation of naphthalene in a fluidized bed

This example serves to illustrate how the operating temperature and the corresponding cooling requirements can be found for a highly exothermic reaction followed by additional exothermic reactions. The catalysed gas phase reaction between naphthalene and oxygen occurs along the following scheme:

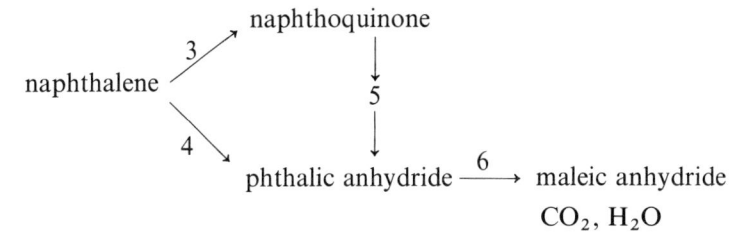

```
                  naphthoquinone
                 ↗ 3              
    naphthalene                  ↓ 5
                 ↘ 4              
                    phthalic anhydride  ──6──→  maleic anhydride
                                                 $CO_2$, $H_2O$
```

Demaria et al. [20] studied the kinetics of these reactions in a fluidized bed with two different catalysts, A and B. On deriving the reaction velocity constants k_3 to k_6 they considered all reactions to be homogeneous and of the first order with respect to the concentrations of the species converted; the concentration of oxygen in the reaction mixture, which was in great excess, was taken as constant. They furthermore assumed the reaction mixture to have a uniform composition in the fluidized bed so that it was considered to behave like a tank reactor. They showed that k_5 was several times greater than k_3, and that k_3 was equal to k_4. Hence we shall further simplify the above reaction scheme to

naphthalene → phthalic anhydride → combustion products,
(A)   (P)   (X)

where P is the desired product. For the first order reaction velocity constants we take

$$k_P = k_3 + k_4 \quad \text{and} \quad k_X = k_6.$$

The values of k_P and k_X as derived from the measurements by Demaria et al. have been plotted in figure VI.46 against $1/T$. On extrapolating the lines beyond the range of the experiments we find the values of T_R and k_R for both catalysts (see Table VI.5); the values of E_X/E_P are found from the ratio of the slopes of k_X and k_P.

In order to obtain a high yield, η_P, k_P should be one or two orders of magnitude greater than k_X; this means that the reaction temperature should be well above T_R

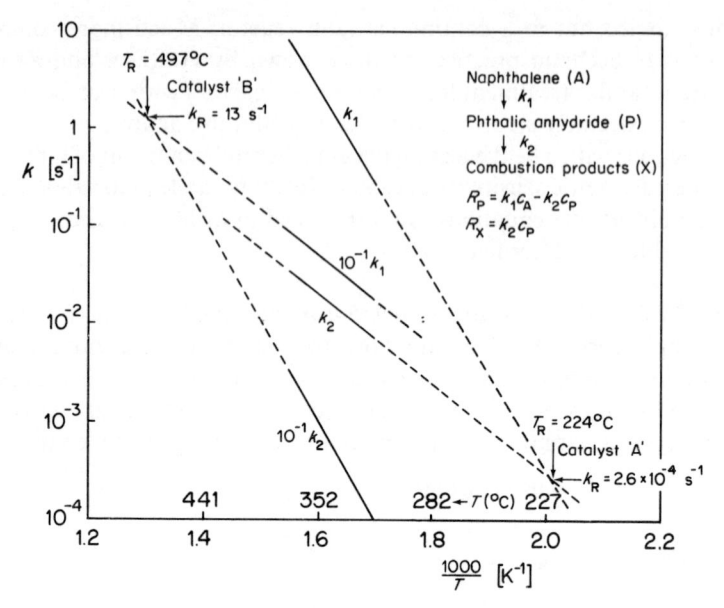

Fig. VI.46 Reaction velocity constants for the oxidation of naphthalene with two different catalysts [19]; Example VI.6

Table VI.5 Data for Example VI.6

Oxidation of naphthalene with air at atmospheric pressure; kinetic data: see figure VI.46.
$(\Delta H_r)_P = -1880 \times 10^6$ J/kmol
$(\Delta H_r)_X = -3280 \times 10^6$ J/kmol
$c_p = 1040$ J/kg°C
$U = 350$ W/m²°C
$c_{Ao}/c = 0.010$
$T_o = 150$ °C

	Catalyst A			Catalyst B
T_R (K)	497			770
k_R (s^{-1})	2.6×10^{-4}			13
E_X/E_P	0.47			2.19
$k_R \tau$	10^{-4}	5×10^{-3}		67
τ (s)	0.4	20		5
$(T_1)_{opt}/T_R$	1.41	1.22		0.83
$(T_1)_{opt}$ (°C)	423	331		365
$\zeta_A = 1 - c_A/c_{Ao}$	0.977	0.941		0.787
$\sigma_P = c_P/(c_{Ao} - c_A)$	0.957	0.814		0.955
$\eta_P = c_P/c_{Ao}$	0.934	0.768		0.752
heat removal line in figure VI.47	(d)	(e)		(f)
A/V_r (m²/m³)	32.8	1.4		7.4
T_c (°C)	348	275		334

(224 °C) for catalyst A, and for catalyst B well below its corresponding T_R value (497 °C). It is seen from figure VI.46 that a reaction temperature in the vicinity of 350 °C will be desirable for both catalysts.

It is now assumed that the residence time distribution in an industrial fluidized bed reactor is very similar to that of the ideal tank reactor. It should be noted that this is not necessarily true; see chapter IV. On this assumption, η_P curves can be calculated as a function of T_1/T_R for various values of the dimensionless residence time $k_R\tau$ in the manner indicated earlier in this section and with a result very similar to the top graph of figure VI.45. For each value of $k_R\tau$ a maximum in η_P occurs at a certain value of T_1/T_R; the corresponding reaction temperature T_1 will be called the optimum temperature $(T_1)_{opt}$ at a given residence time τ. For further calculations, we adopt two residence times (0.4 and 20 s) for catalyst A, with corresponding reaction temperatures $(T_1)_{opt}$ of 423 and 331 °C, respectively; a residence time of 5 s is selected for catalyst B with $(T_1)_{opt} = 365$ °C. These figures are shown in Table VI.5 together with the product distribution obtained. It is seen that with catalyst A the yield of phthalic anhydride increases with increasing reaction temperature, provided the contact time is reduced in order to keep to the maximum within the yield curve. A comparison between catalysts A and B at the longer residence times shows that a higher yield is obtained with A, but that B gives a better selectivity under the conditions indicated.

For investigating the requirements of heat removal, the relative heat production curves (compare figure VI.45, bottom graph) are constructed for the three cases mentioned above. These are shown in figure VI.47, where the operating points have been indicated by S_{A1}, S_{A2} and S_B. The data used for the calculation of the curves have been collected in Table VI.5. The heat removal lines which must pass through the operating points have the formula (see equations (VI.56.b) and (VI.57.b))

$$\frac{\rho c_p (T_1 - T_o) + \rho U A (T_1 - T_c)/\Phi_m}{-c_{Ao}[(\Delta H_r)_P + (\Delta H_r)_X]} = f(T_1)$$

If it is desired to run the reactor adiabatically with a given feed temperature T_o (say, 150 °C), the indicated points of operation could only be realized by proper adjustment of the naphthalene feed concentration c_{Ao}. However, stable operation might result only in the case of catalyst A with $\tau = 0.4$ s (line a in figure VI.47). The feed mole fraction c_{Ao}/c would then have to be 0.00438, but if it became higher than 0.00456 (line b) a nearly complete combustion of the desired product would result at $T_1/T_R = 2.36$, or $T_1 = 900$ °C. Below a mole fraction of 0.00423 (line c), on the other hand, the reaction is extinguished. It will be clear that this range of feed concentration is too narrow for safe operation; moreover, these concentrations are rather low. It is necessary, therefore, to work with much steeper heat removal lines, which means that the reactor has to be cooled with a medium having a relatively high cooling temperature T_c.

Figure VI.47 contains the heat removal lines d, e and f for the three different cases treated here and for $c_{Ao}/c = 0.010$. Only one stable operating condition is possible with these lines, so that the possibility of excessive combustion of the

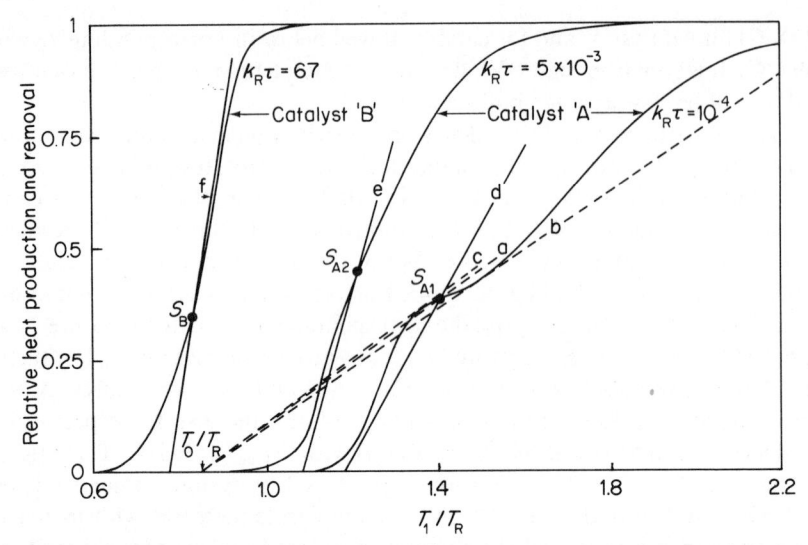

Fig. VI.47 Heat production and heat removal for the oxidation of naphthalene; Example VI.6 [From Westerterp [19]. Reproduced by permission of Pergamon Press Ltd]

phthalic anhydride is excluded. The lines have the character of a compromise: when the slope of the lines through the operating points is diminished, the flexibility of the reactor is greatly reduced; an increase of the slope means that more heat transfer area has to be built in and that a higher coolant temperature has to be used. The values of UA/Φ_m and T_c associated with the heat removal lines d, e and f—as indicated in Table VI.5—must therefore be regarded as minimum values.

It is seen from the tabulated results that a high yield of phthalic anhydride would be obtained with catalyst A at $\tau = 0.4$ s, but that an extremely high heat exchange area per unit reactor volume would be required. This can only be achieved by rapid circulation of catalyst through an external heat exchanger. For calculating such a system, the last term in equation (VI.56) should be replaced by $\Phi_s c_{ps} (T_1 - T'_o)/\Phi_m$, where Φ_s is the mass flow, c_{ps} the specific heat and T'_o the entrance temperature of the circulating catalyst. The design of the catalyst heat exchanger then becomes a separate problem. A reasonably small heat exchange area is required with $\tau = 20$ s, and a once-through yield of about 77% is obtainable. The same yield but a better selectivity results with catalyst B and $\tau = 5$ s; the cooling area is rather large and the temperature of the coolant must be high.

This illustration should be regarded merely as a rough approximation on the basis of rather crude simplifications and extrapolations. It shows, however, in a semi-quantitative manner that the maximum permissible reaction temperature is an elastic concept and that very special cooling conditions are required.

Example VI.6 ended)

The theory of this section was further developed for *tank reactors* by, among others, Luss and Chen [38], Pikios and Luss [39] and Tsotsis, Haderi and

Schmitz [40]: they derived uniqueness and multiplicity criteria, also for endothermic reactions and for Langmuir-Hinshelwood kinetics [40]. A unique solution is found e.g. for the conditions (VI.53). Westerterp and Ptasinski [41] studied maximum allowable temperatures and hot spots for multiple reactions in cooled *tubular reactors* and give design criteria for safe operation at higher selectivities; they indicate that the design criteria can be met in general without entering into the danger area of high parametric sensitivity in a tubular reactor.

VI.7 THE DYNAMIC BEHAVIOUR OF MODEL REACTORS

Only the steady state characteristics of continuous flow reactors have been considered thus far in this chapter. In practice, non-stationary performance due to disturbances is equally important. Intentional disturbances are applied in starting and stopping operations and when a change is made in the operating conditions; the designer and the operator have to be aware of the resulting transient conditions. Unintentional disturbances such as variations in feed conditions, in steam pressures, etc., have to be counteracted by automatic control; the dynamic response of the plant to such disturbances largely determines the control system to be used and the dynamic characteristics of the required controllers.

With *isothermal* reactor operation, a change in feed rate or in feed composition will result in a gradual change of the conversion from one value to another. The time interval in which the transition occurs will then be of main interest. Due to the introduction of time as a new independent variable, ordinary differential equations now have to be solved for tank reactors (which in the steady state were described by means of algebraic equations). Mason and Piret [35, 36] applied these to the starting-up of isothermal cascades of tank reactors; on the basis of their study they recommend procedures for rapidly attaining on-stream conditions in a cascade. Partial differential equations have to be solved for the treatment of the transient behaviour of isothermal tubular reactors (which in the steady state were described by means of ordinary differential equations). Solutions for particular cases can be readily obtained by means of modern calculation aids, even when the conversion rate expressions are non-linear functions of the concentrations.

With *non-isothermal* operation, the transient behaviour of a reactor may show temperature excursions which temporarily or permanently lead to undesired reactor conditions. This is a result of coupling between the material and energy balances. The simultaneous solution of two ordinary non-linear differential equations presents difficulties with respect to the treatment of the steady state operation of tubular reactors (compare the discussion on the parametric sensitivity of tubular reactors in Section VI.3); it can be imagined that the corresponding non-stationary problems must be rather intractable. On the other hand, the dynamic behaviour of an autothermally operating tank reactor belongs to a class of less complicated problems. Aris and Amundson [22] did pioneering work in this field. Uppal, Ray and Poore [23] improved the basic mathematics on the basis of bifurcation principles. In 1972 Perlmutter [24] published a text book on reactor stability. A review of most of the publications in this field of investigations up to 1974 was given by Schmitz [25]. The salient points of the theory of the unsteady autothermal tank reactor will be outlined below, and something will also be said on plug flow reactors.

VI.7.1 The autothermal tank reactor

The equations governing the dynamic behaviour of a tank reactor are essentially the material balance

$$\frac{d(\rho V_r w_{J1})}{dt} = \Phi_{mo} w_{Jo} - \Phi_{m1} w_{J1} + M_J R_{J1} V_r \tag{VI.58}$$

and the energy balance

$$\frac{d(\rho V_r u_1)}{dt} = \Phi_{mo} h_o - \Phi_{m1} h_1 + UA(T_c - T_1). \tag{VI.59}$$

In the energy balance the heat capacity of the vessel itself and the agitator has been neglected; this is permissible only for large industrial reactors, not for small laboratory equipment. It is now assumed that the density, ρ, and the reaction volume, V_r, remain constant so that $\Phi_{mo} = \Phi_{m1} = \Phi_m$. We furthermore consider a single reaction the extent of which is expressed by the relative degree of conversion of a reactant A: $\zeta \equiv (w_{Ao} - w_{A1})/w_{Ao}$. The index 1, indicating the conditions in the reactor and at its outlet, will be omitted. Upon introduction of the average residence time, $\tau = \rho V_r/\Phi_m$, molar concentrations and $\theta = t/\tau$, we find for equation (VI.58):

$$\frac{d\zeta}{d\theta} = -\zeta + \frac{\tau R_A}{c_{Ao}}. \tag{VI.60}$$

Equation (VI.59) can be worked out by using the expressions (VI.5) and (VI.6) for u and h. If we further assume that $u \simeq h$ (liquid) and that c_p and $(\Delta h_r)_A$ are constant, we may write for equation (VI.59)

$$\frac{dT}{d\theta} - \frac{(\Delta H_r)_A c_{Ao}}{\rho c_p} \frac{d\zeta}{d\theta} = T_o - T + \frac{(-\Delta H_r)_A c_{Ao}}{\rho c_p} \zeta$$
$$+ \frac{UA\tau}{\rho c_p V_r}(T_c - T) \tag{VI.61}$$

A further simplification can be obtained by introducing the adiabatic temperature rise (for complete conversion of A):

$$\Delta T_{ad} = \frac{-(\Delta H_r)_A c_{Ao}}{\rho c_p}$$

and the Stanton number

$$St' = \frac{UA\tau}{\rho c_p V_r},$$

and by eliminating the term with $d\zeta/d\theta$ in equation (VI.61) by means of equation (VI.60). The result is

$$\frac{dT}{d\theta} = T_o - T - \Delta T_{ad}\frac{R_A\tau}{c_{Ao}} + St'(T_c - T). \qquad (VI.62)$$

As explained in section VI.5, not more than three solutions of equations (VI.50) and (VI.62) are possible in the steady state. These solutions can be visualized by plotting the function ζ and the straight line $[T - T_o + (T - T_c)St']$ in a ζ–T graph and looking for the intersections (cf. figures VI.28 and VI.29). The intermediate solution is unstable and the other two are said to be statically stable. The upper stable solution is of main interest since it is associated with the highest degree of conversion; this point will be given the index s.

In describing non-stationary behaviour use may be made of a ζ–T plane for a study of what happens when the reactor conditions deviate from a stable solution. Provided the conversion rate is a known function of ζ and T, the values of $d\zeta/d\theta$ can be calculated from equations (VI.60) and (VI.62) for each point (ζ, T); hence, it is possible in principle to approximate a sequence of (ζ, T) combinations and corresponding times along which a stable solution is eventually reached from an arbitrary starting point. If we are interested only in the ζ–T path, the time can be eliminated from equations (VI.60) and (VI.62), which results in:

$$\frac{d\zeta}{dT} = \frac{-\zeta + \dfrac{R_A\tau}{c_{Ao}}}{T_o - T - \Delta T_{ad}\dfrac{R_A\tau}{c_{Ao}} + St'(T_c - T)}. \qquad (VI.63)$$

We shall now examine more closely the possible ζ–T paths after a change in operating conditions. It is accordingly assumed that at a certain moment the reactor works in a point (ζ, T) which is not a statically stable state corresponding to the new set of conditions. If the new conditions do not change, one of the stable states will ultimately be attained. We shall concern ourselves mainly with the manner in which the new upper stable point (ζ_s, T_s) is reached. For this discussion, we shall adopt the conversion rate expression for a first order reaction:

$$R_A = kc_{Ao}(1 - \zeta).$$

As in the approximation used in section VI.3, we now write for k:

$$k = k_\infty \exp(-E/RT)$$
$$= k_\infty \exp(-E/RT_s)\exp\left[\frac{E}{RTT_s}(T - T_s)\right]$$
$$\simeq k_s \exp\left[\frac{E}{RT_s^2}(T - T_s)\right] = k_s e^{\Delta\vartheta}.$$

The above approximation is permissible in the vicinity of the upper stable solution belonging to the new operating conditions after a change has been brought about. At its temperature T_s the reaction rate constant is k_s; the quantity $\Delta\vartheta$ is the dimensionless deviation from this temperature. If we furthermore introduce

$$\Delta\vartheta_o = (T_o - T_s)E/RT_s^2,$$
$$\Delta\vartheta_c = (T_c - T_s)E/RT_s^2,$$
$$\Delta\vartheta_{ad} = \Delta T_{ad}E/RT_s^2,$$

then equation (VI.63) becomes

$$\frac{d\zeta}{d\Delta\vartheta} = \frac{-\zeta + k_s\tau(1-\zeta)e^{\Delta\vartheta}}{-(1+St')\Delta\vartheta + \Delta\vartheta_{ad}k_s\tau(1-\zeta)e^{\Delta\vartheta} + \Delta\vartheta_o + St'\Delta\vartheta_c}. \quad (VI.64)$$

In the upper stable solution ($\Delta\vartheta = 0$), both $d\zeta/dt$ (equation (VI.50)) and dT/dt (equation (VI.62)) are zero, and consequently the nominator and the denominator of equation (VI.64) must vanish. This gives the well-known relations

$$k_s\tau = \zeta_s/(1-\zeta_s)$$

and

$$\Delta\vartheta_o + St'\Delta\vartheta_c = -\Delta\vartheta_{ad}k_s\tau(1-\zeta_s) = -\Delta\vartheta_{ad}\zeta_s.$$

Substitution of these into equation (VI.64) yields the general formula for first order reactions:

$$\frac{d\zeta}{d\Delta\vartheta} = \frac{-\zeta + \dfrac{\zeta_s}{1-\zeta_s}(1-\zeta)e^{\Delta\vartheta}}{-(1+St')\Delta\vartheta + \dfrac{\Delta\vartheta_{ad}\zeta_s}{1-\zeta_s}(1-\zeta)e^{\Delta\vartheta} - \Delta\vartheta_{ad}\zeta_s}. \quad (VI.65)$$

The slope of the $\zeta-\Delta\vartheta$ path can be calculated from equation (VI.65) for each point in a $\zeta-\Delta\vartheta$ plane, as shown in figure VI.48. The locus of all points where $d\zeta/d\Delta\vartheta = 0$ is found when the nominator of equation (VI.65) is zero. In figure VI.48, this is Curve 1, which corresponds to the heat production curve in figure VI.23. Curve 2 is found by putting the denominator equal to zero; it is the locus of all points for which $d\zeta/d\Delta\vartheta = \infty$. The heat removal line corresponding to figure VI.24 is given in this system by the equation (Line 3)

$$\zeta = \zeta_s + (1+St')\frac{\Delta\vartheta}{\Delta\vartheta_{ad}}. \quad (VI.66)$$

This line has a slope of arctan $[(1+St')/\Delta\vartheta_{ad}]$. For points of the $\zeta-\Delta\vartheta$ path lying on this line, we have $d\zeta/d\Delta\vartheta = 1/\Delta\vartheta_{ad}$, as can be derived from equation (VI.65). The value of $d\zeta/d\Delta\vartheta$ can also be calculated at several other points. Some results have been indicated by the arrows in figure VI.48 for a chosen combination of ζ_s,

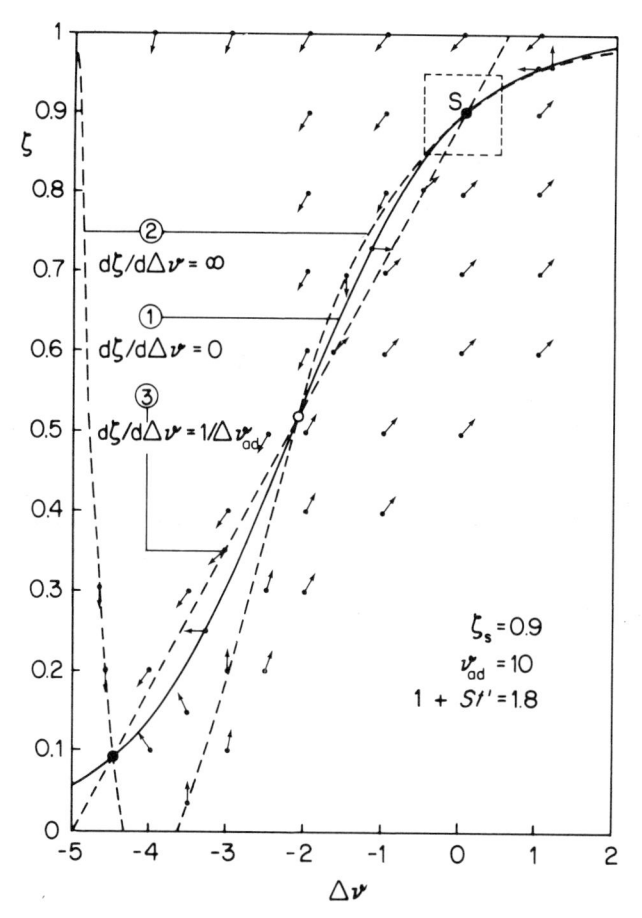

Fig. VI.48 The ζ–$\Delta\vartheta$ plane for a first order exothermic reaction in a cooled tank reactor

$\Delta\vartheta_{ad}$ and $(1 + St')$. An impression can be obtained from these regarding the manner in which a stable situation is reached after a disturbance.

It is seen from figure VI.48 that, starting from a high degree of conversion on the left of Curve 2, both ζ and $\Delta\vartheta$ will invariably decrease. If the starting point is relatively far from Curve 2 the reaction will be extinguished; if it is sufficiently near, the ζ–$\Delta\vartheta$ path will vertically cross Curve 2 above the unstable steady solution, and it will then pass through Curve 1 (horizontally) and through Curve 3 (with a slope $1/\Delta\vartheta_{ad}$). In this case, the conversion and the temperature will again rise, cross Curves 2 and 1 to the right of the upper stable operating point, and again turn to the left to reach this point S. Apparently, if one starts from a moderate degree of conversion and from a $\Delta\vartheta$-value to the right of Curve 2, the reactor temperature will pass through a maximum value which may be much higher than T_s. This feature may be undesirable in particular cases (such as when an exothermic side reaction will occur at this high temperature) and special precautions must be taken in the starting-up of such a system. On the other hand, with a low conversion and a temperature in the vicinity of Curve 2 as a start, the conversion will temporarily rise, pass through a maximum and eventually reach the lower stable point; in such a case the ignition of the reaction has not succeeded or the reaction is extinguished.

It is possible to divide the ζ–$\Delta\vartheta$ plane into a part from which only the upper stable point is reached. The dividing line lies slightly to the left of Curve 2 below the unstable point. It cannot easily be constructed with the method given above. In fact, the best way to trace the ζ–$\Delta\vartheta$ paths for a given problem is by using a computer. This was done by Bilous and Amundsen, whose paper [26] contains a number of results in the form of graphs like figure VI.48.

It can be deduced from figure VI.48 that the stable point S is approached by a turning movement. The region around S has been enlarged in figure VI.49 to allow closer inspection of this phenomenon. Curves 1, 2 and 3 are identical with those in figure VI.48. However, three possible combinations of $\Delta\vartheta_{ad}$ and St' are considered in figure VI.49. It is seen that, with adiabatic operation ($St' = 0$), $d\zeta/d\Delta\vartheta$ for points lying on Line 3 coincides with this line. Consequently, with an adiabatic autothermal reactor the point S will always be reached along Line 3, either from below or from above. With the cooled reactor ($St' > 0$), Line 3 can be crossed and the ζ–$\Delta\vartheta$ path can approach S by a spiral movement (e.g., $St' = 0.8$). However, when the reactor is very strongly cooled (e.g., $St' = 2.6$), figure VI.49 shows that S cannot be reached although it is a *statically* stable point. If the arrows for this case are followed, it is found that the ζ–$\Delta\vartheta$ path will ultimately

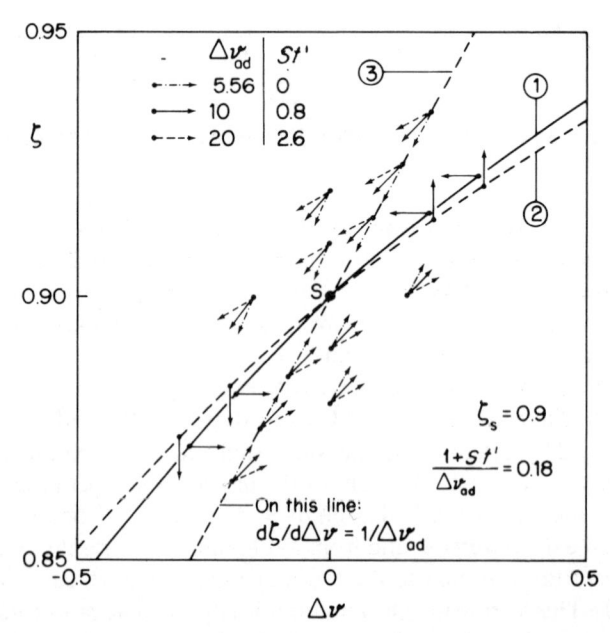

Fig. VI.49 The region around the statically stable operating point S in figure VI.48

turn around S in a limit cycle; this results in a steady oscillation of the reactor conditions.

A more precise examination of the stability conditions of S can be made by considering the region so close to S that equations (VI.60) and (VI.62) can be expressed as linear functions of $(\zeta - \zeta_s)$ and $\Delta\vartheta$. If we put:

$$\zeta - \zeta_s \equiv \Delta\zeta, \quad e^{\Delta\vartheta} = 1 + \Delta\vartheta,$$

and neglect small second-order terms, we find, from equation (VI.60),

$$\frac{d\Delta\zeta}{d\theta} = -\frac{1}{1 - \zeta_s}\Delta\zeta + \zeta_s\Delta\vartheta \equiv a_{11}\Delta\zeta + a_{12}\Delta\vartheta, \qquad \text{(VI.67)}$$

and from equation VI.62

$$\frac{d\Delta\vartheta}{d\theta} = \frac{\Delta\vartheta_{ad}\zeta_s}{1 - \zeta_s}\Delta\zeta + [\Delta\vartheta_{ad}\zeta_s - (1 + St')]\Delta\vartheta \equiv a_{21}\Delta\zeta + a_{22}\Delta\vartheta. \qquad \text{(VI.68)}$$

The solution of these equations has the form:

$$\Delta\zeta = b_{11}e^{p_1\theta} + b_{12}e^{p_2\theta}$$
$$\Delta\vartheta = b_{21}e^{p_1\theta} + b_{22}e^{p_2\theta}$$

in which p_1 and p_2 are defined by

$$p_{1,2} = \frac{a_{11} + a_{22}}{2}\left[1 \pm \sqrt{\left(1 - \frac{4(a_{11}a_{22} - a_{12}a_{21})}{(a_{11} + a_{22})^2}\right)}\right].$$

The solution is stable if $\Delta\zeta$ and $\Delta\vartheta$ approach zero with increasing time. This only occurs when the real parts of p_1 and p_2 are negative. Hence, the stability conditions are

$$a_{11}a_{22} - a_{12}a_{21} > 0 \qquad \text{(VI.69)}$$

and

$$-a_{11} - a_{22} > 0. \qquad \text{(VI.70)}$$

Inserting the values for the coefficients a from equations (VI.67) and (VI.69), we find, from equation (VI.70),

$$-\frac{1}{1 - \zeta_s}[\Delta\vartheta_{ad}\zeta_s - (1 + St')] + \Delta\vartheta_{ad}\frac{\zeta_s^2}{1 - \zeta_s} > 0$$

or

$$\frac{1 + St'}{\Delta\vartheta_{ad}} > \zeta_s(1 - \zeta_s), \qquad \text{(VI.71)}$$

which is equivalent to

$$\frac{1 + St'}{\Delta T_{ad}} > \frac{E}{RT_s^2}\frac{k\tau}{(1 + k\tau)^2}. \qquad \text{(VI.71.a)}$$

This condition states that at the point S the slope of the heat removal line (line 3 in figure VI.48) must be greater than that of the heat production line. Apparently, equation (VI.71) is the condition for *static stability*, which was already mentioned in Section VI.5.

The other condition, (VI.69), becomes

$$\frac{1}{1-\zeta_s} - \Delta\vartheta_{ad}\zeta_s + 1 + St' > 0$$

or

$$\frac{1}{\Delta\vartheta_{ad}\zeta_s} > (1-\zeta_s) - \frac{1-\zeta_s}{\Delta\vartheta_{ad}\zeta_s}(1+St'). \qquad (VI.72)$$

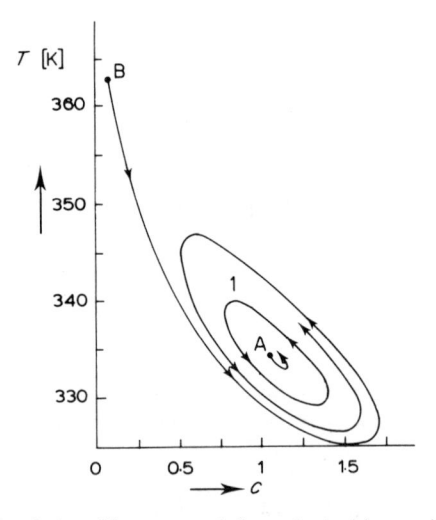

Fig. VI.50 The limit cycle 1 and its approach from the inside starting point A and outside starting point B; Heemskerk et al. [27]

The right-hand side of equation (VI.72) is the tangent to Curve 2 at the point S— compare equation (VI.66). The left-hand term appears to be the value of $d\zeta/d\Delta\vartheta$ when the $\zeta-\Delta\vartheta$ path crosses the line $\Delta\vartheta = 0$; this can be verified by substituting $\Delta\vartheta = 0$ into equation (VI.65). This second condition initially caused some confusion, Gilles and Hofmann [31] were the first to give a physico-chemical interpretation to it. It is the condition for *dynamic stability*; if it is met, the reactor also works dynamically stable at its stable operating point ζ_s, T_s. However, if this condition is not met, the reactor is dynamically unstable and does not work at the stable point, but operates in a fixed *limit cycle* around the stable point ζ_s, T_s. This is shown in figure VI.50, where Curve 1 is the limit cycle. The reactor temperature now cycles at a fixed frequency and with a fixed amplitude around the statically stable operating point. Now if a reactor works under dynamically unstable conditions, whereas condition (VI.71) is met, and if the conditions before a change were represented by a point A in figure VI.50 and after the change, the new conditions would lead to the limit cycle 1 in figure VI.50: the reactor temperature starts to cycle with ever-increasing amplitudes till the limit cycle is reached.

Once the limit cycle conditions are reached, the following process is repeated again and again. Assume that the reaction mixture conditions are represented by point 1 in figure VI.51: here T and c_A are higher then corresponds to the stable operating point at the prevailing τ and ΔT_{ad} of the system. So the reaction rate R_A is too high, the reaction mixture is heated up, the temperature increases and simultaneously c_A is reduced. When point 2 in figure VI.51 is reached, R_A has been reduced already that much that the reaction mixture starts to be cooled by the cold feed, but R_A is still so high that the consumption of reactant A is still higher than the supply of fresh A with the feed. In point 3 c_A reaches a minimum, R_A is now so low that build-up of A in the reaction mixture, which is still being cooled by the cold feed and the cooling medium, starts again. When point 4 is reached sufficient A has been built up to increase the reaction rate so much that the reaction heat evolution is now so high that the cooling of the reaction mixture stops and the heating up is renewed again. The build-up of A continues till point 1 is reached, then the whole cycle starts again.

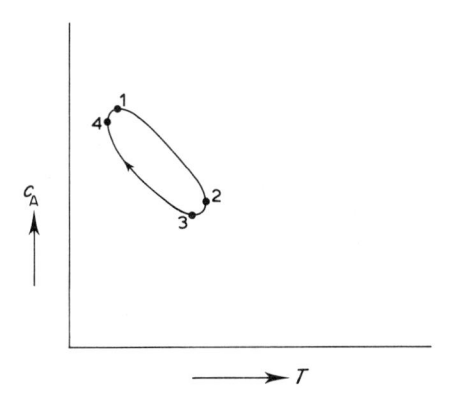

Fig. VI.51 The dynamic behaviour in a limit cycle in a c_A–T plot

The approach to a limit cycle in a temperature–time plot is shown in figure VI.52.a. Conversely, if initial conditions were represented by point B, after a change reactor temperature and concentration start to cycle with ever-decreasing amplitudes as shown in figures VI.50 and VI.52.b, till again the same limit cycle is reached. In figure VI.52 the lines represent the curves calculated from the theory and the points measured by Heemskerk et al. [27]. Much work has been done on a laboratory scale to demonstrate the existence of limit cycles. We mention the work of Chang and Schmitz [28] and Wirges and Hugo [29, 30]. Genuine liquid phase experiments, however, were first carried out by Heemskerk et al. [27]. Both Wirges [30] and Heemskerk et al. [27] demonstrated that in small laboratory reactors the heat capacity of the reactor has also to be taken into account: this can easily be done by replacing the term $d(\rho V_r u)/dt = \rho c_p V_r dT/dt$ in equation (VI.59) by $(\rho V_r c_p + c_p' M) dT/dt$, where c_p' is the specific heat capacity and M the mass of the

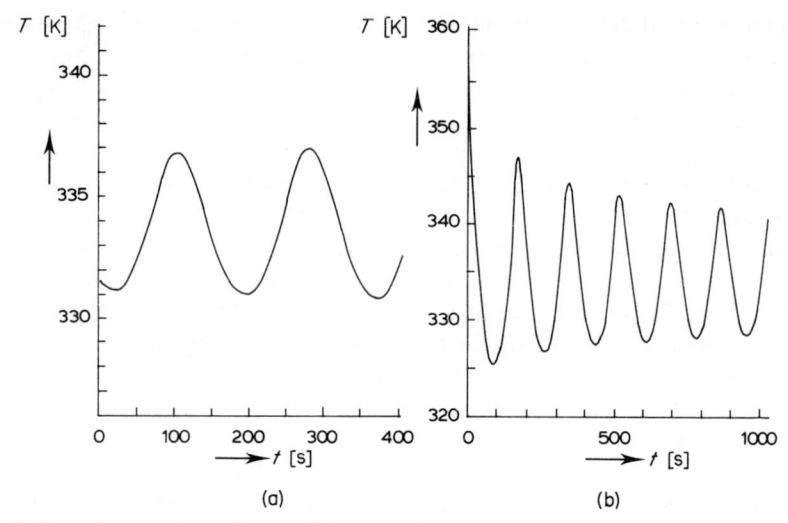

Fig. VI.52 Approach to a limit cycle in a T–t plot; Heemskerk et al. [27]: (a) from inside; (b) from outside

reactor (walls, baffles, agitator and other fittings). In very delicate experiments Heemskerk et al. [27] proved the theory of this section to be correct by comparing the relevant parameters (St', ΔT_{ad}, $k_{1\infty}$, τ, $c'_p M/\rho c_p V_r$, etc.) derived from the best fit of their dynamic experiments with those calculated independently from measured physical data: their results differ only by 0.7–4.9%.

For an adiabatic reactor, the conditions (VI.71) and (VI.72) are identical; statically stable points will always be dynamically stable too, as was also shown in figure VI.49. Apparently, the interaction between the reactor and the cooler is responsible for the occurrence of limit cycles. Applying equation (VI.72) to the two other cases illustrated in figure VI.49, we calculate

$$\Delta \vartheta_{ad} = 10 \to \frac{1}{9} > \frac{0.72}{9}$$

and

$$\Delta \vartheta_{ad} = 20 \to \frac{1}{18} \not> \frac{1.54}{18}.$$

Hence it follows that the system is dynamically unstable in the latter case. Such a situation should be avoided in practice. For example, it can be easily verified that the last two tank reactors of the cascade treated in Example VI.4 and VI.5.b are stable not only statically but in the dynamic sense as well. Condition VI.72 can be rewritten in the following way:

$$\frac{UA\tau}{\rho c_p V_r} > \frac{\dfrac{E\Delta T_{ad}}{RT_s^2}\zeta_{As} + \zeta_{As} - 2}{1 - \zeta_{As}}. \tag{VI.72.a}$$

So the Stanton number for a CISTR must surpass a certain value before dynamic instability—stable cycling around a stable operating point—can occur. In general it can be said that limit cycles are possible, if in a *high conversion* reactor a combination of the following factors coincide:

— a weak temperature dependence of the reaction rate (E/R low)
— a low reactant concentration in the feed (ΔT_{ad} low)
— a low heat effect (ΔT_{ad} low)
— a high reactor temperature (T_s high)
— a not too high conversion ($1 - \zeta_{As}$ relatively large)
— a gas phase reaction (ρc_p low)
— a long residence time (τ high)
— a small (laboratory) reactor (A/V_r high)
— good heat transfer coefficients (U high)

Conversely dynamic instability (limit cycles) can be prevented by adjusting these parameters in the opposite direction.

In a *low conversion* reactor condition (VI.72) simplifies to $St' > \Delta\vartheta_{ad}\zeta_{As} - 2$, or

$$\frac{\Delta\vartheta_{ad}\zeta_A - 2}{St'} < 1. \tag{VI.72.b}$$

Because St' is always positive, limit cycles will definitely occur, when $\Delta\vartheta_{ad}\zeta_{As} - 2 = 0$ or when

$$\Delta T_{ad} \leq \frac{2RT_s^2}{E\zeta_{As}}.$$

This condition is much more easily fulfilled, especially in a long cascade with low conversions per tank. To prevent limit cycles in low conversion reactors, concentration of the reactants in the feed must be low and the reaction temperature high. Dubil and Gaube [32] described an industrial tank reactor for the hydroformylation of propylene (with $\Delta\vartheta_{ad} = 30$ or $E/R = 13\,100$ K, $T_s = 415$ K and $\Delta T_{ad} = 395$ K), which had to be run for selectivity reasons at low propylene conversions and which operated continuously under cycling conditions.

The theory outlined in this section has been developed for first order reactions. It will be clear that reactions with orders differing from one do not present new aspects: they show a similar behaviour, although the shape of the heat production curves is somewhat different. An exception has to be made for (a) single zero order reactions, where the heat production curve shows an abrupt change at $\zeta_A = 1$ and forms a plateau; (b) reactions with Langmuir–Hinshelwood kinetics, Monod kinetics, etc., where more than three operating points are possible; and (c) multiple reactions with also more than three possible operating points. These cases will not be treated here; more will be said on multiplicity and reactor dynamics in Chapter IX, when we discuss heat effects in heterogenous reactors.

The design engineer and the plant operator are less interested in a multitude of plots with $\zeta_A - T$ trajectories for unsteady reactor conditions. They just want to prevent instabilities and to design for and operate at statically stable conditions. However, the question still remains of the dynamic stability, which cannot be abstracted from a heat production–heat withdrawal plot. Will limit cycles occur, and if so, what conditions have to be chosen to prevent them? What will happen if the residence time in the reactor is changed? Will the reactor start to cycle? We will elucidate this in the following Examples and try to answer these questions by using the technique first put forward by Heemskerk et al. [27].

Example VI.7.a *Will a tank reactor be stable?*
A tank reactor has to be designed for a fast gas phase first order reaction with $k = 4.71 \times 10^8 \exp(-8000/T) \text{s}^{-1}$ and $\Delta T_{ad} = 200 \text{ K}$. The feed is introduced at $T_o = 400 \text{ K}$ and the reactor is cooled with water with an average temperature of $T_c = 300 \text{ K}$. Furthermore preliminary design studies indicate very short residence times between 0.2 and 1.5 seconds. For control purposes a quick response of the cooling is required, so that $UA/\rho c_p V_r = 1.0 \text{ s}^{-1}$ has been chosen. We will design a completely stable reactor and draft instructions for the plant operator how to prevent instability and cycling.

For this reaction the heat production curves and heat withdrawal lines are shown in figure VI.29. We will now draft a diagram of the relevant temperatures versus the reference temperature T_R, which is defined as $T_R = (T_o + St'T_c)/(1 + St')$ with $St' = UA\tau/\rho c_p V_r$. For all operating points heat production rate and heat withdrawal rate are in balance. This gives, for the relation between ζ_s and T_s,

$$\zeta_s = \frac{(1 + St')}{\Delta T_{ad}} (T_s - T_R). \tag{a}$$

Rewriting this gives

$$T_s - T_R = \frac{\Delta T_{ad}}{1 + St'} \frac{k_s \tau}{1 + k_s \tau}, \tag{b}$$

with $k_s = k_\infty \exp(-E/RT_s)$. In figure VI.53 this is line 3. At zero conversion we find

$$T_s - T_R = 0. \tag{c}$$

This is line 1 in the same Figure. Similarly we find line 2 for full conversion ($\zeta_s = 1$):

$$T_s - T_R = \frac{\Delta T_{ad}}{1 + St'}. \tag{d}$$

Equation (VI.71) gives the condition for static stability. For the border line case both terms in this equation are equal. If we eliminate ζ_s from equation (VI.71) by substitution of equation (a) into it we have a direct relation between T_s and T_R. By putting both terms in equation (VI.71) equal we find, for the locus of the border line,

$$(T_s - T_R)^2 - \frac{\Delta T_{ad}}{1 + St'}(T_s - T_R) + \frac{RT_s^2}{E} \times \frac{\Delta T_{ad}}{1 + St'} = 0. \qquad (e)$$

This border line, represented by curve 4 in figure VI.53, envelopes the area where only unstable operating points—point 0 in figure VI.29—are possible (the area of multiplicity).

Condition (VI.72)—following the same procedure as for condition (VI.71)—gives the border line which envelopes the area, where limit cycles occur:

$$(T_s - T_R)^2 - \left[\frac{\Delta T_{ad}}{1 + St'} + \frac{RT_s^2}{E}\right](T_s - T_R) + \frac{RT_s^2}{E}\frac{2 + St'}{1 + St'}\frac{\Delta T_{ad}}{1 + St'} = 0. \qquad (f)$$

This is curve 5 in figure VI.53. Above point B in figure VI.53 statically stable operation is possible, but in the range between points A and B limit cycles will occur. The farther away from A, the larger the amplitudes of the limit cycles will be. The design point of figure VI.29 is also given, so this *design* with $\tau = 0.5$ s is *satisfactory*.

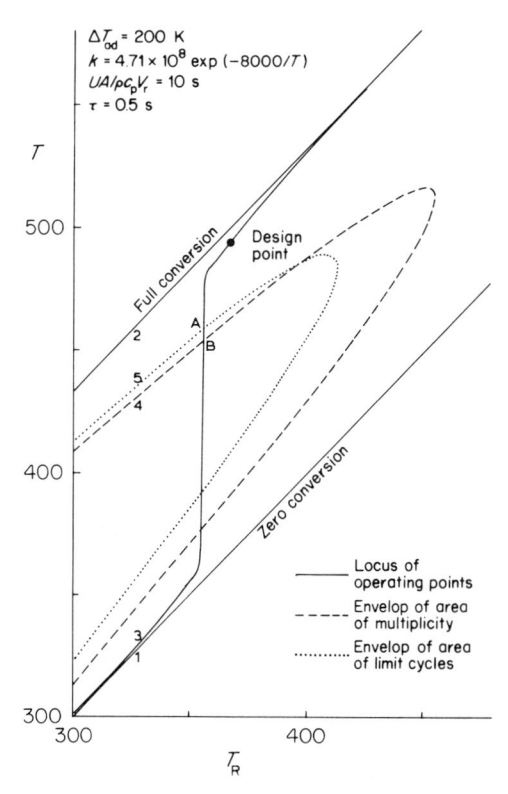

Fig. VI.53 Stability diagram for a tank reactor; Example VI.7.a

A larger residence time will make the design easier and is possible as indicated in figure VI.29. The same diagram, but now for $\tau = 1.0$ s, is drawn in figure VI.54. The design point of figure VI.29 is also given. Although statically stable, the design point is within the limit cycle area, so this *design is dynamically unstable*. We therefore choose for the time being the design with $\tau = 0.5$ s.

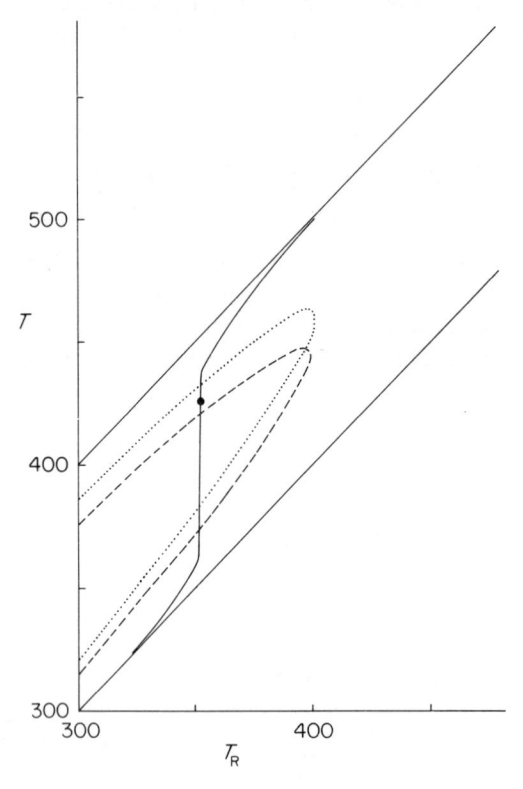

Fig. VI.54 Stability diagram for a tank reactor; Example VI.7.a. Same meaning of lines as in figure VI.53, $\tau = 1.0$ s

Once the reactor has been designed, built and installed, the operator has to keep it in good steady conditions. During operation the reactor load often changes. Both heat withdrawal rate and heat production rate are a function of the residence time; the same holds true for the reference temperature T_R. For the design conditions $\Delta T_{ad} = 200$ K, $T_o = 400$ K and $T_c = 300$ K the same relations as before are given in figure VI.55 as a function of τ for the reactor designed for $\tau = 0.5$ s. We can read from this diagram that cycling starts from $\tau = 0.92$ s and for longer residence times. So the operator has to load the reactor always at a throughput higher than $0.50/0.92 = 54\%$ of the design rate in order to prevent cycling. We also see from figure VI.55 that at $\tau = 0.98$ s the reaction will be extinguished.

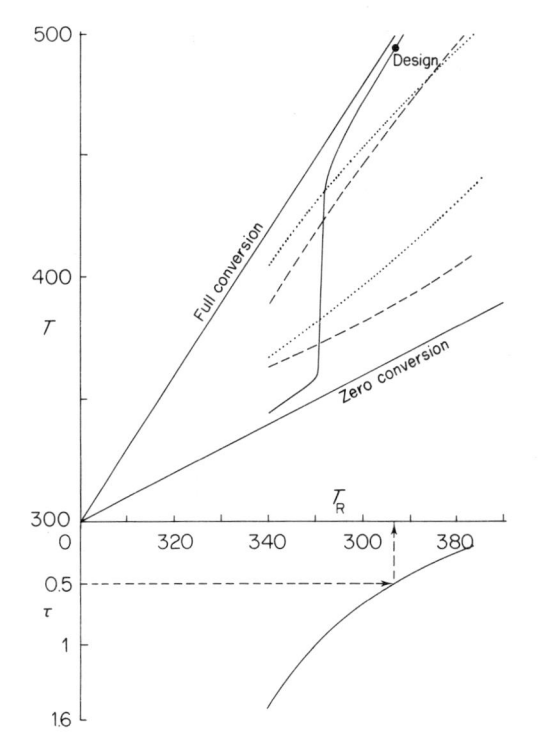

Fig. VI.55 Influence of varying throughputs on the stability of the reactor of figure VI.53. Same meaning of lines as in figure VI.53

We see from the diagram that as soon as a limit cycle would start at a high conversion level the operator has to *reduce the rate of flow of the cooling medium.* Evidently T_R has to be increased to avoid dynamic instabilities at high conversions. Reducing the flow rate increases the average temperature T_c of the cooling medium and consequently also T_R. This is in contrast to the first impulse of the operator who normally would increase the cooling medium flow rate to suppress temperature fluctuations. Also we can see from figure VI.55 that overloading the reactor—decreasing the residence time—would do no harm, in as far as reactor construction materials can stand the higher reaction temperatures.

The same calculations can be repeated for residence times between 0.5 and 1.0 seconds in order to investigate whether another, higher residence time than $\tau = 0.5$ s will also give a satisfactory reactor performance and a somewhat easier design. The design engineer can also manipulate ΔT_{ad} (reducing the feed concentration by dilution), T_o (preheating or precooling the reactor feed) and T_c (choosing another cooling medium instead of water); all these measures are expensive in investment in separation and cooling/heating equipment and increase the energy consumption of the plant.

Example VI.7.b *Stability of a liquid phase reactor*

The reactor described in Example VI.5.c will be checked on its dynamic and static stability at varying throughputs. We will assume that for the design case (iii) in Example VI.5.c has been chosen. The pertinent data on the reaction and the reactor can be found in Example VI.5.c. In this case $UA/\rho c_p V_r = 500 \times 1.45/3.5 \times 10^6 = 4.64 \times 10^{-4} \text{s}^{-1}$. The results of the stability calculations have been plotted directly versus the residence time in the reactor; see figure VI.56. We see that we can underload the reactor at least up to $\tau = 5600 \text{ s}$ without limit cycles to occur. At $\tau = 400 \text{ s}$ the reaction is extinguished. Static instability can occur up to $\tau = 4200 \text{ s}$. So the reactor will work satisfactory at least from $1500/5600 = 27\%$ to $1500/400 = 375\%$ of its design capacity.

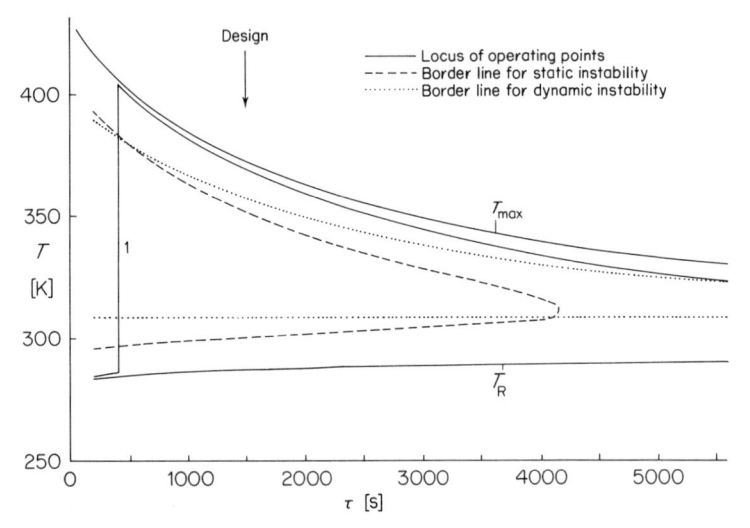

Fig. VI.56 Influence of varying throughput on the stability of the reactor of Example VI.7.b

VI.7.2 Tubular reactor

A CISTR, either alone or in a cascade, can exhibit instabilities as has been explained above. In contrast, a PFR will not exhibit dynamically unstable characteristics, because by definition it does not exhibit backmixing of heat. In a PFR after a change in conditions the new steady state will pass as a plug through the reactor and after a period of the residence time τ_L the new steady state conditions will have been established. However, this situation will change when by means of a heat exchanger the reaction heat is mixed back to the reactor feed stream. Then all phenomena of multiplicity, hysteresis and limit cycles can occur as outlined in Section VI.5. In principle, this will not bring any new, unsuspected aspects.

For a cooled PFR the wall, if its heat capacity is not negligible with respect to the heat capacity of the reaction mixture—which means that in a tube with ρ', c'_p

and a wall thickness d_w the factor $4d_w\rho'c_p'/d_t\rho c_p$ is not negligible—will act as a heat buffer and have a dampening effect on sudden changes either by absorbing or supplying heat.

There are in principle three different ways of cooling a PFR: (a) with a boiling liquid, in which case T_c is constant; (b) with the cooling medium in cocurrent flow, in which case T_c continuously increases; (c) with the cooling medium in countercurrent flow. In the last case the cooling medium is heated up, but after passing the hot spot in the direction of the reactor inlet in principle the situation might reverse, so that the cooling fluid is cooled by the cold reactor feed; see figure VI.57. In this case backmixing of reaction heat to the reactor feed by means of the cooling medium occurs and all phenomena of hysteresis, multiplicity and cycling might occur. Among others, Eigenberger [33] has discussed a first order reaction with $k = 10^{13} \exp(-11\,000/T)\,\text{s}^{-1}$ and $\Delta T_{ad} = 91\,\text{K}$ in a tubular reactor with $UA/\rho c_p V_r = 11.1 \times 10^{-3}\,\text{s}^{-1}$, where hysteresis occurred. To ignite the reaction the cooling medium had to be heated up.

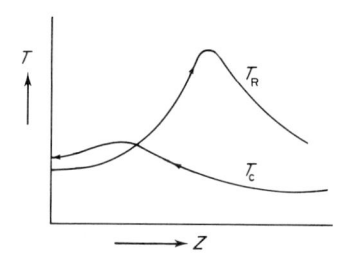

Fig. VI.57 A cooled plug flow reactor. Reaction mixture and cooling medium flow counter-currently

As soon as multiple reactions come into play, multiplicity, hysteresis and cycling can also occur in a PFR. For example, Bush [34] described experiments in a PFR on the chlorination of methylchloride

$$Cl_2 + CH_3Cl \rightarrow CH_2Cl_2 \rightarrow CHCl_3 \rightarrow CCl_4 + HCl,$$

where steady oscillations—limit cycles—persisted. We will not discuss further multiple reactions in this section.

REFERENCES

1. Hougen, O. A., Watson, K. M., and Ragatz, R. A.: *Chemical Process Principles*, part I, *Material and Energy Balances*, 2nd Edition, Wiley, New York, 1958; part II, *Thermodynamics*, 2nd edition, Wiley, New York, 1959
2. Hugo, P.: *Chem. Ing. Techn.* **52**, 712 (1980)
3. Hugo, P., Konczalla, M., and Mauser, H.: *Chem. Ing. Techn.* **52**, 761 (1980)
4. Perona, J. J. and Thodos, G.: *A.I.Ch.E. Journal* **3**, 230 (1957)
5. Jahnke, E., Embe, F., and Lösch, P.: *Tables of Higher Functions*, McGraw-Hill, New York (1960)
6. Almasy, G.: *Acta Chim. Hung.* **24**, 197, 243 (1960)
7. Douglas, J. M. and Eagleton, L. C.: *Ind. Eng. Chem. Fund.* **1** (2), 116 (1962)

8. Barkelew, C. H.: *Chem. Eng. Progr. Symp.*, Ser. No. 25, **55,** 37 (1959)
9. van Welsenaere, R. J. and Froment, G. F.: *Chem. Eng. Sci.* **25,** 1503 (1970)
10. Dente, M. and Collina, A.: *Chim. et Industrie* **46,** 752 (1964)
11. Hlavacek, V., Marek, M., and John, T. M.: *Coll. Czechoslov. Chem. Comm.* **34,** 3868 (1969)
12. Bilous, O. and Amundson, N. R.: *A.I.Ch.E. Journal* **2,** 117 (1956)
13. Amundson, N. R.: *De Ingenieur* **67,** Ch 73 (1955)
14. Wagner, C.: *Chem. Ing. Techn.* **18,** 28 (1945)
15. van Heerden, C.: *Ind. Eng. Chem.* **45,** 1242 (1953)
16. van Heerden, C.: *Chem. Eng. Sci.* **8,** 133 (1958)
17. Shah, M. J.: *Ind. Eng. Chem.* **59,** 72 (1967)
18. Baddour, R. F., Brian, P. L. T., Logeais, B. A., and Emery, J. P.: *Chem. Eng. Sci.* **20,** 281 (1965)
19. Westerterp, K. R.: *Chem. Eng. Sci.* **17,** 423 (1962)
20. Demaria, F., Longfield, J. E., and Butler, C.: *Ind. Eng. Chem.* **53,** 259 (1961)
21. Hoftijzer, P. J. and Zwietering, T. N.: *Chem. Eng. Sci.* **14,** 241 (1961)
22. Aris, R. and Amundson, N. R.: *Chem. Eng. Sci.* **7,** 121, 132, 148 (1958)
23. Uppal, A., Ray, W. H., and Poore, A. B.: *Chem. Eng. Sci.* **29,** 967 (1974); **31,** 205 (1976)
24. Perlmutter, D. D.: *Stability of Chemical Reactors*, Prentice-Hall, Englewood Cliffs (1972)
25. Schmitz, R. A.: *Chem. React. Eng. Reviews* **148,** 156 (1975)
26. Bilous, O. and Amundson, N. R.: *A.I.Ch.E. Journal* **1,** 513 (1955)
27. Heemskerk, A. H., Dammers, W. R., and Fortuin, J. M. H.: *Chem. Eng. Sci.* **35,** 439 (1980)
28. Chang, M. and Schmitz, R. A.: *Chem. Eng. Sci.* **30,** 21, 837 (1975)
29. Hugo, P. and Wirges, M. P.: *A.C.S. series* 65, Houston, 498 (1978)
30. Wirges, H. P.: *Chem. Eng. Sci.* **35,** 2141 (1980)
31. Gilles, E. D. and Hofmann, H.: *Chem. Eng. Sci.* **15,** 328 (1961)
32. Dubil, H. and Gaube, J.: *Chem. Ing. Techn.* **45,** 529 (1973)
33. Eigenberger, G.: *Chem. Ing. Techn.* **46,** 11 (1974)
34. Bush, S. F.: *Proc. Roy. Soc.* **A309,** 1 (1969)
35. Mason, D. R. and Piret, E. L.: *Ind. Eng. Chem.* **42,** 817 (1950)
36. Mason, D. R. and Piret, E. L.: *Ind. Eng. Chem.* **43,** 1210 (1951)
37. Slinko, M. G. and Muler, A. L.: *Kin. i Kat.* (in Russian) **2,** 467 (1961)
38. Luss, D. and Chen, G. T.: *Chem. Eng. Sci.* **30,** 1483 (1975)
39. Pikios, C. A. and Luss, D.: *Chem. Eng. Sci.* **34,** 919 (1979)
40. Tsotsis, T. T., Haderi, A. E., and Schmitz, R. A.: *Chem. Eng. Sci.* **37,** 1235 (1982)
41. Westerterp, K. R. and Ptasinski, K. J.: *Chem. Eng. Sci.* **38,** (1983)

Chapter VII
Multiphase reactors, single reactions

VII.1 THE ROLE OF MASS TRANSFER

In the preceding chapters the conversion rate was used as a quantity to be employed for purposes of calculation. It expressed the number of mass or molar units of a certain component converted per unit time and per unit volume (or per unit mass of catalyst), and it was supposed to be a known function of the temperature, the pressure and the composition of the reaction mixture. No attention was paid to the question whether the reaction was proceeding uniformly or whether the conversion rate would turn out to vary locally when inspected on a smaller scale. In fact, the material balances were applied on a scale which was generally large with respect to the scale of possible local variations.

As we discussed in Chapter V, such local variations in the nature and composition of the reaction mixture will occur even in systems consisting of one phase only, if mixing on a molecular scale is not complete. Above all, however, marked variations will be encountered in systems made up of more than one phase; these will be called heterogeneous systems.

Contrary to reactions in a single phase mixed on a molecular scale, not all reactive molecules present in multi-phase systems may be available for chemical reaction. Thus, to give an example, a *homogeneous reaction* between two reactants A and B proceeds in one phase, although A is originally present in a different phase. Consequently, conversion can only take place when A is transferred to the phase in which the reaction proceeds. The rate of this mass transfer may therefore influence the conversion rate, especially if the rate of the chemical reaction proper is high compared to the rate of supply of A. Another example is a *heterogeneous reaction* at the interface between two phases, e.g. at a catalyst surface. Here the reactants have to be transported by a diffusional process to the interface where the reaction takes place; in this case mass transfer to the reaction surface may considerably influence the overall conversion rate. Hence, the course of a chemical reaction in a multi-phase system is, in principle, described by a combination of chemical kinetics and mass transfer phenomena.

The design of a reactor for a multi-phase system is in full agreement with the general theory developed in the preceding chapters, as long as an overall (or macroscopic) conversion rate is used. The difference between the conversion rate expressions for a homogeneous well-mixed system and a multi-phase system lies in the mathematical formulation. In the latter case it has to account for the interaction between chemical reaction and mass transfer. The main purpose of Chapters VII and VIII is to analyse this interaction and to develop expressions for the overall conversion rate from this analysis. For each species participating in a reaction in a multi-phase system, a formal material balance can be written down for each phase separately. The general expression (I.1) will, of course, still be valid. However, in specifying the term $-\Delta\Phi_{mJ}$, which denotes the net supply of J, a distinction must be made between a convective part, $-\Delta(\Phi_m w_J)$ and a part representing the mass flow of J from the interface to the bulk of the phase considered, $J_J M_J a$. Such balances become, for a total volume V_r (large with respect to the scale of dispersion, but so small as to be governed by uniform conditions) containing two phases L and G, with a reaction going on in phase L only (see also figure VII.1):

$$\frac{dm_{JL}}{dt} = -\Delta(\Phi_{mL} w_{JL}) + J_{JL} M_J a V_r + R_J M_J (1 - \varepsilon) V_r \quad [\text{kg/s}] \quad \text{(VII.1)}$$

and

$$\frac{dm_{JG}}{dt} = -\Delta(\Phi_{mG} w_{JG}) + J_{JG} M_J a V_r \quad [\text{kg/s}]. \quad \text{(VII.2)}$$

The term $R_J M_J (1 - \varepsilon) V_r$ is the production rate by homogeneous chemical reaction in phase L with $1 - \varepsilon$ being the volume of reaction phase per unit volume of reactor. Since the composition may vary on a much smaller scale than the volume V_r considered, an average space value has to be used. The term $J_J M_J a V_r$ is the rate of mass transfer of J in the volume under consideration, $V_r (= S \Delta h)$; see figure VII.1

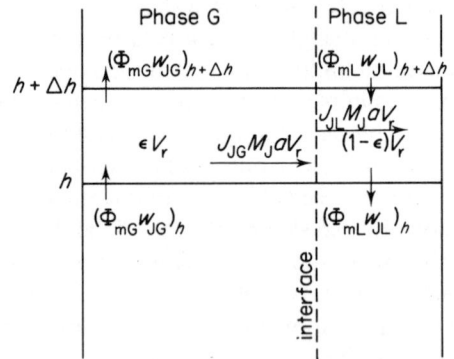

Fig. VII.1. Schematic representation of a two phase column reactor counter-current operation

with a = interfacial area per unit volume of reactor (specific contact area) [m²/m³]
J_J = molar rate of transfer of J per unit interfacial area [kmol/m²s] (its numerical value is taken positive if component J is transferred to the phase and negative if transferred from the phase under consideration).

The overall average production rate $\bar{R}_J$ on the basis of the total volume V_r is defined by means of a balance over both phases together:

$$\frac{dm_J}{dt} = -\Delta(\Phi_{mL}w_{JL} + \Phi_{mG}w_{JG}) + \langle R_J \rangle M_J V_r \tag{VII.3}$$

A comparison between equation (VII.3) and the sum of equations (VII.1) and (VII.2) yields, for the overall conversion rate,

$$\langle R_J \rangle = R_J(1 - \varepsilon) + (J_{JL} + J_{JG})a. \tag{VII.4}$$

With *homogeneous reactions*, no reaction takes place at the interface, and hence

$$J_{JL} = -J_{JG}.$$

With purely *heterogeneous reactions*, a homogeneous reaction does not occur; hence in a liquid–solid system,

$$R_{JS} = R_{JL} = 0.$$

On the other hand, since there is a net production or consumption at the interface,

$$J_{JL} + J_{JS} \neq 0,$$

and accordingly the overall production rate per m³ reactor volume becomes

$$\langle R_J \rangle = (J_{JL} + J_{JS})a \quad [\text{kmol/m}^3\text{s}].$$

Mostly, one of the phases, e.g. phase S, is a material reacting with, or catalysing the reaction of, components in a fluid L, in which case $J_{JS} = 0$. The above equation then states that the rate of production of J is equal to the mass flow of J from the interface into the fluid L. The value of this mass flow can only be found by considering the combination of mass transfer and chemical reaction at the interface on a smaller scale (see Section VII.12).

Besides information on J, also information on the values of ε and a is necessary for design purposes (see equations (VII.1), (VII.2)). Therefore, we will give entries for data on mass transfer coefficients, hold-up and interfacial area at the end of this chapter. The main object of this chapter is to derive and summarize relations for J in case only one reaction occurs.

The next chapter deals with J in case of more reactions. Then the relation between mass transfer and selectivity calls for attention.

VII.2 A QUALITATIVE DISCUSSION ON MASS TRANSFER WITH HOMOGENEOUS REACTION

In this section we discuss qualitatively a homogeneous reaction between the species A and B proceeding in one phase, the reaction phase, while A is supplied by mass transfer from another, adjoining phase. Analysis of the interaction between mass transfer and chemical reaction will ultimately give us expressions for the flux of reactant A across the interface and for the overall conversion rates of A and B to be used in material balances of the type of equation (VII.3), by means of which a reactor calculation can be carried out in the manner indicated in Chapter II.

The reaction phase is supposed to be a flowing medium and the other phase can be either a non-miscible fluid or a solid. In the present analysis, we arbitrarily assume that the reaction phase is a liquid and that the phase providing the reactant A is a gas. In the following, therefore, we consider the absorption of a gas A by a liquid in which A reacts; however, the results are likewise applicable to other operations which are entirely similar, such as liquid–liquid extraction with chemical reaction, and catalytic conversion of gases in porous particles.

Now, with regard to gas absorption, we shall consider a heterogeneous system with a gas–liquid interfacial area per total two-phase volume of a (m^2/m^3) and with a fractional volume of $1 - \varepsilon$ of (liquid) reaction phase. We shall further assume that the bulk concentration of B in the liquid, $\bar{c}_B$, and the concentration of A in the liquid at the interface, c_{Ai}, are known. The latter value can often be found by conventional methods from the gas composition, the mass transfer coefficient of A in the gas phase, k_G, the flux of A and the physical solubility of A in the liquid. Therefore, we focus on the processes in the reaction phase so that no special index, indicating the phase, has to be attached to the concentrations and the various physical properties of this phase.

VII.2.1 Concentration distribution in the reaction phase

A qualitative picture of the various possible concentration distributions of A and B in the boundary region of the liquid is shown in figure VII.2. The curves 1 refer to a case where the reaction between A and B is relatively slow. Most of the conversion takes place in the bulk of the liquid, and the drop in concentration of

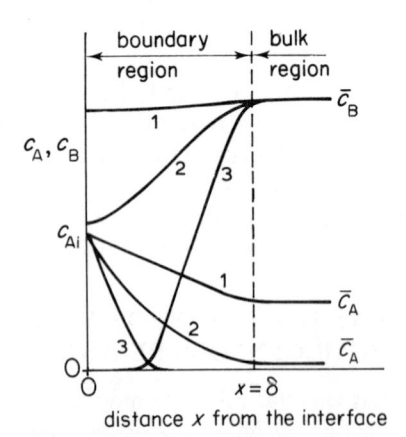

Fig. VII.2. Possible concentration profiles in the boundary region for a homogeneous reaction between a component A transferred through the interface $x = 0$ and a component B present in the bulk of the reaction phase

A over the boundary region is mainly due to a diffusional resistance. Should the ratio of interfacial area and bulk volume be decreased, this concentration drop would become greater since less A would then be supplied per unit volume of reaction phase, resulting in a lower bulk concentration $\bar{c}_A$.

The curves 2 give qualitative concentration distributions for a moderately high chemical reaction velocity. Here, a relatively great fraction of A that diffuses through the interface is already converted within the boundary region; this can be seen, for example, from the concentration distribution of B which apparently is consumed in the film. In stationary situations the rate of consumption equals the rate of supply by diffusion from the bulk. In comparison with the preceding case, the *surface area* of the interface has become more important for the total conversion rate relative to the *volume* of the reaction phase.

Finally, distributions of type 3 will be found when the reaction is so fast that it occurs entirely within the boundary region in a rather narrow reaction zone towards which A and B have to be transported by diffusion. No reaction takes place in the bulk region, and the total conversion rate is proportional to the interfacial area. The position of this reaction zone greatly depends on the ratio $\bar{c}_B/c_{Ai}$.

The above qualitative discussion makes it clear that at least three important parameters play a part in a mass transfer process with a reaction between A and B. They are:

— the surface volume ratio of the reaction phase;
— the ratio of chemical reaction velocity and maximum mass transfer rate;
— the ratio $\bar{c}_B/c_{Ai}$.

A more quantitative study of the influence of these parameters would require a detailed knowledge of the mechanism of mass transfer without chemical reaction. Fluid flow along solid boundaries in a dispersed system mainly involves steady state diffusion through a developing boundary layer [1]. Simple non-steady diffusion models — more appropriate in the case of flow along mobile surfaces — have become known under the name 'penetration or surface renewal theories'. More sophisticated, but therefore also more complicated mass transfer models are available as well. For prediction of the interaction of the chemical reaction with the rate of mass transfer however, simple mass transfer models are in general sufficient. This is because many studies in this field indicate that in a given reaction system the ratio of the mass transfer rate with chemical reaction and the physical mass transfer rate is fairly independent of the mass transfer mechanism itself [2]. This makes possible an almost quantitative discussion of the influence of chemical reaction on mass transfer based on the simplest but otherwise unrealistic model: the stagnant film model. Another consequence is that it is often unnecessary to match the flow conditions in laboratory experiments with those in large-scale equipment. Hence, laboratory reactors may be chosen on the grounds of simplicity and convenience. Experience in interpreting the results however, is required [3]. See further Section VII.15. We will start our quantitative analysis with the general material balance for mass transfer with reaction (next section).

From this general equation we will firstly discuss the simplest mass transfer models, film and penetration (Section VII.4), and secondly the influence of reaction on the mass transfer rate (Sections VII.5–VII.12).

VII.3 GENERAL MATERIAL BALANCE FOR MASS TRANSFER WITH REACTION

To analyse the influence of reaction on the rate of mass transfer, we start from a molar balance for component A over a stationary volume element $\Delta x \, \Delta y \, \Delta z$. It follows from equation (I.1) that

$$\frac{\partial c_A}{\partial t} \Delta x \, \Delta y \, \Delta z = -\frac{\Delta \Phi_{mA}}{M_A} + R_A \Delta x \, \Delta y \, \Delta z \quad \left[\frac{\text{kmol}}{\text{s}}\right]. \quad (\text{VII.5})$$

Contrary to the derivation of equation (VII.1), the volume $\Delta x \, \Delta y \, \Delta z$ is considered to be small with respect to the scale of dispersion.

Transport of A is caused by *fluid flow* and by *diffusional processes* superimposed on the net fluid flow. In case only molecular diffusion plays a role, the latter component is given by the first law of Fick [1, 4]. For only one species A diffusing through stationary boundaries, Fick's law in the x-direction is as follows:

$$(J_A)_x = \frac{-c}{1 - x_A} D_A \frac{\partial x_A}{\partial x},$$

in which x_A is the molar fraction of species A and c the total molar concentration.

In the majority of mass transfer with reaction applications the molar density does not vary substantially within the zone where the mass transfer process takes place, while also x_A is often much smaller than one. Then Fick's law reduces to

$$(J_A)_x = -D_A \frac{\partial c_A}{\partial x}. \quad (\text{VII.6})$$

In our analysis we will generally use the simple expression (VII.6), but in practical applications its validity should be checked, especially when temperature gradients are also present; see Section VII.14.

Thus

$$\frac{-\Delta \Phi_{mA}}{M_A} = \left[\left(v_x c_A - D_A \frac{\partial c_A}{\partial x}\right)_x - \left(v_x c_A - D_A \frac{\partial c_A}{\partial x}\right)_{x+\Delta x}\right] \Delta y \, \Delta z$$

$$+ \left[\left(v_y c_A - D_A \frac{\partial c_A}{\partial y}\right)_y - \left(v_y c_A - D_A \frac{\partial c_A}{\partial y}\right)_{y+\Delta y}\right] \Delta x \, \Delta z$$

$$+ \left[\left(v_z c_A - D_A \frac{\partial c_A}{\partial z}\right)_z - \left(v_z c_A - D_A \frac{\partial c_A}{\partial z}\right)_{z+\Delta z}\right] \Delta x \, \Delta y \quad (\text{VII.7})$$

Substituting equation (VII.7) into equation (VII.5), dividing by $\Delta x \, \Delta y \, \Delta z$ and taking the limit for Δx, Δy and $\Delta z \to 0$ results, for a constant D_A, in

$$\frac{\partial c_A}{\partial t} = -\frac{\partial v_x c_A}{\partial x} - \frac{\partial v_y c_A}{\partial y} - \frac{\partial v_z c_A}{\partial z}$$

$$+ D_A \left(\frac{\partial^2 c_A}{\partial x^2} + \frac{\partial^2 c_A}{\partial y^2} + \frac{\partial^2 c_A}{\partial z^2} \right) + R_A. \quad \text{(VII.8)}$$

In constant density problems, equation (VII.8) reduces to

$$\frac{\partial c_A}{\partial t} = -v_x \frac{\partial c_A}{\partial x} - v_y \frac{\partial c_A}{\partial y} - v_z \frac{\partial c_A}{\partial z}$$

$$+ D_A \left(\frac{\partial^2 c_A}{\partial x^2} + \frac{\partial^2 c_A}{\partial y^2} + \frac{\partial^2 c_A}{\partial z^2} \right) + R_A. \quad \text{(VII.9)}$$

Further simplifications are often possible. For stationary problems $\partial c_A/\partial t = 0$. For mass transfer without reaction $R_A = 0$. For mass transfer in stagnant elements

$$v_x \frac{\partial c_A}{\partial x} = v_y \frac{\partial c_A}{\partial y} = v_z \frac{\partial c_A}{\partial z} = 0. \quad \text{(VII.10)}$$

Examples in which equation (VII.10) is applicable are mass transfer in stagnant drops and mass transfer in porous solids; see Section VII.12.2. Equation (VII.10), however, is also often used in fluid–fluid and fluid–solid mass transfer problems: both the stagnant film model and the penetration models of Higbie and Danckwerts are based on it. These useful mass transfer models will be discussed in the next section.

VII.4 MASS TRANSFER WITHOUT REACTION

In this section we briefly summarize the simple models describing the rate of mass transfer to or from an interface. Theoretically, this rate can be calculated for each particular case by solving equation (VII.8). The solution gives the concentration profile near the interface. Differentiation of the solution gives $(\partial c_A/\partial x)_{x=0}$, with $x =$ the coordinate perpendicular to the interface. Substitution of the value of this quantity in Fick's law gives the diffusional component of the mass transfer rate:

$$J_A = -D_A \left(\frac{\partial c_A}{\partial x} \right)_{x=0}. \quad \text{(VII.11)}$$

In practice, however, this procedure is not applicable in most cases; exceptions are transfer into and from quiescent media and laminar flow. The reason is a lack of knowledge on the exact flow profile near the interface. Therefore, *mass transfer coefficients*, defined as

$$k_{LA} \text{ (or } k_{GA}) \equiv \left| \frac{(J_A)_{\text{phs}}}{c_{Ai} - \bar{c}_A} \right| \quad [\text{m/s}] \quad \text{(VII.12)}$$

are mostly obtained by experiment, see Section VII.16.

In equation (VII.12), J_A is the transfer rate of A through the interface. The subscript 'phs' is added to stress that the definition is restricted to the case of no reaction of A, that is to physical mass transfer of A. Defined this way, k_L and k_G are independent of c_A for all cases in which the logarithmic mean of x_{Ai} and $\bar{x}_A \ll 1$ [5]. Their numerical values mainly depend on the geometry of the system, the physical constants, η, ρ, D of the fluid, and on the local hydrodynamics. Therefore, both theoretical and empirical relations on mass transfer are commonly presented in the dimensionless form $Sh = f(Re, Sc, Fo)$. In the literature, the gas phase mass transfer coefficient is also defined as

$$k'_{GA} \equiv \left| \frac{(J_A)_{phs}}{p_{Ai} - \bar{p}_A} \right|. \tag{VII.13}$$

We prefer the use of k_G to k'_G because the first quantity has the same dimension as k_L. For an ideal gas

$$k_G = k'_G RT. \tag{VII.14}$$

Mass transfer coefficients are slightly different for different components, but in the following it will usually not be given an index for reference to the component under consideration. In calculating the rate of mass transfer of component A from one phase to another, the mass transfer coefficients of both phases have to be taken into account, as well as the distribution coefficient of A between the two phases. Because of the conditions of continuity we have, for example, the absorption of a gas into a liquid (see figure VII.3)

$$J_A = k_G(\bar{c}_{AG} - c_{AiG}) = k_L(c_{AiL} - \bar{c}_{AL}). \tag{VII.15}$$

It can be assumed for nearly all practical cases that equilibrium exists between the two phases at the interface. Therefore c_{AL} and c_{AG} at the interface are connected by an equilibrium relation

$$c_{AiL} = m_A c_{AiG} \tag{VII.16}$$

in which the dimensionless distribution coefficient m_A for species A generally depends on the composition of the two phases and on the temperature and the pressure. If m_A is independent of the concentration of A under otherwise constant conditions, equation (VII.16) indicates a proportionality similar to Henry's law for liquid–gas equilibrium and to Nernst's law for liquid–liquid equilibrium. At a constant value of m_A, the concentration at the interface can be easily eliminated from equation (VII.15), giving for the molar flux:

$$\begin{aligned} J_A &= \left(\frac{1}{k_L} + \frac{m_A}{k_G} \right)^{-1} (m_A \bar{c}_{AG} - \bar{c}_{AL}) \\ &= \left(\frac{1}{m_A k_L} + \frac{1}{k_G} \right)^{-1} \left(\bar{c}_{AG} - \frac{\bar{c}_{AL}}{m_A} \right). \end{aligned} \tag{VII.17}$$

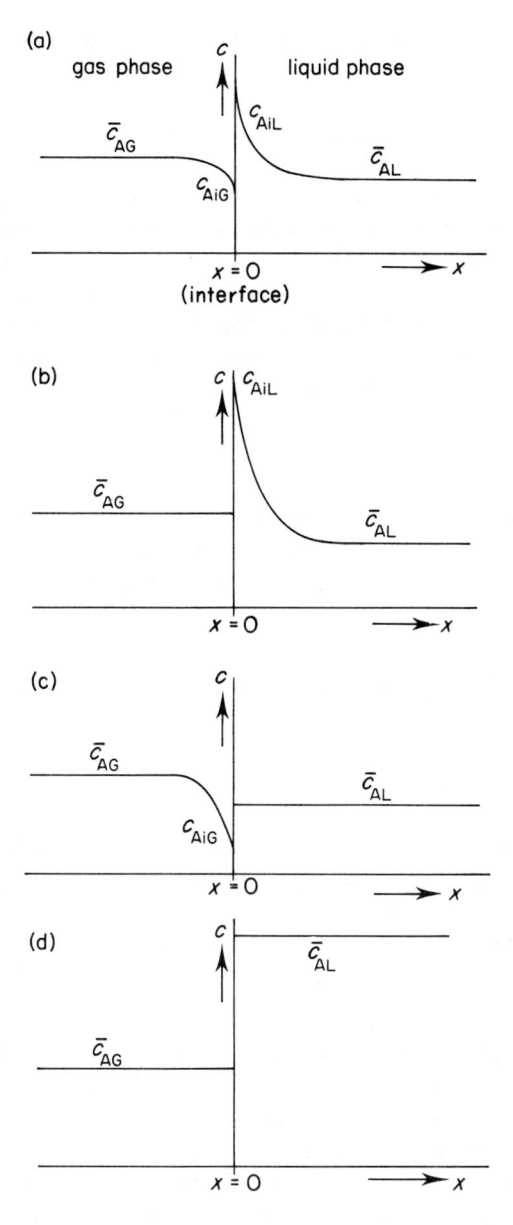

Fig. VII.3. Concentration profiles near gas–liquid interface in case of absorption: (a) general case; (b) no gas phase resistance; (c) no liquid phase resistance; (d) equilibrium

For $k_G/mk_L \gg 1$, gas phase resistance against mass transfer is negligible and equation (VII.17) reduces to

$$J_A = k_L(m_A \bar{c}_{AG} - \bar{c}_{AL}). \tag{VII.18}$$

For $k_G/mk_L \ll 1$, liquid phase resistance against mass transfer is negligible and equation (VII.17) reduces to

$$J_A = k_G(\bar{c}_{AG} - \bar{c}_{AL}/m_A). \tag{VII.19}$$

Equilibrium is obtained for $\bar{c}_{AL} = m_A \bar{c}_{AG}$. Figure VII.3 qualitatively shows concentration profiles for the general case and for the asymptotic cases. Because of the difficulties met in describing mass transfer to and through interfaces from first principles, simple models have been developed which are thought to approximate real hydrodynamics in the vicinity of the interface more or less accurately. The most simple ones are the stagnant film model and the penetration model. Both models are widely and successfully applied in analysing mass transfer with chemical reaction problems. Therefore we shall discuss them in Sections VII.4.1 and VII.4.2.

Example VII.4 *Chemical absorption of NH_3 from air in a packed column*
NH_3 is scrubbed from air at 15 bars pressure by a 1M H_2SO_4 solution, flowing in countercurrent, through a packed column. It is required to calculate the height L of the column, when the following data are given:

Mean temperature:	20°C
Air flow rate, 50 000 m³/day (NTP), Φ_{vG}:	0.0414 m³/s
Liquid flow rate, 2.5 m³/h, Φ_{vL}:	0.694×10^{-3} m³/s
NH_3 content in air feed p_{Ao}/p:	1 vol %
NH_3 content at outlet, p_{AL}/p:	5×10^{-3} vol %
Column diameter:	0.52 m
Cross-sectional area of column, S:	0.212 m²
Diameter of Raschig ring packing material:	0.025 m
Specific surfaces of Raschig rings:	220 m²/m³
Effective specific surface for mass transfer, a, estimated at 50%:	110 m²/m³

The gas is assumed to behave ideally. The NH_3 (component A) dissolves in the acid solution, diffuses inwards and at the same time reacts rapidly with the H^+ ions according to $NH_3 + H^+ \rightarrow NH_4^+$. The mass transfer coefficients for component A are $k_G = 15.7 \times 10^{-3}$ m/s and $k_L = 0.37 \times 10^{-3}$ m/s. The solubility of NH_3 in aqueous solutions is very high, $m \simeq 10^3$, and therefore the resistance against mass transfer is completely in the gas phase, so that $k_G/mk_L \ll 1$. Heat effects may be neglected. The change in NH_3 content of the gas at the height z above the inlet is given by the material balance for A over the gas phase:

$$0 = -\Phi_{vG} d\left(\frac{p_A}{RT}\right) - J_A a S \, dz. \tag{a}$$

It has been assumed here that longitudinal dispersion in the gas flow can be neglected.

We'll first investigate whether we have sufficient acid available to convert the required amount of ammonia. Ammonia to be converted:

$$\Phi_{vG} \frac{p_{Ao} - p_{AL}}{RT} = 0.0414 \times \frac{(10^{-2} - 5 \times 10^{-5})}{8317 \times 298} \times 15 \times 10^5 = 250 \times 10^{-6} \text{ kmol/s}$$

Feed rate of H^+ ions: $\Phi_{vL}(c_{H^+})_o = 0.694 \times 10^{-3} \times 2 = 1.388 \times 10^{-3}$ kmol/s so that, at complete absorption of NH_3,

$$\zeta_{H^+} = 0.18.$$

Hence there is sufficient acid fed to convert the required amount of NH_3.

We'll see in Section VII.5 that equation (VII.17) cannot usually be applied to solve this type of problem because the reaction may influence the rate of mass transfer in the reaction phase. However, the resistance against mass transfer in this example is completely in the gas phase. Therefore, it is no longer important what exactly happens in the liquid reaction phase, as long as $\bar{c}_{AL}$ remains 0. The latter is very likely as we have both a rapid reaction and sufficient acid in the reaction phase. Consequently, we can apply the asymptotic solution of equation (VII.17):

$$J = k_G(\bar{c}_{AG} - 0) \quad \text{or} \quad J = k_G p_A/RT. \tag{b}$$

Substitution of equation (b) in (a) gives

$$\ln \frac{p_{AL}}{p_{Ao}} = -\frac{k_G aSL}{\Phi_{vG}} \tag{c}$$

or

$$L = -\frac{0.0414}{15.7 \times 10^{-3} \times 110 \times 0.212} \ln(5 \times 10^{-3}) = 0.6 \text{ m}.$$

In practice the column is made somewhat larger to compensate for the effect of axial dispersion in the gas phase.

Note that $-\ln(p_{AL}/p_{Ao})$ is known as the number of transfer units required (*NTU*) and $\Phi_{vG}/k_G aS(= u/k_G a)$ as the height of a transfer unit (*HTU*). So equation (c) is equivalent to

$$NTU = \frac{L}{HTU}.$$

VII.4.1 Stagnant film model

This model originates from Nernst and is based on the following assumptions:

— The fluid can be divided into two zones: a stagnant film of thickness δ near the interface and a well-mixed bulk behind it in which no concentration gradients occur.
— The mass transfer process is a stationary process (steady state).

With these assumptions, and neglecting the drift flow caused by diffusion of A, equation (VII.9) reduces to

$$D_A \left(\frac{\partial^2 c_A}{\partial z^2} + \frac{\partial^2 c_A}{\partial y^2} + \frac{\partial^2 c_A}{\partial z^2} \right) = 0, \qquad 0 \leq x \leq \delta. \tag{VII.20}$$

In case the extension and the radius of curvature of the interface are large compared to the thickness of the stagnant layer δ, equation (VII.20) reduces to

$$D_A \frac{d^2 c_A}{dx^2} = 0. \tag{VII.21}$$

The boundary conditions are $c_A = c_{Ai}$ for $x = 0$ and $c_A = \bar{c}_A$ for $x = \delta$. Solving this equation gives

$$D_A \frac{dc_A}{dx} = \text{constant} = -D_A(c_{Ai} - \bar{c}_A)/\delta. \tag{VII.22}$$

Substitution of equation (VII.22) into equation (VII.11) gives

$$J_A = -D_A \left(\frac{dc_A}{dx} \right)_{x=0} = \frac{D_A}{\delta}(c_{Ai} - \bar{c}_A). \tag{VII.23}$$

From equation (VII.12) it follows that

$$\frac{D_A}{\delta} = k_L \quad (\text{or } k_G). \tag{VII.24}$$

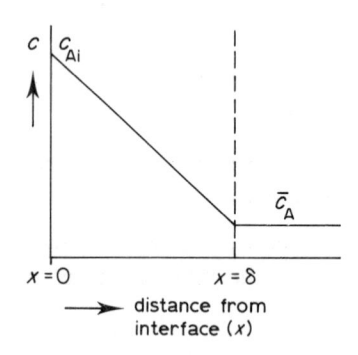

Fig. VII.4. Concentration profile according to stagnant film model

In fact, equation (VII.24) is the definition equation of the film thickness δ, which cannot be measured, because it does not exist in reality. From the physical point of view, the film model is often not very realistic. The advantage of the model is its simplicity and mediative power. Besides it describes mass transfer through an interface combined with chemical reaction sufficiently accurate for most practical cases.

VII.4.2 Penetration models of Higbie and Danckwerts

The penetration model originates from Higbie [6]. It describes the mass transfer process as follows. The interface is considered to be covered with small stagnant fluid elements. After a contact time τ with the interface, the elements return into the well-mixed bulk of the fluid while new elements originating from the bulk take their place at the interface. Coming from the bulk, all new stagnant elements have a uniform initial concentration, $\bar{c}_A$. At the interface, mass transfer into or from the stagnant elements takes place by diffusion.

Contrary to the film theory, this diffusion process is considered to be essentially non-stationary. Figure VII.5 shows qualitatively concentration profiles as a function of time, for the case of diffusion into the elements.

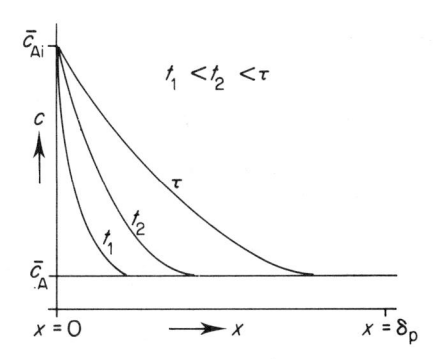

Fig. VII.5. Concentration profiles in stagnant element as a function of contact time with the interface (penetration theory)

The depth of penetration is assumed to be small with respect to the thickness of the element, δ_p. Mathematically this assumption means that the mass transfer process can be described by non-stationary diffusion into a semi-infinite continuum. Neglecting the drift flow caused by diffusion of A, equation (VII.8) reduces to Fick's second law of diffusion:

$$\frac{\partial c_A}{\partial t} = D_A \frac{\partial^2 c_A}{\partial x^2} \qquad (VII.25)$$

with boundary conditions:

$$t = 0, \quad x > 0 : \quad c_A = \bar{c}_A$$
$$t > 0, \quad x = 0 : \quad c_A = c_{Ai} \qquad (VII.26)$$
$$t > 0, \quad x = \infty : \quad c_A = \bar{c}_A$$

The solution is [4]

$$\frac{c_{Ai} - c_A}{c_{Ai} - \bar{c}_A} = \mathrm{erf}\left(\frac{x}{2\sqrt{(D_A t)}}\right) = \frac{2}{\sqrt{\pi}} \int^{x/2\sqrt{(D_A t)}} e^{-z^2}\,dz. \qquad (VII.27)$$

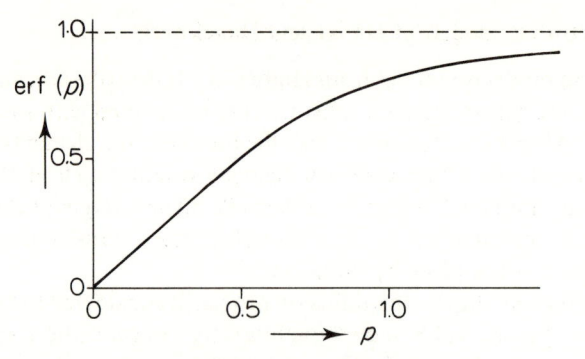

Fig. VII.6. erf (p) as a function of p

Differentiation of equation (VII.27) gives, for $x = 0$,

$$-\left(\frac{\partial c_A}{\partial x}\right)_{x=0} = \frac{1}{\sqrt{(\pi D_A t)}}(c_{Ai} - \bar{c}_A). \tag{VII.28}$$

Combination with equation (VII.11) results in the expression for the time-dependent flux through the interface:

$$J_A = -D_A \left(\frac{\partial c_A}{\partial x}\right)_{x=0} = \sqrt{\left(\frac{D_A}{\pi t}\right)}(c_{Ai} - \bar{c}_A). \tag{VII.29}$$

Equation (VII.29) shows how the mass transfer rate decreases with time. This decrease is caused by a decrease in time of $\partial c_A/\partial x$ at the interface, as shown in figure VII.5. The time average molar flux of A through the interface follows from

$$\bar{J}_A = \frac{1}{\tau}\int_0^\tau J_A \, dt = 2\sqrt{\left(\frac{D_A}{\pi \tau}\right)}(c_{Ai} - \bar{c}_A). \tag{VII.30}$$

Combining with equation (VII.12) gives

$$k_{LA} \quad (\text{or } k_{GA}) = 2\sqrt{\left(\frac{D_A}{\pi \tau}\right)}. \tag{VII.31}$$

Originally, the penetration theory was developed for absorption of a gas into a liquid in bubble columns and in packed columns. In those cases equation (VII.31) in fact is an equation from which the penetration model parameter τ can be calculated from experimentally measured k_L. Occasionally τ can be calculated from first principles, e.g. in a laminar jet (see Section VII.15), a laminar film reactor and a spray column, while for completely mobile bubbles the value

$$\tau \simeq \frac{d_b}{v_b} \quad [s] \tag{VII.32}$$

has been suggested [6].

A modification of Higbie's penetration model has been developed by Danckwerts [7]. Instead of a constant contact time τ, he assumes that the chance of a surface element being replaced within a given time is independent of its age. Thus, each surface element has an equal chance to be replaced during the next time unit. Defining s as the fraction of elements of any age group that is replaced per time unit [s^{-1}], he derived

$$k_L = \sqrt{(D_A s)}. \tag{VII.33}$$

Both the stagnant film model and the penetration models of Higbie and Danckwerts (the latter being also known as surface renewal models) are applied in calculating the influence of chemical reaction on the rate of mass transfer (this chapter) and for calculating the influence of mass transport limitations on selectivity (next chapter). The models have in common that the hydrodynamic conditions near the interface are described by one parameter: either the film thickness, δ, or the contact time, τ, or the probability of replacement, s. Mostly this model parameter is obtained via the definition in equations (VII.24), (VII.31) and (VII.33), with k_L or k_G obtained either from experiment or from the theory of hydrodynamics. The choice of the model is rather arbitrary because the results obtained with them are often much the same (see Section VII.5.3). Exceptions are found for a large difference in the value of reactant diffusion coefficients, see Section VII.7.3, and for certain complex reactions, see Section VIII.6.2.

Because of the similarity of results, application of more complicated and possibly more realistic mass transfer models does not result in substantially better predictions of the influence of reaction on the rate of mass transfer [2]. In particular this is because our knowledge of the hydrodynamical, physical and kinetical data necessary for the calculations is often inaccurate; this is due partly to the nature of the data — for example, there is spread in the history and diameter of drops and bubbles (fluid–fluid systems) and spread in catalyst particle properties from particle to particle and from day to day (fluid–solid systems). This explains why the simple film theory is often also applied in mass transfer with reaction problems in which the film theory is clearly an oversimplification with respect to the hydrodynamics of the system.

A different question is the suitability of equations (VII.24), (VII.31) and (VII.33), for predicting the influence of D_A on k_L or k_G. Apparently, this influence depends on the hydrodynamics of the system.

VII.5 MASS TRANSFER WITH HOMOGENEOUS IRREVERSIBLE FIRST ORDER REACTION

VII.5.1 Penetration models

The Higbie model for bulk concentration of the transferred reactant being zero in the reaction phase

Penetration models start from equation (VII.25) with the extension of a reaction term:

$$\frac{\partial c_A}{\partial t} = D_A \frac{\partial^2 c_A}{\partial x^2} + R_A. \tag{VII.34}$$

In case A disappears by a first order reaction, R_A is negative and given by

$$R_A = -k_1 c_A. \qquad (\text{VII.35})$$

Substitution of equation (VII.35) in equation (VII.34), and solving with the boundary conditions (VII.26) gives, for bulk concentration $\bar{c}_A = 0$ [8],

$$\frac{c_A}{c_{Ai}} = \tfrac{1}{2}e^{-x\sqrt{(k_1/D_A)}}\left[1 - \operatorname{erf}\left(\frac{x}{2\sqrt{(D_A t)}} - \sqrt{(k_1 t)}\right)\right]$$

$$+ \tfrac{1}{2}e^{x\sqrt{(k_1/D_A)}}\left[1 - \operatorname{erf}\left(\frac{x}{2\sqrt{(D_A t)}} + \sqrt{(k_1 t)}\right)\right]. \qquad (\text{VII.36})$$

Differentiation gives the flux through the interface:

$$J_{At} = -D_A\left(\frac{\partial c_A}{\partial x}\right)_{t,x=0} = \sqrt{(k_1 D_A)}\left\{\operatorname{erf}[\sqrt{(k_1 t)}] + \frac{e^{-k_1 t}}{\sqrt{(\pi k_1 t)}}\right\} c_{Ai}. \qquad (\text{VII.37})$$

Average mass transfer rate over the first τ seconds follows from

$$\bar{J}_{A\tau} = \frac{1}{\tau}\int_0^\tau J_{At}\,dt.$$

With equation (VII.37),

$$\bar{J}_{A\tau} = \sqrt{(k_1 D_A)}\left\{\left(1 + \frac{1}{2k_1\tau}\right)\operatorname{erf}[\sqrt{(k_1\tau)}] + \frac{1}{\sqrt{(\pi k_1\tau)}}e^{-k_1\tau}\right\} c_{Ai}. \qquad (\text{VII.38})$$

Equation (VII.38) gives the mass transfer rate according to the Higbie model for $\bar{c}_A = 0$.

Figure VII.7 shows how the mass transfer rate J_{At} decreases with time t and the average transfer rate $\bar{J}_{A\tau}$ with contact time τ. Simple asymptotic solutions are obtained for $k_1 t < 0.5$ and $k_1 t > 2$. For $k_1 t > 2$, the mass transfer rate

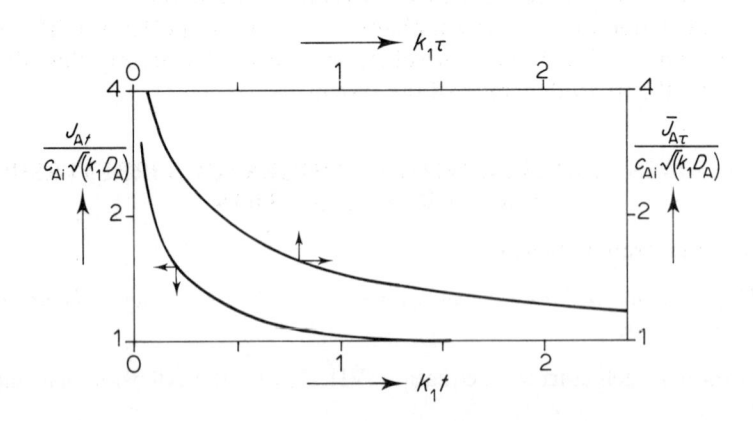

Fig. VII.7. Mass transfer rate (J_{At}) as a function of time (t), and average transfer rate $(\bar{J}_{A\tau})$ as a function of contact time (τ)

approaches a stationary minimum value

$$J_{At} = \sqrt{(k_1 D_A)} c_{Ai} \qquad (VII.39)$$

and

$$\bar{J}_{A\tau} = \sqrt{(k_1 D_A)} \left(1 + \frac{1}{2k_1\tau}\right) c_{Ai}. \qquad (VII.40)$$

The latter rate becomes independent of τ for $k_1\tau > 5$ (within 10%):

$$\bar{J}_{A\tau} = \sqrt{(k_1 D_A)} c_{Ai}. \qquad (VII.41)$$

The fact that the mass transfer rate approaches a constant value, being larger than zero, is an essential difference from physical absorption. In the latter case J_{At} decreases continuously to zero for t approaching infinity. Then the semi-infinite element is saturated with A. Due to chemical reaction, a stationary concentration profile of A is reached for $k_1 t > 2$. This can be understood from the following reasoning. Figure VII.8 shows two concentration profiles for t_1 and t_2, with $t_2 > t_1$. The transfer rate (J_A) for $t = t_1$ is larger than for $t = t_2$, because

$$-\left(\frac{\partial c_A}{\partial x}\right)_{\substack{x=0,\\t=t_1}} > -\left(\frac{\partial c_A}{\partial x}\right)_{\substack{x=0,\\t=t_2}}.$$

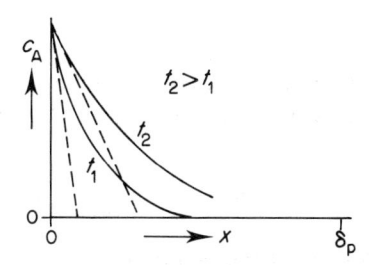

Fig. VII.8. Change of concentration gradient at the interface with time (penetration theory)

Total conversion rate of A, per unit area interface, $|R_A''|$, increases in time, because of increasing $c_A(x)$:

$$|R_{At}''| = \frac{1}{\delta_p} \left\{\int_0^{\delta_p} k_1 c_A \, dx\right\}_t.$$

For $t = 0$, J_A approaches to infinity while R_A'' is zero. The equilibrium profile and the stationary mass transfer rate will be obtained for such a decrease of J_A and such an increase of $(-R_A'')$ that the condition

$$J_{At} = |R_{At}''| \qquad (VII.42)$$

is fulfilled. This is approximately true from $k_1 t > 2$ or $t > 2/k_1$. For $k_1\tau > 5$, equation (VII.42) is, on average, nearly fulfilled over the whole contact time.

Therefore, as long as $\tau > 5/k_1$, the exact value of τ (i.e. the mass transfer coefficient), does not influence the mass transfer rate. That is why equation (VII.41) does not contain the parameter τ any longer. A second consequence of the fact that equation (VII.42) is nearly fulfilled over the whole contact time τ, for $k_1\tau > 5$, is that the amount of unreacted A that is transported to the bulk at $t = \tau$ is relatively small: practically all of A is already converted near the interface; nearly no A reaches the bulk unconverted.

The group $k_1\tau$ is dimensionless. It is the ratio between the rate of reaction at unit concentration (k_1), and the rate of element refreshment $(1/\tau)$, which is related to the rate of mass transfer without reaction. Thus, the higher the value of $k_1\tau$ is, the higher the rate of reaction is, compared to the rate of mass transfer without reaction. Apparently, for $k_1/(1/\tau) > 5$, the rate of reaction is so high that nearly all reactant is converted near the interface, while the mass transfer rate is independent of the mass transfer parameter τ, because of a stationary concentration profile already realized at $t \ll \tau$.

From equation (VII.31),

$$k_L \text{ (or } k_G) = 2\sqrt{\left(\frac{D_A}{\pi\tau}\right)}, \qquad \text{(VII.31)}$$

it follows that

$$k_1\tau = \frac{4k_1 D_A}{\pi k_L^2} \quad \text{or} \quad \frac{4k_1 D_A}{\pi k_G^2}. \qquad \text{(VII.43)}$$

The group $\sqrt{(k_1 D_A)}/k_L$ is known as the Hatta number, Ha, for first order reactions. Substitution of equation (VII.43) into equation (VII.40) gives (for $Ha > \sqrt{(\pi/2)}$)

$$\bar{J}_{A\tau} = \sqrt{(k_1 D_A)}\left(1 + \frac{\pi}{8Ha^2}\right)c_{Ai}. \qquad \text{(VII.44)}$$

For $Ha > 2$, this expression further simplifies to

$$\bar{J}_{A\tau} = \sqrt{(k_1 D_A)}c_{Ai}. \qquad \text{(VII.41)}$$

The stationary concentration profile for $k_1\tau > 2$ follows from equation (VII.36), which reduces to

$$c_A/c_{Ai} = e^{-x\sqrt{(k_1/D_A)}}. \qquad \text{(VII.45)}$$

For $x = 2\sqrt{(D_A/k_1)}$, $c_A/c_{Ai} = 1/e$. Thus, depth of penetration reduces with increasing reaction rate and decreasing diffusion rate.

Simplified expressions for J_{At} (equation (VII.37)) and $\bar{J}_{A\tau}$ (equation (VII.38)) are obtained for $k_1 t < 0.5$ as well as for $k_1 t > 2$. In the former case equation (VII.37) reduces to

$$J_{At} \simeq \sqrt{\left(\frac{D_A}{\pi t}\right)}(1 + k_1 t)c_{Ai} \qquad \text{(VII.46)}$$

and equation (VII.38) to

$$\bar{J}_{A\tau} \simeq 2\sqrt{\left(\frac{D_A}{\pi\tau}\right)}\left(1 + \frac{k_1\tau}{3}\right)c_{Ai} \qquad (VII.47)$$

for $k_1\tau < 0.5$. For $k_1\tau < 0.1$ ($Ha < 0.3$), equation (VII.47) further reduces to the equation for mass transfer without reaction:

$$\bar{J}_{A\tau} \simeq 2\sqrt{\left(\frac{D_A}{\pi\tau}\right)}c_{Ai} = k_L c_{Ai}. \qquad (VII.48)$$

In this asymptotic case the reaction rate is so slow with respect to the rate of mass transfer that the reaction does not have any influence on the rate of mass transfer. Under the same conditions (that is for $k_1\tau < 0.1$ or $Ha < 0.3$), this conclusion also holds for $\bar{c}_A \neq 0$ in the reaction phase. Thus, for $Ha < 0.3$, equation (VII.48) can be extended to

$$\bar{J}_{A\tau} \simeq 2\sqrt{\left(\frac{D_A}{\pi\tau}\right)}(c_{Ai} - \bar{c}_A) = k_L(c_{Ai} - \bar{c}_A) \quad \text{or} \quad k_G(c_{Ai} - \bar{c}_A). \qquad (VII.49)$$

In all other cases (that is for $k_1\tau > 0.1$ or $Ha > 0.3$) due to the reaction occurring, the rate of mass transfer is enhanced compared to the rate obtained in absence of reaction. This is caused by an increase of $\partial c_A/\partial x$ at the interface, due to an extra decrease of A by reaction. A term of current usage is the so-called *enhancement factor*, defined as

$$E_A = \frac{\bar{J}_{A\tau} \text{ with reaction}}{\bar{J}_{A\tau} \text{ without reaction}}. \qquad (VII.50)$$

with both fluxes J based on the same c_{Ai}.

Thus it follows from the above analysis that the enhancement factor E_A varies, according to the Higbie penetration model, from $E_A = 1$ (for $k_1\tau < 0.1$) to $E_A = \frac{1}{2}\sqrt{(\pi k_1\tau)}$ (for $k_1\tau > 5$) or from $E_A = 1$ (for $Ha < 0.3$) to $E_A = Ha$ (for $Ha > 2$), provided the reaction is first order and provided that the bulk concentration of A in the reaction phase is zero.

The Danckwerts model

Starting from the Danckwerts modification of the penetration theory, Danckwerts [9] derived for mass transfer with first order reaction in e.g. a liquid

$$\bar{J}_A = k_L\left(c_{Ai} - \frac{\bar{c}_A}{1 + Ha^2}\right)\sqrt{(1 + Ha^2)}. \qquad (VII.51)$$

This solution is relatively simple, compared to the Higbie solution (equation (VII.38)). This is the more remarkable because the Danckwerts solution also covers the case that the bulk concentration differs from zero. Realizing that, in case of mass transfer to the reaction phase, $\bar{c}_A \leq c_{Ai}$, it follows that for $Ha > 2$ the term $\bar{c}_A/(1 + Ha^2)$ can be neglected in equation (VII.51). For slow reaction

($Ha < 0.3$) and for fast reaction ($Ha > 2$), the same asymptotic solutions are found as with the Higbie model. The enhancement factor E_A, defined (equation (VII.50)) as

$$E_A = \frac{\bar{J}_A}{k_L(c_{Ai} - \bar{c}_A)} \qquad \text{(VII.52)}$$

follows from equations (VII.51) and (VII.52):

$$E_A = [\sqrt{(1 + Ha^2)}]\frac{c_{Ai} - \dfrac{\bar{c}_A}{1 + Ha^2}}{c_{Ai} - \bar{c}_A} \qquad \text{(VII.53)}$$

Thus, E_A depends on concentrations for the regime $0.3 < Ha < 2$, providing $\bar{c}_A \neq 0$. In that case, $\bar{J}_A$ is not proportional to the driving force ($c_{Ai} - \bar{c}_A$), as is the case in absorption without reaction. In practice, however, this situation will be seldom met, because the range of k_1 values for which this condition holds is relatively small.

Example VII.5.1 *Laminar jet model reactor*

As will be discussed in Section VII.15, a so-called laminar liquid jet is a useful model reactor for measuring the kinetics of fast gas–liquid reactions (see figure VII.46). This is because the hydrodynamics are accurately known.

Suppose we have a laminar liquid jet of 10 cm length. Liquid flow rate is $\Phi_L = 2.6 \times 10^{-6}\,\text{m}^3/\text{s}$. The gas phase consists of pure A at 1 bar. The rate of absorption of A in the liquid jet is measured as $\Phi_G = 10 \times 10^{-9}\,\text{kmol/s}$. A does not chemically react with the liquid. The solubility of A is $0.1\,\text{kmol}/(\text{m}^3\,\text{bar})$. The entering liquid does not contain any A. If we can assume that the liquid velocity is constant all over the jet, what then is the diffusion coefficient of A in the liquid?

A reactant B is added to the liquid. It is known that A reacts with B according to $R_A = -k_1 c_A$. By adding B, the rate of absorption of A appears to have increased by a factor of 4. What is the rate constant k_1, if we know that the average diameter of the jet is $1.5 \times 10^{-3}\,\text{m}$?

The answers to these questions can be obtained as follows:

(a) the liquid flow rate and the gas absorption rate are known. Hence the average outlet concentration of A in the liquid is

$$\bar{c}_{A1} = \Phi_G/\Phi_L = 3.8 \times 10^{-3}\,\text{kmol/m}^3.$$

With $c_{Ai} = 0.1\,\text{kmol/m}^3$, it follows that the centre of the jet, even near the outlet, can hardly contain any A. We will first neglect the curvature of the area and approximate the problem by physical absorption in a semi-infinite continuum:

$$\bar{J}_\tau = 2\sqrt{\left(\frac{D}{\pi\tau}\right)} c_{Ai},$$

$$\Phi_G = \bar{J}_\tau A = 2\sqrt{\left(\frac{D}{\pi\tau}\right)} c_{Ai}\pi dl \qquad \text{(a)}$$

$$\tau = \pi d^2 l/(4\Phi_L) \qquad \text{(b)}$$

in which d is the diameter and l is the length of the jet. From equations (a) and (b) it follows that

$$\Phi_G = 4c_{Ai}\sqrt{(\Phi_L l D_A)}$$

or

$$D_A = \Phi_G^2/(16c_{Ai}^2 \Phi_L l),$$

and we find

$$D_A = 2.4 \times 10^{-9} \, \text{m}^2/\text{s}.$$

Let's now check the validity of our neglect of the curvature of the surface area. This is correct as long as the depth of penetration (δ) is small compared to the jet radius. Defining δ by

$$\bar{J}_\tau = \frac{\Phi_G}{\pi d l} = 2\sqrt{\left(\frac{D_A}{\pi \tau}\right)} c_{Ai} \equiv \frac{D_A}{\delta} c_{Ai},$$

we find that

$$\frac{\delta}{r} = \frac{\pi D_A l}{\Phi_G} = 0.15.$$

The assumption therefore is valid.

(b) For $k_1 \tau > 5$ it follows from equations (VII.40) and (VII.50) that

$$E_A = \tfrac{1}{2}\sqrt{(\pi k_1 \tau)}\left(1 + \frac{1}{2k_1 \tau}\right),$$

with $E_A = 4$ we find $k_1 \tau = 20.5$. Hence, the assumption $k_1 \tau > 5$ is correct. We can now calculate the value of k_1.

With $\tau = \pi d^2 l/(4\Phi_L) = 0.0675\,\text{s}$ we calculate $k_1 = 300\,\text{s}^{-1}$.

This illustration shows the importance of the laminar jet as a model reactor for measuring fast gas–liquid kinetics; under homogeneous conditions it would be impossible to measure such fast reaction rates. Note that we started with the knowledge that the reaction was first order in A and zero order in B. In practice this information is often unknown. However, we can determine the kinetic rate expression by systematically varying both c_{Ai} and $\bar{c}_B$ in the experiment.

VII.5.2 Stagnant film model

Recalling from Section VII.4 the basic assumptions underlying the stagnant film model, it follows that with homogeneous reaction, equation (VII.21) must be extended by a chemical production term:

$$D_A \frac{d^2 c_A}{dx^2} + R_A = 0. \tag{VII.54}$$

The boundary conditions are the same as without reaction:

$$c_A = c_{Ai} \quad \text{for} \quad x = 0$$
$$c_A = \bar{c}_A \quad \text{for} \quad x = \delta.$$

Solution of equation (VII.54) for first order kinetics $R_A = -k_1 c_A$ gives, for $0 \leq x \leq \delta$,

$$\frac{c_A}{c_{Ai}} = \frac{1}{\sinh \phi} \left\{ \sinh\left[\phi - x\sqrt{\left(\frac{k_1}{D_A}\right)}\right] + \frac{\bar{c}_A}{c_{Ai}} \sinh\left[x\sqrt{\left(\frac{k_1}{D_A}\right)}\right] \right\} \quad \text{(VII.55)}$$

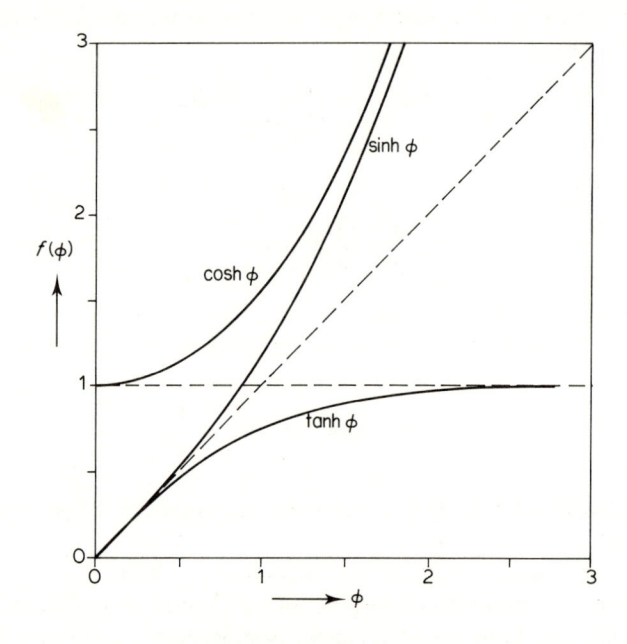

Fig. VII.9. Hyperbolic functions:
$\sinh \phi = (e^{\phi} - e^{-\phi})/2$;
$\cosh \phi = (e^{\phi} + e^{-\phi})/2$;
$\tanh \phi = \sinh \phi / \cosh \phi$;
$\phi < 0.3$: $\sinh \phi \simeq \tanh \phi \simeq \phi$,
   $\cosh \phi \simeq 1$;
$\phi > 2$: $\tanh \phi \simeq 1$;
for any ϕ: $(\cosh \phi)^2 - (\sinh \phi)^2 = 1$

with the dimensionless parameter ϕ defined as

$$\phi \equiv \delta \sqrt{\left(\frac{k_1}{D_A}\right)}. \quad \text{(VII.56)}$$

The physical meaning of the important parameter ϕ follows from rewriting equation (VII.56) in the form

$$\phi^2 = \frac{k_1 c_{Ai} \delta}{\frac{D_A}{\delta}(c_{Ai} - 0)}. \qquad \text{(VII.57)}$$

Thus, ϕ^2 is equal to the ratio of the maximum conversion rate of A in the film per unit area interface and the maximum diffusional transport through the film in case of absence of reaction. When $\phi \ll 1$, hardly any reaction takes place in the film, and when $\phi \gg 1$ the reactant A coming from the interface is converted entirely within the film. In fluid phases δ follows from

$$\delta = \frac{D_A}{k_L} \quad \left(\text{or } \frac{D_A}{k_G}\right). \qquad \text{(VII.24)}$$

In that case:

$$\phi = \sqrt{\left(\frac{k_1 D_A}{k_L^2}\right)} = Ha. \qquad \text{(VII.58)}$$

We note that the dimensionless group Ha is the reaction modulus ϕ for fluid phases, for which the concept of a mass transfer coefficient has been developed. Equations (VII.55) and (VII.56) are also applicable for reactions in stagnant films or porous pellets. In those cases, δ has a direct physical meaning, e.g. (half of) the thickness of the film or, in case of pellets of volume V and area A, $\delta = V/A$; see Section VII.5.4.

By differentiation of equation (VII.55), the mass transfer rate through the interface is obtained:

$$J_A = -D_A \left(\frac{dc_A}{dx}\right)_{x=0} = \frac{D_A}{\delta}\left(c_{Ai} - \frac{\bar{c}_A}{\cosh\phi}\right)\frac{\phi}{\tanh\phi}. \qquad \text{(VII.59)}$$

Also of interest is the flux of A through the plane $x = \delta$ into the bulk of the reaction phase:

$$-D_A \left(\frac{dc_A}{dx}\right)_{x=\delta} \equiv D_A \gamma. \qquad \text{(VII.60)}$$

From differentiation of equation (VII.55), it follows that

$$\gamma = -\left(\frac{dc_A}{dx}\right)_{x=\delta} = \sqrt{\left(\frac{k_1}{D_A}\right)}\left(\frac{c_{Ai}}{\sinh\phi} - \frac{\bar{c}_A}{\tanh\phi}\right). \qquad \text{(VII.61)}$$

The asymptotic solutions for $\phi > 2$ and $\phi < 0.3$, respectively, are equal to those found with the penetration models:

Reaction rate of A is fast compared to mass transfer rate of A ($\phi > 2$).

Realizing that $\bar{c}_A \leq c_{Ai}$ (A is transferred to the reaction phase), it follows that for $\phi > 2$, $\bar{c}_A/\cosh\phi \ll c_{Ai}$. Further: $\tanh\phi \simeq 1$ for $\phi > 2$. Therefore, equation

(VII.59) reduces for $\phi > 2$ to

$$J_A = \frac{D}{\delta} c_{Ai}\phi = \sqrt{(k_1 D_A)} c_{Ai}. \qquad (VII.62)$$

Hence $E_A = \phi$.

This result is identical to that obtained for the penetration theory (equation (VII.41)). Keep in mind that $\phi > 2$ implies that the reactivity of A is so high that A can't reach the bulk of the reaction phase. Therefore the bulk of the reaction phase is not effective in case of fast reactions, because of transport limitations.

Reaction rate of A is slow compared to mass transfer rate of A ($\phi < 0.3$).

Equation (VII.59) reduces to

$$\phi < 0.3: \qquad J_A = \frac{D_A}{\delta}(c_{Ai} - \bar{c}_A)$$

or

$$Ha < 0.3: \qquad J_A = k_L(c_{Ai} - \bar{c}_A). \qquad (VII.23)$$

The same asymptotic solution has been derived for the penetration model (equation (VII.49)): the rate of mass transfer is not enhanced by reaction, as long as $\phi < 0.3$ ($E_A = 1$).

VII.5.3 General conclusion on mass transfer with homogeneous irreversible first order reaction

In the previous two sections, analytical expressions have been presented for mass transfer followed by irreversible first order reactions according to both the film model and the penetration models of Higbie and Danckwerts. The film model mass transfer concept differs essentially from the two penetration models. Also the relations obtained for the rate of mass transfer in case of first order reactions look quite different for each of the three models (see Table VII.1). Numerically, however, the differences are less pronounced. All three models predict the same asymptotic solutions for $Ha < 0.3$ and $Ha > 2$, respectively. The differences are maximal for Ha of the order of one. Even in that case, however, the enhancement factors, as calculated from the three models, agree within 10% for $\bar{c}_A = 0$ and within 20% for $\bar{c}_A \neq 0$, see Table VII.2. This agreement is very satisfactory for most engineering purposes.

For 'fast' reactions, that is for $\phi = Ha > 2$, or $k_1 \tau > 5$, all models predict

$$J_A = \sqrt{(k_1 D_A)} c_{Ai} = k_L c_{Ai} \phi. \qquad (VII.41)$$

Thus, for $\phi > 2$, the rate of mass transfer becomes independent of the mass transfer coefficient. A second important conclusion is that for such a fast reaction, the bulk concentration $\bar{c}_A$ must be zero in the reaction phase. This result is far reaching. It implies that the bulk of the reaction phase behaves as an inert phase;

Table VII.1 The rate of mass transfer through an interface in mass transfer with homogeneous irreversible first order reaction, according to different mass transfer models. For reasons of convenience all relations are rewritten in terms of k_L and Ha by elimination of τ and $k_1\tau$ by means of equation (VII.43) and by substitution of $\phi = Ha$

$\bar{c}_A$	Model	J_A	
0	Stagnant film	$J_A = k_L c_{Ai} \dfrac{Ha}{\tanh Ha}$	(VII.59)
	Higbie (penetration)	$J_A = k_L c_{Ai} Ha \left[\left(1 + \dfrac{\pi}{8Ha^2}\right) \text{erf}\left(\dfrac{2Ha}{\sqrt{\pi}}\right) + \dfrac{1}{2Ha} e^{-(4/\pi)Ha^2} \right]$	(VII.38)
	Danckwerts (penetration)	$J_A = k_L c_{Ai} \sqrt{1 + Ha^2}$	(VII.51)
$\neq 0$	Stagnant film	$J_A = k_L \left(c_{Ai} - \dfrac{\bar{c}_A}{\cosh Ha} \right) \left(\dfrac{Ha}{\tanh Ha} \right)$	(VII.59)
	Danckwerts (penetration)	$J_A = k_L \left[c_{Ai} - \dfrac{\bar{c}_A}{1 + Ha^2} \right] \sqrt{(1 + Ha^2)}$	(VII.51)

this means that information on the exact flow pattern in the bulk of the reaction phase is not required for reactor design purposes. This may substantially simplify (experiments necessary for) the design of the reactor. For 'slow' reactions, that is for $\phi = Ha < 0.3$ or $k_1\tau < 0.1$, all models predict

$$J_A = k_L(c_{Ai} - \bar{c}_A). \tag{VII.49}$$

The reaction does not enhance the rate of mass transfer in this case. The derived expressions for the flux J_A contain the interfacial concentration c_{Ai}. Unless the resistance against mass transfer in the phase where A originates from is negligible, e.g. if that phase contains only pure A, c_{Ai} itself is a function of J_A. Provided m is known, c_{Ai} can easily be eliminated. We deal with this problem in Example VII.5.3.

Table VII.2. Enhancement factor for mass transfer with first order, homogeneous irreversible reaction, according to different models for $Ha = \phi = 1$

$\bar{c}_A$	Model	$(E_A)_{Ha=1}$
0	Stagnant film	1.31
0	Penetration; Higbie	1.38
0	Penetration; Danckwerts	1.41
$\tfrac{1}{2}c_{Ai}$	Stagnant film	1.78
$\tfrac{1}{2}c_{Ai}$	Penetration; Danckwerts	2.12

Example VII.5.3 *Influence of the gas phase resistance on the flux through an interface; first order reaction*

Gas phase G contains a mixture of inert nitrogen and a reactant A; $\bar{c}_{AG} = 0.01 \, \text{kmol/m}^3$.

The gas is bubbled through a liquid in which A absorbs and is then converted according to a first order reaction. The fraction absorbed is rather small, so that $\bar{c}_{AG} = \text{constant} = 10^{-2} \, \text{kmol/m}^3$. Furthermore, the bulk concentration of A in the liquid is zero. The solubility $m = 1$, where $m = c_L/c_G$ at equilibrium. Further data are $k_G = 10^{-2} \, \text{m/s}$, $k_L = 10^{-4} \, \text{m/s}$ and $D_{AL} = 10^{-9} \, \text{m}^2/\text{s}$. The value of k_1 can vary between 4×10^{-3} and $40 \times 10^{-6} \, \text{s}^{-1}$, depending on the concentration of a catalyst. We are asked to:

— derive an analytical expression for the flux of A through the interface, J_A, containing only known parameters and k_1
— construct a graph of both J_A and E_A as a function of $\phi = \sqrt{(k_1 D_{AL})}/k_L$.

The basic equations are

$$J_A = k_L(c_{iL} - 0)E_A,$$

$$J_A = k_G(\bar{c}_G - c_{iG}),$$

$$c_{iL}/c_{iG} = m.$$

Elimination of the unknown concentrations gives

$$J_A = \frac{k_G \bar{c}_{AG}}{1 + k_G/(mk_L E_A)} = \frac{k_L m \bar{c}_{AG} E_A}{1 + \frac{mk_L}{k_G} E_A}$$

with $E_A = \phi/\tanh \phi$ according to the film theory.
ϕ varies from 0.02 to 2000. For J_A two asymptotic solutions are found:

(a) Resistance in gas phase is negligible

This is so for $k_G/mk_L E_A \gg 1$, hence for $E_A \ll \dfrac{k_G}{mk_L} = 100$; see the profiles with $\phi = 0.1$ and 1 in Figure VII.10

(b) Gas phase resistance is rate controlling

This is so for $k_G/mk_L E_A \ll 1$ or $E_A \gg k_G/mk_L = 100$, hence for $\phi \gg 100$; see the profile for $\phi = 1000$ in figure VII.10. Figure VII.10 gives E_A and $J_A/k_L m \bar{c}_{AG}$ as a function of ϕ. Notice that $J_A/k_L m \bar{c}_{AG} = E_A$ as long as $k_G/mk_L E_A \gg 1$, but that with increasing E_A gas phase resistance becomes controlling, thus preventing a further enhancement of the rate of mass transfer with ever increasing values of ϕ. The maximum possible rate of mass transfer is

$$J = k_G(\bar{c}_{AG} - 0)$$

or the maximum enhancement relative to the maximum rate without reaction:

$$\left(\frac{J}{mk_L \bar{c}_{AG}}\right)_{max} = \frac{k_G}{mk_L}.$$

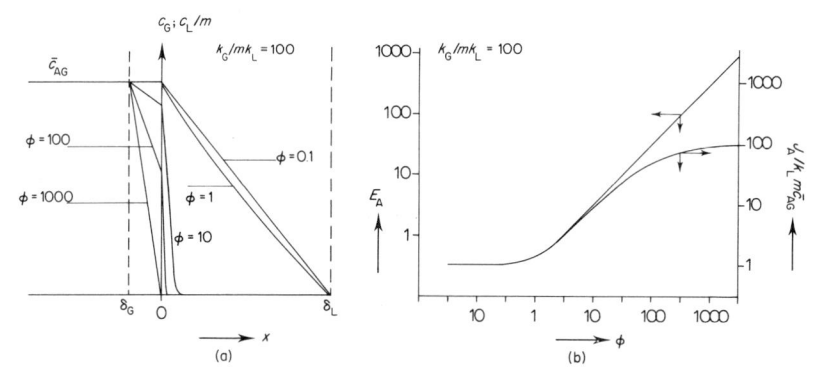

Figure VII.10. Mass transfer with a first order reaction and resistances in two phases: (a) concentration profiles near the interface; (b) enhancement factors and dimensionless flux through the interface. The film thicknesses δ_L and δ_G are not to scale

VII.5.4 Applications

(a) The reaction phase is completely stagnant (films, spheres, pellets)

For mass transfer followed by reaction in stagnant liquid films or solid sheets, the theory from the preceding section is applicable, by substituting for δ the half of the thickness of the film or the sheet, respectively. The equations in the preceding sections have been derived with the assumption that the radius of the film was large compared to the film thickness.

If the reaction phase is a sphere or a pellet of arbitrary form, the curvature, in principle, plays a role. Aris [10], however, showed that the equations, derived for flat films, remain valid to a good approximation if 'film thickness' δ is defined as

$$\delta = V/A \tag{VII.63}$$

where V is the volume of the pellet and A, its surface area. In the middle of the film, sphere or cylindrical pellet, a plane, point or line of symmetry is found. Therefore the flow through the centre is zero. This means that

$$-\frac{dc_A}{dx} \equiv \gamma = 0 \quad \text{for} \quad x = \delta \tag{VII.64}$$

Substitution of equation (VII.64) in equation (VII.61) gives

$$\frac{\bar{c}_A}{c_{Ai}} = \frac{1}{\cosh \phi} \tag{VII.65}$$

Substitution of equation (VII.65) in equation (VII.59) gives the rate of mass transport into the pellet or sheet:

$$J_A = \frac{c_{Ai} D_A}{\delta} \phi \tanh \phi \tag{VII.66}$$

with: $\bar{c}_A$ = the concentration of A in the middle of the sheet or pellet;
$\delta = V/A$ for pellets and $d/2$ for stagnant sheets of thickness d.

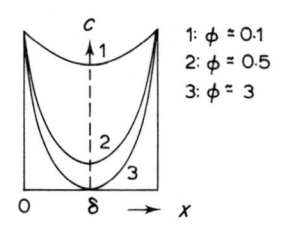

Fig. VII.11. Concentration profiles in symmetrical bodies for mass transfer with irreversible reaction

To show how mass transfer may limit the conversion rate, equation (VII.66) is written as

$$J_A = k_1 c_{Ai} \delta \frac{\tanh \phi}{\phi} \equiv k_1 c_{Ai} \delta \eta \qquad (VII.67)$$

where η is a degree of utilization of the reaction phase, called the effectiveness factor, with $0 < \eta \leq 1$.

For slow reactions ($\phi < 0.3$), $\eta = 1$. That is, the reaction phase is 'saturated' with reactant. The rate of reactant supply is completely determined by the kinetics of the reaction. Then, reactor capacity becomes, for c_{Ai} constant all over the reactor,

$$-\Phi_A = J_A a V_r = k_1 c_{Ai}(1 - \varepsilon) V_r \quad [\text{kmol/s}] \qquad (VII.68)$$

with $1 - \varepsilon =$ fraction reaction phase (hold-up) and $a =$ the interfacial area per unit volume reaction mixture. The whole reaction phase is maximally effective. With increasing ϕ, the reaction becomes faster and the effectiveness factor decreases to $1/\phi$ for $\phi > 2$. Then, only part of the reaction phase is effective. For constant c_{Ai}, the reactor capacity approaches

$$-\Phi_A = JaV_r = \sqrt{(k_1 D)} c_{Ai} a V_r \quad [\text{kmol/s}] \qquad (VII.69)$$

Realizing that aV_r is the total interfacial area, equations (VII.68) and (VII.69) show that for fast reactions, reactor capacity is determined by total interfacial area, instead of total volume of the reaction phase, while for slow reactions the reverse conclusion holds.

In column reactors, c_{Ai} may vary over the height. Then equations like (VII.1) and (VII.2) are required to design the reactor. The two equations are coupled via $J_{JL} = -J_{JG}$ with J given by equation (VII.66), at any height.

Equation (VII.67) shows how the conversion rate is limited by a diffusion limitation in transport of A. Equation (VII.66) shows how the rate of mass transfer is enhanced by reaction of A. Thus, enhanced mass transport and diffusion limited conversion rate are essentially complementary concepts. In Example VII.12.2.a we will derive an exact solution for a sphere and show that the approximate solutions presented here are accurate within 15%.

(b) *Bulk concentration in reaction phase and interface concentration are constant all over the reactor*

In case of a fast reaction ($\phi > 2$), the bulk concentration of reactant A is zero in the reaction phase. Then, A cannot reach the bulk and the problem reduces to the previous one with $\phi > 2$. For $\phi < 2$, at least part of the reaction takes place in the bulk of the reaction phase. Now, two mass balances for A are required to calculate the system: one over the interfacial diffusion film and one over the bulk of the reaction phase. Both balances are coupled via the exchange flux between the two regions

$$(J_A)_{x=\delta} = -D_A \left(\frac{dc_A}{dx}\right)_{x=\delta} \equiv D_A \gamma.$$

In the following, we will consider a continuous two phase fluid–fluid tank reactor, say a gas–liquid tank reactor, with reactant A diffusing from the gas phase into the liquid phase, the latter being the reaction phase. Both phases are perfectly mixed on a macro scale. This means that c_{Ai} and $\bar{c}_A$ are constant in both phases all over the reactor. Figure VII.12.a gives a sketch of the reactor.

An important variable in two phase reactors is the specific contact area a. There is a relation between the value of a, the fraction reaction phase $(1 - \varepsilon)$ and the fineness of the dispersion. Suppose that the non-reaction phase consists of N spherical particles of different diameters d. Then

$$a = N\pi \langle d^2 \rangle / V_r$$

where $\langle d^2 \rangle$ is the average of the square of the diameters. Similarly the volume of the dispersed phase is

$$\varepsilon V_r = N \frac{\pi}{6} \langle d^3 \rangle.$$

Hence, the specific contact area is

$$a = \frac{6\varepsilon \langle d^2 \rangle}{\langle d^3 \rangle} = \frac{6\varepsilon}{d_s},$$

where d_s is the so-called Sauter mean diameter of the particles:

$$d_s = \frac{\sum d^3}{\sum d^2}.$$

This shows that a is inversely proportional to the average size of the dispersed particles. If these are bubbles or drops, their size does not only depend on the fluid properties of both phases but also, to a large extent, on the interplay between the mechanisms responsible for their breaking up and mutual coalescence.

A model describing the two phase reactor under consideration may start from a mass balance for A over the bulk of the reaction phase (see figure VII.12)

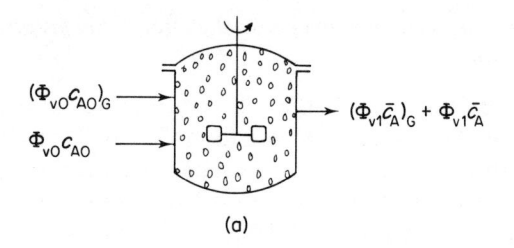

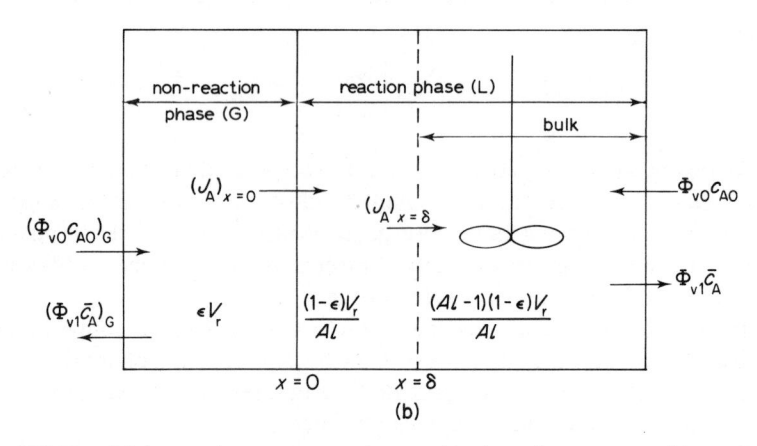

Fig. VII.12. (a) A continuous two phase stirred tank reactor; (b) a schematic representation of mass transfer with chemical reaction in such a reactor

according to accumulation = in by diffusion through the plane $x = \delta$, + in by flow, − out by flow and + produced by reaction, which becomes

$$V_{bL}\frac{\partial \bar{c}_A}{\partial t} = -D_A\left(\frac{\partial c_A}{\partial x}\right)_{x=\delta} aV_r + \Phi_{vo}c_{Ao} - \Phi_{v1}\bar{c}_A - k_1\bar{c}_A V_{bL} \quad \text{(VII.70)}$$

with V_{bL} = bulk volume of reaction phase (m³).

Defining the ratio between the total reaction phase volume and the reaction phase film volume as the 'Hinterland ratio' Al:

$$Al = \frac{1-\varepsilon}{a\delta} = \frac{(1-\varepsilon)k_L}{aD_A} \quad \text{(VII.71)}$$

we obtain, after rewriting,

$$\left(\frac{\partial \bar{c}_A}{\partial t} + k_1\bar{c}_A\right)\left(\frac{Al-1}{Al}\right) = \frac{D_A \gamma a}{(1-\varepsilon)} + \frac{\Phi_{vo}c_{Ao}}{(1-\varepsilon)V_r} - \frac{\bar{c}_A}{\tau_L} \quad \text{(VII.72)}$$

with τ_L, the residence time of the reaction phase, given by $\tau_L = (1-\varepsilon)V_r/\Phi_{V1}$. In equation (VII.72), γ and $\bar{c}_A$ are unknown. In Section VII.5.2 relations have been derived for γ as $f(\bar{c}_A)$ in equation (VII.61) and J_A as $f(\bar{c}_A)$ in equation (VII.59). From these three equations J_A and $\bar{c}_A$ can be obtained. At stationary conditions,

$\partial \bar{c}_A/\partial t = 0$. Then, substitution of equation (VII.61) into equation (VII.72) gives

$$\frac{\bar{c}_A}{c_{Ai}} = \frac{\dfrac{1}{\cosh \phi} + \dfrac{\Phi_{vo} c_{Ao}}{\Phi_{v1} c_{Ai}} \dfrac{1}{k_1 \tau_L} Al\phi \tanh \phi}{\left\{\dfrac{1}{k_1 \tau_L} + \dfrac{Al-1}{Al}\right\} Al\phi \tanh \phi + 1} \qquad \text{(VII.73)}$$

with equation (VII.59) it follows that

$$J_A = \frac{D_A c_{Ai}}{\delta} \left[\frac{\phi}{\tanh \phi} - \frac{\dfrac{\phi}{\sinh \phi \cosh \phi} + \dfrac{\Phi_{vo} c_{Ao}}{\Phi_{v1} c_{Ai}} \dfrac{1}{k_1 \tau_L} \dfrac{Al\phi^2}{\cosh \phi}}{\left\{\dfrac{1}{k_1 \tau_L} + \dfrac{Al-1}{Al}\right\} Al\phi \tanh \phi + 1}\right]. \qquad \text{(VII.74)}$$

Part of A that reaches the bulk of the reaction phase is converted in the bulk. The remaining part leaves the reactor unconverted. The ratio is

$$\frac{k_1 \bar{c}_A V_{bL}}{\Phi_{v1} \bar{c}_A} = \frac{(Al-1)/Al}{1/(k_1 \tau_L)}.$$

For an efficient reactor, this ratio will be high. In case this ratio is low, the reactor acts more as an absorber or extractor. Therefore, in the following we will assume:

$$\frac{1}{k_1 \tau_L} \ll \frac{Al-1}{Al}. \qquad \text{(VII.75)}$$

For

$$k_L c_{Ai} a V_r \gg \Phi_{vo} c_{Ao}, \qquad \text{(VII.76)}$$

the amount of A transferred into the reaction phase will be much higher than the amount of A introduced into the reactor via the feed of the reaction phase. This will be often the case. It can be shown that condition (VII.76) is equivalent to

$$\frac{\phi}{\sinh \phi \cosh \phi} \gg \frac{\Phi_{vo} c_{Ao}}{\Phi_{v1} c_{Ai}} \frac{1}{k_1 \tau_L} \frac{Al\phi^2}{\cosh \phi}. \qquad \text{(VII.77)}$$

Thus, in case the reaction phase fed to and drawn from the reactor contains only minor amounts of A with respect to the amount transferred through the interface, equations (VII.73) and (VII.74) reduce to

$$\frac{\bar{c}_A}{c_{Ai}} = \frac{1}{(Al-1)\phi \sinh \phi + \cosh \phi} \qquad \text{(VII.78)}$$

and

$$J_A = c_{Ai} \sqrt{(k_1 D_A)} \frac{(Al-1)\phi + \tanh \phi}{(Al-1)\phi \tanh \phi + 1} \qquad \text{(VII.79)}$$

From equations (VII.78) and (VII.79), the enhancement factor E_A can be calculated:

$$E_A = \frac{J_A}{k_L(c_{Ai} - \bar{c}_A)} = \phi \left[\frac{(Al-1)\phi + \tanh\phi}{(Al-1)\phi \tanh\phi + 1 - 1/\cosh\phi} \right]. \quad \text{(VII.80)}$$

The unknown interface concentration c_{Ai} is eliminated via the balance $-J_{AG} = J_{AL}$ with $-J_{AG} = k_G(\bar{c}_{AG} - c_{AiG})$ and $c_{Ai} = mc_{AiG}$. Hence

$$J_A = \frac{k_G \bar{c}_{AG}}{1 + k_G/(k_L m E_A)}. \quad \text{(VII.81)}$$

The capacity of the reactor follows from

$$-\Phi_A = J_A a V_r. \quad [\text{kmol/s}] \quad \text{(VII.82)}$$

The mole balance of A over the non-reaction phase G is

$$(\Phi_{vo} c_{Ao})_G - (\Phi_{v1} \bar{c}_A)_G = -\Phi_A. \quad \text{(VII.83)}$$

With the set equations (VII.74), (VII.81), (VII.82) and (VII.83), the reactor is completely described. Below, the asymptotic solutions will be discussed.

Fast reaction ($\phi > 2$) For fast reactions ($\phi > 2$) equations (VII.78), (VII.79) and (VII.80) reduce to

$$J_A = \sqrt{(k_1 D)} c_{Ai}; \quad E_A = \phi; \quad \bar{c}_A = 0. \quad \text{(VII.84)}$$

Reactor capacity is proportional to total interfacial area (aV_r) and becomes independent of the film thickness δ and the mass transfer coefficient k_L. Reactor capacity is also independent of the bulk volume of the reaction phase, because this bulk is not effective, due to the absence of reactant in the bulk.

Slow reaction For slow reactions ($\phi < 0.3$) equations (VII.78), (VII.79) and (VII.80) reduce to

$$J_A = \frac{D}{\delta} c_{Ai} \frac{Al\phi^2}{(1 + Al\phi^2)}; \quad E_A = 1; \quad \frac{\bar{c}_A}{c_{Ai}} = \frac{1}{1 + Al\phi^2}. \quad \text{(VII.85)}$$

Mass transfer is not enhanced; depending on $Al\phi^2$ the bulk is saturated with A ($Al\phi^2 \ll 1$) or contains A only in a relatively low concentration ($Al\phi^2 \gg 1$). Figure VII.13 shows concentration profiles for several combinations of ϕ and $Al\phi^2$.

When A is supplied to the reaction phase by mass transfer only, the product $J_A a$ is equal to the overall conversion rate of A. It is of interest to compare this quantity with the greatest possible conversion rate which would occur if the entire reaction phase were in equilibrium with the interface ($= k_1 c_{Ai}(1 - \varepsilon)$). The ratio of these two conversion rates, which might be called the degree of utilization of the reaction phase, results from equation (VII.79).

$$\eta = \frac{J_A a}{k_1 c_{Ai}(1 - \varepsilon)} = \frac{1}{\phi Al} \frac{(Al-1)\phi + \tanh\phi}{(Al-1)\phi \tanh\phi + 1}. \quad \text{(VII.86)}$$

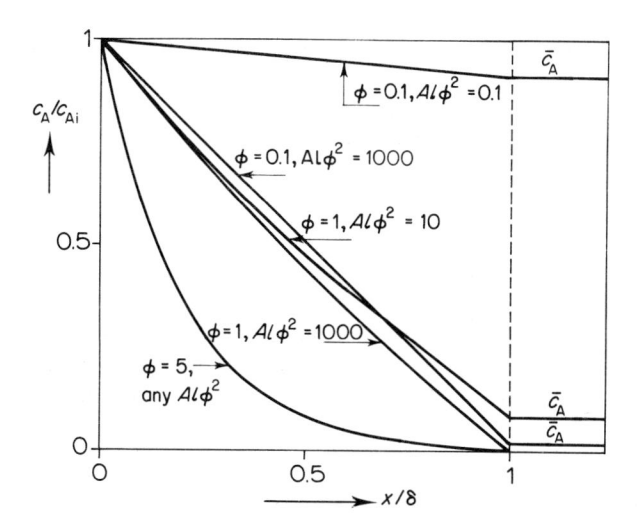

Fig. VII.13. Concentration distribution in a hypothetical mass transfer film for a first order reaction in A

For slow reaction ($\phi < 0.3$) and $Al \gg 1$, the degree of utilization becomes approximately equal to $1/(1 + Al\phi^2)$ or $\bar{c}_A/c_{Ai}$. Hence it follows that the degree of utilization approaches unity for $Al\phi^2 \ll 1$ and $\phi < 0.3$. For a maximum utilization of the reaction phase, the optimal Hinterland ratio will decrease with increasing ϕ; in other words, the faster the reaction is the larger the value of the specific contact area must be to ensure efficient use of the bulk of the reaction

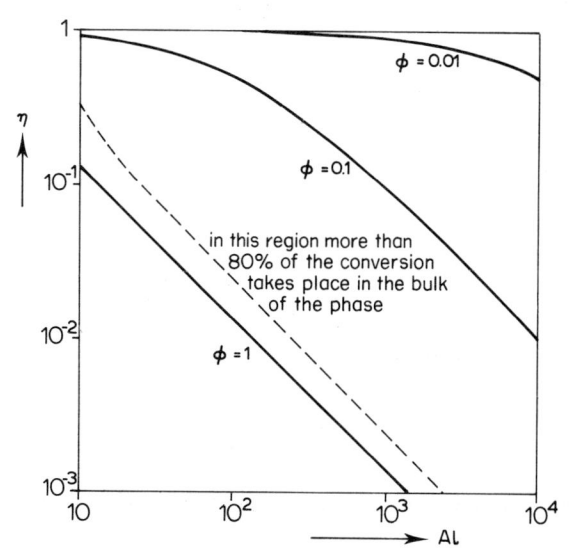

Fig. VII.14. Degree of utilization of reaction phase volume for mass transfer of A with reaction

phase. The Hinterland ratio strongly depends on the energy input per unit volume reactor through the stirrer. There is an optimization problem here. The cost of intensive stirring usually exceeds the cost of a larger reactor by far. Therefore a larger reactor with poor utilization of the bulk of the reaction phase is often a cheaper option than a smaller unit which consumes much more energy by intensive stirring.

Conclusion In case the degree of utilization approaches one ($\phi < 0.3$; $Al\phi^2 \ll 1$), the reactor capacity depends on the volume of the reaction phase and is independent of the value of the specific contact area. In case the degree of utilization is much smaller than one ($\phi > 2$ or $Al\phi^2 \gg 1$), reactor capacity is proportional to the interfacial area and independent of the volume of the reaction phase.

VII.6 MASS TRANSFER WITH HOMOGENEOUS IRREVERSIBLE REACTION OF COMPLEX KINETICS

In case the kinetics are complex, but still a function of c_A only, that is, for $R_A = R(c_A)$, a solution is obtained in the following way (film theory):

$$D_A \frac{d^2 c_A}{dx^2} + R_A = 0, \qquad (VII.87)$$

with boundary conditions

$$c_A = c_{Ai} \qquad \text{for } x = 0,$$

$$-\frac{dc_A}{dx} = \gamma \qquad \text{for } c_A = \bar{c}_A.$$

Introducing a new variable,

$$p \equiv \frac{dc_A}{dx},$$

and realizing that

$$\frac{d^2 c_A}{dx^2} = \frac{dp}{dx} = \frac{dp}{dc_A} \frac{dc_A}{dx} = p \frac{dp}{dc_A},$$

it follows that equation (VII.87) can be written as

$$p \frac{dp}{dc_A} = -\frac{R_A}{D_A}. \qquad (VII.88)$$

Since R_A is only a function of c_A, equation (VII.88) can be integrated. The result is

$$\left(\frac{dc_A}{dx}\right)_{x=0} = -\sqrt{\left(2 \int_{\bar{c}_A}^{c_{Ai}} \left(\frac{-R_A}{D_A}\right) dc_A + \gamma^2\right)} \qquad (VII.89)$$

while at any point in the film we have

$$\frac{dc_A}{dx} = -\sqrt{\left(2\int_{\bar{c}_A}^{c_A}\left(\frac{-R_A}{D_A}\right)dc_A + \gamma^2\right)} \equiv -F_A. \qquad (VII.90)$$

Equation (VII.90) contains two parameters that are still unknown: $\bar{c}_A$ and γ. The latter one, γ, follows from a balance over the bulk. In case the reaction phase fed to and drawn from the bulk contains only minor amounts of unconverted A, this balance is

$$aD_A\gamma = -R(\bar{c}_A)(1-\varepsilon)\left(\frac{Al-1}{Al}\right). \qquad (VII.91)$$

$\bar{c}_A$ follows theoretically from equation (VII.90) by integration:

$$\int_0^\delta dx = \delta = \int_{\bar{c}_A}^{c_{Ai}}\frac{dc_A}{F_A}. \qquad (VII.92)$$

Thus, the solution is obtained by solving the set equations (VII.90) and (VII.92) and a bulk balance such as equation (VII.91). The solution is not usually easy to obtain because equation (VII.92) is difficult to integrate. For fast reaction, however, the reactant cannot reach the plane $x = \delta$ and to a good approximation it can then be assumed that both $\bar{c}_A$ and γ approach zero. Hence, equation (VII.89) reduces to

$$\left(\frac{dc_A}{dx}\right)_{x=0} = \frac{-J_A}{D_A} = -\sqrt{\left(2\int_0^{c_{Ai}}\left(\frac{-R_A}{D_A}\right)dc_A\right)} \qquad (VII.93)$$

The integral in equation (VII.93) is usually easily obtained. Therefore, equation (VII.93) is very useful for arbitrary 'fast' kinetics of type $R_A = R(c_A)$. However, we still need a definition of what 'fast' means in arbitrary kinetics. For fast reactions ($\phi > 2$) and first order kinetics, it was found in Section VII.5.3 that

$$J_A = -D_A\left(\frac{dc_A}{dx}\right)_{x=0} = \frac{D_A}{\delta}c_{Ai}\phi, \qquad (VII.41)$$

suggesting a general definition for ϕ according to

$$\phi = \frac{-\delta}{c_{Ai}}\left(\frac{dc_A}{dx}\right)_{x=0}. \qquad (VII.94)$$

Combination of equations (VII.93) and (VII.94) results in

$$\phi = \frac{\delta}{c_{Ai}}\sqrt{\left(\frac{2}{D_A}\int_0^{c_{Ai}}-R_A\,dc_A\right)}. \qquad (VII.95)$$

Bischoff [11] showed that, in case a general ϕ is defined according to equation (VII.95),

$$J_A = \frac{D_A}{\delta}c_{Ai}\phi/\tanh\phi \qquad (VII.59)$$

holds for $\phi > 2$, whatever the kinetic rate equation may be. For the case of no bulk present (stagnant foam, stagnant drops and pellets for which $\gamma = 0$), an approximate solution is found, which holds for any ϕ. The starting point for this case is the effectiveness factor η (equation (VII.67)), which is defined in a general form as

$$\eta = \frac{-J_A}{\delta R(c_{Ai})}. \tag{VII.96}$$

From equation (VII.93), η becomes

$$\eta = \frac{-D_A}{\delta R(c_{Ai})} \sqrt{\left(2 \int_0^{c_{Ai}} \left(\frac{-R_A}{D_A}\right) dc_A\right)} \tag{VII.97}$$

Bischoff showed that equation (VII.97) holds for both fast and slow reactions, provided that $\gamma = 0$ [12]. That condition is always fulfilled in case of no bulk present, because then the plane $x = \delta$ becomes a plane of symmetry. Equation (VII.67) suggests, for fast reaction ($\phi > 2$)

$$\eta = \frac{1}{\phi}. \tag{VII.98}$$

To satisfy equation (VII.98), a modulus ϕ' is defined as the reciprocal of η, defined in equation (VII.97):

$$\phi' = \frac{-\delta R(c_{Ai})}{D_A} \left\{2 \int_0^{c_{Ai}} \left(\frac{-R_A}{D_A}\right) dc_A\right\}^{-1/2}. \tag{VII.99}$$

With this definition, equation (VII.67) holds in the form

$$J_A = -R(c_{Ai})\delta \frac{\tanh \phi'}{\phi'} = -R(c_{Ai})\delta\eta. \tag{VII.100}$$

for any value of ϕ' within 15%, provided that the kinetic order is not less than a half [12].

For nth order kinetics: $R_A = -k_n c_A^n$, it follows from equations (VII.95) and (VII.99) that

$$\phi = \delta \sqrt{\left(\frac{2}{n+1}\frac{k_n}{D_A}c_{Ai}^{n-1}\right)} \tag{VII.101}$$

and

$$\phi' = \delta \sqrt{\left(\frac{n+1}{2}\frac{k_n}{D_A}c_{Ai}^{n-1}\right)} = \frac{n+1}{2}\phi, \tag{VII.102}$$

while equation (VII.100), for these kinetics and $\phi \gg 2$, reduces to $J_A = \frac{D_A}{\delta} c_{Ai}\phi$. A plot of η against ϕ' for nth order kinetics with $n = 0, \frac{1}{2}, 1, 2$ and 3, is presented in figure VII.15.

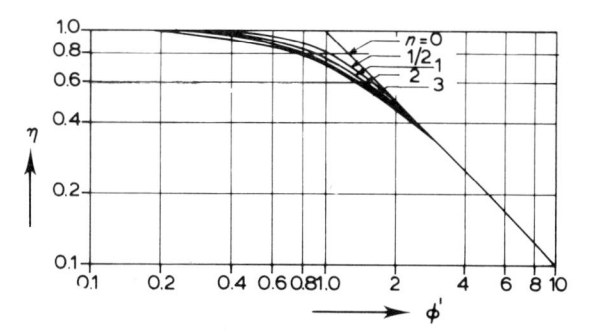

Fig. VII.15. General plot of the degree of utilization (η) against the reaction modulus (ϕ') for $R_A = -k_n C_A^n$ and no mixed bulk present. (From Bischoff [12].) (Reproduced by permission of the American Institute of Chemical Engineers)

Conclusion In case of no mixed bulk present, J_A is given by equation (VII.100), provided that ϕ' is defined according to equation (VII.99). In case a well-mixed bulk is present, J_A is given by equation (VII.59), provided that ϕ is defined according to equation (VII.95) and that $\phi > 2$. In case of a mixed bulk present and $0.3 < \phi < 2$, the solution can, in general, only be obtained by numerical procedures. For $\phi < 0.2$, the reaction is essentially a bulk reaction and a bulk balance in combination with the equation for physical mass transfer (VII.17) is sufficient to solve the problem.

Concerning more complex kinetics, Bischoff [12] discusses the Langmuir–Hinschelwood type of adsorption kinetics:

$$R_A = \frac{-c_A}{\dfrac{1}{k_1} + \dfrac{c_A}{k_0}}$$

and the case of a first order reaction with volume change. Sada and Kumazawa [13] and Bischoff [11] discuss the case of mass transfer with reaction of kinetics

$$R_A = \frac{-c_A^n}{\dfrac{1}{k_m} + \dfrac{c_A^n}{k_n}}.$$

VII.7 MASS TRANSFER WITH HOMOGENEOUS IRREVERSIBLE REACTION OF ORDER (1,1) WITH $Al \gg 1$

In the preceding three sections, equations were derived for the rate of mass transfer through an interface, under occurrence of reaction of general kinetics $R_A = R(c_A)$. In case a second reactant is involved, e.g. in: $A + v_B B \rightarrow v_P P$, the equations derived in Sections VII.5 and VII.6 hold as well, as long as c_B is constant all over the reaction zone. For instance, in kinetics $R_A = -kc_A c_B^n$, the group kc_B^n may be taken as a constant (k'), thus reducing the problem to (pseudo) first order kinetics of form $R_A = -k'c_A$. In general, however, c_B will not be a constant in the reaction zone. In this section we will discuss the system

$$A(1) \to A(2)$$

$$A(2) + v_B B(2) \to v_P P(2) + v_Q Q(2); \qquad R_A = -k_{1,1} c_A c_B$$

with (1) and (2) phase (1) and (2) respectively. Three limiting cases are possible:

Slow reaction There is mention of 'slow reaction' in case the rate of reaction is slow with respect to the rate of mass transfer. Characteristic of this regime is that the rate of mass transfer is not enhanced by the occurring reaction and that the reaction mainly proceeds in the bulk of the reaction phase. The degree of utilization is among other things dependent on the Hinterland ratio (*Al*), see Fig. VII.16.

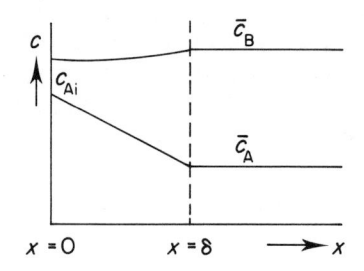

Fig. VII.16. Concentration profiles in slow reaction (film theory)

Fast reaction There is mention of 'fast reaction' in case the rate of reaction is so high, with respect to the rate of mass transfer, that A is completely converted near the interface. Then the rate of mass transfer is enhanced by the chemical reaction and $\bar{c}_A = 0$, see figure VII.17.

Instantaneous reaction This in fact is an asymptotic solution of the previous case. Now, the reaction is so high, that the conversion rate is completely limited by diffusion of both A and B into the reaction zone. A and B do not occur at the same place. The place where A and B meet each other is known as the reaction plane

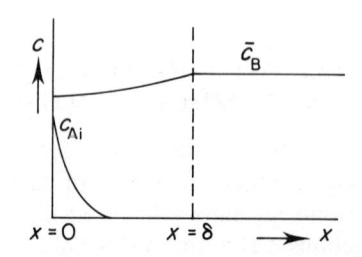

Fig. VII.17. Concentration profiles in fast reaction (film theory)

(δ_r) in which both c_A and c_B are zero. In this case the enhancement factor E_A reaches its maximal value; see figure VII.18. Below, each case will be analysed in detail.

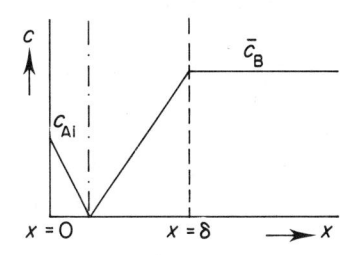

Fig. VII.18. Concentration profiles in instantaneous reaction (film theory)

VII.7.1 Slow reaction

The rate of mass transfer through the interface is not enhanced by reaction, which mainly takes place in the bulk of the reaction phase. This regime is identical to the case $\phi < 0.3$, for reactions of order (1,0) (Section VII.5). However, there is one difference. As the (pseudo) first order reaction rate constant k_1 is given by

$$k_1 = k_{1,1}\bar{c}_B, \qquad (\text{VII.103})$$

ϕ is now defined as

$$\phi = \delta\sqrt{(k_{1,1}\bar{c}_B/D_A)}. \qquad (\text{VII.104})$$

This implies that in reactors where the bulk composition of B depends on place, ϕ will be place dependent as well. For such reactors, it must be checked that the condition $\phi < 0.3$ holds at any place in the reactor.

In line with results obtained in Section VII.5 (application 2, $\phi < 0.3$), the degree of utilization is given by equation (VII.86), provided negligible amounts of A enter or leave the reactor with the reaction phase. Hence, for $Al \gg 1$ and constant bulk concentration,

$$\eta \simeq \frac{\bar{c}_A}{c_{Ai}} = \frac{1}{1 + Al\phi^2}. \qquad (\text{VII.105})$$

Notice that slow reaction does not exclude a concentration gradient of B in the film (speaking in terms of film theory). However, as long as $Al \gg 1$, possible concentration gradients of B in the film are not of interest in calculating the total conversion rate because the reaction mainly takes place in the bulk, even at a low value of $\bar{c}_A/c_{Ai}$. This is a consequence of the value of E_A which approaches one, indicating that

$$-D_A\left(\frac{\partial c_A}{\partial x}\right)_{x=0} \simeq -D_A\left(\frac{\partial c_A}{\partial x}\right)_{x=\delta}.$$

VII.7.2 Fast reaction

For ϕ, as defined in equation (VII.104), larger than 2, the reaction exclusively takes place near the interface. In terms of film theory, all reaction occurs in the film. So far, the situation is identical with the previously treated case of first order reactions. A complicating factor now is that the value of the enhancement factor, E_A, is influenced by whether diffusion limitation of B occurs in the reaction zone. Absence of diffusion limitation of B means in this case that, though the reaction is fast with respect to mass transfer of A, it is slow with respect to mass transfer of B. If this condition is fulfilled, c_B equals $\bar{c}_B$ all over the film. Then, the reaction can be treated as a pseudo first order reaction of order (1,0) with k_1 given by

$$k_1 = k_{1,1}\bar{c}_B. \quad \text{(VII.103)}$$

The criterion determining when this approach is allowed and when not will be derived below:

The assumption of a constant c_B ($=\bar{c}_B$), all over the film, is allowed if

$$\frac{\bar{c}_B - c_{Bi}}{\bar{c}_B} \ll 1; \quad \text{(VII.106)}$$

see figure VII.19. The reaction mainly takes place near the interface in a small zone of thickness δ_r, with

$$\delta_r \equiv -\frac{c_{Ai}}{(dc_A/dx)_{x=0}}; \quad \text{(VII.107)}$$

see figure VII.19. The conversion rate of B in the reaction zone must equal the transport rate of B from the bulk to the reaction zone $0 < x < \delta_r$.

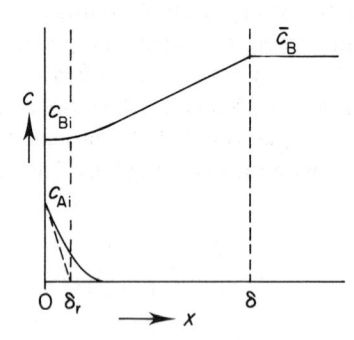

Fig. VII.19. General concentration profile for $\phi > 2$ (film theory)

Assume, as a conservative simplification, that all B must diffuse to the interface:

$$J_B \simeq \frac{D_B}{\delta}(\bar{c}_B - c_{Bi}) \quad \text{(VII.108)}$$

Approximating further, all over the reaction zone $c_B \simeq c_{Bi}$ gives

$$J_A = \frac{D_A}{\delta}c_{Ai}\phi_i, \quad \text{(VII.109)}$$

with $\phi_i = \delta\sqrt{(k_1 c_{Bi}/D_A)}$. Combining equations (VII.108) and (VII.109) via $J_A = J_B/v_B$ gives

$$\frac{D_A}{\delta} c_{Ai} \phi_i \simeq \frac{D_B}{\delta v_B}(\bar{c}_B - c_{Bi}). \qquad (VII.110)$$

From this result, it follows that

$$\frac{\bar{c}_B - c_{Bi}}{\bar{c}_B} \ll 1 \quad \text{for} \quad \phi \ll \frac{D_B \bar{c}_B}{v_B D_A c_{Ai}}. \qquad (VII.111)$$

Conclusion:

$$E_A = \phi \left(= \delta \sqrt{\left(\frac{k_{1,1} \bar{c}_B}{D_A}\right)} \right) \quad \text{for} \quad 2 < \phi \ll \frac{D_B \bar{c}_B}{v_B D_A c_{Ai}}. \qquad (VII.112)$$

With $\delta = D_A/k_L$, ϕ is called the Hatta number for reactions of order (1,1).

In case ϕ is not much smaller than $D_B \bar{c}_B/v_B D_A c_{Ai}$, diffusion limitation of B occurs. Then, maximum transport rate of B, due to diffusion ($D_B \bar{c}_B/\delta$), is slow compared to maximum conversion rate of B in the reaction zone ($v_B D_A c_{Ai} \phi/\delta$) and c_{Bi} is substantially smaller than $\bar{c}_B$. With increasing value of the ratio $\phi/(D_B \bar{c}_B/v_B D_A c_{Ai})$, the concentration profile picture changes from figure VII.20a to VII.20d, the latter being the asymptotic picture for instantaneous reaction. In

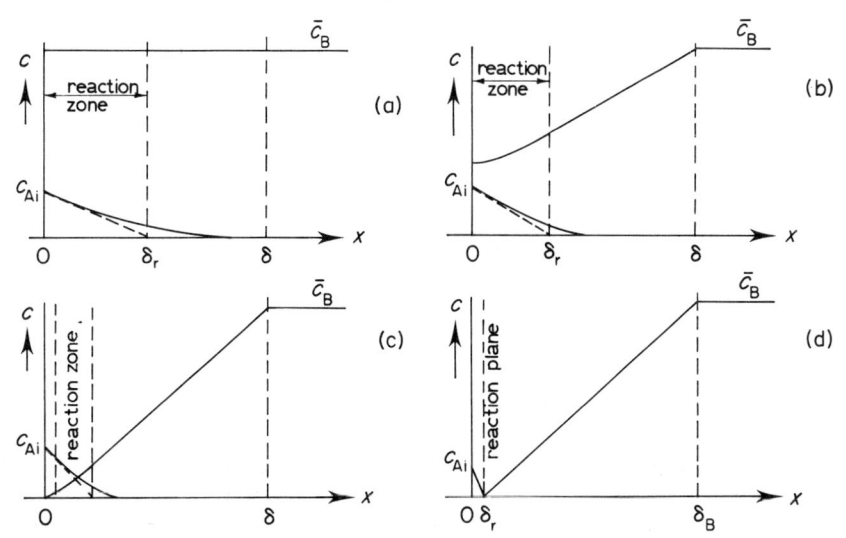

Fig. VII.20. Mass transfer with irreversible homogeneous chemical reaction of order (1,1):

(a) $2 < \phi \ll \dfrac{D_B \bar{c}_B}{v_B D_A c_{Ai}}$, $E_A = \phi$; (b) $2 < \phi \simeq \dfrac{D_B \bar{c}_B}{v_B D_A c_{Ai}}$, E_A from equations (VII.136) and (VII.138) or from Figure VII.22; (c) $2 < \phi \simeq \dfrac{5 D_B \bar{c}_B}{v_B D_A c_{Ai}}$, E_A from equations (VII.36) and (VII.138) or from Figure VII.22; (d) $\phi > 2$, $\phi > \dfrac{10 D_B \bar{c}_B}{v_B D_A c_{Ai}}$, instantaneous reaction,

$$E_A = 1 + \frac{D_B \bar{c}_B}{v_B D_A c_{Ai}} \quad \text{(equation (VII.116))}$$

case diffusion limitation of B occurs, the observed conversion rate will be lower than without diffusion limitation of B. Therefore, in case of diffusion limitation of B, E_A will be smaller than ϕ. The exact expression for E_A in that case will be presented in Section VII.7.4. To which extent diffusion limitation of B occurs depends on the mass transfer coefficient of B (D_B/δ or k_{LB}). This implies that, in contrast to the case $2 < \phi \ll D_B \bar{c}_B / v_B D_A c_{Ai}$, J_A is no longer independent of k_L. For $\phi \gg D_B \bar{c}_B / v_B D_A c_{Ai}$, the third limiting case, that of instantaneous reaction, occurs (see Section VII.7.3).

VII.7.3 Instantaneous reaction

In case the rate of reaction is fast compared to the rate of mass transport of both A and B, the reaction is said to be 'instantaneous'. In the preceding section it has been derived that this case occurs for both $\phi > 2$ and $\phi \gg D_B \bar{c}_B / v_B D_A c_{Ai}$. Then, the reaction zone reduces to a plane at $x = \delta_r$: A and B react so fast that they are not found at a same place. Transport of both A and B to the plane $x = \delta_r$ is diffusion driven. In terms of film theory (see figure VII.20d)

$$J_A = \frac{D_A}{\delta_r} c_{Ai},$$

$$J_B = \frac{D_B}{\delta - \delta_r} \bar{c}_B \tag{VII.113}$$

$$J_A = \frac{J_B}{v_B}. \tag{VII.114}$$

Elimination of δ_r gives

$$J_A = \frac{D_A}{\delta} c_{Ai} \left(1 + \frac{D_B \bar{c}_B}{v_B D_A c_{Ai}}\right). \tag{VII.115}$$

Thus

$$E_A = 1 + \frac{D_B \bar{c}_B}{v_B D_A c_{Ai}} \equiv E_{A\infty}, \tag{VII.116}$$

where $E_{A\infty}$ is the maximum possible enhancement factor. Notice that the rate of mass transfer through the interface is no longer a function of the reaction rate. This is a consequence of the fact that the transfer rate is completely limited by mass transfer. In contrast with the regime $2 < \phi \ll D_B \bar{c}_B / v_B D_A c_{Ai}$, J_A is again a function of $\delta = D/k_L$, that is, J_A is dependent on the hydrodynamics of the system.

Conclusion

Due to chemical reaction, the rate of mass transfer through the interface may be enhanced from $J_A = (D_A/\delta)c_{Ai}$ (for $\phi < 0.3$) to (according to film theory)

$$J_A = \frac{D_A}{\delta} c_{Ai} + \frac{D_B}{v_B \delta} \bar{c}_B \quad \text{for} \quad 2 < \phi \gg \frac{D_B \bar{c}_B}{v_B D_A c_{Ai}}. \tag{VII.117}$$

It has been shown [14] that, for $D_B = D_A$, equation (VII.117) holds for any geometry, any fluid flow model and any reaction fast enough to deplete the surface of reactant B. Differences occur, however, for $D_A \neq D_B$. Higbie's penetration model assumes that the contact time τ is only a function of hydrodynamics. With this assumption it follows $k_L \sim \sqrt{D}$ (equation (VII.31)). In terms of film theory ($\delta = D/k_L$), this means $\delta \sim \sqrt{D}$. Thus

$$\delta_B = \delta_A \sqrt{\left(\frac{D_B}{D_A}\right)} \tag{VII.118}$$

The relevant δ in equation (VII.113) is δ_B; see figure VII.20d. Hence, equation (VII.115) must be written as

$$J_A = \frac{D_A}{\delta_B} c_{Ai} \left(1 + \frac{D_B \bar{c}_B}{v_B D_A c_{Ai}}\right). \tag{VII.119}$$

Combining equations (VII.118) and (VII.119) gives

$$J_A = \frac{D_A c_{Ai}}{\delta_A} \left(1 + \frac{D_B \bar{c}_B}{v_B D_A c_{Ai}}\right) \left(\frac{D_A}{D_B}\right)^{0.5} \tag{VII.120}$$

and

$$E_A = \left(1 + \frac{D_B \bar{c}_B}{v_B D_A c_{Ai}}\right) \left(\frac{D_A}{D_B}\right)^{0.5}. \tag{VII.121}$$

This, indeed is the solution for instantaneous reaction, according to the penetration model. In cases where the mass transfer coefficient is described by the equation

$$Sh = c Re^m Sc^n, \tag{VII.122}$$

it can be generally stated [15, 16], that the enhancement factor at instantaneous reaction is, at least in good approximation:

$$E_{A\infty} \simeq \left(1 + \frac{D_B \bar{c}_B}{v_B D_A c_{Ai}}\right) \left(\frac{D_A}{D_B}\right)^n \tag{VII.123}$$

$n = \frac{1}{2}$ in the penetration model,
$n = \frac{1}{3}$ for boundary layer flow
$n = \frac{1}{3} - \frac{1}{4}$ for fully developed turbulent flow.

In case $\bar{c}_B/v_B c_{Ai}$ is not much larger than one, the approximation

$$E_{A\infty} \simeq 1 + \frac{D_B \bar{c}_B}{v_B D_A c_{Ai}} \left(\frac{D_A}{D_B}\right)^n \tag{VII.124}$$

is preferred [15].

Especially at instantaneous reactions, high mass transfer rates are possible. Theoretically, Fick's law, which is the basis of all previous derivations, refers to a plane through which no net mass flux occurs [1]. Therefore, the effect of the drift flow (also called the convective term) is neglected in the previous analyses. At high mass fluxes, however, this drift flow can be quite substantial. Stepanek and Shilimkan [17] studied the effect of the drift flow on the enhancement factor in absorption with instantaneous reaction, on the basis of film theory and obtained (for $v \neq 0$)

$$E_{A\infty} = 1 - \frac{D_B}{vD_A} \frac{\ln[1 + (v/v_B)\bar{x}_B]}{\ln[1 - x_{Ai}]} \qquad \text{(VII.125)}$$

with $v = v_P + v_Q - v_B$. \hfill (VII.126)

For $A + B \to P$, $v = 0$. In that case

$$E_{A\infty} = 1 - \frac{D_B}{v_B D_A} \frac{\bar{x}_B}{\ln(1 - x_{Ai})}. \qquad \text{(VII.127)}$$

For $x_{Ai} \ll 1$, equation (VII.127) reduces to

$$E_{A\infty} = 1 + \frac{D_B}{v_B D_A} \frac{\bar{x}_B}{x_{Ai}},$$

which is in agreement with equation (VII.119), provided that the molar density is constant. For $v \geq 0$, the enhancement factor is lowered by the occurring drift flow; for $v < 0$ the direction of the effect depends on $\bar{x}_B$ and x_{Ai}.

VII.7.4 General approximated solution

The general solution follows from solving the differential equations describing the reaction and mass transport process near the interface. For the stagnant film model,

$$D_A \frac{d^2 c_A}{dx^2} - k_{1,1} c_A c_B = 0, \qquad \text{(VII.128)}$$

$$D_B \frac{d^2 c_B}{dx^2} - v_B k_{1,1} c_A c_B = 0, \qquad \text{(VII.129)}$$

with boundary conditions $c_A = c_{Ai}$ for $x = 0$
$c_A = \bar{c}_A$ for $x = \delta$
$c_B = \bar{c}_B$ for $x = \delta$
$dc_B/dx = 0$ for $x = 0$.

The latter boundary condition states that no B is transferred to the non-reaction phase, that is, that B is not volatile (in gas–liquid systems) or not soluble in the non-reaction phase (in liquid–liquid systems). If B is transferred to the non-reaction phase, a mass balance for B over the non-reaction phase is necessary as well, with both B-balances coupled via $J_B = -D_B(dc_B/dx)_{x=0}$. The approximated solution has been presented by van Krevelen and Hoftijzer [18]. In recent years, their method of approximation has been successfully applied in solving many complex problems in mass transfer with reaction. Therefore we will deal with their method in detail. The essence of the van Krevelen and Hoftijzer method is the approximation of the concentration profile of component B by a constant c_{Bi} all over the reaction zone; see figure VII.21. The higher the value of δ/δ_r, the better the approximation is. With this approximation, the reaction becomes essentially pseudo first order with $k_1' = k_{1,1} c_{Bi}$. The solution is given by equation (VII.59):

$$J_A = \frac{D_A}{\delta} c_{Ai} \frac{\phi_i}{\tanh \phi_i} \left(1 - \frac{\bar{c}_A}{c_{Ai} \cosh \phi_i}\right), \qquad (VII.130)$$

with ϕ_i the value of ϕ at interface concentration c_{Bi}:

$$\phi_i = \delta \sqrt{\left(\frac{k_{1,1} c_{Bi}}{D_A}\right)} = \phi \sqrt{\left(\frac{c_{Bi}}{\bar{c}_B}\right)}. \qquad (VII.131)$$

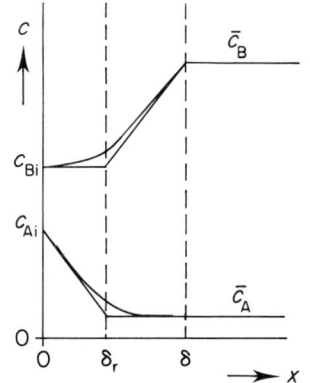

Fig. VII.21. Concentration profile approximation according to van Krevelen and Hoftijzer [18]

An enhancement factor, E_{A0}, is defined according to

$$J_A = -D_A \left(\frac{dc_A}{dx}\right)_{x=0} \equiv \frac{D_A}{\delta} c_{Ai} E_{A0}. \qquad (VII.132)$$

Notice that E_{A0} only equals the enhancement factor E_A in case $\bar{c}_A = 0$. The reaction zone is defined in the usual way:

$$\delta_r = -\frac{c_{Ai}}{(dc_A/dx)_{x=0}}. \qquad (VII.107)$$

Combination of equations (VII.107) and (VII.132) gives

$$E_{A0} = \frac{\delta}{\delta_r}. \qquad (VII.133)$$

The solution of the problem, equations (VII.130) and (VII.131), contains the still unknown concentration of B at the interface (c_{Bi}). This concentration is determined from the mass balance:

$$J_A = \frac{J_B}{v_B}. \qquad (VII.134)$$

Transport of B to the reaction zone proceeds by diffusion:

$$J_B = \frac{D_B}{\delta - \delta_r}(\bar{c}_B - c_{Bi}); \qquad (VII.135)$$

see figure VII.21. Substitution of equations (VII.132) with (VII.133) and (VII.135), in equation (VII.134), gives

$$\frac{D_A}{\delta_r}(c_{Ai} - \bar{c}_A) = \frac{D_B}{v_B(\delta - \delta_r)}(\bar{c}_B - c_{Bi}),$$

so that

$$\frac{c_{Bi}}{\bar{c}_B} = 1 - \left(\frac{\delta - \delta_r}{\delta_r}\right)\frac{v_B D_A(c_{Ai} - \bar{c}_A)}{D_B \bar{c}_B} \simeq \left(1 - (E_{A0} - 1)\frac{v_B D_A c_{Ai}}{D_B \bar{c}_B}\right)$$

or

$$\frac{c_{Bi}}{\bar{c}_B} \simeq \frac{E_{A\infty} - E_{A0}}{E_{A\infty} - 1}. \qquad (VII.136)$$

Substitution in equation (VII.131) gives ϕ_i as a function of known parameters:

$$\phi_i = \phi\sqrt{\left(\frac{E_{A\infty} - E_{A0}}{E_{A\infty} - 1}\right)}. \qquad (VII.137)$$

Combination of equations (VII.130), (VII.132) and (VII.137) gives

$$E_{A0} = \frac{\phi\sqrt{[(E_{A\infty} - E_{A0})/(E_{A\infty} - 1)]}}{\tanh \phi\sqrt{[(E_{A\infty} - E_{A0})/(E_{A\infty} - 1)]}}$$

$$\times \left(1 - \frac{\bar{c}_A}{c_{Ai} \cosh \phi\sqrt{[(E_{A\infty} - E_{A0})/(E_{A\infty} - 1)]}}\right). \qquad (VII.138)$$

From equation (VII.138), the enhancement factor E_{A0}, and hence J_A, is obtained by trial and error.

The three limiting solutions for slow (Section VII.7.1), fast (Section VII.7.2) and instantaneous reaction (Section VII.7.3) can all be derived from equation (VII.138). The accuracy of the approximated solution is remarkably good: the deviation from the exact numerical solution is less than 3% [19].

Figure VII.22 presents the solution for $\bar{c}_A = 0$, that is for $Al\phi^2 \gg 1$. As shown by Brian et al. [20], the Higbie penetration model solution is nearly identical to

403

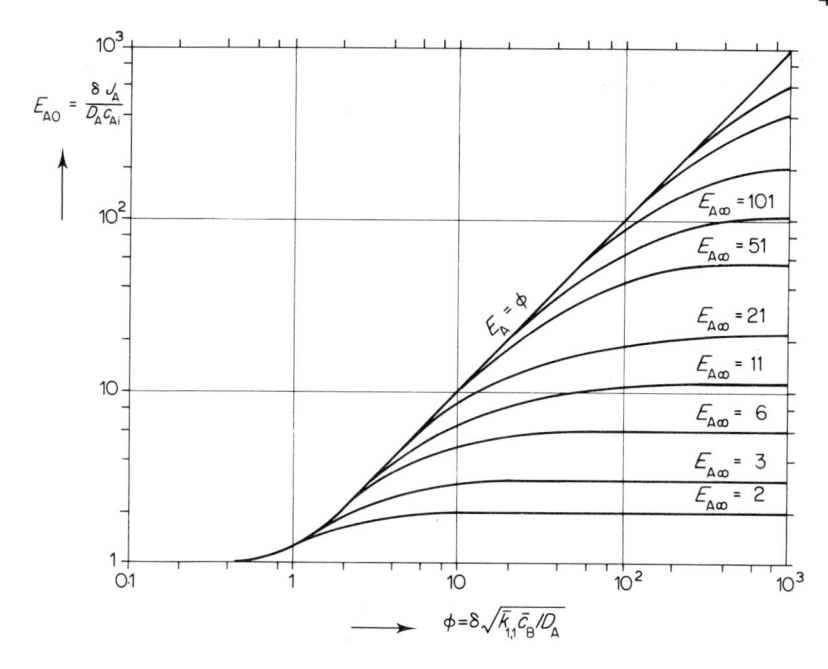

Fig. VII.22. Enhancement factor E_A for (1,1) order kinetics with $\bar{c}_A = 0$ [From van Krevelen and Hoftijzer [18]. Reproduced by permission of Recueil des travaux chimiques des Pays-Bas]

the film model solution, provided that for the instantaneous enhancement factor ($E_{A\infty}$), the corresponding expression for the penetration theory is used (equation (VII.121)). The expressions derived for the flux J_A and for $E_{A\infty}$ contain the interfacial concentration c_{Ai}. Unless the resistance against mass transfer in the phase where A originates from is negligible (e.g. if that phase contains only pure A) c_{Ai} itself is a function of J_A. Then c_{Ai} has to be eliminated with a flux expression for the non-reaction phase. In general, this only can be done numerically; see Example VII.7.4.a.

Example VII.7.4.a *The influence of a gas phase resistance on the flux through an interface for a reaction of order* (1,1)

Gas phase G contains a mixture of inert nitrogen and a reactive component A with concentration $c_A = 0.01 \text{ kmol/m}^3$. The gas is bubbled through a liquid in which A absorbs and is converted by reaction with B according to 1,1-kinetics. The fraction absorbed is rather small, so that $\bar{c}_{AG}$ is constant $= 0.01 \text{ kmol/m}^3$. It is assumed that the bulk concentration of A in the liquid is zero. The solubility of A is $m = 1$. Further data are: $k_G = 10^{-2} \text{ m/s}$, $k_{LA} = k_{LB} = 10^{-4} \text{ m/s}$, $D_{BL} = D_{AL} = 10^{-9} \text{ m}^2/\text{s}$ and $v_B = 1$. If the reaction is an instantaneous acid–base reaction under all practical circumstances we will derive expressions for J as a function of the input data and of $\bar{c}_B$ which is varied from 10^{-3} kmol/m^3 up to

10 kmol/m³. Further, we will construct a graph of $J_A/mk_L\bar{c}_{AG}$ as a function of $\bar{c}_B/\bar{c}_{AG}$, and also for values of k_G/k_L of 10, 2 and 1.

The partial fluxes of A in the liquid and gas phase are, respectively,

$$J_A = k_L\left\{mc_{AiG} + \frac{D_B}{v_B D_A}\bar{c}_B\right\} \tag{a}$$

and

$$J_A = k_G(\bar{c}_{AG} - c_{AiG}). \tag{b}$$

Elimination of c_{AiG} gives

$$\frac{J_A}{mk_L\bar{c}_{AG}} = \left\{1 + \frac{D_B}{v_B D_A}\frac{\bar{c}_B}{m\bar{c}_{AG}}\right\}\frac{1}{(1 + mk_L/k_G)}. \tag{c}$$

However, equation (c) holds only as long as

$$J_A < k_G(\bar{c}_{AG} - 0), \tag{d}$$

being the maximum possible flux through the interface. This limitation cannot be found by equation (c) because that equation has been derived from equation (a), in which we assumed that both A and B are transported to a reaction plane where both concentrations are zero. However, for $c_{AiG} = 0$, the reaction plane has reached the interface. In that case

$$k_G\bar{c}_{AG} = k_L\frac{D_B}{v_B D_A}\bar{c}_B,$$

so that for

$$\frac{D_B\bar{c}_B}{v_B D_A\bar{c}_{AG}} > \frac{k_G}{k_L}$$

c_{Bi} is no longer zero for complete gas phase resistance. Figure VII.23 gives concentration profiles for $k_{GA} = k_{LA} = k_{LB}$ and $v_B = m = 1$. From the above, it follows that J_A is given either by equation (c) or by

$$J_A = k_G\bar{c}_{AG}, \tag{d}$$

whichever gives the lower value for J_A. The result is presented in figure VII.24 as a dimensionless flux, $J_A/(k_L m\bar{c}_{AG})$, as a function of the values of the dimensionless groups that govern this problem: $D_B\bar{c}_B/(v_B D_A m\bar{c}_{AG})$ and k_G/mk_L. Note that if the rate of reaction is no longer instantaneous, equation (b) must be solved simultaneously with equation (VII.138). This can only be done numerically.

Figure VII.25 gives results for mass transfer with reaction of order (1,1) for $mk_L/k_G = 0.01$, $D_B\bar{c}_B/v_B D_A m\bar{c}_{AG} = 200$ and $\bar{c}_{AL} = 0$ with ϕ varying from 0.1 to 1000. From the theory presented in the next section it follows that the figure will be also approximately valid for kinetics of order (n, m). Notice that for $\phi > 100$, E_A becomes smaller than ϕ because diffusion limitation of B becomes important. The difference between E_A and $J_A/k_L m\bar{c}_{AG}$ is caused by gas phase mass transfer

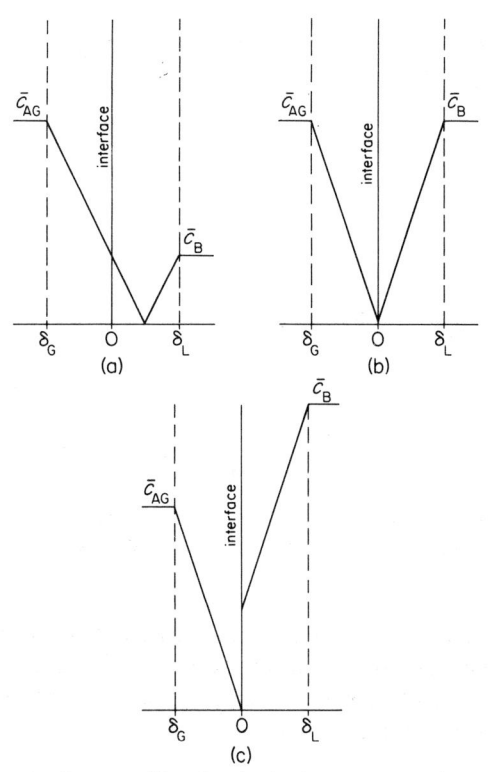

Fig. VII.23. Concentration profiles for instantaneous reaction with mass transfer resistance in two phases (G and L): (a) $D_B \bar{c}_B / v_B D_A m \bar{c}_{AG} < k_G / m k_L$; (b) $D_B \bar{c}_B / v_B D_A m \bar{c}_{AG} = k_G / m k_L$; (c) $D_B \bar{c}_B / v_B D_A m \bar{c}_{AG} > k_G / m k_L$. In the drawings, $k_G = k_{LA} = k_{LB}$; $v_B = m = 1$

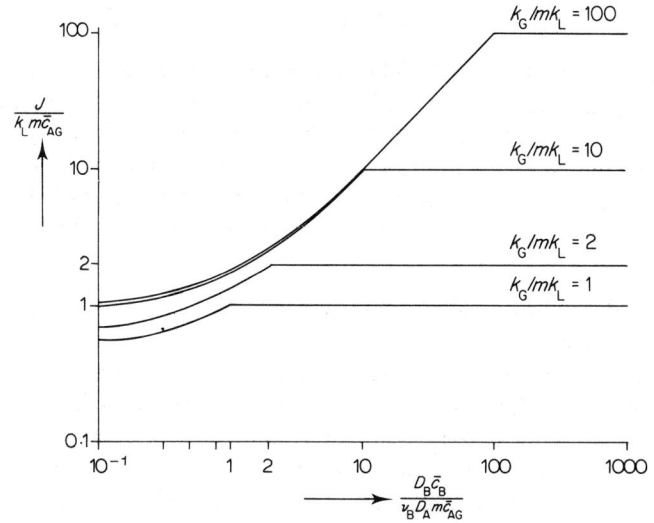

Fig. VII.24. The dimensionless flux $J/(k_L m \bar{c}_{AG})$ as a function of $D_B \bar{c}_B/(v_B D_A m \bar{c}_{AG})$ for instantaneous reaction with resistance against mass transfer in both phases

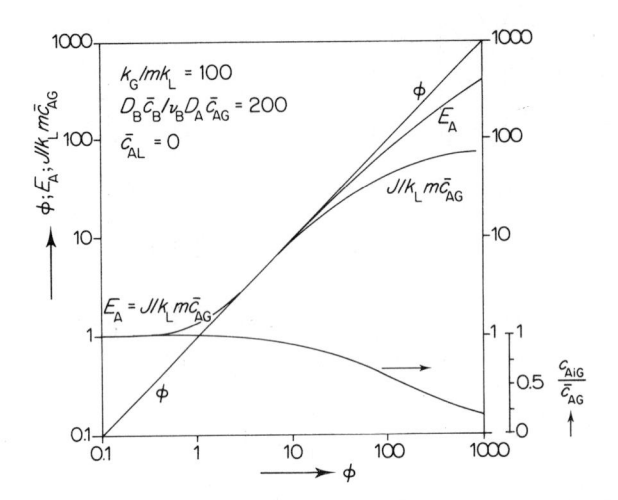

Fig. VII.25. Reaction of order (1,1) with mass transfer resistance in both phases. The influence of ϕ on E_A, $J/k_L m \bar{c}_{AG}$, and $c_{AiG}/\bar{c}_{AG}$

resistance. The increase of E_A for $\phi > 500$ may suggest that in that region the reaction is still not instantaneous: in fact the increase in E_A in this region is merely caused by the decrease of c_{Ai} with increasing ϕ, due to increasing dominance of gas phase resistance with increasing ϕ.

Example VII.7.4.b *Soap Ltd, soap boilers since 1775*
A traditional house produces an organic sulfonic acid P by sulphonation of an aromatic component B with liquid SO_3 (A) according to

$$A(L) + B(L) \rightarrow P(L), \quad R_A = -k_{1,1} c_A c_B.$$

The reaction is carried out in a CISTR. SO_3 is added in excess (5%) and is present as a separate phase. The reactor product flows through a cyclone for separation of the SO_3 excess while unconverted B (0.1 kmol/m³) remains in the product as a pollutant. Both B and P are insoluble in A. Solubility of A in the product phase is 3 kmol/m³. The droplet diameter is assumed to be independent of SO_3 hold-up; further $D_B/D_A = 1$. The reactor is operated with $\phi = 10$ and a Hinterland ratio $Al = 100$.

How much production capacity could be gained if the excess of SO_3 is doubled, keeping ζ_B the same?
Contradictory publications from the early sixties on the possible toxicity of B got the attention of the Ministry of Health in the early seventies. The Government proposed to reduce the amount of unreacted B in the surfactant by a factor of 25. Soap Ltd protested, arguing that only negligible profits were made, while its significance for local employment opportunities was still substantial. All this was in vain and a law became effective by January 1980. As it turned out, the company could still produce the same amount of product for the same price while meeting

the new specifications. How can this be explained? Because of this remarkable success, the Government forced a new reduction of unconverted B in P, just 2 years later, now by an extra factor of 400, without easing its strict product price control regulations. Soap Ltd operated at a substantial loss in 1982. Why?

All these questions can be answered as follows:
(a) We have $\phi = 10$ and $E_\infty = 1 + \bar{c}_B/c_{Ai} = 1.033$, so that the reaction is instantaneous. Doubling SO_3 excess doubles a_i and, in this case, also doubles the production capacity.

(b) Lowering $\bar{c}_B$ by a factor of 25 lowers ϕ by $\sqrt{25}$, so that now $\phi = 2$. The reaction is still instantaneous with $E_\infty = 1.001$. As a result, the production capacity is only 3% less than previously.

(c) After the last demand of the Government ϕ is reduced to 0.1, resulting in a slow reaction in the bulk of the liquid phase. The production rate is consequently reduced from

$$JaV_r = k_L(c_{Ai} - 0) E_\infty a V_r \text{ to } JaV_r = k_L(c_{Ai} - \bar{c}_A)aV_r$$

with

$$\frac{\bar{c}_A}{c_{Ai}} = \frac{1}{1 + Al\phi^2} = 0.5.$$

Hence the production capacity is reduced by a factor of 2. A second tank in series with the first one is necessary.

Soap had to borrow for the additional investments, but the additional costs of depreciation, maintenance, interest, etc. could not be compensated for by a price increase. Consequently the low profits turned into a loss in 1982.

Note that now doubling the excess of SO_3 will not result in doubling of the production capacity: with Al roughly proportional to $1/a$, the gain is only a factor 1.5. Note furthermore that the assumption of insolubility of B and P in A is often not realistic, particularly in sulphonation reactions.

Example VII.7.4.c *Chemical absorption of CO_2 from air in a packed column*
CO_2, (A), is scrubbed from air at 15 bars pressure by a 1 M NaOH solution flowing counter-currently through a packed column. Mass transfer coefficients, gas and liquid flow rates, temperature and column characteristics are the same as in Example VII.4. Again, relative gas inlet and outlet concentrations of the component to be stripped are 1 and 5.10^{-3} vol %, respectively. The absorbed CO_2 reacts in the liquid phase with OH^- ions (species B) according to a second order reaction $R_A = -k_{1,1}c_A c_B$. The relevant properties of this system can be found in the literature [21–23], which also contains sufficient information on the influence of the ionic strength of the solution. Further data are distribution coefficient, $m = c_{Ai}/(p_{Ai}/RT) = 0.535$, diffusivity of A in the liquid, $D_A = 1.77 \times 10^{-9} \text{ m}^2/\text{s}$, reaction rate constant at 20°C: $k_{1,1} = 5700 \text{ m}^3/\text{s.kmol}$. What is the height, L, of the column to meet the CO_2 outlet specification?

It appears that under the conditions given above, the reaction occurs entirely in the boundary region next to the liquid–gas interface. Hence, the bulk concentration of A in the liquid is zero. Taking into account the gas-side mass transfer coefficient and the liquid-side mass transfer coefficient multiplied by the chemical acceleration factor E_A, we obtain

$$J_A = \frac{p_A}{RT} \frac{mE_A k_L}{1 + mE_A k_L/k_G}. \tag{a}$$

The change in CO_2 content of the gas at the height z above the inlet is given by the material balance for A over the gas phase:

$$0 = -\Phi_{vG} d\left(\frac{p_A}{RT}\right) - J_A a S\, dz. \tag{b}$$

It has been assumed here that longitudinal dispersion in the gas flow can be neglected. The required length L of the column can be found by integrating (b) after (a) has been substituted:

$$\frac{SL}{\Phi_{vG}} = -\int_{p_{Ao}}^{p_{AL}} \frac{1 + mE_A k_L/k_G}{mE_A k_L} \frac{dp_A}{a p_A}. \tag{c}$$

On integration, the values of m, k_L, k_G and a may be assumed to be constant. The chemical acceleration factor E_A, however, depends on the values of $\phi = \sqrt{(k_{1,1} \bar{c}_B D_A / k_L^2)}$ and $\bar{c}_B/c_{Ai}$, both of which change along the column height. The value of c_{Ai} follows from p_{Ai}, which is found from

$$J_A = k_G \frac{p_A - p_{Ai}}{RT} \tag{d}$$

and $\bar{c}_B$ is calculated from the stoichiometric condition that for the conversion of each mole of CO_2 two moles of OH^- are needed:

$$\Phi_{vG} d\left(\frac{p_A}{RT}\right) = \tfrac{1}{2}\Phi_{vL}\, d\bar{c}_B \tag{e}$$

with the boundary condition that at the top of the column $\bar{c}_B = 1 \text{ kmol/m}^3$ and $p_{AL}/p = 50 \times 10^{-6}$. In equation (e), longitudinal dispersion in the liquid phase has been neglected.

E_A can now be found for various values of p_A by the simultaneous solution of equations (a), (d) and (e) and by using the theory of the preceding section. The results are shown in the Table VII.3.

It is seen from this table by comparing p_A and p_{Ai} that only 5 to 10% of the mass transfer resistance lies on the gas side. Moreover, the values of $\bar{c}_B/c_{Ai}$ are so high that the theory for first order reaction in A can be applied and that $E_A = \phi$. The values in the last column can be plotted against p_A and graphic or numerical integration according to equation (c) yields

$$\frac{SL}{\Phi_{vG}} = 34.4 \text{ s} \quad \text{or} \quad L = \frac{34.4 \times 0.0414}{0.212} = 6.75 \text{ m}.$$

Table VII.3. Conditions in CO_2 absorption column; Example VII.7.4.c

p_A	$\bar{c}_B$	p_{Ai}	$\sqrt{\dfrac{k_{1,1}\bar{c}_B D_A}{k_L^2}}$	$\bar{c}_B/c_{Ai}$	$E_A = \phi\,\dfrac{1 + mE_A k_L/k_G}{mE_A k_L a p_A}$
(N/m²)	(kmol/m³)	(N/m²)	$= \phi$			(10³ × m² s/N)
15 000	0.25	14 200	4.3	80	4.3	0.75
10 000	0.50	9 300	6.1	245	6.1	0.81
5 000	0.75	4 600	7.4	740	7.4	1.36
2 000	0.90	1 800	8.1	2 280	8.1	3.13
1 000	0.95	900	8.4	4 800	8.4	6.05
500	0.975	450	8.4	9 900	8.4	12.1
200	0.99	180	8.5	25 000| 8.5	29.9
100	0.995	90	8.5	51 000| 8.5	59.8
75	1.00	68	8.5	67 000| 8.5	79.7

It still has to be ascertained whether the assumptions of isothermal operation and of the absence of longitudinal dispersion in both phases are correct. The heat liberated by the reaction is approximately 90×10^6 J/kmol of CO_2 converted. This heat is mainly taken up by the liquid stream and it can be calculated that its temperature increase would be 8 °C. This would result in only a slight increase in absorption rate; although $k_{1,1}$ rapidly rises with increasing temperature, this effect is offset partly by a decrease of the solubility of CO_2.

With regard to longitudinal dispersion, Stemerding [24] mentions a value of $D_1 \approx 1.5 \times 10^{-3}$ m²/s for the flow of liquid over 0.013 m Raschig rings. For the liquid, therefore, the value of $N' = \langle v \rangle L/D_1$ will be of the order of 15. Longitudinal dispersion will also occur to some extent in the gas stream. It will therefore be safe to compensate for longitudinal dispersion in both phases by making the column about 20% longer than calculated.

VII.8 MASS TRANSFER WITH IRREVERSIBLE HOMOGENEOUS REACTION OF ARBITRARY KINETICS WITH $Al \gg 1$

For a system: $A(1) \to A(2)$

$$A(2) + v_B B(2) \to v_P P + v_Q Q \tag{VII.139}$$

with arbitrary kinetics: $R_A = R(c_A, c_B)$, an enhancement factor for mass transfer can, in principle, be found by applying the van Krevelen–Hoftijzer approximation. With this approximation the rate equation reduces to

$$R_A = R(c_A, c_B) \simeq R(c_A, c_{Bi}) = R(c_A). \tag{VII.140}$$

This way the problem has been reduced to the system treated in Section VII.6.

The solution will contain c_{Bi} as a parameter. This unknown interfacial concentration can always be eliminated via

$$\frac{c_{Bi}}{\bar{c}_B} \simeq \frac{E_{A\infty} - E_{A0}}{E_{A\infty} - 1}, \qquad \text{(VII.136)}$$

which has been derived without using the rate equation. Therefore the relation holds for any kinetics, as long as B does not react in the zone $\delta_r - \delta$. As an example, we will discuss the kinetics of order (n, m):

$$R_A = -k_{n,m} c_A^n c_B^m. \qquad \text{(VII.141)}$$

According to van Krevelen and Hoftijzer, the rate equation reduces to

$$R_A \simeq -k_{n,m} c_A^n c_{Bi}^m = R(c_A).$$

For $\phi > 2$, the solution is (see Section VII.6, equation VII.59)

$$E_{A0} = \phi_i / \tanh \phi_i, \qquad \text{(VII.142)}$$

with ϕ_i obtained by solving equation (VII.95)

$$\phi_i = \delta \sqrt{\left(\frac{2}{n+1} \frac{k_{n,m}}{D_A} c_{Ai}^{n-1} c_{Bi}^m\right)}. \qquad \text{(VII.143)}$$

Elimination of c_{Bi} by equation (VII.136) and substituting equation (VII.143) in equation (VII.142) results in

$$E_{A0} = \phi \left\{\frac{E_{A\infty} - E_{A0}}{E_{A\infty} - 1}\right\}^{m/2} \Big/ \tanh\left[\phi \left\{\frac{E_{A\infty} - E_{A0}}{E_{A\infty} - 1}\right\}^{m/2}\right] \qquad \text{(VII.144)}$$

with ϕ, analogously to equation (VII.101), the ϕ or Ha number for (n,m) order kinetics:

$$\phi = \delta \sqrt{\left(\frac{2}{n+1} \frac{k_{n,m}}{D_A} c_{Ai}^{n-1} \bar{c}_B^m\right)} \qquad \text{(VII.145)}$$

or

$$Ha = \frac{1}{k_L} \sqrt{\left(\frac{2}{n+1} k_{n,m} D_A c_{Ai}^{n-1} \bar{c}_B^m\right)}. \qquad \text{(VII.146)}$$

From equation (VII.144), which is closely analogous to the solution for (1,1) order kinetics (equation (VII.138)), E_{A0} is obtained by trial and error. Equation (VII,144) was first derived by Hikita and Asai [25]. They showed that the equation remains valid, to a good approximation, for $\phi < 2$, provided that $\bar{c}_A = 0$. For $\phi \ll E_{A\infty}$, there is no diffusion limitation of B and equation (VII.142) reduces to

$$E_{A0} = \frac{\phi}{\tanh \phi}, \qquad \text{(VII.147)}$$

which is identical to the previously derived enhancement factor for (pseudo) first order reactions with $\bar{c}_A = 0$ (equation (VII.59)). Contrary to the latter equation, however, equation (VII.147) is an approximated solution (accuracy within 10%). For fast reactions and no diffusion limitation of B ($2 < \phi \ll E_{A\infty}$) equation (VII.147) reduces to the well-known relation

$$E_{A0} = E_A = \phi.$$

In this case, the conversion rate per unit reactor volume follows from:

$$J_A a = \frac{D_A a}{\delta} c_{Ai} \phi \quad \text{for} \quad 2 < \phi \ll E_{A\infty}, \tag{VII.148}$$

or

$$J_A a = a \sqrt{\left(\frac{2 D_A k_{n,m}}{n+1}\right)} c_{Ai}^{(n+1)/2} \bar{c}_B^{m/2}. \tag{VII.149}$$

Equation (VII.149) shows how, due to diffusion limitation of A, the apparent overall kinetic reaction order changes to a value closer to one: order (0,0) changes to ($\frac{1}{2}$,0); order (1,1) to (1,$\frac{1}{2}$) and order (2,2) to ($\frac{3}{2}$,1).

For $2 < \phi \gg D_B \bar{c}_B / v_B D_A c_{Ai}$, the reaction becomes instantaneous. Then the general, kinetics independent solution for instantaneous reaction holds. Provided ϕ is defined according to equation (VII.145), the solution of equation (VII.144), presented in a van Krevelen–Hoftijzer plot E_{A0} as $f(\phi, E_{A\infty})$, is nearly independent of the order m and not very dependent on the order n, provided $0 \leq n \leq 2$ (maximum deviation for $n = 1$ in the order of 25% for $\phi \simeq E_{A\infty}$). Approximate solutions for (n, m) order kinetics according to penetration models have been presented as well. Hikita and Asai [25] analysed the problem for the Higbie and De Coursey [26] for the Danckwerts model. The latter is the more practical one because it presents E_{A0} in an explicit form:

$$E_{A0} = -\frac{\phi^2}{2(E_A - 1)} + \sqrt{\left(\frac{\phi^4}{4(E_{A\infty} - 1)^2} + \frac{E_{A\infty} \phi^2}{(E_{A\infty} - 1)} + 1\right)}. \tag{VII.150}$$

If $\bar{c}_A = 0$ and D_A and D_B are not very different, equation (VII.150) is accurate within 5%. Differences between film and penetration model solutions are practically negligible.

Conclusion The van Krevelen–Hoftijzer approximation forms, in combination with the theory on arbitrary kinetics $R_A = R(c_A)$, presented in Section VII.6, a valuable method for analysing mass transfer with arbitrary kinetics $R_A = R(c_A, c_B)$.

For reference purposes it is mentioned that the case of mass transfer with autocatalytic reaction

$$A(1) \to A(2), \quad A(2) + \dot{R} \to \dot{R} + \dot{R}$$

has recently been analysed both numerically [27] and analytically [28]. In the latter reference, an approximate solution is obtained by a profile substitution in the film of type:

$$\frac{c_A}{c_{Ai}} = \frac{\sinh[m(1-x)]}{\sinh m},$$

where m is a collocation parameter. The problem is of importance in gas–liquid oxidation and chlorination reactions.

VII.9 MASS TRANSFER WITH IRREVERSIBLE REACTION OF ORDER (1,1) FOR A SMALL HINTERLAND COEFFICIENT

With reaction kinetics $R_A = R(c_A)$, the theory from Sections VII.5 and 6 is, in principle, applicable for any value of Al. The situation may be different for kinetics $R_A = R(c_A, c_B)$. The main reason is that film and penetration theory assume a constant $\bar{c}_B$ either at $x = \delta$ (film theory) or at infinity (penetration model). For small values of Al and at high conversion of B, deviations may be expected due to depletion of reactant B: for example, in the centre of a drop.

The problem is of interest in liquid–liquid reactions and gas–liquid reactions carried out in spray towers and turbulent contact absorbers where the liquid phase is the dispersed phase. Experimental studies are scarce. Recently, however, Wellek and Brunson [29] showed that for circulating drops ($d_p v_r/D_A \simeq 10^6$; $Re \simeq 600$) the film theory gives reasonable results. Concerning the deviations that occur at a large conversion of B, Brunson and Wellek [30] present a numerical analysis with the Fourier number, a type of Peclet number ($d_p v_r \mu_c/D_A(\mu_c + \mu_d)$), $E_{A\infty}$ and the Hatta number as variables. The case $Pe = 0$ has been included as well. Their conclusion is that, in practice, the deviations are important only in case the reaction phase is a gas bubble. In liquid drops, the depth of penetration in general is small compared to the diameter, so that the deviations are small as well.

Recently, a fundamental numerical analysis for drops with $Re < 1$ [31] and a numerical and analytical solution for a stagnant film [32] have been published. The latter is for instantaneous reaction only.

VII.10 MASS TRANSFER WITH REVERSIBLE HOMOGENEOUS REACTIONS

An important industrial example of mass transfer with reversible reaction is the removal of hydrogen sulphide from natural gas, gasoline, kerosine, etc., with an aqueous hot carbonate solution or an aqueous amine solution. Other examples are the absorption of SO_3 in sulphuric acid and of SO_2 or Cl_2 in aqueous solutions. The most simple case is mass transfer of A followed by a simple reaction:

$$A(1) \to A(2), \qquad A(2) \rightleftarrows P(2).$$

We will discuss it for porous solids in Section VII.12.2. Here we will treat the more complex case with two reactants involved:

$$A(1) \rightarrow A(2)$$

$$A(2) + v_B B(2) \rightleftarrows v_P P(2) + v_Q Q(2)$$

and with the general kinetics

$$R_A = -k_{n,m} c_A^n c_B^m + k_{n',m'} c_P^{n'} c_Q^{m'}. \tag{VII.151}$$

Based on the van Krevelen–Hoftijzer approximation Onda et al. [33] derived an approximated solution according to the film model. Their assumptions are:

— B, P and Q are not transferred to phase (1)
— the composition in the bulk and at $x = \delta$ is in equilibrium.

Material balances in the film are

$$D_A \frac{d^2 c_A}{dx^2} = k_{n,m} c_A^n c_B^m - k_{n'm'} c_P^{n'} c_Q^{m'} \equiv -R_A \tag{VII.152}$$

$$D_B \frac{d^2 c_B}{dx^2} = -v_B R_A \tag{VII.153}$$

$$D_P \frac{d^2 c_P}{dx^2} = v_P R_A \tag{VII.154}$$

$$D_Q \frac{d^2 c_Q}{dx^2} = v_Q R_A \tag{VII.155}$$

with boundary conditions

at $x = 0$: $c_A = c_{Ai}$; $c_B = c_{Bi}$; $c_P = c_{Pi}$; $c_Q = c_{Qi}$ \hfill (VII.156)

at $x = \delta$: $c_A = \bar{c}_A$; $c_B = \bar{c}_B$; $c_P = \bar{c}_P$; $c_Q = \bar{c}_Q$ \hfill (VII.157)

From equations (VII.152) and (VII.154), it follows that

$$D_A \frac{d^2 c_A}{dx^2} + \frac{D_P}{v_P} \frac{d^2 c_P}{dx^2} = 0. \tag{VII.158}$$

Integration of equation (VII.158), subject to boundary conditions (VII.156) and (VII.157), gives the relation between c_A and c_P all over the film:

$$c_P = \bar{c}_P - \left[\frac{v_P}{D_P} D_A \left(\frac{dc_A}{dx} \right)_{x=0} - \left(\frac{dc_P}{dx} \right)_{x=0} \right] (\delta - x) - \frac{v_P}{D_P} D_A (c_A - \bar{c}_A) \tag{VII.159}$$

which reduces at the interface to

$$\left[-D_A \left(\frac{dc_A}{dx} \right)_{x=0} + \frac{D_P}{v_P} \left(\frac{dc_P}{dx} \right)_{x=0} \right] \delta = D_A (c_{Ai} - \bar{c}_A) + \frac{D_P}{v_P} (c_{Pi} - \bar{c}_P) \tag{VII.160}$$

Analogously to equation (VII.160), equations for c_{Bi} and c_{Qi} as a function of c_{Pi} are obtained:

$$c_{Bi} = \bar{c}_B - \frac{v_B D_P}{v_P D_B}(c_{Pi} - \bar{c}_P) - \frac{v_B}{D_B}\left[\frac{D_B}{v_B}\left(\frac{dc_B}{dx}\right)_{x=0} + \frac{D_P}{v_P}\left(\frac{dc_P}{dx}\right)_{x=0}\right]\delta \quad \text{(VII.161)}$$

$$c_{Qi} = \bar{c}_Q + \frac{v_Q D_P}{v_P D_Q}(c_{Pi} - \bar{c}_P) - \frac{v_Q}{D_Q}\left[\frac{D_Q}{v_Q}\left(\frac{dc_Q}{dx}\right)_{x=0} - \frac{D_P}{v_P}\left(\frac{dc_P}{dx}\right)_{x=0}\right]\delta. \quad \text{(VII.162)}$$

At this point, the boundary conditions at the interface call for attention. Equations (VII.160), (VII.161) and (VII.162) are general in the sense that the equations also hold for desorption of B, P and Q. In case B, P and Q are not transferred to phase (1), the boundary conditions at the interface follow from:

$$J_A + D_A\left(\frac{dc_A}{dx}\right)_{\Delta x} = -R_A \Delta x, \quad \text{(VII.163)}$$

$$D_B\left(\frac{dc_B}{dx}\right)_{\Delta x} = -v_B R_A \Delta x, \quad \text{(VII.164)}$$

$$D_P\left(\frac{dc_P}{dx}\right)_{\Delta x} = v_P R_A \Delta x, \quad \text{(VII.165)}$$

$$D_Q\left(\frac{dc_Q}{dx}\right)_{\Delta x} = v_Q R_A \Delta x. \quad \text{(VII.166)}$$

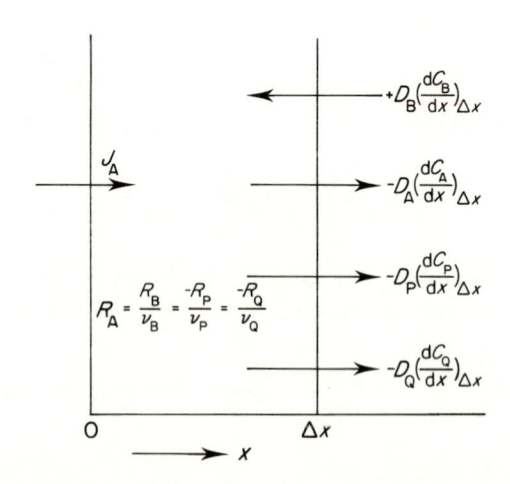

Fig. VII.26. Fluxes at the interface for the system:
$A(1) \rightarrow A(2)$
$A(2) + v_B B(2) \rightarrow v_P P(2) + v_Q Q(2)$

For finite values of $k_{n,m}$ and $k_{n',m'}$ these relations reduce for $\Delta x = 0$ to the well known set:

$$J_A = -D_A \left(\frac{dc_A}{dx}\right)_{x=0}$$

$$\left(\frac{dc_B}{dx}\right)_{x=0} = \left(\frac{dc_P}{dx}\right)_{x=0} = \left(\frac{dc_Q}{dx}\right)_{x=0} = 0. \quad (VII.167)$$

For infinite values of $k_{n,m}$ and $k_{n',m'}$ however, the right-hand side of equations (VII.163)–(VII.166) becomes undefined. Then, the boundary conditions follow from eliminating the reaction terms, resulting in

$$J_A = -\left[D_A\left(\frac{dc_A}{dx}\right)_{x=0} + \frac{D_P}{v_P}\left(\frac{dc_P}{dx}\right)_{x=0}\right] \quad (VII.168)$$

$$\frac{D_B}{v_B}\left(\frac{dc_B}{dx}\right)_{x=0} + \frac{D_P}{v_P}\left(\frac{dc_P}{dx}\right)_{x=0} = 0 \quad (VII.169)$$

$$\frac{D_Q}{v_Q}\left(\frac{dc_Q}{dx}\right)_{x=0} - \frac{D_P}{v_P}\left(\frac{dc_P}{dx}\right)_{x=0} = 0. \quad (VII.170)$$

Hence, equations (VII.160)–(VII.162) reduce to

$$J_A = \frac{D_A}{\delta}(c_{Ai} - \bar{c}_A) + \frac{D_P}{v_P \delta}(c_{Pi} - \bar{c}_P) \quad (VII.171)$$

$$c_{Bi} = \bar{c}_B - \frac{v_B D_P}{v_P D_B}(c_{Pi} - \bar{c}_P) \quad (VII.172)$$

$$c_{Qi} = \bar{c}_Q + \frac{v_Q}{v_P}\frac{D_P}{D_Q}(c_{Pi} - \bar{c}_P). \quad (VII.173)$$

E_A follows from equation (VII.171), with $E_A = J_A \delta / D_A(c_{Ai} - \bar{c}_A)$:

$$E_A = 1 + \frac{D_P}{v_P D_A}\left(\frac{c_{Ai} - \bar{c}_P}{c_{Ai} - \bar{c}_A}\right). \quad (VII.174)$$

Onda et al. [33] obtain the same set of equations (VII.172)–(VII.174), but with the assumption

$$\left(\frac{dc_B}{dx}\right)_{x=0} = \left(\frac{dc_P}{dx}\right)_{x=0} = \left(\frac{dc_Q}{dx}\right)_{x=0} = 0.$$

For instantaneous reaction however, these assumptions are not justified, as will be shown at the end of this section. Onda et al. use the symbol Φ which is equal to $E_A(c_{Ai} - \bar{c}_A)/c_{Ai}$; i.e. $\Phi = E_{A0}$. In the literature on both reversible and irreversible reactions, the use of E_A is more conventional than E_{A0}.

A numerical solution is obtained in the following way. Analogously to equation

(VII.159), expressions are derived for c_B and c_Q as a function of c_A. These relations are substituted in equation (VII.152), resulting in

$$D_A \frac{d^2 c_A}{dx^2} = R(c_A, x),$$

which can be solved numerically. An approximated solution has been obtained by Onda et al., by applying the van Krevelen–Hoftijzer substitution for both c_B and c_Q. That is, $c_B = c_{Bi}$ and $c_Q = c_{Qi}$, over the whole reaction zone.

Moreover, they linearize c_A^n and $c_P^{n'}$ according to $2c_{Ai}^{n-1} c_A/(n+1)$ and $2c_{Pi}^{n'-1} \times c_P/(n'+1)$ respectively. Argument for this linearization is found in the solution for irreversible nth order kinetics in A which, to a good approximation, equals the solution for first order kinetics in A, provided that k_n is replaced by $2k_n c_{Ai}^{n-1}/(n+1)$ [25] see previous section. The approximation $c_Q = c_{Qi}$ seems the most risky one because there is a net formation rate of Q in the film which means that the Q profile resembles the P profile rather than the B profile.

With these linearizations, equation (VII.152) reduces to

$$D_A \frac{d^2 c_A}{dx^2} = \frac{2}{n+1} k_{n,m} c_{Ai}^{n-1} c_{Bi}^m c_A - \frac{2}{n'+1} k_{n',m'} c_{Qi}^{m'} c_P. \quad (\text{VII}.175)$$

Elimination of c_P with equation (VII.159), and solving the resulting differential equation, subject to the boundary conditions (VII.156) and (VII.157), gives

$$E_A = \frac{1 + \dfrac{(n'+1) D_P}{(n+1) v_P D_A \Gamma} + \dfrac{D_P}{v_P D_A} \left\{ \dfrac{(n'+1) \bar{c}_A}{(n+1)(c_{Ai} - \bar{c}_A)\Gamma} - \dfrac{\bar{c}_P}{(c_{Ai} - \bar{c}_A)} \right\} \left\{ 1 - \dfrac{1}{\cosh \phi_i} \right\}}{1 + \dfrac{(n'+1)}{(n+1)} \dfrac{D_P}{v_P D_A \Gamma} \dfrac{\tanh \phi_i}{\phi_i}}$$

(VII.176)

with

$$\phi_i = \phi \sqrt{\left[\left(\frac{c_{Bi}}{\bar{c}_B}\right)^m \left(1 + \frac{n+1}{n'+1} \frac{v_P D_A}{D_P} \Gamma\right) \right]}$$

(VII.177)

and

$$\Gamma = \frac{k_{n',m'} c_{Pi}^{n'-1} c_{Qi}^{m'}}{k_{n,m} c_{Ai}^{n-1} c_{Bi}^m} \quad (\text{VII}.178)$$

and ϕ the Hatta number for irreversible reactions of order (n, m) as defined by equations (VII.141) and (VII.145).

So far, the analysis is straightforward. An asymptotic enhancement factor $E_{A\infty}$ is obtained for instantaneous reaction, that is for ϕ approaching infinity. Then, equation (VII.176) reduces to

$$E_{A\infty} = 1 + \frac{D_P}{v_P D_A} \left\{ \frac{n'+1}{n+1} \frac{c_{Ai}}{(c_{Ai} - \bar{c}_A)\Gamma} - \frac{\bar{c}_P}{(c_{Ai} - \bar{c}_A)} \right\}. \quad (\text{VII}.179)$$

Elimination of $E_{A\infty}$ by means of equation (VII.174) results in

$$\frac{k_{n',m'}\, c_{Pi}^{n'}\, c_{Qi}^{m'}}{k_{n,m}\, c_{Ai}^{n}\, c_{Bi}^{m}} = \frac{n'+1}{n+1}. \quad \text{(VII.180)}$$

At instantaneous reaction, however, there is always equilibrium indicating that the value of the left-hand side expression of equation (VII.180) must be equal to one. Onda et al. therefore suggest replacing the group $(n'+1)/(n+1)$ in equations (VII.167), (VII.177) and (VII.179) by unity. Hence

$$E_A = \frac{\left(1 + \dfrac{D_P}{v_P D_A \Gamma}\right) + \dfrac{D_P}{v_P D_A}\left(\dfrac{\bar{c}_A}{(c_{Ai}-\bar{c}_A)\Gamma} - \dfrac{\bar{c}_P}{(c_{Ai}-\bar{c}_A)}\right)\left(1 - \dfrac{1}{\cosh\phi_i}\right)}{1 + \dfrac{D_P \tanh\phi_i}{v_P D_A \Gamma \phi_i}} \quad \text{(VII.181)}$$

with

$$\phi_i = \phi \sqrt{\left[\left(\frac{c_{Bi}}{\bar{c}_B}\right)^m \left(1 + \frac{v_P D_A}{D_P}\Gamma\right)\right]} \quad \text{(VII.182)}$$

and

$$E_{A\infty} = 1 + \frac{D_P}{v_P D_A}\left\{\frac{1}{\Gamma} - \frac{\bar{c}_P}{c_{Ai}}\right\}\frac{c_{Ai}}{(c_{Ai}-\bar{c}_A)}. \quad \text{(VII.183)}$$

For instantaneous reaction, equilibrium is obtained instantaneously, which means that Γ in equation (VII.183) reduces to $\Gamma = c_{Ai}/c_{Pi}$. Hence

$$E_{A\infty} = \left(1 + \frac{D_P}{v_P D_A}\frac{c_{Pi}-\bar{c}_P}{c_{Ai}-\bar{c}_A}\right), \quad \text{(VII.184)}$$

which is identical to equation (VII.174). For irreversible instantaneous reaction, equation (VII.184) reduces to the previously derived relation:

$$(E_{A\infty})_{k_{n',m'}=0} = 1 + \frac{D_B \bar{c}_B}{v_B D_A c_{Ai}}. \quad \text{(VII.116)}$$

This may be checked by realizing that for an irreversible instantaneous reaction $\bar{c}_A = 0$ and $J_B/v_B = J_P/v_P$, or

$$\frac{D_B \bar{c}_B}{v_B(\delta_m - \delta_r)} = \frac{D_P}{v_P}\frac{(c_{Pi}-\bar{c}_P)}{(\delta_m - \delta_r)}.$$

The last equation states that the amount of B transported to the reaction zone equals the amount of P transported from the reaction zone with respect to the stoichiometric coefficients.

For the general reversible case, the enhancement factor E_A can be calculated from equations (VII.172), (VII.173), (VII.178), (VII.181) and (VII.182) by trial and error, starting with the assumption of an appropriate value of c_{Pi}. In case of instantaneous reaction, E_A follows from equations (VII.172), (VII.173) and (VII.184) and the equilibrium relation at the interface:

$$\frac{c_{Pi}^{n'} c_{Qi}^{m'}}{c_{Ai}^{n} c_{Bi}^{m}} = \frac{k_{n,m}}{k_{n'm'}} = K. \tag{VII.185}$$

Figures VII.27 and VII.28 show the enhancement factor E_A as a function of ϕ and a dimensionless equilibrium constant K' defined as

$$K' = \frac{k_{n,m}}{k_{n',m'}} c_{Ai}^{m+n-m'-n'}. \tag{VII.186}$$

Figure VII.27 shows how the enhancement factor E_A is suppressed by a decrease of K'. The figure also shows that the approximation is very good. Onda *et al.* state that their approximation is accurate within 10%.

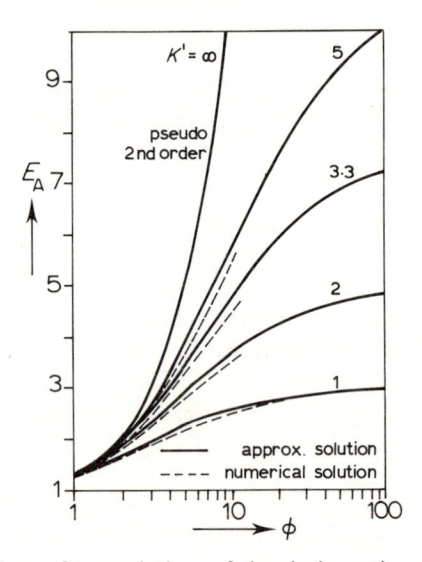

Fig. VII.27. Comparison of two solutions of chemical reaction $A \rightleftharpoons \frac{1}{2}P$, with
$$R_A = -k_{2,0} c_A^2 + k_{1,0} c_P$$
for $D_A = D_P$, $\bar{c}_A/c_{Ai} = 10^{-2}$ and $\bar{c}_P/c_{Ai} = 0.5 \times 10^{-3}$ with $K' = k_{2,0} c_{Ai}/k_{1,0}$. [From Onda *et al.* [33]. Reproduced by permission Pergamon Press Ltd]

In figure VII.28, E_A is plotted against ϕ for various values of K' and $\bar{c}_B/c_{Ai}$, but for the same asymptotic enhancement factor $E_{A\infty}$ as given by equation (VII.184). The enhancement factor appears to be insensitive to the values of K' and $\bar{c}_B/c_{Ai}$ as long as $E_{A\infty}$ does not change. This means that, at least for kinetics in which $m = n = m' = n' = 1$, it is permissible to use the classical van Krevelen–Hoftijzer graph (figure VII.22) to determine E_A for an equilibrium reaction, provided that $E_{A\infty}$ is determined by equations (VII.184), (VII.185), (VII.172) and (VII.173).

Additional data of Onda *et al.* suggest that this statement can be extended to the general (n,m)–(n',m') order kinetics, analogous to the Hikita–Asai extension

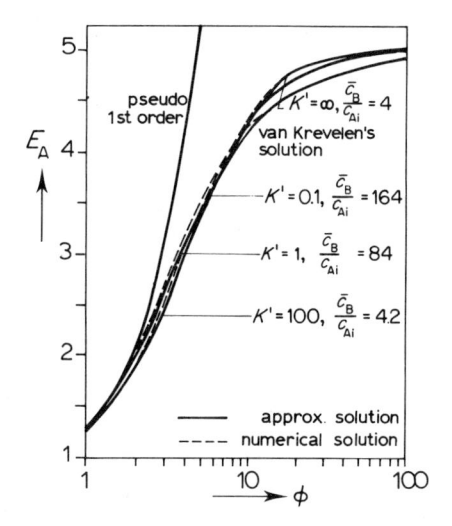

Fig. VII.28 Comparison of two solutions of chemical reactions $A + B \rightleftharpoons P + Q$ for equal diffusivities, $\bar{c}_A = \bar{c}_P = \bar{c}_Q = 0$ and $E_{A\infty} = 5$ [From Onda et al. [33] Reproduced by permission of Pergamon Press Ltd]

of the van Krevelen–Hoftijzer graph, though the solution is somewhat sensitive for the total order $(m' + n')$ of the reverse reaction.

Related readings in the field are Secor and Beutler [34] for a numerical solution of the general reversible kinetics, based on the penetration theory of Higbie; Danckwerts [35] and Olander [36] for instantaneous reversible reactions and Sada and Kumazawa [37] for the complex two-step reaction $2A \rightleftharpoons A^* + A$ and $A^* + B \rightarrow P$ with general power law kinetics. Both a numerical solution and an approximated analytical solution, based on the van Krevelen–Hoftijzer approximation, have been presented for the latter case.

Note that with instantaneous reversible reactions, the boundary conditions at the interface are of interest because they differ from those met with irreversible reactions or with reversible reactions of finite rate. Consider as an example the following simple system:

$$A(1) \rightarrow A(2)$$
$$A(2) + B(2) \rightleftharpoons P(2) \qquad R_A = k_1 c_P - k_{1,1} c_A c_B \qquad (VII.187)$$

For instantaneous reaction, there is always equilibrium, at any point in the film. Hence

$$\frac{c_{Pi}}{c_{Ai} c_{Bi}} = K \qquad (VII.188)$$

or

$$c_{Bi} \left(\frac{dc_A}{dx} \right)_{x=0} + c_{Ai} \left(\frac{dc_B}{dx} \right)_{x=0} = \frac{1}{K} \left(\frac{dc_P}{dx} \right)_{x=0} \qquad (VII.189)$$

Combination of equations (VII.168), (VII.169), (VII.188) and (VII.189) gives

$$\frac{-D_P(dc_P/dx)_{x=0}}{J_A} = \left[1 + \frac{D_A}{D_B}\frac{Kc_{Ai}^2}{c_{Pi}} + \frac{D_A}{D_P}\frac{c_{Ai}}{c_{Pi}}\right]^{-1}. \quad \text{(VII.190)}$$

It follows from equation (VII.190) that dc_P/dx is zero only for $K = \infty$ (irreversible reactions) and for $K \to 0$ ($c_{Ai}/c_{Pi} \to \infty$). Figure VII.29 shows concentration profiles for different values of Kc_{Ai}.

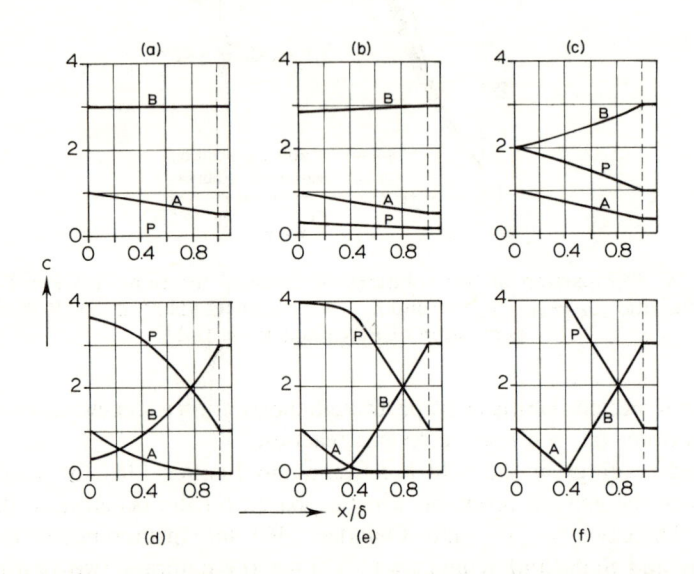

Fig. VII.29 Effect of the equilibrium constant K on the concentration profiles for mass transfer with an instantaneous equilibrium reaction of type $A(1) \to A(2)$; $A(2) + B(2) \rightleftharpoons P(2)$; $\bar{c}_B/c_{Ai} = 3$, $D_B/D_A = 0.5$, $D_P = D_B$: (a) $Kc_{Ai} = 10^{-3}$; (b) $Kc_{Ai} = 10^{-1}$; (c) $Kc_{Ai} = 1$; (d) $Kc_{Ai} = 10$; (e) $Kc_{Ai} = 100$; (f) $Kc_{Ai} = 10^4$ (From Olander [36]. Reproduced by permission of the American Institute of Chemical Engineers)

VII.11 REACTION IN A FLUID–FLUID SYSTEM WITH SIMULTANEOUS MASS TRANSFER TO THE NON-REACTION PHASE (DESORPTION)

Desorption of reactants and reaction products is commonly encountered in industrial practice. It may occur inevitably if a reactant or a product is volatile (gas–liquid systems), or soluble in the non-reaction phase (liquid–liquid systems). It may also be applied purposely to achieve a special goal. Such applications are:

— desorption in order to reuse a chemical absorbent. An example is the sulfinol process from Shell by which H_2S is removed from natural gas or refinery gases with an amine solution which acts as chemical absorbent. The liquid is reused after stripping the H_2S in a second column at a lower pressure and a higher temperature, according to [38]:

$$R_2NH_2^+ + HS^- \rightleftharpoons R_2NH + H_2S\uparrow;$$

— desorption of a reaction product in order to get an equilibrium reaction to completion. Examples are manifold, e.g. in esterification and polymerization reactions as in the caprolactam polymerization to Nylon-6 [39]:

$$(-NH_2-) + (-COOH) \rightleftharpoons (-NHCO-) + H_2O\uparrow;$$

— desorption of a desired (intermediate) product to realize a higher selectivity of the process. An example is the manufacturing of vinyl stearate according to

$$CH_3COOC_2H_3 + CH_3(CH_2)_{16}COOH \rightleftharpoons CH_3(CH_2)_{16}COOC_2H_3 + CH_3COOH.$$

vinyl acetate   stearic acid   vinyl stearate   acetic acid

An unwanted secondary reaction is

$$CH_3COOC_2H_3 + CH_3COOH \rightleftharpoons (CH_3COO)_2CHCH_3.$$

ethylidene acetate

Geelen and Wijffels [40] showed that, from the selectivity point of view, it is advantageous to use a distillation column as a chemical reactor for this process. In this reactor acetic acid evaporates as soon as it is formed, thus preventing the unwanted secondary reaction. Other examples are given in Section VIII.6.

Physically, mass transfer into and from a phase proceeds by exactly the same mechanism: the value of the mass transfer coefficient is independent of the direction of mass transfer. Also desorption with reaction is not different in principle from absorption with reaction. Therefore desorption with reaction problems may be analysed successfully with the stagnant film and penetration models as well. The solutions for reaction with ab- and desorption generally will be different. This difference is caused by the difference in boundary conditions at the interface. If B is transferred to the non-reaction phase, the boundary condition for B at the interface becomes

$$-D_B(dc_B/dx)_{x=0} = J_B \qquad (VII.191)$$

instead of $(dc_B/dx)_{x=0} = 0$ for the cases where B is not volatile (G/L systems) or not soluble in the non reaction phase (L/L systems). For instantaneous reversible reaction as described in the previous section,

$$A(1) \rightarrow A(2)$$
$$A(2) + v_B B(2) \rightleftharpoons v_P P(2) + v_Q Q(2) \qquad (VII.192)$$

with desorption of B: B(2) → B(1), the appropriate boundary condition for B at the interface is no longer given by equation (VII.169). The correct boundary condition becomes

$$-\left\{D_B\left(\frac{dc_B}{dx}\right)_{x=0} + \frac{D_P v_B}{v_P}\left(\frac{dc_P}{dx}\right)_{x=0}\right\} = J_B. \quad \text{(VII.193)}$$

As B desorbs, J_B will have a negative sign. As an illustration, we will discuss a simple case, based on the film theory.

Consider the simple instantaneous equilibrium reaction (see figure VII.30); A(2) ⇌ P(2) with, at any place:

$$c_P/c_A = K \quad \text{(VII.194)}$$

Assume stripping of P: P(2) → P(1). According to film theory,

$$D_A \frac{d^2 c_A}{dx^2} = -R_A \quad \text{(VII.54)}$$

$$D_P \frac{d^2 c_P}{dx^2} = R_A \quad \text{(VII.195)}$$

with boundary conditions

$$x = 0: c_P = c_{Pi}; \; c_A = \frac{c_{Pi}}{K}$$

$$x = \delta: c_P = \bar{c}_P; \; c_A = \frac{\bar{c}_P}{K}.$$

From equations (VII.54) and (VII.195), it follows that

$$D_A \frac{d^2 c_A}{dx^2} + D_P \frac{d^2 c_P}{dx^2} = 0. \quad \text{(VII.196)}$$

The solution is

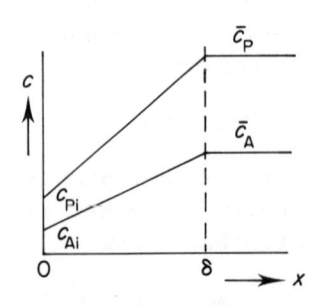

$$D_A c_A + D_P c_P = c_1 x + c_2. \quad \text{(VII.197)}$$

Fig. VII.30. Concentration profiles for instantaneous reversible reaction A⇌P with $K = 2$ and with desorption of component P

The desorption rate of P is equal to the total flux of P to the interface either in unconverted form (as P) or in converted form (as A):

$$J_P = -\left[D_A \left(\frac{dc_A}{dx}\right)_{x=0} + D_P \left(\frac{dc_P}{dx}\right)_{x=0} \right]. \qquad (\text{VII.198})$$

From equations (VII.197) and (VII.198) in combination with the boundary conditions, we obtain the desorption rate of P as

$$J_P = \frac{D_P}{\delta}\left(1 + \frac{D_A}{D_P K}\right)(c_{Pi} - \bar{c}_P). \qquad (\text{VII.199})$$

For the analogous absorption with instantaneous reaction,

$$P(1) \rightarrow P(2)$$

$$P(2) \rightleftharpoons A(2) \quad \text{with} \quad K = \frac{c_P}{c_A},$$

the reader may derive the same results as equation (VII.199). Hence, the enhancement factors for absorption and desorption are equal:

$$E_A = 1 + \frac{D_A}{D_P}\frac{1}{K}.$$

Whether absorption or desorption occurs depends on the sign of $(c_{Pi} - \bar{c}_P)$. Asymptotic solutions of equation (VII.199) are

$$K \rightarrow \infty \quad J_P = \frac{D_P}{\delta}(c_{Pi} - \bar{c}_P)$$

$$K \rightarrow 0 \quad J_P = \frac{D_A}{\delta}(c_{Ai} - \bar{c}_A).$$

For further information on reaction with desorption (both gas–liquid and liquid–liquid), the reader is referred to the comprehensive survey published by Shah and Sharma [41].

Related to the subject of this section is the problem of mass transfer with reaction in two phases. This may occur in liquid–liquid systems where two reactive liquids have some solubility in either phase, as in nitration, hydrolysis and alkylation processes, see Figure VII.31. The problem has been analysed on the basis of film theory by Rod [42, 43], Mashkar and Sharma [44], Sada et al. [45, 46] and Merchuk and Farina [47].

For $\phi \gg E_{A\infty}$, the reaction always takes place in one phase only. This conclusion also holds approximately for ϕ and $E_{A\infty}$ of same order in one of the two phases and for great differences in solubility.

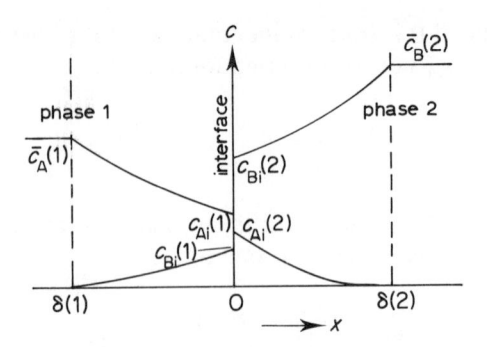

Fig. VII.31. Mass transfer with reaction in two phases

VII.12 THE INFLUENCE OF MASS TRANSFER ON HETEROGENEOUS REACTIONS

As discussed in Section VII.1, a heterogeneous reaction takes place at the interface between two adjacent phases. In the case of heterogeneous catalysis, this interface is the surface of a phase of catalyst, and the reactants are present in the fluid phase surrounding the catalyst. With a non-catalysed heterogeneous reaction, the interface is the surface of a more or less pure material which itself reacts with components of the surrounding fluid phase. In both cases the chemical conversion at the surface is, in principle, influenced by the fact that the reactants have to be transferred to the interface and the products away from it.*
Accordingly, it is of interest to investigate the interplay between the true *chemical reaction rate* and the physical transport phenomena in order to arrive at an expression for the *overall* conversion rate (per unit reactor volume or per unit mass of catalyst). The latter expression can then be used for reactor calculations in the manner indicated in Chapters II to VI.

To this end, let us consider the general case of a reaction at the surface of a particle of porous catalyst along which flows a fluid containing the reactant(s).†
It is generally recognized that the following steps occur successively in such a system:

(i) mass transfer of reactant(s) from the main flow to the external surface of the particle;
(ii) transport of reactant(s) by a diffusional process through the pores into the particle;
(iii) adsorption of reactant(s) on internal catalyst surface;
(iv) chemical conversion in the adsorbed state;
(v) desorption of product(s);

* The treatment of a heterogeneous reaction between a fluid and a solid is entirely similar but for the fact that the solid is consumed whereas in general a solid catalyst retains its geometry.

† In fluid–solid reactions it may occur that the reaction product forms a new solid phase, e.g. in the thermal decomposition of calcium carbonate [48] and in the oxidation of zinc sulphide in an air stream [49].

(vi) transport of product(s) by a diffusional process through the pores out of the particle;
(vii) mass transfer of product(s) from the external surface into the main flow.

The steps (i), (ii), (vi) and (vii) are entirely of a physical nature. If the particle is non-porous, the steps (ii) and (vi) do not occur.

The maximum use is made of a porous catalyst when the entire internal surface is readily accessible to the reactants; this will be the case when the chemical conversion rate is slow with respect to the rates of transport according to the mechanisms (i) and (ii), (vi) and (vii). Such a situation may also be realized for moderately fast reactions, provided the external mass transfer is made sufficiently high (by a high fluid velocity along the particle) and if the diffusional resistance inside the particle is made sufficiently small (by a high internal porosity and a small particle diameter). If the chemical reaction is very fast with respect to the greatest possible physical transport, it will occur in the outer part of the particle. In the extreme case, only the external particle surface will be of use to the reaction; the internal surface then plays no part and may equally well be absent.

A quantitative treatment of the conversion rate along the lines given above for a given reaction system is very difficult to undertake. No sufficient general data are available for an *a priori* evaluation of steps (ii)–(vi). The internal diffusion greatly depends on the pore texture of the particle of catalyst (pore size distribution and the connection between the pores) and on the fact whether the transport is brought about by normal molecular diffusion, Knudsen diffusion (with gases, when the mean free path of the molecules exceeds the average pore size) or migration in the adsorbed state.

In the first case the coefficient of internal diffusion D_i may be 10 to 100 times smaller than the molecular diffusivity in the free fluid. For these matters, the reader is referred to the more general papers by Thiele [50], Zeldowitch [51], Wagner [52], Wicke [53], van Krevelen [54] and Satterfield [55].

A vast amount of work has been done on the kinetics of the heterogeneous reaction itself, particularly on the basis of kinetic measurements where the inference by physical transport resistances was eliminated as much as possible. In general, the expression for the kinetics of the surface conversion rate contains all three effects (iii), (iv) and (v), since it is almost impossible to study these separately. This is reflected by the complexity of many experimental rate expressions for heterogeneous catalytic reactions; see e.g. Hougen and Watson [56].

Because heterogeneous reactions are essentially surface reactions, production rates may be expressed per unit surface, e.g. for first order reaction

$$R_A'' = -k_1'' c_{Ai}, \qquad \text{(VII.200)}$$

with

R_A'' = the production rate per unit surface [kmol/m^{2}s];
c_{Ai} = the concentration of A in the direct vicinity of the surface, in equilibrium with the concentration of A at the surface [kmol m^3]:

k_1'' = surface reaction rate constant [m/s]. For nth order kinetics the dimension of the surface rate constant k_n'' is [m^{3n-2}/kmol^{n-1}s].

It is convenient to distinguish between the following two cases:

(1) the whole surface is equally easy accessible for the reactants, e.g. in catalytic reaction on the surface of massive (non-porous) pellets as in oxidation of ammonia on platinum wire and in many fluid–solid reactions in which the particle dimension decreases in time due to reaction;
(2) reactions in porous catalysts in which the accessibility of the surface decreases with increasing distance from the external surface.

The first case is an asymptotic situation of the latter one. We will discuss both cases and will show that the theory derived for mass transfer with homogeneous kinetics is equally well applicable for heterogeneous kinetics. As an example, mass transfer with an equilibrium reaction will be discussed in detail.

VII.12.1 Heterogeneous reaction at an external surface

In mass transfer with homogeneous reaction, the transport and the reaction process occur at the same place (in the film near the interface): mass transfer and reaction are *parallel*, each other interacting processes. Heterogeneous reaction at an external surface is an example of mass transfer and reaction in *series*: the transport and the reaction process occur at different places. Therefore the reaction does not interfere directly with the mass transport process. As a consequence, enhanced mass transfer will not occur in general. An exception to this rule is possible in fluid–fluid–solid systems (to be discussed in the next note).

Irreversible first order reaction

In case a component A is transferred to a solid surface where it is converted according to first order kinetics, the stationary conversion rate follows from

$$J_A = \frac{D_A}{\delta}(\bar{c}_A - c_{Ai}) \tag{VII.201}$$

$$R_A'' = -k_1'' c_{Ai} \tag{VII.200}$$

and

$$J_A = -R_A'', \tag{VII.202}$$

with c_{Ai} the local concentration of A near the surface in equilibrium with the concentration at the surface. Elimination of this unknown concentration with equations (VII.200)–(VII.202) results in (see figure VII.32)

$$J_A = \bar{c}_A / \left\{ \frac{\delta}{D_A} + \frac{1}{k_1''} \right\} \tag{VII.203}$$

or, for a gas–solid reaction,

$$J_A = \bar{c}_A / \left\{ \frac{1}{k_G} + \frac{1}{k_1''} \right\}. \tag{VII.204}$$

By analogy with the treatment of homogeneous reactions, we may define an effectiveness factor η'' as

$$\eta'' = \frac{\text{actual concentration near reactive surface}}{\text{maximal concentration near reactive surface}}. \tag{VII.205}$$

Hence $\eta'' = c_{Ai}/\bar{c}_A$ \hfill (VII.206)

From equations (VII.201)–(VII.203), it follows that

$$\eta'' = \frac{1}{1 + (k_1''\delta/D_A)} \tag{VII.207}$$

or, for example, for a gas–solid reaction,

$$\eta'' = \frac{k_G}{k_G + k_1''} \tag{VII.208}$$

and

$$J_A = k_1'' \bar{c}_A \eta''. \tag{VII.209}$$

Two asymptotic solutions are (see figure VII.33)

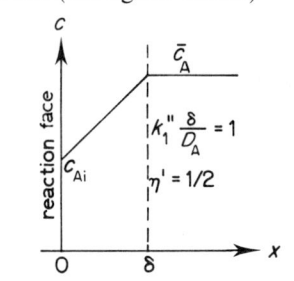

Fig. VII.32. Concentration profile for mass transfer with heterogeneous reaction $R_A'' = -k_1'' c_{Ai}$ (film theory)

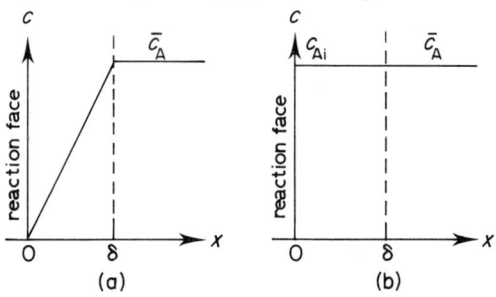

Fig. VII.33. Asymptotic concentration profiles for mass transfer with heterogeneous reaction $R_A'' = -k_1'' c_{Ai}$ (film theory): (a) $k_1''\delta/D_A \gg 1$; (b) $k_1''\delta/D_A \ll 1$

Fast reaction regime For $k_1''/k_G \gg 1$, the effectiveness factor η' approaches zero. The rate of mass transfer completely limits the conversion rate which follows from $J_A = k_G(\bar{c}_A - 0) = k_G \bar{c}_A$

Slow reaction regime For $k_1''/k_G \ll 1$, the effectiveness factor approaches one. The kinetic reaction rate completely limits the conversion rate $J_A = k_1'' \bar{c}_A$.

The capacity of a reactor follows from

$$\Phi_A = \int_{V_r} J_A a \, dV_r. \tag{VII.210}$$

From the above analysis, it follows that increase of reactor capacity can be realized by an increase of reaction surface and by

— improvement of the mass transfer coefficient for the case $k_G/k_1'' \ll 1$;
— increase of the reaction rate for the case $k_G/k_1'' \gg 1$.

The system analysed above is easily extended to a three phase reactor with mass transfer and reaction in series: a gas absorbs in a liquid in which non-porous particles are suspended. At the surface of the particles, reaction takes place. Examples are the hydrogenation of organic liquids with nickel as a catalyst and the alkylation of a liquid reactant with, for example, gaseous ethylene and with $AlCl_3$ as a catalyst.

The concentration profile may look as shown in figure VII.34. At the gas–liquid interface and at the reaction surface equilibrium is assumed: $c_{AiL} = mc_{AiG}$. If a_p is the specific catalyst surface area per unit volume reactor, a the specific gas–liquid surface area per unit volume reactor, and k_L and k_{LS} the liquid phase mass transfer coefficients for transfer from gas to liquid and liquid to solid, respectively, one can derive, analogously to the derivation of equation (VII.204),

$$J_A a = \bar{c}_{AG} \bigg/ \left\{ \frac{1}{k_G a} + \frac{1}{m k_L a} + \frac{1}{m k_{LS} a_p} + \frac{1}{m k_1'' a_p} \right\}. \tag{VII.211}$$

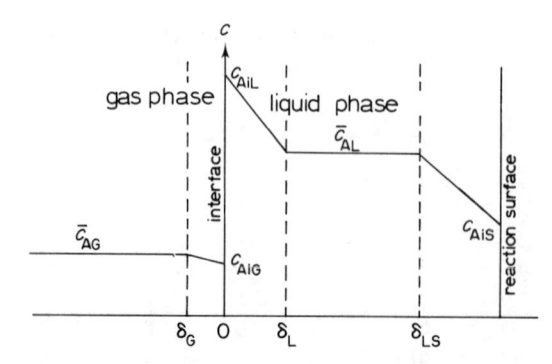

Fig. VII.34. Concentration profiles in mass transfer and reaction in series in a gas–liquid–solid system (film theory)

Reactor capacity per unit volume appears to depend on four resistances in series: the gas phase transfer resistance, two liquid phase transfer resistances and the kinetic resistance. The highest resistance will limit the capacity of the reactor.

Implicitly it is assumed in the derivation of equation (VII.211) that all particles are in the bulk of the liquid phase. However, for

$$\frac{1}{k_{LS}a_p} + \frac{1}{k_1'' a_p} \ll \frac{1}{k_L a} \tag{VII.212}$$

the concentration of A in the bulk of the liquid phase approaches zero. Then it is possible that only the particles in the liquid film near the gas–liquid interface are active, and mass transfer and reaction become parallel steps resulting in a chemical enhanced absorption of A into the liquid. The problem has been studied both theoretically and experimentally by Sada et al. [57, 58]. It may be of interest in the removal of SO_2 from off-gases with aqueous slurries of hydroxide. The analysis has been extended to the absorption of a gas into a liquid containing small porous particles with diameter d_p much smaller than the film diameter δ_L. In the liquid-filled pores the gas may react catalytically at the catalyst surface [58]. The criterion for the applicability of equation (VII.211) is the value of $\bar{c}_{AL}$ calculated from it. The equation is applicable for

$$\frac{\bar{c}_{AL}}{(c_{AiL} + \bar{c}_{AL})/2} Al \gg 1 \tag{VII.213}$$

because the left term of this equation approximates the ratio of reaction taking place in the bulk and in the film.

Arbitrary kinetics

The balance

$$J_A = \frac{D_A}{\delta}(\bar{c}_A - c_{Ai}) = -R_A'' \tag{VII.214}$$

holds for any kinetics (film model). From equation (VII.214), the unknown interface concentration c_{Ai} is easily eliminated; e.g. for nth order kinetics $R_A'' = -k_n'' c_{Ai}^n$, J_A follows from the implicit equation

$$J_A = \frac{D_A}{\delta}\left(\bar{c}_A - \left(\frac{J_A}{k_n''}\right)^{1/n}\right). \tag{VII.215}$$

Equation (VII.215) shows that only for first order kinetics, the overall conversion rate J_A remains first order with respect to $\bar{c}_A$. In all other cases the apparent macro kinetics become more complex.

In the event of two components taking part in the surface reaction, an extra balance for the second component must be considered as well. For $A + \nu_B B \to \nu_P P + \nu_Q Q$, we have

$$J_A = \frac{D_A}{\delta}(\bar{c}_A - c_{Ai}) = k_{n,m}'' c_{Ai}^n c_{Bi}^m = -R_A'', \tag{VII.216}$$

$$J_B = \frac{D_B}{\delta}(\bar{c}_B - c_{Bi}) = -v_B R''_A. \tag{VII.217}$$

Both balances are coupled via

$$J_A = J_B/v_B \tag{VII.218}$$

From equations (VII.216) and (VII.218) the unknown concentrations c_{Bi} and c_{Ai} can be eliminated, resulting in J_A as $f(D_A, D_B, \delta, k''_{n,m}, \bar{c}_A, \bar{c}_B)$.

The most prevalent model for heterogeneous kinetics is the Langmuir–Hinshelwood model for the conversion of A into P. That model has been successfully applied in many heterogeneous reactions e.g. the dehydrogenation of secondary alcohols on nickel [59], the reduction of iron oxides with hydrogen [60, 61] and the reaction of carbon with various gases, as e.g. CO_2 [62]. According to this model, the conversion rate at the surface is proportional to the fraction of active surface sites covered by A. Calling this fraction θ one gets: $-R''_A = k\theta$. The adsorption rate is assumed to be proportional to the concentration of A near the interface (c_{Ai}) and to the fraction of uncovered surface sites ($\alpha c_{Ai}(1-\theta)$); the desorption rate is assumed to be proportional to the fraction of covered active surface sites ($\beta\theta$). With $\alpha/\beta = K =$ the Langmuir adsorption coefficient [m³/kmol] and $kK = k''_1$, the surface reaction rate constant, it follows for the stationary situation in which adsorption and desorption rates are equal, after elimination of θ that

$$J_A = -R''_A = \frac{k''_1 c_{Ai}}{1 + K c_{Ai}} \tag{VII.219}$$

Combination of equations (VII.219) and (VII.214) gives J_A as a function of the observable $\bar{c}_A$.

Example VII.12.1.a *Hydrogenation of an aldehyde in a gas-liquid slurry continuous tank reactor*

A small company produces an alcohol by hydrogenation of an aldehyde. Pure liquid aldehyde, B ($M = 80$, $\rho = 800\,\text{kg/m}^3$, $c_{Ao} = 10\,\text{kmol/m}^3$) is continuously fed to a CISTR. By continuously bubbling pure hydrogen through the tank the aldehyde is hydrogenated according to B + $H_2 \rightarrow$ alcohol. The liquid density is independent of the conversion, ζ_B, no by-products are formed and the vapour pressure of both reactant B and the product are negligible. Finely dispersed nickel is used as a catalyst. Further data are: $p = 100$ bars, conversion $\zeta_B = 0.88$, average space time $V_r/\Phi_L = 600\,\text{s}$, average particle diameter $d_p = 10^{-5}\,\text{m}$, external particle surface area $a_p = 30 \times 10^3\,\text{m}^2/\text{m}^3$ reactor and the gas hold up $\varepsilon_G = 0.1$. The process is quite profitable and the company looks for opportunities to double the production capacity of the existing plant without substantial new investments or operating expenses. Suppose you, as a reaction engineering consultant, get the contract to advise the company. How do you proceed and what will be your advice?

As no key reaction engineering data are available you can't give straightforward advice in this stage. Diffusivities and solubilities may often be reasonably accurately estimated from the literature. Suppose you find $D_{H2,L} \simeq 4 \times 10^{-9} \, m^2/s$ and the solubility of H_2 at process conditions $\simeq 0.5 \, kmol/m^3$ liquid. You could guess a value for $k_L a$ as well (see Section VII.17), but it is safer to measure it. This could be done with a transient physical absorption experiment without any catalyst present. From $t = 0$, the relevant H_2 flow is introduced into a batch of reaction mixture of a volume as present under actual operating conditions. The liquid phase H_2 concentration build-up is measured continuously. Suppose you find

$$\ln(1 - \bar{c}_{H2}/c_{H2,i}) = -0.03t;$$

then $k_L a = 0.03 \, s^{-1}$. This relation can easily be derived.

We notice that 50% saturation is reached in 23 seconds. This means there is not much time for a good measurement as it also takes time for the liquid dispersion to be established. Therefore a steady state experiment with a continuous liquid flow will give more accurate data but probably is also more expensive. Derive the appropriate equations for such an experiment.

For small particles the mass transfer coefficient around the particles k_{LS} follows from the conservative estimation $Sh = 2$, so that $k_{LS} d_p = 2D_{H2,L}$ or $k_{LS} = 0.8 \times 10^{-3} \, m/s$. This gives $k_{LS} a_p = 24 \, s^{-1}$. Reaction kinetics and internal catalyst surface area are unknown but you may formally write $J_A a_p = \bar{R}_A'' a_p$, where $\bar{R}_A''$ is an effective reaction rate per m^2 external particle surface. With porous particles and no internal and external diffusion limitation the observed effective $\bar{R}_A''$ will be proportional to d_p. Hence, $\bar{R}_A''$ is not the true reaction rate per m^3 internal surface! You now may treat the problem as mass transfer with reaction in series with two mass transfer resistances and one reaction resistance:

$$J_A a = \frac{c_{AiL}}{\dfrac{1}{k_L a} + \dfrac{1}{k_{LS} a_p} + \dfrac{1}{(\bar{R}_A''/c_{AiS}) a_p}} \tag{a}$$

and $J_A a V_r = \Phi_L c_{Bo} \zeta_B$, or

$$J_A a = c_{Bo} \zeta_B \Phi_L / V_r \tag{b}$$

from (a) and (b) by substituting the data, you get $c_{ASi}/\bar{R}_A'' a_p = 0.75$. Comparing with the other two resistances: $1/k_L a = 33.3 \, s$ and $1/k_{LS} a_p = 0.042 \, s$, it follows that the liquid side resistance around the gas bubbles completely controls the conversion rate. Adding more catalyst or introducing a more reactive catalyst will not increase the capacity appreciably. On the contrary, catalyst concentration can be substantially reduced without much loss of capacity. Also an increase of temperature will not have much effect as $k_L a$ is only weakly dependent on temperature. The only way to increase the conversion rate is to increase the pressure or the product $k_L a$. Doubling p is usually not possible because most probably the reactor is designed for the normal operating pressure. With a guess for k_L in the order of $0.5 \times 10^{-3} \, m/s$ we get for a an estimate of $600 \, m^{-1}$. This is already a rather high value so installing a more powerful stirrer will not be either of much help nor cheap. Consequently you had better advise the company to consider installing a second reactor parallel to the first one. Predictably the company will not be deeply impressed by such advice, but maybe your suggestion to try and reduce catalyst concentration by a factor of 5 or 10 may readily pay back your consulting expenses, if catalyst losses are reduced accordingly.

Note that you arrived at your advice without the slightest idea on the actual reaction kinetics. It is possible to obtain kinetic information from small scale experiments at lower temperatures where the resistance against reaction is rate controlling. By extrapolation according to the Arrhenius relation for the temperature dependence you then learn on the possible kinetics at actual operating conditions. But such an expensive experimental programme will not pay off as long as the conversion rate is not controlled by the chemical reaction rate.

***Example VII.12.1.b** Dimerization in a tube reactor*
A liquid mixture flows turbulently through a tube. The wall of the tube acts catalytically on the dimerization of one of the components of the mixture according to a surface reaction $A \to \frac{1}{2}P$ with $R''_A = -k''_2 c^2_{Ai}$ in which c_{Ai} is the concentration close to the wall. We will derive for plug flow the differential equation for the conversion ζ as a function of the following parameters: c_{Ao} (feed concentration) Φ_L (liquid flow rate) d_t (tube diameter) k_L (mass transfer coefficient to the wall) k''_2 (surface reaction rate constant [m^4/kmol.s]) and z (coordinate in direction of flow).

The basic equations for a reactor volume element $\pi d_t \, dz$ are:

$$-\Phi_L \frac{d\bar{c}_A}{dz} = \pi d_t k''_2 c^2_{Ai} = \pi d_t k_L (\bar{c}_A - c_{Ai}),$$

in which $\bar{c}_A$ and c_{Ai} are functions of z. The local concentration near the wall can be eliminated by substituting

$$c_{Ai} = \frac{-k_L + \sqrt{(k_L^2 + 4k_L k''_2 \bar{c}_A)}}{2k''_2},$$

and $\bar{c}_A$ can be replaced by $c_{Ao}(1 - \zeta)$, so that

$$\Phi_L c_{Ao} \frac{d\zeta}{dz} = \frac{\pi d_t k_L^2}{4k''_2} \left\{ \sqrt{\left(1 + \frac{4k''_2 c_{Ao}(1-\zeta)}{k_L}\right)} - 1 \right\}^2 \quad \text{(a)}$$

For $k''_2 c_{Ao}/k_L \ll 1$, the conversion rate is kinetically controlled and (a) reduces to

$$\Phi_L \frac{d\zeta}{dz} = \pi d_t k''_2 c_{Ao}(1 - \zeta)^2.$$

For $k''_2 c_{Ao}/k_L \gg 1$ mass transfer controls the conversion rate and equation (a) reduces to

$$\Phi_L \frac{d\zeta}{dz} = k_L \pi d_t (1 - \zeta).$$

VII.12.2 Reactions in porous solids

In this extensive field, a distinction can be made between systems in which the solid reacts and in which it only acts as a carrier–catalyst system. In the first group a second distinction is possible between systems in which the solid is completely gasified or liquefied and systems in which the overall particle size remains unchanged. For each category somewhat modified models on mass transfer with reaction have been developed. Within a category, models may differ for fast and slow reaction compared to the rate of mass transfer. For example in reactions of Type II (see Table VII.4), that is in gasification and liquefaction of porous solids, a model in which all pores increase in diameter with increasing time without changing the overall dimension or a model in which the particle dimension decreases in time may be realistic. The first model will be applicable for slow reaction, taking place throughout the pellet. The latter model gets physical significance in case the rate of mass transfer into the particles is much slower than the rate of reaction. Then, a small reaction zone near the external surface exists and in an asymptotic situation (complete resistance against mass transfer outside the particle) the model becomes identical to mass transfer and reaction at non-porous solids.

Reactions of Type III (see Table VII.4) differ from gasification and liquefaction in the fact that the porous structure of the solid continues to exist, though it may be affected by the reaction.

For reactions of type II and III, that is for a solid being converted, figure VII.35 shows possible concentration profiles at different reaction times and for different reaction rates. The pictures are analogous to those for mass transfer with homogeneous reaction with $D_A \gg D_B$ according to penetration theory. Indeed it is often possible to treat these heterogeneous reactions as (pseudo) homogeneous reactions, especially for reactions of type I of Table VII.4. Often, the concentration of A can be assumed to be nearly constant on the scale of the pore diameter. Then a 'homogeneous' production rate may be defined according to

$$R_A = a_s R_A'' \tag{VII.220}$$

Table VII.4. Reactions in porous solids

Type	Examples
I. Solid phase remains unchanged	Many catalytic gas–solid and liquid–solid reactions, e.g. catalytic cracking, catalytic desulphurization, catalytic hydrogenation
II. Solid reacts to a gas or a liquid	Gasification of coal: $2C + O_2 \rightarrow 2CO$; water–gas reaction $C + H_2O \rightarrow CO + H_2$
III. Solid reacts but particle overall size remains unchanged	Burning coke from spent cracking catalyst; roasting of metal sulphides; burning of cement

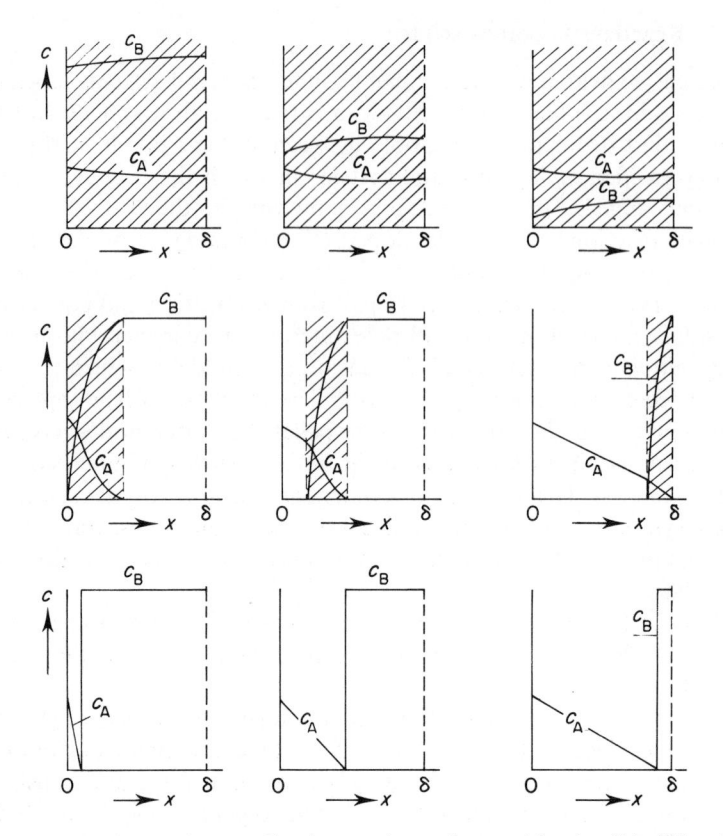

Fig. VII.35. Concentration profiles in reactions of type A(gas) + B(solid) → P(solid) + Q(gas). From left to right, the reaction time increases. From top to bottom, the reaction rate increases from slow, via fast to instantaneous. The reaction zone has been hatched

e.g. for first order kinetics

$$-R_A = k_1'' a_s c_A \qquad \text{(VII.221)}$$

with c_A the local concentration in the pore, which is assumed to be a unique function of the radial coordinate (in spherical pellets), a_s the specific internal surface area of the particle per unit porous particle volume, R_A the local production rate per m³ particle, which is a function of the radial coordinate. This way, the particle is considered as a continuum in which the reaction proceeds according to homogeneous kinetics with, in case of say a first order reaction, a rate constant k_1 given by

$$k_1 = k_1'' a_s \; [\text{s}^{-1}]. \qquad \text{(VII.222)}$$

By that, the formal description becomes identical to mass transfer with homogeneous reaction with $Al = 1$. This means that, for a catalytic reaction of arbitrary kinetics $R_A'' = f(c_A)$, the conversion rate can be calculated by equations

(VII.99), (VII.100) and (VII.220). In case of Langmuir–Hinshelwood kinetics, for instance (see equation VII.219)

$$-R_A = -R_A'' a_s = \frac{k_1'' a_s c_A}{1 + K c_A}.$$

ϕ' follows from equation (VII.99):

$$\phi' = \frac{\delta k_1'' a_s c_{Ai}}{D_i(1 + K c_{Ai})} \left\{ 2 \int_0^{c_{Ai}} \frac{k_1'' a_s c_A}{D_i(1 + K c_A)} dc_A \right\}^{-1/2} \quad (VII.223)$$

with δ = volume of the pellet/external surface area (i.e. $d_p/6$ for a sphere) and D_i = coefficient of internal diffusion of A in the pellet. The solution is

$$\phi' = \frac{\delta \sqrt{\left(\frac{k_1'' a_s}{2 K D_i}\right)} \frac{K c_{Ai}}{1 + K c_{Ai}}}{\sqrt{\left[c_{Ai} - \frac{1}{K} \ln(1 + K c_{Ai}) \right]}} \quad (VII.224)$$

and from equation (VII.100)

$$J_A \simeq -R''(c_{Ai}) a_s \delta \frac{\tanh \phi'}{\phi'} \quad (VII.225)$$

or

$$J_A = k_1'' a_s \delta \frac{c_{Ai}}{1 + K c_{Ai}} \frac{\tanh \phi'}{\phi'} \quad (VII.226)$$

with J_A the conversion rate of A per m² external particle surface. Figure VII.36 shows that the accuracy of equation (VII.226) is within 15% [12]. Other kinetics may be treated in a similar way.

In case a reactive solid B is present (Table VII.4, reactions of Type II and III), a transient penetration description must be applied with $D_A \gg D_B$ and with corrections for the finite depth of the particle as briefly discussed in Section VII.9. Extra complications may arise, however, mostly caused by structural changes

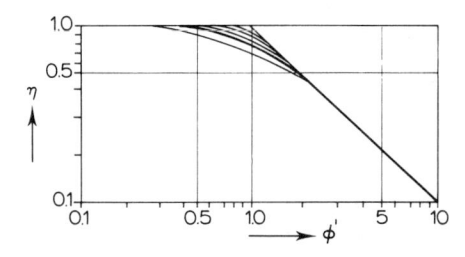

Fig. VII.36. Degree of utilization (η) as a function of the reaction modulus (ϕ') in Langmuir–Hinshelwood kinetics, for different values of c_{Ai} and K. The solid line is the relation $\eta = (\tanh \phi')/\phi'$ [From: K. B. Bischoff [12]. Reproduced by permission of the American Institute of Chemical Engineers]

due to e.g. sintering, particle cracking, enlarging of the pores and changing of the reactive area a_s with time. Most of these refinements fall outside the scope of this general textbook. We refer to a specialized work in the field, namely the handbook of Szekely et al. on gas–solid reactions [63].

The concentration profiles presented in the three bottom figures are typical for instantaneous gas–solid reactions where the particles keep their original shape. This is usually the case if one of the products is also a solid as in the reduction of iron ore:

$$CO + \frac{1}{3}Fe_2O_3 \rightarrow \frac{2}{3}Fe + CO_2$$

or if an inert solid structure is maintained as in coal gasification and combustion if the ash structure is not destroyed:

$$O_2 + C + ash \rightarrow CO_2 + ash \quad \text{or} \quad H_2O + C + ash \rightarrow CO + H_2 + ash.$$

As long as the reaction is instantaneous, the reaction zone is reduced to a reaction plane that moves from the surface of the particle to the centre as time progresses. Models based on this phenomenon are called unreacted shrinking core models. Because of the existence of a reaction plane, mass transfer and reaction are in series. Therefore, derivation of expressions for the conversion rate as a function of time or with solid conversion are relatively simple and straightforward for any kinetics; see Example VII.12.2.c.

Shrinking core models are also relevant for non-instantaneous reactions provided the diffusivity of the gaseous reactant in the unreacted core is much lower than in the converted shell, or more precisely for $D_{ic}/r_c \ll k''$, where D_{ic} is the diffusivity of A in, and r_c the radius of, the unreacted core.

We conclude our survey on mass transfer with a single reaction with the treatment of some integrated examples, in which most of the previously discussed asymptotic effects are once more met. The first case is a reversible reaction in a porous catalyst particle allowing for an external resistance in mass transfer (k_G) and allowing for the curvature of the external surface.

Example VII.12.2.a *Catalytic equilibrium reaction in a spherical porous particle*
An equilibrium isomerization reaction $A \rightleftharpoons P$ is carried out with a porous catalyst in the form of spherical particles with radius r_p. The following assumptions will be made:

— The chemical rate of the forward reaction is given by $k_1'' c_A$ and that of the reversed reaction by $k_2'' c_P$ (both in kmol/m² s); the ratio k_1''/k_2'' is equal to the chemical equilibrium constant K.
— The temperature in the particle is uniform.
— The specific internal surface area a_s (m²/m³) is distributed uniformly over the particle.
— The effective internal coefficient of diffusion D_i and the external mass transfer coefficient k_G are constants and have the same value for A and for P.
— The composition of the fluid outside the particle is given by $\bar{c}_A$ and $\bar{c}_P$.

The material balance for species A in the catalyst particle gives the differential equation for spherical symmetry and in the steady state:

$$0 = \frac{d}{dr}\left[r^2 D_i \frac{dc_A}{dr}\right] - r^2 a_s(k_1'' c_A - k_2'' c_P), \tag{a}$$

where c_A and c_P are the concentrations of A and P in the pores and functions of the distance r from the centre of the particles; see figure VII.37. The boundary conditions are

$$r = 0: \quad \frac{dc_A}{dr} = 0$$

$$r = r_p: \quad D_i \frac{dc_A}{dr} = k_G(\bar{c}_A - c_A) - k_1'' c_A + k_2'' c_P. \tag{b}$$

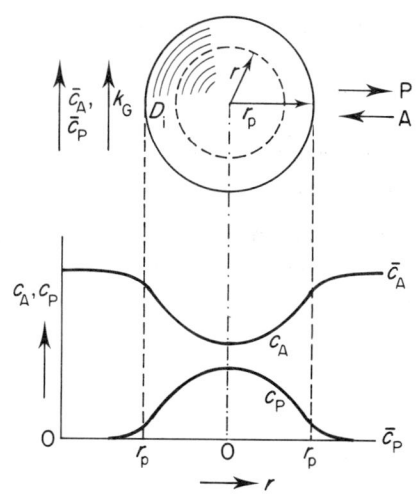

Fig. VII.37. Heterogeneous equilibrium reaction in a spherical porous particle. Concentration distributions are calculated from equation (c) in Example VII.12.2.a with $K = 5$, $\phi = 0.66$, $D_i/k_G r_p = 0.1$ and $a_s r_p = 10^5$

In the latter condition the fact is taken into account that A and P may react at the external surface of the particle which is assumed to be equal to its geometric surface.

If we write the corresponding equations for component P, it is found that $c_A + c_P = \bar{c}_A + \bar{c}_P$, so that c_P can be expressed in terms of c_A and equations (a) and (b) can be solved. The concentration distribution in the particle is found to be

$$c_A = \bar{c}_A - \frac{K\bar{c}_A - \bar{c}_P}{K+1}\left[1 - \frac{\frac{r_p}{r}\sinh\left(\frac{r}{r_p}3\phi\right)}{\frac{D_i 3\phi}{k_G r_p}\cosh 3\phi + \left[1 + \frac{D_i}{k_G r_p}\left(\frac{9\phi^2}{a_s r_p} - 1\right)\right]\sinh 3\phi}\right], \tag{c}$$

where

$$\phi = \frac{V}{A}\sqrt{\left(\frac{k_1'' a_s(K+1)}{D_i K}\right)} = \frac{r_p}{3}\sqrt{\left(\frac{k_1'' a_s(K+1)}{D_i K}\right)}.$$

It is seen that the following quantities appear as parameters in this complicated expression:

K = the equilibrium constant
$D_i/k_G r_p$ = a measure of the ratio of external mass transfer resistance and internal diffusional resistance
ϕ = the reaction modulus for the equilibrium reaction $A \rightleftharpoons P$. Notice that for $K \to \infty$ and $k_1'' a_s = k_1$, equation (VII.222),

$$\phi = \frac{V}{A}\sqrt{\frac{k_1}{D_i}},$$

the known reaction modulus for first order irreversible reaction where $\delta = V/A$, equation (VII.63).
$a_s r_p$ = a measure of the ratio of internal and external surface area.

Figure VII.37 gives an example of a concentration distribution according to equation (c). Equation (c) can be greatly simplified for a few combinations of extreme values of these parameters, e.g.:

— for a very slow reaction in a porous particle, $\phi \ll 1$ and $c_A \simeq \bar{c}_A$;
— for a rapid irreversible reaction in a porous particle with a negligible external mass transfer resistance, $\phi \gg 1$, $K = \infty$ and $D_i/k_G r_p \ll 1$:

$$c_A \simeq \bar{c}_A \frac{r_p}{r} \exp(-3\phi(1 - r/r_p)),$$

which indicates a rapid and almost exponential decrease of c_A just below the outer surface of the particle.
— when no internal surface area is present, $\phi = 0$ and $a_s r_p = 0$; see also, for example, Frank-Kamenetzki [64]:

$$c_A|_{r=r_p} = \frac{k_G K \bar{c}_A + k_1''(\bar{c}_A + \bar{c}_p)}{k_G K + k_1''(1+K)}.$$

We can now obtain from equation (c), the overall production rate to be used in reactor calculations. If the production rate of A is expressed in molar units per unit time and per unit volume of the solid, it is equal to the molar flow of A away from the particle divided by its volume:

$$R_A = -k_G(\bar{c}_A - c_A|_{r=r_p})4\pi r_p^2 / \tfrac{4}{3}\pi r_p^3,$$

and with equation (c)

$$R_A = -\frac{3k_G}{r_p}\left[\frac{3\phi + \left(\frac{3\phi^2}{a_s r_p} - 1\right)\tanh 3\phi}{3\phi + \left(\frac{k_G r_p}{D_i} + \frac{3\phi^2}{a_s r_p} - 1\right)\tanh 3\phi}\right]\frac{K\bar{c}_A - \bar{c}_P}{K+1}. \quad (d)$$

Table VII.5 shows a few special forms of this equation. When $\phi \ll 1$, the concentration in the particle is uniform at a level determined by the relative importance of the mass transfer coefficient k_G. If k_G is sufficiently large, neither external mass transfer nor internal diffusion limits the rate of conversion, and the surface area is utilized to the greatest possible extent, equation (d2).

The reaction occurs almost entirely near the periphery of the particle if $\phi > 1$. With a sufficiently high mass transfer coefficient, the conversion rate is proportional to $\sqrt{(k_1'' a_s D_i)}$ (equation (d5)); this result is analogous to the

Table VII.5. Special forms of Equation (d), Example VII.12.2.a

Slow reaction in a porous particle

$\phi \ll 1$:
$$R_A \approx - \frac{k_G k_1''(K+1)(a_s r_p + 3)}{r_p[k_G K + k_1''(K+1)(1 + a_s r_p/3)]} \frac{K\bar{c}_A - \bar{c}_p}{K+1} \quad (d1)$$

$\tanh 3\phi \approx 3\phi - 9\phi^3$

special cases:
$$R_A \approx - \frac{k_1''(K+1)(a_s r_p + 3)}{r_p K} \frac{K\bar{c}_A - \bar{c}_p}{K+1} \quad (d2)$$

$k_G K/k_1''(K+1) \gg 1$
(greatest possible conversion rate)

$k_G K/k_1''(K+1) \ll 1$
$$R_A \approx - \frac{3k_G}{r_p} \frac{K\bar{c}_A - \bar{c}_p}{K+1} \quad (d3)$$

Fast reaction in a porous particle

$\phi > 1$ and $a_s r_p \gg \phi$:
$$R_A \approx - \frac{3k_G(3\phi - 1)}{r_p(3\phi - 1 + k_G r_p/D_i)} \frac{K\bar{c}_A - \bar{c}_p}{K+1} \quad (d4)$$

$(\tanh 3\phi \simeq 1)$

special cases:
$k_G r_p/D_i \gg 3\phi$
$\phi \gg 1$
$$R_A \approx - \frac{3}{r_p} \sqrt{\left(\frac{k_1'' a_s D_i (K+1)}{K}\right)} \frac{K\bar{c}_A - \bar{c}_p}{K+1} \quad (d5)$$

$k_G r_p/D_i \ll 3\phi$
$$R_A \approx - \frac{3k_G}{r_p} \frac{K\bar{c}_A - \bar{c}_p}{K+1} \quad (d6)$$

Reaction at the outer surface of a non-porous particle

$a_s = 0$:
$$R_A \approx - \frac{3k_G k_1''(K+1)}{r_p[K + k_1'(K+1)]} \frac{K\bar{c}_A - \bar{c}_P}{K+1} \quad (d7)$$

special cases:
$Kk_G/k_1''(K+1) \gg 1$
$$R_A \approx - \frac{3k_1''(K+1)}{r_p K} \frac{K\bar{c}_A - \bar{c}_P}{K+1} \quad (d8)$$
(greatest possible conversion rate)

$Kk_G/k_1''(K+1) \ll 1$
$$R_A \approx - \frac{3k_G}{r_p} \frac{K\bar{c}_A - \bar{c}_P}{K+1} \quad (d9)$$

expression describing the mass flux for diffusion combined with a fast reaction; see equation (VII.41). With non-porous particles, two extreme cases may occur where either the chemical réaction rate or the mass transfer rate is entirely in control (equation (d8) and (d9) respectively). The expression (d9) is identical with (d3) and (d6), since in all three cases it was assumed that the external mass transfer would be the main limiting factor in the conversion rate. In practice, however, the conditions for which (d3) and (d6) apply are rarely encountered.

From equation (d) we also may derive the exact equation for mass transfer with first order irreversible reaction ($K = \infty$) in a sphere. With no gas phase resistance ($k_G \to \infty$) and with $k_1'' a_s = k_1$ we find

$$R_A = -\frac{3D_i}{r_p^2}\left(\frac{3\phi - \tanh 3\phi}{\tanh 3\phi}\right) c_{Ai}$$

or, with $R_A \equiv -k_1 c_{Ai} \eta$,

$$\eta_{exact} = \frac{1}{3\phi^2}\left(\frac{3\phi - \tanh 3\phi}{\tanh 3\phi}\right).$$

From equation (VII.67) we have as an approximate solution

$$\eta_{approx} = (\tanh \phi)/\phi.$$

Both solutions give the same asymptotic values for $\phi \ll 1$ and $\phi \gg 1$ while for $\phi = 1$ the exact and approximate η are 0.67 and 0.76 respectively. Hence, the accuracy of the approximation is within 15%. For the more general case with k_G and K not infinite, the effectiveness factor defined as the ratio of the actual conversion rate and the greatest possible conversion rate follows from

$$\eta = \frac{R_A \text{ (equation (d))}}{R_A \text{ (equation (d2), Table VII.5)}}.$$

Figure VII.38 shows a few curves of η as a function of the parameter ϕ for a porous catalyst of which $a_s r_p \gg 1$. The curves show that the degree of utilization of the internal catalyst surface decreases as the chemical reaction rate and the physical resistance to the transport of reactant become greater. It is seen that, with a given combination of reaction system and catalyst, the variables available for improving the effectiveness factor are the particle size and, to a lesser extent, the mass transfer coefficient k_G (which is roughly proportional to $(v/r_p)^{1/2}$. Inversely, a variation of these two parameters in conversion experiments makes it possible to ascertain to what extent physical transport influences the overall conversion rate. Thus, if it is found that the conversion rate is not influenced by the fluid velocity, it may be concluded that the effect of external mass transfer is negligible; internal diffusion may, however, still be the controlling factor. Whether this is true can be decided by investigating the influence of the particle diameter.

The effect of temperature on the conversion rate greatly depends on the relative importance of the physical transport phenomena, which are much less

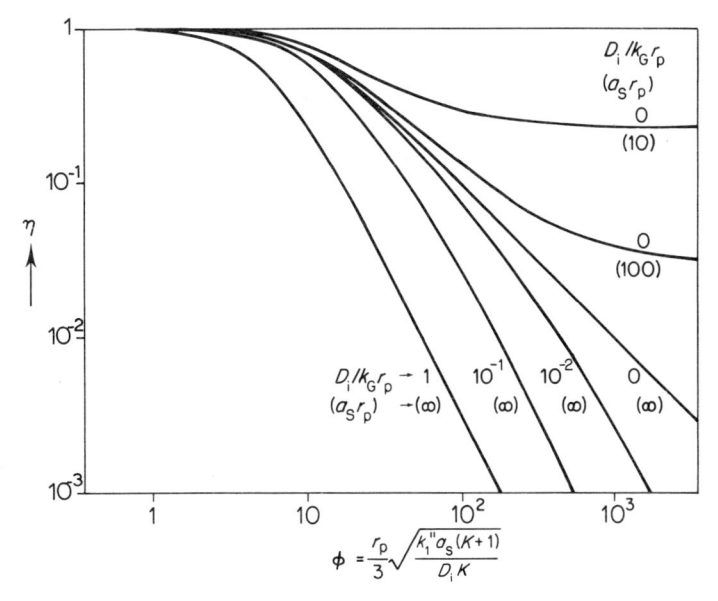

Fig. VII.38. Degree of utilization or effectiveness factor (η) of a porous catalyst for a first order equilibrium reaction

$$\phi = \frac{r_p}{3}\sqrt{\frac{k_1'' a_s (K+1)}{D_i K}}$$

temperature-dependent than the chemical reaction velocity itself. This is qualitatively shown in figure VII.39, where the logarithm of $|R|_A$ has been plotted as a function of $1/T$ for various cases; the slope of a curve in this diagram is a measure of the 'activation energy' of the overall conversion rate. The slope will indicate the activation energy of the chemical reaction only at such a low temperature that equation (d2) in Table VII.5 applies. On increasing the temperature, the external mass transfer may directly become rate-determining (Curve 1, for a relatively small value of $k_G r_p/D_i$) or eventually, after a temperature region has been passed where equation (d5) of Table VII.5 applies approximately (Curve 2). In this particular region, the apparent activation energy of the overall conversion rate is approximately one half of the chemical activation energy. With catalyst particles having a relatively small internal surface area, it is possible that a second intermediate region occurs with the chemical activation energy (Curve 3, equation (d8)). In this situation, the internal surface is no longer active but the chemical reaction at the external surface still determines the conversion rate. It may be concluded that experimentally determined activation energies of heterogeneous reactions (catalysed or not) may often not be equal to the activation energy of the reaction proper, and that extrapolation of experimental results of this nature to other temperatures may be hazardous. This will also be true for extrapolation to other particle sizes. A curve analogous to Curve 2 of figure VII.39 is found for the reaction between oxygen and coal (Wicke [65, 66]). Figure VII.40 shows, as an example, the overall conversion rate of oxygen (A) in a bed of porous coke particles as determined by Hedden [67].

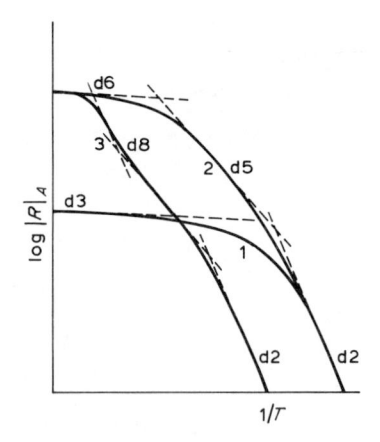

Fig. VII.39. Schematic representation of the effect of temperature on the conversion rate with a porous catalyst. $a_s r_p$ of curves 1 and 2 > $a_s r_p$ of curve 3; $k_G r_p / D_i$ of curves 2 and 3 > $k_G r_p / D_i$ of curve 1

Two effects that may modify the results given above call for attention. Firstly, in gas reactions it may occur that the number of moles changes upon conversion. In such a case, a net molar flow in radial direction is established; this flow sensibly affects the diffusive transport of reactants into the catalyst particle and of products out of it if this transport is brought about by free molecular diffusion in the pores. If, on the other hand, the mean free path of the gas molecules is much longer than the diameter of the pores, Knudsen diffusion prevails and the various species migrate independently of each other; this effect needs not be considered in this case.

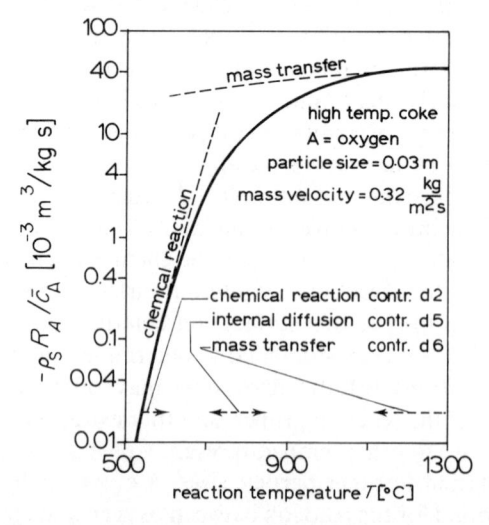

Fig. VII.40. Overall conversion rate for first order reaction between air oxygen (A) and porous coke [From Hedden [67]. Reproduced by permission of Pergamon Press Ltd]

A second effect not taken into account relates to the possibility that the temperature in a porous particle is not uniform. This may occur in relatively large and poorly conducting particles of catalyst in which a reaction with a high heat effect takes place. An early paper by Carberry [68] indicates that for first order kinetics the factor of catalytic effectiveness is significantly affected when

$$\left[\frac{E}{RT}\frac{\bar{c}_A(\Delta H_r)_A D_i}{\lambda_s T}\right] > 1.$$

Schilson and Amundson [69] developed an elegant method for calculating the factor of catalytic effectiveness for exothermic and endothermic reactions with Knudsen diffusion in the pores. In Sections VII.14 and IX.2.1 we will discuss this temperature effect in more detail.

Example VII.12.2.b   *Packed bed reactor versus riser*
For carrying out the equilibrium reaction of Example VII.12.2.a we have the following options:

— a packed bed reactor where relatively large catalyst particles form the stationary reaction phase; average $d_p = 0.02$ m and interparticle porosity $\varepsilon = 0.5$;
— a riser where small catalyst particles are cocurrently entrained by the gas flow, are separated in a cyclone and then recycled to the reactor inlet; $d_p = 0.2 \times 10^{-3}$ m and interparticle porosity $\varepsilon = 0.97$.

What is the ratio, CAP^*, of the two conversion rates per m³ reactor, as a function of ϕ?

Let us first assume that gas phase resistance is absent. With subscript ri for the riser and pb for the packed bed reactor we define CAP^* as

$$CAP^* = \frac{(1-\varepsilon)_{pb}(R_A)_{pb}}{(1-\varepsilon)_{ri}(R_A)_{ri}} \tag{a}$$

From equation (d) of Example VII.12.2.a, we derive, neglecting the external surface compared to the internal surface ($3\phi^2/a_s r_p \ll 1$ as is nearly always the case for porous catalysts):

$$R_A = k_1 \frac{K\bar{c}_A - \bar{c}_P}{(K+1)} \eta, \tag{b}$$

with

$$\eta = \frac{1}{3\phi^2}\left(\frac{3\phi - \tanh 3\phi}{\tanh 3\phi}\right) \simeq \frac{\tanh \phi}{\phi}. \tag{c}$$

we further know that

$$\phi_{ri} = \phi_{pb}\frac{(d_p)_{ri}}{(d_p)_{pb}}.$$

Substitution of these data into the relation for CAP^* gives

$$CAP^* = \frac{(1-\varepsilon)_{pb}}{(1-\varepsilon)_{ri}} \frac{(d_p)_{ri}}{(d_p)_{pb}} \frac{\tanh \phi_{pb}}{\tanh\left\{\phi_{pb} \frac{(d_p)_{ri}}{(d_p)_{pb}}\right\}}. \tag{d}$$

For $\phi_{pb} < 0.2$,

$$CAP^* = \frac{(1-\varepsilon)_{pb}}{(1-\varepsilon)_{ri}},$$

which is the ratio of catalyst volumes in the two reactors. For $\phi_{pb} > 200$,

$$CAP^* = \frac{(1-\varepsilon)_{pb}}{(1-\varepsilon)_{ri}} \frac{(d_p)_{ri}}{(d_p)_{pb}},$$

which is the ratio of the external catalyst surfaces in the two reactors. From equation (d) and with the data given, the value of CAP^* as a function of ϕ_{pb} is plotted in figure VII.41.

How would a gas phase resistance influence this picture?

Typical values for k_G can be found from Section VII.17.1. Large particles would probably have a lower k_G than small particles. Let us assume the following values: $(k_G)_{pb} = 0.05$ m/s, $(k_G)_{ri} = 0.2$ m/s and $D_i = 10^{-5}$ m²/s. Gas phase resistance will be substantial for $\phi > k_G \delta / D_i = \frac{k_G d_p}{6 D_i}$. This condition is fulfilled in the packed bed for $\phi_{pb} > 17$ and in the riser for $\phi_{ri} > 0.66$ or $\phi_{pb} > 66$. Hence, the ultimate production rate ratio reduces to $CAP^* = \frac{(k_G a_p)_{pb}}{(k_G a_p)_{ri}} = 0.04$.

If both external and internal resistances play a role the relation for the production rate ratio CAP^* follows by using equation (d) of Example VII.12.2.a:

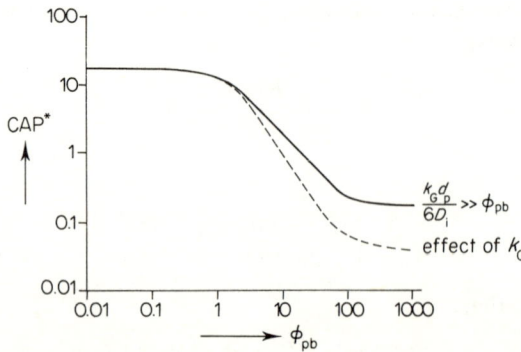

$$R_A = -\frac{6}{d_p} \frac{K \bar{c}_A - \bar{c}_P}{K+1} \frac{1}{\dfrac{1}{k_G} + \dfrac{d_p}{2D_i} \dfrac{\tanh 3\phi}{3\phi - \tanh 3\phi}}, \tag{d}$$

Fig. VII.41. The conversion rate ratio for a packed bed and a riser reactor (CAP^*) as a function of ϕ in a packed bed, ϕ_{pb}, with data as given in Example VII.12.2.b

where

$$\frac{\tanh 3\phi}{3\phi - \tanh 3\phi} \simeq 1/(3\phi \tanh \phi),$$

and by substituting this for both reactors in equation (a) for CAP^*.

This illustration clearly shows that for fast reactions small catalyst particles must be used (risers, fluidized beds). On the other hand, for slow reactions packed beds with larger particles may be the best choice, because they have a larger effective catalyst volume per m³ reactor. Whatever reactor is used, the conversion will be limited by the equilibrium conversion. For example, for plug flow we have

$$-u\frac{dc_A}{dz} = (1-\varepsilon)R_A$$

with equation (d), and after substitution of $\bar{c}_A = c_{Ao}(1-\zeta)$ and $\bar{c}_P = \zeta c_{Ao}$ it follows that

$$1 - \frac{(K+1)\zeta}{K} = \exp(-L/HTU),$$

where

$$HTU = \frac{u_S d_p}{6(1-\varepsilon)}\left(\frac{1}{k_G} + \frac{d_p}{6D_i \phi \tanh \phi}\right).$$

For a sufficiently long reactor $L \to \infty$ and $\zeta_e = K/(K+1)$.

Especially for $K < 1$, expensive technical provisions are necessary for the separation of the products and the recycling of the unconverted reactants. A new solution coping with this problem is the gas–solid trickle flow reactor where the catalyst particles 'rain' through the column and where the catalyst material is selected in such a way that the products remain adsorbed at the surface. This method of countercurrent operation allows for complete conversion in one cycle. In a second vessel, e.g. a fluidized bed, the products are desorbed continuously at a higher temperature and/or a lower pressure level after which the catalyst is recycled to the top of the reactor.

Example VII.12.2.c *Conversion rate of a solid according to a sharp interface shrinking unreacted core model*
A solid metal oxide is reduced with carbon monoxide according to the equilibrium reaction:

$$A(G) + v_B B(S) \to v_P P(G) + v_Q Q(S)$$

with

$$R_A'' = -k_S''\left(c_A - \frac{c_P}{K}\right)$$

where the species P is CO_2 and Q the pure metal.

The reaction takes place at the sharp boundary between the porous metal product layer and the shrinking unreacted non-porous metal oxide core. The reaction is assumed to occur as a sequence of the following steps:

— diffusion of the reducing gas A through the external film and the porous product layer of the pure metal;
— chemical reaction at the sharp interface between A and B;
— diffusion of the product gas P through the product layer and the external gas film into the bulk of the gas.

We will derive an expression for ζ_B as a function of time and discuss the several possible rate limiting cases. The rate of transport of A to the particle equals the diffusion rate of A in the reacted shell and the conversion rate of A at the core surface so that:

$$J_A 4\pi r_p^2 = k_{GA} 4\pi r_p^2 (\bar{c}_A - c_{Ai}) = \frac{D_{iA} 4\pi r_p r_c}{r_p - r_c}(c_{Ai} - c_{Ac})$$

$$= k_S'' 4\pi r_c^2 \left(c_{Ac} - \frac{c_{Pc}}{K}\right), \tag{a}$$

where subscripts i and c refer to particle surface and unreacted core interface respectively (see figure VII.42). The same holds for P:

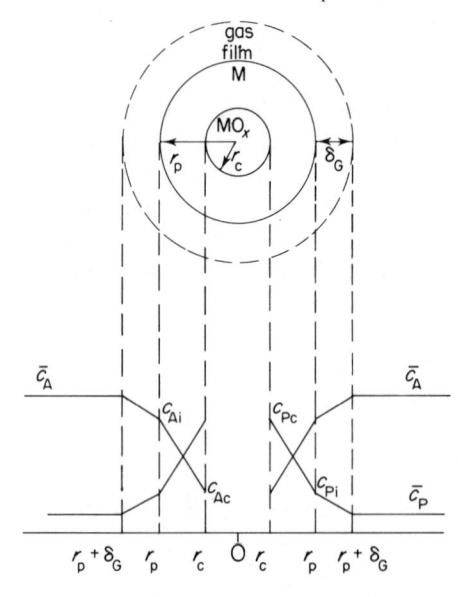

$$J_P 4\pi r_p^2 = k_{GP} 4\pi r_p^2 (c_{Pi} - \bar{c}_P) = \frac{D_{iP} 4\pi r_p r_c}{r_p - r_c}(c_{Pc} - c_{Pi}) \tag{b}$$

Fig. VII.42. Unreacted shrinking core model with equilibrium reaction:
$$CO + \frac{1}{x}MO_x \rightleftharpoons CO_2 + \frac{1}{x}M$$

and further $J_A = J_P/v_P$. \hfill (c)

From equations (a), (b) and (c) the local concentrations c_{Ai}, c_{Ac}, c_{Pi} and c_{Pc} can be eliminated in the usual way; we then get

$$J_A = \frac{D_A}{r_p}\left(\bar{c}_A - \frac{\bar{c}_p}{K}\right)\frac{1}{A^*}, \qquad (d)$$

with

$$A^* = \frac{2}{Sh_A} + \frac{D_A}{D_{iA}}\left(\frac{1-r^*}{r^*}\right) + \frac{D_A}{k_s''r_p}\frac{1}{(r^*)^2} + \frac{D_A}{KD_{iP}}\left(\frac{1-r^*}{r^*}\right) + \frac{2}{Sh_P}\frac{D_A}{KD_P} \qquad (e)$$

and $r^* = r_c/r_p$. \hfill (f)

Equation (d) gives an expression for J_A as a function of r^* which will change from 1 to 0 during the reduction process. If $\bar{c}_{Bo}$ is the concentration of metal oxide in the virgin particle we find the rate of change of the core diameter dr^*/dt from

$$J_A 4\pi r_p^2 = -4\pi r_i^2 \frac{dr_i}{dt}\frac{\bar{c}_{Bo}}{v_B} \qquad (g)$$

and, after substitution of equation (d),

$$\frac{dr^*}{dt} = -\frac{D_A}{r_p^2}\frac{v_B}{\bar{c}_{Bo}}\left(\bar{c}_A - \frac{\bar{c}_p}{K}\right)\frac{1}{r^{*2}}\frac{1}{A^*}. \qquad (h)$$

For constant values of $\bar{c}_A$ and $\bar{c}_P$ during the whole conversion process equation (h) can easily be integrated. With $r^{*3} = 1 - \zeta_B$ it follows that

$$\left(\frac{2}{Sh_A} + \frac{2}{Sh_P}\frac{D_A}{KD_p}\right)\frac{\zeta_B}{3} + \left(\frac{D_A}{D_{iA}} + \frac{D_A}{KD_{iP}}\right)\left[\frac{1}{6} - \frac{1}{2}(1-\zeta_B)^{2/3} + \frac{1}{3}(1-\zeta_B)\right]$$
$$+ \frac{D_A}{k_s''r_p}[1 - (1-\zeta_B)^{1/3}] = \frac{D_A}{r_p^2}\frac{v_B}{\bar{c}_{Bo}}\left(\bar{c}_A - \frac{\bar{c}_p}{K}\right)t. \qquad (i)$$

Equation (i) e.g. is very useful in conversion experiments with a thermal balance where $\bar{c}_P$ and $\bar{c}_A$ are kept constant. In countercurrent flow blast furnaces or in moving bed gasifiers, $\bar{c}_A$ and $\bar{c}_P$ vary with ζ_B, so equation (h) must be applied in stepwise calculations.

Asymptotic cases of (i) for irreversible reactions are given in Table VII.6 where τ is the time required for complete conversion and t/τ is the time required to reach a conversion ζ_B. Figure VII.43 shows $d\zeta_B/d(t/\tau)$ as a function of ζ_B for the various asymptotic solutions. Notice that τ is different for each regime (see Table VII.6). Notice also that in practice the rate controlling step may depend on ζ. For low conversions either external diffusion or kinetics will control the conversion rate. For large conversions and not too small particles often the internal diffusion resistance controls the rate of conversion.

Table VII.6. Converstion–time relations according to the shrinking core model for various controlling regimes

Controlling resistance	t/τ	τ
External mass transfer	ζ_B	$\dfrac{r_p \bar{c}_{Bo}}{3 k_G v_B \bar{c}_A}$
Diffusion through converted shell	$1 - 3(1 - \zeta_B)^{2/3} + 2(1 - \zeta_B)$	$\dfrac{r_p^2 \bar{c}_{Bo}}{6 D_{iA} v_B \bar{c}_A}$
Chemical kinetics	$1 - (1 - \zeta_B)^{1/3}$	$\dfrac{r_p \bar{c}_{Bo}}{k_S'' v_B \bar{c}_A}$

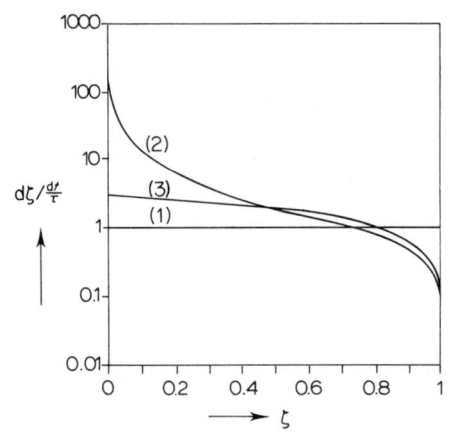

Fig. VII.43. Conversion rate as a function of conversion according to shrinking core model: (1) external diffusion controls; (2) diffusion in converted shell controls; (3) chemical kinetics control

VII.13 GENERAL CRITERION FOR ABSENCE OF MASS TRANSPORT LIMITATION

As discussed in previous sections, the conversion rate is not limited by mass transport for $\phi < 0.2$ and $Al\phi^2 \ll 1$. Then $\bar{c}_A = c_{Ai}$. Particularly in kinetic research, it is of importance to know whether or not the experiment has been carried out in the kinetic regime, that is under absence of mass transport limitations. As long as the kinetics are unknown, a criterion as $\phi < 0.2$ and $Al\phi^2 \ll 1$ is not of much help because calculation of ϕ asks for the kinetics of the reaction. Fortunately there exists a criterion that can be applied without knowing the actual kinetics. Knowledge of the observed conversion rate $(-\Phi_A)$ is sufficient. We can write the generally valid relation:

$$J_A a V_r = -\Phi_A = \frac{D_A}{\delta} \Delta c_A E_A a V_r \qquad \text{(VII.227)}$$

with $\Delta c_A = c_{Ai} - \bar{c}_A$, so that

$$\frac{\Delta c_A}{c_{Ai}} E_A = \frac{-\Phi_A}{V_r} \frac{\delta}{D_A a c_{Ai}} \qquad \text{(VII.228)}$$

In case of negligible (defined as less than 5%) mass transport limitation

$$\frac{\Delta c_A}{c_{Ai}} \leq 0.05 \text{ and } E_A = 1 \text{ so that we have } \frac{\Delta c_A}{c_{Ai}} E_A \leq 0.05.$$

In case of complete mass transfer limitation

$$\frac{\Delta c_A}{c_{Ai}} \simeq 1 \text{ and } E_A \geq 1 \text{ so that we have } \frac{\Delta c_A}{c_{Ai}} E_A \geq 1.$$

Hence, no mass transfer limitation has occurred for

$$\phi'' \equiv \frac{-\Phi_A}{V_r} \frac{\delta}{D_A a c_{Ai}} \leq 0.05 \qquad \text{(VII.229)}$$

and complete mass transfer limitation has occurred for

$$\phi'' \equiv \frac{-\Phi_A}{V_r} \frac{\delta}{D_A a c_{Ai}} \geq 1 \qquad \text{(VII.230)}$$

In gas–liquid reactions, criterion (VII.229) may be written as

$$\phi'' \equiv \frac{-\Phi_A}{V_r} \frac{1}{k_L a c_{Ai}} \leq 0.05,$$

while in stagnant drops or porous pellets the criterion reduces to

$$\phi'' \equiv \frac{-\Phi_A}{(1-\varepsilon)V_r} \frac{\delta^2}{D_i c_{Ai}} \leq 0.05 \qquad \text{(VII.231)}$$

with δ = volume/external surface ($=d_p/6$ for spherical particles), $\Phi_A/[(1-\varepsilon)V_r]$ = the observed conversion rate per m³ particle or drop. For first order kinetics, it follows from equation (VII.231) that $\phi'' = \phi^2 \eta$, with limits $\phi'' = \phi^2$ (for $\eta = 1$, that is for $\phi < 0.2$) and $\phi'' = \phi$ (for $\eta = 1/\phi$, that is for $\phi > 2$).

A somewhat more sophisticated analysis shows that the criterion (VII.231) is conservative. Dependent on the kinetics it may be less stringent [70]:

$$\frac{-\Phi_A}{(1-\varepsilon)V_r} \frac{\delta^2}{D_i c_{Ai}} \leq 0.03 - 0.7. \qquad \text{(VII.232)}$$

On the other hand, for a numerical value exceeding 0.7, definite existence of diffusion effects is indicated. For refinements with respect to the criterion (VII.232), the reader is referred to the review of Mears [71], and to Section IX.2.1.

Example VII.13 *Gasification of wood char according to a grain model*
One of the many gas–solid reaction models is the so-called grain model. Here the porous particle is thought to consist of grains of reactive material which are kept together by an inert matrix and are surrounded by macro pores. For relatively low reaction rates with respect to mass transfer around and in the particles, the gas composition close to the grain surface equals the bulk gas composition. If the Thiele modulus of the grains in the particle (ϕ_{grain}) is much larger than one, these grains can be assumed to react according to a kinetically controlled shrinking core. Such a model is only realistic if the diffusivity in the grains is much smaller than the diffusivity in the macro pores. We will now:

(a) derive an expression for the solid conversion as a function of time for a constant surface conversion rate;
(b) check the grain model for gasification rates of wood char according to
$$CO_2 + C \rightarrow 2 CO$$
(A) (B)

for the following conditions: $T = 801\ °C$, $\bar{c}_{AG} = 1.75\ mol/m^3$ and $p = 1\ bar$. Moreover, $M_B = 12$, the true carbon density $\rho_{BS} = 2200\ kg/m^3$, the initial internal specific contact area $a_{So} = 140 \times 10^6\ m^2/m^3$ particle, the initial internal porosity $\varepsilon_o = 0.68$ and $d_p = 10^{-3}\ m$. The observed conversions as a function of time are given in Table VII.7 (measured by F. G. van den Aarsen, to be published);
(c) check on external and internal diffusion limitations.

With the number of grains in a particle remaining constant during the conversion process, the grain diameter will decrease in time and so will a_S. With n being the number of grains/m³ we have $n\dfrac{\pi}{6}d_g^3 = (1 - \varepsilon)$, for which it follows for the interfacial area that

$$\frac{a_S}{a_{So}} = \frac{(1 - \varepsilon)^{2/3}}{(1 - \varepsilon_o)^{2/3}} = (1 - \zeta_B)^{2/3}. \qquad (a)$$

Table VII.7. Char conversion as a function of time for conditions given in Example VII.13

ζ_B (%)	t (s)	ζ_B (%)	t (s)
0.995	79	41.138	3740
7.435	606	46.192	4264
13.551	1130	51.041	4785
19.412	1652	55.683	5305
24.795	2147	68.218	6870
30.204	2659	85.011	10002
35.930	3217	90.509	12084

Equation (a) gives the change of a_S as a function of ζ_B and initial area a_{S_o}. Assuming a reaction:

$$A(G) + v_B B(S) \rightarrow v_B P(G)$$

with rate

$$R_A = R_A'' a_S, \quad (b)$$

the mass balance for B becomes

$$R_A = -\frac{1}{v_B} \frac{\rho_B}{M_B} \frac{d(1-\varepsilon)}{dt}. \quad (c)$$

Combination of (a), (b) and (c) gives, after integration, with the boundary conditions $\zeta = 0$ for $t = 0$,

$$(1 - \zeta_B)^{1/3} = 1 - \frac{t}{\tau} \quad (d)$$

in which

$$\tau = \frac{3\rho_B(1-\varepsilon_o)}{M_B v_B a_{S_o} R_A''}. \quad (e)$$

Moreover, for (pseudo) first order holds: $R_A'' = k_1'' \bar{c}_{AG}$. Correlating $(1-\zeta_B)^{1/3}$ with t by a straight line gives, as the best fit,

$$(1 - \zeta_B)^{1/3} = 1.00641 - \frac{t}{2.164 \times 10^4}$$

with a correlation coefficient 0.9988. This is very high and therefore the approach seems promising. But it is not a proof. Equation (d) is exactly the same as obtained in the previous example for the regime of a kinetically controlled shrinking core. If only the product $k_1'' a_S$ is measured there is no way to distinguish between the two models. We know however that in char with $\varepsilon_o = 0.68$, there are macro pores throughout the particle, and therefore we prefer the grain model.

With respect to the last question, negligible external mass transfer resistance can be expected for

$$\frac{k_1 d_p}{6 k_G} = \frac{-R_A \pi d_p^3}{6 k_G \bar{c}_A \pi d_p^2} \ll 1,$$

in which R_A has to be taken as the maximum conversion rate per unit particle volume; hence $R_A = R_A'' a_{S_o}$. With $\tau = 2.164 \times 10^4$ s, it follows from equation (e) that $R_A = -8.13 \times 10^{-3}$ kmol/m³ s. Assuming $Sh = 2$ and $D_{CO_2} = 8.7 \times 10^{-5}$ m²/s, it follows that

$$\frac{k_1 d_p}{6 k_G} = \frac{-R_A d_p^2}{12 D \bar{c}_A} = 4.5 \times 10^{-3},$$

which implies that there is no external diffusion limitation. To check for the absence of internal diffusion limitation we can use equation (VII.232). With $\delta = d_p/6$ this results in the condition

$$\frac{R_A d_p^2}{D_i \bar{c}_A} < 1.$$

From the data given it follows that this condition is fulfilled for $D_i/D > 5.4 \times 10^{-2}$; this is true because of the high internal porosity of the char particle with $\varepsilon > 0.68$.

VII.14 HEAT EFFECTS IN MASS TRANSFER WITH REACTION

So far, we have not considered possible local temperature effects caused by the chemical reaction. In principle, however, it is possible that the temperature in the reaction zone differs substantially from the bulk temperature. Temperature gradients also cause problems with the real driving forces for mass transfer. We have based our calculations on the concentration gradient as the driving force for mass transfer, e.g. in

$$J_A = \frac{D_A}{\delta}(\bar{c}_A - c_{Ai}) = -D_A \frac{dc_A}{dx}.$$

However, it must be realized that these equations in principle are not correct if a temperature gradient exists. The temperature gradient is accompanied by a density gradient. This may result in a concentration gradient, even without giving a mass flux, e.g. the concentration difference $px_{A1}/(RT_1) - px_{A2}/(RT_2)$ will not result in a mass flux if $x_{A1} = x_{A2}$, although the concentrations are different due to the temperature difference. Correct equations have to be based on

$$J_A = -D_A \frac{c}{(1-x_A)} \frac{dx_A}{dx},$$

in which c is the total concentration and x_A the molar fraction of A. Fortunately, in many processes where mass transfer with reaction causes a local temperature gradient, this rigorous approach is not necessary and we will also in this section use the concentration as the driving force for mass transfer. However, the accuracy of this simplified approach should be checked for each particular problem and, if deemed to be necessary, our analysis should be based on molar fraction gradients as driving forces for mass transfer.

VII.14.1 Mass transfer with reaction in series

Consider the case of mass transfer followed by reaction at the external surface of a non-porous particle. Then

$$J_A = \frac{D_A}{\delta_m}(\bar{c}_A - c_{Ai}) \qquad (VII.233)$$

where δ_m is the stagnant film thickness for mass transfer. The temperature at the surface follows from

$$J_h = J_A(\Delta H_r)_A = \frac{\lambda}{\delta_h}(\bar{T} - T_i), \tag{VII.234}$$

where δ_h is the film thickness for heat transfer. From equations (VII.233) and (VII.234) it follows that

$$\frac{\bar{T} - T_i}{\bar{c}_A - c_{Ai}} = \frac{D_A}{\lambda}\frac{\delta_h}{\delta_m}(\Delta H_r)_A. \tag{VII.235}$$

In case the Chilton–Colburn analogy holds,

$$\frac{Nu}{Re\,Pr^{1/3}} = \frac{Sh}{Re\,Sc^{1/3}} \tag{VII.236}$$

it follows ($\alpha \equiv \lambda/\delta_h$) that

$$\frac{\delta_h}{\delta_m} = Le^{1/3} \tag{VII.237}$$

where Le is the Lewis number ($= Sc/Pr = \lambda/(\rho c_p D_A)$). Experimentally the value $0.7 Le^{1/3}$ has been observed in packed beds [72]. Combination of equations (VII.235) and (VII.237) gives

$$\frac{\bar{T} - T_i}{\bar{c}_A - c_{Ai}} = \frac{(\Delta H_r)_A}{\rho c_p} Le^{-2/3}. \tag{VII.238}$$

The difference between equation (VII.238) and the relation for the wet bulb temperature is that in the latter case there is equilibrium at the interface. In mass transfer with reaction the relation is more complex. In case of a first order reaction

$$J_A = k_1'' c_{Ai} = \frac{D_A}{\delta_m}(c_A - c_{Ai}), \tag{VII.239}$$

where

$$k_1'' = k_\infty'' e^{-E/RT}, \tag{VII.240}$$

it follows from equations (VII.238) and (VII.240), after elimination of c_{Ai} by means of equation (VII.239), that

$$(T_i - \bar{T}) = \frac{-(\Delta H_r)_A}{\rho c_p}\frac{\bar{c}_A}{Le^{2/3}}\frac{1}{[1 + D_A e^{E/RT_i}/(\delta_m k_\infty'')]}. \tag{VII.241}$$

From equation (VII.241), T_i can be calculated. The maximum temperature difference is found for complete mass transfer limitation. In that case $D_A/\delta_m k_1'' \ll 1$. Hence

$$(T_i - \bar{T})_{max} = -\frac{(\Delta H_r)_A \bar{c}_A}{\rho c_p Le^{2/3}} = \frac{\Delta T_{ad}}{Le^{2/3}}. \tag{VII.242}$$

It is evident that equation (VII.242) is independent of the kinetics. For many liquids Lewis is of the order of 10–100, while Lewis for gases is of the order of one. Therefore in gas–solid reactions considerable temperature differences between gases and solids occur more frequently than in gas–liquid reactions. In case of very soluble gases with fast reactions, however, large temperature gradients are not exceptional. This case will be treated below.

VII.14.2 Mass transfer with simultaneous reaction in a gas–liquid system

We first postulate that

— Considerable heat effects will mainly occur in fast reactions. Therefore we will assume that $\bar{c}_{AL} = 0$.
— Unless $E_A < 3$, the reaction zone is small with respect to the diffusion zone. Hence in general only a small error is introduced by assuming that all heat is released at the interface. This statement is safe because $\delta_h > \delta_m$ in liquids.
— When evaporation does not play an important role, the resistance against heat transfer will be in the gas phase. Therefore we assume that all heat will be transported to or from the bulk of the reaction phase.

With these assumptions, it follows (by analogy with equation (VII.238)), that

$$T_i - \bar{T} \simeq - \frac{[(\Delta H_r)_A + (\Delta H_a)_A]}{\rho c_p Le^{2/3}} c_{Ai} E_A \qquad (VII.243)$$

where $(\Delta H_a)_A$ is the molar heat of absorption of A. c_{Ai} can be eliminated via

$$J_A = k_L c_{Ai} E_A = k_G \left(\bar{c}_{AG} - \frac{c_{Ai}}{m} \right). \qquad (VII.244)$$

Combination of equations (VII.243) and (VII.244) gives

$$T_i - \bar{T} = \frac{-[(\Delta H_r)_A + (\Delta H_a)_A]}{\rho c_p Le^{2/3}} \frac{m \bar{c}_{AG} E_A}{(1 + m k_L E_A / k_G)}. \qquad (VII.245)$$

In general both m and E_A depend on temperature. Usually m will decrease with temperature, while for $E_A \ll E_{A\infty}$, E_A will increase. For m and E_A the values at surface temperature should be taken. Therefore $T_i - \bar{T}$ can be calculated from equation (VII.245) by trial and error only. We will further elaborate on this phenomenon in Section (IX.1.1).

Penetration models on simultaneous heat and mass transfer with reaction have been discussed by Danckwerts [73, 74], Shah [75] and Clegg and Mann [76]. Refinements as the influence of liquid bulk flow, the effect of evaporation and gas phase heat transfer have been discussed by Tamir et al. [77].

In analysing practical problems, however, lack of accurate data is often more limiting than lack of theory. This has been shown by Beenackers and van Swaaij [78, 79] and by Mann and Moyes [80]. These authors sulphonated benzene and dodecyl benzene respectively and reported an interfacial temperature rise up to

60 °C. Beenackers and van Swaaij found evidence for a large viscosity increase and a large benzene sulphonic acid accumulation in the reaction zone at the interface. Moreover, free convection induced by the reaction was found to be the rate controlling mass transfer mechanism in a stirred cell reactor (described in section VII.15.1). Under such conditions, a limitation for accurate analysis may be found in a lack of data on solubility, diffusivity and reaction rate which cannot be measured but which will be influenced by both surface temperature and reactor product concentrations at the surface. Some of these difficulties are also reported by Mann and Moyes [80], who interpreted their experimental findings in terms of a solubility correlation calculated from Raoult's law. Large heat effects are also reported for the absorption of HCl in ethylene glycol [81] and the absorption of Cl_2 in toluene [82].

VII.14.3 Mass transfer with simultaneous reaction in a porous pellet

To give an introduction in this area, we will follow Weisz and Hicks [83] for a first order reaction. We start from the mass balance for A in a porous spherical pellet (stationary case)

$$D_i \frac{d^2 c_A}{dr^2} + \frac{2}{r}\frac{dc_A}{dr} = k_1 c_A. \tag{VII.246}$$

For temperature gradients within the pellet, k_1 becomes place dependent. Assume that k_1 depends on T according to Arrhenius:

$$k_1 = k_\infty e^{-E/RT}. \tag{VII.247}$$

Define k_{1i} as the rate constant at the particle surface temperature T_i. Then

$$k_1 = k_{1i} \exp\left(\frac{E}{RT_i} \frac{(T - T_i)}{T_i\left\{1 + \frac{(T - T_i)}{T_i}\right\}}\right). \tag{VII.248}$$

A relation between the local temperature $T(r)$ and the local concentration $c_A(r)$ follows from the consideration that the heat transport is related to the mass transport via:

$$J_h = J_A \Delta H_r$$

Further is postulated:

$$\lambda_s \frac{dT}{dr} = \Delta H_r D_i \frac{dc_A}{dr} \tag{VII.249}$$

or

$$T - T_i = \frac{\Delta H_r D_i}{\lambda_s}(c_A - c_{Ai}). \tag{VII.250}$$

Combination of equations (VII.248) and (VII.250) gives

$$k_1 = k_{1i} \exp\left[\frac{E}{RT_i} \frac{\Delta H_r D_i (c_A - c_{Ai})}{\lambda_s T_i \left\{1 + \frac{\Delta H_r D_i}{\lambda_s T_i}(c_A - c_{Ai})\right\}}\right]. \quad (VII.251)$$

Defining $\beta \equiv E/RT_i$
and

$$\gamma \equiv \frac{-\Delta H_r D_i c_{Ai}}{\lambda_s T_i} = \left(\frac{T - T_i}{T_i}\right)_{max} \quad (VII.252)$$

we obtain, with equation (VII.246),

$$D_i \frac{d^2 c_A}{dr^2} + \frac{2}{r}\frac{dc_A}{dr} = k_{1i} c_A \exp\left[\gamma \beta \frac{1 - \frac{c_A}{c_{Ai}}}{1 + \gamma\left(1 - \frac{c_A}{c_{Ai}}\right)}\right]. \quad (VII.253)$$

This equation gives c_A as $f(r)$. The equation has been solved numerically by Weisz and Hicks [83]. They present the solution in the form $\eta = f(\phi, \beta, \gamma)$. The effectiveness factor η is defined in the usual way but is referred to the surface temperature T_i:

$$\eta = \frac{-J_A}{\delta R_A(T_i)}. \quad (VII.254)$$

ϕ is the Thiele modulus at surface temperature T_i: $\phi = \delta\sqrt{(k_{1i}/D_i)}$; β is a dimensionless parameter that indicates how sensitive the reaction rate is for a change in temperature; γ is the dimensionless maximum possible temperature difference between the centre of the pellet and the surface. A high value of γ is obtained for poor conductivity (λ_s low) combined with fast mass transport into the pellet (D_i high) and a reaction proceeding with a large heat effect. For endothermic reactions γ will be negative. Table VII.8 gives values for β and γ for some commercial processes. Figure VII.44 presents some η–ϕ curves for different values of γ and β. It appears that for strong exothermic reactions (γ large) whose reaction rate increases rapidly with temperature (β large), the temperature effect becomes so important that effectiveness factors much larger than 1.0 are possible. Then the conversion rate is increased greatly due to the local high temperature in the pellet. Notice also the possibility of multiple solutions: at one particular ϕ-value, at most 3 η-values are possible, each at a different average particle temperature. This effect is quite analogous to the existence of multiple solutions in the autothermal reactor (discussed in Section VI.5).

We need a simple criterion depending only on observables which determines whether interparticle behaviour will be isothermal or not. Weisz derived that a particle is free of both mass and heat transport limitation for

$$\frac{\Phi_A}{(1-\varepsilon)V_r} \frac{\delta^2}{D_i c_{Ai}} e^{\gamma\beta/(1+\gamma)} < 0.1.$$

Table VII.8. Parameter values for some heterogeneous reactions. [From V. Hlaváček and M. Kubíček [84], *Chem. Eng. Sci.*, **25,** 1537 (1970)]

Reaction	γ	β	$\dfrac{Sh}{Nu} \dfrac{D_A}{D_i} \dfrac{\lambda_S}{\lambda_G}$
Ethylene hydrogenation	0.0029	13.2	580
Benzene hydrogenation	0.03	15.7	150
Methane oxidation	0.0056	21.5	310
Ethylene oxidation	0.068	—	45
Naphtalene oxidation	0.015	22.2	2.7
$SO_2 \rightarrow SO_3$	0.0031	—	105
Acrylonitrile synthesis	0.0019	9.9	4260

More generally applicable criteria have been reviewed by Mears [71]. From this reference we report as a criterion for isothermal operation:

$$|\gamma\beta| < 0.05n,$$

with n the order of the kinetics. In view of the great importance we will further elaborate on this in Section IX.2.1.

Intraparticle heat effects are only part of the problem. They must be coupled with the external (interparticle) heat and mass transfer discussed in the beginning of this section. This has been done by Hlaváček and Kubíček [84] and by McGreavy and Cresswell [85]. Figures VII.45 shows how for γ, β combinations

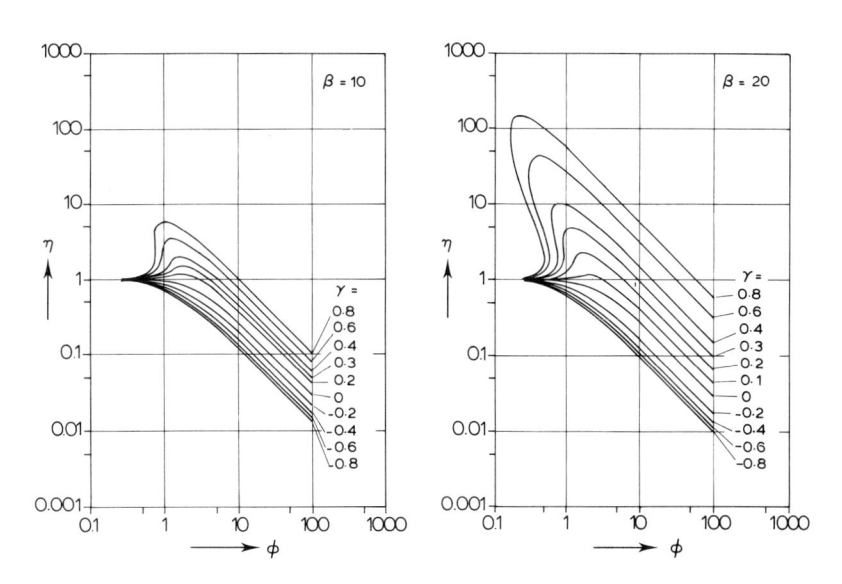

Fig. VII.44. The effectiveness factor (η) as a function of ϕ and γ for two values of β (10 and 20 respectively) [From Weisz and Hicks [83]. Reproduced by permission of Pergamon Press Ltd]

by which no large intraparticle heat effects are expected (see figure VII.44), large heat effects may already occur due to poor external heat transfer. ShD_A/D_i and $Nu\lambda_G/\lambda_S$ (called Sh and Nu by Hlaváček) are used as parameters in figures VII.45. The numerical values of their ratio being 20 and 5 respectively are not very high, compared to the values reported in Table VII.8. Hence, many catalytic reactions can be described by assuming particles of a uniform temperature. That temperature often will differ appreciably from the bulk gas phase temperature. This model has been discussed by Hlaváček and Kubiček [86].

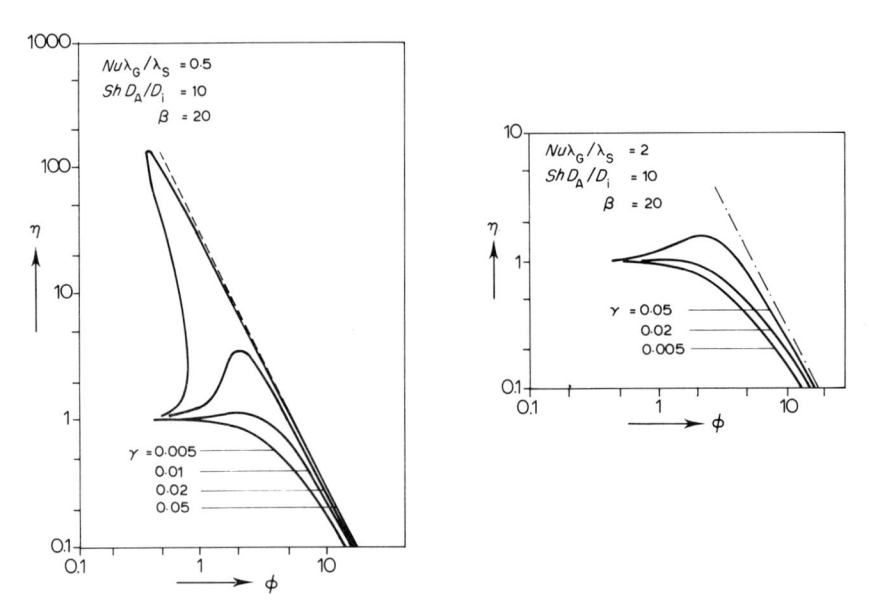

Fig. VII.45. The effectiveness factor (η) as a function of ϕ for two values of $Nu\lambda_G/\lambda_S$. First order reaction. Sphere. Dashed line: asymptotic solutions [From Hlavácek and Kubiček [84]. Reproduced by permission of Pergamon Press Ltd]

VII.15 MODEL REACTORS FOR STUDYING MASS TRANSFER WITH CHEMICAL REACTION IN HETEROGENEOUS SYSTEMS

Inherent to fast kinetics is the impossibility to carry out kinetic experiments in such a way that the reactants can be considered as being premixed. As long as the rate of the reaction is high with respect to the rate of mass transfer, the reaction comes to completion before well-premixed conditions have been established. Hence, fast kinetics must be studied essentially under conditions of diffusion limitation of at least one reactant. Therefore special two-phase reactors have been designed to study fast reactions under controlled conditions of mass transfer. These reactors are called model reactors. They have in common that the value of the interfacial area is accurately known. This makes it possible to calculate J_A from an observed mass transfer rate:

$$J_A = \frac{-\Phi_A}{aV_r} \text{ kmol/m}^2\text{ s} \qquad \text{(VII.255)}$$

The reactors also have in common that the mass transfer coefficient(s) can be varied in a controlled way.

The practical procedure for finding the reaction kinetics usually is as follows: J_A is measured as a function of the mass transfer coefficient(s), the reaction temperature and the bulk concentrations of reactants, products and possibly catalysts. Mostly the observed transfer rate J_A can be compared to the rate that would have occurred in case of no reaction $(J_A)_{phs}$. This way the enhancement factor E_A is obtained. The kinetic orders follow from the concentration dependency of J_A, except for $\phi \gg E_{A\infty}$. In the latter case, however, the reaction is instantaneous, which means that the actual kinetics are not relevant for reactor design for the particular conditions under which the experiment has been carried out. Once the kinetic order(s) have been determined, the rate constants follow from the actual values of E_A.

VII.15.1 Model reactors for gas–liquid reactions

For gas–liquid reactions the most important model reactors are the laminar film reactor (or wetted wall column), the laminar jet reactor and the stirred cell reactor, see figure VII.46. Often, the depth of penetration of the gas into the laminar film and laminar jet is so small that the absorption can be considered to occur in a semi infinite medium. This means that the liquid mass transfer coefficient follows from

$$k_L = 2\sqrt{\left(\frac{D_A}{\pi\tau}\right)} \qquad \text{(VII.256)}$$

with $\tau = L/v$, L = length of jet or film and v = surface liquid velocity. v can be calculated from first principles (see e.g. [1]). Thus, k_L can be varied by changing L and/or v. In the stirred cell reactor, the stirrer speed is so low that the liquid surface remains flat. Here k_L must be measured by a physical absorption experiment. A relation of the form

$$Sh = cRe^n Sc^{1/3} \qquad \text{(VII.257)}$$

is found with $n = 0.8$–1. In the stirred cell reactor the residence time of the liquid is an independent variable, which makes the reactor especially suitable for studying the kinetics as a function of conversion. The film and jet reactors are considered to be more accurate but also more difficult to operate properly. The most important difference, however, is the different contact time range that each reactor covers (see Table VII.9 [87]).

With the laminar film and jet reactors, no proper measurements are possible for $\phi < 1$ because then the enhancement is too low and the absorption rate

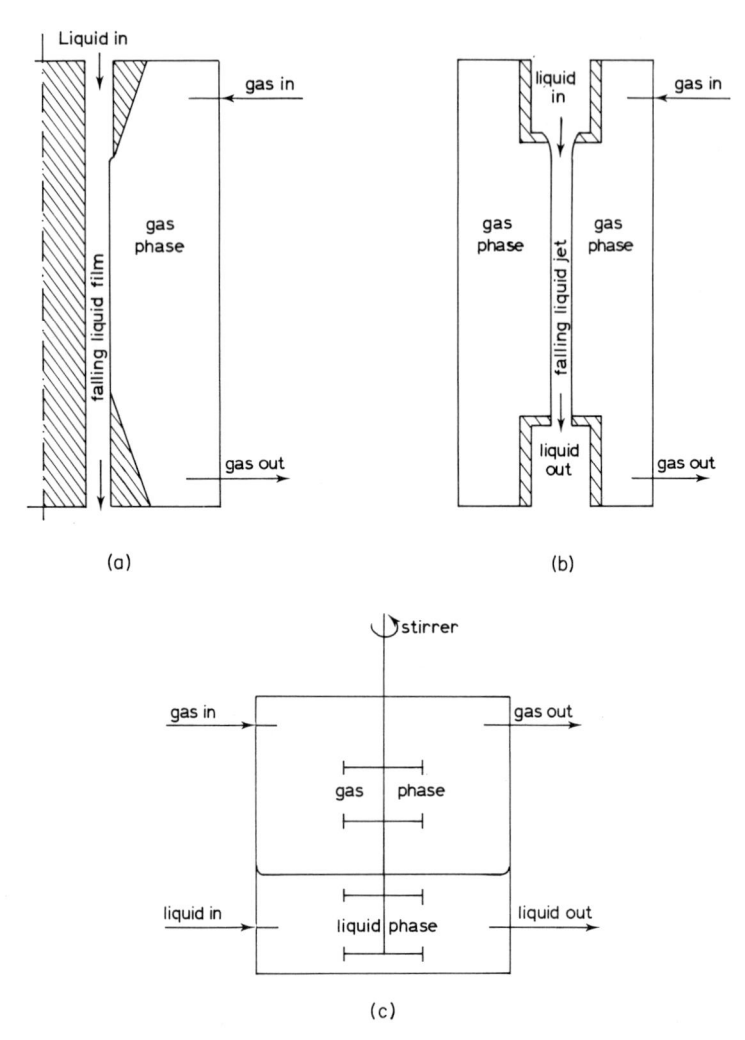

Fig. VII.46. Model reactors for gas–liquid reactions: (a) laminar film (wetted wall); (b) laminar jet; (c) stirred cell

becomes independent of kinetics. Preferably, experimental conditions are such that $\phi > 2$. The same holds for the stirred cell, though for very slow reactions $(Al\phi^2 \ll 1)$ the reactor may be used as well. In that case, measurements are carried out in the kinetic regime. Also in the instantaneous regime no kinetic information is obtained (except the information that the reaction is instantaneous). Therefore proper kinetic experiments are usually carried out in such a way that

$$2 < \phi < 0.1 \frac{D_B \bar{c}_B}{v_B D_A c_{Ai}}. \tag{VII.258}$$

Table VII.9. Contact times in gas–liquid model reactors. [From Groothuis [87]]

Reactor	τ [s]
Laminar jet	0.001–0.1
Laminar film	0.1–2
Stirred cell	1–10

As ϕ depends on the contact time τ ($\phi \propto \sqrt{\tau}$), proper experimental conditions differ in each reactor. The jet is particularly suitable for kinetic research on very fast reactions while the stirred cell is suitable for kinetic research on intermediate fast reactions. The experimental techniques contain many pitfalls and should be carried out with a thorough understanding of the theory on mass transfer with reaction and a knowledge of possible deviations such as inlet and end effects, temperature rise at the interface and gas phase resistance. Notice that criterion (VII.258) can only be checked afterwards when the kinetics are already known. General introductions to the field are the book of Danckwerts [9] and a review from Laurent et al. [88]. Specific references are: for the laminar jet [23, 89–92], for the laminar film [23, 89, 93] and for the stirred cell [94–98]. Levenspiel and Godfrey [98] suggest some modifications, one being the independent stirring of gas and liquid phase. They also present an experimental scheme by which the kinetics can be obtained. Related to the jet and the film reactor is the rotating drum contactor developed by Danckwerts and Kennedy [99]. Applied contact times varied between 0.01 and 0.25 seconds.

Example VII.15.1.a *Kinetic data from absorption experiments in a stirred cell model reactor*

A chemical reaction of type $A(G) \rightarrow A(L)$ and $A(L) + B(L) \rightarrow P(L)$ is investigated in a small laboratory reactor where a liquid L, in which B is dissolved, is brought into contact with a gas consisting of pure A. Then the liquid is slowly stirred at a constant stirrer speed in such a way that the liquid surface always remains flat, so that the value of the gas–liquid interfacial area remains accurately known. The solvent itself is inert. Immediately after starting the stirrer the initial absorption rate is measured by recording the initial rate of pressure decrease with time.

The following introductory experiment is done: The reactor is filled with 10^{-3} m^3 inert solvent and with pure gaseous A above the liquid with $V_G = 10^{-3}$ m^3 and $p_A = 1$ bar. Immediately thereafter the stirrer is started and an initial absorption rate of 5×10^{-9} kmol/s is recorded. The absorption process reaches equilibrium at $p_A = 0.62$ bar. If the experiment is carried out correctly, you safely may assume that the liquid did not yet contain an appreciable amount of A when stirring began. For no stirring the absorption of A is caused by diffusion in a semi-infinite liquid; this is a very slow process, provided free convection can be neglected. We are asked to solve the following problems:

(a) What is the solubility of A in the liquid?
(b) What value of k_{LA} can be calculated from the experimental results presented in Table VII.10 for a stirred cell diameter of 0.1 m?
(c) What is the kinetic order in A and B respectively and what is the value of the reaction rate constant?

Table VII.10. Experimental results, Example VII.15.1.a

exp	$\bar{c}_B$ (kmol/m³)	p_A (bar)	initial absorption rate of A (kmol/s)
1	0	2	10×10^{-9}
2	1	1	20×10^{-9}
3	1	2	40×10^{-9}
4	2	1	28.3×10^{-9}

Additional data are $D_A = D_B = 2 \times 10^{-9}$ m²/s and further m and k_{LA} are independent of $\bar{c}_B$.

The answers to the questions are

(a) $m = c_{AL}/c_{AG}$ at equilibrium. The fraction of A in the liquid phase at equilibrium is $1 - 0.62 = 0.38$. Hence

$$m = \frac{0.38/V_L}{0.62/V_G} = 0.61$$

from which follows that c_{Ai} at 20°C and 1 bar $= 0.61 p_{Ai}/RT = 0.025$ kmol/m³.

(b) As no B is present, the absorption is essentially physical. So

$$\Phi_A = k_L S(c_{Ai} - 0)$$

$$10 \times 10^{-9} = k_L \times \frac{\pi}{4} \times (0.1)^2 \times 0.05$$

which gives $k_L = 25 \times 10^{-6}$ m/s.

(c) Comparing experiment (a) with the introductory experiment shows that J_A is proportional to p_A, from which follows that m is a constant, independent of p_A. Comparing experiments 2 and 3 with 1 and with the introductory experiment shows that $E_A = 4$ while we know that $E_{A\infty} = 1 + D_B \bar{c}_B/(D_A c_{Ai}) = 41$ and 21 respectively. Hence $2 < E_A \ll E_{A\infty}$, which means $E_A = \phi = 4$.

Suppose we have (n, m) kinetics; then ϕ is proportional to $\sqrt{(c_{Ai}^{n-1} \bar{c}_B^m)}$. As ϕ is independent of c_{Ai}, apparently $n = 1$. Comparing experiment 4 with experiment 2 shows $E = \phi \propto \sqrt{\bar{c}_B}$ and therefore also $m = 1$. The value of $k_{1,1}$ now follows from experiment 2, for example,

$$E = \phi = 4 = \sqrt{\left(\frac{k_{1,1} \times 2 \times 10^{-9} \times 1}{(25 \times 10^{-6})^2}\right)}$$

which gives $k_{1,1} = 5 \,[\text{m}^3/\text{kmol.s}]$.

This example shows the usefulness of this model reactor for the measurement of reaction kinetics. In practice of course a series of data over a range of $\bar{c}_B$, c_{Ai} will be collected and fitted mathematically with n and m as parameters. As long as gas phase resistance is negligible and $E_A \ll E_{A\infty}$, this can be done graphically by plotting $\log J_A$ versus $\log (mc_{AG})$ and $\log (\bar{c}_B)$, the slopes being $(n-1)/2$ and $m/2$ respectively.

A numerical optimization allowing for both gas phase resistance and diffusion limitation of B is more generally applicable. Then the flux is fitted by varying n and m in the equation

$$J_A = \frac{c_{AG}}{\dfrac{1}{k_G} + \dfrac{1}{mk_L E_A}},$$

with E_A from equations (VII.144) and (VII.145) respectively. With increasing diffusion limitation either for A in the non-reaction phase or for B in the reaction phase, J_A becomes less sensitive to the actual kinetics, resulting in decreased accuracy of the method. Ultimately, for $k_G \ll mk_L E_A$ or $E_A = E_{A\infty}$ the absorption flux J_A becomes independent of the kinetics.

Please note that the assumption that k_L is independent of $\bar{c}_B$ is valid only as long as neither B nor P influence the (local) viscosity near the surface very much. Corrections are possible by $Sh = cRe^n Sc^{1/3}$, where the value of the exponent n slightly depends on the stirred cell reactor geometry (see [94-8]).

Finally we have to remark that many reactions are not of the order (n, m). Therefore we always have to check how good our correlation is.

Example VII.15.1.b *Kinetic data from experiments in a wetted wall reactor*
Carbon dioxide is physically absorbed in water at 20°C in a laminar film reactor of length L and circumference πd_t (see figure VII.46). The water entering the reactor is free of carbon dioxide; the liquid bulk concentration of CO_2 is negligible all over the reactor and the partial pressure of CO_2 in the gas feed is 0.05 bar. The total pressure is 1 bar. The bulk of the gas phase may be assumed to be ideally mixed.

The thickness of the liquid film (d_f), which is not the film thickness according to the film theory, follows from [1]:

$$d_f = \left[\frac{3\mu_L \Phi_{vL}}{\pi d_t g \rho_L}\right]^{1/3} \quad [\text{m}] \tag{a}$$

and the interfacial liquid velocity (v_i) from

$$v_i = \frac{3}{2}\left(\frac{\Phi_{vL}}{\pi d_t}\right)^{2/3} \left(\frac{\rho_L g}{3\mu_L}\right)^{1/3} \quad [\text{m/s}]. \tag{b}$$

The contact time for both gas and liquid is defined as

$$\tau = \frac{L}{v_i} \quad [\text{s}]. \tag{c}$$

In the equations Φ_{vL} is the liquid flow rate [m³/s]. Further data are: $g = 9.8$ m/s², $\mu_L = 10^{-3}$ N.s/m², $\rho_L = 1000$ kg/m³, $\pi d_t = 0.05$ m and $L = 0.1$ m. Further, it is known that

$$m_{CO_2} = \{c_{CO_2,L}/c_{CO_2,G}\}_{\text{equil.}} = 0.8, \qquad D_{CO_2,L} = 1.5 \times 10^{-9}\,\text{m}^2/\text{s}$$

and

$$D_{CO_2,G} = 15 \times 10^{-6}\,\text{m}^2/\text{s}.$$

We are now asked to

(a) prove that the resistance against mass transfer is mainly situated in the liquid phase;
(b) calculate the relative decrease of the CO_2 concentration over the gas phase between in- and outlet $\bar{c}_{A1G}/\bar{c}_{AoG}$ for $\Phi_{vL} = \Phi_{vG} = 10^{-6}$ m³/s,
(c) derive the relationship between δ in the film theory and τ in the penetration theory for an identical average molar flux through the interface;
(d) derive an expression for the Hinterland coefficient

$$Al = f(\Phi_{vL}, \mu_L, \rho_L, g, L, \pi d_t, D_{CO_2,L})$$

and discuss the influence of Φ_{vL} on Al;
(e) calculate the Hinterland coefficient Al for the flow rates under (b).

After the execution of those introductory physical absorption experiments, an amine (B) is dissolved into the water. This amine reacts with carbon dioxide according to $CO_2 + 2B \rightarrow P$. From experiments similar to those described in the previous Example it is known that this reaction is first order in both CO_2 and amine. The diffusion coefficient of amine in the solution is $D_B = 0.5 \times 10^{-9}$ m²/s. As also discussed in the previous Example, the accuracy of the kinetic measurements decreases with increasing gas phase resistance.

(f) If we can assume that all physical parameters are independent of $\bar{c}_B$ and equal to those in water, what is then the maximum allowable ratio $\bar{c}_{BL}/\bar{c}_{CO_2,G}$ to be sure that less than 10% of the resistance will be located in the gas phase?
(g) To measure the rate constant $k_{1,1}$, the following experimental conditions have been selected: $\bar{c}_B = 2$ kmol/m³, pure CO_2 is used and the system is operated at $p = 0.5$ bar and $\Phi_{vL} = 10^{-5}$ m³/s. After closing the gas outlet, a gas flow of $\Phi_{vG} = 4.38 \times 10^{-6}$ m³/s must still be supplied to keep the pressure constant. How can we calculate the rate constant $k_{1,1}$ from these data?

The answers to the various questions are:

(a) The gas phase resistance is negligible if $k_G/mk_L \gg 1$. Applying the penetration theory gives

$$k_G = 2\sqrt{\left(\frac{D_G}{\pi\tau}\right)} \quad \text{and} \quad k_L = 2\sqrt{\left(\frac{D_L}{\pi\tau}\right)}.$$

Further, with $\tau_G = \tau_L$, we derive

$$k_G/k_L = \sqrt{(D_G/D_L)} \quad \text{or} \quad k_G/mk_L = \frac{1}{m}\sqrt{(D_G/D_L)} = 125,$$

so there is no gas phase resistance.

(b) From $k_L = 2\sqrt{(D_L/\pi\tau)}$ with $\tau = L/v_i$ and v_i from equation (b) it follows that $k_L = 0.56 \times 10^{-6}$ m/s. Further, with $J_A A = k_L m(\bar{c}_{AG})_1 \pi d_t L$ and $\Phi_{vG}(\bar{c}_{AGo} - \bar{c}_{AG1}) = J_A A$ we find

$$\frac{\bar{c}_{AGo}}{\bar{c}_{AG1}} = 1 + \frac{k_L m \pi d_t L}{\Phi_{vG}} = 1.22.$$

(c) If we use

$$k_L = \frac{D_L}{\delta} = 2\sqrt{\left(\frac{D_L}{\pi\tau}\right)}$$

we find $\delta = \frac{1}{2}\sqrt{(\pi D_L \tau)}$.

(d) From the definition of Al (equation (VII.71)) it follows that $Al = d_f/\delta$. From the answer to (c) we know that $\delta = \frac{1}{2}\sqrt{(\pi D_L \tau)} = \frac{1}{2}\sqrt{(\pi D_L L/v_i)}$. From equations (a) and (b) it now follows that

$$Al = 0.79\left\{\frac{gD_L^2 L^3}{(\Phi_{vL}/d_t)^4}\right\}^{-1/6} \left(\frac{\rho_L D_L}{\mu_L}\right)^{-1/6}.$$

This relation describes how Al increases with increasing Φ_{vL}.

(e) Substitution of the data gives directly: $Al = 2.4$.

(f) The ratio of the gas phase resistance to the total resistance is

$$\frac{1/k_G}{1/k_G + 1/mk_L E_A}.$$

As $E_A \leq E_{A\infty}$, a safe condition is

$$\frac{1}{1 + k_G/mk_L E_{A\infty}} < 0.1.$$

This equals $E_{A\infty} < k_G/(9mk_L)$, so that

$$E_{A\infty} = 1 + \frac{1}{2}\frac{D_B \bar{c}_B}{D_A c_{AiL}} < 13.9.$$

or

$$\frac{\bar{c}_B}{mc_{AiG}} \leq 77.4.$$

With less than 10% gas phase limitation or $c_{AiG} \geq 0.9\bar{c}_{AG}$ we then have $\bar{c}_B/\bar{c}_{AG} \leq 77.4 \times 0.9\, m = 55.7$.

(g) With pure CO_2, the gas phase resistance is non existent. With the relations $J_A \pi d_t L = \Phi_{vG} p_A/RT$ [kmol/s], $J_A = k_L m p_A E_A/RT$ and $E_A = \Phi_G/k_L m\pi d_t L$ we can recalculate k_L as under (b). The result is $k_L = 0.12 \times 10^{-3}$ m/s. (Remember that k_L is proportional to $\Phi_{vL}^{1/3}$). It now follows that $E_A = 9.13$ and

$$E_{A\infty} = 1 + \frac{D_B \bar{c}_B RT}{2 D_A m p_A} = 21.3.$$

Because the observed value of E_A is larger than 2 and smaller than $E_{A\infty}$ it follows that E_A is close to the value of ϕ. From figure VII.22 we estimate $\phi = 10$. In fact, this problem is a typical penetration theory problem. With this theory,

$$E_{A\infty} = 1 + \tfrac{1}{2}\sqrt{\left(\frac{D_B}{D_A}\right)} \frac{\bar{c}_B}{m} \frac{RT}{p_A} = 36.5.$$

Now, from equation (VII.150) we find $\phi = 10.3$. With $\phi = \sqrt{(k_1 D_A \bar{c}_B)}/k_L$ it follows that $k_1 = 509$ kmol/m³ s.

Notice that the method of measuring the extra CO_2 supply, necessary to keep the pressure constant, calls for very pure CO_2 to prevent accumulation of inerts in the reactor.

VII.15.2 Model reactors for liquid–liquid reactions

The stirred cell reactor is equally applicable for liquid–liquid reactions as for gas–liquid reactions. Sharma and co-workers [100–3] reported results, based on the design described by Danckwerts and Gilham [94]. Landau and Chin [104] modified this reactor (starting from a design proposed by Procházka and Bulička [105]) to maintain an undisturbed flat interfacial area at higher stirrer speeds.

VII.15.3 Model reactors for fluid–solid reactions

The number of available fluid–fluid model reactor types is limited by the requirement that the value of the interfacial area must be accurately known. Fluid–fluid reactors in which one of the phases is dispersed usually do not meet this requirement. This is different for fluid–solid reactors because here both the internal and external specific contact areas are solid phase properties. Hence, the interfacial areas in such reactors are determined by the type and the weight of solid phase present in the reactor which are often independent of the hydrodynamics in the reactor, so here, the required information on the value of the interfacial areas does not limit the number of suitable model reactors for kinetic research. However, there are other limitations:

— For every phase the residence time distribution at least must be known, but to keep the analysis of the experimental data simple, the model reactor should preferably be of the plug flow or of the ideal mixer type.
— The temperature distribution at least must be known, but an isothermal fluid phase, isothermal pellets and if possible no temperature difference between the pellets and the fluid phase are preferred.

Table VII.11 gives the main model reactors for fluid–solid kinetic measurements characterized by the residence time distribution. The fixed bed reactor may be operated at very low conversions (differential reactor) or at higher conversions (integral reactor). In the differential reactor, the bulk fluid flow composition is uniform and the conversion rate per unit area external surface follows directly from the measured conversion via

$$J_A = \frac{\Phi_v c_{A_0} \zeta}{a V_r}. \qquad (VII.259)$$

Assuming particular kinetics (R_A), J_A can be further elaborated with the theory on mass transfer with reaction as treated in this chapter

$$J_A = f(k_G, D_i, a_s, R_A). \qquad (VII.260)$$

Table VII.11. Fluid–solid model reactors, classified according to RTD-characteristics

Solid phase \ Fluid phase	Plug flow (or batch)	Ideally mixed flow on macro scale
Batch	Fixed bed reactor; single pellet diffusion reactor; thermo balance Slurry batch reactor	Fixed bed recycle reactor (with high recycle ratio) Basket type mixed reactor
Ideally mixed on macro scale	Riser reactors with recycle of solid phase	Stirred slurry tank reactor Riser with recycle of both phases

By varying Φ_v and d_p, k_G varies, so that information can be obtained on the location of the rate controlling resistance: in the pellet or in the fluid phase. By varying the catalyst dimension d_p, ϕ is varied ($\Phi::d_p$), so that we can determine whether the reaction is slow or fast with respect to the internal mass transfer. Once the relative resistances are determined, variation of the reactant concentrations and of the temperature gives the information necessary to decide whether a particular kinetic equation R_A (c_A, c_P) fits the conversion data.

A disadvantage of the differential reactor can be the occurrence of inaccuracies in the measurement of conversion and selectivity due to the small concentration variations applied. Moreover, it is difficult to obtain kinetic data at higher conversions especially with autocatalytic reactions where kinetics depend on conversion and on back-mixing.

With the integral reactor these difficulties can be eliminated because in this reactor ζ can be varied by changing the residence time or the mass of solid phase. Now, other disadvantages are introduced. The most serious one is caused by the poor heat transfer of a fixed bed which prevents establishing isothermal conditions for reactions with moderate or high heat effects. Further the interpretation of the data is somewhat more complicated because $\bar{c}_A$, and consequently J_A and R_A are place dependent, with

$$\Phi_v c_{Ao} \zeta_A = \int_0^L J_A a V_r \, dz. \qquad (VII.261)$$

Especially for an unknown complicated kinetic function R_A derivation of the correct R_A from the experimental results by use of equations (VII.260) and (VII.261) is more cumbersome than fitting equations (VII.259) and (VII.260). This is even more true for complex reaction networks e.g. for isomerization and cracking reactions of crude oil fractions, where the integral method is very laborious to derive the individual rate constants. But also in this case, experiments in integral packed bed, fluid bed or riser reactors may be required to test the process under realistic conditions. The kinetic information that can be obtained from such experiments will greatly benefit from measuring the concentration simultaneously at several points in the reactor.

Model reactors with perfect mixing of the fluid phase on a macro scale and a uniform composition may combine the advantage of both the differential contactor — isothermal operation on a macro scale and differential measurement technique — and the integral reactor with its arbitrary variable fluid conversion. These reactors are shown in the right-hand column of Table VII.11. In case the solid phase is stationary the fluid phase mass transfer coefficient can be varied too, either by changing the fluid load in the fixed bed reactor or by varying the stirrer speed in the basket type mixed reactors. The latter reactor type was developed by Carberry [106] and consists of a cell in which the catalyst is placed in the large stirrer sheets, held together by fine screens. In rotating the stirrer the fluid phase is pressed through the catalyst layers, thus realizing good solid–fluid contact. Modified types are described in the review of Bennett et al. [107]. The recycle reactors are good mixers with respect to the fluid phase only as long as the recycle ratio is much larger than one, preferably larger than 25.

Reactors in which also the solid phase is perfectly mixed on a macro scale, such as the stirred slurry tank and the riser reactor with recycle of both phases, are particularly useful for a fast catalyst deactivation process. Notice that the residence time of both phases can be varied independently by introducing an extra recycle flow of one of the phases after the fluid solid separator. Regretfully the fluid phase mass transfer coefficient in these slurry reactors is difficult to vary. Some variation is possible by changing d_p, but then ϕ also changes.

Especially for analysing reaction mechanisms in complex catalytic (pseudo) monomolecular reaction networks, transient methods are often used by application of micro pulse techniques in small catalytic packed beds. These catalytic bed reactors either simultaneously act as a chromatographic separation column or are followed by such a column. After this, the product gas is led through a flame ionization or conductivity detector to determine the product distribution. An introduction to these techniques is the comprehensive review on laboratory reactors for heterogeneous catalytic processes by Christoffel [108].

The single pellet diffusion reactor [109] allows us to measure the gas phase composition as actually present in the centre of a catalyst pellet. In this laboratory reactor (see figure VII.47) the gas flows along the top surface of a pellet

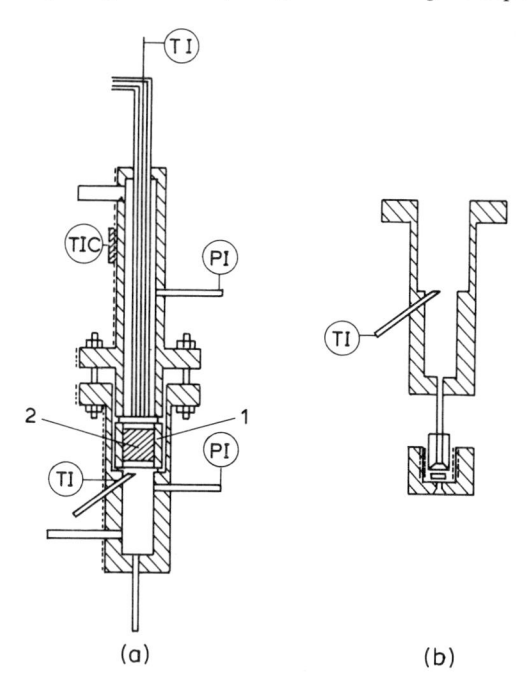

Fig. VII.47. Single pellet diffusion reactor: (a) set up for permeability and internal diffusion measurements; (b) part of the cell for measuring kinetics, with sample device: 1. catalyst container; 2. pellet (From Christoffel and Schuler [109]. Reproduced by permission of Verlag Chemie GmbH)

of thickness δ. Reactive components diffuse into the pellet and are converted. Reaction products diffuse upwards and then leave the pellet. The space below the pellet is closed and, once stationary conditions have been established, there is no net flow into this space. Then $dc/dx = 0$ holds at the bottom surface of the pellet, so that concentrations measured in the space below the pellet represent the concentrations in the centre of a pellet of characteristic length δ that is accessible on all sides. Under isothermal conditions the concentration curve follows by solving

$$\frac{d^2 c_A}{dx^2} = -R_A.$$

For first order kinetics, the solution is

$$\frac{\bar{c}_A}{c_{Ai}} = \frac{1}{\cosh \phi}$$

where $\bar{c}_A$ is the concentration in the space below and c_{Ai} the concentration at the top surface of the pellet. Hence, ϕ follows directly from one concentration measurement provided the order of the reaction is known. The latter can be determined by changing the partial pressure of the reactant in the feed.

The method is useful for the measurement of kinetics under diffusion controlled conditions, for measuring actual effective internal diffusivities under reaction conditions, provided the kinetics are known, and, above all, to study catalyst de-activation mechanisms and to establish what type of poisoning occurs (e.g. pore mouth poisoning or uniform poisoning). Also interparticle non-isothermal effects and the existence of multiple steady states in the pellet can be studied in this way.

For gas–solid reactions or solid–solid reactions where the solid mass changes with conversion as in pyrolysis, gasification, metal ore reduction or sulphidation of calcium oxide, commercially available thermo balances can be used to follow the conversion rate with time by measuring the weight of the sample continuously. This can be done at one single temperature or with a temperature–time program to simulate the temperature history of a particle in an actual reactor. Also the net heat production or consumption during the conversion process can be measured simultaneously. Preferably, the measurements will be carried out in the absence of diffusion limitation, although diffusion controlled single particle conversion rates may provide relevant information as well.

Kinetic research on fluid–solid catalysed reactions exhibits many pitfalls. The worker in the field therefore must be fully versed with both the theory on mass transfer with reaction and the characteristics of each available model reactor. Good entrances for the latter subject are the reviews of Christoffel [108] and Weekman [110]. We further mention applications of the basket type mixed reactor to liquid–solid systems [111, 112] and the development of a laminar tube flow liquid–solid model reactor [113]. In the latter reactor the wall consists of a dissolving solid or catalyst.

Once the model reactor has been selected, the correct rate equation must be found in an experimental programme which, because of the high costs involved, should be as efficient as possible. To this end see the review of Froment [114].

Example VII.15.3 *The measurement of reaction kinetics and internal pellet diffusivity with a single pellet diffusion reactor*

The isomerization of n-hexane at 453°C and total pressure of 1.6 bars has been investigated by Christoffel and Schuler [109]. They used a single pellet diffusion reactor with pellet thickness δ of 6.22×10^{-3} m. The porous pellet material contained 0.35 wt % Pt on Al_2O_3. The flow rate along the pellet was so high that the observed conversion rate was independent of the flow rate. The partial pressure of n-hexane in the feed was varied between 15 and 160 mbar. The ratio of the n-hexane concentration in the space below the pellet ($\bar{c}_A$) and in the gas mixture above the pellet (c_{Ai}) was found to be independent of c_{Ai}: $\bar{c}_A/c_{Ai} = 0.191$. The apparent conversion rate was found to be proportional to the n-hexane concentration in the feed and to be represented by: $-R_A = 0.184 c_{Ai}$ [kmol/m³ pellet.s]. From these results the following information can be obtained:

— the order of the reaction;
— the value of the kinetic rate constant;
— the actual effective diffusivity in the pellet.

Because the conversion rate is independent of the external flow rate, external diffusion limitation can be neglected, $k_G \gg (D_i/\delta)E_A$. Both from the proportionality of the conversion rate with the feed concentration and the independence of $\bar{c}_A/c_{Ai}$ from c_{Ai} it follows that the reaction is first order, so that $\bar{c}_A/c_{Ai} = 1/\cosh\phi$, $-R_A = k_1 c_{Ai} (\tanh\phi)/\phi$, $D_i = \delta^2 k_1/\phi^2$. From the data given we obtain $\phi = 2.34$, $k_1 = 0.439 \text{ s}^{-1}$ and $D_i = 3.1 \times 10^{-6} \text{ m}^2/\text{s}$. Notice that the effective diffusivity has been measured during actual reaction conditions. The advantage of this is somewhat limited by the fact that the pellet is often artificially prepared by grinding and pelletizing a commercial catalyst, so that the measured diffusivity may differ from that in the commercial catalyst.

VII.16 MEASUREMENT TECHNIQUES FOR MASS TRANSFER COEFFICIENTS AND SPECIFIC CONTACT AREAS IN MULTI-PHASE REACTORS

The capacity of a multi-phase reactor is characterized by

1. the residence time distribution of each phase;
2. the flow directions: co-current, counter-current or cross flow;
3. the hold-up of each phase;
4. the value of the specific contact area;
5. the values of the mass transfer coefficients.

The latter three parameters strongly depend on the hydrodynamics and influence the reactor capacity per unit volume ($J_A a$) at given bulk concentrations, i.e. for a gas–liquid reaction:

— for $\phi < 0.2$ and $Al\phi^2 \ll 1$, the capacity per m³ depends on ε:

$$J_A a = -R(\bar{c}_A, \bar{c}_B, \bar{c}_P, \bar{c}_Q)(1 - \varepsilon); \qquad (VII.262)$$

— for $\phi < 0.2$ and $Al\phi^2 \gg 1$, the capacity per m³ depends on the product $k_L a$:

$$J_A a = k_L a c_{Ai};$$

— for $2 < \phi \ll 1 + D_B \bar{c}_B / v_B D_A c_{Ai}$, the capacity depends on the value of the contact area:

$$J_A a = k_L \phi a c_{Ai}, \qquad (VII.263)$$

in which the product $k_L \phi$ is independent of k_L;

— for $\phi \gg 1 + D_B \bar{c}_B / v_B D_B c_{Ai}$, the capacity depends again on the product $k_L a$:

$$J_A a = k_L a c_{Ai} E_{A\infty}; \qquad (VII.264)$$

— in all cases the capacity depends on k_G unless

$$k_G \gg m k_L E_A.$$

So we can conclude that unless the reaction is 'very slow', k_G, k_L and a must be known for the design of a multi-phase reactor, whereas to conclude whether the reaction is 'very slow' the order of magnitude of k_L, k_G and a must be known.

VII.16.1 Measurement of the specific contact area a

The measurement of ε is straightforward and does not need further discussion. The same holds for a in case of solid particles of a regular shape. For spherical particles $a = \sum n_i \pi d_p^2$, in which n_i is the number of particles per unit volume with a diameter between $d_p - \frac{1}{2}\Delta d_p$ and $d_p + \frac{1}{2}\Delta d_p$. For the measurement of a in fluid–fluid reactors several methods are available. A survey is given by Reith [115]. The most important methods are:

(a) *The photographic technique* (see van Dierendonck et al. [116] and Towell, Strand and Ackerman [117]). The Sauter mean dispersed phase diameter $\bar{d}_p$ is measured photographically:

$$\bar{d}_p = \sqrt{\left(\sum n_i d_p^2 / \sum n_i \right)}.$$

Now the specific contact area follows from $a = 6\varepsilon/\bar{d}_p$. Disadvantages of the method are that only local values of a are obtained, that it is time consuming and that the dispersion may be disturbed by the sample fittings.

(b) *The light scattering technique* (see Calderbank [118]). A parallel beam of light with an intensity I_0 is sent through the dispersion over a distance l and the remaining light intensity I is measured. Then, for $l = 3$ cm and $a < 600 \, \text{m}^2/\text{m}^3$, we have $a = \frac{4}{l} \ln \frac{I_0}{I}$. Disadvantages of this technique are [115] that only local values of a are obtained, that the measurement fittings may disturb the dispersion and that the method is limited to values of a up to $600 \, \text{m}^{-1}$, while values of up to $2000 \, \text{m}^{-1}$ are not uncommon in industrial reactors. The latter advantage can be overcome by modifications introduced by Landau et al. [119, 120].

Both the photographic and the light scattering technique are restricted to systems in which the dispersed phase is present as bubbles or drops. They fail in slug flow or in gas–liquid packed colums.

(c) *The chemical absorption technique.* This technique has been introduced by Westerterp [121, 122] and Yoshida and Yoshiharu [123] and has found a wide application since then. General review articles have been published by Linek and Vacek [124], Sharma et al. [125-7] and by Laurent et al. [88]. The principle of the method is simple: if a component is transferred from a non-reacting phase into a reacting phase in which it reacts, chemically enhanced, according to $J_A a = k_L \phi a c_{Ai}$, then the product $k_L \phi$ is a function only of kinetics and is independent of k_L. Under these conditions a follows straightforwardly from the observed conversion rate and the kinetics, e.g., for a mixed tank reactor,

$$a = \frac{J_A a V_r}{k_L \phi c_{Ai} V_r} = \frac{-\Phi_A}{k_L \phi c_{Ai} V_r}.$$

For kinetics of order (1,1): $k_L \phi = \sqrt{(k_{1,1} D_A \bar{c}_B)}$ and we have:

$$a = \frac{-\Phi_A}{\sqrt{(k_{1,1} D_A \bar{c}_B)} c_{Ai} V_r}.$$

For a stationary column reactor a can be determined by solving

$$\frac{d u \bar{c}_A}{dz} - D_1 \varepsilon \frac{d^2 \bar{c}_A}{dz^2} + a \sqrt{(k_{1,1} D_A \bar{c}_B)} c_{Ai} = 0$$

with the appropriate boundary conditions. In this relation $\bar{c}_A$, u, ε and D_1 refer to the (dispersed) non-reaction phase. a is found by trial and error. In practice many complications may occur, e.g.

— the reaction influences the area [128];
— gas phase resistance is not negligible;
— the condition $2 < \phi \ll E_{A\infty}$ is not fulfilled in a part of the reactor;
— the measured area depends on the ratio

$$\frac{\text{factor by which reaction increases the capacity of the reaction phase}}{\text{enhancement factor}}.$$

The latter effect has been observed in packed columns [129].

Most common model reactions are the absorption of CO_2 in hydroxide solution (see Danckwerts [9]) and the absorption of oxygen in aqueous sodium sulphite solutions catálysed by Co^{2+}, see Reith and Beek [130], and Linek and Vacek [124]. The latter system is sensitive to the sulphite kinetics, the cobalt and the water quality while possibly an induction period before the reaction starts cannot be excluded [126]. The suitability of a great number of gas–liquid reactions as model reactions has been discussed by Sharma et al. [126, 127]. Sharma and Danckwerts [126] also discuss suitable systems for liquid–liquid reactors.

VII.16.2 Measurement of the product $k_L a$

The product $k_L a$ is measured in three different ways:

(a) by mass transfer without reaction, e.g. for a CISTR

$$k_L a = \frac{-\Phi_A}{(c_{Ai} - \bar{c}_{AL})V_r};$$

(b) by mass transfer with a slow reaction with $\phi < 0.2$, $Al\phi^2 \gg 1$, e.g. for a CISTR

$$k_L a = \frac{-\Phi_A}{c_{Ai} V_r};$$

(c) with an instantaneous reaction, e.g. for a CISTR

$$k_L a = \frac{-\Phi_A}{\left(c_{Ai} + \dfrac{D_B \bar{c}_B}{v_B D_A}\right) V_r},$$

in which Φ_A is the amount of A transferred from the non-reacting phase per unit of time.

Because the liquid must not be saturated, method (a) is limited to $k_L a \tau_L \ll 1$. Moreover, for $\bar{c}_{AL}$ not much smaller than c_{Ai}, the residence time distribution of the reaction phase must be known. Methods (b) and (c) have a wider range of applicability because of the higher capacity of the reaction phase for component A. The residence time distribution of the reaction phase does not influence the experiment in method (b) and also not in method (c) as long as $\bar{c}_B$ is constant. In case $D_B \bar{c}_B / (v_B D_A c_{Ai}) \gg 1$ the result obtained with method (c) is also independent of the RTD of the non-reaction phase and of c_{Ai}, because then

$$k_L a \simeq \frac{-\Phi_A}{V_r} \frac{v_B D_A}{D_B \bar{c}_B}.$$

However, care must be taken to avoid gas phase resistance; this can be done by applying pure gases.

Laurent et al. [88] and Sharma et al. [126, 127] discuss the methods and review reaction systems which are suitable for gas–liquid reactors. Sharma and Danckwerts [126] review reactions for liquid–liquid reactors as well.

In liquid–solid systems external mass transfer and reaction are in series. Here $k_L a$ can be determined by method (a). In case of a fast reaction c_{Ai} goes to zero and the equation from method (b) holds in the form $k_L a = -\Phi_A/(\bar{c}_A V_r)$. A proven technique in slurry reactors is the dissolving of solid particles and the measurement of the liquid concentration of dissolved material in the outlet of the reactor. Small-scale experiments to determine k_L, k_G and a are preferably carried out with the same G/L/S system as will be used in the large-scale reactor.

VII.16.3 Measurement of the product $k_G a$

Measurement techniques for $k_G a$ are reviewed by Sharma et al. [125–7] and by Laurent et al. [88]. With systems in which the gas phase resistance prevails, $k_G a$ is determined analogously to the determination of $k_L a$ in a slurry reactor. In case the gas phase is perfectly mixed,

$$k_G a = \frac{-\Phi_A}{\bar{c}_{AG} V_r}.$$

This relation holds for both gas–liquid and gas–solid systems provided the gas phase resistance is completely controlling the rate of the transfer process. Measurement of gas phase resistance and of $k_G a$ are often difficult because of the high value of k_G as compared to k_L. For this reason in gas–solid systems the solid phase is often diluted with inert particles, thus retarding the process by lowering a. Methods used in gas–solid reactors are the evaporation of water from porous particles, the sublimation of organic solids like naphthalene and the adsorption of organic gases on porous particles (see Kunii and Levenspiel [131]).

In gas–liquid systems also a chemical reaction can be chosen for which $k_G/mk_L E_A$ is of the order of one. From the relation

$$J_A a = \frac{\bar{c}_{AG}}{\dfrac{1}{k_G a} + \dfrac{1}{mk_L a E}}$$

it follows, e.g. for a CISTR, that

$$\frac{1}{k_G a} + \frac{1}{mk_L a E_A} = \frac{\bar{c}_{AG} V_r}{-\Phi_A}.$$

By varying the rate constant with a catalyst or in case the kinetics depend on $\bar{c}_B$, by varying $\bar{c}_B$, E_A can be changed such that $k_G/(mk_L E_A)$ is changed from

$$k_G/(mk_L E_A) < 1 \quad \text{to} \quad k_G/(mk_L E_A) > 1.$$

For $E_A = \phi$, the values of $k_G a$ and a can be found simultaneously by plotting $-V_r \bar{c}_{AG}/\Phi_A$ against $1/k_L \phi$.

For $E_A = E_{A\infty}$, $k_G a$ and $k_L a$ can be found simultaneously by plotting $-V_r \bar{c}_{AG}/\Phi_A$ against $1/E_{A\infty}$.

However, these methods are time consuming because a series of experiments with varying ϕ is necessary to obtain one set of $k_G a$, a or $k_G a$, $k_L a$ respectively, that is at one particular combination of gas and liquid load! Therefore we suggest the simultaneous absorption of two gases, one gas reacting chemically enhanced giving $k_L a$ (for $E_A = 1$ or $E_A = E_{A\infty}$) or a (for $E_A = \phi$) and the other gas giving $k_G a$ at exactly the same hydrodynamic conditions (see the next section).

VII.16.4 Measurement of mass transfer coefficients k_L, k_G

The mass transfer coefficients k_L and k_G can be derived from the previously discussed measurements of a and of the products $k_L a$ and $k_G a$ via $k_G = k_G a/a$ and $k_L = k_L a/a$. For fluid–solid reactions this method gives no problems. However, in fluid–fluid reactions the hydrodynamics of the system may be influenced by the chemical reaction, causing a change in both k_L and a with a change of the reaction rate. This has been shown by Linek [128, 132, 133] and Brian et al. [134]. This disadvantage can be overcome by using the so-called Danckwerts method [9] in which carefully selected chemical reactions of a moderate fast rate are applied with $\phi \simeq 1$. Sharma and Danckwerts review suitable reaction systems for both gas–liquid and liquid–liquid reactors [126]. According to the Danckwerts penetration model for kinetics (1,1), no diffusion limitation of B and $\bar{c}_A$ zero, J_A is given by

$$J_A = c_{Ai}\sqrt{(D_A k_{1,1} \bar{c}_B + k_L^2)}$$

so that, e.g. for a CISTR,

$$\frac{-\Phi_A}{c_{Ai} V_r} = a\sqrt{(D_A k_{1,1} \bar{c}_B + k_L^2)}.$$

Now $k_{1,1} \bar{c}_B$ is varied from $\phi \ll 1$ to $\phi \gg 1$ by changing $\bar{c}_B$, or preferably by changing $k_{1,1}$ by variation of the catalyst concentration. In case k_L and a do not change with the reaction rate a straight line is obtained in a plot of $(-\Phi_A/c_{Ai} V_r)^2$ against $k_{1,1} \bar{c}_B$. From the slope, $a^2 D_A$, and the intercept, $(k_L a)^2$, both k_L and a are obtained. Regretfully, this method is very time consuming because every (k_L, a)-combination calls for a whole series of experiments. Therefore Robinson and Wilke [135] and Beenackers and Van Swaaij [136] introduced the method of simultaneous mass transfer of two components. One component reacts chemically enhanced according to $E_A = \phi$ and gives information on the value of the interfacial area, the other component is inert and gives information on the value of the product $k_L a$ under exactly the same hydrodynamical conditions. In this way k_L and a are obtained by one experiment only.

In principle this method is equally well suited for measuring k_G and a in one single experiment. A possible system is the simultaneous absorption of H_2S and CO_2 in aqueous solutions of amines. For CO_2, the conditions can be choosen such that $E_A = \phi$. With this reaction the area, a, is then obtained. The absorption rate of H_2S is, under certain conditions, limited by k_G. In that case, the product $k_G a$ will be obtained from the observed H_2S conversion in the same experiment and k_G follows from $k_G = k_G a/a$.

Example VII.16.4 *Simultaneous measurement of k_L and a in a cyclone reactor*
Liquid is introduced tangentially in a cyclone reactor whereas the gas is introduced through a porous stainless steel wall [136]. Gas that is not absorbed leaves the cyclone via the top (vortex) while most of the liquid leaves the cyclone via the bottom (apex) (see figure VII.48). Around 25% of the liquid is entrained with the gas through the vortex. Further $\Phi_L = 100 \times 10^{-6}$ m³/s and $\Phi_G = 125 \times 10^{-6}$ m³/s. The effective reactor volume, V_r, is located in the gas–liquid dispersion; this is the porous cylindrical section with $V_r = 3.5 \times 10^{-6}$ m³. To measure in one experiment both k_L and a simultaneously, a degasified aqueous solution of 2.07 M NaOH is used while the gas consists of a mixture of oxygen and carbon dioxide, 90 and 10 vol % respectively. The reaction temperature is $T = 20°C$. The reaction of CO_2 with OH^- takes place according to

$$CO_2 + 2\,OH^- = CO_3^{2-} + H_2O$$

with $R_A = -k_{1,1} c_{OH^-} c_{CO_2}$. The kinetics and the product $k_{1,1} D_{CO_2}$ were measured in a laminar jet with a degasified NaOH solution and an H_2O saturated CO_2 gas at a total pressure of 0.2 bar, and $k_{1,1} D_{CO_2} = 1.32 \times 10^{-5}$ m⁵/kmol s² was found at 20°C. The average solubility of CO_2 at process conditions is $(c_{CO_2})_i = 2.95 \times 10^{-3}$ kmol/m³. We take an average, because there is a pressure drop between the wall and the central gas core and the solubility follows Henry's

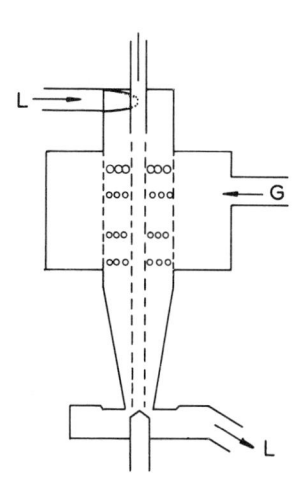

Fig. VII.48. Cyclone reactor; liquid is introduced tangentially; gas via porous wall [From Beenackers and van Swaaij [136]. Reproduced by permission of DECHEMA Deutsche Ges. f. chem. Apparateweesen, Frankfurt/M, Fed. Rep. Germany]

law. The CO_3^{2-} and O_2 concentrations in the liquid leaving through the apex were measured with results: $\bar{c}_{CO_3^{2-}} = 2.05 \times 10^{-3}\,\text{kmol/m}^3$ and the degree of oxygen saturation of the liquid $\bar{c}_{O_2}/(c_{O_2})_i = 0.26$, in which $(c_{O_2})_i$ was taken at the actual average pressure in the reactor section of 1.5 bars. We will derive both k_L and a from this experiment.

Assuming the reactor to behave as a perfect mixer for the liquid phase, we can apply for the physical absorption of oxygen:

$$J_{O_2} a V_r = \Phi_L \bar{c}_{O_2} = k_L a V_r ((c_{O_2})_i - \bar{c}_{O_2})$$

or

$$k_L a = \frac{\Phi_L}{V_r} \frac{(\bar{c}_{O_2})/(c_{O_2})_i}{1 - \bar{c}_{O_2}/(c_{O_2})_i} \quad \text{(a)}$$

which leads to $k_L a = 10\,\text{s}^{-1}$. In case the absorption of CO_2 is chemically enhanced with $\phi > 2$, we can get the specific contact area from

$$J_{CO_2} a V_r = \Phi_L \bar{c}_{CO_3^{2-}} = \sqrt{(k_{1,1} D_{CO_2} \bar{c}_{OH^-})} a V_R (c_{CO_2})_i$$

or

$$a = \frac{\Phi_L c_{CO_3^{2-}}}{\sqrt{(k_{1,1} D_{CO_2} \bar{c}_{OH^-})} V_r (c_{CO_2})_i} \quad \text{(b)}$$

and find $a = 3800\,\text{m}^2/\text{m}^3$ reactor section. Now k_L follows from $k_L = k_L a/a$ and we have $k_L = 2.6 \times 10^{-3}$ m/s. We did not know ϕ in advance because k_L is not known beforehand. However, we can check the assumption afterwards:

$$\phi = \frac{\sqrt{(k_{1,1} D_{CO_2} \bar{c}_{OH^-})}}{k_L} = 2.01.$$

Hence, our method has been justified.

In practice, the method can be extended for still higher values of k_L up to $\phi = 1$ provided $\sqrt{(k_{1,1} D_{CO_2} \bar{c}_{OH^-})}$ in equation (b) is replaced by $k_L \phi/\tanh \phi$. Then k_L and a can be found from equations (a) and (b) with a root finding procedure. Note, however, that the experimental accuracy will decrease rapidly with ϕ decreasing below 2. We further want to remark:

— The assumption of a perfectly mixed liquid bulk, necessary for the $k_L a$ measurement, is questionable but does not affect the result very much as long as the degree of saturation with oxygen does not become too high. In principle the method can be extended to higher values of $k_L a V_r$ by applying a slow oxygen consuming reactant in the liquid phase ($\phi < 0.2$) which acts as an increased oxygen capacity for the liquid, thus preventing saturation effects, e.g. sodium sulphite can be used to this end.
— Due to the very low oxygen solubility, only a minor fraction of the oxygen introduced is absorbed in the liquid, while a substantial fraction of CO_2 is absorbed, about 36%. This must be taken into account by calculating the average effective CO_2 interface concentration in the reactor.
— We implicitly assumed that $\phi \ll E_\infty$ and $k_G \ll m k_L E$. This remains to be checked.

VII.17 NUMERICAL VALUES OF MASS TRANSFER COEFFICIENTS AND SPECIFIC CONTACT AREAS IN MULTI-PHASE REACTORS

During the past decades a considerable amount of data has been accumulated on physical mass transfer in dispersed systems. The theoretical and experimental results can most conveniently be expressed in terms of a dimensionless Sherwood number for physical mass transfer:

$$Sh = k_L d/D_A \quad \text{or} \quad k_G d/D_A, \quad \text{(VII.265)}$$

where d is a characteristic dimension of the dispersion (e.g. bubble size, drop size, liquid film thickness, particle size) and D is the coefficient of molecular diffusion of the species considered. Orders of magnitude of the Sherwood number are given in Table VII.12. For mass transfer between the interior of a small bubble or drop and its surface we have roughly:

$$10 < Sh < 25;$$

the lower figure applies approximately when the inside fluid is stagnant and the higher figure can be used in case of free internal circulation [137–140]. For liquid films flowing by gravity over a packing, the Sh number, based on film thickness, lies in the same range. A general entry on this topic is the comprehensive review of van Landeghem [141].

Table VII.12. Characteristic data of equipment for gas absorption with chemical reaction in aqueous systems

Type of gas–liquid contactor	Specific surface, a (m^{-1})	Volume fraction liquid phase, $(1-\varepsilon)$	$Sh = k_L d/D_A$	$(1-\varepsilon)/a\delta = ((1-\varepsilon)/ad)Sh$
spray column	60	0.05	10–25	2–10
plate column	150	0.15	200–600	40–100
packed column	100	0.08	10–100	10–100
wetted-wall column	50	0.05	10–50	10–50
bubble column	20	0.98	400–1000	4000–10^4
agitated bubble contactor	200	0.90	100–500	150–800

VII.17.1 Fluid–solid reactors

Beek [142] correlated the literature data on mass transfer coefficients, to particles via the Stanton number St $(=\varepsilon k_G/u$ or $\varepsilon k_L/u)$ by

$$St Sc^{2/3} = 0.81 \left(\frac{u d_p}{v}\right)^{-0.5} \quad \text{for} \quad 5 < \frac{u d_p}{v} < 500 \quad \text{(VII.266)}$$

and

$$StSc^{2/3} = 0.6 \left(\frac{ud_p}{v}\right)^{-0.43} \quad \text{for} \quad 50 < \frac{ud_p}{v} < 2000, \quad \text{(VII.267)}$$

where ε is the fraction of fluid phase.

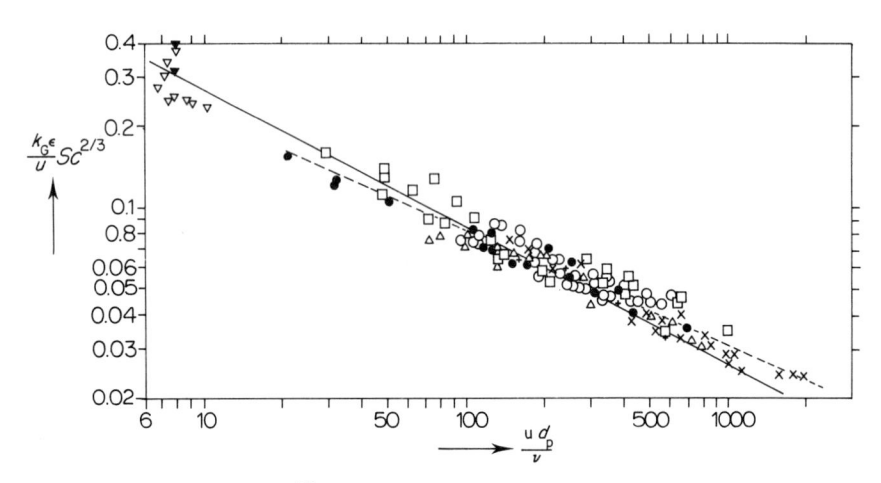

Fig. VII.49. Values of $StSc^{2/3}$ as a function of ud_p/v for packed and fluidized beds. [From Beek [142]. Reproduced by permission of Elsevier Science Publishers B.V.]

Equation (VII.266), which frequently is presented as $Sh = (0.81/\sqrt{\varepsilon})Re_p^{0.5}Sc^{1/3}$, was originally obtained by Thoenes and Kramers [143] for packed beds of spherical particles. The equation holds equally well for liquid–solid fluidized beds with $100 < Sc < 1000$ and $\varepsilon_0 < \varepsilon < 0.63$, in which ε_0 is the porosity of the packed bed, $\varepsilon_0 \simeq 0.43$. Equation (VII.267) holds for both gas–solid and liquid–solid packed and fluidized beds for $0.6 < Sc < 2000$ and $\varepsilon_0 < \varepsilon < 0.75$.

For a single particle a lower limit is found:

$$Sh_p = 2 \quad \text{for} \quad Re_p \ll 1 \quad \text{(VII.268)}$$

Thus, for Reynolds numbers below the lower limit of equation (VII.266) the correlation

$$Sh_p = 2 + \frac{0.81}{\sqrt{\varepsilon}} Re_p^{0.5} Sc^{1/3} \quad \text{(VII.269)}$$

may be suggested. However, it is known that the Sherwood number for particle systems becomes very much smaller than 2 at low Reynolds values. Thus, equation (VII.269) is not valid for $Re_p < 10$. Figure VII.50 gives Sh and Nu numbers for particles in packed beds at low Péclet numbers. Notice that for gases $Sc \simeq 1$, so that then $Pe \simeq Re$. The dramatic decrease of Sh and Nu at low Reynolds numbers has been explained by Schlünder [144, 145] from the non-

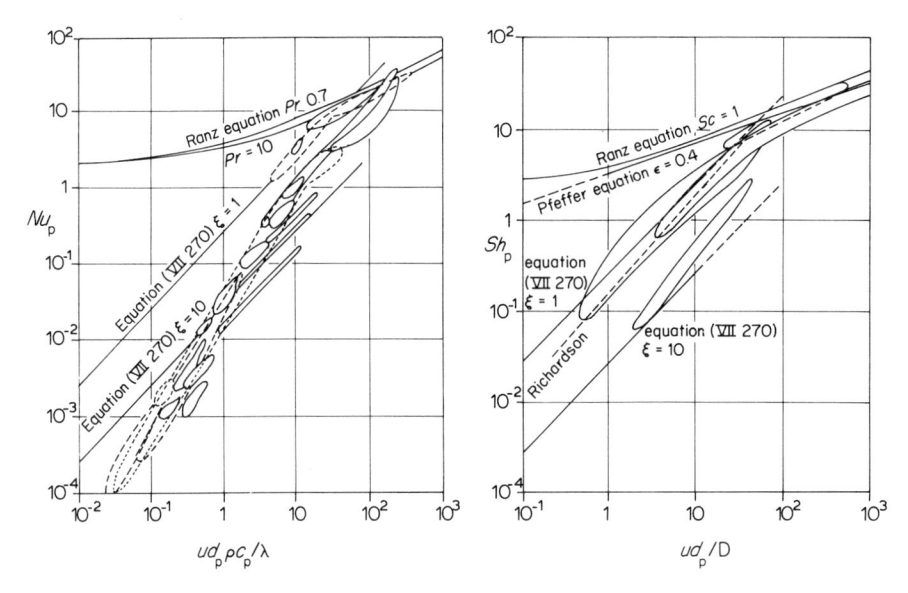

Fig. VII.50. Heat and mass transfer in packed beds at low Reynolds numbers [From Kunii and Suzuki [148]. Reproduced by permission of Pergamon Press Ltd]

uniform size distribution of surface and passage elements in a two-phase reactor. At low flow rates, only the channels with the largest diameters are effective while the smaller ones no longer contribute. Striking results can be observed as a decrease in conversion with decreasing flow rate in a certain regime of low Péclet numbers. Similar phenomena have been observed for fluid bed reactors (see figure VII.51) and for gas–liquid column reactors. Alternative explanations have been presented by Nelson and Galloway [146] and by Rowe [147].

The enclosed areas in figure VII.50 indicate experimentally observed heat and mass transfer data. Also shown in this figure is the Ranz equation which is similar to equation (VII.269). Kunii and Suzuki [148] correlated the experimental data for low Reynolds numbers by

$$Sh_p = \frac{\phi_s}{6(1-\varepsilon)\xi} \frac{ud_p}{D} \qquad (VII.270)$$

in which ϕ_s is the particle shape factor and ξ a channelling factor. Unfortunately, the value of ξ cannot be predicted *a priori*. Therefore the uncertainty in the value of Sh and Nu may be up to a factor of 5 at low Re_p.

Equations (VII.266) and (VII.267) also are not obeyed in fluidized beds at high superficial velocities where bubbles are found. Especially in gas–solid fluidized beds the phases tend to segregate and large gas bubbles exist. In this case the overall mass transfer from gas to solid will be much lower than predicted by equation (VII.267). The process is very complex in the bubbling regime. Gas is transferred from the bubbles into the so-called 'dense phase'; a more or less

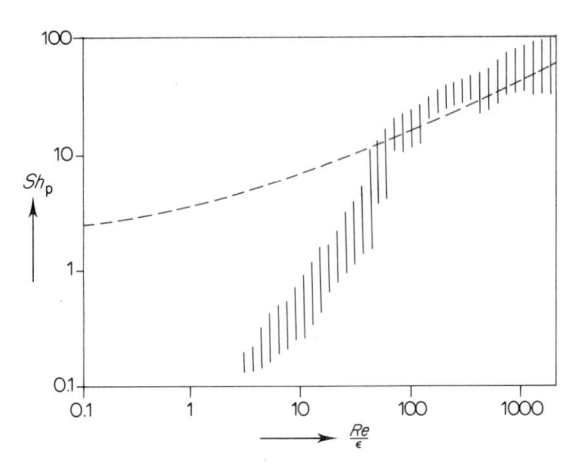

Fig. VII.51. Experimentally observed mass transfer coefficients in a fluidized bed; compared with relation (VII.269); $Sc = 2.35$ [From Schlünder [145], Reproduced by permission of Krausskopf Verlag für Wirtschaft GmbH, Mainz]

homogeneous dense gas–solid mixture (see figure VII.52). This is followed by mass transfer of gas from the dense phase to the particles, which again obeys the mass transfer relations mentioned above.

The rate of mass transfer between the bubbles and the dense phase depends on the concentration difference between the 'bubble gas' and the gas in the dense phase. Just as in gas–liquid systems, this can be expressed by a mass transfer coefficient and an interfacial area.

Recent reviews of models and relations are given by van Swaaij [149] and Myauchi et al. [150]. The empirical relations of van Swaaij and Zuiderweg [151, 152] give the product $k_G a$ directly for the cracking catalyst:

$$\frac{u}{k_G a} = \left(1.8 - \frac{1.06}{d_t^{0.25}}\right)\left(3.5 - \frac{2.5}{L^{0.25}}\right) \tag{VII.271}$$

where d_t and L are the diameter and the height of the fluidized bed expressed in metres. Also the empirical relation of Kobayashi et al. [153] is often used:

$$k_G a = 0.11 \varepsilon_b / d_b \quad [\text{s}^{-1}] \tag{VII.272}$$

where ε_b is the bubble hold-up and d_b is the equivalent bubble diameter (in metres). This relation requires the knowledge of d_b, see e.g. Kato and Wen [154].

For the mass transfer coefficient governing gas exchange from bubbles to the dense phase, both theoretical [155, 156] and empirical [157, 158] relations have been suggested. An early example is the relation of Davidson and Harrison [155]:

$$k_G = \tfrac{3}{4} u_{mf} + \left(\frac{gD^2}{d_b}\right)^{0.25} \tag{VII.273}$$

where u_{mf} is the superficial gas velocity at incipient fluidization. In deriving this relation the mass transfer coefficient is supposed to be situated only at the bubble side of the interfacial area. This has been challenged by recent experimental results. For example Drinkenburg and Rietema [159] found that under full fluidizing conditions no influence of the molecular diffusion coefficient on k_G could be observed. Many more mechanisms are involved [149] such as bubble coalescence [160] and adsorption on particles [161].

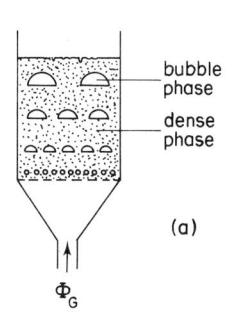

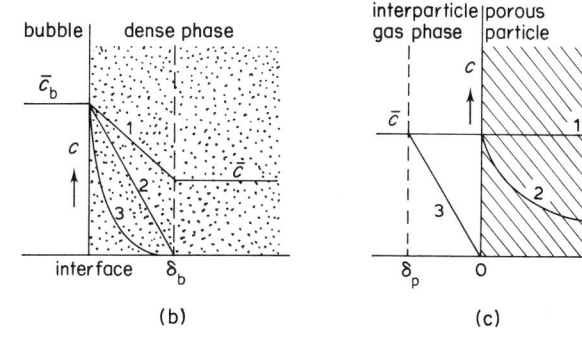

Fig. VII.52. Mass transfer in fluidized beds: (a) bubbling fluidized bed; (b) mass transfer from bubble to dense phase. Explanation of the profiles:
1. slow reaction with respect to mass transfer from bubbles; $(k_G a)_b \simeq k_1$;
2. the same with $(k_G a)_b \ll k_1$;
3. $\sqrt{(k_1 D)/k_{Gb}} \equiv \phi_b > 2$

(c) mass transfer in the dense phase. Explanation of the profiles:
1. $\left(\dfrac{V}{A}\right)_p \sqrt{\dfrac{k_1'' a_s}{D_i}} \equiv \phi_p < 0.2$
   $k_1 = (1 - \varepsilon_d) k_1'' a_s$;
2. $k_1 = (1 - \varepsilon_d) k_1'' a_s (\tanh \phi_p)/\phi_p$;
3. $k_1 = k_{Gp} a_p$

Definitions: k_1 = apparent rate constant per unit volume of the dense phase (s^{-1}). ε_d = interparticle gas volume in dense phase divided by the dense phase volume.
Subscripts d, p and b refer to dense phase, particles and bubbles respectively

For design purpose however, much experimental evidence has been obtained that k_G is a constant, only depending on the type of solid applied, e.g.

$$\text{for sand, } k_G = 0.008 \text{ m/s, for silica, } k_G = 0.017 \text{ m/s} \quad \text{(VII.274)}$$

(see van Swaaij [162]). The value of the interfacial bubble area can be found by dividing the values of $k_G a$ and k_G as predicted by the relations given above. This way, an average value for a is obtained that is independent of the bed height. Recent investigations have shown, however, that with good distributor plates very small bubbles can be obtained which rapidly coalesce to larger bubbles, resulting in a sharp decrease in interfacial area with increasing distance from the distributor plate. This has been described by Darton et al. [163, 164], Werther [157, 165] and van Swaaij [162]. According to Darton the equivalent bubble diameter decreases with z according to:

$$d_b = 0.54(u - u_{mf})^{2/5}(z + 4\sqrt{A_0})^{4/5}/g^{1/5} \quad \text{(VII.275)}$$

until a certain bed height z^* is reached. For $z > z^*$ bubble coalescence is in equilibrium with bubble break-up and d_b remains constant:

$$d_b = 0.54(u - u_{mf})^{2/5}(z^* + 4\sqrt{A_0})^{4/5}/g^{1/5} \quad \text{(VII.276)}$$

for $z > z^*$, with all units in S.I. A_0 is the so-called catchment area, a measure for the initial bubble size. For an ideal plate $A_0 = 0$. The value of z^* depends on the fraction of fines, which are particles with $d_p < 40 \times 10^{-6}$ m. Typical values are between $z^* = 1$ m for 30% fines and 2 m for 5% fines [165].

The value of the interfacial area follows from

$$a = \frac{1}{L} \int_0^L \varepsilon_b \frac{6}{d_b} dz \quad \text{(VII.277)}$$

where ε_b is the local bubble hold-up. If we neglect the dense phase through-flow we may write, for ε_b

$$\varepsilon_b = u/v_b, \quad \text{(VII.278)}$$

where v_b = the absolute bubble rising velocity. Werther showed [165]

$$v_b = \psi\sqrt{(gd_b)}, \quad \text{(VII.279)}$$

in which $\psi = 0.64$ for $d_t \leq 0.1$ m
$\phantom{\text{in which }\psi} = 1.6 d_t^{0.4}$ for $0.1 \text{ m} \leq d_t < 1$ m
$\phantom{\text{in which }\psi} = 1.6$ for $d_t > 1$ m,

provided sufficient fines are present in the bed. From relations (VII.276) to (VII.279), the specific bubble area can be calculated as a function of the bed height.

Similarly to gas-slurry reactors, also in fluidized beds the mass transfer from the bubbles may be enhanced by chemical reaction on the solids in the diffusion film around the bubbles. This has been described by Werther [165, 166] and van Swaaij [162].

Example VII.17.1 *Design of a fluid bed reactor*

We will design two fluidized bed reactors. In the first reactor gas A is catalytically converted in the porous particles by a relatively slow first order reaction A → P with $k_1 = 0.6 \text{ s}^{-1}$. The second reactor is for a similar reaction but with a much higher rate: $k_1 = 60 \text{ s}^{-1}$. Both reactors should operate at 300°C and the conversion should be 98% in both cases. The selected silica catalyst is of the cracking catalyst type with a settled bed density of 1000 kg/m³ and an average particle diameter of $d_p \simeq 70 \times 10^{-6}$ m. The actual gas flow velocity through the dense phase equals the minimum fluidization velocity and is $u_{mf} \simeq 0.01$ m/s. To prevent excessive attrition of the catalyst the superficial gas velocity is limited to 0.15 m/s. From the desired capacity a bed diameter of 3 m is derived. The remaining important question is the required bed height to obtain 98% conversion for each case. For a quick estimate of $k_G a$ we use the equation of van Swaaij and Zuiderweg [151, 152], relation (VII.271).

If the reactor capacity is completely limited by the rate of mass transfer from the bubbles to the dense phase and if we neglect the gas flow through the dense phase because $u_{mf} \ll u$, the conversion equation is simply

$$\frac{c_1}{c_o} = e^{-N_\alpha}, \tag{a}$$

in which N_α is the number of transfer units:

$$N_\alpha = k_G a L / u \tag{b}$$

For $c_1/c_o = 0.02$, N_α should be 4.

In case also the reaction rate still influences the conversion, the conversion can simply be derived from the van Deemter model or directly from the mass balance neglecting axial mixing of the dense phase. The conversion equation for that case reads

$$\frac{c_1}{c_o} = \exp\left(\frac{-1}{\frac{1}{N_r} + \frac{1}{N_\alpha}}\right)$$

N_r is the number of reaction units, given by $N_r = k_1(1 - \varepsilon_b)L/u$. We estimate the bubble fraction ε_b to be about 0.1. Table VII.13 gives the results of the calculations. It is clear that the reaction is almost mass transfer limited. To obtain the required conversion, 8 m bed height is sufficient for the fast reaction and 9–10 m for the slow reaction. Possibly the mass transfer could be chemically enhanced, but this can only be checked if we are able to predict the interfacial area and the mass transfer coefficient separately. This can be done with a bubble growth model, such as has been published by Werther [165], Krishna et al. [167] and van Swaaij [162]. We use the last one here. For k_G we use relation (VII.274): $k_G = 0.017$ m/s. In theory k_G should weakly depend on the diameter of the bubbles (relation (VII.273)) but, as mentioned above, such a relation has not been observed in freely flowing beds.

Table VII.13. The conversion as a function of the fluid bed height for two reaction rate constants; data from Example VII.17.1

L [m]	N_α	H_α	\multicolumn{2}{c}{$k_1 = 0.6\,\text{s}^{-1}$}	\multicolumn{2}{c}{$k_1 = 60\,\text{s}^{-1}$}
			N_r	c_1/c_0	N_r	c_1/c_0
7	3.58	1.95	25.2	0.044	2520	0.028
8	4	2	28.8	0.030	2880	0.018
9	4.4	2.04	32.4	0.0207	3240	0.012
10	4.8	2.08	36	0.0145	3600	0.008

The interfacial area can be calculated with the set relations (VII.276) to (VII.279), where we will assume sufficient fines to be present to have $z^* = 1$ m, and a nearly ideal porous plate distributor with $A_0 = 56 \times 10^{-6}\,\text{m}^2$. Notice that technical distributors may be much less ideal, resulting in much larger values of A_0. The result of these calculations is presented in figure VII.53. Both the local value of a (a_z) and the average value are presented as a function of z and L respectively. For $z > 1$ m, a_z remains constant because a constant bubble size is reached. The average value still decreases with L due to the decreasing influence of a_z for $z < z^*$. From the values of a and k_G it is found that e.g. at $L = 10$ m:

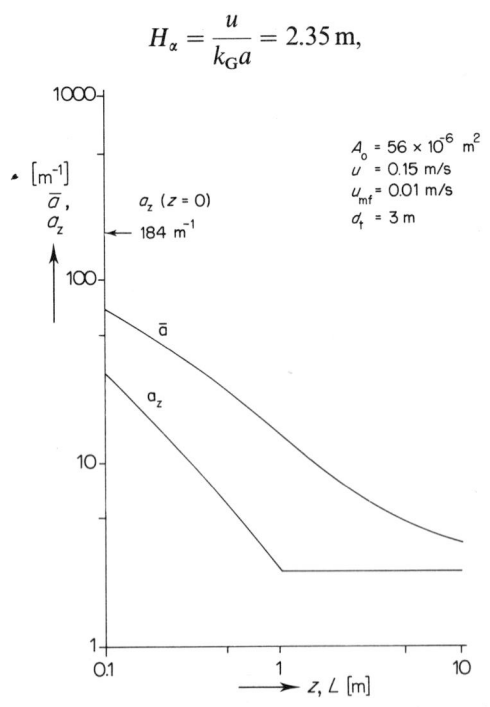

$$H_\alpha = \frac{u}{k_G a} = 2.35\,\text{m},$$

Fig. VII.53. The local (a_z) and the average ($\bar{a}$) specific bubble area in a fluidized bed as a function of distance from the distributor (z) and the bed height (L), respectively (Example VII.17.1)

which is slightly higher than calculated with the relation of van Swaaij and Zuiderweg (see Table VII.13).

We may now calculate the Hatta number if we estimate the diffusion coefficient in the dense phase: $D_{\text{eff}} \simeq 20 \times 10^{-6}\,\text{m}^2/\text{s}$

$$Ha = \frac{\sqrt{(k_1 D_{\text{eff}})}}{k_G}.$$

The result is $Ha \simeq 0.2$ for the slow reaction and $Ha \simeq 2$ for the fast reaction. For the fast reaction a much lower bed height is required because the conversion equation taking into account the enhancement factor will be

$$\frac{c_1}{c_o} = e^{-k_G E_A aL/u}$$

If we take $E_A \simeq Ha$ the value of $\dfrac{N_a k_G aL}{u}$ need only to be 2 to obtain a conversion of 98%. This means that L needs only to be ± 2–$5\,\text{m}$. Chemical enhancement in fluid beds has been observed experimentally. This is shown in figure VII.54. The experimental results of van Swaaij and Zuiderweg [151, 152] can be explained with chemical enhancement, especially if also non-sphericity of the bubbles is taken into account. Werther proposed a spherical factor of 1.6 for bubbles in a fluid bed, which seems to fit well.

VII.17.2 Fluid–fluid (–solid) reactors

In Table VII.14 we give the order of magnitude of k_L in gas–liquid contactors with aqueous systems. In all conventional contactors such as stirred tank reactors, bubble columns, tray columns, packed columns and spray columns the mass

Table VII.14. Order of magnitude of liquid side mass transfer coefficients in different gas–liquid reactors (aqueous systems)

Reactor	$10^4 k_L$ [m/s]	Reference
stirred cell	0.2–1	[97]
co-current gas–liquid tube reactor	2–10	[168]
cyclone	10–40	[136]
power jet	2–4	[169]
stirred tank	1–4	[170]
bubble column	1–4	[171]
bubble cap plate column	3–4	[172]
packed column, countercurrent	0.5–2	[173]
packed column, co-current	1–5	[174]
spray column	1–3	[175]

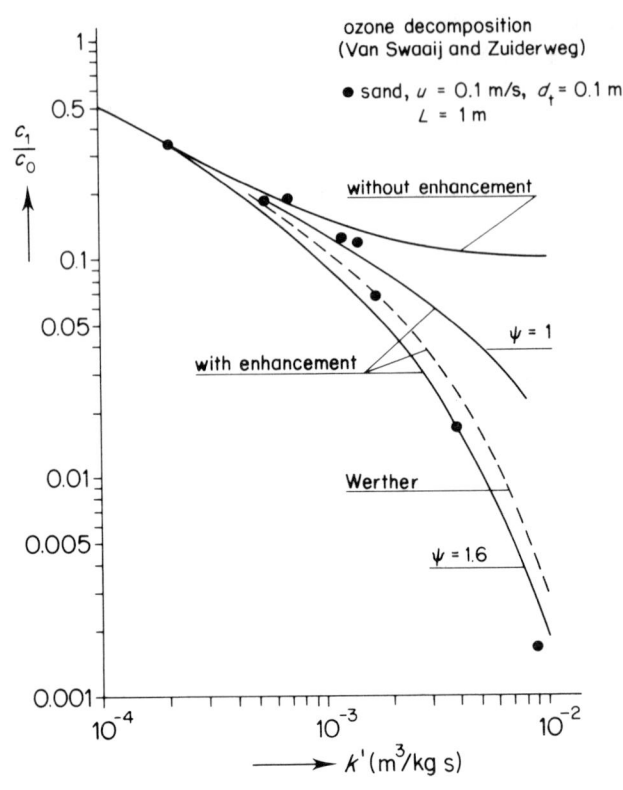

Fig. VII.54. Calculated and observed ozone decomposition for a fluidized sand bed. The drawn lines are calculated according to the method of Example VII.17.1, with and without chemical enhancement. In the Werther model the bubble form factor has already been taken into account ($\psi = 1.6$). Porous plate distributor, $A_0 = 56 \times 10^{-6}$ m^2, $k_G = 0.008$ m/s, $D_{\text{eff}} = D_G$

transfer coefficients are of a comparable order of magnitude. Considerably higher values of k_L are obtained only under extreme conditions such as in the cyclone gas–liquid reactor [136]; see also Example VII.16.4. Here the centripetal acceleration of up to 25 000 m/s^2 enhances the mass transfer process.

The *specific contact area* is often related to the power consumption. The relation between these two parameters depends on the reactor type as shown in figure VII.55. Reviews covering conventional gas–liquid reactors have been presented by Laurent and Charpentier [176], Charpentier [177] and van Landeghem [141]. Mass transfer in tray columns has been reviewed by Sharma *et al.* [178], in spray columns by Mehta and Sharma [175], in packed beds and in trickle bed reactors by Charpentier [179], Satterfield [180] and Blok *et al.* [181]. For bubble columns we refer to the reviews of Gestrich *et al.* for liquid side mass transfer coefficients [182], specific contact area [183] and hold up [184] and further to Mashelkar [185]. The review of van de Vusse and Wesselingh also includes gas–liquid–solid bubble columns and stirred tanks [186]. For

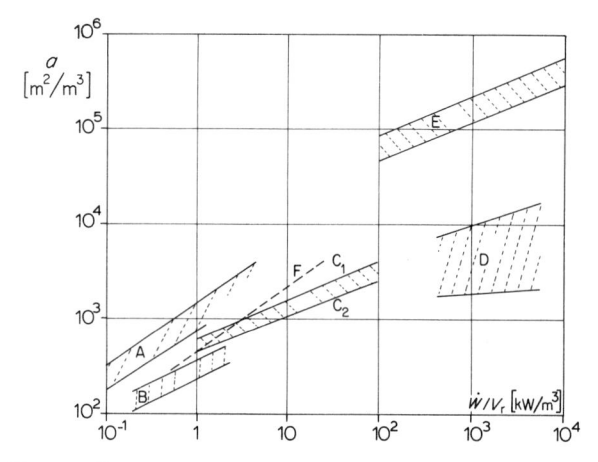

Fig. VII.55. The specific contact area, a, as a function of the specific energy consumption for different gas–liquid reactors. A = power jet reactor; B = bubble column; C = packed column; C1: with Raschig rings, C2 with spheres, D: venturi scrubber, E: Power jet tube reactor, F: stirred tank reactor [From Nagel, O., Kurten, H., and Sinn, R., *Chem. Ing. Tech.* **44**, 899 (1972). Reproduced by permission of Verlag Chemie GmbH]

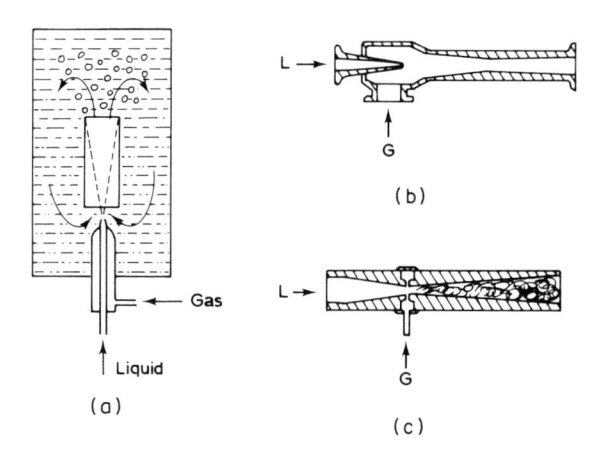

Fig. VII.56. (a) Power jet reactor; (b) Power jet tube reactor; and (c) venturi scrubber [From Laurent, A. and Charpentier, J. C., [176] *Chem. Eng. J.* (Lausanne) **8**, 85 (1974). Reproduced by permission of Elsevier Sequoia S. A.]

gas–liquid stirred tank reactors we refer to the reviews of Sideman *et al.* [187], Reith [115], Reith and Beek [188] and van Dierendonck, Fortuin and Venderbos [116]. For liquid–liquid systems data can be found in Hanson [189]. Data on gas–liquid–solid reactors are still relatively scarce. For both gas–liquid trickle flow reactors and gas–slurry reactors Shah [190] gives an introduction. See further van de Vusse and Wesselingh [186].

REFERENCES

1. Bird, R. B., Stewart, W. E., Lightfoot, E. N.; *Transport Phenomena*, Wiley, New York, 1960
2. Den Hartog, H. J. and Beek, W. J., *Appl. Sci. Res.* **19**, 311 (1968)
3. Porter, K. E. and Roberts, D.; *Chem. Eng. Sci.* **24**, 695 (1969)
4. Beek, W. J. and Muttzall, K. M. K.; *Transport Phenomena*, Wiley, London, 1975
5. Sherwood, T. K., Pigford, R. L., and Wilke, C. R.; *Mass Transfer*, McGraw-Hill, New York, 1975
6. Higbie, R.; *Trans. Am. Inst. Chem. Eng.*, **31**, 365 (1935)
7. Danckwerts, P. V.; *Ind. Eng. Chem.*, **43**, 1460 (1951)
8. Danckwerts, P. V.; *Trans. Faraday Soc.*, **46**, 300 (1950)
9. Danckwerts, P. V.; *Gas–Liquid Reactions*, McGraw-Hill, London, 1970
10. Aris, R.; Elementary Chemical Reactor Analysis, Prentice-Hall, Englewood Cliffs, 1969
11. Bischoff, K. B.; *Chem. Eng. Sci.* **29**, 1348 (1974)
12. Bischoff, K. B.; *AIChE J* **11**, 351 (1965)
13. Sada, E. and Kumazawa, H.; *Chem. Eng. Sci.* **28**, 1903 (1973)
14. Brunson, R. J. and Wellek, R. M.; *Chem. Eng. Sci.* **25**, 904 (1970)
15. Lightfoot, E. N.; *Chem. Eng. Sci.* **17**, 1007 (1962)
16. Baldi, G. and Specchia, V.; *Chem. Eng. J.* (Lausanne) **11**, 81 (1976).
17. Stepanek, J. B. and Shilimkan, R. V.; *Trans. Inst. Chem. Eng.* **52**, 313 (1974)
18. van Krevelen, D. W. and Hoftijzer, P. J.; *Rec. Trav. Chim. Pays-Bas* **67**, 563 (1948)
19. De Santiago, M. and Farina, I. H.: *Chem. Eng. Sci.* **25**, 744 (1970)
20. Brian, P. L. T., Hurley, J. F., and Hasseltine, E. A.: *A.I.Ch.E. J.* **7**, 226 (1961)
21. Van Krevelen, D. W. and Hoftijzer, P. J.: *Chim. Ind. XXIème Congrès Int. Chim. Ind.*, 168 (1948)
22. Nijsing, R. A. T. O. and Kramers, H., *Chem. Eng. Sci.* **8**, 81 (1958)
23. Nijsing, R. A. T. O., Hendriksz, R. H., and Kramers, H.: *Chem. Eng. Sci.* **10**, 88 (1959)
24. Stemerding, S.: *Chem. Eng. Sci.* **14**, 209 (1961)
25. Hikita, H. and Asai, S.: *Int. Chem. Eng.* **4**, 332 (1964)
26. De Coursey, W. J.: *Chem. Eng. Sci.* **29**, 1867 (1974)
27. Sim, M. T. and Mann, R.: *Chem. Eng. Sci.* **30**, 1215 (1975)
28. Juvekar, V. A.: *Chem. Eng. Sci.* **31**, 91 (1976)
29. Wellek, R. M. and Brunson, R. J., *Can. J. Chem. Eng.* **53**, 150 (1975)
30. Brunson, R. J. and Wellek, R. M.: *A.I.Ch.E. J.* **17**, 1123 (1971)
31. Brounshtein, B. I., Fishbein, G. A. and Rivikind, V. Ya: *Int. J. Heat Mass Transfer* **19**, 193 (1976)
32. Sada, E. and Ameno, T.: *J. Chem. Eng. Jpn* **6**, 247 (1973)
33. Onda, K., Sada, E., Kobayashi, T. and Fujine, M.: *Chem. Eng. Sci.* **25**, 753 (1970)
34. Secor, R. M. and Beutler, J. A.: *A.I.Ch.E. J.* **13**, 365 (1967)
35. Danckwerts, P. V.: *Chem. Eng. Sci.* **23**, 1045 (1968)
36. Olander, D. R.: *A.I.Ch.E. J.* **6**, 233 (1960)
37. Sada, E. and Kumazawa, A.: *Chem. Eng. Sci.* **29**, 335 (1974)
38. Dun, C. L., Freitas, E. R., Goodenbour, J. W., Henderson, W. T., and Papadopoulos, M. N.: *Hydrocarbon Process.* **43**, 150 (March 1964)
39. Hoftijzer, P. J. and Van Krevelen, D. W.: *Chem. React. Eng., Proc. Eur. Symp.* 4th, 1968, Pergamon, Oxford, 1971, p. 134
40. Geelen, H. and Wijffels, J. B.: *Chem. React. Eng., Proc. Eur. Symp.*, 3rd, 1964, Pergamon, Oxford, p. 125 (1965)
41. Shah, Y. T. and Sharma, M. M.: *Trans. Inst. Chem. Eng.* **54**, 1 (1976)
42. Rod, V.: *Collect Czech. Chem. Commun.* **38**, 3228 (1973)
43. Rod, V.: *Chem. Eng. J.* (Lausanne) **7**, 137 (1974)
44. Mhaskar, R. D. and Sharma, M. M.: *Chem. Eng. Sci.*, **30**, 811 (1975)

45. Sada, E., Kumazawa, H., and Butt, M. A.: *Can. J. Chem. Eng.* **54**, 97 (1976)
46. Sada, E., Kumazawa, H., and Butt, M. A.: *Can. J. Chem. Eng.* **55**, 475 (1977)
47. Merchuk, J. C. and Farina, I. H.: *Chem. Eng. Sci.* **31**, 645 (1976)
48. Narsimkan, G., *Chem. Eng. Sci.:* **16**, 7 (1961)
49. Denbigh, K. G. and Beveridge, G. S. G.: *Trans. Inst. Chem. Eng.* **40**, 23 (1962)
50. Thiele, E. W., *Ind. Eng. Chem.:* **31**, 916 (1939)
51. Zeldowitch, J. B.: *Acta Physicochim. URSS* **10**, 583 (1939)
52. Wagner, C.: *Z. Phys. Chem.* Abt. A **193**, 1 (1943)
53. Wicke, E.: *Angew Chem.* **19**, 57, 94 (1947)
54. Van Krevelen, D. W.: *Chem. Weekbl.* **47**, 427 (1951)
55. Satterfield, C. N.: *Mass Transfer in Heterogeneous Catalysis*, M.I.T. Press, Cambridge, Mass., 1970
56. Hougen, O. A. and Watson, K. M.: *Chemical Process Principles*, III, *Kinetics and Catalysis*, Wiley, New York, 1947
57. Sada, E., Kumazawa, H. and Butt, M. A.: *Chem. Eng. Sci.* **32**, 972 (1977)
58. Sada, E., Kumazawa, H., Butt, M. A. and Sumi, T.: *Chem. Eng. Sci.* **32**, 970 (1977)
59. Thonon, C. and Jungers, J. C.: *Bull. Soc. Chim. Belg.* **58**, 331 (1949)
60. McKewan, W. M.: *Trans. Metal. Soc. AIME* **224**, 387 (1962)
61. Walker, Jr., P. L., Rusinko, Jr., F., and Austin, L. G.: *Adv. Catal.* **11**, 133 (1959)
62. Rao, Y. K.: *Chem. Eng. Sci.* **29**, 1935 (1974)
63. Szekely, J., Evans, J. W., and Sohn, H. Y.: *Gas–Solid Reactions*, Academic Press, New York, 1976
64. Frank-Kamenetzki, D. A.: *Stoff und Wärme Übertragung in der Chemischen Kinetik*, Springer, Berlin, 1959
65. Wicke, E.: *Z. Electrochem.* **60**, 774 (1956)
66. Wicke, E. and Hedden, K., *Z. Electrochem.* **57**, 636 (1953)
67. Hedden, K.: *Chem. Eng. Sci.* **14**, 317 (1961)
68. Carberry, J. J.: *A.I.Ch.E. J.* **7**, 350 (1961)
69. Schilson, R. E. and Amundson, N. R.: *Chem. Eng. Sci.* **13**, 226 (1961)
70. Weisz, P. B.: *Z. Phys. Chem.* (Frankfurt am Main) **11**, 1 (1957)
71. Mears, D. A.: *Ind. Eng. Chem., Process Des. Dev.* **10**, 541 (1971)
72. De Acetis, J. and Thodos, G. T.: *Ind. Eng. Chem.* **52**, 1003 (1960)
73. Danckwerts, P. V.: *Appl. Sci. Res.*, Sect. A **3**, 385 (1953)
74. Danckwerts, P. V.: *Chem. Eng. Sci.* **22**, 472 (1967)
75. Shah, Y. T.: *Chem. Eng. Sci.* **27**, 1469 (1972)
76. Clegg, G. T. and Mann, R.: *Chem. Eng. Sci.* **24**, 321 (1969)
77. Tamir, A., Danckwerts, P. V. and Virkar, P. D.: *Chem. Eng. Sci.* **30**, 1243 (1975)
78. Beenackers, A. A. C. M. and van Swaaij, W. P. M.: *Chem. Eng. J.* (Lausanne) **15**, 25 (1978)
79. Beenackers, A. A. C. M. and van Swaaij, W. P. M.: *Chem. Eng. J.* (Lausanne) **15**, 39 (1978)
80. Mann, R. and Moyes, H.: *A.I.Ch.E. J.* **23**, 17 (1977)
81. Clegg, G. T. and Kilgannon, R. B. F.: *Chem. Eng. Sci.* **26**, 669 (1971)
82. Mann, R. and Clegg, G. T.: *Chem. Eng. Sci.* **30**, 97 (1975)
83. Weisz, P. B. and Hicks, J. S.: *Chem. Eng. Sci.* **17**, 265 (1962)
84. Hlaváček, V. and Kubiček, M.: *Chem. Eng. Sci.* **25**, 1537 (1970)
85. McGreavy, C. and Cresswell, D. L.: *Chem. Eng. Sci.* **24**, 608 (1969)
86. Hlaváček, V. and Kubiček, M.: *Chem. Eng. Sci.* **25**, 1761 (1970)
87. Groothuis, H.: *Ingenieur* (The Hague) **78**, Ch. 9 (1966)
88. Laurent, A., Prost, C. and Charpentier, J. C.: *J. Chim. Phys.* **72**, 236 (1975)
89. Nijsing, R. A. T. O.: *Ph.D. Thesis*, Delft (1957)
90. Scriven, L. E. and Pigford, R. L.: *A.I.Ch.E. J.* **4**, 439 (1958)
91. Scriven, L. E. and Pigford, R. L.: *A.I.Ch.E. J.* **5**, 397 (1959)
92. Sharma, M. M. and Danckwerts, P. V.: *Chem. Eng. Sci.* **18**, 729 (1963)

93. Roberts, D. and Danckwerts, P. V.: *Chem. Eng. Sci.* **17,** 961 (1962)
94. Danckwerts, P. V. and Gilham, A. J.: *Trans. Inst. Chem. Eng.* **44,** T42 (1966)
95. Hikita, H. and Ishikawa, H.: *Bull. Univ. Osaka Prefect.* Ser. A **18,** 427 (1969)
96. Hikita, H., Ishikawa, H., and Murakami, Y.: *Bull. Univ. Osaka Prefect.* Ser. A **19,** 34 (1970)
97. Jhaveri, A. S. and Sharma, M. M.: *Chem. Eng. Sci.* **22,** 1 (1967)
98. Levenspiel, O. and Godfrey, J. H.: *Chem. Eng. Sci.* **29,** 1723 (1974)
99. Danckwerts, P. V. and Kennedy, A. M.: *Chem. Eng. Sci.* **8,** 201 (1958)
100. Kothari, P. J. and Sharma, M. M.: *Chem. Eng. Sci.* **21,** 391 (1966)
101. Nanda, A. K. and Sharma, M. M.: *Chem. Eng. Sci.* **21,** 701 (1966)
102. Nanda, A. K. and Sharma, M. M.: *Chem. Eng. Sci.* **22,** 769 (1967)
103. Sharma, M. M. and Nanda, A. K.: *Trans. Inst. Chem. Eng.* **46,** T44 (1968)
104. Landau, J. and Chin, M.: *Can. J. Chem. Eng.* **55,** 161 (1977)
105. Procházka, J. and Bulička, J.: *Solvent Extr., Proc. Int. Solvent Extr. Conf.*, The Hague (1971), Society of Chemical Industry, London, 1971, p. 823
106. Carberry, J. J.: *Ind. Eng. Chem.* **56,** 39 (1964)
107. Bennett, C. O., Cutlip, M. B., and Yang, C. C.: *Chem. Eng. Sci.* **27,** 2255 (1972)
108. Christoffel, E. G.: in *Catal. Rev. Sci. Eng.* (eds. J. J. Carberry and H. Heinemann) **24,** 159 (1982)
109. Christoffel, E. G. and Schuler, J. C.: *Chem. Ing. Tech.* **52,** 844 (1980)
110. Weekman, Jr., V. W.: *A.I.Ch.E. J.* **20,** 833 (1974)
111. Suzuki, M. and Kawazo, K., *J. Chem. Eng. Jpn* **8,** 79 (1975)
112. Teshima, H. and Ohashi, Y.: *J. Chem. Eng. Jpn*, **10,** 70 (1977)
113. Schmalzer, D. K., Hoelscher, H. E., and Cowherd, C. (Jr.): *Can. J. Chem. Eng.* **48,** 135 (1970)
114. Froment, G. F.: *Chem. React. Eng., Proc. Int. 4th, Eur. 6th Symp.*, Heidelberg (1976), Dechema, Frankfurt (M), 1976, p. 421
115. Reith, T.: *Br. Chem. Eng.* **15,** 1559 (1970)
116. Van Dierendonck, L. L., Fortuin, J. M. H., and Venderbos, D.: *Chem. React. Eng., Proc. Eur. Symp., 4th.*, 1968, Pergamon, 1971, p. 205
117. Towell, G. D., Strand, C. P. and Ackerman, G. H.: *A.I.Ch.E.–I. Chem. E., Symp. Ser.*, No. 10, Institute of Chemical Engineers, London, 1965, p. 97
118. Calderbank, P. H.: *Trans. Inst. Chem. Eng.* **36,** 443 (1958)
119. Landau, J., Boyle, J., Gomaa, H. G., and Al Taweel, A. M.: *Can. J. Chem. Eng.* **55,** 13 (1977)
120. Landau, J., Gomaa, H. G., and Al Taweel, A. M.: *Trans. Inst. Chem. Eng.* **55,** 212 (1977)
121. Westerterp, K. R.: *Eng. Dr. Thesis*, Delft, 1962
122. Westerterp, K. R., van Dierendonck, L. L., and de Kraa, J. A.: *Chem. Eng. Sci.* **18,** 157 (1963)
123. Yoshida, F. and Yoshiharu, M.: *Ind. Eng. Chem., Process Des. Dev.* **2,** 263 (1963)
124. Linek, V. and Vacek, V.: *Chem. Eng. Sci.* **36,** 1747 (1981)
125. Sharma, M. M. and Mashelkar, R. A.: *I. Chem. Eng. Symp. Ser.*, Institute of Chemical Engineers, London, **28,** 10 (1968)
126. Sharma, M. M. and Danckwerts, P. V.: *Br. Chem. Eng.* **15,** 522 (1970)
127. Sridharan, K. and Sharma, M. M.: *Chem. Eng. Sci.* **3** , 767 (1976)
128. Linek, V.: *Collect. Czech. Chem. Commun.* **34,** 1299 (1969)
129. Joosten, G. E. H. and Danckwerts, P. V.: **28,** 453 (1973)
130. Reith, T. and Beek, W. J.: *Chem. Eng. Sci.* **28,** 1331 (1973)
131. Kunii, D. and Levenspiel, O.: *Fluidization Engineering*, Wiley, New York, 1969
132. Linek, V.: *Chem. Eng. Sci.* **27,** 627 (1972)
133. Linek, V. and Mayrhoferova, J.: *Chem. Eng. Sci.* **24,** 481 (1969)
134. Brian, P. L. T., Vivian, J. E. and Matiatos, D. C.: *A.I.Ch.E. J.* **13,** 28 (1967)
135. Robinson, C. W. and Wilke, C. R.: *A.I.Ch.E. J.* **20,** 285 (1974)

136. Beenackers, A. A. C. M. and van Swaaij, W. P. M.: *Chem. React. Eng., Proc. Eur., 6th., Intern. Symp., 4th*, 1976, Dechema, Frankfurt (M), 1976, VI-260
137. Bond, W. N. and Newton, D. A.: *Phil. Mag.* **4**, 24, 898 (1927)
138. Bond, W. N. and Newton, D. A.: *Phil. Mag.* **5**, 794 (1928)
139. Kronig, R. and Brink, J. C.: *Appl. Sci. Res.* **A2**, 143 (1950)
140. Garner, F. H. and Hammerton, D.: *Chem. Eng. Sci.* **3**, 1 (1954)
141. van Landeghem, H.: *Chem. Eng. Sci.*, **35**, 1912 (1980)
142. Beek, W. J., in A. A. H. Drinkenburg (ed.): *Proc. Int. Symp. Fluidization*, Neth. Univ. Press, Amsterdam, 507 (1967)
143. Thoenes, D. and Kramers, H.: *Chem. Eng. Sci.* **8**, 271 (1958)
144. Schlünder, E. U.: *Chem. Eng. Sci.* **32**, 845 (1977)
145. Schlünder, E. U.: *Verfahrenstechnik*, **10**, 645 (1976)
146. Nelson, P. A. and Galloway, T. R.: *Chem. Eng. Sci.* **30**, 1 (1975)
147. Rowe, P. N.: *Chem. Eng. Sci.* **30**, 7 (1975)
148. Kunii, D. and Suzuki, M.: *Int. J. Heat and Mass Transfer* **10**, 845 (1967)
149. van Swaaij, W. P. M.: in Luss, D. and Weekman (Jr.), V. W. (eds.), *A.C.S. Symp. Ser.* **72**, *Chem. React. Engng Revs*, Houston, 193 (1978)
150. Myauchi, T., Furusaki, S., Morooka, S. and Ikeda, Y.: in *Advances in Chemical Engineering*, Vol. 11, Academic Press, New York, 1981
151. van Swaaij, W. P. M. and Zuiderweg, F. J.: in *Proc. 5th Europ. Symp. on Chem. React. Engng.* B 9-25, Elsevier, Amsterdam, 1972
152. van Swaaij, W. P. M. and Zuiderweg, F. J.: in Angelino, H. et al., *Proc. Int. Symp. Fluidization and its Appl.*, Cepadues edit., Toulouse, 454 (1973)
153. Kobayashi, H., Arai, F. and Sunagawa, T.: *Chem. Eng.* (Tokyo) **31**, 239 (1967)
154. Kato, H. and Wen, C. Y.: *Chem. Eng. Sci.* **24**, 1351 (1969)
155. Davidson, J. F. and Harrison, D.: *Fluidised Particles*, Cambridge Univ. Press, 1963
156. Kunii, D. and Levenspiel, O.: *Fluidization Engineering*, Wiley, New York, 1969
157. Werther, J.: *Chem. Eng. Sci.* **35**, 372 (1980)
158. Werther, J.: *German Chem. Eng.* **1**, 243 (1978)
159. Drinkenburg, A. A. H. and Rietema, K.: *Chem. Eng. Sci.* **28**, 259 (1973)
160. Chiba, T. and Kobayashi, H.: *Chem. Eng. Sci.* **25**, 1375 (1970)
161. Bohle, W. and van Swaaij, W. P. M.: in Davidson, J. F. and Kearns, D. L., *Proc. Second Eng. Found. Conf.*, Cambridge University Press, 1978, p. 167
162. van Swaaij, W. P. M.: in Davidson, J. F. and Harrison, D., *Fluidization*, 2nd ed., London, Academic Press, 1983
163. Darton, R. C.: La Nauze, R. D., Davidson, J. F. and Harrison, D., *Trans. Inst. Chem. Engrs.* **55**, 274 (1977)
164. Darton, R. C.: *Trans. Inst. Chem. Engrs.* **57**, 134 (1979)
165. Werther, J.: *Chem. Ing. Tech.* **50**, 850 (1978)
166. Werther, J. and Hegner, B.: *Chem. Ing. Tech.* **52**, 106 (1980)
167. Krishna, R.: in Rodrigues, Calo, J. M. and Sweed, N. H. (eds.), *Nato Advanced Study Inst.*, series E, no 52, Sijthoff and Noordhoff, Leiden, 1981
168. Gregory, G. A. and Scott, D. S., in Rhodes, E. and Scott, D. S. (eds.): *Cocurrent Gas–liquid Flow*, Plenum Press, New York, 1969
169. Nagel, O., Kurten, H., and Sinn, R.: *Chem. Ing. Tech.* **42**, 474 (1970)
170. Koetsier, W. T. and Thoenes, D.: *Chem. Eng. J.* (Lausanne) **5**, 71 (1973)
171. Calderbank, P. H. and Moo Young, M. B.: *Chem. Eng. Sci.* **16**, 39 (1961)
172. Porter, K. E., King, M. B., and Varshney, K. C.: *Trans. Inst. Chem. Eng.* **44**, T274 (1966)
173. Danckwerts, P. V. and Sharma, M. M.: *Chem. Eng.* (London) **44**, CE 244 (1966)
174. Gianetto, A., Spechia, V. and Baldi, G.: *A.I.Ch.E. J.* **19**, 916 (1973)
175. Mehta, K. L. and Sharma, M. M.: *Brit. Chem. Eng.* **15**, 1440 (1970)
176. Laurent, A. and Charpentier, J. C.: *Chem. Eng. J.* (Lausanne) **8**, 85 (1974)

177. Charpentier, J. C.: in Luss, D. and Weekman (Jr.) V. W. (eds) *A.C.S. Symp. Ser.* **79,** 193 (1978)
178. Sharma, M. M., Mashelkar, R. A. and Mehta, V. D.: *Brit. Chem. Eng.* **14,** 70 (1969)
179. Charpentier, J. C.: in *CHISA, Proc 1975*, Section G, Prague, 1975
180. Satterfield, C. N.: *A.I.Ch.E. J.* **21,** 209 (1975)
181. Blok, J. R., de Wilt, G. J. and Drinkenburg, A. A. H.: *Polytech. Tijdschr., Procestechniek* (The Hague) **31,** 681 (1976)
182. Gestrich, W., Esenwein, H., and Kraus, W.: *Chem. Ing. Tech.* **48,** 399 (1976)
183. Gestrich, W., and Krauss, W.: *Chem. Ing. Tech.* **47,** 360 (1975)
184. Gestrich, W. and Räse, W.: *Chem. Ing. Tech.* **47,** 8 (1975)
185. Mashelkar, R. A.: *Brit. Chem. Eng.* **15,** 1297 (1970)
186. Van de Vusse, J. G. and Wesselingh, J. A.: *Chem. React. Eng. Proc. Eur. 6th, Int. 4th. Symp.*, Dechema, Frankfurt (M) 1976
187. Sideman, S., Hortacsu, O., and Fulton, J. W.: *Ind. Eng. Chem.* **33**(7), 58 (1966)
188. Reith, T. and Beek, W. J.: *Chem. React. Eng., Proc. Eur. 4th.*, Brussels (1968); Pergamon, 1971, p. 191
189. Hanson, C. (ed.): *Recent Advances in Liquid–Liquid Extraction*, Pergamon, Oxford, 1971
190. Shah, Y. T.: *Gas–Liquid–Solid Reactor Design*, McGraw-Hill, 1979

Chapter VIII

Multi-phase reactors, multiple reactions

VIII.1 INTRODUCTION

The previous chapter deals with mass transfer of one reactant in series or parallel with one reaction only. In this chapter we mainly discuss mass transfer with at least two reactions. More often than not, only one of the products is desired commercially and therefore the reactions have to be carried out in such a way that an optimum conversion to desired product will be obtained. Therefore this chapter mainly deals with the problem of selectivity and methods of optimizing selectivities. For single phase reactors the subject has been treated in Chapter III. This chapter extends the treatment to multiphase reactors; that is, it includes mass transfer effects. Hence, this chapter mainly deals with the influence of mass transfer on the selectivity of chemical processes.

In case all reactions are slow with respect to the rate of mass transfer, the reactions proceed either in the bulk of the reaction phase (in fluid–fluid reactions) or at concentrations which are in equilibrium with the bulk concentrations in the non-reaction phase (heterogeneous reactions). Basically, cases of slow reaction can be solved with the theory of Chapter III. Yet the possibility to improve the selectivity of a process by continuous extraction of the wanted intermediate is so interesting that we will deal with it explicitly (Section VIII.6.2). There is really a new problem in case at least one of the reactions is fast with respect to the rate of mass transfer. It will, therefore, be the main subject of this chapter.

Consequences of mass transfer limitation are (see figure VIII.1):

(a) The concentration of reactants transferred from a non-reaction phase into the reaction zone is lower in that reaction zone than at the interface ($c_A \leq c_{Ai}$).
(b) The concentration of reactants, originally present in the reaction phase, is lower in the reaction zone than in the remaining reaction phase ($c_B \leq \bar{c}_B$).
(c) The concentration of reaction products in the reaction zone is higher than in the remaining reaction phase ($c_P \geq \bar{c}_P$).
(d) The magnitude of the effects mentioned under (b) and (c) depends on ϕ and E_∞.

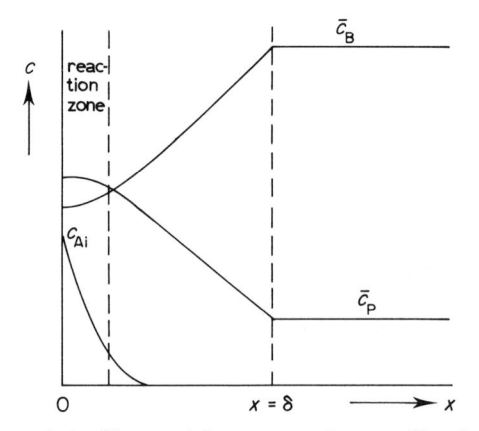

Fig. VIII.1. Characteristic film model concentration profiles for a fast fluid–fluid reaction;

$$A(1) \rightarrow A(2)$$
$$A(2) + B(2) \rightarrow P(2)$$

(e) The concentration of species that are both reaction product and reactant (e.g. intermediate products) depends on the relative reaction rates and on the rate of mass transfer.

Except for zero order reactions, reaction rates are a function of at least the reactant concentrations. The local differential selectivity of a process depends on the relative rates of the desired and the undesired reactions respectively. Therefore, in general, the selectivity realized depends on the local concentrations in the reaction zone. The latter are influenced by the value of the mass transfer coefficient. Hence, the selectivity obtained may depend on the value of the mass transfer coefficient(s).

The relation between mass transfer and selectivity depends on the reaction kinetics of the system. Wheeler [1], who did pioneering work in this field, distinguishes the following basic types:

— independent parallel reactions

$$A \rightarrow P$$
$$A' \rightarrow X;$$

— dependent parallel reactions

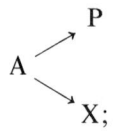

$$A \begin{array}{c} \nearrow P \\ \searrow X; \end{array}$$

— consecutive reactions $A \rightarrow P \rightarrow X$;

These used to be called systems of Type I, II and III selectivity, respectively. We will discuss each type in detail.

In addition to this, we will pay attention to systems in which more than one reactant is transferred in series or parallel with one or more reactions, that is to the topic of simultaneous mass transfer with reaction. Table VIII.1 gives the structure of this chapter.

Table VIII.1. Structure of Chapter VIII

Mass transfer	Reactions	Mass transfer reaction	Section
of one reactant only (A)	Parallel $A(+B) \to P$ $A(+B,B') \to X$	Series	VIII.3.1
		Parallel	VIII.3.2
	Consecutive $A \to P \to X$	Series	VIII.6.1
		Parallel	VIII.6.2
	Mixed parallel–consecutive $A(2) + B(2) \to P(2)$ $P(2) + B(2) \to X(2)$		VIII.7.1
	Mixed parallel–consecutive $A(2) + B(2) \to P(2)$ $A(2) + P(2) \to X(2)$		VIII.7.2
	Complex		VIII.7.3
of two reactants (A and A')	Parallel, independent: $A \to P$; $A' \to X$	Series	VIII.2.1
		Parallel	VIII.2.2
	Parallel, dependent $A + B \to P$ $A' + B \to X$		VIII.4
	$A + A' \to P$		VIII.5

VIII.2 SIMULTANEOUS MASS TRANSFER OF TWO REACTANTS A AND A' WITH INDEPENDENT PARALLEL REACTIONS A → P AND A' → X (TYPE I SELECTIVITY)

Especially in the petrochemical industry the feed to a reactor often consists of a mixture of reactants (e.g. aromatics, olefines or hydrocarbons with boiling points between T_1 and T_2) which all are subject to a certain conversion process (e.g. cracking, isomerization, hydrogenation). In such a system, a number of parallel reactions occur, all with a specific reaction rate. We assume in this section that all reactions are independent of each other. In the case of homogeneous single phase reactors, the selectivity ratio of the process, defined as the ratio between desired and undesired reactions, depends on the kinetics and the state of mixing only. In case mass transfer plays a role, it may affect the selectivity as well. It is convenient to distinguish between systems in which the mass transfer process and the reaction process are either parallel or in series.

VIII.2.1 Mass transfer and reaction in series

Hydrogenation, dehydrogenation and cracking of a mixture of olefines and hydrocarbons, respectively, may proceed at a catalyst surface according to

$$A \to A \text{ (cat)} \to P$$

$$A' \to A' \text{ (cat)} \to X$$

with P the desired and X the undesired product. In case the catalyst is non-porous, mass transfer and reaction are in series and the adsorption rates follow from simultaneous solving of the set equations (VIII.1–3):

$$J_A = k_{GA}(\bar{c}_A - c_{Ai}) \tag{VIII.1}$$

$$R''_A = f(c_{Ai}) \tag{VIII.2}$$

$$J_A = -R''_A \tag{VIII.3}$$

Hence, for first order kinetics, according to

$$-R''_A = R''_P = k''_P c_{Ai}, \tag{VIII.4}$$

elimination of c_{Ai} results in

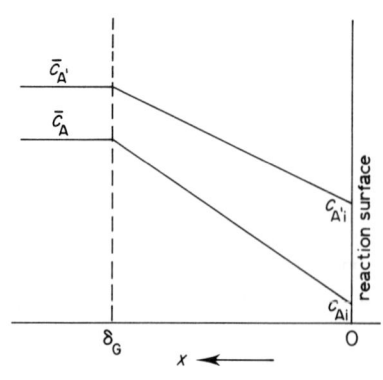

$$R''_P = J_A = \frac{\bar{c}_A}{(1/k_{GA}) + (1/k''_P)}. \tag{VIII.5}$$

Fig. VIII.2. Simultaneous mass transfer of two reactants A and A' in series with independent parallel reactions $A \to P$ and $A' \to X$;
 Concentration profiles according to film theory for $k_{GA} = k_{GA'}$.

The ratio R_P''/R_X'' is a direct measure for the differential selectivity σ_P' ($R_P''/R_X'' = \sigma_P'/(1 - \sigma_P') = \sigma_P'/\sigma_X'$). The goal is a maximum differential selectivity; hence a maximum value for R_P''/R_X''. From equation (VIII.5) it follows that

$$\frac{\sigma_P'}{\sigma_X'} = \frac{R_P''}{R_X''} = \frac{J_A}{J_{A'}} = \frac{\left(\dfrac{1}{k_{GA'}} + \dfrac{1}{k_X''}\right)\bar{c}_A}{\left(\dfrac{1}{k_{GA}} + \dfrac{1}{k_P''}\right)\bar{c}_{A'}}. \quad \text{(VIII.6)}$$

Figure VIII.3 shows some typical characteristics of equation (VIII.6). Two asymptotic cases can be distinguished:

(a) $k_G \gg k_P''$ and k_X''. There is no mass transfer limitation. The concentrations of A and A' on the catalyst are in equilibrium with the bulk concentrations. The formation rate ratio approaches

$$\frac{\sigma_P'}{\sigma_X'} = \frac{R_P''}{R_X''} = \frac{k_P'' \bar{c}_A}{k_X'' \bar{c}_{A'}}. \quad \text{(VIII.7)}$$

(b) $k_G \ll k_P''$ and k_X''. There is complete mass transfer limitation. The concentrations on the catalyst of both A and A' approach zero and the formation rate ratio approaches

$$\frac{\sigma_P'}{\sigma_X'} = \frac{R_P''}{R_X''} = \frac{k_{GA} \bar{c}_A}{k_{GA'} \bar{c}_{A'}}. \quad \text{(VIII.8)}$$

Note that the selectivity in this case becomes independent of the ratio k_P''/k_X''.

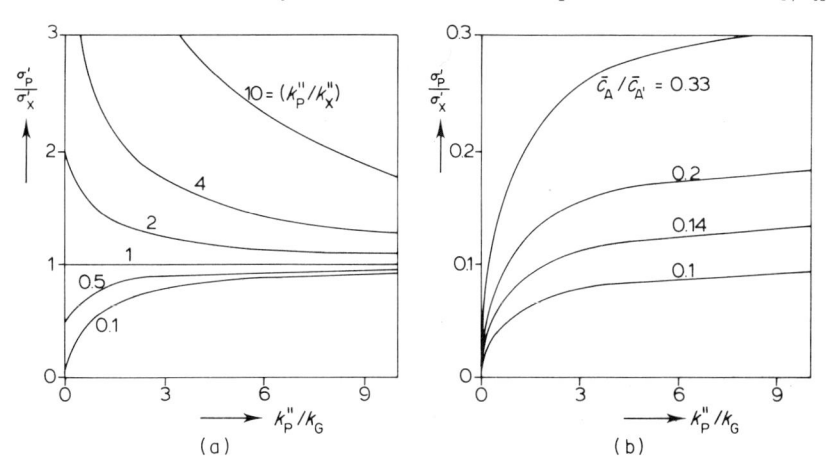

Fig. VIII.3. Selectivity change in a system of two independent first order reactions due to the reaction regime change from the kinetic region to the mass transfer controlled region for $k_{GA'} = k_{GA}$

(a) $\bar{c}_A = \bar{c}_{A'}$, various k_P''/k_X''; (b) $k_P''/k_X'' = 0.1$, various $\bar{c}_A/c_{A'}$ [From J. Horak and F. Jiráček, Collect. Czech. Chem. Commun., **39**, 2532 (1974). Reproduced by permission of Academia, Publishing House of the Czechoslovakian Academy of Sciences]

For $k_P'' > k_X''$ and $k_{GA} = k_{GA'}$, case (a) gives a better selectivity (equation (VIII.7)) than case (b) (Equation (VIII.8)). Then, case (a) is the most favourable one. Theoretically, it can be realized by sufficiently low temperatures and high mass transfer coefficients. The latter asks for high fluid velocities relative to the catalyst and for small catalyst particles. For $k_P'' < k_X''$, the reverse is true and mass transfer limitation favours selectivity. For different kinetics, qualitatively the same results are obtained. The general conclusion for any kinetics

$$R_P'' = f(c_{Ai})$$

$$R_X'' = F(c_{A'i})$$

is that the formation rate ratio varies from

$$\frac{R_P''}{R_X''} = \frac{f(\bar{c}_A)}{F(\bar{c}_{A'})} \tag{VIII.9}$$

for the kinetic regime, to

$$\frac{R_P''}{R_X''} = \frac{k_{GA}\bar{c}_A}{k_{GA'}\bar{c}_{A'}} \tag{VIII.8}$$

for the diffusion controlled regime.

VIII.2.2 Mass transfer and reaction in parallel

For the same reaction system as in the previous section

$$A \rightarrow P$$

$$A' \rightarrow X$$

but now with porous pellets instead of non-porous ones or with a fluid–fluid reaction system, the theory can be extended to the case of mass transfer both in series and in parallel with reaction. Figure VIII.4 gives a qualitative picture of the

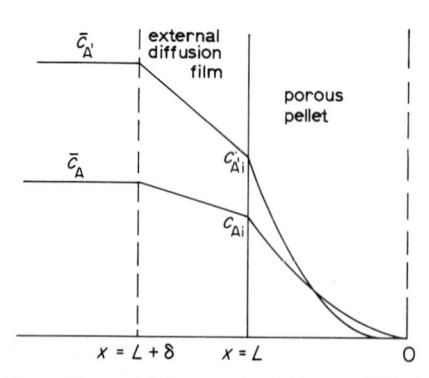

Fig. VIII.4. Characteristic film model concentration profiles for A and A' in mass transfer to and in a porous pellet, followed by independent reactions

concentration profiles of A and A'. Both A and A' are transported to the interface according to, e.g. for A,

$$J_A = k_{GA}(\bar{c}_A - c_{Ai}) \tag{VIII.10}$$

In the porous pellet reaction and diffusion are parallel and, in case of first order reactions, the flux through the interface follows from equation (VII.66):

$$J_A = \frac{D_{iA}}{L} c_{Ai} \phi_A \tanh \phi_A, \tag{VIII.11}$$

with L = volume of the pellet divided by the external surface of the pellet,

$$\phi_A = L\sqrt{(k_P'' a_S / D_{iA})},$$

D_{iA} = coefficient of internal diffusion of species A. If the molecule dimensions of A and A' differ greatly, D_{iA} may differ substantially from $D_{iA'}$.

Elimination of the unknown concentration near the external interface by combination of equations (VIII.10) and (VIII.11) gives, for A and A' respectively,

$$J_A = \bar{c}_A \Big/ \left\{ \frac{1}{k_{GA}} + \frac{1}{\sqrt{(k_P'' a_S D_{iA})}\tanh \phi_A} \right\}$$

$$J_{A'} = \bar{c}_{A'} \Big/ \left\{ \frac{1}{k_{GA'}} + \frac{1}{\sqrt{(k_X'' a_S D_{iA'})}\tanh \phi_{A'}} \right\}. \tag{VIII.12}$$

Hence, the formation rate ratio of P and X is given by

$$\frac{R_P''}{R_X''} = \frac{J_A}{J_{A'}} = \frac{\bar{c}_A \left\{ \dfrac{1}{k_{GA'}} + \dfrac{1}{\sqrt{(k_X'' a_S D_{iA'})}\tanh \phi_{A'}} \right\}}{\bar{c}_{A'} \left\{ \dfrac{1}{k_{GA}} + \dfrac{1}{\sqrt{(k_P'' a_S D_{iA})}\tanh \phi_A} \right\}}.$$

(VIII.13)

Asymptotic solutions are

(a) $k_G \gg \sqrt{(k_P'' a_S D_{iA})}\tanh \phi_A$ and $\sqrt{(k_X'' a_S D_{iA'})}\tanh \phi_{A'}$. With these conditions satisfied, external diffusion limitations are absent, and c_{Ai} and $c_{A'i}$ equal $\bar{c}_A$ and $\bar{c}_{A'}$ respectively. Equation (VIII.13) reduces to

$$\frac{R_P''}{R_X''} = \frac{\bar{c}_A \sqrt{(k_P'' D_{iA})}\tanh \phi_A}{\bar{c}_{A'} \sqrt{(k_X'' D_{iA'})}\tanh \phi_{A'}}. \tag{VIII.14}$$

For both ϕ_A and $\phi_{A'} < 0.2$, also internal diffusion limitations are absent and equation (VIII.14) further reduces to

$$\frac{R_P''}{R_X''} = \frac{k_P'' \bar{c}_A}{k_X'' \bar{c}_{A'}}, \tag{VIII.7}$$

being the equation for a completely kinetically controlled system. For both ϕ_A and $\phi_{A'} > 2$, internal diffusion limitation is complete and equation (VIII.14) reduces to

$$\frac{R_P''}{R_X''} = \sqrt{\left(\frac{k_P'' D_{iA}}{k_X'' D_{iA'}}\right) \frac{\bar{c}_A}{\bar{c}_{A'}}}. \tag{VIII.15}$$

(b) $k_G \ll \sqrt{(k_P'' a_s D_{iA})} \tanh \phi_A$ and $\sqrt{(k_X'' a_s D_{iA'})} \tanh \phi_{A'}$.

Now, external diffusion limitation is complete and c_{Ai} and $c_{A'i}$ approach zero. Equation (VIII.13) reduces to equation (VIII.8):

$$\frac{R_P''}{R_X''} = k_{GA} \bar{c}_A / k_{GA'} \bar{c}_{A'}. \tag{VIII.8}$$

For $k_P''/k_X'' > 1$ and equal diffusivities, selectivity will increase in the direction as given by the equations (VIII.8), (VIII.15) and (VIII.7). Then, maximum selectivities will be obtained for a high relative fluid velocity (k_G large), small pellets (k_G large and ϕ small) and large pores (D_i large and therefore ϕ small). If necessary, ϕ can be further reduced by lowering the temperature. The latter measure however may give an unfavourable result in case $E_A > E_{A'}$, due to a relative decrease of k_P''/k_X'' with decreasing temperature.

The *general conclusion* from this and the previous section is that for a system of independent, first order, parallel reactions, and equal diffusivities, the selectivity ratio changes from

$$R_P''/R_X'' = k_P'' \bar{c}_A / k_X'' \bar{c}_{A'} \tag{VIII.7}$$

for no external nor internal diffusion limitation, via

$$R_P''/R_X'' = \sqrt{(k_P''/k_X'')\bar{c}_A/\bar{c}_{A'}} \tag{VIII.15}$$

for complete internal diffusion limitation only to

$$R_P''/R_X'' = \bar{c}_A/\bar{c}_{A'} \tag{VIII.8}$$

for complete external diffusion limitation. A shift in this direction is desired only for $k_P'' < k_X''$. Different kinetics will affect the conclusion only qualitatively. Independently of the kinetics, the formation rate ratio will be between the value for kinetic limitation (equation (VIII.9)) and that for complete mass transfer limitation (equation (VIII.8)).

The theory can be extended to fluid–fluid reactions with $Al \neq 0$ by using the appropriate expressions for J (equation (VII.59) instead of equation (VIII.11)). For both ϕ_A and $\phi_{A'}$ larger than 2, the results are the same as for $Al = 0$. In all other cases, a balance over the bulk is necessary to calculate the concentration of the reactants in the bulk of the reaction phase. Such an analysis is straightforward. An important example in the field of fluid–fluid reactions is the removal of H_2S from natural gas, refinery gas or gas mixtures obtained from the gasification of coal or oil. Widely applied is the absorption of H_2S in amines. In case CO_2 is also present, which is often the case, this component is absorbed as well. The reaction between H_2S and amines is mostly instantaneous and the H_2S

absorption rate is usually completely limited by gas phase resistance. Hence $J_{H_2S} \propto k_G$. Absorption of carbon dioxide is usually chemically enhanced if gas phase resistance is absent. Hence $J_{CO_2} \neq f(k_G, k_L)$. Absorption of CO_2 lowers the effectiveness of the process by consuming amine, which is consequently no longer available for binding of H_2S. Therefore a high value of the mass transfer coefficient k_G may favour the selectivity of the process, via

$$\frac{J_{H_2S}}{J_{CO_2}} \propto k_G.$$

A problem related to the theory of this section is found in the field of catalyst deactivation. Poisoning of a catalyst may affect both k_P'' and k_X'', and, by that, spoil the selectivity of the process. A different effect occurs if, as in hydrodesulphurization [2], the poisons are preferentially absorbed at the pore mouth. This way an extra resistance against mass transfer may be built up in time. In case this resistance becomes rate controlling, the relative formation rate may change from the original value

$$\frac{R_P''}{R_X''} = \frac{\bar{c}_A \sqrt{(k_P'' D_{iA})} \tanh \phi_A}{\bar{c}_{A'} \sqrt{(k_X'' D_{iA'})} \tanh \phi_{A'}}$$

(in case of first order kinetics and no external resistance) to

$$\frac{R_P''}{R_X''} = \frac{D_{iA} \bar{c}_A}{D_{iA'} \bar{c}_{A'}}$$

with a simultaneous remarkable reduction of the overall conversion rate. Figure VIII.5 illustrates this mechanism.

VIII.3 MASS TRANSFER OF ONE REACTANT (A) FOLLOWED BY TWO DEPENDENT PARALLEL REACTIONS

$$A(+B) \to P$$

$$A(+B, B') \to X$$

(TYPE II SELECTIVITY)

If all parallel reactions are independent (previous section) the enhancement factor can be calculated for each reaction separately with the theory of mass transfer with one reaction only (Chapter VII). With that, the selectivity follows directly from the ratio of the different calculated mass transfer rates.

If the parallel reactions are dependent however, the model description becomes more complicated. The simplest example is

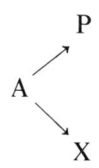

$$A \begin{array}{c} \nearrow P \\ \searrow X \end{array}$$

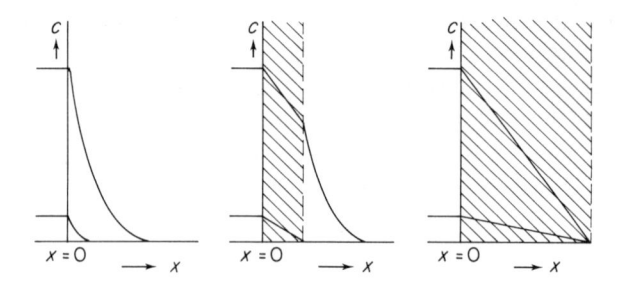

Fig. VIII.5. Schematic concentration profiles with increasing mass transfer resistance at the catalyst surface due to adsorption of poisons, starting at the pore mouth ($x \neq 0$). The shaded area is thought to be completely poisoned

If both reactions are of the same order, say of order m, the problem can be reduced to mass transfer with one chemical reaction of rate

$$R_A = -(k_P + k_X)c_A^m,$$

as has been described in Section VII.6. The local formation rate ratio

$$\frac{R_P}{R_X} = \frac{k_P c_A^m}{k_X c_A^m} = \frac{k_P}{k_X}$$

is independent of c_A and therefore independent of any mass transfer or diffusional effect. This is no longer true when both reactions are of different orders, say m and n. Lowering of the reactant concentration relatively favours the rate of the reaction of lowest order, according to

$$\frac{R_P}{R_X} = \frac{k_P}{k_X} c_A^{m-n}. \tag{VIII.16}$$

Mass transfer limitation lowers the concentration of the reactant in the reaction zone. Hence, mass transfer limitation must be avoided in case the desired reaction is of highest order. A quantitative analysis follows below.

VIII.3.1. Mass transfer and reaction in series

$$A \begin{array}{c} \nearrow P \quad R_P'' = k_P'' c_A^m \\ \searrow X \quad R_X'' = k_X'' c_A^n \end{array}$$

In case of heterogeneous reaction at a non-porous pellet, the conversion rate of A (J_A) and c_{Ai} follow from solution of the set

$$J_A = k_G(\bar{c}_A - c_{Ai}) \tag{VIII.17}$$

$$J_A = k_P'' c_{Ai}^m + k_X'' c_{Ai}^n. \tag{VIII.18}$$

Then the formation rate ratio and consequently the differential selectivity ratio follow from

$$\frac{R_P''}{R_X''} = \frac{\sigma_P'}{\sigma_X'} = \frac{k_P''}{k_X''} \bar{c}_{Ai}^{(m-n)}. \tag{VIII.19}$$

In case the desired reaction is of the highest order, a high $\bar{c}_A$ and a high k_G will favour the selectivity. The latter asks for small particles and a relatively high fluid velocity.
The criterion for absence of diffusion limitation is

$$\frac{k_G}{k_P'' \bar{c}_A^{m-1} + k_X'' \bar{c}_A^{n-1}} \gg 1. \tag{VIII.20}$$

In case of large heat effects, the theory must be extended by adding the heat balance (equation (VII.238), provided that radiation may be neglected), and the two rate equations as a function of temperature to the set of equations (VIII.17, VIII.18) in order to calculate the actual reaction temperature at the particle surface by solving these five equations simultaneously. Though this is left to the reader, we point out that in case of heat effects, mass transfer may affect selectivity even in case both reactions are of the same order, because k_P''/k_X'' depends on the surface temperature in case $E_P \neq E_X$.

Example VIII.3.1 *Mass transfer with parallel first order reactions*
A problem of mass transfer in series with the parallel reactions:

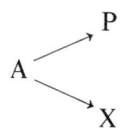

$$A \begin{array}{c} \nearrow P \\ \searrow X \end{array}$$

is discussed and for the case that both reactions are of the first order, it is derived from equation (VIII.6) that the differential selectivity ratio is given by

$$\frac{R_P''}{R_X''} = \frac{(1/k_G) + (1/k_X'')}{(1/k_G) + (1/k_P'')}.$$

Why is this result not correct?
The solution to this is as follows. Equation (VIII.6) has been derived for two different reactants A and A', A can be converted to P only and A' to X only. The proposed solution therefore is only correct if we can trace two types of reactant A in the gas phase, one capable of being converted to P, the other to X, which is not

possible because every molecule of A has the same chance, $k_P''/(k_P'' + k_X'')$, of being converted to P, and therefore

$$\frac{R_P''}{R_X''} = \frac{k_P''}{k_X''}$$

as also follows from equation (VIII.19) for $m = n$.

VIII.3.2 Mass transfer and reaction in parallel

Both for porous particles and for fluid–fluid systems, the rate of mass transfer follows from solving

$$D_A \frac{d^2 c_A}{dx^2} + R_A = 0, \qquad (VII.87)$$

with boundary conditions: $c_A = c_{Ai}$ for $x = 0$ and $-dc_A/dx = \gamma$ for $c_A = \bar{c}_A$. For fluid–fluid systems, where a bulk is present, an additional balance is required to find γ. For a mixer this balance is:

$$aD_A \gamma = -R(\bar{c}_A)(1-\varepsilon)\left(\frac{Al-1}{Al}\right), \qquad (VII.91)$$

with $R(\bar{c}_A) = R_A$ at bulk concentrations. Notice that in case of porous solids, for D_A the effective internal diffusion coefficient, D_{iA}, of A must be taken, while then $\gamma = 0$ for $c_A = \bar{c}_A$ ($\equiv (c_A)_{x=\delta}$). The method of solution is explained in Section VII.6.

From equation (VII.90) it follows for the (m, n) order system A → P, A → X, that

$$\frac{dc_A}{dx} = -\sqrt{\left\{\frac{2}{D_A}\left[\frac{k_P}{m+1}(c_A^{m+1} - \bar{c}_A^{m+1}) + \frac{k_X}{n+1}(c_A^{n+1} - \bar{c}_A^{n+1})\right]\right.}$$
$$\left. + \left[(k_P \bar{c}_A^m + k_X \bar{c}_A^n)\frac{(1-\varepsilon)(Al-1)}{aD_A Al}\right]^2\right\}. \qquad (VIII.21)$$

Numerical integration of this equation from $\bar{c}_A$ to c_{Ai} gives an expression for $\bar{c}_A$ (equation (VII.92)), and from $\bar{c}_A$ to c_A gives c_A as a function of x. Total absorption rate follows from

$$J_A = -D_A \left(\frac{dc_A}{dx}\right)_{x=0}$$

and the differential selectivity ratio from

$$\frac{\sigma_P'}{\sigma_X'} = \left(\frac{1}{\delta}\int_0^\delta \frac{k_P}{k_X} c_A^{m-n} dx\right)\frac{1}{Al} + \left(\frac{k_P}{k_X} \bar{c}_A^{m-n}\right)\frac{Al-1}{Al}. \qquad (VIII.22)$$

For *slow reactions* in fluid–fluid systems with $Al \gg 1$, the reactions are essentially bulk reactions and Equation (VIII.22) reduces to

$$\frac{\sigma'_P}{\sigma'_X} = \frac{k_P \bar{c}_A^{m-n}}{k_X} \quad \text{for } Al \gg 1 \text{ and } \phi < 0.2.$$

The criterion for slow reaction is $\phi < 0.2$, with ϕ defined by its general definition according to equation (VII.95).

In general the bulk concentration $\bar{c}_A$ depends on the value of the interfacial area a and on k_L. For example, in a mixer in which all of A is transferred to the reaction phase and reacts slowly in the reaction phase,

$$k_L a(c_{Ai} - \bar{c}_A) = (k_P \bar{c}_A^m + k_X \bar{c}_A^n)(1 - \varepsilon).$$

In case the desired reaction is of the highest order, $\bar{c}_A$ must be as close to c_{Ai} as possible to favour a good selectivity. This calls for a sufficiently large value of the product $k_L a$. For example, for $(c_{Ai} - \bar{c}_A)/c_{Ai} < 10^{-1}$, the criterion is

$$\frac{k_L a}{k_P c_{Ai}^{m-1} + k_X c_{Ai}^{n-1}} > 10.$$

In case no bulk is present, as in *fluid–porous solid reactions*, equations (VIII.21, VIII.22) simplify by substituting $Al = 1$ and $\gamma = 0$. Roberts [3] gives exact solutions for (m, n) combinations $(1, 0)$, $(2, 0)$ and $(2, 1)$, all for $\gamma = 0$. He presents his results as a function of the modulus ϕ'' (called ϕ by Roberts) and defined as

$$\phi'' = \frac{|\Phi_A|\delta^2}{(1 - \varepsilon)V_r D_A c_{Ai}}. \tag{VII.235}$$

As discussed in Section VII.13, this modulus can be calculated directly from experimental data without knowing the kinetics, while its numerical value indicates whether the reaction proceeds in the diffusional regime ($\phi'' > 1$), the kinetic regime ($\phi'' < 0.05$) or the intermediate regime.

Figures VIII.6–VIII.9 present effectiveness factors (η) and relative differential

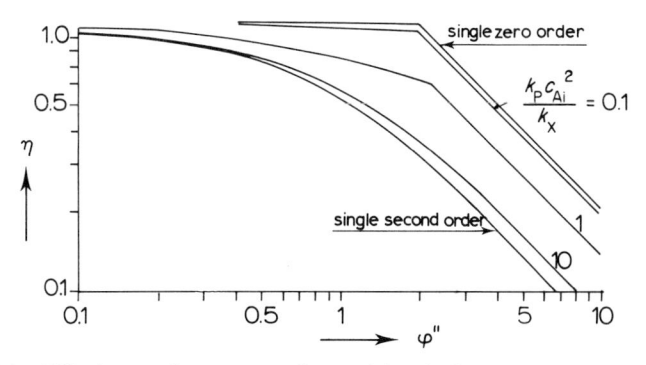

Fig. VIII.6. Effectiveness factor, η, vs. the modulus ϕ'' for various values of $k_P c_{Ai}^2/k_X$ for mass transfer with second ($m = 2$, A → P) and zero ($n = 0$, A → X) order parallel reactions [From Roberts [3]. Reproduced by permission of Pergamon Press Ltd]

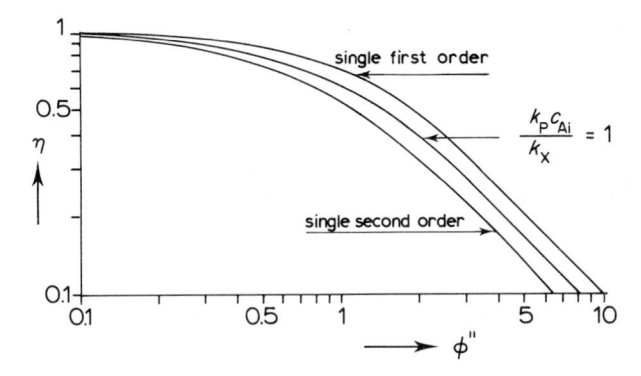

Fig. VIII.7. Effectiveness factor, η, vs. the modulus ϕ'' for various values of $k_P c_{Ai}/k_X$ for mass transfer with second ($m = 2$, A → P) and first ($n = 1$, A → X) order parallel reactions [From Roberts [3]. Reproduced by permission of Pergamon Press Ltd]

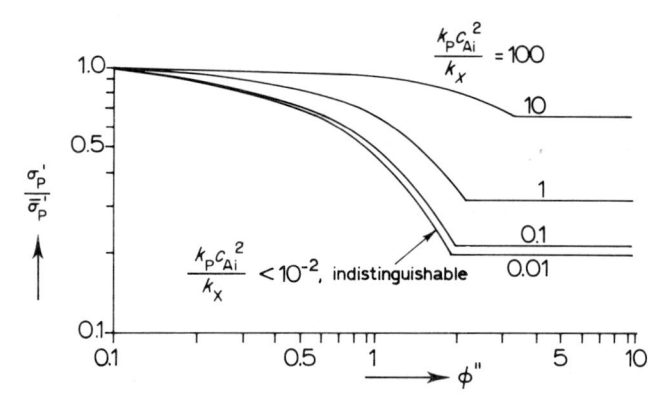

Fig. VIII.8 Relative selectivity, $\sigma'_P/\bar{\sigma}'_P$, vs. the modulus ϕ'' for various values of $k_P c_{Ai}^2/k_X$ for mass transfer with second ($m = 2$, A → P) and zero ($n = 0$, A → X) order parallel reactions [From Roberts [3]. Reproduced by permission of Pergamon Press Ltd]

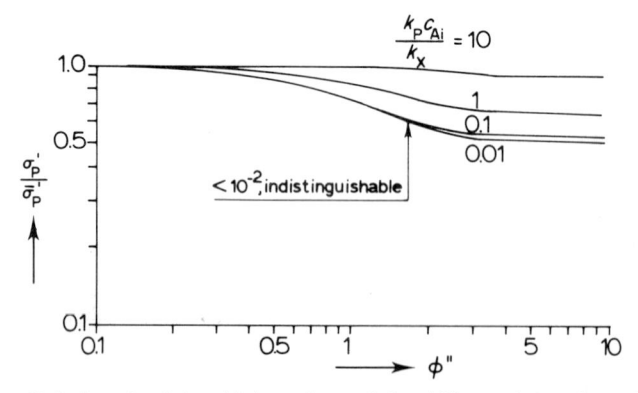

Fig. VIII.9. Relative selectivity $\sigma'_P/\bar{\sigma}'_P$, vs. the modulus ϕ'' for various values of $k_P c_{Ai}/k_X$ for mass transfer with second ($m = 2$, A → P) and first ($n = 1$, A → X) order parallel reactions [From Roberts [3]. Reproduced by permission of Pergamon Press Ltd]

selectivities as a function of ϕ'' for the cases that the desired reaction is second order ($m = 2$) and the undesired reaction is zero ($n = 0$) and first order ($n = 1$) respectively. The $\eta-\phi''$ curves show a familiar pattern, with η approaching the value one for $\phi'' \ll 1$ and decreasing η for increasing ϕ'' for $\phi'' > 0.05$. In contrast to $\eta-\phi'$ graphs (see Section VII.6), the curves vary somewhat for $\phi'' > 2$, depending on the relative reaction rate ratio $k_P c_{Ai}^{m-n}/k_X$. This is caused by the fact that ϕ'' is defined slightly differently from ϕ'. The asymptotic solution for $\phi'' \gg 1$ is [3]:

$$\eta = \sqrt{\left[2\left(\frac{1}{m+1} + \frac{k_x}{(n+1)k_P c_{Ai}^{m-n}}\right)\right]} / \left(\delta \sqrt{\left[\frac{k_P c_{Ai}^{m-1}}{D_A}\left(1 + \frac{k_x}{k_P c_{Ai}^{m-n}}\right)\right]}\right) \quad \text{(VIII.23)}$$

For the reaction A → P much faster than the reaction A → X ($k_P c_{Ai}^{m-n}/k_X \gg 1$) this reduces to

$$\eta = 1 / \left(\delta \sqrt{\left[\frac{(m+1)k_P c_{Ai}^{m-1}}{2D_A}\right]}\right) = \frac{1}{\phi'},$$

being the known expression for mth order kinetics for $\phi > 2$ (Section VII.6). In a similar way one finds $\eta = 1/\phi'$ for $\phi'' \gg 1$ and $k_P c_{Ai}^{m-n}/k_X \ll 1$ with ϕ' the expression for nth order kinetics of the reaction A → X.

Concerning the selectivity, it is apparent that for $\phi'' > 0.05$ increasing diffusion effects occur with increasing ϕ''. Because the desired reaction is of the highest order, there will be a decrease in selectivity after $\phi'' > 0.05$. In figures (VIII.8) and (VIII.9) the differential selectivity σ'_P has been divided by $\bar{\sigma}'_P$, being the differential selectivity under absence of diffusional effects. Notice that $\sigma'_P/\bar{\sigma}'_P$ decreases to a constant minimum level obtained for $\phi'' > 5$. The decrease in selectivity due to mass transfer limitation is larger for larger values of the difference in reaction order ($m - n$) and for smaller values of the formation rate ratio at exterior surface conditions $R_P(c_{Ai})/R_X(c_{Ai})$:

$$\frac{R_P(c_{Ai})}{R_X(c_{Ai})} = \frac{k_P}{k_X} c_{Ai}^{m-n}. \quad \text{(VIII.16)}$$

For $k_P c_{Ai}^{m-n}/k_X \ll 1$, a maximum value for the decrease in selectivity due to mass transfer limitation is found. Then equation (VIII.21) reduces to

$$\frac{dc_A}{dx} \simeq -\sqrt{\left(\frac{2k_X}{D_A(n+1)}\right)} c_A^{(n+1)/2} \quad \text{(VIII.24)}$$

(because $\bar{c}_A = \gamma = 0$ for complete mass transfer limitation). The differential selectivity is given by

$$\sigma'_P = \frac{\int_0^\delta k_P c_A^m \, dx}{-D_A(dc_A/dx)_{x=0}}. \quad \text{(VIII.25)}$$

Combination of equations (VIII.24) and (VIII.25) gives

$$\sigma'_P = \frac{\int_0^{c_{Ai}} k_P \sqrt{\left(\frac{(n+1)D_A}{2k_X}\right) c_A^{m-(n+1)/2}}\, dc_A}{\sqrt{\left(\frac{2k_X D_A}{n+1}\right) c_{Ai}^{(n+1)/2}}}$$

or
$$\sigma'_P = \frac{k_P}{k_X} \frac{(n+1)}{(2m-n+1)} c_{Ai}^{m-n} \qquad (VIII.26)$$

Because we assumed $k_P c_{Ai}^m \ll k_X c_{Ai}^n$, $\bar{\sigma}'_P$ approaches

$$\bar{\sigma}'_P \simeq \frac{k_P}{k_X} c_{Ai}^{m-n}. \qquad (VIII.27)$$

Hence, the maximum possible relative decrease in differential selectivity is found for $\phi'' \gg 1$ and $k_P c_{Ai}^{m-n} \ll k_X$. Then:

$$\left(\frac{\sigma'_P}{\bar{\sigma}'_P}\right) = \left(\frac{\sigma'_P}{\bar{\sigma}'_P}\right)_{min} = \frac{n+1}{2m-n+1}. \qquad (VIII.28)$$

Table VIII.2 shows the maximum effect according to equation (VIII.28). The equation holds only for $m - n > 0$. In case P is the desired product, $(\sigma'_P/\bar{\sigma}'_P)_{min}$ is the ultimate decrease in differential selectivity due to mass transfer limitation obtained at low selectivities ($k_P c_{Ai}^{m-n} \ll k_X$) and $\phi'' \gg 1$. In case P is the undesired product, $(\sigma'_P/\bar{\sigma}'_P)_{min}$ is the ultimate decrease in by-product formation obtained at

Table VIII.2. Maximum decrease in selectivity, $(\sigma'_P/\bar{\sigma}'_P)_{min}$, due to mass transfer limitation for the system

$$A(1) \rightarrow A(2)$$

$$A(2) \begin{array}{l} \nearrow P(2) \quad R_P = k_P c_A^m \\ \searrow X(2) \quad R_X = k_X c_A^n \end{array}$$

for different values of m and n with $m > n$.

m	n	$(\sigma'_P/\bar{\sigma}'_P)_{min}$
$\frac{1}{2}$	0	0.50
1	0	0.33
2	0	0.20
1	$\frac{1}{2}$	0.6
2	$\frac{1}{2}$	0.33
2	1	0.50

high selectivities ($k_P c_{Ai}^{m-1} \ll k_X$) and $\phi'' \gg 1$. As Table VIII.2 shows, mass transfer limitation may influence the selectivity substantially, up to a factor of 5 for kinetics of order (2, 0).

It is mentioned here that, in case one of the reactions is of zero order, the boundary conditions must be carefully defined in order to avoid physically unrealistic results. An analysis of this boundary value problem is given by Blackmore *et al.* [4] for both parallel and consecutive systems.

So far, the reactions we discussed have been monomolecular. Of at least equal importance are bimolecular reactions, such as the system:

$$A(1) \to A(2)$$
$$A(2) + B(2) \to P(2)$$
$$A(2) + B'(2) \to X(2)$$

with (1) and (2), non-reaction phase and reaction phase, respectively. Onda *et al.* [5, 6] obtained approximated solutions for this system with general (m,m'), (n,n') order kinetics by using an approximation method similar to that used by van Krevelen and extended by Hikita and Asai (Sections VII.7.4 and VII.8). Simple solutions are readily obtained for asymptotic situations. Such a situation is the case for which both reactions are instantaneous with respect to mass transfer. For that case, Figure VIII.10 shows the concentration profiles according to the film theory. Assuming $v_A = v_B = v'_B = 1$, it follows from the equations:

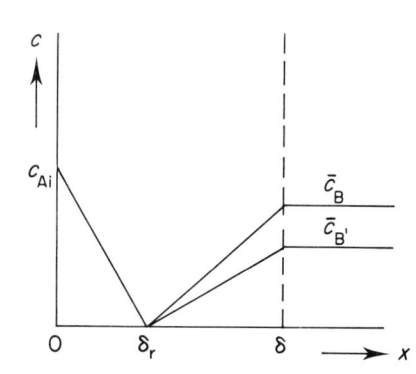

$$J_A = \frac{D_A}{\delta_r} c_{Ai}, \quad J_B = \frac{D_B}{\delta - \delta_r} \bar{c}_B, \quad J_{B'} = \frac{D_{B'}}{\delta - \delta_r} \bar{c}_{B'} \quad \text{and} \quad J_A = J_B + J_{B'},$$

Fig. VIII.10. Concentration profiles (film model) for mass transfer with instantaneous reaction of type

$$A(1) \to A(2)$$
$$A(2) + B(2) \to P(2)$$
$$A(2) + B'(2) \to X(2)$$

and after elimination of δ_r, that:

$$J_A = \frac{D_A}{\delta} c_{Ai} E_{A\infty}$$

where

$$E_{A\infty} = 1 + \frac{D_B \bar{c}_B}{D_A c_{Ai}} + \frac{D_{B'} \bar{c}_{B'}}{D_A c_{Ai}}. \qquad (\text{VIII.29})$$

The differential selectivity σ'_B, for instantaneous reactions called $\sigma'_{B\infty}$, is given by

$$\sigma'_{B\infty} = \frac{D_B \bar{c}_B}{D_B \bar{c}_B + D_{B'} \bar{c}_{B'}}. \qquad (\text{VIII.30})$$

This result is identical to that of completely mass transfer limited independent parallel reactions (Section VIII.2.2).

A second asymptotic solution is obtained for enhanced mass transfer of A without diffusion limitation of both B and B'. Figure VIII.11 shows the concentration profiles. On substitution of $k'_P = k_P \bar{c}_B^{m'}$ and $k'_X = k_X \bar{c}_{B'}^{n'}$ the system becomes equivalent to the elementary case $A \to P$.
$\qquad\qquad\qquad\qquad\qquad\qquad\qquad\qquad\qquad\qquad\searrow$
$\qquad\qquad\qquad\qquad\qquad\qquad\qquad\qquad\qquad\qquad\quad X$

A third asymptotic solution is obtained for one reaction instantaneous and the other slow, without diffusion limitation of B'. Figure VIII.12 shows the concentration profiles. For $0 < x < \delta_r$,

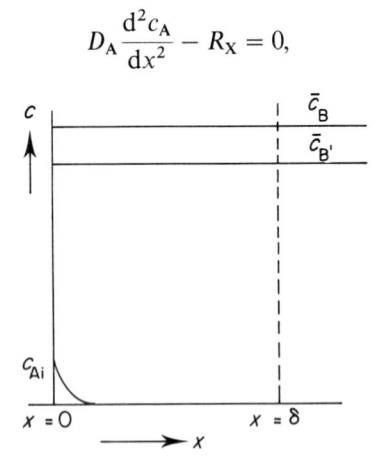

$$D_A \frac{d^2 c_A}{dx^2} - R_X = 0,$$

Fig. VIII.11. Concentration profiles for mass transfer with fast reaction ($\phi > 2$) and both $(1 + D_B \bar{c}_B / D_A c_{Ai})$ and $(1 + D_{B'} \bar{c}_{B'} / D_A c_{Ai}) \gg \phi$, for the system:

$$A(1) \to A(2)$$
$$A(2) + B(2) \to P(2)$$
$$A(2) + B'(2) \to X(2)$$

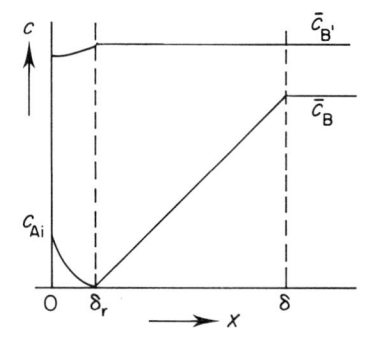

Fig. VIII.12. Concentration profiles (film model) for mass transfer followed by both an instantaneous and a slow reaction, for the system:

$$A(1) \rightarrow A(2)$$
$$A(2) + B(2) \rightarrow P(2)$$
$$A(2) + B'(2) \rightarrow X(2)$$

with R_X the formation rate of X, according to $A + B' \rightarrow X$. In general, R_X is a function of both c_A and $c_{B'}$. As long as $c_{B'}$ in the reaction zone is constant ($\bar{c}_{B'}$), R_X can be written as

$$R_X = R_X(c_A, \bar{c}_{B'}) = R'_X(c_A).$$

Hence, dc_A/dx can be obtained by the theory of mass transfer with complex kinetics $R(c_A)$ as treated in Section VII.6. Boundary conditions are $c_A = c_{Ai}$ for $x = 0$ and $c_A = 0$ for $x = \delta_r$. A second equation is necessary to find δ_r. It is the balance at $x = \delta_r$, stating that all A that reaches $x = \delta_r$ will react with component B:

$$-D_A \left(\frac{dc_A}{dx}\right)_{x=\delta_r} = \frac{D_B}{(\delta - \delta_r)} \bar{c}_B.$$

Solution of the equations gives dc_A/dx. The rate of mass transfer follows from $J_A = -D_A (dc_A/dx)_{x=0}$ and the differential selectivity ratio from

$$\frac{\sigma'_X}{\sigma'_P} = \frac{-(dc_A/dx)_{x=0} + (dc_A/dx)_{x=\delta_r}}{-(dc_A/dx)_{x=\delta_r}}. \tag{VIII.31}$$

In case the reaction $A + B' \rightarrow X$ is of such a slow rate that, even for the case that the production rate of P is completely mass transfer limited, $\sigma'_X \ll \sigma'_P$ holds, it follows from equation (VIII.31) that

$$(dc_A/dx)_{x=\delta_r}/(dc_A/dx)_{x=0} \simeq 1,$$

which means that the c_A-curve approaches a straight line. In that case, the total mass transfer rate approaches the value for a single instantaneous reaction $A + B \rightarrow P$:

$$J_A = \frac{D_A}{\delta_r} c_{Ai} = \frac{D_A}{\delta} c_{Ai} E_{P\infty} \qquad \text{(VIII.32)}$$

where

$$E_{P\infty} \equiv 1 + \frac{D_B \bar{c}_B}{D_A c_{Ai}}, \qquad \text{(VIII.33)}$$

while the differential selectivity can be calculated straightforwardly via:

$$\sigma'_X = \frac{\int_0^{\delta_r} R'_X(c_A) \, dx}{J_A}.$$

That is, for reaction $A + B' \rightarrow X$ of order $(n - n')$,

$$\sigma'_X \simeq \frac{\int_0^{\delta_r} k_X \bar{c}_{B'}^{n'} c_{Ai}^n \left(1 - \frac{x}{\delta_r}\right)^n dx}{J_A}.$$

By equations (VIII.32), (VIII.33) it follows that

$$\sigma'_X \simeq \tfrac{1}{2} \phi_X^2 / E_{P\infty}^2, \qquad \text{(VIII.34)}$$

where ϕ_X is the modulus for the reaction $A + B' \rightarrow X$:

$$\phi_X = \delta \sqrt{\left(\frac{2 k_X c_{Ai}^{n-1} \bar{c}_{B'}^{n'}}{(n+1) D_A}\right)}. \qquad \text{(VIII.35)}$$

In case the reaction $A + B' \rightarrow X$ is undesired, this regime is highly unfavourable for a good selectivity which can be lowered by decades due to mass transfer limitation. Improving the mass transfer will improve the selectivity drastically, because ϕ_X is proportional to δ, that is to $1/k_L$, thus:

$$\sigma'_X \propto \frac{1}{k_L^2}.$$

Such systems call for special reactors with very high mass transfer coefficients, i.e. cyclone reactors [7–9] or centrifuges.

The following example is different. During oxidation of organic materials with air in the liquid phase, peroxides usually occur as intermediates. These peroxides may react to undesired termination products according to a second order reaction. In case the desired reaction is first order, we have the system.

$O_2(G) \to O_2(L)$

$O_2(L) + B(L) \to I(L)$, any kinetics

$$I(L) \begin{array}{c} \overset{k_P}{\nearrow} P(L) \quad R_P = k_P c_I \\ \underset{k_X}{\searrow} X(L) \quad R_X = k_X c_I^2 \end{array}$$

in which I is the peroxide intermediate product. Figure VII.13 shows concentration profiles at the gas–liquid interface. Because the undesired side reaction is of the highest order, it follows from

$$\frac{R_P}{R_X} = \frac{k_P}{k_X c_I}$$

at any place, that the peroxide concentration must be kept as low as possible. This means that, for maximal selectivity, the accumulation of peroxide at the interface must be eliminated. Two ways are possible:

(a) Increasing the transport rate of peroxide to the bulk of the liquid by increasing the liquid side mass transfer coefficient. In conventional reactors, however, mass transfer coefficients are all of the same order of magnitude (see Table VII.14). Only in special reactors, as cyclones, substantially higher k_L values can be realized. Such reactors, however, are expensive.

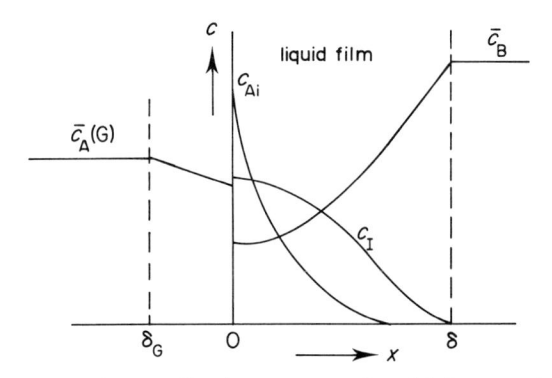

Fig. VIII.13. Concentration profiles for mass transfer with fast reaction of type:

$$A(1) \to A(2)$$
$$A(2) \to I(2)$$
$$I(2) \begin{array}{c} \nearrow P(2) \\ \searrow X(2) \end{array}$$

(b) Decreasing the production rate of peroxide per unit interfacial area. This can be done by operating at a lower partial oxygen pressure and by a lower k_G. To some extent the latter might be possible by realizing large bubbles instead of small ones. This should mean that stirring must be avoided, which is a surprising conclusion. Notice that by lowering of k_G and c_{Ai} the production capacity will decrease. Selectivity is improved at the expense of capacity.

On more complex systems, literature is still scarce. Mass transfer followed by reversible parallel reactions has been studied by Li et al. [10]. Golikeri and Luss [11] analysed mass transfer followed by a mixture of many parallel nth order reactions and showed how it is possible to estimate the effectiveness factor η without knowing the distribution function of the Thiele modulus among the various species. The value of the two observables, overall conversion rate and its relative dispersion, prove to give sufficient information for calculating η. Bailey and Horn [12] analysed the influence of cyclic changing of the reactant concentrations on the selectivity of a process with complex reaction kinetics. The influence of intraparticle heat effects on selectivity have been analysed by Østergaard [13] for the case of both reactions being of the first order.

Example VIII.3.2 *The absorption of CO_2 in amines aided by a shuttle*

Removal of carbon dioxide from natural gas is necessary before the gas can be transported as LNG. This can be done by absorbing CO_2 in aqueous solutions of amines. The tertiary amine methyl diethanol amine (MDEA) has a large capacity for binding CO_2 at ambient temperature according to:

$$CO_2 + R_3N + H_2O \rightarrow R_3NH^+ + HCO_3^-.$$

Unfortunately the reaction rate is not very high, so that relatively large columns are necessary. The secondary amine methyl ethanol amine (MEA) reacts much faster with CO_2. However, the equilibrium

$$CO_2 + 2R_2NH \rightleftarrows R_2NCOO^- + R_2NH_2^+$$

is not favourable for a high CO_2 absorption capacity of the liquid.

We will therefore use R_3N as the main chemical for binding CO_2 while some MEA is added to enhance the absorption rate of CO_2. This may work successfully if the reactions

$$R_2NH_2^+ + R_3N \rightleftarrows R_3NH^+ + R_2NH$$
$$R_2COO^- + H_2O \rightleftarrows HCO_3^- + R_2NH$$

proceed sufficiently fast in the bulk of the liquid in order to produce the R_2NH, which is then re-used to pick up CO_2 again at the interface. Hence, the secondary amine shuttles between interface and bulk to transport the CO_2, whereas the CO_2 is bound in the bulk by the tertiary amine under reproduction of the secondary amine. We will use an aqueous solution of 2.2 kmol/m³ amine. The concentration of MEA in this solution $\bar{c}_B$ is an operating variable and can be chosen between 0 and 0.6 kmol/m³ and the concentration of MDEA, $\bar{c}_C$, consequently between 2.2 and 1.6 kmol/m³. The CO_2 concentration at the interface, c_{Ai}, varies with the column height from 0.25 to 0.0025 kmol/m³. Additional data are $T = 300$ K, $k_B = 33\,412$ m³/kmol s, $k_C = 11.5$ m³/kmol s, $k_L = 100 \times 10^{-6}$ m/s, $D_A = 1.9 \times 10^{-9}$ m²/s, $D_B = 1 \times 10^{-9}$ m²/s and $D_C = 0.8 \times 10^{-9}$ m²/s. We will investigate the influence of the concentration of the secondary amine for various values of c_{Ai} on the enhancement factor for the CO_2 absorption.

From component material balances in the film it follows that:

$$D_A \frac{d^2 c_A}{dx^2} = k_B c_A c_B + k_C c_A c_C \tag{a}$$

$$D_B \frac{d^2 c_B}{dx^2} = 2 k_B c_A c_B \tag{b}$$

$$D_C \frac{d^2 c_C}{dx^2} = k_C c_A c_C. \tag{c}$$

To solve these equations we will use the approximated method developed by Onda et al [5], who express the enhancement factor (for $\bar{c}_{AL} = 0$ only) as

$$E_A = 1 + \frac{D_B(\bar{c}_B - c_{Bi})}{v_B D_A c_{Ai}} + \frac{D_C}{v_C D_A} \frac{(\bar{c}_C - \bar{c}_{Ci})}{c_{Ai}}, \tag{d}$$

while also

$$E_A = \frac{\phi_e}{\tanh \phi_e}, \tag{e}$$

in which ϕ_e is an effective Hatta number given by

$$\phi_e = \sqrt{\frac{k_B D_A}{k_L^2} \left(c_{Bi} + \frac{k_C}{k_B} c_{Ci} \right)}. \tag{f}$$

In our case $v_B = 2$ and $v_C = 1$. We now assume quadratic concentration profiles in the liquid film at the interface:

$$\frac{c_B}{\bar{c}_B} = \left(1 - \frac{c_{Bi}}{\bar{c}_B} \right) x^2 + \frac{c_{Bi}}{\bar{c}_B} \tag{g}$$

$$\frac{c_C}{\bar{c}_B} = \left(\frac{\bar{c}_C}{\bar{c}_B} - \frac{c_{Ci}}{c_{Bi}} \right) x^2 + \frac{c_{Ci}}{c_{Bi}}. \tag{h}$$

Eliminating c_A by dividing equation (b) by equation (c) gives

$$\frac{D_C}{D_A} \frac{d^2 c_C}{dx^2} = \frac{k_C c_C}{2 k_B c_B} \frac{d^2 c_B}{dx^2} \tag{i}$$

Substitution of equations (g) and (h) into (i) gives

$$\frac{c_{Ci}}{\bar{c}_B} = \frac{1 - \dfrac{D_B}{12 D_C}\dfrac{k_C}{k_B}\left(1 - \dfrac{c_{Bi}}{\bar{c}_B}\right)}{1 + \dfrac{D_B}{12 D_C}\dfrac{k_C}{k_B}\dfrac{(1 - c_{Bi}/\bar{c}_B)}{c_{Bi}/\bar{c}_B}}\frac{\bar{c}_C}{\bar{c}_B}. \tag{j}$$

We now have four equations (d), (e), (f) and (j) with four unknown parameters c_{Ci}, c_{Bi}, ϕ_e and E_A, so that E_A can be found by trial and error.

Figure VIII.14 gives some results. It is clear that the shuttle is particularly effective in the top of the column where the CO_2 concentration is low. This is caused by the change in

$$1 + \frac{D_B \bar{c}_B}{\nu_B D_A c_{Ai}} (\equiv E_{B\infty})$$

from 1.63 to 163, if c_{Ai} decreases from 0.25 to 0.0025 kmol/m^3.

To design the column, balances over the bulk have to be set up to calculate the liquid bulk concentrations as a function of the column height. We assumed them constant in our Example; this, of course, is not realistic.

VIII.4 SIMULTANEOUS MASS TRANSFER OF TWO REACTANTS (A AND A') FOLLOWED BY DEPENDENT PARALLEL REACTIONS WITH A THIRD REACTANT: A + B → P, A' + B → X

Simultaneous mass transfer followed by reaction is of importance in both fluid–solid and fluid–fluid reactions. We have already mentioned the

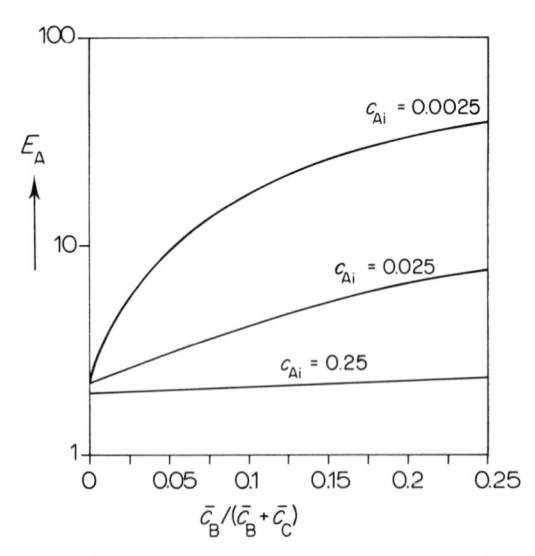

Fig. VIII.14. The effect of the concentration of a shuttle compound on the enhancement of the CO_2 absorption for various CO_2 interface concentrations; data as given in Example VIII.3.2

dehydrogenation and cracking of hydrocarbons as gas–solid examples (Section VIII.2.1) and the absorption of H_2S and CO_2 in amines and carbonate solutions as gas–liquid examples (Section VIII.2.2). More examples are found in Table VIII.3. Additional examples are the simultaneous absorption of ethylene or propylene with chlorine in the chlorohydrin synthesis [36], the selective absorption of chlorine from gas containing chlorine and carbon dioxide [37], the selective absorption of phosgene from gas containing carbon dioxide [38] and the simultaneous absorption of Cl_2 and SO_2 in acid solutions [39].

Simultaneous mass transfer with independent reactions has been discussed in Section VIII.2. We now therefore discuss simultaneous mass transfer followed by dependent parallel reactions. The basic form of such a system is:

$$A(1) \to A(2)$$

$$A'(1) \to A'(2)$$

$$A(2) + B(2) \to P(2)$$

$$A'(2) + B(2) \to X(2).$$

Although phase (2) can be a porous solid phase (e.g. gasification of coal in which oxygen, water, carbon monoxide and hydrogen may compete for reaction with carbon) we will analyse the problem in particular with reference to fluid–fluid systems.

Figure VIII.16 shows concentration profiles for the general case and for several interesting asymptotic regimes. The general case is the one in which both reactions have a final rate with some diffusion limitation in the transport of B to the interface. Goettler and Pigford [40] analysed it numerically, starting from the penetration theory. Shklyar et al. [41] did the same but started from the film theory. An approximate solution for (m, m'), (n, n') kinetics has been obtained by Onda et al. [5]. Based on film theory and with an approximation method, similar to that used by Van Krevelen and extended by Hikita and Asai (Sections VII.7.4 and VII.8), they derived a set of equations from which the enhancement factors can be obtained via a trial and error procedure. A similar approximation method starting from the penetration and the surface renewal theory is available as well [6]. For equal diffusivities, the results are found to be insensitive to the theory applied. Figure VIII.15 shows enhancement factors for each of the two components A and A' with increasing Hatta number of one reaction for different values of E_∞ for that reaction. (Both reactions are of order (1, 1) and Ha_A and $E_{A\infty}$ are the Hatta number and the instantaneous enhancement factor for reaction species A only, that is, in absence of the parallel reaction).

Notice how the more reactive component may suppress the rate of mass transfer of the less reactive one. Both A and A' compete for reactant B and the faster one reaction is, the less B is available for the competing reaction. If there is noticeable diffusion limitation of B and both reactions are of comparable rate, the enhancement factors for both reactions are lower than if each reaction occurs alone. As in the previous section, we will discuss the different asymptotic solutions in more detail.

Table VIII.3. Some industrially important examples of simultaneous absorption of more than one gas in a liquid [Reproduced with permission from P. A. Ramachandran and M. M. Sharma, [43] *Trans. Inst. Chem. Eng.*, **49**, 253 (1971)]

A. Physical absorption:

No.	Gases being absorbed	Solvent used	Reference
1.	Mixture of aliphatic hydrocarbons	Light oil	Nelson [14]
2.	CO_2, H_2S	Sulpholane, propylene carbonate, N-methyl pyrrolidone, or methanol	Swaim [15]
3.	NH_3, CH_3NH_2	Water	Sittig [16]
4.	Acetone, isopropanol	Water	Goldstein and Waddams [17]
5.	Ethanol, acetaldehyde	Water	Goldstein and Waddams [17]
6.	CS_2, CO, CO_2, H_2S	Light oil	Kirk and Othmer [18]
7.	C_2H_2, CO_2, CO	Dimethylformamide or N-methyl-pyrrolidone	Sittig [16]

B. Absorption accompanied by chemical reaction:

No.	Gases being absorbed	Solvent used	Reference
1.	CO_2, H_2S and CO_2, COS, H_2S	(a) Aqueous potassium carbonate solutions	Garnell *et al.* [19] Gioia [20]
		(b) Aqueous potassium phosphate solutions	Gioia and Marrucci [21]

	(c) Aqueous sodium hydroxide solution	Astarita and Gioia [22]
	(d) Aqueous triethanolamine	Kohl and Riesenfeld [23]
	(e) Aqueous ammonia	Kohl and Riesenfeld [23]
	(f) Aqueous monoethanolamine	Danckwerts and Sharma [24]
	(g) Aqueous diethanolamine	Danckwerts and Sharma [24]
	(h) Aqueous diglycolamine	Dingman and Moore [25]
2. Isobutylene, butenes	Aqueous sulphuric acid	Davis et al. [26, 27]
3. SO_2, O_2	Water containing manganese sulphate as a catalyst	Gehlawat and Sharma [28] Coughanour and Krause [29] Gunn and Saleem [30]
4. $COCl_2$, CO_2	Water, aqueous carbonate solution or aqueous caustic soda	Manogue and Pigford [31]
5. Cl_2, CO_2	Water, aqueous carbonate solution or aqueous caustic soda	
6. CO, CO_2	Ammoniacal cuprous chloride	Van Krevelen and Baans [32]
7. HCN, CO_2	Alkalis	Taylor [33]
8. SO_2, CO_2	Aqueous carbonate solution or aqueous caustic soda	Goettler and Pigford [34]
9. Mercaptans, CO_2	Alkalis	Lites, Dilman and Tagintsev [35]

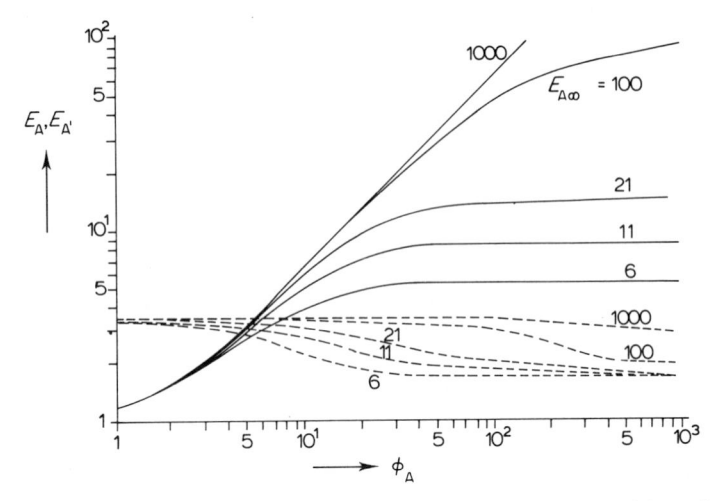

Fig. VIII.15. Enhancement factors, E_A and $E_{A'}$ (dashed), as a function of ϕ_A and $E_{A\infty}$ for $\phi_{A'} = 5$ and $E_{A'\infty} = 6$. Reaction system:

$$A(1) \rightarrow A(2)$$
$$A'(1) \rightarrow A'(2)$$
$$A(2) + B(2) \rightarrow P(2)$$
$$A'(2) + B(2) \rightarrow X(2).$$

Reactions of order (1,1) [From Shklyar et al. [41]. Reproduced by permission of Plenum Publishing Corporation]

VIII.4.1 Complete mass transfer limitation in non-reaction phase

Figure VIII.16(b) gives the concentration profiles for this case. The mass transfer rates are $J_A = k_G \bar{c}_{AG}$ and $J_{A'} = k_G \bar{c}_{A'G}$ and the differential selectivity ratio is

$$\frac{\sigma'_A}{\sigma'_{A'}} = \frac{\bar{c}_{AG}}{\bar{c}_{A'G}}, \qquad \text{(VIII.36)}$$

which is independent of the reaction kinetics. Whether or not this situation is favourable for a good selectivity depends on the kinetics of the reactions. An example is the absorption of CO_2 and H_2S from air in a concentrated aqueous ammonia solution [42].

VIII.4.2 One reactant mass transfer limited in non-reaction phase

This case will occur for $\phi_A \leq 1 + D_B \bar{c}_B / D_A m_A \bar{c}_{AG}$ and A' very soluble or reactive such that

(a) $\quad k_G \bar{c}_{A'G} \ll k_L \bar{c}_B \qquad$ for $\quad \phi_{A'} \gg 1 + \dfrac{D_B \bar{c}_B}{D_{A'} m_{A'} \bar{c}_{A'G}}$

or

(b) $k_G \ll m_{A'} k_L \phi_{A'}$ for $1 < \phi_{A'} \ll 1 + \dfrac{D_B \bar{c}_B}{D_{A'} m_{A'} \bar{c}_{A'G}}$

or

(c) $k_G \ll m_{A'} k_L$ for $\phi_{A'} < 1$ and $\bar{c}_{A'L} \ll m\bar{c}_{A'G}$

In case of good selectivity, that is for $J_{A'} \ll J_A$, the mass transfer rate of A is not much influenced by the mass transfer of A' and J_A equals the van Krevelen solution:

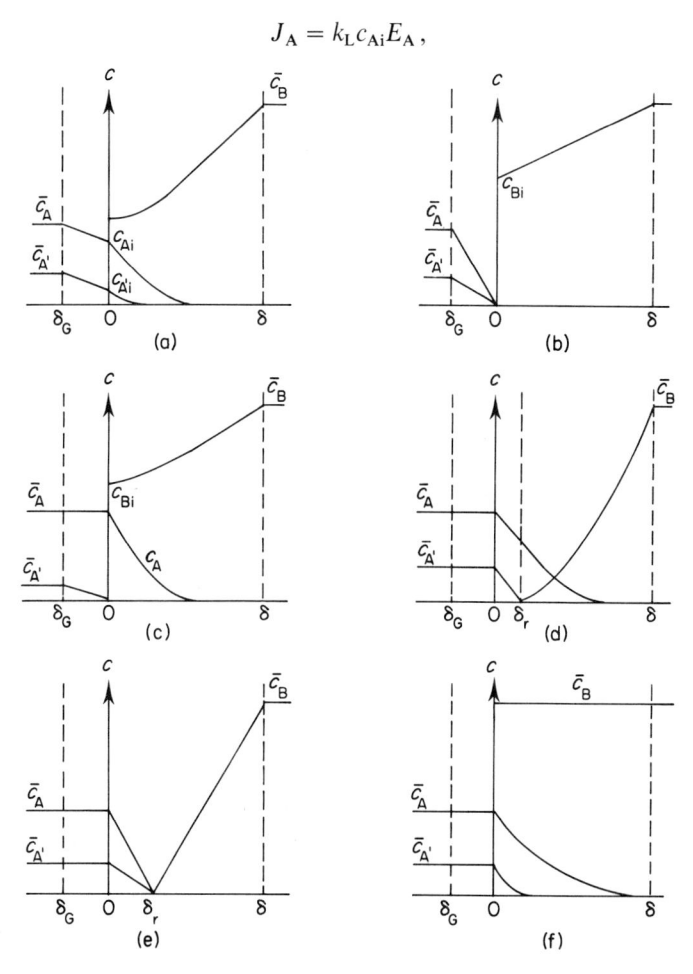

$$J_A = k_L c_{Ai} E_A,$$

Fig. VIII.16. Concentration profiles for simultaneous mass transfer of two reactants followed by reaction with a component, originally present in the reaction phase (irreversible reactions): (a) general case; (b) two components mass transfer limited in non-reaction phase; (c) one reactant limited in non-reaction phase; (d) one reaction instantaneous; (e) two reactions instantaneous, (f) no diffusion limitation of component B

where E_A is as in equation (VII.138), while

$$J_{A'} = k_G \bar{c}_{A'G}$$

and the by-product formation rate ratio is

$$\frac{\sigma'_{A'}}{\sigma'_A} = \frac{k_G \bar{c}_{A'G}}{k_L c_{Ai} E_A}. \qquad (VIII.37)$$

If $\sigma'_{A'}/\sigma'_A$ found in this way is much smaller than one, there is indeed good selectivity, which justifies the use of equation (VII.138) for estimating J_A. If the value $\sigma'_{A'}/\sigma'_A$ found in this way is not much smaller than one, a more elaborate analysis is necessary, which takes into account the influence of the side reaction on the concentration profile of B. Analogously to the van Krevelen method, this can be done by taking c_{Bi} = constant in the reaction zone in solving the equation

$$D_A \frac{d^2 c_A}{dx^2} = k_P R(c_A, c_B). \qquad (VIII.38)$$

Hence

$$D_A \frac{d^2 c_A}{dx^2} = k_P R(c_A, c_{Bi})$$

with c_{Bi} from the mass balance $J_A + J_{A'} = J_B$. This has been done by Ramachandran and Sharma [43]. With decreasing selectivity, their method will give greater deviations compared to the exact solution. The reason is that with an instantaneous surface reaction the boundary condition $-D_B(dc_B/dx)_{x=0} = 0$ no longer holds. The correct boundary condition for this case is:

$$D_B \left(\frac{dc_B}{dx} \right)_{x=0} = J_{A'}. \qquad (VIII.39)$$

Therefore, a more accurate approximated solution can be obtained by substituting in equation (VIII.38) for c_B the expression

$$c_B = c_{Bi} + \frac{J_{A'}}{D_B} x.$$

This has been done by Ouwerkerk [44]; see figure VIII.17. Asymptotic solutions are obtained for no diffusion limitation of B, thus for $\phi_A \ll 1 + D_B \bar{c}_B / D_A m_A \bar{c}_{AG}$ in combination with one of the three conditions formulated at the beginning of this section. Then equation (VIII.37) holds with:

$$\frac{\sigma'_{A'}}{\sigma'_A} = \frac{k_G \bar{c}_{A'G}}{k_L c_{Ai} \phi_A} \qquad \text{for} \quad 2 < \phi_A \ll 1 + \frac{D_B \bar{c}_B}{D_A m_A \bar{c}_{AG}} \qquad (VIII.40)$$

and

$$\frac{\sigma'_{A'}}{\sigma'_A} = \frac{k_G \bar{c}_{A'G}}{k_L \{c_{Ai} - \bar{c}_{AL}\}} \qquad \text{for} \quad \phi_A < 0.2 \qquad (VIII.41)$$

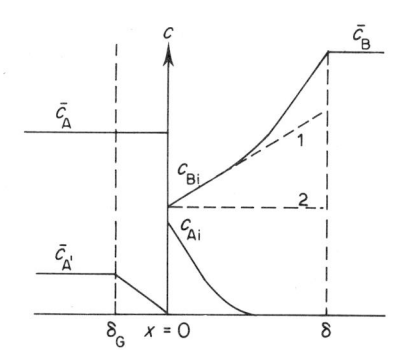

Fig. VIII.17. Concentration profiles for simultaneous mass transfer, with one reaction $(A' + B \rightarrow X)$ mass transfer controlled in the non-reaction phase: (1) Ouwerkerk approximation of c_B profile; (2) Ramachandran and Sharma approximation of c_B profile

provided all A reaching the bulk is converted in the reactor. Figure VIII.18 shows the concentration profiles for these two asymptotic solutions. For fast reaction of A (figure VIII.18(a)), the by-product formation rate ratio, $\sigma'_{A'}/\sigma'_A$ is proportional to k_G and independent of k_L (because ϕ_A in equation (VIII.40) is proportional to $1/k_L$). For slow reaction of A, the by-product formation rate ratio $\sigma'_{A'}/\sigma'_A$ turns out to be

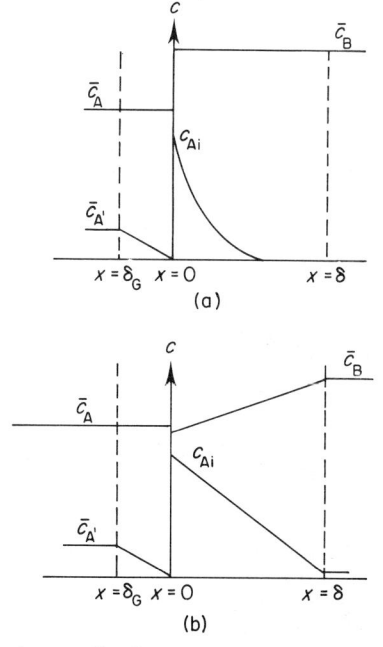

Fig. VIII.18. Concentration profiles for asymptotic cases of simultaneous mass transfer, with one reaction $(A' + B \rightarrow X)$ mass transfer controlled in the non-reaction phase and the other reaction: (a) fast with no mass transfer limitation of B; (b) slow.

proportional to the ratio k_G/k_L. This result is evident because the problem is reduced to simultaneous physical mass transfer with resistances in different phases.

VIII.4.3 One reaction instantaneous

For $\phi_{A'} \gg E_{A'\infty}$ without complete mass transfer limitation of A' in the non-reaction phase, the reaction plane for the reaction $A' + B \to X$ moves from the interface (previous section) into the reaction phase. Figure VIII.19 shows typical profiles. Mathematical formulation of the problem, according to the film model, is straightforward, e.g. for the reaction $A + B \to P$ of order $(1,1)$:

For $0 < x < \delta_r$, $\quad D_A \dfrac{d^2 c_A}{dx^2} = 0$ \hfill (VIII.42)

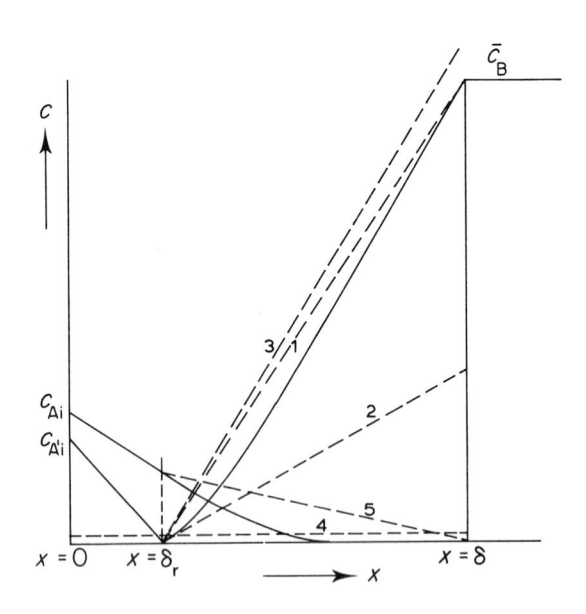

$$D_{A'} \dfrac{d^2 c_{A'}}{dx^2} = 0 \quad \text{(VIII.43)}$$

Fig. VIII.19. Film model of the simultaneous mass transfer of two components (A and A') in a reactive liquid, containing B, with one fast and one instantaneous parallel reaction. Dashed: approximate profiles for c_B in solving the corresponding mathematical model: (1) Goettler and Pigford [34] and Ramachandran and Sharma [43]; (2) Ouwerkerk [44]; (3) Cornelisse et al. [45]; (4) Onda et al. [5]; (5) Assumed c_A profile by Ramachandran and Sharma [43] [From Cornelisse et al. [45]. Reproduced by permission of Pergamon Press Ltd]

For $\delta_r < x < \delta$, $\quad D_A \dfrac{d^2 c_A}{dx^2} = k_p c_A c_B$ \hfill (VIII.44)

$$D_B \dfrac{d^2 c_B}{dx^2} = k_p c_A c_B,$$ (VIII.45)

with boundary conditions:

for $x = 0$: $\quad c_A = c_{Ai}; \quad c_{A'} = c_{A'i}$,

for $x = \delta_r$: $\quad c_A = (c_A)_{\delta_r}; \quad c_{A'} = 0; \quad c_B = 0$,

for $x = \delta$: $\quad c_A = \bar{c}_A$ or $c_A = 0; \quad c_{A'} = 0; \quad c_B = \bar{c}_B$.

(a) *One reaction instantaneous and one reaction slow*

For $\phi_A < 0.2$ conversion of A mainly proceeds in the bulk. This problem can be treated as two independent parallel reactions with the result, see also Figure VIII.20:

$$J_A = k_L(c_{Ai} - \bar{c}_A)$$
$$J_{A'} = k_L c_{A'i} E_{A'\infty}$$

and:

$$\frac{\sigma'_{A'}}{\sigma'_A} = \frac{c_{A'i} E_{A'\infty}}{(c_{Ai} - \bar{c}_A)}.$$

Hence the selectivity ratio is independent both of the mass transfer coefficient and of the kinetics (except indirectly via the value of $\bar{c}_A$), provided all A in the bulk is converted.

(b) *One reaction instantaneous and one reaction fast*

To solve the set (VIII.42)–(VIII.45), the following additional equations are used:

$$J_A = -D_A(dc_A/dx)_{x=0}$$
$$J_{A'} = -D_{A'}(dc_{A'}/dx)_{x=0}$$
$$J_B = D_B(dc_B/dx)_{x=\delta}$$
$$J_B = J_A + J_{A'}$$

and at $x = \delta_r$: $-D_{A'}(dc_{A'}/dx)_{x=\delta_r} = D_B(dc_B/dx)_{x=\delta_r}$.

Numerical solutions have been presented by Goettler and Pigford [40] and Sada et al. [46]. Approximate solutions have been presented by Goettler and Pigford [34], Ouwerkerk [44], Onda et al. [5], Ramachandran and Sharma [43] and Cornelisse et al. [45]. All approximations are based on linearizations of the B-profile in the zone $\delta_r < x < \delta$, while Ramachandran and Sharma linearized the A profile in that zone as well. Figure VIII.19 shows the different approximations used by the different authors.

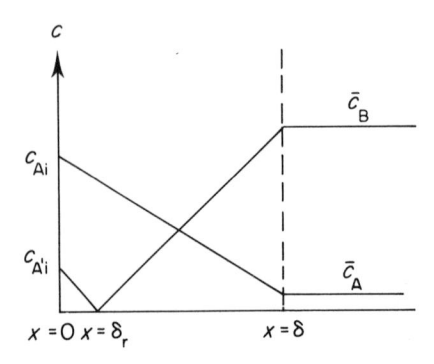

Fig. VIII.20. Film model concentration profiles for simultaneous mass transfer with one instantaneous (A' + B → X) and one slow (A + B → P) reaction

As shown in figure VIII.21, the solutions of Goettler and Pigford [34], Ouwerkerk [44] and Cornelisse et al. [45] closely agree with each other. Differences with the numerical solutions are small [34]. Only Cornelisse et al. present an explicit solution.

For $\phi_A > 5$:

$$E_A = \frac{1 + a_i + a_i'}{a_i} - \frac{a_i'}{a_i} \left[\frac{F - S}{2} + \sqrt{\left(\frac{F - S}{2}\right)^2 + \frac{S a_i'}{1 + a_i + a_i'}} \right] \quad \text{(VIII.46)}$$

and

$$E_{A'} = 1 - \frac{a_i}{a_i'}(E_A - 1) + \frac{1}{a_i'} \quad \text{(VIII.47)}$$

with $a_i = D_A c_{Ai}/D_B \bar{c}_B$

$a_i' = D_{A'} c_{A'i}/D_B \bar{c}_B$

$F = (a_i + a_i')/(1 + a_i + a_i') \quad \text{(VIII.48)}$

and $S = 1.37/[(1 + a_i + a_i')\phi_A^2]^{1/3}$.

It can be shown that F (equation (VIII.48)) equals δ_r/δ if both reactions are instantaneous. Notice from figure VIII.21 that the enhancement factor for the reaction A + B → P is lowered by the occurrence of the competing side reaction. In the absence of the side reaction the enhancement factor is $E_A = \phi_A$ for $\phi_A \ll E_{A\infty}$. Due to the side reaction, which competes for the consumption of B, it is found that $E_A < \phi_A$. For $\phi_A \gg E_{A\infty}$, both reactions become instantaneous. That case will be treated in the next section. The by-product formation rate ratio follows from

$$\frac{\sigma_{A'}'}{\sigma_A'} = \frac{c_{A'i} E_{A'}}{c_{Ai} E_A}, \quad \text{(VIII.49)}$$

with E_A and $E_{A'}$ from equations (VIII.46) and (VIII.47) respectively.

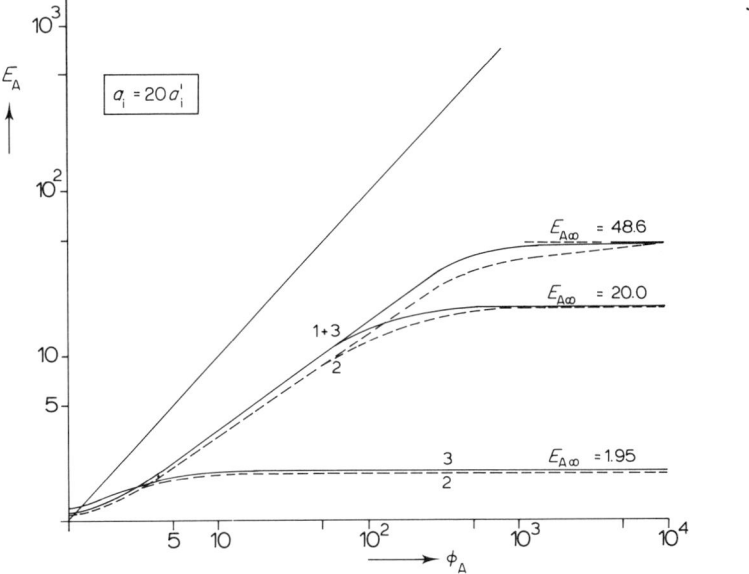

Fig. VIII.21. Plot of the enhancement factor of A (E_A) vs. ϕ_A for simultaneous mass transfer with reactions

$$A(1) \to A(2)$$
$$A'(1) \to A'(2)$$
$$A(2) + B(2) \to P(2)$$
$$A'(2) + B(2) \to X(2),$$

of which one is instantaneous (A' + B → X) and the other is of order (1,1), for $a_i = 20a_i'$ and three values of $E_{A\infty}$: (1) approximation of Goettler et al. [34]; (2) approximation of Ouwerkerk [44]; (3) approximation of Cornelisse et al. [45]. [From Cornelisse et al. [45]. Reproduced by permission of Pergamon Press Ltd]

In a particular process, different regimes may occur at different places in the reactor. For example, in absorbing H_2S and CO_2 with ethanol amine, the H_2S concentration will be lower in the top of the column than in the bottom (we assume a proper operation!). The reaction between H_2S and amines is instantaneous. Whether the prevailing mass transfer resistance is in the liquid or in the gas phase depends on the ratio $k_G/mk_L E_{A'\infty}$, where

$$\frac{k_G}{mk_L E_{A'\infty}} \simeq \frac{k_G c_{A'i}}{mk_L \bar{c}_B}.$$

This ratio decreases with decreasing $c_{A'i}$. Hence, it is possible that gas phase resistance prevails at the top of the column (previous section) while it is much less important in the bottom of the column. Figure VIII.22 shows selectivity ratios as a function of the concentration of A' in the gas phase for a typical ratio of k_G/mk_L. With decreasing $c_{A'G}$, the reaction plane $x = \delta_r$ moves towards the interface. At

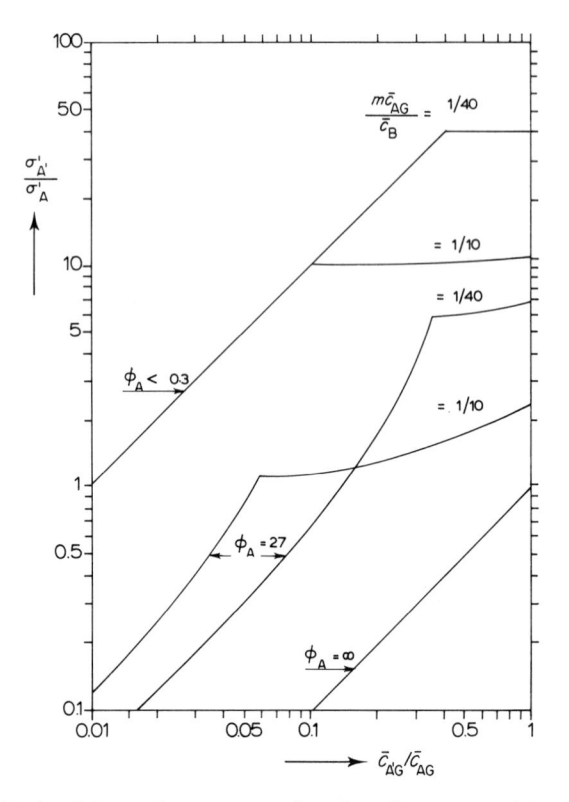

Fig. VIII.22. Ratio of absorption rates as a function of gas composition (parameters ϕ_A and $c_{A'G}/c_{AG}$) for simultaneous mass transfer of A and A' with parallel reactions of which one reaction (A' + B → X) is instantaneous and $k_G/mk_L = 100$ for both gases [From Ouwerkerk [44]. Reproduced by permission of C. Ouwerkerk]

the concentration at which the reaction phase coincides with the interface, the curves show a discontinuity.

Asymptotic solutions are:

For $\phi_A < 0.3$, the reaction A + B → P is slow. Then, for $E_{A'\infty} \gg 1$ and for $c_{A'G}$ sufficiently high to establish a reaction plane away from the interface, the selectivity ratio becomes

$$\frac{\sigma'_{A'}}{\sigma'_A} = \frac{k_L c_{A'i} E_{A'\infty}}{k_L c_{Ai}} \simeq \frac{\bar{c}_B}{c_{Ai}},$$

a value independent of both $c_{A'G}$ and k_G. For $\phi_A < 0.3$ and a low value of $\bar{c}_{A'G}$, the interface is the reaction plane for A' + B → X. Then k_G limits this reaction and $\sigma'_{A'}/\sigma'_A$ equals:

$$\frac{\sigma'_{A'}}{\sigma'_A} = \frac{k_G \bar{c}_{A'G}}{k_L c_{Ai}},$$

which is proportional to $\bar{c}_{A'G}$. The line $\phi_A = \infty$ means both reactions are instantaneous. This case will have our attention in the next section.

Experimentally examined examples of simultaneous absorption, followed by reactions, are the simultaneous absorption of H_2S and CO_2 in amines [47] and the simultaneous absorption of SO_2 and CO_2 in aqueous solutions of hydroxide [34]. Care must be taken, however, because not all reactions involved can be considered as being irreversible under all conditions. The latter effect is subject of a recent study of Cornelisse et al. [48].

VIII.4.4 Both reactions instantaneous

In case both reactions are so fast that they can be considered as instantaneous with respect to mass transfer, the equations become very simple again. Figure VIII.16(e) shows the concentration profiles (film theory). From

$$J_A = D_A c_{Ai}/\delta_r, \quad J_{A'} = D_{A'} c_{A'i}/\delta_r \quad \text{and} \quad J_B = D_B \bar{c}_B/(\delta - \delta_r)$$

it follows that

$$\frac{1}{E_A} = \frac{1}{E_{A'}} = \frac{\delta_r}{\delta} = \frac{D_A c_{Ai} + D_{A'} c_{A'i}}{D_A c_{Ai} + D_{A'} c_{A'i} + D_B \bar{c}_B} \quad \text{(VIII.50)}$$

and

$$\frac{\sigma'_{A'}}{\sigma'_A} = \frac{D_{A'} c_{A'i}}{D_A c_{Ai}}. \quad \text{(VIII.51)}$$

E_A decreases continuously with increasing $c_{A'i}$ and vice versa. Hence, again it appears that one reactant may suppress the mass transfer of the other one by competing for reactant B. The penetration theory gives the same result, provided that D_j is replaced by $\sqrt{D_j}$ in equations (VIII.50), (VIII.51). The solution can be easily extended to simultaneous absorption of more than two reactants and to stoichiometric reaction coefficients different from one.

VIII.4.5 No diffusion limitation of reactant originally present in reaction phase

Figure VIII.16(f) shows the concentration profiles for this case. The criterion is $\phi_A + \phi_B \ll E_A$ with E_A from equation (VIII.50). By putting $k_P \bar{c}_B \equiv k'_P$ and $k_X \bar{c}_B \equiv k'_X$, the system reduces to $A \to P$ and $A' \to X$, which has been treated in Section VIII.2.

VIII.4.6 More complex systems

Multi-component mass transfer followed by reaction is still a subject of research. Literature data are scarce, especially on experimental studies. The removal of CO_2 and H_2S by hydroxide solutions has been described by a set of instantaneous reactions in which the product of one reaction may be reactant for the other reaction [22, 49, 50]:

$$CO_2(G) \rightarrow CO_2(L)$$

$$H_2S(G) \rightarrow H_2S(L)$$

$$CO_2 + 2OH^- \rightarrow CO_3^{2-} + H_2O \quad \text{(VIII.52)}$$

$$H_2S + OH^- \rightarrow HS^- + H_2O \quad \text{(VIII.53)}$$

$$H_2S + CO_3^{2-} \rightarrow HS^- + HCO_3^- \quad \text{(VIII.54)}$$

$$HCO_3^- + OH^- \rightarrow CO_3^{2-} + H_2O. \quad \text{(VIII.55)}$$

As shown in figure VIII.23, two reaction planes are found, known as the primary and secondary reaction planes. Without reaction (VIII.54), both H_2S and CO_2 would react at the same reaction plane. Due to the reactivity of H_2S, not only with OH^- but also with CO_3^{2-}, the mass transfer rate of H_2S is further enhanced, resulting in a reaction plane δ_{r1} that is nearer to the interface than the plane for the CO_2 reaction. Consequently, the reaction between H_2S and OH^- will no longer occur because these two reactants will not meet each other. The differential selectivity ratio, defined as $\sigma'_{H_2S}/\sigma'_{CO_2}$, is greater than found with the theory of two coupled simple instantaneous reactions (Section VIII.4.4). This has been confirmed experimentally in a wetted wall column [49]. Both theory and experiment show that the selectivity for H_2S increases with a decrease in OH^- concentration. In practice an optimization problem results because lowering of the hydroxide concentration lowers the absorption capacity of the liquid.

A different system is the simultaneous absorption of a mixture of $COCl_2(A)$ and

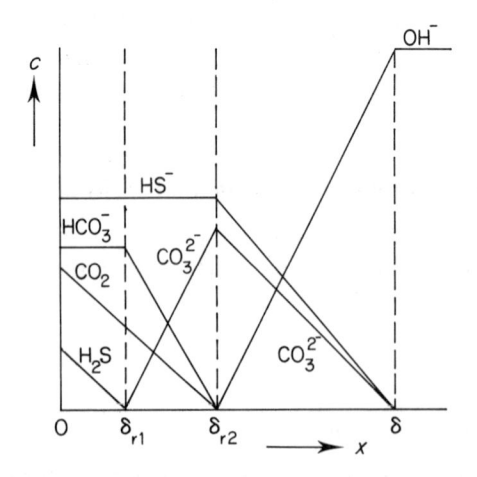

Fig. VIII.23. Simultaneous mass transfer of H_2S and CO_2 in hydroxide solutions according to the film model. [From Astarita [50, p. 171]. Reproduced by permission of Elsevier Publishing Company]

$CO_2(A')$ in water with the aim to remove the obnoxious compound phosgene. The scheme is

$$A(G) \to A(L)$$
$$A'(G) \to A'(L)$$
$$A(L) + B(L) \to A'(L)$$
$$A'(L) + B(L) \to X(L).$$

Here, the hydrolysis reaction of phosgene results in the formation of CO_2:

$$COCl_2 + H_2O \to CO_2 + 2H^+ + 2Cl^-.$$

The formation of CO_2 in the film may greatly reduce the absorption rate of CO_2. Under certain conditions no carbon dioxide will be absorbed at all, even at zero bulk CO_2 concentration in the liquid. This system has been analysed by Ramachandran and Sharma [38]. Simultaneous absorption followed by a complex consecutive reaction of type

$$A(G) \to A(L)$$
$$A'(G) \to A'(L)$$
$$A(G) + B(L) \to I(L) + P$$
$$A'(G) + I(L) \to B(L)$$

has been analysed by Onda et al. [51] and by Ramachandran and Sharma [52]. The system seems to be relevant for the absorption of C_2H_4 and O_2 in a $CuCl_2$ solution (oxochlorination process) [43].

The simultaneous absorption of two gases followed by equilibrium reactions of type

$$A + B \rightleftharpoons P$$
$$A' + B \rightleftharpoons X$$

has been analysed only for the case where both reactions are instantaneous [53, 54]. Broader of scope is the contribution of Cornelisse et al. [48], who presented a numerical method for a system of coupled partial differential equations, describing the rate of mass transfer for absorption and/or desorption of gases in a liquid accompanied by chemical reactions of complex reversible nature. The method gives stable results under all physically important conditions.

Mass transfer of two gases to or into a (porous) catalyst is also a case of simultaneous mass transfer. From the selectivity point of view, the scheme

$$A + A' \to P$$
$$A + P \to X$$

is of particular importance. We will deal with it in Section VIII.7.

Example VIII.4.6.a *The influence of mass transfer on the selective absorption of H_2S from a mixture of CO_2 and H_2S*

We discussed the suitability of stirred cell model reactors for investigating reactive gas–liquid systems in Example VII.15.1.a. One of our coworkers, G. F. Versteeg, recently measured in a stirred cell k_L and k_G as a function of agitator frequency ω and found, for the mass transfer to a 2M aqueous diisopropanol amine (B) solution at 302 K, $k_L = 16.1 \times 10^{-6}\omega$ and $k_G = 6.54 \times 10^{-3}\omega$ for $\omega < 1.7$ rev/s.

He also absorbed CO_2 (A) and H_2S (A') simultaneously in this solution and found the selectivity ratio to depend on ω according to

$$\frac{\sigma'_{A'}}{\sigma'_A} = 11.22\omega \frac{\bar{c}_{A'G}}{\bar{c}_{AG}}$$

for $0.6 < \omega < 1$ rev/s, $p_{H_2S} < 2 \times 10^{-3}$ bar and $p_{CO_2} \simeq 2\text{–}5 \times 10^{-2}$ bar. Both A and A' react with B. The pseudo first order rate constant k_1 for $CO_2 + B \to X$ is k_1(2M DIPA, $T = 302$ K) $= 700\,\text{s}^{-1}$. The reaction between H_2S and B is instantaneous. Additional data are $D_{A'G} = 1.5 \times 10^{-6}\,\text{m}^2/\text{s}$, $D_{AG} = 1.6 \times 10^{-6}\,\text{m}^2/\text{s}$, $D_B = 0.193 \times 10^{-9}\,\text{m}^2/\text{s}$, $D_{A'L} = 0.59 \times 10^{-9}\,\text{m}^2/\text{s}$, $D_{AL} = 0.77 \times 10^{-9}\,\text{m}^2/\text{s}$, $m_{A'} = 2.13$ and $m_A = 0.672$. We will investigate the observed selectivity ratio function.

Because $D_{AG} \simeq D_{A'G}$ we can conclude that $k_{GA} \simeq k_{GA'}$. For an average value of $\omega = 0.8\,\text{s}^{-1}$, we get $k_L = 12.8 \times 10^{-6}\,\text{m/s}$ and $k_G = 5.23 \times 10^{-3}\,\text{m/s}$.

We further calculate for the reaction $CO_2 + B \to X$ that $\phi = 57.3$, $E_\infty \simeq 10^3$ and $k_G/mk_L\phi \gg 1$. In the same way we find for $H_2S + B \to P$ that $E'_\infty > 10^4$ and $k_G/mk_L E'_\infty \ll 1$. Hence, H_2S absorbs according to the relation

$$J_{A'} = k_G(\bar{c}_{A'G} - 0)$$

and CO_2 according to

$$J_A = \sqrt{(k_1 D_{AL})}\, m \bar{c}_{AG}.$$

Hence

$$\frac{\sigma'_{A'}}{\sigma'_A} = \frac{J_{A'}}{J_A} = \frac{k_G}{m_A \sqrt{(k_1 D_{AL})}} \cdot \frac{\bar{c}_{A'G}}{\bar{c}_{AG}}.$$

Substitution of the data given leads to

$$\frac{\sigma'_{A'}}{\sigma'_A} = 13.2\omega \frac{\bar{c}_{A'G}}{\bar{c}_{AG}},$$

which agrees to within 20% with experimental results.

Example VIII.4.6.b *The negative enhancement factor*

The two reactions discussed in Example VIII.4.6.a are, in fact, equilibrium reactions:

$$H_2S + B \rightleftarrows BH^+ + HS^- \quad (a)$$

$$CO_2 + 2B \rightleftarrows BH^+ + BCOO^- \quad (b)$$

The following experiment was done on this system. A stirred cell reactor was partly filled with an aqueous diisopropanol solution (B) loaded with CO_2. After some time the system reached equilibrium, so that the CO_2 and H_2O partial pressures above the solution approached constant values, the equilibrium pressures.

Starting from this equilibrium situation, a continuous flow of H_2S was added to the gas phase in the cell over a period of 2.000 seconds. Both CO_2 and H_2S partial pressure above the amine solution were continuously recorded until a new equilibrium was approached, about an hour after the H_2S feed had stopped.

Figure VIII.24 shows the results. Notice that during the introduction of H_2S the CO_2 started to move away from the initial equilibrium by being stripped out of the solution, even up to a partial pressure twice as high as the original equilibrium pressure. After the H_2S feed was stopped, it took an hour before equilibrium was established again. During this period, nearly all CO_2 which had desorbed during the supply of H_2S was reabsorbed into the liquid. We will explain this experiment qualitatively.

The absorption of H_2S increases the bulk concentration of BH^+, and thus causes reaction (b) to shift to the left, resulting in a desorption of CO_2. However, from the fact that after the experiment, nearly all CO_2 has returned into the bulk of the solution again, we conclude that this shift in the bulk concentrations caused

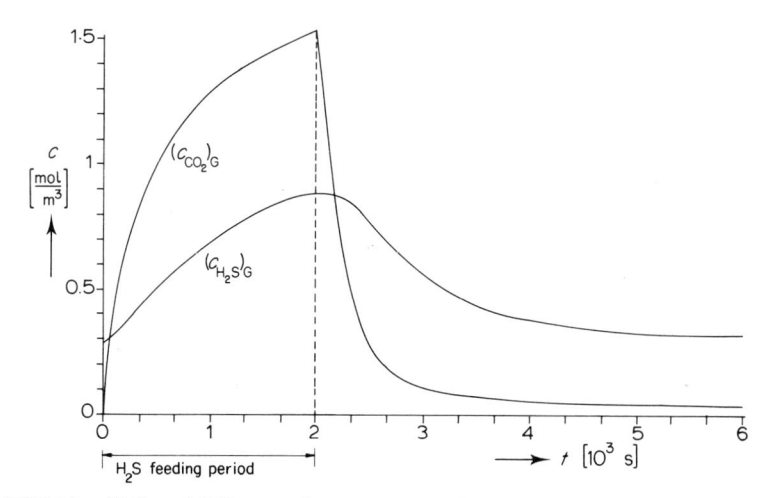

Fig. VIII.24. H_2S and CO_2 gas phase concentration as a function of time during and after introduction of H_2S to the gas phase above a CO_2 amine equilibrium solution in a stirred cell reactor, for data presented in Example VIII.4.6.b

only a minor shift in reaction (b), so that this cannot explain the observed temporary desorption. The explanation must therefore be found in the film near the interface. The absorption of H_2S is instantaneous and causes a local excess of BH^+ at the interface. This causes reaction (b) to shift locally to the left, thus starting a local production and accumulation of free CO_2 at the interface. As a result CO_2 starts to desorb and continues to do so until the concentration of CO_2 in the gas phase is in equilibrium with the increased local concentration in the liquid at the interface. The desorption may be enhanced and therefore this experiment proves the existence of a negative enhancement factor $E = J/(k_L(c_i - \bar{c}))$. A negative E means that desorption occurs while the overall driving force for absorption $(c_i - \bar{c})$ is positive. The experiment shows that the effect can be very significant. In the example desorption occurs up to a partial CO_2 pressure twice as high as the pressure in equilibrium with the bulk concentrations. Once the H_2S absorption rate starts to decrease because of approaching equilibrium, then also the local excess of BH^+ at the interface will decrease. This will cause reaction (b) at the interface to shift back to the right and reabsorption of CO_2 will start.

Figure VIII.25 gives typical concentration profiles. Notice the positive slope for CO_2 at the interface, indicating desorption under conditions of a substantial overall driving force for absorption $(c_i - \bar{c})/\bar{c} \simeq 2$. A quantitative model is presented by Cornelisse et al. [48] while additional experimental evidence will be published in the near future (P.M.M. Blauwhoff and van Swaaij).

VIII.5 SIMULTANEOUS MASS TRANSFER OF TWO REACTANTS (A AND A') WHICH REACT WITH EACH OTHER

The basic type of this system is:

$$A(1) \to A(2)$$
$$A'(1) \to A'(2)$$
$$A(2) + A'(2) \to P(2) \tag{VIII.56}$$

A major part of heterogeneous catalytic gas phase processes and many fluid–fluid processes belong to this type. Table VIII.4 gives examples of the latter class. With only one reaction proceeding, there is no selectivity problem. The mass transfer rate of one reactant may be enhanced by the presence of the other, and vice versa. Numerical calculation of J_A and $J_{A'}$ can be done by solving the equations (film theory)

$$D_A \frac{d^2 c_A}{dx^2} = D_{A'} \frac{d^2 c_{A'}}{dx^2} = -R(c_A, c_{A'}),$$

with

$$c_A = c_{Ai}, \quad c_{A'} = c_{A'i} \quad \text{for} \quad x = 0$$
$$c_A = \bar{c}_A, \quad c_{A'} = \bar{c}_{A'} \quad \text{for} \quad x = \delta.$$

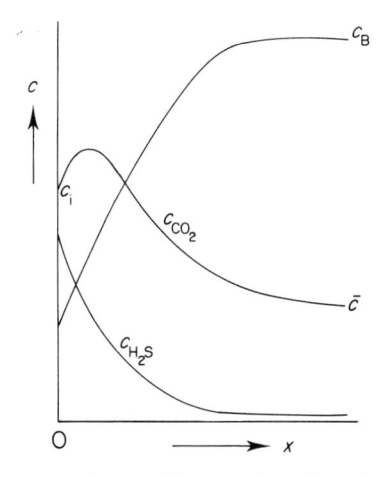

Fig. VIII.25. Liquid concentration profiles near interface showing desorption of CO_2 while $(c_i - \bar{c})$ is positive

A numerical solution based on the penetration model has been presented by Roper et al. [73]. Approximated solutions have been obtained by assuming linear [43] and exponential [74] concentration profiles of the reactants transferred. Inaccuracies are substantial (up to 30 and 45% respectively). Juvekar [75] substitutes for c_A a profile

$$\frac{c_A}{c_{Ai}} = \frac{\sinh[m(1 - x/\delta)]}{\sinh m}$$

with m a constant. For $c_{A'}$ = constant all over the film, this concentration profile is exact, with $m = \phi_A \, (\equiv \delta\sqrt{(kc_{A'}/D_A)})$. His solution is (for $\bar{c}_A = 0$ and 1,1 order kinetics)

$$E_A = m/\tanh m,$$

$$E_{A'} = \frac{m}{(E_{A\infty} - 1)\tanh m} + \frac{E_{A\infty} - 2}{E_{A\infty} - 1}$$

with $E_{A\infty} = 1 + D_{A'}c_{A'i}/D_A c_{Ai}$ and m found by trial and error from the equation

$$\frac{m}{\phi_A \sinh m} - \frac{m}{\phi_A \tanh m} - \frac{\bar{c}_{A'}/c_{A'i}}{m \sinh m}$$

$$+ \frac{1}{m \tanh m}\left(\frac{2E_{A\infty} - 3}{2E_{A\infty} - 2}\right) - \frac{1}{(2E_{A\infty} - 2)\sinh^2 m}$$

$$- \frac{1}{m^2}\left(\frac{E_{A\infty} - 2}{E_{A\infty} - 1} - \frac{\bar{c}_{A'}}{c_{A'i}}\right) = 0.$$

For $\phi_A > 1$, the accuracy is within 10%.

Table VIII.4. Some industrially important examples of simultaneous absorption of gases which react with each other in a liquid phase [Reproduced with permission from Ramachandran and Sharma [43]]

No.	Gases being absorbed	Product formed	Liquid medium employed	Reference
1	NH_3, CO_2	Ammonium carbamate	Water, aqueous monoethanol-amine,	Hatch and Pigford [55]
			Aqueous urea nitrate	Schmidt [56]
2.	C_2H_4, Cl_2	Dichloroethane	Dichloroethane	Balasubramanian et al. [57]
3.	Vinyl chloride, chlorine	Trichloroethane	Trichloroethane	Goldstein and Waddams [17]
4	Acetylene, Cl_2	Tetrachloroethane	Tetrachloroethane	FIAT-843 [58]
5	C_2H_4, HCl	Ethyl chloride	Ethyl chloride	CIOS XXXII.31 [59]
6	C_2H_4, HBr	Ethyl bromide	Ethyl bromide	Goldstein and Waddams [17]
7	SO_2, Cl_2	Sulphuryl chloride	Sulphuryl chloride	Kirk and Othmer [18]
8	SO_3, HCl	Chlorosulphonic acid	Chlorosulphonic acid	Kirk and Othmer [18]
9	SO_2, H_2S	Sulphur	(a) Molten sulphur	Chem. Eng. News [60] Wieworowski [61] (US Patent assigned to Freeport Sulphur Co.)
			(b) Triethylene glycol solution	Goar [62]

10	NO, H_2	Hydroxylamine salts	Sulphuric acid solution	Jockers et al. [63] (German patent)
11	CO_2, H_2S	Thiourea	Calcium cyanamide solution	Hahn [64]
12	Dimethyl sulphide, NO_2	Dimethyl sulphoxide	Dimethyl sulphoxide	Sittig [16]
13	Dimethyl ether, SO_3	Dimethyl sulphate	Dimethyl sulphate	Kirk and Othmer [18]
14	C_2H_4, O_2	Acetaldehyde	Solution of $CuCl_2$ containing $PdCl_2$ catalyst	Chandalia [65] Hatch [66]
15	C_3H_6, CO, H_2	Butyraldehyde and butanol	Butanol	Goldstein and Waddams [17] Oliver and Booth [67]
16	Heptene or octene, CO, H_2	Octanols or nonanols	Octanols or nonanols	Oliver and Booth [67]
17	C_2H_4, Cl_2,	Ethylene chlorohydrin	Water	Miller [68] Domask and Kobe [69]
18	HCl, O_2	Chlorine		Oblad [70]
19	C_2H_4, HCl, O_2	Dichloroethane	Aqueous $CuCl_2$	Friend et al. [71]
20	Ethylene, propylene	Co-polymer	Non-Newtonian solution of the copolymer	Jordan [72]
21	PCl_3, O_2	Phosphorus oxychloride	Phosphorus oxychloride	Kirk and Othmer [18]

Somewhat more accurate is the solution obtained by Sada et al. [76]. They substitute for c_A/c_{Ai} the profile

$$c_A/c_{Ai} = (1 - x/\delta)^{E_A},$$

which is exact for mass transfer without reaction for $\bar{c}_A = 0$ ($E_A = 1$). Implicit equations for E_A and $E_{A'}$ are obtained this way:

$$E_A = \frac{\phi_A^2 D_A}{(E_{A\infty} - 1)D_{A'}} \left[\frac{1}{2(E_A + 1)} + \frac{E_{A\infty} - 2}{(E_A + 3)} + \frac{\bar{c}_{A'}(E_{A\infty} - 1)}{c_{A'i}(E_A + 2)(E_A + 3)} \right] + 1$$

$$E_{A'} = E_A \left(\frac{1}{(1 - \bar{c}_{A'}/c_{A'i})} \frac{1}{(E_{A\infty} - 1)} \right).$$

Table VIII.5 shows results for $1 < \phi_A < 3$. For $c_{Ai} = c_{A'i}$ and bulk concentrations zero, the solution reduces to

$$E_A = E_{A'} = \sqrt{(1 + \tfrac{1}{2}\phi_A^2)}$$

For $c_{Ai}/c_{A'i}$ very different from one, the problem often simplifies by the fact that the concentration of the reactant that is in excess may be assumed to be constant in the reaction zone.

The main conclusion is that profile substitution is a powerful tool to get approximate solutions in complex mass transfer with reaction problems, but that substantial errors are easily introduced. A similar conclusion is drawn from comparing different approximation methods in the simultaneous absorption of two gases in a reactive liquid [45]. Recently Hikita et al. [77] showed that a linear profile substitution for one of the reactants according to Ouwerkerk [44] (Section VIII.4.3),

$$c_A = c_{Ai} + \left(\frac{dc_A}{dx}\right)_{x=0} x,$$

also results in accurate approximate analytical solutions.

Table VIII.5. Accuracy of approximate solutions for simultaneous mass transfer of two reactants (A and A') which react with each other; $\bar{c}_{A'} = 0$ and $E_{A\infty} = 4$ [From Sada et al. [76]. Reproduced by permission of Canadian Journal of Chemical Engineering]

ϕ_A	Numerical solution E_A	Sada et al. solution E_A	Juvekar solution E_A	Chaudhari–Doraiswamy solution E_A	Ramachandran–Sharma solution E_A
1.00	1.239	1.23	—	0.65	—
1.50	1.511	1.49	1.45	1.08	0.84
2.00	1.853	1.80	1.76	1.50	1.20
2.50	2.245	2.14	2.05	1.93	1.52
3.00	2.667	2.52	2.48	2.40	1.87

Concerning more complex studies we mention a study of Sada et al. [78] on the simultaneous absorption of three reacting gases, a study of simultaneous mass transfer followed by three parallel reactions [79] and a study of the simultaneous absorption of chlorine and olefins in aqueous solutions to produce chlorohydrins [36].

VIII.6 MASS TRANSFER WITH CONSECUTIVE REACTIONS A → P → X (TYPE III SELECTIVITY)

Probably, most complex reactions are of a mixed parallel consecutive type, as will be discussed in Section VIII.7. Systems with mass transfer followed by an elementary consecutive reaction of type A(1) → A(2) and A(2) → P(2) → X(2) are not uncommon, however. In gas–liquid systems the oxidation of hydrocarbons can be an example, with P the desired unstable peroxide, while catalytic reactions in porous catalysts of mixed parallel consecutive type, A + B → P and P + B → X, reduce to A → P → X by absence of diffusion limitation of B in the particles, e.g. because of excess of B. We will treat the cases of mass transfer and reactions both in series and parallel.

VIII.6.1 Mass transfer and reaction in series

Selectivity calculations for mass transfer and reaction in series, as for catalytic reactions at non-porous surfaces, are always straightforward, for any kinetics. We will give the equations (for both reactions first order) in terms of k_G for A → P → X:

$$J_A = k_{GA}(\bar{c}_A - c_{Ai})$$

$$J_P = k_{GP}(\bar{c}_P - c_{Pi})$$

$$R''_A = -k''_P c_{Ai}$$

$$R''_P = k''_P c_{Ai} - k''_X c_{Pi}$$

$$J_A = -R''_A$$

$$J_P = -R''_P.$$

Elimination of the unknown concentrations at the interface gives, for the differential selectivity,

$$\sigma'_P = -\frac{J_P}{J_A} = \frac{1 - \frac{\bar{c}_P}{\bar{c}_A}\frac{k''_X}{k_{GP}}\left(1 + \frac{k_{GA}}{k''_P}\right)\frac{k_{GP}}{k_{GA}}}{1 + \frac{k''_X}{k_{GP}}}. \qquad \text{(VIII.57)}$$

Before discussing this solution in detail, we will look at the concentration profiles of A and P near the interface, for two asymptotic solutions: the kinetic regime (figure VIII.26(a)) and the completely mass transfer controlled regime (figure

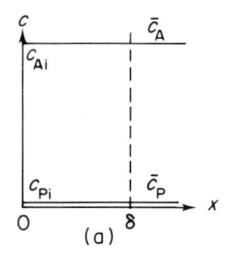

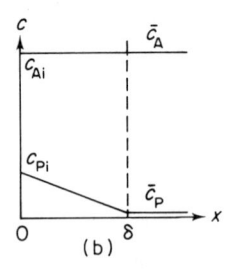

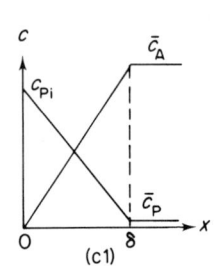

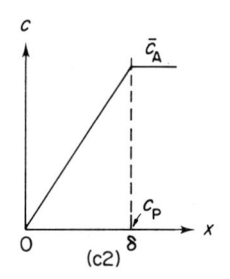

Fig. VIII.26. Concentration profiles of A and P for mass transfer in series with consecutive first order reactions A → P → X with $k_{GA} = k_{GB}$

(a) Kinetic regime

$k_{GA}/k_P'' \gg 1; k_{GP}/k_X'' \gg 1$

$$\sigma_P' = 1 - \frac{k_X'' \bar{c}_P}{k_P'' \bar{c}_A} \tag{VIII.58}$$

(b) no diffusion limitation of A:

$k_{GA}/k_P'' \gg 1$

$$\sigma_P' = \left(1 - \frac{\bar{c}_P k_X''}{\bar{c}_A k_P''}\right) \bigg/ \left(1 + \frac{k_X''}{k_{GP}}\right) \tag{VIII.59}$$

(c) complete diffusion limitation of A:

$k_{GA}/k_P'' \ll 1$

$$\sigma_P' = \left(1 - \frac{\bar{c}_P}{\bar{c}_A} \frac{k_X''}{k_{GA}}\right) \bigg/ \left(1 + \frac{k_X''}{k_{GP}}\right) \tag{VIII.60}$$

(c1) $k_X''/k_{GA} \ll 1$

$$\sigma_P' = 1 - \frac{\bar{c}_P}{\bar{c}_A} \frac{k_X''}{k_{GA}} \tag{VIII.61}$$

(c2) $k_X''/k_{GP} \gg 1$

$$\sigma_P' = \frac{k_{GP}}{k_X''} - \frac{\bar{c}_P}{\bar{c}_A} \frac{k_{GP}}{k_{GA}} \tag{VIII.62}$$

VIII.26(c)). We see that with increasing mass transfer limitation, $c_{Pi}/\bar{c}_P$ increases and $c_{Ai}/\bar{c}_A$ decreases. Differential selectivity, in terms of 'near-surface' concentrations, is

$$\sigma'_P = 1 - \frac{k''_P \, c_{Pi}}{k''_X \, c_{Ai}}$$

or

$$\sigma'_P = 1 - \frac{k''_P \, c_{Pi}}{k''_X \, \bar{c}_P} \cdot \frac{\bar{c}_A}{c_{Ai}} \cdot \frac{\bar{c}_P}{\bar{c}_A}.$$

Due to mass transfer limitation the product $\dfrac{c_{Pi}}{\bar{c}_P} \cdot \dfrac{\bar{c}_A}{c_{Ai}}$ may increase from 1 (kinetic regime) to infinity (completely mass transfer controlled). Hence, selectivity decreases tremendously with increasing mass transfer limitation. Consequently, mass transfer limitations must be avoided with heterogeneous consecutive reactions $A \to P \to X$. From this, it follows that a sufficient increase of temperature to enhance the reaction rates always eventually spoils the selectivity, even at equal activation energies of the two reactions. This is because reaction rates usually increase much faster with temperature than mass transfer rates. Similar conclusions are valid for mass transfer and a consecutive reaction in parallel in porous solids, but not necessarily for systems in which P cannot escape from the reaction phase, e.g. in gas–liquid reactions with P not volatile.

The asymptotic solutions, derived from equation (VIII.57) and presented in figure VIII.26, are self-explanatory. Some aspects are noteworthy. In homogeneous single phase consecutive kinetics, initial selectivity for $\bar{c}_P = 0$ always equals one, because then $R_X = 0$ because of lack of P. When mass transfer plays a role, this is no longer true. It follows from equation (VIII.57) that, for $\bar{c}_P = 0$,

$$\sigma'_P = 1/(1 + k''_X/k_{GP}),$$

which is only one for $k''_X/k_{GP} \ll 1$, that is, for the kinetic regime. Again, the reason for this decrease in selectivity is accumulation of P at the interface.

For the conversion rate of A completely controlled by the rate of mass transfer, selectivity highly depends on k''_X/k_G (figures VIII.26(c1, c2)). With increasing k''_X/k_G, selectivity decreases rapidly to $\sigma'_P = 0$ for $k''_X/k_G \gg 1$ with $\bar{c}_P = 0$ and to negative values for $\bar{c}_P > 0$ (equation (VIII.62)). Figure VIII.27 shows how differential selectivity, σ'_P, decreases dramatically with increasing mass transfer limitation of the first reaction. For $k''_P/k_G \simeq 10^{-2}$, there is no mass transfer limitation and selectivities are very high due to the selected values of k''_P/k''_X and $\bar{c}_P/\bar{c}_A$. Negative selectivities indicate a net consumption of P.

Similar to the results obtained for homogeneous single phase reactions, selectivity decreases with increasing $\bar{c}_P/\bar{c}_A$, that is with increasing conversion. Hence, also these reaction systems are preferentially carried out in plug flow reactors (e.g. packed beds) at low conversions. Starting with a feed, free of P, the

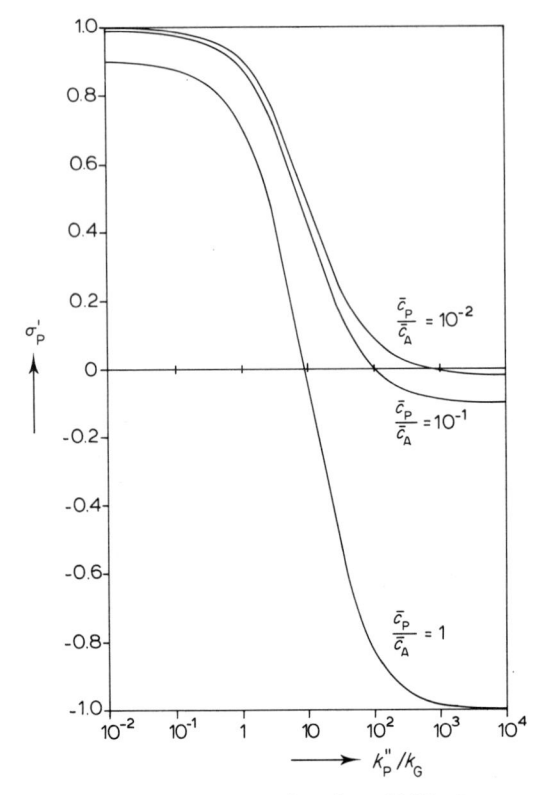

Fig. VIII.27. Differential selectivity, σ'_P, as a function of k''_P/k_G for mass transfer in series with irreversible first order reactions $A \to P \to X$, with $k_{GA}/k_{GP} = 1$ and $k''_X/k''_P = 0.1$ for different values of $\bar{c}_P/\bar{c}_A$

maximal concentration ratio $(\bar{c}_P/\bar{c}_A)_{max}$ that can be obtained is found for that conversion for which differential selectivity, σ'_P, has decreased to zero. Hence, from equation (VIII.57), it follows that

$$\left[\frac{\bar{c}_P}{\bar{c}_A}\right]_{max} = \frac{1}{\dfrac{k''_X}{k''_P}\left(1 + \dfrac{k''_P}{k_{GA}}\right)}. \qquad (VIII.63)$$

Figure VIII.28 shows the decrease in maximal obtainable $\bar{c}_P/\bar{c}_A$ as a function of k''_P/k_{GA} for different values of the reaction rate constants ratio. This figure too, shows the enormous spoiling effect of diffusion limitation on selectivity.

For reaction orders different from one, the conclusions may change quantitatively. Horak and Chuchvalec [80] analysed this system for reaction orders m and n of the first and the consecutive reaction respectively. They showed that for m or n different from one, σ'_P not only depends on the ratio $\bar{c}_P/\bar{c}_A$, but also on the absolute value of $\bar{c}_A$ itself. The influence of $\bar{c}_A$ on σ'_P at a constant $\bar{c}_P/\bar{c}_A$ may be

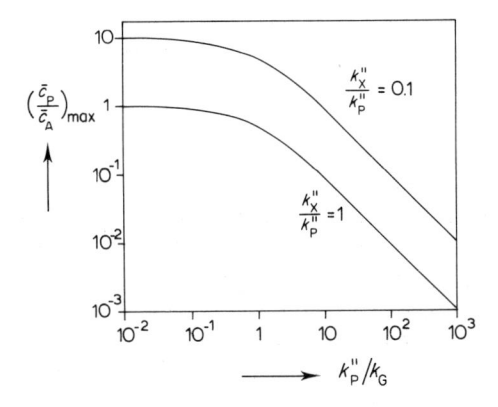

Fig. VIII.28. Maximum value of $\bar{c}_P/\bar{c}_A$ as a function of k_P''/k_G for mass transfer in series with irreversible first order reactions A → P → X with $k_{GA}/k_{GP} = 1$ for two values of k_X''/k_P''

different for the kinetic and for the diffusion regime. Table VIII.6 shows for which conditions it is advantageous for a good selectivity to operate at low or high concentration of $\bar{c}_A$.

VIII.6.2 Mass transfer and reaction in parallel

The most general case for mass transfer and consecutive reactions parallel is a fluid–fluid (e.g. a gas–liquid) system with $Al > 1$ and P transferable to the non-reaction phase (e.g. volatile). Figure VIII.29(a) qualitatively shows concentration profiles for A and P.

Table VIII.6. Effect of increase of concentration $\bar{c}_A$ on selectivity at constant hydrodynamic conditions and $\bar{c}_P/\bar{c}_A = 0.1$, for mass transfer in series with a consecutive reaction system A → P → X of the order m, n respectively. [From Horak and Chuchvalec [80]. Reproduced by permission of Academia, Publishing House of the Czechoslovakian Academy of Sciences]

m	n	$\dfrac{m}{n}$	Kinetic regime	Mass transfer regime
>1	>1	1	no influence	decreases
<1	<1	1	no influence	increases
1	1	1	no influence	no influence
>1	≥1	>1	increases	increases or decreases
≥1	<1	<1	decreases	decreases
≤1	<1	>1	increases	increases
<1	<1	<1	decreases	increases or decreases
<1	>1	<1	decreases	decreases
>1	<1	>1	increases	increases

A special case is obtained for no bulk (Figure VII.29(c)). Again it appears that mass transfer and reaction in a pellet can be treated as a special case of mass transfer and reaction in a liquid. From the point of selectivity, however, it makes an essential difference whether P can be transferred to a non-reaction phase (e.g. outside the pellet or drop, figure VIII.29(c), where P is safe for degradation, or that P can only be transferred to the bulk of the reaction phase (figure VIII.29(b)), where it is still subject to the degradation reaction P → X. It will be our approach to present the general two-film theory solution (figure VIII.29(a)) for first order kinetics, according to Bridgwater [81] and to derive from it the asymptotic solution for mass transfer with consecutive reaction in a porous pellet (figure VIII.29(c)), while we will conclude with the case of figure VIII.29(b).

With Bridgwater [81], we assume a stationary operated mixer as reactor in which the amounts of A and P, which leave the reactor unconverted with the reaction phase, are respectively small compared to the amounts of A transferred to the reaction phase and of P transferred from (or to) the reaction phase. Then $\sigma'_P = -J_P/J_A$. In case this condition is not satisfied, bulk concentrations of either A and/or P will be substantial, which means that mass transfer and at least the consecutive reaction are in series, which is a straightforward problem, to be discussed as an example at the end of this section. The transport equation for A in the film is

$$D_A \frac{d^2 c_A}{dx^2} = k_P c_A,$$

with boundary conditions

at $x = 0$: $c_A = c_{Ai}$

at $x = \delta$: $-aD_A \left(\frac{dc_A}{dx}\right)_{x=\delta} = (1-\varepsilon)\left(\frac{Al-1}{Al}\right) k_P \bar{c}_{AL}.$

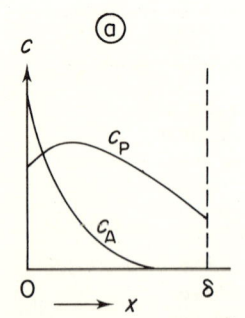

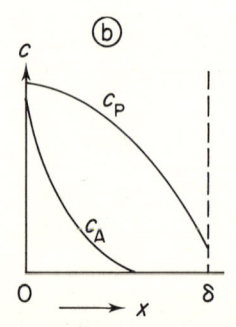

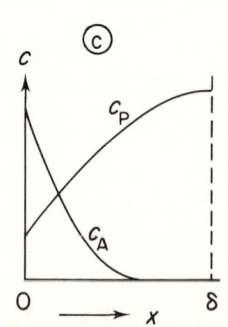

Fig. VIII.29. Mass transfer parallel with consecutive reaction A → P → X: (a) general case; $Al > 1$, P transferable to non-reaction phase; (b) $Al > 1$, P not transferable to non-reaction phase; (c) $Al = 1$, P transferable to non-reaction phase (Case A, p. 548)

The solution is given in Chapter VII:

$$J_A = c_{Ai}\sqrt{(k_P D_A)}\frac{(Al-1)\phi + \tanh\phi}{(Al-1)\phi\tanh\phi + 1} \qquad \text{(VII.79)}$$

or, with the definition equation for E_{A0},

$$J_A \equiv \frac{D_A}{\delta}c_{Ai}E_{A0} \qquad \text{(VII.132)}$$

$$E_{A0} = \phi\left[\frac{(Al-1)\phi + \tanh\phi}{(Al-1)\phi\tanh\phi + 1}\right]. \qquad \text{(VIII.64)}$$

Combination of equation (VII.132) with a balance over the gas film gives c_{Ai} as a function of $\bar{c}_{AG}$:

$$\frac{c_{Ai}}{m_A \bar{c}_{AG}} = \frac{1}{1 + \dfrac{D_A E_{A0} m_A}{\delta k_{GA}}}. \qquad \text{(VIII.65)}$$

The transport equation for P in the film is

$$D_P \frac{d^2 c_P}{dx^2} = k_X c_P - k_P c_A \qquad \text{(VIII.66)}$$

with boundary conditions

$$x = 0: \quad -D_P\left(\frac{dc_P}{dx}\right)_{x=0} = k_{GP}\left(\bar{c}_{PG} - \frac{c_{Pi}}{m_P}\right)$$

$$x = \delta: \quad -aD_P\left(\frac{dc_P}{dx}\right)_{x=\delta} = (1-\varepsilon)\left(\frac{Al-1}{Al}\right)(k_X \bar{c}_{PL} - k_P \bar{c}_{AL}).$$

Solving equation (VIII.66) subject to the boundary conditions gives [81]

$$\sigma_P' = -\frac{J_P}{J_A} = \left[\frac{1}{1 - \dfrac{k_X D_A}{k_P D_P}}\right] \times$$

$$\left[1 - \frac{\phi_P\left[\dfrac{D_P}{D_A \phi}\sqrt{\dfrac{k_X D_A}{k_P D_P}}\right]\tanh\phi_P + [Al-1]\dfrac{k_X}{k_P}\right]\left[\dfrac{m_P k_{LP}}{k_{GP}} + \dfrac{1 + \dfrac{m_P \bar{c}_{PG}}{m_A \bar{c}_{AG}}\left(\dfrac{D_P}{D_A} - \dfrac{k_X}{k_P}\right)}{E_{A0}}\right]}{\dfrac{D_P}{D_A \phi}\sqrt{\dfrac{k_X D_A}{k_P D_P}} + \dfrac{m_P k_{LP}}{k_{GP}}\phi(Al-1)\dfrac{k_X}{k_P}\sqrt{\dfrac{k_X D_A}{k_P D_P}} + \dfrac{k_X}{k_P}\left[Al - 1 + \dfrac{m_P k_{LP}}{k_{GP}}\right]\tanh\phi_P}\right]$$

(VIII.67)

where ϕ_P is the reaction modulus for the consecutive reaction:

$$\phi_P = \delta\sqrt{\left(\frac{k_X}{D_P}\right)} = \phi\sqrt{\left(\frac{k_X D_A}{k_P D_P}\right)}. \qquad \text{(VIII.68)}$$

Case A   Reaction in a pellet or drop ($Al = 1$):

This specific problem has been treated initially by Wheeler. Equation (VIII.67) simplifies to

$$\sigma_P' = \left[\frac{1}{1-\frac{k_X D_A}{k_P D_P}}\right]\left[1 - \frac{\tanh \phi_P \left[\frac{m_P k_{LP}}{k_{GP}} + \frac{1}{E_{A0}}\left[1 + \frac{m_P \bar{c}_{PG}}{m_A \bar{c}_{AG}}\left(\frac{D_P}{D_A} - \frac{k_X}{k_P}\right)\right]\right]}{\frac{1}{\phi_P} + \frac{m_P k_{LP}}{k_{GP}} \tanh \phi_P}\right]$$
(VIII.69)

with E_{A0} from equation (VIII.64): $E_{A0} = \phi \tanh \phi$. For reactions in porous pellets $m_P = m_A = 1$. In case all the mass transfer resistance is inside the pellet, equation (VIII.69) further simplifies to

$$\sigma_P' = \left[\frac{1}{1-\frac{k_X D_A}{k_P D_P}}\right]\left[1 - \frac{\phi_P \tanh \phi_P}{\phi \tanh \phi}\left[1 + \frac{\bar{c}_{PG}}{\bar{c}_{AG}}\left(\frac{D_P}{D_A} - \frac{k_X}{k_P}\right)\right]\right]. \quad \text{(VIII.69)}$$

As characteristic for any consecutive reaction, selectivity decreases with increasing conversion. Maximum differential selectivity is found for $\bar{c}_{PG} = 0$:

$$(\sigma_P')_{\bar{c}_{PG}=0} = \frac{1}{\left(1-\frac{k_X D_A}{k_P D_P}\right)}\left[1 - \frac{\sqrt{\left(\frac{k_X D_A}{k_P D_P}\right)} \tanh\left[\phi\sqrt{\left(\frac{k_X D_A}{k_P D_P}\right)}\right]}{\tanh \phi}\right]. \quad \text{(VIII.70)}$$

Figure VIII.30 shows differential selectivities as a function of the reaction modulus ϕ of the first reaction, for various values of k_X/k_P and $\bar{c}_P/\bar{c}_A$. The general conclusion is that diffusion limitation in the pellet spoils the selectivity and therefore has to be avoided. Decrease in selectivity due to mass transfer limitation is less pronounced than with mass transfer and reaction in series (previous section). The reason is that with increasing ϕ (e.g. by temperature increase) the reaction zone comes closer to the outer surface of the pellet, thus somewhat facilitating the transport of P out of the pellet.

Possible ways of avoiding diffusion limitation are smaller pellets, lower temperature and special pellets with two types of pores: large (macro) pores to ensure a fast transport into and out of the pellet and small (micro) pores for a large reaction surface. Carberry gives a mass transfer with reaction model for this type of catalyst [82]. Van de Vusse [83], Blackmore et al. [4] and Krishna [84] discussed the scheme for reaction orders different from one. Langmuir–Hinshelwood kinetics has been discussed by Komiyama and Inoue [85], Butt [86] extended the theory to non-isothermal particles and Smith [87] gives estimation methods for both interface and intraparticle temperature rise from observables. We concluded in section VII.14 that temperature gradients within the pellet are often negligible while gradients outside the pellet are more common. Experimental results have been discussed by Weisz and Swegler [88] and by Komiyama and Inoue [89].

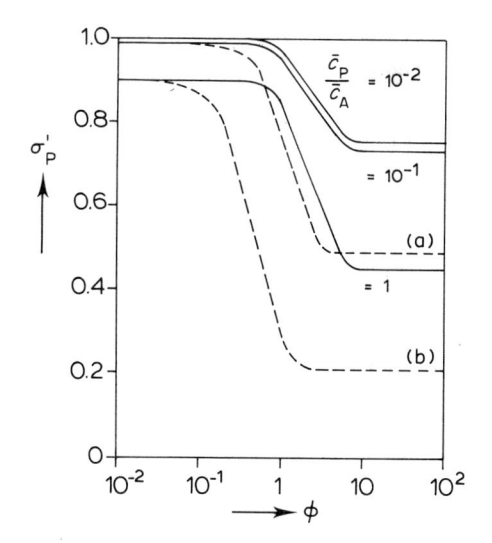

Fig. VIII.30. Differential selectivity, σ'_P, vs. ϕ for mass transfer parallel with first order reactions $A \to P \to X$ with $D_A = D_P$ and $k_X/k_P = 0.1$ for different values of $\bar{c}_P/\bar{c}_A$.
Dashed curves: $\bar{c}_P/\bar{c}_A = 10^{-2}$, $k_X/k_P = $ (a)1 and (b)10

Case B Both reactions fast (ϕ and $\phi_P > 2$)

In case both reactions are fast compared to mass transfer ($\tanh \phi_P = \tanh \phi = 1$ and $E_{A0} = \phi$) equation (VIII.67) reduces to

$$\sigma'_P = \frac{1}{1 - \dfrac{k_X D_A}{k_P D_P}} \left[\frac{1 - \sqrt{\dfrac{k_X D_A}{k_P D_P}} \left[1 + \dfrac{m_P \bar{c}_{PG}}{m_A \bar{c}_{AG}} \left(\dfrac{D_P}{D_A} - \dfrac{k_X}{k_P} \right) \right]}{1 + \dfrac{k_{LP} m_P}{k_{GP}} \phi \sqrt{\dfrac{k_X D_A}{k_P D_P}}} \right]. \quad \text{(VIII.71)}$$

Both reactions occur exclusively in the film near the interface. Neither A nor P can reach the bulk. Hence, Al has no influence on selectivity and the equation holds for both $Al = 1$ (porous pellet) and $Al \gg 1$; σ'_P is high for no gas phase resistance in transport of P ($k_{LP} m_P/k_{GP}$ small), for low concentrations of P ($m_P \bar{c}_{PG}/m_A \bar{c}_{AG}$ small) and decreases with increasing ϕ and k_X/k_P.

Case C First reaction fast, second reaction slow ($\phi > 2$; $\phi_P < 0.2$ and ($k_X D_A/k_P D_P \ll 1$))

Equation (VIII.67) reduces to

$$-\frac{J_P}{J_A} = \frac{\dfrac{D_P}{D_A} + Al\phi \dfrac{k_X}{k_P} \left[\phi - 1 - \dfrac{m_P \bar{c}_{PG} D_P}{m_A \bar{c}_{AG} D_A} \right]}{\dfrac{D_P}{D_A} + Al\phi^2 \dfrac{k_X}{k_P} \left(1 + \dfrac{k_{LP} m_P}{k_{GP}} \right)}. \quad \text{(VIII.72)}$$

We know from equation (VII.85) that for all P diffusing to the reactive bulk, $\bar{c}_{PL}$ will be zero if $Al\phi_P^2 \gg 1$, In the case under consideration, however, part of P diffuses into the gas phase. As P is formed close to the interface, the fraction that diffuses into the gas phase is, approximately, in the order of

$$\frac{k_{GP}}{k_{GP} + m_P k_{LP}}.$$

The criterion for $\bar{c}_{PL} = 0$ may therefore be modified to

$$\bar{c}_{PL} = 0 \quad \text{for} \quad Al\phi_P^2 \gg \frac{k_{GP}}{k_{GP} + m_P k_{LP}}.$$

Let us call this Case CI. Except for very soluble species, $k_{LP}m_P/k_{GP}$ is in the order of one or smaller. Equation (VIII.72) then reduces to

$$\sigma_P' = -\frac{J_P}{J_A} = \frac{\phi - 1 - \dfrac{m_P \bar{c}_{PG}}{m_A \bar{c}_{AG}} \dfrac{D_P}{D_A}}{\phi(1 + k_{LP}m_P/k_{GP})}. \tag{VIII.73}$$

Now, selectivity is, as expected, independent of Al. It is independent of k_X/k_P too. P is exclusively formed in the film; the fraction that reaches the gas phase is saved, the fraction that reaches the bulk is converted. Hence, selectivity increases with decreasing $m_P \bar{c}_{PG} D_P / m_A \bar{c}_{AG} D_A$ and decreasing $k_{LP}m_P/k_{GP}$.

Increase of ϕ may improve selectivity to an asymptotic value by moving the reaction zone closer to the interface, thus facilitating escape of P to the gas phase. The asymptotic value is

$$\sigma_P' = \frac{k_{GP}}{k_{GP} + m_P k_{LP}} \quad \text{for} \quad \phi \gg 1 + \frac{m_P \bar{c}_{PG} D_P}{m_A \bar{c}_{AG} D_A}.$$

If $Al\phi_P^2 \ll 1/\left(1 + \dfrac{k_{LP}m_P}{k_{GP}}\right)$, the bulk is saturated with P (Case CII) and

$$-\frac{J_P}{J_A} = 1 - Al \frac{k_X}{k_P} \phi \frac{m_P \bar{c}_{PG}}{m_A \bar{c}_{AG}}. \tag{VIII.74}$$

In this case, preferably a reactor is selected for which Al is as low as possible (e.g. a spray tower, a packed column or a chromatography column with an adsorbed liquid phase. Selectivity is not necessarily $-J_P/J_A$ because the outgoing liquid may also contain substantial amounts of P.

Case D   First reaction slow, second reaction fast

$$(\phi < 0.2, \phi_P > 2, k_X D_A / k_P D_P \gg 1)$$

A diffuses to the bulk before it reacts to P. P is converted to X before it can reach the gas phase. Consequently, σ'_P is low. Equation (VIII.67) reduces to

$$\sigma'_P = -\frac{J_P}{J_A} = \frac{D_P k_P}{D_A k_X}\left[\frac{-1 + \phi\sqrt{\left(\frac{k_X D_A}{k_P D_P}\right)}\left(1 - \frac{m_P \bar{c}_{PG} k_X}{m_A \bar{c}_{AG} k_P}\left(1 + \frac{1}{Al\phi^2}\right)\right)}{1 + \frac{k_{LP} m_P}{k_{GP}}\phi\sqrt{\left(\frac{k_X D_A}{k_P D_P}\right)}}\right].$$

Case E Both reactions slow ($\phi < 0.2$, $\phi_P < 0.2$)

Equation (VIII.67) reduces to

$$-\frac{J_P}{J_A} = \frac{\frac{D_P}{D_A}\left[1 - \frac{m_P \bar{c}_{PG}}{m_A \bar{c}_{AG}}\frac{k_X}{k_P}(1 + Al\phi^2)\right]}{\frac{D_P}{D_A} + Al\phi^2\frac{k_X}{k_P}\left(1 + \frac{k_{LP} m_P}{k_{GP}}\right)}.$$

Mass transfer and reaction are in series and, as expected, a high ratio $-J_P/J_A$ asks for low values of $m_P\bar{c}_{PG}/m_A\bar{c}_{AG}$, $Al\phi^2$, $k_{LP}m_P/k_G$ and k_X/k_P. According to Table VII.12 a spray column or a packed column look favourable from the selectivity point of view. Because liquid concentrations of A and P are substantial, $-J_P/J_A$ does not necessarily equal σ'_P. The latter may be higher, as discussed, at the end of this section (example VIII.6.2).

Case F P not volatile (figure VIII.29.b)

In case at least one of the reactions is slow, selectivity σ'_P is determined by bulk conditions only:

$$\sigma'_P = \frac{k_P \bar{c}_A - k_X \bar{c}_P}{k_P \bar{c}_A}$$

with $\bar{c}_A$ and $\bar{c}_P$ determined by mass balances over the gas and liquid phase (see the example at the end of this section). In case both reactions are fast, selectivity is zero.

A special, interesting situation exists if P is not the desired product itself, but either X_1 or X_2 in the system

$$A \rightarrow P \begin{array}{c} \nearrow X_1 \\ \searrow X_2 \end{array}$$

This system has already been discussed in Section VIII.3.2. Table VIII.7 shows simplified solutions for $D_A = D_P$, $\bar{c}_{PG} = 0$ and $k_{GP}/k_{LP}m_P \gg 1$ (no gas phase resistance). Figures VIII.31 and VIII.32 show the influence of Al and of $k_{LP}m_P/k_{GP}$ on $-J_P/J_A$ for $\phi < 0.2$ and > 2 respectively.

Table VIII.7. $-J_P/J_A$ for mass transfer with the consecutive first order reaction A(1) → A(2), A(2) → P(2) → X(2) and P(2) → P(1) and with $D_A = D_P$, $k_G = \infty$ and $\bar{c}_{PG} = 0$

Case	$-J_P/J_A$
A ($Al = 1$)	$\dfrac{1}{\left(1 - \dfrac{k_X}{k_P}\right)} \left[1 - \dfrac{\sqrt{\left(\dfrac{k_X}{k_P}\right)} \tanh \phi \sqrt{\left(\dfrac{k_X}{k_P}\right)}}{\tanh \phi}\right]$
B ($\phi > 2$, $\phi_P > 2$)	$\dfrac{1 - \sqrt{\left(\dfrac{k_X}{k_P}\right)}}{1 - \dfrac{k_X}{k_P}}$
C ($\phi > 2$, $\phi_P < 0.2$)	$1 - \dfrac{Al\phi k_X/k_P}{1 + Al\phi^2 k_X/k_P}$
CI ($Al\phi_P^2 \gg 1$)	$1 - 1/\phi$
CII ($Al\phi_P^2 \ll 1$)	$1 - Al\phi_P^2/\phi \simeq 1$
D ($\phi < 0.2$, $\phi_P > 2$)	$\dfrac{k_P}{k_X}\left[\phi\sqrt{\left(\dfrac{k_X}{k_P}\right)}\left(1 + \dfrac{1}{Al\phi^2}\right) - 1\right]$
E ($\phi < 0.2$, $\phi_P < 0.2$)	$\dfrac{1}{1 + Al\phi^2(k_X/k_P)}$
F ($k_G = 0$)	0

In fluid–fluid systems the penetration model is thought to be more realistic than the film theory. Szekely and Bridgwater [90] discuss mass transfer with a consecutive reaction A → P → X according to the penetration theory. Results may differ from the film theory by as much as 21.5%, even for $D_A/D_P = 1$ (see figure VIII.33). Kuo [91] discusses the system for a two parameter unsteady state mass transfer model (the film penetration theory). We conclude this section with an example in which balances over the bulk are necessary to calculate the yield.

Example VIII.6.2 *Mass transfer with a slow consecutive reaction in a stirred tank reactor*

$$B(2) \to P(2) \to X(2)$$
$$P(2) \to P(1)$$

Suppose that the reactions occur homogeneously in the phase L of an emulsion. The other fluid phase, G, is used to extract the desired product P from the reaction phase in order to improve the reaction yield. It is assumed that the reaction is carried out in one stirred continuous tank reactor in the steady state and at constant temperature, and furthermore that:

— B and X are insoluble in phase G;
— the distribution coefficient $m = c_{PL}/c_{PG}$ is constant;

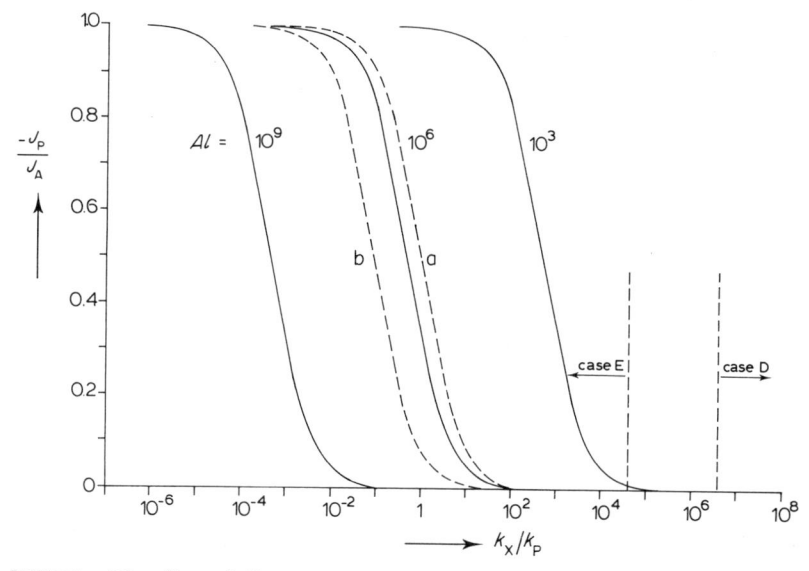

Fig. VIII.31. The effect of Al on $-J_P/J_A$ for mass transfer with consecutive irreversible first order reactions with $\phi < 0.2$ and $k_{LP} m_P/k_{GP} = 1$ ($\phi = 10^{-3}$, $\bar{c}_{PG} = 0$, $D_A = D_P$). The dashed curves show the influence of $k_{LP} m_P/k_{GP}$ for $Al = 10^6$; for curves (a) and (b), $k_{LP} m_P/k_{GP} = 0.1$ and 10 respectively [From Bridgwater [81]. Reproduced by permission of Pergamon Press Ltd.]

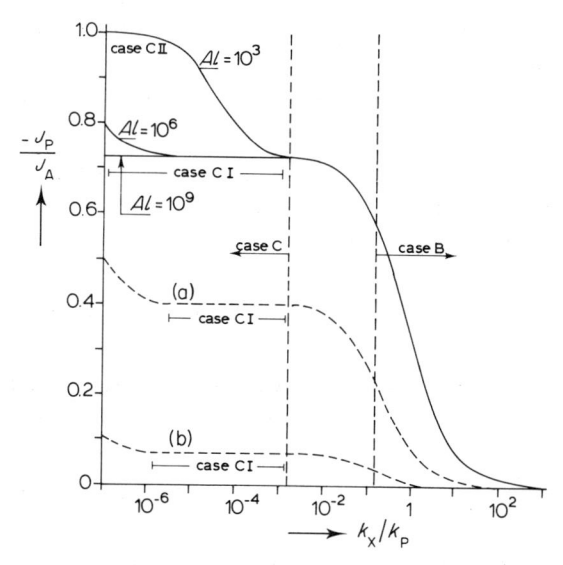

Figure VIII.32. The effect of Al on $-J_P/J_A$ for mass transfer with consecutive irreversible first order reactions with $\phi > 2$ and $k_{LP} m_P/k_{GP} = 0.1$ ($\phi = 5$, $\bar{c}_{PG} = 0$, $D_A = D_P$). The dashed curves show the influence of $k_{LP} m_P/k_{GP}$ for $Al = 10^6$: for curves (a) and (b), $k_{LP} m_P/k_{GP} = 1$ and 10 respectively [From Bridgwater [81]. Reproduced by permission of Pergamon Press Ltd.]

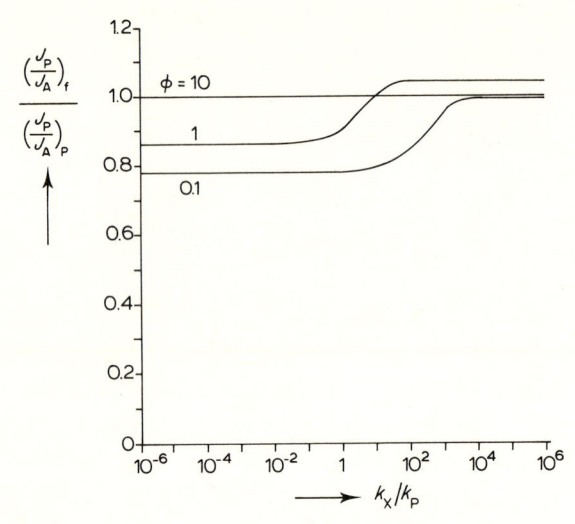

Fig. VIII.33. The ratios of J_P/J_A for mass transfer with consecutive irreversible first order reactions as predicted by film and penetration theory as a function of k_X/k_P and ϕ, ($D_A = D_P$) [From Szekely and Bridgwater [90]. Reproduced by permission of Pergamon Press Ltd]

— the composition in phase L is uniform;
— the conversion rate is small with respect to the rate of mass transfer of product P, so that the two phases are in equilibrium with respect to this component;
— the densities of the two feed streams are equal to the corresponding densities at the outlet.

Reference is made to figure VIII.34 for the symbols used. In view of the above assumptions, the material balance for species B (which occurs only in the phase L) is, according to equation (VII.1),

$$V_L \frac{dc_B}{dt} = 0 = \Phi_{vL}(c_{Bo} - c_{B1}) - k_P c_{B1} V_L \quad \text{[kmol/s]}. \quad \text{(VIII.75)}$$

Since for P equilibrium is assumed between both phases, the following material balance can be drawn up for the entire reactor contents:

$$V_L \frac{dc_P}{dt} = 0 = -\Phi_{vL} c_{P1} - \Phi_{vG} c_{P1}/m + k_P c_{B1} V_L - k_X c_{P1} V_L. \quad \text{(VIII.76)}$$

c_{B1} and c_{P1}, upon introduction of the average residence time of the reaction phase $\tau_L = V_L/\Phi_{vL}$ and the extraction factor $E_P = \Phi_{vG}/m\Phi_{vL}$ are found to be

$$\frac{c_{B1}}{c_{Bo}} = \frac{1}{1 + k_P \tau_L}$$

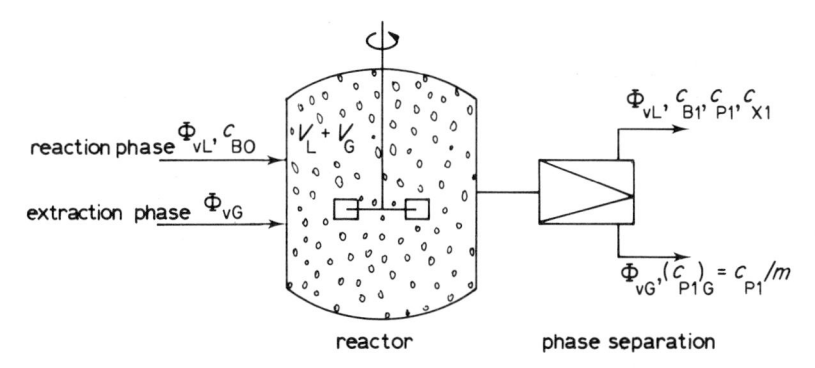

Fig. VIII.34. Consecutive reactions in a liquid system with simultaneous extraction of the intermediate product P by phase G

and

$$\frac{c_{P1}}{c_{Bo}} = \frac{k_P \tau_L}{(1 + k_P \tau_L)(1 + E_P + k_X \tau_L)}.$$

The yield η_P is the ratio of the moles of P produced per mole of A fed:

$$\eta_P = \frac{\Phi_{vL} c_{P1}(1 + E_P)}{\Phi_{VL} c_{Bo}} = (1 + E_P)\frac{c_{P1}}{c_{Bo}}. \quad (\text{VIII.77})$$

It is seen that the yield has a maximum as a function of τ_L when k_P, k_X and E_P remain constant. From the condition:

$$\frac{\partial \eta_P}{\partial \tau_L} = 0,$$

we find that this maximum occurs at

$$k_P \tau_L = \sqrt{[(1 + E_P)k_P/k_X]},$$

at which

$$\eta_P = \frac{(1 + E_P)k_P/k_X}{[1 + \sqrt{((1 + E_P)k_P/k_X)}]^2}. \quad (\text{VIII.78})$$

Apparently, $k_P \tau_L$ and η_P depend on a single parameter $(1 + E_P)k_P/k_X$ for a maximum yield. These functions are shown in figure VIII.35. Since the value of E_P would be zero without extraction of P, it can be reasoned that the optimum result of the extraction is an apparent increase, by the factor $(1 + E_P)$, of the ratio of the desired and the undesired conversion rates. The above method of calculation can be extended to a cascade of tank reactors, as was shown by Hofmann [92]. The improvement in yield by extraction of P entails the need for a greater reaction volume. It is seen from equation (VIII.78) that, with respect to the case $E_P = 0$, the average residence time of the reaction phase must be increased by the factor $\sqrt{(1 + E_P)}$. Furthermore, the extraction fluid G has to be accommodated, which

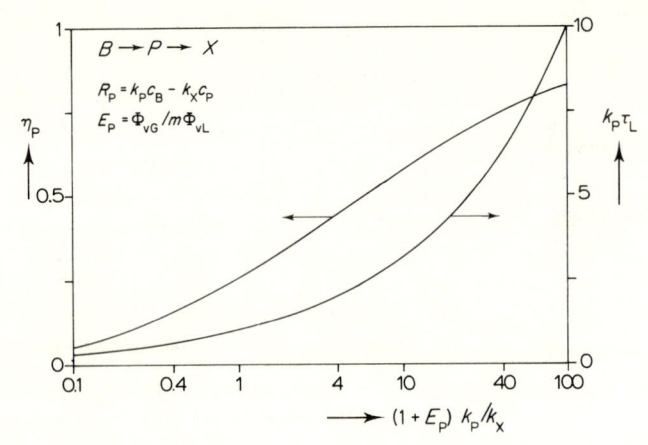

Fig. VIII.35 Maximum yield of P and corresponding dimensionless residence time (equation (VIII.78)) for consecutive first order reactions in a tank reactor with simultaneous extraction

implies that a further factor of $(V_L + V_G)/V_L = V_r/V_L$ must be applied for obtaining the total reaction volume. As to the volume ratio V_G/V_L, its value is determined mainly by the requirements of obtaining a sufficiently large mass transfer surface per unit volume, and to a lesser extent by the value of E_P to be applied. It was shown, e.g. by Trambouze [93], that by taking special measures at the outlet of an agitated emulsion reactor the value of V_G/V_L can be made different from the ratio of the volumetric flow rates Φ_{vG}/Φ_{vL}.

The theory of this example was successfully applied by Schoenemann and Hofmann [94–97] to the preparation of furfural from xylose in water containing hydrochloric acid. They found that the reaction is relatively slow and proceeds according to the scheme:

$$\begin{array}{c} \text{xylose (B)} \\ \downarrow \\ \text{intermediate product} \\ \downarrow \\ \text{furfural (P)} \\ \downarrow \\ \text{resin (X)} \end{array} \Bigg\} \to \text{condensation product (Y)}$$

The production rate of furfural could be satisfactorily described by

$$R_P = k_P c_B - k_X c_B c_P - k_Y c_P,$$

in which the rate constants depend on the temperature and the HCl concentration. It is seen that it is desirable to keep the value of c_P in the reaction phase low so as to suppress the side reactions leading to the products X and Y. This can be achieved by extracting the furfural from the reaction mixture. This problem was extensively studied by Schoenemann and Hofmann, who calculated — on the basis of the experimental rate data and distribution ratios m — the furfural yield for various reactor types and process variables (xylose concentration in the feed, HCl concentration and reaction temperature). They verified experimentally that a yield η_P of 0.63 could be obtained in a cascade of three identical tank reactors with tetraline as solvent phase and with a value of $\Phi_{vG}/\Phi_{vL} = 10$. In a batch process carried out under the same conditions, but without extraction, the maximum yield would have been only 0.10.

For cocurrent and counter current column reactors this type of yield calculations have been carried out by Hofmann [98], van de Vusse [99] and by Ohki and Inoue [100]. Depending on E_P and $k_P\tau_L$ either a counter-current or a co-current column are preferable for obtaining the highest yield. Another application is the chlorination of benzene in a distillation column [101]:

$$Cl_2 + C_6H_6 \rightarrow C_6H_5Cl + HCl$$

$$Cl_2 + C_6H_5Cl \rightarrow C_6H_4Cl_2 + HCl$$

$$Cl_2 + C_6H_4Cl_2 \rightarrow C_6H_5Cl_3 + HCl, \text{ etc.}$$

A good yield of C_6H_5Cl could be obtained by continuous removal of C_6H_5Cl as a bottom product. A similar example is the chlorination of alkanes in the gas phase with the products separating off as a liquid phase through condensation [99].

VIII.7 MASS TRANSFER WITH MIXED CONSECUTIVE PARALLEL REACTIONS

The reaction scheme

$$A + B \rightarrow P$$

$$A + P \rightarrow X$$

is consecutive with respect to B:

$$B \xrightarrow{A} P \xrightarrow{A} X$$

and parallel with respect to A:

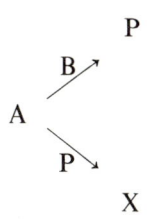

$$A \begin{array}{c} \xrightarrow{B} P \\ \searrow_P X \end{array}$$

In catalytic reactions, both A and B are transferred into the catalyst from the environment of the pellet. Hence, this is a case of simultaneous mass transfer followed by a complex chemical reaction. Several gas phase chlorinations, hydrogenations and oxidations belong to this type. Some results from analog simulations have been published by Wirges and Rähse [102], while Hell et al. published numerical results [103]. General characteristics are similar to those found for the system $A \rightarrow P \rightarrow X$. High yields are favoured by avoiding diffusion limitation and by keeping $\bar{c}_P$ as low as possible. The influence of pore diffusion on the (liquid phase) hydrogenation of poly unsaturated fatty acid glycerides has been described with this mixed consecutive parallel reaction scheme [103]. In the following two sub-sections we shall discuss systems in which A is transferred to and B is already present in the reaction phase with special reference to fluid–fluid systems

VIII.7.1 The system: $A(1) \to A(2)$; $A(2) + B(2) \to P(2)$; $P(2) + B(2) \to X(2)$

In case of no diffusion limitation of B, the system becomes similar to $A \to P \to X$ (section VIII.6). In case of diffusion limitation of B, there are some new aspects. One is that the enhancement factor for the absorption of A may depend on both reaction rates because both reactions compete for B. Enhancement factor calculations have been carried out by Brian and Beaverstock [104] and Ramachandran [105] while Hikita et al. extended the theory for the case in which the consecutive step is reversible [106, 107].

Figure VIII.36 shows enhancement factors as a function of Ha for

$$1 + \frac{D_B \bar{c}_B}{D_A c_{Ai}} \equiv E_{A\infty} = 5$$

and for various values of $k_X D_A/(k_P D_P)$. With increasing values of $k_X D_A/(k_P D_P)$, the instantaneous enhancement factor decreases from the value $1 + D_B \bar{c}_B/(D_A c_{Ai})$ (belonging to $A + B \to P$) to the value $1 + D_B \bar{c}_B/(2D_A c_{Ai})$ (belonging to $A + 2B \to P$).

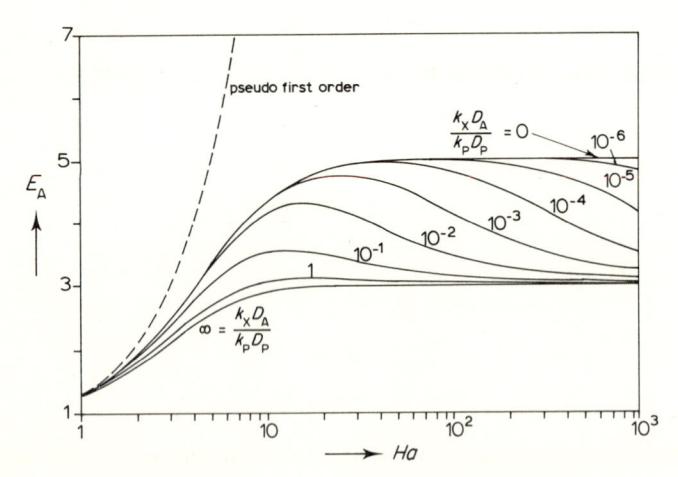

Fig. VIII.36. Enhancement factor, E_A, vs. $\sqrt{(k_P \bar{c}_B)}/k_L$ for the system $A(1) \to A(2)$:
$$A(2) + B(2) \to P(2)$$
$$B(2) + P(2) \to X(2),$$
with both reactions of order (1,1) and $D_B \bar{c}_B/D_A c_{Ai} = 4$ [From Brian and Beaverstock [104]. Reproduced by permission of Pergamon Press Ltd]

Systems which, under certain conditions, can be described according to this theory, are the absorption of chlorine in ferrous chloride [104], carbon dioxide in mono-ethanol amine [104], both chlorine and sulphur dioxide in hydroxide solutions [108] and carbon dioxide in sodium hydroxide or carbonate solutions [109]. The selectivity problem has not received much attention in the literature until now.

For fluid–fluid reaction systems we may distinguish between cases where P is transferable to the non-reaction phase and where it is not. Only if the first reaction A + B → P is fast, with ϕ at least in the order of $E_{A\infty}$, does the problem differ from that treated in the previous section. For P not transferable to the non-reaction phase, if the consecutive reaction is fast as well, selectivity is practically zero. For P not transferable to the non-reaction phase, if the consecutive reaction is slow, mass transfer and the consecutive reaction are in series and treatment of the problem goes along lines developed in the example of the previous section.

There remains mass transfer with both reactions fast, with ϕ_A at least in the order of $E_{A\infty}$ and P volatile. We will give a qualitative discussion only. Figure VIII.37 shows typical concentration profiles for $\phi_A \gg E_{A\infty}$. Assuming that all P that diffuses from $x = \delta_r$ in the direction of the bulk will ultimately be converted into X gives

$$\sigma_P' = -\left(\frac{dc_P}{dx}\right)_{x\uparrow\delta_r} \bigg/ \left[\left(\frac{dc_P}{dx}\right)_{x\downarrow\delta_r} + \left(\frac{dc_P}{dx}\right)_{x\uparrow\delta_r}\right]$$

where $x\uparrow\delta_r$ and $x\downarrow\delta_r$ denote x approaching δ_r from the interface and from the bulk respectively. In case the consecutive reaction mainly proceeds in the bulk, that is for

$$\frac{\delta}{(\delta - \delta_r)}\phi_P < 0.2,$$

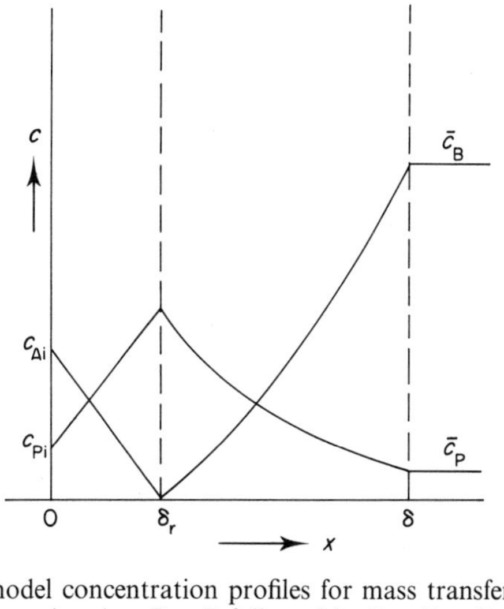

Fig. VIII.37. Film model concentration profiles for mass transfer with instantaneous reaction A + B → P followed by B + P → X

dc_P/dx = constant for $\delta_r < x < \delta$ too. In case, moreover, both c_{Pi} and $\bar{c}_P$ are zero, it follows from figure VIII.35 that

$$\left(\frac{dc_P}{dx}\right)_{x\uparrow\delta_r} \bigg/ \left(\frac{dc_P}{dx}\right)_{x\downarrow\delta_r} = \frac{(\delta - \delta_r)}{\delta_r}.$$

Thus

$$\sigma'_P = 1 - 1/E_{A\infty} \tag{VIII.79}$$

σ'_P decreases with increasing c_{Pi} (e.g. due to gas phase resistance). Increase of the bulk concentration $\bar{c}_P$, on the other hand, may improve selectivity by lowering $(dc_P/dx)_{x\downarrow\delta_r}$. A high $\bar{c}_P$ can be obtained for

$$(1-\varepsilon)k_1(c_P)_{x=\delta_r} \ll \frac{Da}{(\delta-\delta_r)}(c_P)_{x=\delta_r};$$

and hence also for

$$Al\phi_P^2 \ll E_{A\infty}/(E_{A\infty} - 1)$$

which means that a spray column (low Al) may be advisable. With increasing ϕ_P above $\phi_P = 0.2$, more and more B is consumed by P in the zone $\delta_r < x < \delta$. Consequently the plane δ_r moves towards the bulk, increasing the value of δ_r and thus decreasing the differential selectivity σ'_P. Hence selectivity will decrease with increasing ϕ_P. Due to the accumulation of P at the plane $x = \delta_r$, supersaturation of P may occur. If supersaturation is followed by bubble nucleation, this may enlarge the mass transfer coefficient substantially. Shah et al. [110] derived criteria for supersaturation in gas–liquid reactions involving a volatile reaction product.

VIII.7.2 The system: A(1) → A(2); A(2) + B(2) → P(2); A(2) + P(2) → X(2)

In the Section VIII.7.1 product P competes with A to react with B, thus slowing down the rate of mass transfer of A for $\phi_P > 2$. In the system

$$A(1) \to A(2)$$
$$A(2) + B(2) \to P(2)$$
$$A(2) + P(2) \to X(2),$$

P competes with B to react with the absorbing component A, thus enhancing the rate of mass transfer of A for $\phi_P > 2$. With increasing k_X/k_P, the instantaneous enhancement factor increases from $1 + D_B\bar{c}_B/D_A c_{Ai}$ (belonging to the reaction A + B → P) to the value $1 + 2D_B\bar{c}_B/D_A c_{Ai}$ (belonging to the reaction 2A + B → X). Figure VIII.38 shows quantitative results. Numerical and approximated film theory enhancement factor calculations for reaction orders (m, n) and (p, q) respectively, have been presented by Onda et al. [51]. Numerical and approximated penetration theory solutions are discussed by Onda et al. [6, 111]

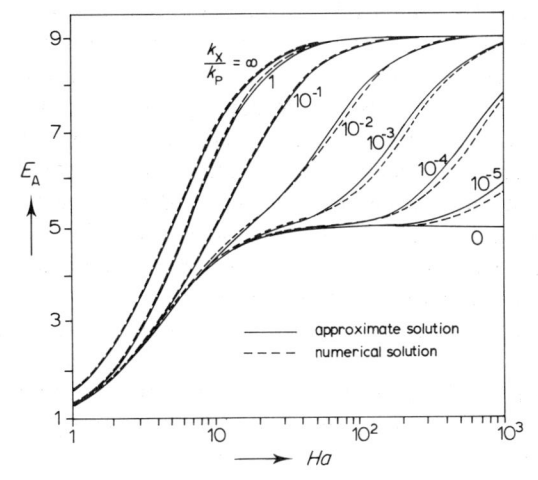

Fig. VIII.38. Enhancement factors, E_A, vs. $\sqrt{(k_P \bar{c}_B)}/k_L$ for the system

$$A(1) \rightarrow A(2)$$
$$A(2) + B(2) \rightarrow P(2)$$
$$A(2) + P(2) \rightarrow X(2)$$

with both reactions of order (1,1) and $D_B \bar{c}_B/(D_A c_{Ai}) = 4$ [From Onda et al. [51]. Reproduced by permission of Pergamon Press Ltd]

and Shah [112, 113]. The approximation methods are similar to those used by van Krevelen and Hoftijzer [114] and by Hikita and Asai [115].

This type of fluid–fluid consecutive reaction is probably the most common one. Many hydrogenations, oxidations, chlorinations, alkylations, sulphonations and nitrations belong to this class. Therefore, the influence of mass transfer on selectivity has received relatively much attention. For P non-volatile this influence, as first pointed out by van de Vusse [116], exists only in case the reaction A + B → P is sufficiently fast. Sufficiently means $\phi > 0.5$ and not much smaller than $E_{A\infty}$ [117], with ϕ and $E_{A\infty}$ referring to the first reaction A + B → P only. The idea is to transport P as fast as possible out of the reaction zone near the interface into the bulk of the reaction phase where it is relatively safe as regards being converted into X due to lack of A which can't reach the bulk for $\phi > 2$. Experimental results are available for chlorination [116, 118–122] and sulphonation [8, 9, 123]. A numerical analysis for (m, n), (p, q)-order kinetics [124], an analog simulation [116] and trial and error procedures [118, 121, 124] by which approximate solutions can be obtained, have been presented. The special case of (0, 1), (0, 1) kinetics has been analysed by Hashimoto et al. [125].

In practice, there is often doubt about the exact values of the relevant parameters c, m, ρ, μ, D and T, mainly because interface conditions may differ from bulk conditions. Moreover, knowledge of k_L may be limited due to free convection induced by interfacial concentration and temperature differences, created by the reactions; the latter effect also influencing k_P/k_X at the interface. Further inaccuracies in the model description may be introduced by evaporation of solvent and/or some products, by interdependence of diffusion coefficients, by spread in c_{Ai} and $\bar{c}_B$ due to back-mixing, etc. Because of these complicating factors, fairly rough approximated relations for σ'_P are often sophisticated enough to discuss experimental results. Harriott [117] derived such a model for the intermediate regime between fast and instantaneous reaction rate of the first step $A \rightarrow P$. We will treat it in a simplified form for $\phi \; (=\sqrt{(k_P \bar{c}_B)}/k_L) > 2$, though Harriott allows for reaction in the bulk as well.

The model is further limited by the assumption that the rate of mass transfer of A is not influenced by the consecutive reaction. Processes with a not too low differential selectivity will satisfy this assumption. With this condition fulfilled, E_A is a function only of ϕ and $E_{A\infty}$ and follows, either from figure VII.21 or from equation (VII.138).

Harriott defines an 'effective' constant concentration of B in the reaction zone, $(\bar{c}_B)_{rz}$, such that

$$E_A = \sqrt{[k_P(\bar{c}_B)_{rz}/D_A]} \qquad \text{(VIII.80)}$$

being the known expression for E_A for $\phi > 2$ in case of a constant concentration of B in the reaction zone (see figure VIII.39). Hence

$$\frac{(\bar{c}_B)_{rz}}{\bar{c}_B} = \left(\frac{E_A}{\phi}\right)^2. \qquad \text{(VIII.81)}$$

For $E_A = \phi$, there is no diffusion limitation of B: $(\bar{c}_B)_{rz} = \bar{c}_B$, while for increasing values of $\phi/E_{A\infty}$, E_A/ϕ becomes increasingly smaller than one, thus predicting

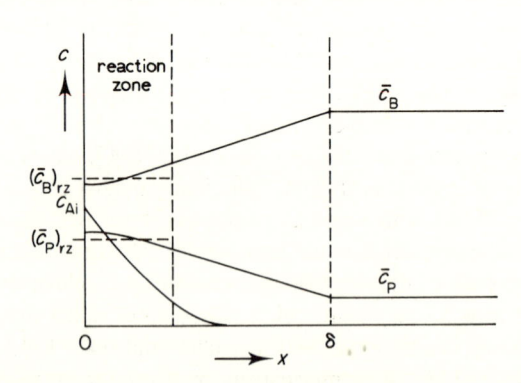

Fig. VIII.39. Schematic film model concentration profiles for the system

$$A(1) \rightarrow A(2)$$
$$A(2) + B(2) \rightarrow P(2)$$
$$A(2) + P(2) \rightarrow X(2)$$

with $\phi \simeq E_{A\infty} \simeq 3$

increasingly lower values of the effective concentration of B in the reaction zone.

As the model is limited to selectivities that are not too low, J_P is of the order of $-J_B$; or

$$D_B(\bar{c}_B - (\bar{c}_B)_{rz}) \simeq D_P((\bar{c}_P)_{rz} - \bar{c}_P), \tag{VIII.82}$$

which is a defining equation for the effective concentration of P in the reaction zone, $(\bar{c}_P)_{rz}$. Because also the consecutive reaction requires A, both reactions can occur in the reaction zone only. Hence, differential selectivity follows from

$$\sigma'_P \simeq 1 - \frac{k_X(\bar{c}_P)_{rz}}{k_P(\bar{c}_B)_{rz}}. \tag{VIII.83}$$

Combination of equations (VIII.81), (VIII.82) and (VIII.83) gives

$$\sigma'_P \simeq 1 - \frac{k_X}{k_P}\left[\frac{\bar{c}_P}{\bar{c}_B}\frac{\phi^2}{E_A^2} + \frac{D_B}{D_P}\left(\frac{\phi^2}{E_A^2} - 1\right)\right]. \tag{VIII.84}$$

For $E_A = \phi$, σ'_P equals the value found for slow reactions in the bulk. With increasing value of ϕ/E_A, $(\bar{c}_B)_{rz}$ decreases and $(\bar{c}_P)_{rz}$ increases, which, according to equation (VIII.83) results in a strong decrease of selectivity. The method becomes increasingly more inaccurate as $(\bar{c}_B)_{rz}$ approaches zero, that is as the rate of the first reaction approaches instantaneity. For that case Beenackers and van Swaaij showed [8] that

$$1 - \sigma'_P \simeq \tfrac{1}{2}\frac{k_X}{k_P}\frac{\phi^2}{E_{A\infty}^2}\left[\frac{\bar{c}_P}{\bar{c}_B} + \frac{D_B}{D_P}\right] \tag{VIII.85}$$

for $\phi \gg E_{A\infty}$ and $(1 - \sigma'_P) \ll 1$.

As expected, maximum selectivity is obtained for minimal $\bar{c}_P/\bar{c}_B$, which asks for low conversions and preferably plug flow of the bulk of the reaction phase. For $\bar{c}_P/\bar{c}_B \ll 1$, $D_A = D_B = D_P$ and $E_{A\infty} \gg 1$, equation (VIII.85) reduces to

$$(1 - \sigma'_P) \simeq \tfrac{1}{2}\frac{k_X D c_{Ai}}{k_L^2}, \tag{VIII.86}$$

which shows the strong increase in selectivity with increasing liquid side mass transfer coefficient. Due to mass transfer limitation, the obtained selectivity can be factors lower than for the kinetic regime. This has been shown experimentally for the sulphonation of benzene with gaseous sulphur trioxide [8, 9, 123]. By applying a cyclone reactor with very large k_L values, selectivity is raised substantially compared to conventional reactors. For σ'_P substantially lower than one, the equations derived in this section may become too inaccurate. Then, more sophisticated approximations can be used [118, 121, 124].

The analysis in Section VIII.7 is based on the interface concentration c_{Ai}. Often some form of diffusion limitation in the non-reaction phase occurs and has to be taken into account as well. We will demonstrate this in the following Example.

Example VIII.7.2 *Selectivity for a consecutive reaction with the controlling resistance completely in the non-reaction phase*

A chlorination reaction proceeds according to

$$Cl_2(G) \to Cl_2(L)$$

$$Cl_2 + B(L) \to P + HCl$$

$$Cl_2 + P \to Q + HCl$$

Both reactions are instantaneous and of the order (1,1). The controlling resistance is completely in the gaseous phase, so that $c_{Ai} = 0$. We will derive a relation for the differential selectivity and demonstrate the influence of $k_G \bar{c}_{AG}/k_L \bar{c}_{AL}$ on the differential selectivity σ'_P in a graph for various k_P/k_X and for $\zeta = 0$. The concentration profiles are shown in figure VIII.40. The process is described by the following set of equations:

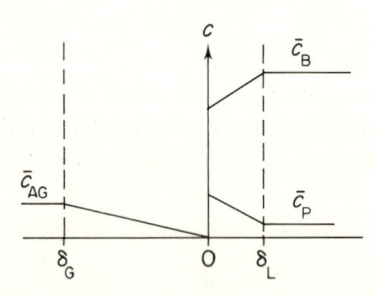

$$\sigma'_P = 1 - \frac{k_X c_{Pi}}{k_P c_{Bi}} \tag{a}$$

$$c_{Pi} = \bar{c}_P + J_P/k_{LP} \tag{b}$$

$$c_{Bi} = \bar{c}_B - J_B/k_{LB} \tag{c}$$

$$J_P = \sigma'_P J_B \tag{d}$$

$$J_B(1 - \sigma'_P) + J_B = J_A \tag{e}$$

$$J_A = k_G \bar{c}_{AG}. \tag{f}$$

Fig. VIII.40. Concentration profiles for consecutive reactions with the controlling resistance in the non-reaction phase

Using relations (b)–(f), relation (a) can be rewritten as follows:

$$\sigma'_P = 1 - \frac{k_X \bar{c}_B}{k_P} \cdot \frac{\dfrac{\bar{c}_P}{\bar{c}_B} + \dfrac{\sigma'_P \beta}{2 - \sigma'_P} \dfrac{k_{LB}}{k_{LP}}}{1 - \dfrac{\beta}{2 - \sigma'_P}},$$

which leads to

$$-\sigma_P'^2 + a\sigma_P' + b = 0, \quad \text{where} \quad \beta = \frac{k_G c_{AG}}{k_{LB} \bar{c}_B},$$

with

$$a = 3 - \beta - \frac{k_X}{k_P}\left(\frac{\bar{c}_P}{\bar{c}_B} - \beta\frac{k_{LP}}{k_{LB}}\right)$$

$$b = \beta - 2 + 2\frac{k_X \bar{c}_P}{k_P \bar{c}_B}.$$

The initial differential selectivity for $\zeta_B = 0$, hence also $\bar{c}_P = 0$, is presented in figure VIII.41 for various values of k_p/k_x. Notice that the mass transfer effects can be disastrous for a good yield. For consecutive reactions, without diffusion limitations we always have $\sigma_P' = 1$ at $\zeta_B = 0$, independent of the value of k_P/k_X. For complete diffusion limitation with both c_{Bi} and c_{Ai} approaching zero, we get $\sigma_P' = 0$, independently of kinetics, even for $\zeta_B = 0$.

Note that the solution presented here is also valid for catalytic reactions at external surfaces if $c_{Ai} = 0$. The theory can be easily extended for reactions at external surfaces where c_{Ai} is not zero.

VIII.7.3 Complex systems

If mass transfer and reaction are in series only, the influence of mass transfer and bulk concentrations on selectivity are always rather easily described. An example

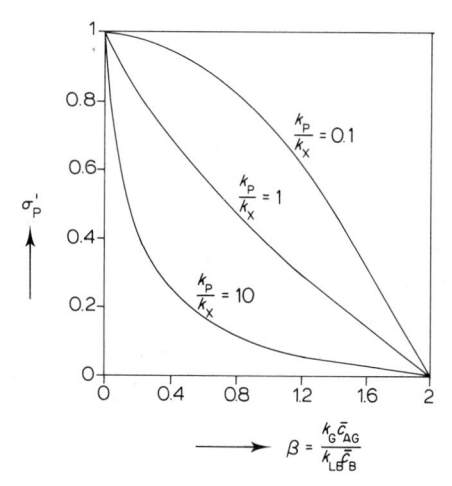

Fig. VIII.41. Initial differential selectivity at $\bar{c}_P = 0$ for mass transfer with reaction for a chlorination reaction where $k_{LP}/k_{LB} = 1$ (Example VIII.72)

is the hydrogenation of triglyceride oils containing di-unsaturated fatty acids at a non-porous catalyst surface [126]. The kinetics are

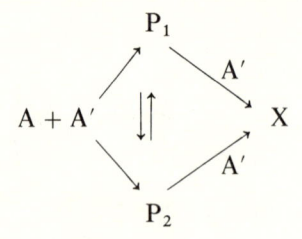

$$A + A' \quad \updownarrow \quad \begin{array}{c} \nearrow P_1 \searrow \\ \\ \searrow P_2 \nearrow \end{array} \begin{array}{c} A' \\ \\ A' \end{array} X$$

with A' is hydrogen, A is di-unsaturated fatty oil, P_1 and P_2 cis- and transmonounsaturated fatty oil respectively and X is saturated fatty oil. The description of mass transfer with consecutive kinetics $A \to P \to X$ can be easily extended to the systems [102]:

$$\begin{array}{c} A \to P \\ \searrow \downarrow \\ X \end{array} \quad \text{and} \quad \begin{array}{c} A \to P \to X_1 \\ \searrow \\ X_2 \end{array}$$

Practical examples are the gas phase air oxidation of ortho-xylene (A) to phthalic acid (P) and CO_2 and H_2O (X) on V_2O_5 catalysts [102], and the catalytic cracking of gas oil (A) to gasoline (P) and to coke and light gases (X) [127].

Based on the film theory, Onda et al. [51] discussed simultaneous absorption of two gases followed by a complex consecutive reaction of type:

$$A(1) \to A(2)$$
$$A'(1) \to A'(2)$$
$$A(2) + B(2) \to P(2)$$
$$A'(2) + P(2) \to B(2)$$

Approximated expressions for the rate of mass transfer based on the van Krevelen–Hoftijzer [114] and Hikita–Asai [115] linearization are available for the penetration and surface renewal theory as well [6].

The hydrogenation of mesityl oxide (A) to methyl isobutylketone (P) and to methyl isobutyl carbinol (X) on copper–chromium oxide catalysts is an example of mass transfer followed by consecutive kinetics, including an equilibrium reaction [128]:

$$A + H_2 \to P$$
$$P + H_2 \rightleftharpoons X$$

More complex reversible mixed parallel-consecutive kinetics have been discussed by Kuo and Huang [129, 130]:

$$A \rightleftharpoons E \rightarrow P \rightarrow \sum X_i$$
$$\searrow \sum X_j$$

and: $\quad A + B \rightleftharpoons E \rightleftharpoons P.$

Mass transfer with Michaelis–Menten kinetics has been studied by Sada et al. [131, 132]:

$$A(1) \rightarrow A(2)$$

$$2A(2) \rightarrow A^*(2) + A(2) \quad -R_A = k_m c_A^m - k_{m'} c_{A*} c_A^{m'-1}$$

$$A^*(2) + B(2) \rightarrow P(2) \quad -R_B = k_n c_{A*} c_B^{n-1}$$

and the overall reaction rate

$$-R_A = \frac{k_m c_A^m}{1 + k_{m'} c_A^{m'-1}/k_n c_B^{n-1}}.$$

Loffler and Schmidt showed that heterogeneous catalytic surfaces having both active and inactive regions may give selectivities different from those predicted by assuming area weighted averages [133]. Numerical examples for the NH_3 oxidation on a Pt gauze and the CO oxidation in the automotive converter indicate that these effects may be important in determining selectivities and conversions in some industrial reactors. Estimates of the interface intraparticle temperature rise for complex reactions have been dealt with by Smith [87].

REFERENCES

1. Wheeler, A.: *Advances in Catalysis*, Vol. III, Academic Press, London, 1951, 249.
2. Dautzenberg, F. M., Van Klinken, J., Pronk, K. M. A., Sie, S. T., and Wijffels, J. B.: in Weekman, V. W. and Luss, D., *Chem. Reaction Eng.* — Houston, A.C.S. Symp. Ser. **65**, 254 (1978).
3. Roberts, G. W.: *Chem. Eng. Sci.* **27**, 1409 (1959)
4. Blackmore, K., Luckett, P. and Thomas, W. J.: *Chem. Eng. Sci.* **30**, 1285 (1975)
5. Onda, K., Sada, E., Kobayashi, T., and Fujine, M.: *Chem. Eng. Sci.* **25**, 1023 (1970)
6. Onda, K., Sada, E., Kobayashi, T., and Fujine, M.: *Chem. Eng. Sci.* **27**, 247 (1972)
7. Beenackers, A. A. C. M. and Van Swaaij, W. P. M.: *Proc. Eur., 6th Intern. Symp., 4th*, 1976, DECHEMA, Frankfurt (M), 1976, VI 260
8. Beenackers, A. A. C. M. and Van Swaaij, W. P. M.: *Chem. Eng. J.* (Lausanne) **15**, 25 (1978)
9. Beenackers, A. A. C. M. and Van Swaaij, W. P. M.: *Chem. Eng. J.* (Lausanne) **15**, 39 (1978)
10. Li, K. Y., Kuo, C. H., and Weeks, J. L. (Jr.): *Can. J. Chem. Eng.* **52**, 569 (1974)
11. Golkeri, S. V. and Luss, D.: *Chem. Eng. Sci.* **26**, 237 (1971)
12. Bailey, J. E. and Horn, F. J. M.: *Chem. Eng. Sci.* **27**, 109 (1972)
13. Østergaard, K.: in Sachtler, W. M. H., Schuit, G. C. A. and Zwietering, P. (eds.), *Third Congress on Catalysis II*, North Holland, Amsterdam, 1965, p. 1348
14. Nelson, W. L.: *Petroleum Refinery Engineering*, McGraw-Hill, New York, 1949

15. Swaim, C. D.: *Hydrocarbon Process.* **49**(3), 127 (1970)
16. Sittig, M.: *Organic Chemical Process Encyclopedia*, Noyes Development Corp., Park Ridge (N.J.), 1967
17. Goldstein, R. F. and Waddams, A. L.: *The Petroleum Chemical Industry*, Spon, London, 1967
18. Kirk, R. E. and Othmer, D. F.: *Encyclopedia of Chemical Technology*, Interscience, New York, 1964
19. Garnell, F. H., Long, R., and Pennell, A.: *J. Appl. Chem.* (London) **8**, 325 (1958)
20. Gioia, F.: *Chim. Ind.* (Milan) **49**, 1287 (1967)
21. Gioia, F. and Marrucci, G.: *Chem. Eng. J.* (Lausanne) **1**, 91 (1970)
22. Astarita, G. and Gioia, F.: *Ind. Eng. Chem., Fundam.* **4**, 317 (1965)
23. Kohl, A. L. and Riesenfeld, F. C.: *Gas Purification*, McGraw-Hill, New York, 1960
24. Danckwerts, P. V. and Sharma, M. M.: *Chem. Eng.* (London) **73**, 244 (1966)
25. Dingman, J. C. and Moore, T. F.: *Hydrocarbon Process.* **47**(7), 138 (1968)
26. Davis, H. S. and Schuler, R.: *J. Am. Chem. Soc.* **52**, 721 (1930)
27. Davis, H. S. and Crandall, S.: *J. Am. Chem. Soc.* **52**, 3769 (1930)
28. Gehlawat, J. K. and Sharma, M. M.: *Chem. Eng. Sci.* **23**, 1173 (1968)
29. Coughanour, D. R. and Krause, F. E.: *Ind. Eng. Chem., Fundam.* **4**, 61 (1965)
30. Gunn, D. J. and Saleem, A.: *Trans. Inst. Chem. Eng.* **48**, T 46 (1970)
31. Manogue, W. H. and Pigford, R. L.: *AIChE J.* **6**, 494 (1960)
32. Van Krevelen, D. W. and Baans, C. M. E.: *J. Phys. Chem.* **54**, 370 (1950)
33. Taylor, M. D.: Paper presented at a Symposium on *Gas Liquid Reactions* by Inst. Chem. Eng. (London), 21 Jan. 1969, at Univ. of Salford
34. Goettler, L. A. and Pigford, R. L.: *Inst. Chem. Eng. Symp. Ser.* **28**, 1 (1968)
35. Lites, I. L., Dilman, V. V., and Tagintsev, B. G.: *Khim. Prom.* **7**, 523 (1968)
36. Chaudhari, R. V., Kulkarni, B. D., Doraiswamy, L. K., and Juvekar, V. A.: *Chem. Eng. Sci.* **30**, 945 (1975)
37. Belg. Patent 815477 (Cl C01 B), 25 November 1974
38. Ramachandran, P. A. and Sharma, M. M.: *Chem. Eng. Sci.* **25**, 1743 (1970)
39. Sadek, S. E., Nawrocky, D. A., Sterbis, E. E., and Vivian, J. E.: *Ind. Eng. Chem., Fundam.*, **16**, 36 (1977)
40. Goettler, L. A. and Pigford, R. L.: *AIChE J.* **17**, 793 (1971)
41. Shklyar, S. L., Aksel'rod, Yu. V., and Sokolinskii, Yu. A.: *Theor. Found. Chem. Eng.* (Engl. Transl.) **10**, 480 (1976)
42. Tamer, A. and Taitel, Y.: *Chem. Eng. Sci.* **29**, 669 (1974)
43. Ramachandran, P. A. and Sharma, M. M., *Trans. Inst. Chem. Eng.* **49**, 253 (1971)
44. Ouwerkerk, C.: *Inst. Chem. Eng., Symp. Ser.* **28**, 39 (1968)
45. Cornelisse, R., Beenackers, A. A. C. M., and Van Swaaij, W. P. M.: *Chem. Eng. Sci.* **32**, 1532 (1977)
46. Sada, E., Ameno, T., and Kondo, M.: *J. Chem. Eng. Jpn.* **9**, 409 (1976)
47. Ouwerkerk, C.: *Hydrocarbon Process.*, **57** (April), 90 (1978).
48. Cornelisse, R., Beenackers, A. A. C. M., van Beckum, F. P. H., and van Swaaij, W. P. M.: *Chem. Eng. Sci.* **35**, 1245 (1980)
49. Onda, K., Takeuchi, H., Kobayashi, T., and Yokota, K.: *J. Chem. Eng. Jpn.* **5**, 27 (1972)
50. Astarita, G.: *Mass Transfer with Chemical Reactions*, Elsevier, Amsterdam, 1967
51. Onda, K., Sada, E., Kobayashi, T., and Fujine, M.: *Chem. Eng. Sci.*, **25**, 761 (1970)
52. Ramachandran, P. A. and Sharma, M. M.: *Chem. Eng. Sci.*, **26**, 1767 (1971)
53. Ramachandran, P. A.: *Chem. Eng. Sci.* **26**, 349 (1971)
54. Astarita, G.: *Chem. Eng. Sci.* **27**, 453 (1972)
55. Hatch, T. F. and Pigford, R. L.: *Ind. Eng. Chem. Fundam.* **1**, 209 (1962)
56. Schmidt, A.: *Chem. Ing. Tech.* **42**, 521 (1970)
57. Balasubramanian, S. N., Rihani, D. N., and Doraiswamy, L. K.: *Ind. Eng. Chem., Fundam.* **5**, 184 (1966)

58. Technical Reports on German Industry: *Chlorinated Hydrocarbons from Acetylene,* F.I.A.T., Final Report No. 843
59. Technical Reports on German Industry, C.I.O.S. XXXII, 31
60. *Chem. Eng. News* **48**(18), 68 (1970)
61. Wieworowski, T. K.: U.S. Patent No. 3 447 903
62. Goar, B. G.: *Hydrocarbon Process.* **47**(9), 248 (1968)
63. Jockers, K., Meier, H., Eberhardt, E., and Taglinger, L.: German Patent No. 1 177 118, Cl.C.01b (1964)
64. Hahn, A. V.: *The Petrochemical Industry-Market and Economics,* McGraw-Hill, New York, 1970
65. Chandalia, S. B.: *Indian J. Technol.* **6**, 88 (1968)
66. Hatch, L. F.: *Hydrocarbon Process.* **49**(3), 101 (1970)
67. Oliver, K. L. and Booth, F. B.: *Hydrocarbon Process.* **49**(4), 112 (1970)
68. Miller, S. A.: *Ethylene and its Industrial Derivitives,* Benn, London, 1969
69. Domask, W. G. and Kobe, K. A.: *Ind. Eng. Chem.* **46**, 680 (1954)
70. Oblad, A. G.: *Ind. Eng. Chem.* **61**(7), 23 (1969)
71. Friend, L., Wender, L., and Yarze, J. C.: *Liquid Phase Oxychlorination of Ethylene to Vinyl Chloride, Adv. Chem. Ser.* **70** (1968)
72. Jordan, D. G.: *Chemical Process Development,* Interscience, New York, 1968
73. Roper, G. H., Hatch, T. F., and Pigford, R. L.: *Ind. Eng. Chem., Fundam.* **1**, 144 (1962)
74. Chaudhari, R. V. and Doraiswamy, L. K.: *Chem. Eng. Sci.* **29**, 675 (1974)
75. Juvekar, V. A.: *Chem. Eng. Sci.* **29**, 1842 (1974)
76. Sada, E., Kumazawa, H., and Butt, M. A.: *Can. J. Chem. Eng.* **54**, 97 (1976)
77. Hikita, H., Asai, S., and Ishikawa, H.: *Ind. Eng. Chem., Fundam.* **16**, 215 (1977)
78. Sada, E., Kumazawa, H., and Butt, M. A.: *Chem. Eng. J.* (Lausanne) **13**, 225 (1977)
79. MHaskar, R. D.: *Chem. Eng. Sci.* **30**, 1441 (1975)
80. Horak, J. and Chuchvalec, J.: *Collect. Czech. Chem. Commun.* **36**, 1740 (1971)
81. Bridgwater, J.: *Chem. Eng. Sci.* **22**, 185 (1967)
82. Carberry, J. J.: *Chem. Eng. Sci.* **17**, 675 (1962)
83. Van de Vusse, J. G.: *Chem. Eng. Sci.* **21**, 654 (1966)
84. Krishna, R.: *Chem. Eng. Sci.* **31**, 399 (1976)
85. Komiyama, H. and Inoue, H.: *J. Chem. Eng. Jpn.* **3**, 117 (1970)
86. Butt, J. B.: *Chem. Eng. Sci.* **21**, 275 (1966)
87. Smith, T. G.: *Chem. Eng. Sci.* **32**, 334 (1977)
88. Weisz, P. B. and Swegler, E. W.: *J. Phys. Chem.* **59**, 823 (1955)
89. Komiyama, H. and Inoue, H.: *J. Chem. Eng. Jpn.* **1**, 142 (1968)
90. Szekely, J. and Bridgwater, J.: *Chem. Eng. Sci.* **22**, 711 (1967)
91. Kuo, C. H.: *AIChE J.* **18**, 644 (1972)
92. Hofmann, H.: *Chem. Eng. Sci.* **14**, 56 (1961)
93. Trambouze, P. J.: *Chem. Eng. Sci.* **14**, 161 (1961)
94. Schoenemann, K. and Hofmann, H.: *Chem. Ing. Tech.* **29**, 665 (1957)
95. Schoenemann, K.: *Chem. Eng. Sci.* **8**, 161 (1957)
96. Hofmann, H.: *Chem. Eng. Sci.* **8**, 113 (1957)
97. Schoenemann, K.: *Chem. Eng. Sci.* **14**, 39 (1961)
98. Hofmann, H.: *Chem. Eng. Sci.* **8**, 113 (1958)
99. Van de Vusse, J. G.: *Chem. Eng. Sci.* **21**, 1239 (1966)
100. Ohki, Y. and Inoue, H.: *Can. J. Chem. Eng.* **47**, 576 (1969)
101. Van den Berg, H.: *Ph.D. Thesis,* Groningen, 1973
102. Wirges, H. P. and Rähse, W.: *Chem. Eng. Sci.* **30**, 647 (1975)
103. Hell, M., Lundqvist, L. E. and Schöön, N. H.: *Acta Polytech. Scand.,* Ch 100 I-V, 59 (1971)
104. Brian, P. L. T. and Beaverstock, M. C.: *Chem. Eng. Sci.* **20**, 47 (1965)
105. Ramachandran, P. A.: *Chem. Eng. Sci.* **27**, 1807 (1972)
106. Hikita, H., Asai, S., and Takatsuka, T.: *Chem. Eng. J.* (Lausanne) **4**, 31 (1972)

107. Hikita, L. and Asai, S.: *Chem. Eng. J.* (Lausanne) **11,** 123 (1976)
108. Hikita, H., Asai, S., Himukashi, Y., and Takatsuka, T.: *Chem. Eng. J.* (Lausanne) **5,** 77 (1973)
109. Hikita, H. and Asai, S.: *Chem. Eng. J.* **11,** 131 (1976)
110. Shah, Y. T., Pangarkar, V. G., and Sharma, M. M.: *Chem. Eng. Sci.* **29,** 1601 (1974)
111. Onda, K., Takeuchi, H., and Fujine, M.: *Chem. Eng. Sci.* **28,** 1637 (1973)
112. Shah, Y. T.: *Chem. Eng. Sci.* **27,** 1171 (1972)
113. Shah, Y. T.: *Chem. Eng. Sci.* **28,** 1639 (1973)
114. Van Krevelen, D. W. and Hoftijzer, P. J.: *Rec. Trav. Chim. Pays-Bas* **67,** 563 (1948)
115. Hikita, H. and Asai, S.: *Int. Chem. Eng.* **4,** 332 (1964)
116. Van de Vusse, J. G.: *Chem. Eng. Sci.* **21,** 631 (1966)
117. Harriott, P.: *Can. J. Chem. Eng.* **48,** 109 (1970)
118. Teramoto, M., Nagayasu, T., Matsui, T., Hashimoto, K., and Nagata, S.: *J. Chem. Eng. Jpn.* **2,** 186 (1969)
119. Teramoto, M., Fujita, S., Kataoka, M., Hashimoto, K., and Nagata, S.: *J. Chem. Eng. Jpn.* **3,** 79 (1970)
120. Inoue, H. and Kobayashi, T.: *Chem. React. Eng. Proc. Eur. Symp.*, 4th, 1968, Pergamon, 1971, p. 147.
121. Pangarkar, V. G. and Sharma, M. M.: *Chem. Eng. Sci.* **29,** 561 (1974)
122. Nakao, K., Hashimoto, K., and Otake, T.: *J. Chem. Eng. Jpn.* **5,** 264 (1972)
123. Beenackers, A. A. C. M.: *ACS Symp. Ser.* **65,** 327 (1978)
124. Teramoto, M., Hashimoto, K., and Nagata, S.: *J. Chem. Eng. Jpn.* **6,** 522 (1973)
125. Hashimoto, K., Teramoto, M., Nagayasu, T., and Nagata, S.: *J. Chem. Eng. Jpn.* **1,** 132 (1968)
126. Hashimoto, K., Teramoto, M., and Nagata, S.: *J. Chem. Eng. Jpn.*, **4,** 150 (1971)
127. Pachovski, R. A. and Wojciechowski, B. W.: *AIChE J.* **19,** 1121 (1973)
128. Hashimoto, K., Teramoto, M., Miyamoto, K., Tada, T., and Nagata, S.: *J. Chem. Eng. Jpn.*, **7,** 116 (1974)
129. Kuo, H. and Huang, C. J.: *Chem. Eng. J.* (Lausanne), **5,** 43 (1973)
130. Kuo, C. H. and Huang, C. J.: *AIChE J.* **16,** 493 (1970)
131. Sada, E. and Kumazawa, H.: *Chem. Eng. Sci.* **28,** 1903 (1973)
132. Sada, E., Kumazawa, H., and Sato, I.: *J. Chem. Eng. Jpn.* **7,** 356 (1974)
133. Loffler, D. G. and Schmidt, L. D.: *AIChE J.* **21,** 786 (1975)

Chapter IX
Heat effects in multi-phase reactors

We now have to synthesize the information provided in all the previous chapters into a realistic model which describes the behaviour of a multiphase reactor sufficiently accurately for design and operation purposes. To all the previous information in this chapter the role of the heat effect will be added; for homogeneous reactors the thermal behaviour has been discussed in Chapter VI. We first will discuss briefly the foregoing chapters so that the reader is aware of all other aspects to be taken into account.

In Chapter VI the conversion rate $\langle R_J \rangle$ was used as a quantity to be employed for purposes of calculation of the role of the heat effect. It expressed the number of mass or molar units of a certain component converted per unit time and per unit volume (or per unit mass of catalyst), and it was supposed to be a known function of the temperature, the pressure and the composition of the reaction mixture. No attention was paid to the question of whether the reaction was proceeding uniformly or whether the conversion rate would turn out to vary locally when inspected on a smaller scale. In fact, the material and energy balances were applied on a scale which was generally small with respect to the reactor dimensions, but still large with respect to the scale of possible local variations.

Such local variations in the nature and composition of the reaction mixture will occur even in systems consisting of one phase only, if mixing on a molecular scale is not complete. Differences in local flow rates in a reactor lead to a residence time distribution for the different volume elements. This was discussed in Chapter IV. The different volume elements can be mixed on a molecular scale or behave as more or less independent agglomerates: this leads to different degrees of micromixing, which under special conditions can profoundly influence the behaviour of a reactor, as was discussed in Chapter V.

Above all, however, marked variations will be encountered in systems made up of more than one phase; these will be called heterogeneous systems.

Contrary to reactions in a single phase mixed on a molecular scale, not all reactants present in heterogeneous systems may be available for chemical reaction. Thus, to give an example, *a homogeneous reaction* between two reactants A and B may have to be carried out in one phase, although A is originally present in a different phase. Apparently, conversion can only take place when A is transferred to the phase in which the reaction can occur. The rate of this mass transfer may therefore influence the conversion rate, especially if the rate of the chemical reaction proper is high with respect to the rate at which A can be supplied.

Another example is a *heterogeneous reaction* at the interface between two phases (e.g., at a catalyst surface). Here the reactants have to be transported by a diffusional process to the interface where the reaction takes place; in this case mass transfer to the reaction surface may considerably influence the overall conversion rate. Hence, chemical reactions in heterogeneous systems are, in principle, combinations of chemical reaction and mass transfer phenomena. The strong interaction between mass transfer and chemical reaction and the formulation of flux and conversion rate expressions has been dealt with in Chapter VII for single and in Chapter VIII for multiple reactions. The choice of the correct model to describe the interactions is of utmost importance. The reader must therefore be fully conversant with the contents of these chapters before he can engage on the synthesis of local heat effects, mixing phenomena and mass transfer and chemical reaction interactions into a realistic overall description of a multiphase reactor.

The influence of temperature on the local conversion rate expressions for multiphase systems has been discussed in Section VII.14; these local variations, however, still have to be translated into the overall reactor performance. As a consequence of the fact that the conversion does not take place in every one of the phases present in a heterogeneous system, also the heat effect accompanying the chemical conversion will be localized in only one (or sometimes two) of the phases or at the interface between two phases. The heat, however, will be transported to the other phases present, and also to the wall surrounding the reaction system. It can be understood that each reactor problem encountered in a heterogeneous system is a special case in itself and that it is hardly possible to give general rules. In view of the great diversity of problems arising in the area of heat effects in multiphase systems we shall limit our discussions to a few types of application frequently encountered in practice, so that all the principal tools needed for handling similar problems have been dealt with.

In Chapters VII and VIII we derived expressions for the molar flux J_A of species A flowing into or to a phase in a heterogeneous system. The total transport across a phase boundary for a given system is equal to the product of the mass or molar flux and the interfacial area A. This quantity per unit reactor volume, the specific surface a, is an extremely important variable in heterogeneous reaction systems and its value is closely related to the degree of dispersion. The value of a can be easily calculated from geometric considerations, if a particle-size analysis is not difficult to obtain. For other situations correlations for a were given in Section VII.17.

There is a relation between the interfacial area A in a system consisting of two phases with a total volume V_r, their volume fractions and the fineness of dispersion. Suppose we have two phases F and G with volumes V_F and V_G. If G is dispersed in F, the porosity of the system is

$$\varepsilon_F = V_F/(V_F + V_G) = V_F/V_r. \tag{IX.1}$$

If confusion is possible we will give ε an index to indicate the phase whose porosity is meant. And as was discussed in Section VII.5.4, we have for the specific surface area per unit volume, a,

$$a = \frac{A}{V_r} = \frac{6(1-\varepsilon)}{d_{av}}. \qquad (IX.2)$$

if ε is the porosity of the continuous phase. This shows that A and a are inversely proportional to the average size of the dispersed particles. If these are bubbles or drops, their size depends not only on the fluid properties of both phases but also to a large extent on the interplay between the mechanisms responsible for their breaking up and mutual coalescence.

The *correct choice* of values for the *properties a and ε* are of utmost importance for the overall performance of a multiphase reactor and will enter both in the mass and the heat balances describing a heterogeneous reaction system. In general mass balances are set up for each relevant species taking part in the reaction and for each phase containing those species. On the other hand, heat balances are set up generally only for the individual phases in the reactor, whereas — if more than one reaction occurs — all the heat effects of each individual reaction have to be taken into account.

Heat will flow across the boundaries of phases. Here the situation is different to mass flow across boundaries. Firstly some boundaries are impermeable for mass flow, such as the reactor wall or the surface of a non-reacting particle. Secondly there is a difference between the distribution laws for mass — such as the Henry or the Nernst laws — and for heat. For heat flow across a boundary, at the interface

$$T_{Fi} = T_{Gi} \qquad (IX.3)$$

will always hold; that is, the temperatures in two phases at their common boundary are equal.

The models used for the description of heat effects in multiphase reactors will always be based on the model reactors discussed in Chapter II. If the reactor exhibits a considerable spread in residence time, this will be accounted for by the axial and radial dispersion model or the cell model. The cell model is equivalent to a cascade of CISTRs with or without back-mixing streams coming from upstream CISTRs. However, the axial dispersion model is the most widely used model.

For a model tank reactor according to the definition of this reactor the concentrations and the temperature in each single phase will be uniform in each phase in the reactor. In plug flow reactors with multiphase systems axial dispersion will be absent for all phases present. Often, however, in multiphase reactors also mixed models have to be used. For example in bubble column gas–liquid reactors sometimes a good approximation of the reactor performance is given by ideal mixing in the liquid phase and by plug flow for the gas flowing through the reactor. But if the gas load is very low and the reactor has a high L/d ratio, the assumption of ideal mixing in the liquid phase is no longer valid, so that a plug flow with axial dispersion model has to be used for the description of the conditions in the liquid phase of the bubble column reactor. Generally it can be stated that besides correct values of ε and a also the *correct choice of a realistic model* determines the practical significance of our reactor calculations.

Before we start to evaluate a reaction in a heterogeneous system and choose a representative model for its description, we first have to make *three checks* in order to find out whether simplifying assumptions, which are often made in practice, are allowable:

— can the reactor be considered as an *isothermally operating* reactor?
— are *local heat effects* on the scale of the dispersion negligible?
— has *axial dispersion* to be taken into account in any of the phases present in the reactor?

If simplifications are allowable the model can become much easier to handle.

The total increase in temperature over the reactor can be deducted from a heat balance over the whole reactor and the heat effect times the total conversion required. Another check is that the Péclet number for heat dispersion is low or $Pe_h < 1$. If the temperature increase is very low the reactor might be treated as an isothermal reactor, provided the local heat effects are also small. For gas–solid systems this can be checked with equations (VII.245) and (VIII.246) and for gas–liquid systems with equation (VII.249), which enable us to determine temperature differences between the bulk and the interface of a phase. If all these temperature differences are small we can regard the reactor as an isothermally operating reactor.

If axial dispersion is so large that complete mixing or so small that plug flow can be assumed, we obtain the simplest models. In practice this is not always the case. A criterion for the neglect of the influence of axial dispersion in a multiphase reactor cannot be derived easily. From the discussion in Section IV.5 it follows that the relevant factor to determine whether axial dispersion can be neglected in homogeneous reactors is $(k\tau)^2/Pe < f$, where f is the allowed fractional deviation of the outlet concentration of the reactant. For multiphase systems more dimensionless groups and these in each separate phase are relevant, e.g. $Pe, Ha, k\tau$ and NTU. Here $NTU = L/HTU$ is the number of transfer units in the reactor, while HTU is the height of a transfer unit.

For isothermal tubular reactors with axial dispersion in both phases many solutions are given for the height of a transfer unit [1–6]. For high values of Pe and an extraction factor $E = 1$ Roes and van Swaaij [7, 8] showed that all the solutions given in the literature for the height of a transfer unit in an isothermal two phase system with axial dispersion and without chemical reaction, reduce to

$$HTU_{ov} = HTU_o + \frac{L}{Pe_F} + \frac{L}{Pe_G},$$

where HTU_{ov} is the overall HTU and HTU_o the same for the system without axial dispersion. Dividing by HTU_o and using the relation $NTU \times HTU = L$, we find

$$\frac{HTU_{ov}}{HTU_o} = 1 + NTU\left(\frac{1}{Pe_F} + \frac{1}{Pe_G}\right).$$

For a certain transfer duty NTU the lengths of columns with and without axial dispersion are related to each other by $L_{ax}/L_o = HTU_{ov}/HTU_o$. Now if we demand that the length of the column is not larger than a fraction f than that of a column with plug flow only, we find as a condition for $E = 1$ and no chemical reaction, that

$$\frac{1}{Pe_F} + \frac{1}{Pe_G} \leq \frac{f}{NTU}, \qquad (IX.4)$$

so the sum of the reciprocal Pe values of the phases flowing through a reactor should be smaller than the fractional increase of the reactor length divided by the required number of transfer units. With equation (IX.4) a first orientation can be obtained of whether axial dispersion has to be taken into account. For a definite decision on the neglect of axial dispersion more information is required.

In this chapter the checks for isothermicity, local heat effects and axial dispersion will be mentioned only if the most suitable model for a multiphase reactor is not known beforehand. We will further limit our discussion of the many possible heterogeneous reactor types to some of the most common types as shown in Table IX.1. As can be seen from this table our main attention will be directed towards the gas–liquid and gas–solid reactors.

IX.1 GAS–LIQUID REACTORS

Gas–liquid reactions are carried out in many different types of apparatus. A survey of the most common types is given in figure IX.1. We will restrict ourselves

Table IX.1 The structure of Chapter IX

Section	Subject	Illustration
IX.1	*Gas–liquid reactors*	
1.1	General	(a) Influence boundary layer temperature on enhancement factor
1.2	Column reactors	(b) Check on isothermicity
		(c) Absorption of H_2S in amine solution
1.3	Bubble column reactors	(d) Reactor with incomplete mixing in liquid phase
1.4	Agitated tank reactors	(e) Oxidation of a hydrocarbon
IX.2	*Gas–solid reactors*	
2.1	Single particle behaviour	(a) NH_3 combustion
		(b) Check for transport resistances
2.2	Catalytic tubular reactors	(c) temperature and concentration profiles
2.3	Moving bed reactors	(d) Reduction of iron ore
2.4	Thermal stability and dynamic behaviour	(e) Decoking of a catalyst
IX.3	*Gas–liquid–solid reactors*	Absorption of H_2S in amine solution

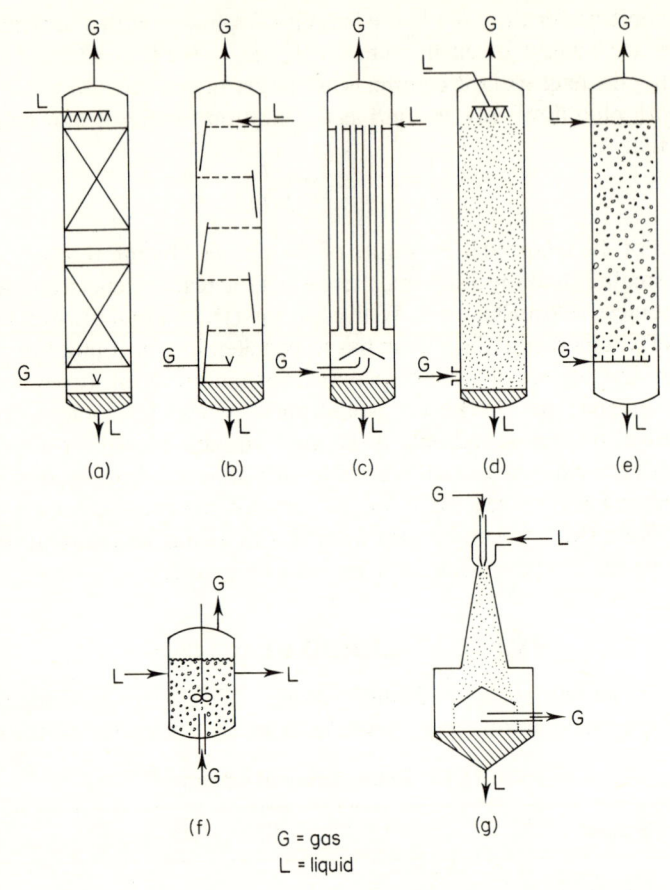

Fig. IX.1 Some common types of gas–liquid reactors: A—packed column; B—tray column; C—falling film reactor; D—spray column; E—bubble column; F—agitated tank reactor; G—venturi reactor

in this section to the discussions of a few examples of these gas–liquid reactors and solve the corresponding simultaneous mass and energy balances. The reader—along the lines developed in this section—can derive the appropriate equations for his individual problem.

A survey of gas–liquid reactors is given by Nagel *et al.* [9]: they give recommendations for the choice of the most appropriate type of reactor. This choice, as has been discussed in Section VII.5, must be based on the required value of the Hinterland coefficient Al ($=\varepsilon Sh/ad$) and the required residence time of the reaction phase as determining factors, but also on yield and selectivity with multiple reactions. Nagel *et al.* [9] correlated the specific interfacial areas obtainable in the various types of gas–liquid reactors with the energy input into the reactor both by the pressure drop of the gas stream and by the agitation. Several good surveys on gas–liquid reactors have been published in the last years, of which we specially want to mention those by Charpentier [10] and van Landeghem [11]. In these surveys also detailed information is given on available correlations for mass and heat transfer coefficients, hold-up of the dispersed phase, interfacial areas and axial dispersion coefficients. A survey on back-mixing in gas–liquid reactors was given by Shah *et al.* [12]. Also two good books exclusively dedicated to gas–liquid reactors by Astarita [13] and by Danckwerts [14] have been published. For a discussion of existing correlations we refer to Section VII.17.

Often also a solid catalyst is used in gas–liquid reactions. The catalyst can either be suspended as free particles in the liquid or packed in a column of rather large particles over which the gas and the liquid flows. In the first case agitated vessels or sparged bubble columns are used, in the second case counter- or cocurrent trickle flow columns. The addition of a solid catalyst to gas–liquid reactions does not present new fundamental problems — the required theory has been treated in Chapters VII and VIII — only the mass transfer and reaction rate equations become more complex. The most difficult problem still remains the finding of reliable correlations for the relevant physical and chemical data of the system under study. See further Section IX.3.

IX.1.1 General

When a gas–liquid reaction is accompanied by a heat effect or is executed over a range of temperatures the change of the properties of the system with temperature also has to be taken into account. Solubilities of gases in liquids as a function of the temperature are reported extensively in the literature. Difficulties arise when the gas reacts with the liquid—such as halogenations or sulphonations of hydrocarbons—because in this case it is not easy to separate chemical and physical phenomena.

The same holds true for the diffusivity in liquids. The temperature $\bar{T}$ in the bulk of the liquid can be calculated with the normal mass and energy balances. With chemically enhanced reactions often steep concentration profiles are present in the boundary layer; therefore temperature profiles can also be expected. This means that the interfacial temperature T_i is higher than $\bar{T}$ for exothermic reactions. A higher temperature increases the reaction rate constant but usually decreases the solubility of the gas; the combined effect cannot be predicted offhand. For physical absorption with a heat of solution, ΔH_a, Danckwerts [15, 16] solved this problem with the penetration theory. The extent of the heat penetration is much greater than the depth of mass penetration, because the thermal diffusivity is much greater than the mass diffusivity. The ratio of the depths of penetration is given by $\sqrt{(\lambda/\rho c_p D)}$; as for many systems $\lambda/\rho c_p \simeq 100D$, the temperature profile for physical absorption extends approximately $10\times$ deeper into the liquid than the concentration profile. It is therefore safe to assume that the absorption takes place at T_i, the interfacial temperature.

Chiang and Toor [17] studied the absorption of NH_3 in water and determined for this gas, which is highly soluble and has a high heat of solution, interfacial temperatures of 15°C above the bulk temperature. Verma and Delancey [47] reported for physical absorption in stagnant liquids interfacial temperature rises of 1.3 K for low and 7–18 K for high absorption heats. Mann et al. [18, 19] studied temperature profiles in gas–liquid reactions using the film theory. Considerable temperature gradients in the boundary layer have been reported in the literature for highly exothermic reactions by, among others, Mann et al. [18, 19] and Beenackers and van Swaaij [48].

The possibility of a temperature profile over the liquid film was discussed in Section VII.14.2., where the following relation for the temperature difference over the film was derived:

$$T_i - \bar{T} = \frac{(-\Delta H_r)_A + (-\Delta H_a)_A}{\rho c_p Le^{0.67}} \frac{m\bar{c}_{AG} E_A}{1 + \dfrac{mk_L E_A}{k_G}}. \tag{VII.249}$$

In this equation the values of m and E_A are still unknown, because they have to be taken at the interface temperature. We now assume that $mk_L E_A \ll k_G$ and introduce the adiabatic temperature rise in the liquid at bulk temperature conditions

$$\Delta T_{adL} = [(-\Delta H_r)_A + (-\Delta H_a)_A] m\bar{c}_{AG}/\rho c_p.$$

Equation (VII.249) can now be written:

$$T_i - \bar{T} = \Delta T_{adL} \frac{E_{A \text{ at } T}}{Le^{0.67}} \frac{(mE_A)_{\text{at } T_i}}{(mE_A)_{\text{at } \bar{T}}} \tag{IX.5}$$

Some data on the value of the parameters are given in Table IX.2. For small values of $T_i - \bar{T}$ we can write

$$(mE_A)_{\text{at } T_i} \simeq (mE_A)_{\text{at } \bar{T}} + (T_i - \bar{T})\left(m\frac{dE_A}{dT} + E_A\frac{dm}{dT}\right)_{\text{at } \bar{T}}$$

which now leads, for equation (IX.5), to

$$T_i - \bar{T} \simeq \Delta T_{adL} \frac{E_A}{Le^{0.67}} \frac{1}{1 - \dfrac{\Delta T_{adL} E_A}{Le^{0.67}}\left(\dfrac{1}{E_A}\dfrac{dE_A}{dT} + \dfrac{1}{m}\dfrac{dm}{dT}\right)}. \tag{IX.6}$$

Table IX.2 Some parameter values for reactions in aqueous systems (pure reactant in the gas phase under atmospheric pressure, $T = 25°C$)

Reactant	$Le_L^{2/3}$	He*	m	c_{AL} (kmol/m³)
NH₃	21.6	0.830	1.20	49.4 × 10⁻³
CO₂	20.4	1.09	0.915	37.5 × 10⁻³
H₂S	22.7	36.1	27.7 × 10⁻³	1.14 × 10⁻³
O₂	15.8	30.0	33.4 × 10⁻³	1.37 × 10⁻³
H₂	11.4	51.1	19.6 × 10⁻³	0.80 × 10⁻³

According to the rule of van 't Hoff we can write for the dimensionless Henry coefficient $He^* = m^{-1} = A \exp(-B/T)$; this gives $dm/m\,dT = -B/\bar{T}^2$. For the enhancement factor E_A the temperature derivative depends on the regime of interaction between mass transfer and chemical kinetics. In case $E_A = Ha$ we have $dE_A/E_A\,dT = E/2R\bar{T}^2$, in which E is the activation energy for the chemical reaction and other properties in Ha are taken as temperature independent. For the diffusion controlled regime E_A is approximately independent of temperature. Substitution into equation (IX.6) leads to

$$T_i - \bar{T} \simeq \Delta T_{adL} \frac{E_A}{Le^{0.67}} \frac{1}{1 + \dfrac{\Delta T_{adL} E_A}{Le^{0.67}}\left(\dfrac{B}{\bar{T}^2} - \dfrac{1}{E_A}\dfrac{dE_A}{dT}\right)} \qquad \text{(IX.7)}$$

Equation (IX.7) shows that the temperature drop over the liquid film is decreased due to the decrease in solubility ($B/\bar{T}^2 > 0$) and increased due to the increase of the reaction rate ($dE_A/dT \geq 0$) at the gas–liquid interface.

Some parameter values for gases reacting in aqueous liquids are given in Table IX.2. Equation (IX.7) can be used to estimate, whether temperature differences over the film have to be taken into account. This will be done in the next Example.

Example IX.1.a *Temperature difference over the liquid film, chlorination of toluene*

Mann and Moyes [18] give data on the reaction of pure chlorine at atmospheric pressure and liquid toluene. $\Delta H_r = -126$ MJ/kmol Cl_2 and $\Delta H_a = -23$ MJ/kmol Cl_2. What will be the temperature difference over the liquid film, if $E_A = 66$ at 25°C?

From data in the literature we find $Le = 120$. From the data of Mann and Moyes we can derive $m = 9.4 \times 10^{-3} \exp(2630/T)$. At 25°C we have $m = 64$. Substituting further literature data gives

$$\Delta T_{adL} = \frac{(126 + 23) \times 10^6 \times 64 \times 10^5}{868 \times 1720 \times 8315 \times 298} = 258 \text{ K}$$

$E_A/Le^{0.67} = 66/24.3 = 2.71$. The value of the activation energy is not known. We now assume that $E/R = 5000$ K and that $E_A = Ha$. We then find the denominator in the correction term in equation (IX.7):

$$1 + 258 \times 2.71\left(2630 - \frac{5000}{2}\right) \times \frac{1}{298^2} = 2.02.$$

This leads to a temperature difference over the film of $T_i - \bar{T} = 345$ K. It is clear that the approximations used to derive equation (IX.7) are no longer valid for such large values of $T_i - \bar{T}$. We conclude that in this case even all the models in Chapter VII, which were derived for constant values of the system properties, have to be revised for non-isothermal conditions over the boundary layer. We refer to Section VII.14.2. Mann and Moyes [18] demonstrated that for very exothermic and rapid reactions under extreme conditions E_A can become lower than 1. If we had assumed $E/R = 10\,000$ K, we would have found a negative value for the denominator of the correction term in equation (IX.7). So either the assumption $E_A = Ha$ or the experiments which led to $E_A = 66$ or both are incorrect!

IX.1.2 Column Reactors

Column reactors can be of the packed, tray or spray type (see figure IX.1). A survey of design procedures was published by Juvekar and Sharma [20] (among others). For packed and spray columns as gas–liquid reactors, usually the following assumptions are made:

— constant pressure in the column;
— constant physical properties of the gas and the liquid;
— constant temperature over a cross-section
  and often as additional assumptions;
— plug flow both in the gas and in the liquid phase;
— isothermal operation;
— negligible gas-side resistance or $k_G = \infty$; the admissibility has to be checked very carefully!

The assumption of plug flow must be carefully checked: for spray columns in general it will not be allowed. The gas-side resistance can be accounted for as demonstrated in Chapter VII; especially for very soluble gases or high enhancement factors, this is necessary. Whether isothermal operation can be assumed follows from an overall energy balance, which states that the reaction and absorption heats evolved must be carried away by the gas and the liquid streams. If only one single reaction takes place in the liquid phase, we have for the overall heat balance of the entire reactor:

$$[(-\Delta H_r) + (-\Delta H_a)]_B \Phi_{vL} c_{Bo} \zeta_B = (\rho c_p \Phi_v)_L (T_{L\,out} - T_{L\,in})$$
$$+ (\rho c_p \Phi_v)_G (T_{G\,out} - T_{G\,in}) \quad \text{(IX.8)}$$

if a reactant A in the gas phase is absorbed in the liquid phase and if all A absorbed has reacted with species B. It is clear that if other heat effects like heat of mixing or dilution occur, these also have to be included in the first term of equation (IX.8). We now introduce the adiabatic temperature rise of the liquid phase, if all reaction heat were taken up by the liquid, ΔT_{adL} and the ratio β of the heat carrying capacity of the gas phase to that of the liquid phase $\beta = (\rho c_p \Phi_v)_G / (\rho c_p \Phi_v)_L$. Then we have, for the equation above,

$$\Delta T_{adL} \zeta_B = (T_{L\,out} - T_{L\,in}) + \beta(T_{G\,out} - T_{G\,in}) \quad \text{(IX.9)}$$

For the situation given in figure IX.2 we can write $T_{top} = T_{G\,out}$ and $T_{bottom} = T_{L\,out}$. For these two unknowns we need a second relation besides IX.9 to solve them. If in figure IX.2 at the top of the reactor the reaction has virtually stopped due to completion, which means that species A has been virtually removed from the gas

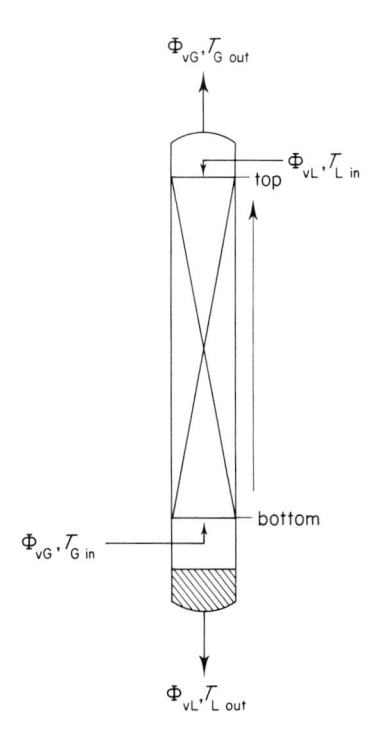

Fig. IX.2 Column gas–liquid reactor with countercurrent operation

stream, we can safely assume for the top cross-section or for the top tray that $T_{top} = T_{L\,in}$. If this is not the case (see Example IX.1.c), a trial and error procedure has to be followed.

For *cocurrent* flow the mixing rule provides the reactor entrance temperature:

$$T_{L\,in} + \beta T_{G\,in} = (1 + \beta)T_{in} \qquad (IX.10)$$

The reactor outlet temperature can be found by equation (IX.8). In general there is a temperature jump at the plane or tray where the reaction starts. These relations enable us to check for isothermicity in the following Example.

Example IX.1.b *Check for isothermicity of a column reactor*
In Example VII.7.4.c the absorption of CO_2 in an NaOH solution in a packed column has been discussed. The inlet temperatures of the gas and the liquid were both 20°C and the conversion of B (NaOH) was $\zeta_B = 0.75$. As one mole of CO_2 reacts with two moles of NaOH in this case $(-\Delta H_r)_B = \frac{1}{2} \times 90\,\text{MJ/kmol B}$. For the gas we find

$$(\rho c_p \Phi_v)_G = 1.31 \times 997 \times 0.579 = 756\,\text{J/K s}$$

and for the liquid:

$$(\rho c_p \Phi_v)_L = 1000 \times 4190 \times 0.694 \times 10^{-3} = 2907 \text{ J/K s}$$

so that $\beta = 0.26$. Further, in this case, $T_{G\,out} = T_{L\,in} = 20°C$, because virtually no reaction takes place in the plane, where the liquid enters into the column. So

$$\Delta T_{adL} = \frac{45 \times 10^6 \times 1 \times 0.75}{1000 \times 4190} = 8°C$$

and $T_{G\,out} = T_{L\,in} = T_{top} = 20°C$ and the outlet temperature of the liquid is $8 + 20 = 28°C$. So the average temperature over the column is 24°C.

A further study on the influence of temperature on the solubility and the reaction rate constant has to reveal whether the assumption of isothermicity was allowed in Example VII.7.4.c. At least the values for the chemical and physical data could have been better taken at the average column temperature of 24°C. If the gas inlet temperature had been $T_{G\,in} = 50°C$, the run-off temperature of the liquid would have been 35.8°C. In all cases we observe a sudden temperature jump for the gas phase at the bottom of the column.

We still have to check for temperature gradients over the boundary layer in the liquid phase. From data in the literature we can derive $He^* = m^{-1} = 2.6 \times 10^{-3} \exp(-2490/T)$. A determination of the values of the dimensionless group $mk_L E_A/k_G$ reveals that it cannot be neglected in comparison to 1; however, in Example VII.7.4.c the pertinent data for p_{Ai} already have been calculated. This enables us to calculate $\Delta T_{adL} = (-\Delta H_r)_A m p_{Ai}/RT(\rho c_p)_L$. From the further data of the same Example we can derive that $\Delta T_{adL} E_A/Le^{0.67}$ is very small. Also $B/\bar{T}^2$ under our condition is much smaller than one. We can also be sure that the factor $dE/E\,dT$ will have no influence on our problem, although data on the temperature dependence of the reaction rate were not given. So from equation (IX.7) it follows that the maximum values of $T_i - \bar{T} = \Delta T_{adL} E_A/Le^{0.67}$. The results are given below:

	T (°C)	$mk_L E_A/k_G$ (—)	E_A (—)	p_{Ai}/RT (kmol/m³)	ΔT_{adL} (K)	$T_i - \bar{T}$ (K)
Bottom	28	0.19	4.3	5.7×10^{-3}	0.228	0.048
Top	20	0.37	8.5	28×10^{-6}	0.0011	0.0005

We see that we can neglect the temperature gradient over the film in the liquid phase.

(Example IX.1.b ended)

Furthermore, plug flow in the gas and the liquid phase cannot always be assumed. The energy balance and the material balance for a small reactor volume element $S\,dz$ and for species J is, if J is transferred from the gas into the liquid phase, for a column reactor, as follows:

for the energy

$$(-\Delta H_2)J_J a \mp [(\varepsilon\lambda_1)_L \mp (\varepsilon\lambda_1)_G]\frac{d^2 T}{dz^2} \pm [(\rho c_p \Phi_v)_L$$

$$\mp (\rho c_p \Phi_v)_G]\frac{1}{5}\frac{dT}{dz} = 0 \quad \text{(IX.11.a)}$$

for J in the gas phase

$$(\varepsilon D_1)_G \frac{d^2 c_{JG}}{dz^2} - \frac{\Phi_{vG}}{S}\frac{dc_{JG}}{dz} - J_J a = 0; \quad \text{(IX.11.b)}$$

and for J in the liquid phase

$$(\varepsilon D_1)_2 \frac{d^2 c_{JL}}{dz^2} - \frac{\Phi_{vL}}{S}\frac{dc_{JL}}{dz} \mp J_J a = 0 \quad \text{(IX.11.c)}$$

In these equations the upper sign in $\pm$ refers to countercurrent and the lower sign to cocurrent operation, if the gas stream flows in the direction of the coordinate z. Similar material balances to (IX.11(b)) and (c) have to be written down for the component in the liquid phase with which species J reacts.

For the energy balance we should have written down two balances each for the gas and for the liquid phase, which are interconnected by a term describing the heat transfer over the interfacial area. This makes the solution of all the simultaneous differential equations very complicated, so that usually a temperature constant and equal in both phases is assumed over a cross-section of the reactor. This is sufficiently accurate in most cases except under very exotic circumstances. The expression $(-\Delta H_r)_J J_J a$ still has to be worked out, because it depends on the region where the heat is liberated or absorbed: in the boundary layer, in the bulk or in both regions of the liquid. Moreover if more than one independent reaction takes place, more than one heat effect has to be taken into account. This will be elaborated in more detail in later Examples.

Generally the gas–liquid reaction column is designed to contact the reaction products only under the desired conditions: the separation of the reactor products from the liquid phase is then undertaken in a separate separation unit.

There are cases where both operations — reaction and separation — are combined in order not only to save on investment and operation costs, but especially to influence chemical or physical equilibria. Such a special case is the distillation column as a reactor, where the reaction takes place in a boiling liquid and where we have a very intensive interchange between reaction and mass transfer. Reaction distillation columns are especially advantageous when (1) we can avoid an azeotropic point by applying a special solvent; (2) the position of the chemical equilibrium is moved into the desired direction by taking reaction products out of the reaction mixture; or (3) the total residence time of the liquid has to be reduced in case of heat sensitive materials. Kaibel et al. [21] describe calculation methods for reaction distillation columns, as do Nelson [22] and Geelen and Wijffels [23] among others. This type of reactor above all is used for equilibrium esterifications, saponifications, transesterifications and polycondensations, where the product H_2O is distilled off. In order to increase the liquid residence times on the trays in this reactor type the overflow weirs are sometimes higher than is usual in normal distillation practice.

In the following Example we will discuss a tray column as a gas–liquid reactor for multiple reactions in the liquid phase. This Example was taken from Blauwhof et al. [44].

Example IX.1.c *Chemical absorption of H_2S from a natural gas stream*

H_2S is removed in a tray column reactor with an aqueous solution containing methyl diethylamine (MDEA) from a natural gas stream of $\Phi_{vGin} = 55.56 \text{ N m}^3/\text{s}$. The gas contains 7 vol. %H_2S and 7 vol. %CO_2. The flow scheme is given in figure IX.3. The reactor pressure is 7 MPa. The amine solution contains 2.0, 3.0 or 3.2 kmol MDEA/m^3 and is introduced into the reactor at 40°C; the gas stream enters at 20°C. The wash liquor feed coming from the regenerator still contains 0.01 kmol H_2S/m^3 and 0.01 kmol CO_2/m^3. Both the H_2S and the CO_2 in the natural gas are absorbed into the liquid and react chemically with the MDEA. The liquid flows countercurrent to the gas and will take up H_2S and CO_2 simultaneously up to 70% of the maximum possible amount. The gas emitted has to contain a maximum of 4×10^{-4} vol. %H_2S. We will calculate the temperature and concentration profiles over the column.

The very complicated reaction scheme is given in figure IX.4, whereas literature references on all the required physical, chemical and kinetic data together with their temperature dependence are given in the study of Blauwhof et al. [44]. The gas phase was described by the Soave modification of the Redlich–Kwong equation of state. Gas–liquid interfacial areas were derived with the correlation of Nonhebel [45] and values for $k_L a$ and $k_G a$ with the correlations of Sharma and Gupta [46].

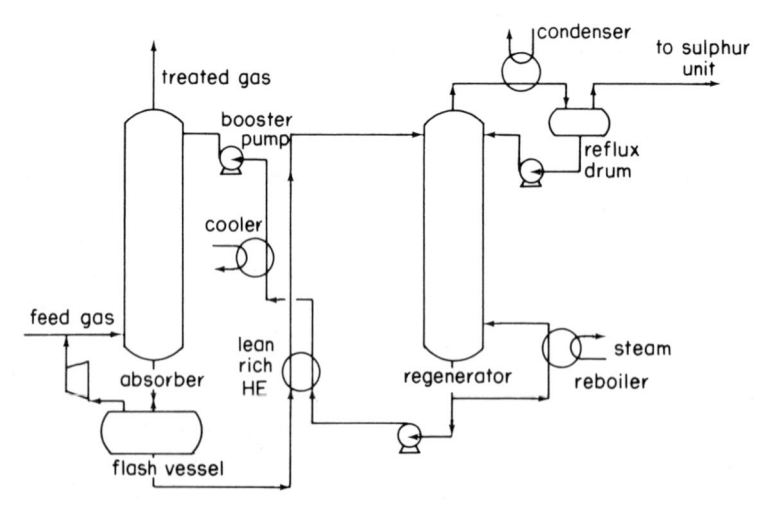

Fig. IX.3 Flow scheme of a natural gas treating plant; Example IX.1.c

585

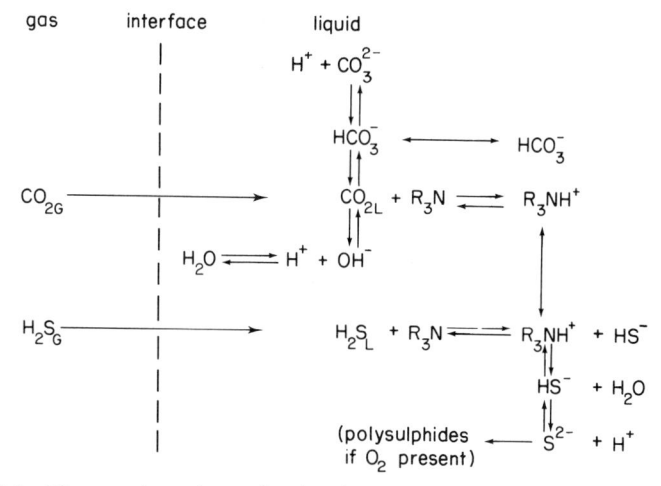

```
gas         interface       liquid
                            H⁺ + CO₃²⁻
                               ↕
                            HCO₃⁻  ⇌  HCO₃⁻
                               ↕
CO₂G ─────────────────────→ CO₂L + R₃N ⇌ R₃NH⁺
                               ↕
                        H₂O ⇌ H⁺ + OH⁻

H₂S_G ────────────────────→ H₂S_L + R₃N ⇌ R₃NH⁺ + HS⁻
                                            ↕
                                          HS⁻ + H₂O
                                            ↕
                          (polysulphides ← S²⁻ + H⁺
                           if O₂ present)
```

Fig. IX.4 The reaction scheme for the absorption of H_2S and CO_2 in MDEA

The tray absorbers are calculated by a tray-to-tray procedure. The absorption column is considered as a series of ideally mixed reactors, each corresponding to an *actual* tray (see figure IX.5). Process conditions and gas and liquid compositions are assumed to be uniform in each reactor. More sophisticated tray models, e.g. with plug flow in the gas phase, are not considered. Gas and liquid phase back-mixing between trays as well as pressure drop over the trays is neglected. The calculation starts with an overall mass balance over the column. For this purpose a CO_2 concentration in the treated gas, $y_{CO_2}^{n+1}$, has to be estimated and the treated gas flow, Φ_{vG}^{n+1}, can be calculated from

$$\Phi_{vG}^{n+1} = \frac{\Phi_{vG}^{1}(1 - y_{H_2S}^{1} - y_{CO_2}^{1})}{1 - y_{H_2S}^{n+1} - y_{CO_2}^{n+1}} \text{ (m}^3/\text{s)} \tag{a}$$

whereas the H_2S concentration in the treated gas is always taken as 4 p.p.m. vol. The superscript numbers refer to the tray numbers. The composition and the flow of the treated gas is fixed now. Next, the liquid volumetric flow rate in the column is calculated in such a way that the rich solution leaving the column at tray 1 is loaded with acid gases to a preset desired value α_{tot}:

$$\alpha_{tot} = \alpha_{H_2S} + \alpha_{CO_2} = \frac{c_{H_2S\,tot\,L}^{1} + c_{CO_2\,tot\,L}^{1}}{c_{Am\,tot\,L}} \tag{b}$$

in which the index 'tot' refers to the total concentration of the respective components in both reacted and unreacted form. The liquid flow is obtained from a total acid balance:

$$\Phi_{vL} = \frac{\Phi_{vG}^{n+1}[c_{H_2SG}^{n+1} + c_{CO_2G}^{n+1}] - \Phi_{vG}^{1}[c_{H_2SG}^{1} + c_{CO_2G}^{1}]}{\alpha_{tot} c_{Am\,tot} - c_{H_2SL}^{n+1} - c_{CO_2L}^{n+1}} \text{ m}^3/\text{s,} \tag{c}$$

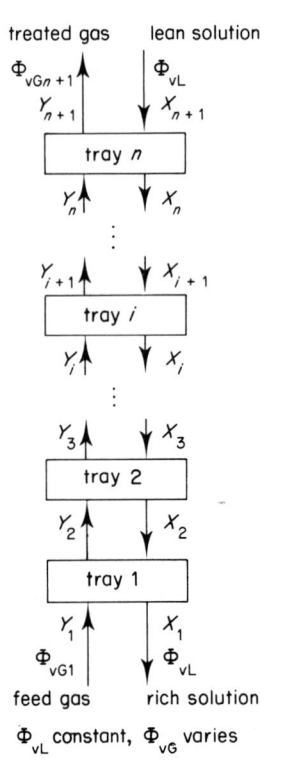

Fig. IX.5 The calculation scheme for the tray column reactor of Example IX.1.c

where all gas phase concentrations are expressed in moles/m^3. We assumed Φ_{vL} to be constant over the column. From the overall mass balances for the individual components the total acid gas concentrations in the rich solution are calculated:

$$c^1_{H_2S\,tot\,L} = \frac{\Phi^1_{vG} c^1_{H_2SG} - \Phi^{n+1}_G c^{n+1}_{H_2SG}}{\Phi_{vL}} + c^{n+1}_{H_2S\,tot\,L} \tag{d}$$

and a similar expression for the CO$_2$ concentration.

The heat carrying capacities of the gas and liquid flows are of the same order of magnitude and, therefore, a heat balance has to be incorporated in the model. The overall heat balance over the absorber is solved assuming:

$$T_G^{n+1} = T_L^{n+1}, \tag{e}$$

which means that the temperature of the top tray is that of the lean amine solution introduced. This enables us to calculate the temperature of the rich solution, provided that gas and liquid inlet temperatures are specified, using equation (IX.8):

$$T_L^1 = \frac{(\Phi_{vG}^1 T_G^1 - \Phi_{vG}^{n+1} T_G^{n+1})(\rho c_p)_G}{(\rho c_p \Phi_v)_L}$$
$$+ \frac{(-\Delta H_r)_{H_2S}[c_{H_2S\,tot\,L}^1 - c_{H_2S\,tot\,L}^{n+1}]}{(\rho c_p)_L}$$
$$+ \frac{(-\Delta H_r)_{CO_2}[c_{CO_2\,tot\,L}^1 - c_{CO_2\,tot\,L}^{n+1}]}{(\rho c_p)_L} + T_L^{n+1}. \tag{f}$$

The gas and liquid compositions and conditions at the absorber bottom are now fixed and provide the starting point for tray-to-tray calculations. In order to solve the tray heat balance, thermal equilibrium is assumed between the gas and liquid flows leaving the tray:

$$T_G^{i+1} = T_L^i. \tag{g}$$

At the absorber top, this equation is contradictory to the assumption for the overall heat balance expressed in equation (e). This discrepancy is, however, negligible because $T_L^{n+1} \simeq T_L^n$, as will be demonstrated later. Now, as the gas temperature on the bottom tray 1 is known from equation (g), the liquid phase equilibrium constants, Henry's coefficients, and gas and liquid phase diffusivities are calculated. The liquid phase composition, more specifically the concentrations of unreacted H_2S, CO_2 and amine, is obtained by means of the equilibrium model described by Blauwhof et al. [44].

Next, H_2S and CO_2 gas phase concentrations at tray 1, that is in the gas leaving tray 1, are estimated. Subsequently gas phase fugacities are calculated by means of the Soave–Redlich–Kwong equation of state. The absorption driving forces are now known. The mole fluxes J_{H_2S} and J_{CO_2} are calculated. After the first estimate of the gas phase concentrations and the subsequent mole flux calculations, iteration proceeds using a Newton–Raphson technique, until the following implicit flux balances are simultaneously satisfied:

$$\Phi_{vG}^2 c_{H_2SG}^2 - \Phi_{vG}^1 c_{H_2SG}^1 = J_{H_2S} a_{tray}$$

and

$$\Phi_{vG}^2 c_{CO_2G}^2 - \Phi_{vG}^1 c_{CO_2G} = J_{CO_2} a_{tray}$$

Using mass and heat balances for tray 1 now $c_{H_2SL}^2$, $c_{CO_2L}^2$ and T_L^2 can be calculated. The tray-to-tray procedure continues until the H_2S specification is met (4 p.p.m.v). If the CO_2 concentration in the treated gas deviates more than 5% from the initially estimated value, the column calculation is repeated right from the start using the calculated CO_2 concentration as a new estimate.

The results of our calculations are shown in figures IX.6–IX.8. With increasing amine concentration the solvent flow and consequently its capacity for heat absorption are reduced, so that the temperature profiles over the column become more pronounced. This is clearly demonstrated in figure IX.6 for three MDEA solutions. With increasing MDEA concentration the absorber temperature level rises and the maximum shifts to higher trays. This change of temperature profiles

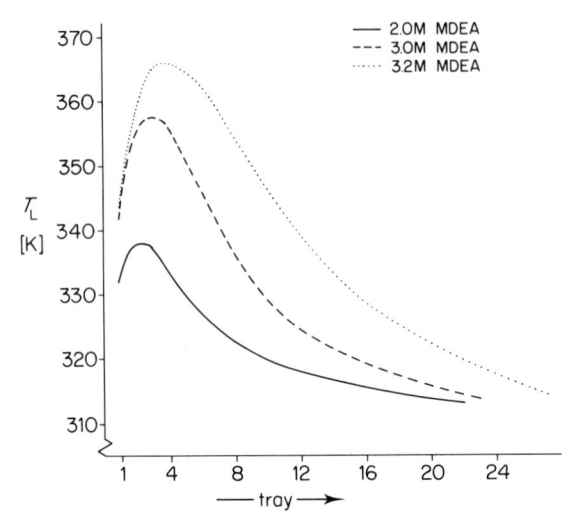

Fig. IX.6 Temperature profiles in the tray column reactor of Example IX.1.c

affects the liquid phase equilibria and thus the absorption driving forces. In figure IX.7 gas phase and unreacted liquid phase concentration c_{H_2SG} and c_{H_2SL}/m_{H_2S} are plotted as a function of the tray number for the three MDEA concentrations. The difference between the gas and liquid phase profiles represents the H$_2$S absorption driving force. Note the logarithmic scale for the concentration. With increasing MDEA concentration c_{H_2SL}/m_{H_2S} increases and the absorption driving forces particularly in the absorber bottom are reduced. Consequently the H$_2$S mole fluxes are reduced and additional trays are required to reach the H$_2$S specification.

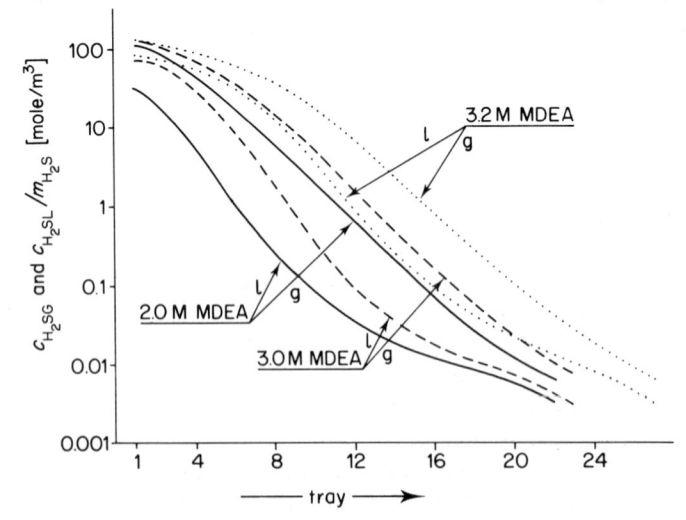

Fig. IX.7 H$_2$S concentration profiles in the tray column reactor of Example IX.1.c

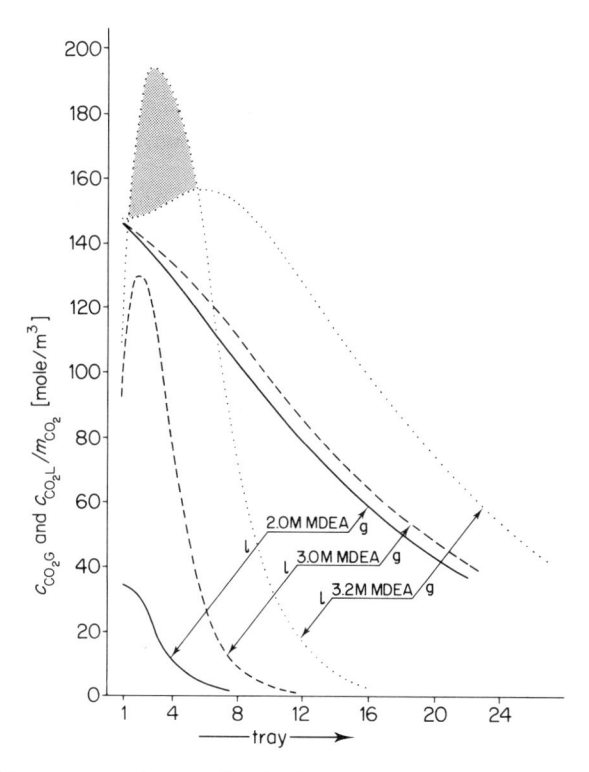

Fig. IX.8 CO_2 concentration profiles in the tray column reactor of Example IX.1.c

The CO_2 concentration profiles show a more significant sensitivity towards the temperature or MDEA concentration (see figure IX.8). As a result the CO_2 liquid concentration rapidly increases with the MDEA molarity in the lower part of the absorber. For the 3.2M MDEA solution this concentration even *exceeds* the gas phase concentration at trays 2–4, resulting in a desorption of CO_2 and an increasing CO_2 gas concentration. With an increase of the amine concentration at a fixed acid gas loading α_{tot}^l, unreacted H_2S and CO_2 liquid phase concentrations rise rapidly and reduce in particular the absorption driving forces at the absorber bottom (see figures IX.7 and 8).

For the conditions specified and a 2.0M MDEA solution the column has a diameter of 2.95 m and a height of 19 m. It contains 22 trays and a total interfacial area of 6120 m². The wash liquor flow is 0.225 m³/s and the CO_2 concentration in the treated gas 1.68 vol %. In the publication of Blauwhof et al. [44] also results for diisopropylamine and for the regeneration — operating at higher temperatures and at 0.15–0.35 MPa — are given. For the stripper–regenerator calculations the same computer program can be used, if the mixture also contains water, which strips the H_2S and CO_2 out of the solution. At regenerator conditions the reaction reverses and H_2S and CO_2 desorb from the liquid phase.

IX.1.3 Bubble column reactors

Bubble column gas–liquid reactors are cylindrical vessels filled with a liquid through which the gas bubbles. The gas inlet system is designed in such a way that small bubbles are formed and the gas is finely dispersed. The liquid flow can be either upward (cocurrent) or downward (countercurrent). The bubble column reactor is often preferred despite its larger volume because its energy consumption is much lower than in the agitated gas–liquid contactor.

In designing a bubble column reactor usually the following assumptions are made:

— the gas flows upward and in plug flow;
— the bulk of the liquid is ideally mixed;
— the pressure is either taken to be constant or to vary linearly with the height L of the reaction mixture as

$$p = p_{\text{top}} + \frac{dp}{dz}(L - z),$$

where z is the coordinate in the direction of the gas flow and L the total liquid height above the gas inlet;
— the mass transfer coefficients k_G and k_L and the specific interfacial area are constant over the whole reaction volume;
— there is no heat transfer resistance between the gas and the liquid.

The assumption of an ideally mixed liquid phase has to be used with care. A rule of thumb is that one gas bubble entrains three times its own volume of liquid: this enables us to estimate whether the liquid is sufficiently mixed to approach ideal mixing. In case of low gas to liquid flow ratios and/or viscous liquids the liquid phase is no longer ideally mixed, so that we should apply an axial dispersion model for the liquid phase: otherwise we would seriously overestimate the required reactor volume. This will be demonstrated in Example IX.1.c.

We will derive now in a dimensionless form the basic equations for the energy and the mass balance for a first order gas–liquid reaction in a bubble column reactor. The reaction is so fast that the concentration of A in the bulk of the liquid is zero ($\phi > 2$). See further Section VII.5.

Component A is transferred from the gas phase into the liquid and reacts in the liquid phase with a chemical rate, described by $R_A = -kc_A$. The material balance for the gas phase is, if $k_G = \infty$ and $Al \gg 1$:

$$\frac{d[\Phi_{vG} p_A/RT]}{d[z/L]} = -J_A a V_r = -k_L E_A c_{Ai} a V_r, \qquad (IX.12)$$

where at $z = 0$, $p_A/RT = p_{Ao}/RT_o$.

Further T is the reactor temperature, L the height of the reaction mixture and T_o the inlet temperature of the gas. For fast first order reactions $E_A = \sqrt{(kD)}/k_L$. We further postulate that the gas pressure is constant over the reactor and that there are no volatile components in the liquid, which can vaporize. If the heat of solution is $-\Delta H_a$, we can write for the solubility coefficient m in $c_A = mp_A/RT$,

$$m = m_o \exp(+\Delta H_a/RT).$$

Now if Φ_{vG} is constant—or A is present in the gas phase only in low concentrations—we can write for equation (IX.12), after dividing by Φ_{vG},

$$\frac{d(p_A/p_{Ao})}{d(z/L)} = -\frac{p_A}{p_{Ao}} \frac{m_o a V_r \sqrt{(k_\infty D)}}{\Phi_{vG}} \exp\left(-\frac{E}{2RT}\right) \exp\left(-\frac{(-\Delta H_a)}{RT}\right)$$

We now introduce a dimensionless temperature difference ϑ, defined by

$$\vartheta = \frac{E + 2(-\Delta H_a)}{2RT_o} \frac{T - T_o}{T_o},$$

so that

$$\frac{T}{T_o} = \frac{2RT_o}{E + 2(-\Delta H_a)} \vartheta + 1 = A\vartheta + 1,$$

and then find, for the gas phase material balance of component A

$$\frac{d(p_A/p_{Ao})}{d(z/L)} = -B \frac{p_A}{p_{Ao}} \exp\left(\frac{\vartheta}{1 + A\vartheta}\right), \tag{IX.13}$$

where $p_A/p_{Ao} = 1$ at $z/L = 0$. The constants are

$$A = \frac{2RT_o}{E + 2(-\Delta H_a)} \quad \text{and} \quad B = \frac{m_o a V_r \sqrt{(k_\infty D)}}{\Phi_{vG}} \exp\left(-\frac{1}{A}\right).$$

Equation (IX.13) can be integrated because ϑ is independent of z/L. Remember that the liquid is ideally mixed, so that the temperature T of the reactor is constant. After integration we obtain

$$p_A/p_{Ao} = \exp\left[-B(z/L)\exp\frac{\vartheta}{1 + A\vartheta}\right]. \tag{IX.14}$$

The total conversion ζ_A of component A in the reactor is now given by

$$\zeta_A = 1 - \frac{p_A}{p_{Ao}} \frac{RT}{RT_o} = 1 - \exp\left(-B\exp\frac{\vartheta}{1 + A\vartheta}\right)/(1 + A\vartheta). \tag{IX.15}$$

The material balance in this case could be solved independently of the energy balance, because of the uniform temperature in the reactor. For solving the energy balance we will assume that the liquid feed enters the reactor at the same temperature T_o as the gas. If this is not the case a mixed entry temperature for the two feed streams can be used as outlined in Section VI.4. Then we find, for the energy balance,

$$[(\rho c_p \Phi_v)_G + (\rho c_p \Phi_v)_L](T - T_o) + UA(T - T_c) =$$
$$(-\Delta H_a - \Delta H_r)\int_0^1 \sqrt{(kD)} a V_r m \frac{p_{Ai}}{RT} d(z/L). \tag{IX.16}$$

If we substitute equation (IX.14) into (IX.16) and divide by $(\rho c_p \Phi_v)_L$, we find that

$$\left[1 + \frac{(\rho c_p \Phi_v)_G}{(\rho c_p \Phi_v)_L}\right]\vartheta + \frac{UA}{(\rho c_p \Phi_v)_L}(\vartheta - \vartheta_c)$$
$$= \frac{-(\Delta H_a + \Delta H_r)aV_r}{(\rho c_p \Phi_v)_L} \frac{p_{A_0}m}{RT_0(1 + A\vartheta)} \sqrt{(kD)} \int_0^1 \frac{p_A}{p_{A_0}} d(z/L).$$

After evaluation of the integral we finally obtain

$$\vartheta + C(\vartheta - \vartheta_c) = \frac{D}{1 + A\vartheta}\left[1 - \exp\left(-B\exp\frac{\vartheta}{1 + A\vartheta}\right)\right]. \quad (IX.17)$$

The constants in this equation are

$$C = \frac{UA}{(\rho c_p \Phi_v)_G + (\rho c_p \Phi_v)_L}$$

and

$$D = \frac{(-\Delta H_a - \Delta H_r)}{(\rho c_p \Phi_v)_G + (\rho c_p \Phi_v)_L} \frac{p_{A_0}}{RT_0} \frac{\Phi_{vG}}{T_0} = \frac{\Delta T_{ad}}{T_0}.$$

Here ΔT_{ad} is the adiabatic temperature rise of the entire reaction mixture. From equation (IX.17) the reactor temperature—that is the value of $\vartheta = (T - T_0)/AT_0$—can be found by trial and error.

Because of the great similarity to a single phase CISTR we can expect multiplicity in bubble columns. This has been demonstrated by Huang and Varma [24]. Depending on the conditions in the reactor as represented by a certain set of values of the parameters A, B, C and D, one or three solutions for ϑ can be found, as explained in Section VI.5. Some results of the work of Huang and Varma [24] for a certain set of parameter values are shown in figure IX.9, where the conversion for a certain reaction, a hydrocarbon chlorination, is shown as a function of the gas residence time, $\tau_G = \varepsilon V_r / \Phi_{vG}$. We see that only a low conversion is possible for gas residence times $0.3 > \tau_G > 11.7$ s. In the range $0.3 < \tau_G < 11.7$ s three solutions are found—of which the intermediate one is unstable—and only in this range a stable operating point with also a high conversion exists. At too high gas rates ($\tau_G < 0.3$ s) the reaction is extinguished by the cold gas; at too low gas rates ($\tau_G > 11.7$ s) the reaction mixture is too strongly cooled by the cooling medium, so that also here the reaction is extinguished.

We will now end our discussion on bubble column reactors with an Example on a reactor, where the liquid is not ideally mixed.

Example IX.1.d *A bubble column reactor with incomplete mixing in the liquid phase*
We will study the performance of a bubble column reactor in which a slow reaction is carried out and where the gas to liquid flow ratio is so low that axial mixing must be taken into account. We will assume that the liquid phase is

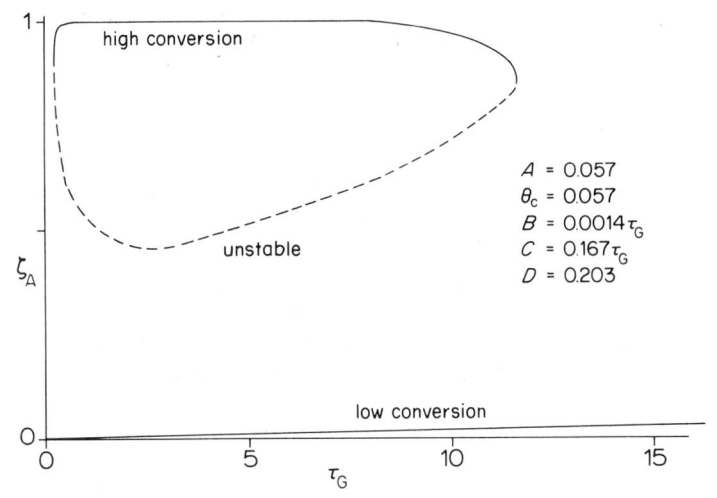

Fig. IX.9 Multiplicity in a bubble column gas–liquid reactor [From Huang and Varma [24]. Reproduced by permission of the American Institute of Chemical Engineers]

uniform in temperature but not in concentration, so that we use the plug flow with axial dispersion model for the liquid phase. In a strong HCl solution in water, cyanogen is absorbed and reacts towards oxamide according to

$$(CN)_2 + 2H_2O \rightarrow (CONH_2)_2.$$

The reaction takes place at a constant pressure of 0.3 MPa and both the gas and the liquid are fed to the column at 20°C. The gas feed consists of air containing 20 vol % cyanogen. Wöhler and Steiner [25] described this reaction and we will use the data they provided or calculate them from the correlations as discussed by van Landeghem [11], Joshi [193] and Shah et al. [194].

It is required to produce 2000 kg oxamide per day at a conversion of 99%. For the first order reaction, $k = 18.6 \times 10^5 \exp(-13\,860/T)$ and $\Delta H_r = -231$ MJ/kmol. The heat of solution of cyanogen into the liquid is $\Delta H_a = -38$ MJ/kmol and for the solubility coefficient m in $mp_A/RT = c_{AL}$ we have $m = 4.55 \times 10^{-6} \exp(+4600/T)$. As operating conditions are chosen $u_G = 5 \times 10^{-3}$ m/s and $u_L = 0.88 \times 10^{-3}$ m/s. From these data it follows that the liquid is pumped around by the gas flow $3 \times 5/0.88 \simeq 17$ times. This is not sufficient to approach ideal mixing, as can be concluded from Example IV.3.b. At these conditions and for our system from the data given in [11] we can derive $\varepsilon_G = 0.017$, $k_L a = 0.5 \times 10^{-3}$ s^{-1}, $a = 26$ m^2/m^3, $k_L = 1.9 \times 10^{-5}$ m/s and $D_1 = 5 \times 10^{-3}$ m^2/s. Moreover $D = 0.6 \times 10^{-9}$ m^2/s. Further data are $\rho_G = 3.50$ kg/m^3, $\rho_L = 1145$ kg/m^3, $c_{pG} = 1018$ J/kg K and $c_{pL} = 4190$ J/kg K. The resistance to mass transfer in the gas phase is negligible ($k_G/mk_L E_A \gg 1$) and the reactor works adiabatically. The temperature in the reactor can now be calculated from an energy balance:

$$[(\rho c_p u)_L + (\rho c_p u)_G](T - 20) = (-\Delta H_r - \Delta H_a)\frac{p_{Ao}}{RT_o}\zeta_A u_G,$$

which leads to $T - 20 = 7.74°C$, so the reactor operates at 28°C. At this temperature $m = 19.7$ and $k = 187 \times 10^{-6}\,\text{s}^{-1}$. The Hatta number now is

$$\phi = \frac{(187 \times 10^{-6} \times 0.6 \times 10^{-9})^{1/2}}{1.9 \times 10^{-5}} = 0.0018$$

At this value of ϕ the reaction takes place in the bulk. The Hinterland coefficient is

$$Al = \frac{1-\varepsilon}{a\delta} = \frac{(1-\varepsilon)k_L}{aD} = \frac{0.983 \times 1.9 \times 10^{-5}}{26 \times 0.6 \times 10^{-9}} = 1200$$

so that $\phi(Al - 1) = 21.5$. There is a considerable concentration difference over the film due to the low specific interfacial area, so that we have to write for our material balances

for the gas phase in plug flow:

$$\frac{dc_G}{dz} + \frac{k_L a}{u_G}(mc_G - c_L) = 0 \tag{a}$$

and for the axially dispersed liquid phase

$$-\frac{dc_L}{dz} + \frac{D_1(1-\varepsilon)}{u_L}\frac{d^2c_L}{dz^2} + \frac{k_L a}{u_L}(mc_G - c_L) - \frac{k(1-\varepsilon)}{u_L}c_L = 0. \tag{b}$$

The values of the constants in our case are

$$\frac{k_L a}{G} = 98.8 \times 10^{-3}\,\text{m}^{-1}, \quad \frac{k_L a}{L} = 0.561\,\text{m}^{-1}$$

$$\frac{D_1(1-\varepsilon)}{u_L} = 5.59\,\text{m}, \quad \frac{k(1-\varepsilon)}{u_L} = 0.209\,\text{m}^{-1}.$$

The boundary conditions for equations (a) and (b) are:

at $z = 0$, $c_G = c_{G\text{in}}$ and $\frac{dc_L}{dz} = 0$

at $z = L$; $c_G = c_{G\text{out}}$ and $u_L c_{L\text{in}} = 0 = u_L c_L - D_1(1-\varepsilon)\frac{dc_L}{dz}$

or $u_L = \frac{D_1(1-\varepsilon)}{c_L}\frac{dc_L}{dz}$

The solution of equations (a) and (b) with these boundary conditions are given by

$$c_L = A_1 e^{\lambda_1 z} + A_2 e^{\lambda_2 z} + A_3 e^{\lambda_3 z} \quad c_G = B_1 e^{\lambda_1 z} + B_2 e^{\lambda_2 z} + B_3 e^{\lambda_3 z}$$

in which $B_i = A_i \left(\frac{\lambda_i u_G}{k_L a} + m\right)^{-1}$ and the values of λ_i can be found by solving:

$$\lambda^3 + \left(\frac{mk_L a}{u_G} - \frac{u_L}{D_1(1-\varepsilon)}\right)\lambda^2 - \frac{1}{D_1(1-\varepsilon)}\left(k(1-\varepsilon) + k_L a + \frac{mk_L a u_L}{u_G}\right)\lambda$$

$$-\frac{mk_L ak}{u_G D_1} = 0.$$

We then find that

$\lambda_1 = 0.106 \qquad \lambda_2 = -0.497 \qquad \lambda_3 = -1.38$
$A_1 = 10^{-6} \qquad A_2 = 1.88 \times 10^{-3} \qquad A_3 = -0.68 \times 10^{-3}$
$B_1 = 20.8 \times 10^{-6} \qquad B_2 = 27.6 \times 10^{-3} \qquad B_3 = -3.89 \times 10^{-3}.$

The units of λ_i are m^{-1} and of A_i and B_i kmol/m^3. With these results concentration profiles for cyanogen in the gas phase and in the bulk of the liquid have been calculated and are given in figure IX.10. For a conversion of $\zeta_A = 0.99$ we find that the column height should be $L = 9.5$ m. We can now check that

$$Pe_L = u_L L/D_1 = 0.88 \times 10^{-3} \times 9.5/(5 \times 10^{-3}) = 1.67.$$

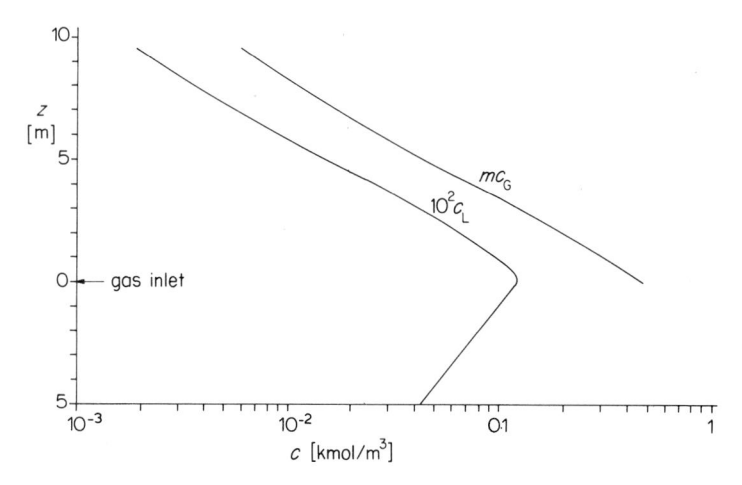

Fig. IX.10 Concentration profiles in the bubble column gas–liquid reactor of Example IX.d

This value is so low that we definitely needed to use the axial dispersion model for the liquid phase.

For a production of 2000 kg/day of oxamide (molecular weight = 88 kg/kmol) the column diameter should be, if P is the daily production in kg,

$$d_t = \left[\frac{4P}{\pi u_G c_{AGin} \zeta_A M_A \times 3600 \times 24}\right]^{0.5}$$

$$= \left[\frac{4 \times 2000}{\pi \times 5 \times 10^{-3} \times 24.4 \times 10^{-3} \times 0.99 \times 88 \times 3600 \times 24}\right]^{1/2}$$

$$= 1.66 \text{ m}.$$

From the concentration profiles we see that there is still non-converted cyanogen dissolved in the liquid phase at the gas inlet. We therefore increase the column length with a reaction zone below the gas inlet, where the dissolved cyanogen can continue to react. If this section is 5 m long the cyanogen concentration is reduced to 4×10^{-4} kmol/m^3. We consider this section to operate in plug flow, because here the liquid is no longer agitated by the gas bubbles. The outlet concentration mentioned represents a loss of 0.6 kg cyanogen per day.

If the liquid were in plug flow the height of the gas–liquid reaction zone would be 0.46 and, if it were ideally mixed, 5.2 times higher than the calculated height of $L = 9.5$ m. Wöhler and Steiner [25] proved experimentally that the model with axial mixing in the liquid phase gave the correct agreement with practical experience in this particular case.

IX.1.4 Agitated gas–liquid reactors

An introduction to the vast literature on agitated gas–liquid reactors and on the estimation of the values of the relevant design and operation variables for this reactor type is given, among others, by Charpentier [10], van Landeghem [11] Joshi, Pandit and Sharma [195] and Barona [26]. We refer to Section VII.17 for a discussion.

For agitated gas–liquid reactors the following assumptions are usually made:

— the gas flow is perfectly mixed;
— the pressure is constant or varies linearly with the liquid height;
— mass transfer coefficients, the gas hold-up and the specific interfacial area are constant over the whole reaction volume;
— the liquid phase is completely mixed.

Hanhart et al. [27], among others, showed that the gas is perfectly mixed at higher agitation rates, so that the assumption of perfect mixing in the gas phase is the more prudent one. Moreover if the gas bubbles do not coalesce, the gas flow is segregated, exhibiting a perfect mixer residence time distribution. Nishikawa et al. [41] also demonstrated that only at very low agitation rates is plug flow of the gas approached. In this region, however, it is better to apply a bubble column reactor, because at too low agitation rates hardly any increase at all of the interfacial area is obtained and the impeller is flooded.

The heat balance of an agitated gas–liquid reactor depends on whether the reactor vessel is cooled by an outside cooling medium or is operated adiabatically. Mostly it is assumed that the liquid does not vaporize. As discussed for bubble columns, the agitated gas–liquid reactor can also exhibit multiplicity. This multiplicity in gas–liquid reactors has attracted much attention. After the first theoretical analyses in the pioneering papers of Schmitz and Amundson [49], Luss and Amundson [50] also developed the theory for two phase segregated systems. For adiabatic gas–liquid CISTRs the theory was further developed by, among others, Hoffman et al. [29] and Raghuram and Shah [51] and for non-adiabatic, cooled reactors by among others, Raghuram et al. [52]

and Huang and Varma [53, 54]. In [51] also cascades of CISTRs with or without back mixing are included. The influence of the gas feed temperature on multiplicity was reported on by Singh et al. [55]. Huang and Varma [54] also studied the dynamic behaviour of a cooled gas–liquid CISTR and indicated the conditions under which limit cycles can occur. These limit cycles were demonstrated experimentally by Ding et al. [28] and by Hancock and Kenney [56].

Multiplicity in gas–liquid CISTRs was demonstrated experimentally by Ding et al. [28]; the phenomena were analysed theoretically by Hoffman et al. [29]. Some of their results are shown in figure IX.11, where the heat production curves and heat removal lines are shown as a function of the reactor temperature for different values of the residence time of the liquid in the reactor. The reaction was a second order one in the liquid phase: we find for $\tau_L = 240$ and 480 s two stable and one unstable and for $\tau_L = 360$ s three stable and two unstable operating points (see Section VI.5). So there are even two more operating points possible than in a single phase CISTR.

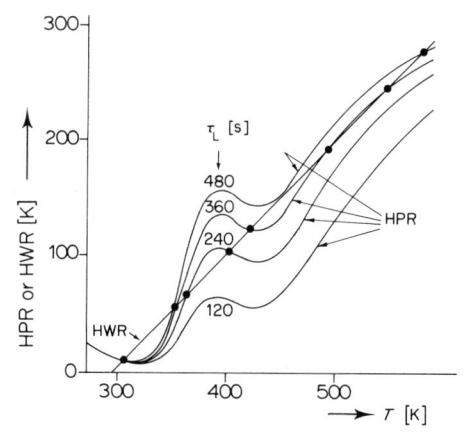

Fig. IX.11 Multiplicity in an agitated gas–liquid reactor [Hoffman et al. [29], Reproduced by permission of the American Institute of Chemical Engineers]

None of the studies mentioned on gas–liquid reactors included evaporation of reactants, products or solvents. Desorption of volatile reaction products and its influence on conversion rates was reviewed by Shah and Sharma [57]. Shah et al. [58] also included the influence of the reaction heat liberation in the boundary layer on the desorption rates. We refer to Section VII.11 for a discussion on desorption phenomena. Sharma et al. [59] were the first to include the influence of the evaporation losses in the heat balance of gas–liquid CISTRs, when they investigated consecutive gas–liquid reactions. They concluded that evaporation losses are an important factor and can induce a large error in the prediction of high temperature steady states. Beskov et al. [30] studied a non-enhanced first order gas–liquid reaction ($\phi < 0.2$) with evaporation losses and with negligible mass transfer resistances in both the gas and the liquid phase ($k_L = \infty$, $k_G = \infty$). They demonstrated the existence of three possible operating points. Because they assumed in their study k_L^{-1} to be zero, they found two operating points less than Hoffman et al. [29]. If they had included finite values for both k_G and k_L they could have found under certain conditions even seven possible operating points, because the 'resistances' k^{-1}, k_L^{-1} and k_G^{-1} each individually exhibit their own temperature behaviour and each one of them can give rise to their own operating points. In general it can be said that, if there are n resistances in series, we can expect $2n + 1$ possible operating points of which $n + 1$ are stable.

There are many different situations in agitated gas–liquid reactors which lead to different design equations, depending on the type of reactions in the liquid phase, the number of resistances involved, the type of cooling and whether the liquid vaporizes or not. We will restrict ourselves here to discussing only one case, as an example of the influence of heat effects on the performance of an agitated gas–liquid reactor. We choose the air oxidation of a dissolved hydrocarbon under conditions of strong evaporation, which leads to rather extensive material and energy balance equations. This was elaborated by Westerterp and Crombeen [135].

A schematic sketch of an air oxidation gas–liquid reactor section is given in figure IX.12. Air and liquid are fed separately into the reactor; the liquid feed consists of solvent and reactant. The reaction heat is partly carried away by the vapours produced which, together with the nitrogen and non-converted oxygen, are flowing to a steam producing cooler–condenser. Here the gas and vapour stream are cooled to a temperature somewhat above the steam temperature. The condensate is separated in a separator and returned to the reactor. The gases and the remainder of the vapours pass to a second cooler–condenser, which is cooled by water, and a second separator. Also this condensate is returned to the reactor; the emitted gases are now fed through a high pressure scrubber, where the last traces of solvents are removed, and discharged into the atmosphere via a pressure control valve. The first condensate can also be cooled in a cooler before being returned to the reactor; the reason for this will be explained in the next Example. The liquid reactor product after separation of the entrained gas bubbles goes to the product separation section, which will not be discussed here.

We will study a partial oxidation reaction of the type

$$B + vO_2 \rightarrow P + (v - 1)H_2O,$$

such as the oxidation of a xylene to a dicarboxylic acid. With this reaction type the liquid reactor product will also contain reaction water besides the solvent. If a molar stream of $\Phi_{vL}c_{B_0}$ of reactant B is fed to the reactor, the molar amount of oxygen consumed for complete conversion of the reactant is equal to $v\Phi_{vL}c_{B_0}$. An excess of oxygen has to be applied in order to maintain a certain partial pressure of oxygen in the gas phase, which is assumed to be perfectly mixed. If we supply an amount of oxygen of X times the required stoichiometric amount, the molar quantity of oxygen fed to the reactor is $v\Phi_{vL}c_{B_0}X$ and the accompanying amount

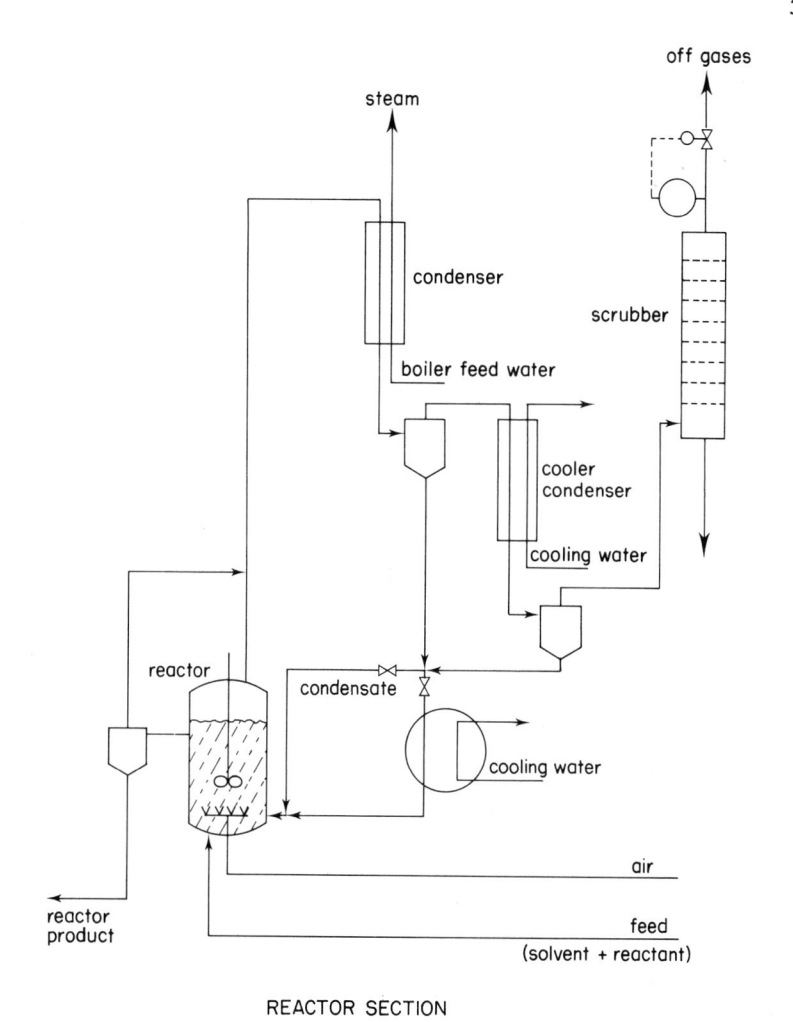

Fig. IX.12 The gas–liquid reactor section in an air oxidation plant

of nitrogen Φ_{N2} then equals $\frac{79}{21} v \Phi_{vL} c_{Bo} X$. The value of X cannot be chosen freely, because an explosive mixture in the gas phase has to be avoided.

The nitrogen passes untouched through the reactor, the cooler–condenser and scrubber to the atmosphere. The total reactor pressure p equals:

$$p = p_W + p_S + p_B + p_P + p_{N2} + p_A, \qquad (IX.18)$$

in which the indices refers to water (W), solvent (S), reactant B (B), product P (P), nitrogen (N2) and oxygen in the reactor outlet (A). We will assume that the vapour pressure of the reactant B is so low that $p_B \simeq 0$ and that product P crystallizes out and exhibits no vapour pressure. For the oxygen the overall material balance is

$$v\Phi_{vL}(c_{Bo} - c_B) = \Phi_{O2\,in} - \Phi_{O2\,out} = \Phi_{O2\,in} - \frac{p_A}{p_{N2}}\Phi_{N2\,out}$$

$$= v\Phi_{vL}c_{Bo}\zeta_B = v\Phi_{vL}c_{Bo}X - 79p_A v\Phi_{vL}c_{Bo}X/21p_{N2}, \quad \text{(IX.19)}$$

which can be rewritten with $p_B = p_P = 0$ as

$$\zeta_B = X \frac{p - \frac{100}{21}p_A - p_W - p_S}{p - p_A - p_W - p_S}. \quad \text{(IX.20)}$$

Here we have used the relation $\dfrac{p_A}{p_{N2}} = \dfrac{p_A}{p - (p_A + p_W + p_S)}$.

From equation (IX.20) we can also derive expressions for the vapour pressures of oxygen and nitrogen in the gas phase; we find

for oxygen:

$$\frac{p_A}{p} = \frac{X - \zeta_B}{\frac{100}{21}X - \zeta_B}\left(1 - \frac{p_W + p_S}{p}\right)$$

and for nitrogen:

$$\frac{p_{N2}}{p} = 1 - \frac{p_A}{p} - \frac{p_W + p_S}{p} = \frac{\frac{79}{21}X}{\frac{100}{21}X - \zeta_B}\left(1 - \frac{p_W + p_S}{p}\right).$$

In these equations the relation between the vapour pressure and the liquid composition is $p_i = p_i^\circ x_i$, in which p_i° is the vapour pressure of pure component i at the reactor temperature and x_i is the mole fraction of i in the liquid. When there is no dissolved product P in the liquid, the mole fractions in the liquid phase are related to the conversion ζ_B of reactant B by

$$x_S = \frac{c_{So}/c_{Bo}}{1 + (v - 2)\zeta_B + c_{So}/c_{Bo}}$$

$$x_W = \frac{(v - 1)\zeta_B}{1 + (v - 2)\zeta_B + c_{So}/c_{Bo}}$$

$$x_B = \frac{1 - \zeta_B}{1 + (v - 2)\zeta_B + c_{So}/c_{Bo}},$$

if c_{Bo} is the molar concentration of the reactant and c_{So} of the solvent in the reactor feed and if the density of the reaction mixture remains constant. At high conversions or at high dilution ratios (c_{So}/c_{Bo} is high) average values of x_S and x_W can be used. For the mole fraction of oxygen in the gas phase is then found

$$\frac{p_A}{p} = \left(\frac{X - \zeta_B}{\frac{100}{21}X - \zeta_B}\right)\left(1 - \frac{p_W + p_S}{p}\right). \tag{IX.21}$$

If the mass transfer resistance in the gas phase is negligible or $k_G = \infty$ and $mp_A/RT = c_A$, the conversion rate of the oxygen in the reactor is given by the mass balance:

$$v\Phi_{vL}c_{Bo}\zeta_B = J_A a V_r = k_L E_A a V_r m p_A/RT. \tag{IX.22}$$

We further assume that the reaction between oxygen and reactant B in the gas phase is of the second order.

Two expressions for the enhancement factor E_A for second order reactions, which cover the whole range of values of ϕ for the chemically enhanced regime or $\phi > 2$, have been developed by Porter [31] and by Kishinevskii et al. [32]. Alper [33] proved that both give very good coverage with the original results of van Krevelen and Hoftijzer [34], so that we will use the simpler one, that of Porter:

$$E_A = 1 + (E_\infty - 1)\left[1 - \exp\left(\frac{1 - \phi}{E_\infty - 1}\right)\right], \tag{IX.23}$$

in which

$$E_\infty = 1 + \frac{vD_A c_B}{D_B c_{Ai}}$$

and

$$\phi = \frac{1}{k_L}\sqrt{(kc_B D_A)}.$$

If the reaction heat is $(-\Delta H_r)_A$ and the heat of absorption $(-\Delta H_a)_A$ per kmol of A reacted and absorbed, the *heat production rate HPR* in the reactor is

$$HPR = v[(-\Delta H_r)_A + (-\Delta H_a)_A]\Phi_{vL}c_{Bo}\zeta_B. \tag{IX.24}$$

For the heat removal four different contributions can be distinguished:

(1) the heat absorbed by the cold liquid feed, which equals $\Phi_{vL}\rho_L c_{pL}(T - T_o)$ if T is the reactor and T_o the liquid feed temperature;
(2) the heat absorbed by the air blown into the reactor, which equals

$$M_{air}c_{pair} \times \frac{100}{21}v\Phi_{vL}c_{Bo}X[(T - T_o) - (T_{oair} - T_o)]$$

if T_{oair} is the air inlet temperature;

(3) the heat carried away by the vapours. If Φ_{molW} and Φ_{molS} are the molar streams and ΔH_{vapW} and ΔH_{vapS} the heats of evaporation of water and solvent respectively, we find with $\Phi_{moli} = p_i \Phi_{N2}/p_{N2}$ for this contribution

$$\frac{[p_W(\Delta H_{vap})_W M_W + p_S(\Delta H_{vap})_S M_S] \times \frac{79}{21} v \Phi_{vL} c_{Bo} X}{p - (p_W + p_S + p_A)};$$

(4) the heat absorbed by the condensate returned to the reactor. If the vapours are completely condensed, this contribution is

$$\frac{(p_W M_W c_{pW} + p_S M_S c_{pS})[(T - T_o) - (T_{cond} - T_o)] \times \frac{79}{21} v \Phi_{vL} c_{Bo} X}{p - (p_W + p_S + p_A)}.$$

Here T_{cond} is the temperature of the condensate entering into the reactor.

The sum of these contributions amounts to the *heat withdrawal rate HWR*. But first we substitute temperature dependent expressions for the physical and chemical parameters:

solvent vapour pressure $\quad : p_S^o = A_S \exp(-B_S/T)$
water vapour pressure $\quad : p_W = A_W \exp(-B_W/T)$
diffusion coefficient $\quad : D_A = A_D \exp(-B_D/T)$
mass transfer coefficient $\quad : k_L = A_L \sqrt{D_A}$
solubility coefficient $\quad : m = A_M \exp(+B_M/T)$
reaction rate constant $\quad : k = A_K \exp(-B_K/T)$

and define a dimensionless reactor temperature by

$$\vartheta = \frac{B_K}{T_o^2}(T - T_o) \qquad (IX.25)$$

so that for the reaction rate constant, if $\beta = T_o/B_K$, the following is now obtained:

$$k = k_o \frac{k}{k_o} = k_o \exp\left(-B_K\left[\frac{1}{T} - \frac{1}{T_o}\right]\right) = k_o \exp\left[\frac{B_K}{T_o^2}(T - T_o)\frac{T_o}{T}\right]$$

$$= k_o \exp\left(\frac{\vartheta}{1 + \beta\vartheta}\right) \qquad (IX.26)$$

and also

$$\frac{T}{T_o} = 1 + \beta\vartheta. \qquad (IX.27)$$

This implies that $k_o = k_\infty \exp(-\beta^{-1})$, which is the value of the reaction rate constant at the temperature T_o. If we now abbreviate the term $\exp[\vartheta/(1 + \beta\vartheta)]$

by using the symbol Θ, then, for example, for the water vapour pressure at the temperature T we can write

$$p_W^o = p_{Wo}^o \Theta^\alpha, \qquad (IX.28)$$

where the index o is used to express 'at the temperature T_o', so that p_{Wo}^o is the vapour pressure of water at T_o, and where $\alpha = B_W/B_K$. This can be derived as follows. At the reactor inlet temperature T_o the vapour pressure is $p_{Wo}^o = A_W \exp(-B_W/T_o)$, and at any temperature T we have $p_W^o = A_W \exp(-B_W/T)$, so that also we can write

$$p_W^o = p_{Wo}^o \exp\left[-B_W\left(\frac{1}{T} - \frac{1}{T_o}\right)\right].$$

The exponent can now be rewritten

$$-B_W\left(\frac{1}{T} - \frac{1}{T_o}\right) = +\frac{B_W}{T_o}\left(1 - \frac{T_o}{T}\right) = \frac{B_W}{T_o}\left(1 - \frac{1}{1+\beta\vartheta}\right)$$

$$= \frac{B_W}{T_o} \times \frac{\beta\vartheta}{1+\beta\vartheta} = \frac{B_W T_o}{B_K T_o} \times \frac{\vartheta}{1+\beta\vartheta} = \alpha \times \frac{\vartheta}{1+\beta\vartheta},$$

which results in equation (IX.28). Similar expressions are obtained for the other parameters.

Dividing both the HPR and the HWR terms by $\rho_L c_{pL} \Phi_{vL}$ and introducing the dimensionless temperature defined by equation (IX.25), the energy balance $HPR = HWR$ becomes, after the addition of the four different contributions to HWR, see Table IX.3 for the symbols used:

$$\vartheta_{ad}\zeta_B = A\vartheta + (1-A)\vartheta_{airo}$$
$$+ \frac{[B + C(\vartheta - \vartheta_{cond})]\Theta^\alpha + [D + F(\vartheta - \vartheta_{cond})]\Theta^\gamma}{\frac{79X}{100X - 21\zeta_B}(1 - P_W H^\alpha - \Theta^\alpha - P_S\Theta^\gamma)} \qquad (IX.29)$$

In equation (IX.29) the conversion ζ_B is still a function of temperature. The value of ζ_B as a function of ϑ can be found by solving the material balance (IX.22) combined with (IX.21), which after introducing ϑ and dividing by $\Phi_{vL} c_{Bo}$ becomes

$$\zeta_B = \frac{P\Theta^{0.5\delta-\varepsilon}}{1+\beta\vartheta} \frac{X - \zeta_B}{\frac{100}{21}X - \zeta_B}(1 - P_W \Theta^\alpha - P_S\Theta^\gamma)E_A, \qquad (IX.30a)$$

in which E_A is given by equation (IX.23), where

$$\phi = \Theta^{0.5}(1 - \zeta_B)^{0.5} R$$

and

$$E_\infty = \frac{(1 - \zeta_B \Theta^\varepsilon(1 + \beta\vartheta))}{\left(\frac{X - \zeta_B}{\frac{100}{21}X - \zeta_B}\right)(1 - P_W\Theta^\alpha - P_S\Theta^\gamma)} \cdot Q + 1 \quad \text{(IX.30.b)}$$

The significance of all the dimensionless constants is given in Table IX.3.

In equation (IX.29) the HWR—the right-hand side—approaches infinity in case $P_W\Theta^\alpha + P_S\Theta^\gamma \to 1$: the HWR is infinitely large when the sum of the partial pressures of the solvent and the water formed equals the total reactor pressure. In this case so many vapours evolve that the mole fraction of oxygen and nitrogen in the gas phase have approached the limit zero. This sets a hypothetical limit to the maximum attainable reactor temperature, which is reached at the boiling point of the reaction mixture.

Due to the vapour pressures of the solvent and the water formed the partial pressure of the oxygen at the interface is being reduced and consequently also the HPR. If the conversion ζ_B and X or the excess air supplied remained constant, the influence of the evaporation would be as shown in figure IX.13, where the absorption rate J is plotted as a function of the reactor temperature. Due to the volatility of the liquid the oxygen partial pressure at the gas–liquid interface is reduced. At high values of E_A gas phase mass transfer limitations will be reached.

Table IX.3 The constants in equations (IX.29) and (IX.30)

$$\vartheta_{ad} = \frac{B_K}{T_o^2} \frac{(-\Delta H_a)_A + (-\Delta H_r)_A}{\rho_L c_{pL}} vc_{Bo}$$

$$\vartheta_{airo} = \frac{B_K}{T_o^2}(T_{oair} - T_o), \quad \vartheta_{cond} = \frac{B_K}{T_o^2}(T_{cond} - T_o)$$

$$\beta = \frac{T_o}{B_K}, \quad \alpha = B_W/B_K, \quad \gamma = B_S/B_K, \quad \delta = B_D/B_K, \quad \varepsilon = B_M/B_K$$

$$A = 1 + \frac{100}{21}\frac{(Mc_p)_{air}vc_{Bo}X}{\rho_L c_{pL}}$$

$$B = \frac{79}{21}\frac{vc_{Bo}X(M\Delta H_{vap})_W}{\rho_L c_{pL}}\frac{B_K}{T_o^2}P_W$$

$$D = \frac{79}{21}\frac{vc_{Bo}X(M\Delta H_{vap})_S}{\rho_L c_{pL}}\frac{B_K}{T_o^2}P_S$$

$$C = \frac{79}{21}\frac{vc_{Bo}X(Mc_p)_W}{\rho_L c_{pL}}P_W \quad F = \frac{79}{21}\frac{vc_{Bo}X(Mc_p)_S}{\rho_L c_{pL}}P_S$$

$$P_W = \frac{p_{wo}^o x_W}{p}, \quad P_S = \frac{p_{So}^o x_S}{p}$$

$$P = \frac{a\tau_L k_{Lo} m_o p}{vc_{Bo}RT_o}, \quad Q = \frac{vc_{Bo}RT_o}{m_o p}, \quad R = \frac{\sqrt{(k_o c_{Bo} D_{Ao})}}{k_{Lo}^2}$$

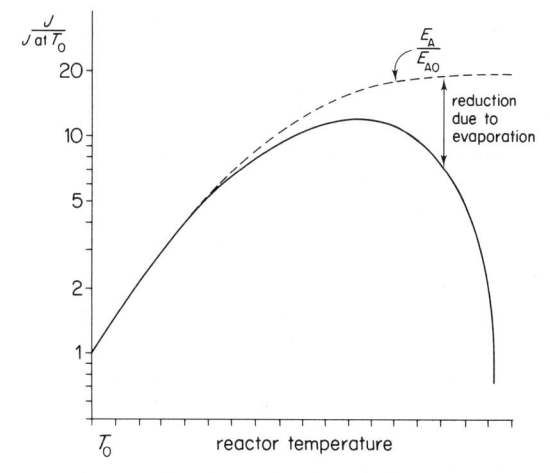

Fig. IX.13 The conversion rate for a second order reaction with (thick line) and without (dotted line) evaporation at constant reactant concentrations as a function of temperature in an agitated gas–liquid reactor

In figure IX.14 the usual plot of the heat production and heat withdrawal rate is given for a certain case and as calculated using equations (IX.29) and (IX.30). We observe the usual S-shape for the HPR at lower reaction temperatures, the same as we met with for the single phase CISTR in Section VI.5. However, at very high reactor temperatures the HPR suddenly drops because here the vapour pressures of water and solvent are extremely high and as a consequence the oxygen partial pressure is greatly reduced, so that no high conversions can be achieved anymore. When $p_W + p_S = p$ the reaction has stopped completely and ζ_B dropped to zero:

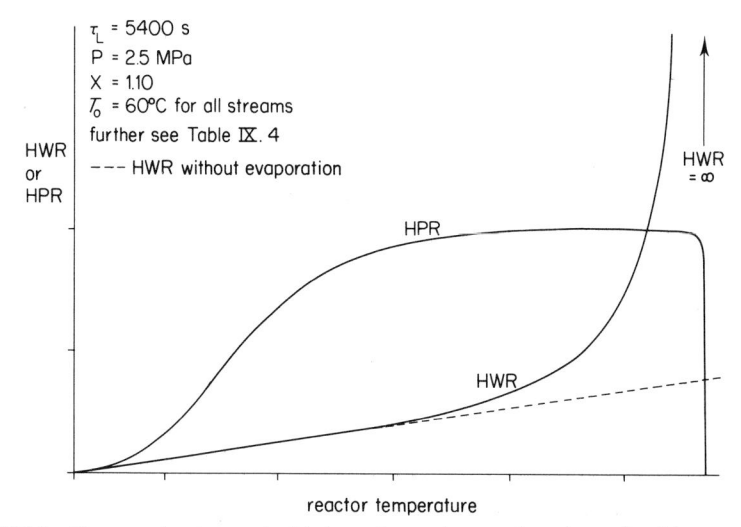

Fig. IX.14 Heat production and withdrawal rates in an agitated gas–liquid reactor with a vaporizing solvent

the gaseous reactant can no longer diffuse into the liquid. For the HWR we observe the usual start with a straight line. As soon as water and solvent start to evaporate, the HWR is enhanced and finally reaches infinity as soon as $p_W + p_S = p$. The upper stable operating point is found when $HWR = HPR$.

More information on the same case of figure IX.14 is given in figures IX.15 and IX.16. The enhancement factor E_A, figure IX.15, increases with reactor temperature and tends to level off to a constant value, the E_∞ value, but as soon as the evaporation really starts, E_A increases again. The curve E_A versus T has an inflexion point and for faster reactions could even have developed a plateau. In figure IX.16 we see that the conversion ζ_B is high over the whole range of reactor temperatures and drops to zero at the maximum attainable temperature. The partial pressure p_A of oxygen, divided by p_{Ao} if no evaporation takes place, is gradually reduced and reaches the limit zero. The mass of vapour produced per unit of mass of liquid feed introduced — the evaporation ratio — increases rapidly as a function of T in the neighbourhood of the upper stable operating point.

For the cooler–condenser design the evaporation ratio is very important. This ratio is given by

$$\frac{\text{vapour}}{\text{liquid}} = \frac{\Phi_{mS} + \Phi_{mW}}{\rho_L \Phi_{vL}} = \frac{M_W p_W + M_S p_S}{p_{N2}} \frac{\Phi_{N2}}{\rho_L \Phi_{vL}}$$

$$= \frac{M_W P_W \Theta^\alpha + M_S P_S \Theta^\gamma}{(1 - P_W \Theta^\alpha - P_S \Theta^\gamma)} \frac{v c_{Bo}}{\rho_L} \left(\frac{100}{21} X - \zeta_B \right), \qquad (IX.31)$$

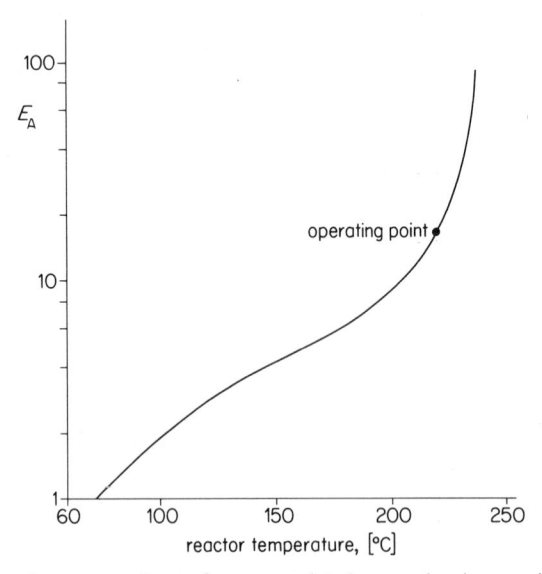

Fig. IX.15 The enhancement factor for a second order reaction in an agitated gas–liquid reactor with a vaporizing solvent as a function of the reactor temperature. Conditions as in Figure IX.14

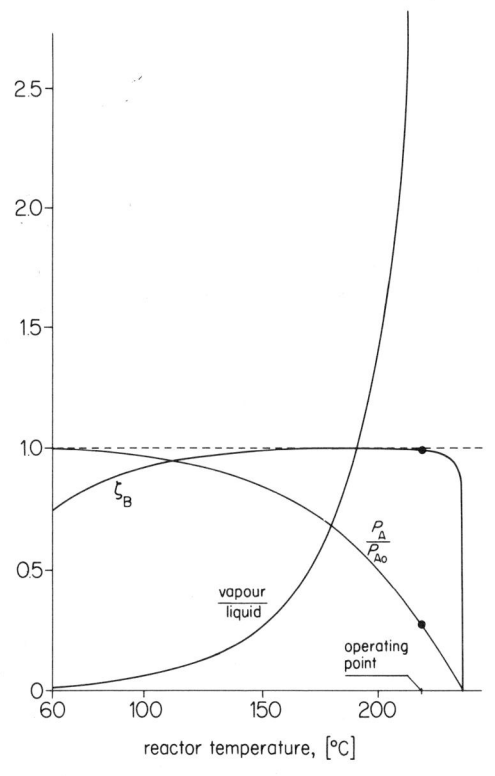

Fig. IX.16 The conversion, oxygen partial pressure reduction and vapour formation in an agitated gas–liquid reactor with a vaporizing solvent as a function of the reactor temperature. Conditions as in figure IX.14

and for the reactor design we have to choose such a reactor diameter d_t that the linear velocity of the gas and vapour is low enough to prevent liquid entrainment. With $\rho_G = p/RT$ kmol/m^3 we can derive for this linear velocity

$$u_o = \frac{\Phi_{vG}}{\frac{1}{4}\pi d_t^2 \rho_G} = \frac{4p\Phi_{N2}}{p_{N2}\pi d_t^2 \rho_G}$$

$$= \frac{v\Phi_{vL}c_{Bo}\left(\frac{100}{21}X - \zeta_B\right)}{(1 - P_W\Theta^\alpha - P_S\Theta^\gamma)} \frac{RT}{p} \frac{4}{\pi d_t^2}. \qquad \text{(IX.32)}$$

The multiplicity characteristics of this reactor type can be investigated by the same method as explained for single phase CISTRs in Section VI.5 by plotting

$$\frac{HWR}{HPR} = f(\vartheta). \qquad \text{(IX.33)}$$

For the values of ϑ where $f(\vartheta) = 1$, operating points are obtained. This is done in figure IX.17 for the same case as in figure IX.14. Here only one stable operating

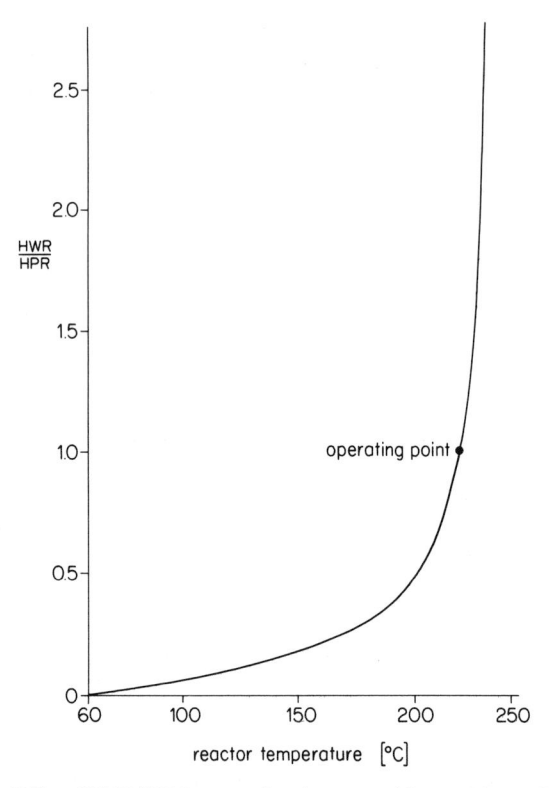

Fig. IX.17 The $f(\vartheta) = HWR/HPR$ curve for the case of figure IX.14. Conditions as in figure IX.14

point is obtained. If the reactor inlet temperature T_o is reduced as shown in figure IX.18 the reactor will develop multiplicity: curve 1 exhibits a plateau but not yet multiplicity, whereas curve 2 with a lower T_o demonstrates three operating points of which the intermediate one is unstable. This is in principal the same behaviour as of the autothermal tank reactor discussed in Section VI.5.

We have to remember that the gas phase mass transfer resistance is negligible as long as $mk_L E_A/k_G \ll 1$. As E_A continues increasing with increasing reaction temperature, sooner or later a temperature level is reached where we have to include also the influence of k_G in our model. We also have to remark that the interfacial area cannot be independent of the evaporation rate and consequently of the reaction temperature. The gas flow through the reaction mixture increases due to the increased vapour formation with increasing reactor temperatures. Now if the bubble diameter is kept constant by the agitation, the number of bubbles increases so that $a = 6\varepsilon_G/\bar{d}_p$ increases linearly with a possible increase in gas hold-up. On the other hand if the number of bubbles remained constant, the bubble diameter would be enlarged due to the uptake of vapour: this would result in an interfacial area decrease. Regretfully nothing is known yet of the influence of vapour formation on the interfacial area in gas–liquid reactors.

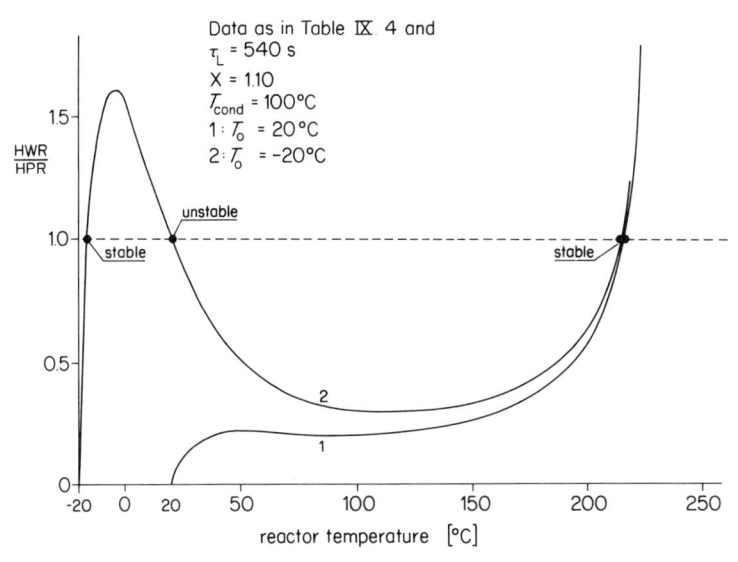

Fig. IX.18 The $f(\vartheta)$ curve for the reaction system of figure IX.17, but at lower reactor inlet temperatures

Example IX.1.e *Oxidation of a dissolved hydrocarbon in an agitated gas–liquid reactor under pressure*
A xylene dissolved in acetic acid is oxidized with air under pressure according to

$$\text{xylene} + 3O_2 \rightarrow \text{dicarboxylic acid} + 2H_2O.$$

The dicarboxylic acid crystallizes out and exerts no vapour pressure. It is now required to design a reactor for a xylene conversion $\zeta_B > 0.98$, with a feed concentration of $c_{Bo} = 2.2\,\text{kmol/m}^3$ and a liquid residence time of one hour. Air under pressure and the xylene solution are available at 20°C.
For the reactor and the reaction system pertinent data are given in Table IX.4, including data obtained in a preliminary design. The estimate of the specific interfacial area needs some explanation. From the early work of, among others, Calderbank [35] and Westerterp *et al.* [36] and later by Reith [37], van Dierendonck *et al.* [38, 39] and Miller [40] and recently by Nishikawa *et al.* [41] the work on correlation of specific interfacial areas still continues. Moreover, very little of the effect of pressure on gas–liquid dispersions is known, but results till now [42, 43] indicate that hardly any influence of pressure has to be expected. We have used the work of Van Dierendonck [38, 39] and found for the agitation level chosen in our preliminary design a gas hold-up of $\varepsilon = 0.33$ and an interfacial area of $2100\,\text{m}^2/\text{m}^3$ reaction volume or $3100\,\text{m}^2/\text{m}^3$ liquid volume at near operating conditions. These data have been used in our calculations on the reactor performance.
Several design variables have been studied more in detail at $\tau_L = 3600\,\text{s}$ by using the equations (IX.29)–(IX.33). The results are shown in Table IX.5 and in

Table IX.4 Data for the xylene oxidation of Example IX.1.e

$p_S^o = 19.9 \times 10^9 \exp(-4605/T) \, N/m^2$

$p_W^o = 29.4 \times 10^9 \exp(-4666/T) \, N/m^2$

$D_A = 7.87 \times 10^{-6} \exp(-2400/T) \, m^2/s$

$k_L = 3.97 \times \sqrt{D_A}$

$m = 443 \times 10^{-6} \exp(+1263/T)$

$k = 1.52 \times 10^{14} \exp(-10\,000/T) \, m^3/kmol\,s$

$\Delta H_{vapS} = 405.4 \times 10^3 \left[\dfrac{594 - T}{202.7}\right]^{0.25} \, J/kg$

$\Delta H_{vapW} = 2257 \times 10^3 \left[\dfrac{647.2 - T}{274.2}\right]^{0.38} \, J/kg$

$\Delta H_r = -1270 \times 10^6 \, J/kmol\,B \quad \Delta H_a = -10.5 \times 10^6 \, J/kmol\,A$

$c_{pair} = 1020 \, J/kg \quad\quad\quad c_{pfeed} = 2100 \, J/kg\,k$

$c_{pW} = 4190 \, J/kg$

$c_{pS} = 1960 \, J/kg\,K \quad\quad\quad \rho_L = 930 \, kg/m^3$

$V_L = 12 \, m^3, \; d_t = 2.35 \, m, \; a_L = 3100 \, m^2/m^3$ liquid

$\Phi_{vL} = 3.4 \times 10^{-3} \, m^3/s \quad\quad c_{Bo} = 2.2 \, kmol/m^3$

$T_{oair} = 20°C \quad T_o = 20°C$

---

figure IX.19, where $f(\vartheta)$ is plotted versus the reactor temperature. We then observe for the design variables studied:

— if the *reactant concentration* is increased to $c_{Bo} = 4.4 \, kmol/m^3$ the conversion drops to too low a value. At the same time air is supplied at such a rate that linear gas velocities are too high and excessive liquid entrainments occur, as can be checked using equation (IX.32).

— a *reactor pressure* of 2.0 MPa gives too low a conversion, and one of 2.5 MPa an adequate conversion. At $p = 3.0$ MPa a conversion higher than desired is obtained. The gain in conversion will probably not offset the higher compression costs for the air.

— an *air excess* of 5% ($X = 1.05$) gives too low a conversion, one of 10% gives an adequate conversion and if $X = 1.20$ the conversion is higher than desired. Also in this case the yield increase probably does not offset the 9% higher air compression costs.

— the installation of a *condensate after cooler* is advantageous because it reduces the reactor temperature and increases the conversion, simultaneously the cooler–condensers can be designed with lower heat transfer areas. Without the after cooler the combined condensate streams will have a temperature of around 170°C. The T_{cond} can become lower if a water cooled

Table IX.5 Influence of design variables on the performance of the reactor of Example IX.1.e

Conditions							
Reactor pressure (MPa)	2.5	2.0	2.5	3.0	2.5	2.5	2.5
Reactant concentrations							
$c_{Bo}/(kmol/m^3)$	4.4	2.2	2.2	2.2	2.2	2.2	2.2
Excess air, ratio	1.10	1.10	1.10	1.10	1.05	1.20	1.10
Condensate temperature (°C)	170	170	170	170	170	170	40
Results							
Value of P							
(see Table IX.3)	5.2	8.3	10.4	12.5	10.4	10.4	10.4
Reactor temperature (°C)	222	211	222	232	223	221	216
Conversion ζ_B	0.940	0.958	0.977	0.988	0.955	0.991	0.984

cooler–condenser is used, but then steam can no longer be produced, and this is an attractive feature of the design of figure IX.12. With an after cooler the total condensate can be cooled to around 40°C, so that per kg liquid evaporated more heat is absorbed. Now the stable operating point is reached at a lower evaporation ratio, which results in a lower partial pressure of vapours in the gases leaving the reactor: therefore the oxygen partial pressure and consequently also the conversion are higher.

To meet the requirements for the reactor performance we recommend, on the basis of the results in figure IX.19 and Table IX.5, an air excess of 10% ($X = 1.10$), a reactant concentration of $c_{Bo} = 2.2$ kmol/m³, a reactor pressure of $p = 2.5$ MPa and the installation of condensate after cooler.

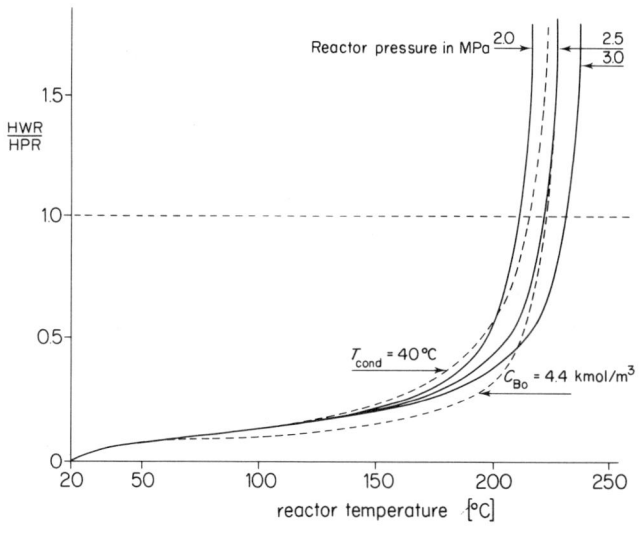

Fig. IX.19 $f(\vartheta)$ curves for different design conditions of the reactor of Example IX.1.e

Multiplicity does not occur, as can be seen from figure IX.19. For the same operating conditions except lower values of P—which implies lower agitation rates, resulting in lower values of $a/(1-\varepsilon)$—multiplicity can occur as demonstrated in figure IX.20, in our case for $P < 0.29$. This corresponds to a value of $a/(1-\varepsilon) = 86\,\text{m}^2/\text{m}^3$ liquid. This value is so low that even without agitator or for an agitator break-down the reactor would maintain a high stable operating temperature, because the air flow itself would already create a larger interfacial area. Therefore we can safely conclude that in our reactor and with our reaction system no multiplicity can occur. Also at start-up only a high operating point can be reached. This does not mean that we can do without the agitator: intense agitation is needed for the high conversion required.

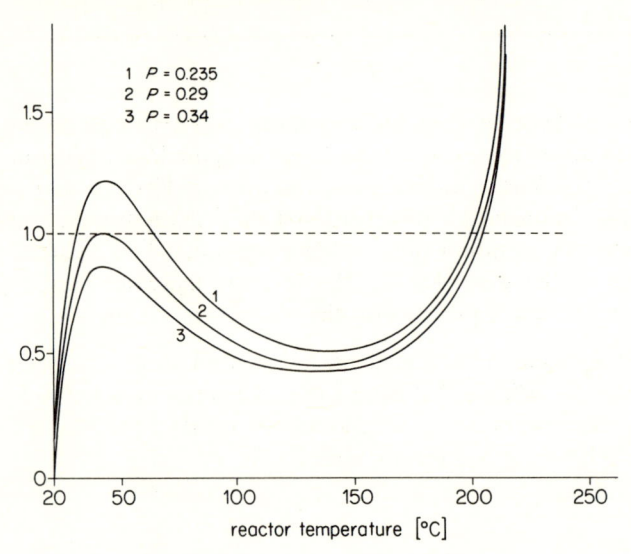

Fig. IX.20 Multiplicity in the reactor of Example IX.1.e

IX.2. GAS–SOLID REACTORS

The effect of temperature on the conversion rate of gas–solid reactions greatly depends on the relative importance of the physical transport phenomena, which are much less temperature-dependent than the chemical reaction velocity itself. This was qualitatively shown in figure VII.35, where the logarithm of the conversion rate had been plotted as a function of $1/T$ for various cases; the slope of a curve in this diagram is a measure of the 'activation energy' of the overall conversion rate. The slope will indicate the activation energy of the chemical reaction only at lower temperatures. On increasing the temperature the external mass transfer may directly become rate determining. This was also demonstrated in figure VII.36. Also other phenomena such as the interaction of the actual chemical kinetics and adsorption equilibria on catalyst surfaces, the change of the number of moles in a gas upon conversion and non-uniform temperatures in porous particles can play a role. We therefore have to be fully aware of all the pitfalls of the interaction of chemical kinetics and physical transport phenomena, and must be familiar with the contents of Chapters VII and VIII before embarking on the study of the behaviour of gas–solid reactors.

IX.2.1 Single particle behaviour

In Section VI.5 we discussed the autothermal operation of an entire reactor system in which an exothermic reaction was carried out. It was shown that this phenomenon is brought about by the feed-back of liberated heat of reaction to the cold feed. A similar situation may occur on a much smaller scale with rapid heterogeneous exothermic reactions; examples were given in Table VII.5. It is encountered in the catalytic combustion of NH_3 with air on a surface of platinum catalyst (see Example IX.2.a) as well as in certain gas–solid reactions, such as the combustion of carbon. The latter process will be used for the following discussion.

Let us consider a particle of lean coal or coke in a fixed bed through which air is flowing. When the coal is burning, the overall reaction between oxygen and carbon may be represented by

$$O_2 + C \to CO_2.$$

Especially at higher temperatures CO is produced at the coal surface, either by oxidation of carbon or by reduction of CO_2, while CO is oxidized to CO_2 very near the surface. However, for the present analysis it may be assumed that the reaction takes place at the coal surface, and accordingly that the net heat of reaction is released at this surface.

Since a burning piece of coal is consumed very slowly, a pseudo-steady state may be assumed to exist in the coal bed, and the temperature of a coal particle may be thought to be constant. In such a case, all of the heat produced by the reaction must be transported from the coal surface to the surroundings. Since, at least in the middle of the burning coal bed, radiation equilibrium will approximately exist, the reaction heat will be transferred by convection to the gas stream. In this way the feed-back of heat is established from the reaction site to the gaseous reactant. Under the above conditions the molar flux J_A of oxygen (which we call the reactant A) to the surface of the particle, multiplied by the heat liberated per molar unit of A converted, $-(\Delta Hr)_A$, must be equal to the heat flux coming from the surface ($J_A(\Delta Hr)_A$, see equation (VII.238)) or

$$HPR = HWR. \tag{IX.34}$$

The value of J_A depends on the texture of the particle and on the chemical and physical rate parameters as indicated in Chapter VII. At a low value of T_i (the temperature of the solid), J_A is mainly determined by the rate of the chemical surface reaction; as T_i increases, a situation is reached where the oxygen is converted at the external particle surface at such a high rate that its concentration at the surface is zero and J_A is entirely determined by mass transfer:

$$J_A = k_G \bar{c}_A. \tag{VII.237}$$

With a constant gas velocity, the mass transfer coefficient k_G does not greatly depend on temperature, so that J_A will be practically constant for a certain value of $\bar{c}_A$ and at high temperatures. Accordingly, the left-hand term of equation (IX.34), i.e. the chemical heat production rate per unit external area, depend on T_i in the manner indicated in figure IX.21, Curves 1a and 1b. The equation of these curves is given by equation (VII.243) after elimination of c_{Ai}, or

$$HPR = \frac{(-\Delta H_r)_A \bar{c}_A}{(1/k_\infty'') e^{E/RT_i} + (1/k_G)} \qquad \text{(IX.35)}$$

The rate of heat transfer to the gas flow can be described by means of a heat transfer coefficient α:

$$HWR = \alpha(T_i - \bar{T}), \qquad \text{(IX.36)}$$

where $\bar{T}$ is the average gas temperature. Since it was assumed that α does not contain a contribution due to thermal radiation, it is fairly independent of temperature. Hence, at a constant gas load and for a given value of $\bar{T}$, HWR is practically a linear function of T_i, as shown in figure IX.21, line 2. The intersections of a curve 1 and a line 2 in a diagram like figure IX.21 represent solutions to equation (IX.34). When there are three solutions, the middle one is unstable for the same reasons as discussed in Section VI.5. The right-hand intersection corresponds with a high conversion rate which can only be obtained after ignition. The corresponding surface temperature lies between 1500 and 2000°C for the burning of coal in atmospheric air. Its value is fairly insensitive to variations in gas velocity and the degree of oxygen conversion in a fixed bed; this is valid not only for one piece of coal but equally well for a large collection of particles.

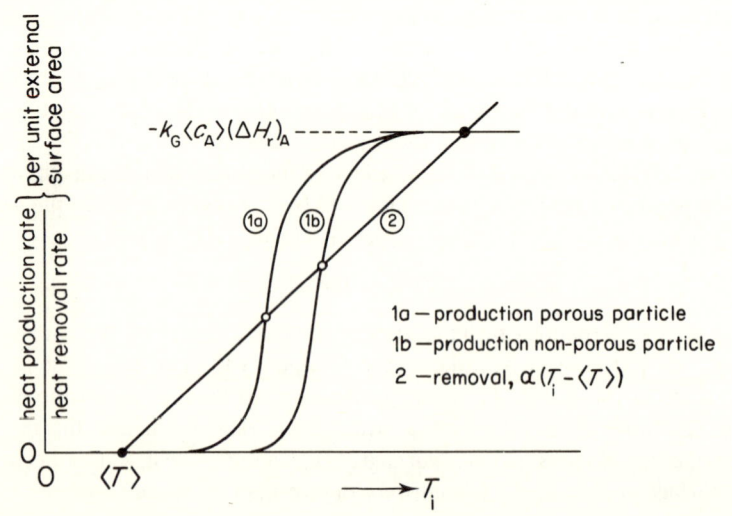

Fig. IX.21 Heterogeneous exothermic reaction; possible solutions to equation (IX.34): ● stable, ○ unstable

To determine the magnitude of the combustion zone in a coal bed, we can start from equation (VII.245) and consider the case where $k''_\infty \exp(-E/RT_i) \gg D_A/\delta_m$:

$$T_i - \bar{T} = \frac{\bar{c}_A}{\rho c_p}(-\Delta H_r)_A Le^{-2/3} \qquad (IX.37)$$

the value of which does not depend on the gas flow rate for constant $\bar{T}$ and $\bar{c}_A$. For an O_2–N_2 mixture the value of Le is about 1.5. For finding the reaction temperature T_i as a function of the degree of conversion, we consider a coal bed as a tubular reactor with a negligible longitudinal dispersion of matter and heat. At the entrance of the combustion zone ($z = 0$), the oxygen concentration and the gas temperature are c_{Ao} and $\bar{T}_o$, respectively. With adiabatic steady operation, the combined heat and material balances give, for a differential length dz of the reactor,

$$0 = \frac{d}{dz}(\Phi_m c_p \bar{T}) + (-\Delta H_r)_A \frac{d}{dz}(\Phi_m \bar{c}_A/\rho)$$

When the increase of the mass flow Φ_m (due to the fact that the product CO_2 is heavier than the reactant O_2) is neglected, after putting $c_A = py_A/RT$ and taking average values for c_p and the molar weight M of the gas over the bed, we obtain on integration

$$\bar{T} - \bar{T}_o = \frac{(-\Delta H_r)_A y_{Ao}}{c_p M}\left(1 - \frac{y_A}{y_{Ao}}\right) = \Delta T_{ad}\left(1 - \frac{y_A}{y_{Ao}}\right). \qquad (IX.38)$$

The following expression for T_i results from equations (IX.38) and (IX.37):

$$T_i - \bar{T}_o = \Delta T_{ad}\left[1 - \frac{y_A}{y_{Ao}}(1 - Le^{-2/3})\right]. \qquad (IX.39)$$

Since the Lewis number for gases is of the order of unity, equation (IX.39) shows that $(T_i - \bar{T}_o)$ will always remain of the order of ΔT_{ad}, irrespective of the degree of conversion (of which $1 - y_A/y_{Ao}$ is a measure). If $Le > 1$, $(T_i - \bar{T}_o)$ will rise to the final value ΔT_{ad} when the reaction is completed. If $Le < 1$, it will drop to this final value; therefore, in the latter case in the reaction zone, $(T_i - \bar{T}_o)$ will be higher than ΔT_{ad}. Moreover, from this reasoning we can conclude that the temperature in the combustion zone of a coal bed is nearly constant, because the gas inlet temperature is constant, of course. This will be demonstrated also for the Pt gauze in Example IX.2.a for NH_3 combustion.

An elegant demonstration of the influence of Le was given by Wicke [60]. When y_A/y_{Ao} is eliminated from equations (IX.38) and (IX.39), the following equation for the upper stable operating temperature T_i as a function of $\bar{T}_o$, ΔT_{ad} and $\bar{T}$ is obtained:

$$\frac{T_i - \bar{T}}{\bar{T}_o + \Delta T_{ad} - \bar{T}} Le^{2/3} = 1. \qquad (IX.40)$$

This equation, as shown in figure IX.22, can be solved graphically by finding the intersection between a horizontal line at a height 1 and an inclined straight line through the point $P(\bar{T}_o + \Delta T_{ad}, Le^{2/3})$. Actually, such a line represents the heat removal rate as a function of $\bar{T}$ in dimensionless form; as the degree of conversion increases, the point of intersection $\bar{T}$ of this line with the abscissa moves from $\bar{T}_o$ to $\bar{T}_o + \Delta T_{ad}$, and the line pivots around the point P. With $Le < 1$, the value of T_i giving a solution to equation (IX.40) moves to the left; with $Le > 1$ it would move to the right with increasing conversion.

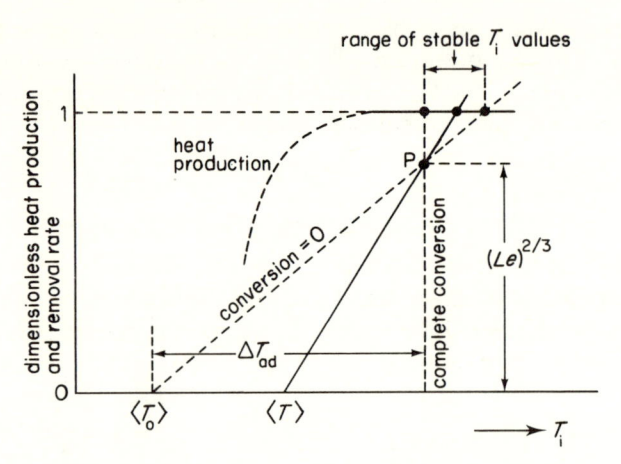

Fig. IX.22 Dimensionless representation of figure IX.21 showing the influence of Le on the course of the solids temperature with the conversion

Finally, the degree of conversion can be calculated as a function of the active bed length z from the material balance:

$$0 = \frac{d}{dz}\left(\Phi_m \frac{y_A}{\bar{M}}\right) + k_G \frac{p}{RT} y_A a_p (1 - \varepsilon_G) S, \qquad (IX.41)$$

in which $a_p(1 - \varepsilon)$ is the external surface area of the particles per unit reactor volume. Simple integration is not possible since not only y_A, but also Φ_m, $\bar{M}$ and k_G depend on z. An approximation of the partial pressure distribution is obtained when Φ_m, $\bar{M}$ and $k_G p/RT$ are assumed to be constant. The solution then becomes

$$\frac{y_A}{y_{Ao}} = \exp\left(-\frac{k_G \bar{M} a_p(1-\varepsilon) S}{RT\Phi_m} z\right) \equiv \exp(-z/z_e).$$

If we use the relation of Thoenes and Kramers [62] ($Sh = 1.9 Re^{0.5} Sc^{0.33}$) for the determination of the gas phase mass transfer coefficient k_G, we find:

$$z_e = \frac{\Phi_m RT}{S\bar{M}p} \frac{1}{a_p(1-\varepsilon)} \frac{1}{k_G} \simeq u \frac{d_p}{6(1-\varepsilon)} \frac{d_p}{1.9 D_A} Re_p^{-0.5} Sc^{-0.33}.$$

With the usual value of $\varepsilon \simeq 0.4$ for fixed beds, z_e becomes

$$z_e \simeq 0.15 d_p \left(\frac{u d_p}{\varepsilon v}\right)^{0.5} \left(\frac{v}{D_A}\right)^{0.67} = 0.15 d_p Re_p^{0.5} Sc^{0.67}.$$

Re_p lies between 10 and 1000 for most practical applications, while Sc is of the order of 1 for gases. Accordingly, z_e varies between 1 and 6 particle diameters for rapid gas–solid reactions of which the conversion rate is entirely determined by mass transfer. Actually, z_e may be somewhat increased by longitudinal dispersion and/or a non-uniform velocity distribution.

As a result, it may be concluded that, with such a reaction as the combustion of coal with atmospheric air, 90 % of the oxygen conversion takes place within a bed length equal to about 3–15 particle diameters, as the value of Re_p goes from 10 to 1000. Consequently, the oxygen in the air passing through a burning coal bed of the height indicated is almost completely converted, particularly if no heat is lost at the end of the reaction zone by radiation to cold surfaces. Because of the consumption of solid fuel, the burning zone moves slowly in the direction of the gas flow, leaving the ashes behind.

In the above we have given only the essential outlines of the theory on temperature effects with rapid heterogeneous exothermic reactions between a gas and a solid. The same principles can be applied to other rapid heterogeneous reactions such as the catalytic recombination of H_2 and O_2 and the catalytic combustion of NH_3 with air. The latter process will be discussed more quantitatively in the next Example.

Example IX.2.a *Catalytic combustion of* NH_3 *with atmospheric air*
Many authors have studied this important reaction, e.g. Roberts and Gillespie [63] and Zharov et al. [64]. Here some of the data published by Oele [65] will be used to analyse the conversion of NH_3 and the temperature distribution in a catalytic burner in which the gas mixture passes at atmospheric pressure through three flat platinum gauzes in succession. Under operating conditions (platinum temperature between 800 and 900°C), the chemical reaction velocity at the platinum surface is so high that the rate of conversion is determined entirely by the rate of mass transfer of the reactant which is not in excess. For several reasons this reaction would become the limiting factor. As a consequence, the system will show autothermal behaviour with a cold feed gas, so that the reaction has to be ignited by preheating temporarily either the gas or the gauze.

The following values of the operating variables are selected for calculating the stable reaction conditions:

Pressure, p	1 bar
Feed temperature, $\bar{T}_o$	60°C
NH_3 in feed, y_{Ao}	0.11
Number of gauzes	3
Gauze wire diameter, d	60×10^{-6} m

Platinum surface/m² gauze,
$a\,\Delta z \quad 1.2\,\text{m}^2/\text{m}^2$
Mass velocity of gas, $\rho u \quad 0.4\,\text{kg/m}^2\,\text{s}$.

Oele's paper supplies the values of the heat of reaction $(\Delta H_r)_A = -226 \times 10^3$ kJ/kmol at 800°C and of the specific molar heat of the feed gas $C_p = 31.8$ kJ/kmol°C between 60 and 800°C with which the rise in adiabatic temperature for complete conversion is calculated:

$$\Delta T_{ad} = \frac{(-\Delta H_r)_A y_{Ao}}{C_p} = 782°C.$$

The mole fraction y_{A1}, y_{A2} and y_{A3} after the first, second and third gauze, respectively, have to be calculated to arrive at the gas and solids temperature. The relation for the concentration profile will be used in the form

$$\frac{y_{A1}}{y_{Ao}} = \exp\left(-\frac{k_G \bar{M} p}{RT}\frac{S}{\Phi_m} a\,\Delta z\right) \tag{a}$$

and similar expressions for y_{A2}/y_{A1} and y_{A3}/y_{A2}. The value of the mass transfer coefficient k_G is calculated by analogy from an empirical heat transfer correlation for gas flow across their wires [66]:

$$Sh = 0.42 Sc^{0.2} + 0.57 Re^{0.5} Sc^{0.33}.$$

Other correlations of Satterfield and Cortez [67] and Shah [68] give nearly the same results. For average film temperatures of about 400°C (first gauze) and 800°C (third gauze), with the physical properties indicated, this produces the results given in Table IX.6.

Table IX.6 Conditions along the platinum gauzes of Example IX.2.a

	400°C	800°C	
μ	31	42	10^{-6} kg/m s
ρ	0.49	0.31	kg/m³
D_B	1.00	2.15	10^{-4} m²/s
Sc	0.63	0.63	—
Re	0.775	0.570	—
Sh	0.82	0.75	—
k_G	1.37	2.7	m/s
$\dfrac{k_G p \bar{M}}{RT}$	0.67	0.83	kg/m² s
$\Delta z/z_e$	2.0	2.5	—

Accordingly we find from equations (a) and (IX.38), if $\bar{T}_1$, etc. is the temperature in the bulk of the gas after passing the first gauze,

$$\frac{y_{A1}}{y_{A0}} \simeq e^{-2.0} = 0.13, \qquad \bar{T}_1 = 740°C.$$

$$\frac{y_{A2}}{y_{A1}} \simeq e^{-2.3} = 0.10, \quad \frac{y_{A2}}{y_{A0}} = 0.013, \qquad \bar{T}_2 = 832°C,$$

$$\frac{y_{A3}}{y_{A2}} \simeq e^{-2.5} = 0.08, \quad \frac{y_{A3}}{y_{A0}} = 0.001, \qquad \bar{T}_3 = 841°C.$$

With $v/D = 0.63$ and $\mu c_p/\lambda = Pr = 0.75$, we have $Le = Sc/Pr = 0.84$. This value is used in equation (IX.40) to calculate the gauze temperature. Because of the large gas temperature difference, particularly over the first gauze, an average temperature of the gas passing through the gauze should be taken for $\bar{T}$, such as the arithmetic average of the temperatures of the incoming and of the outgoing gas. The results are shown in Table IX.7.

Table IX.7

Gauze no.	1	2	3	
$\bar{T}$	400	786	837	°C
$T_i - \bar{T}$	495	63	6	°C
T_i	895	849	843	°C

It is seen that, since $Le < 1$, the catalyst temperature is higher than corresponds with the rise in adiabatic temperature. It decreases towards this value as the reaction is completed. This example furthermore shows that, with 3 gauzes at atmospheric pressure and with $\rho u = 0.4 \text{ kg/m}^2 \text{ s}$, the final conversion is nearly complete. Since the value of $k_G p/RT$ is fairly insensitive to pressure changes, it appears from equation (a) that the conversion will be reduced when ρu is increased by increasing either the velocity or the pressure. A larger number of gauzes must be used for obtaining a complete conversion under these circumstances; burners containing more than 20 gauzes exist [65].

(*Example IX.2.a ended*)

Now still the question remains of when the various diffusional effects really influence the conversion rate in gas–solid reactions. This subject has been discussed in Sections VII.13 and VII.14, and laboratory test reactors for gas–solid reactors in VII.15.3. Many criteria have been developed in the past for the determination of the absence of diffusional resistances. In using the many criteria no more information is required than the diffusion coefficient D_A for gas phase diffusion and D_i for internal diffusion in a porous pellet, heat of reaction and the physical properties of the gas and the solid or catalyst, together with an experimental value of the observed global reaction rate $\langle R_J \rangle$ per unit volume or weight of solid or catalyst. For the time being we recommend the following criteria.

The *intraparticle* concentration and temperature gradients in a porous particle can always be neglected, when the pore effectiveness factor η is practically equal to 1. Assuming that η should be 1 ± 0.05, Mears [69] derived for simple irreversible reactions of the nth order the following criterion: if

$$\frac{\langle R_J \rangle d_p^2}{4\bar{c}_i D_i} < \frac{1}{n - \gamma\beta}, \qquad (IX.42)$$

then the internal resistances can be neglected. However, if $n - \gamma\beta$ approaches 0, the asymptotic method of Petersen [70] has to be used. To check if the particle can be considered as isothermal—whether there are important concentration gradients or not—the criterion of Anderson [71] can be used: if

$$\frac{(-\Delta H_r)_J \langle R_J \rangle d_p^2}{4\lambda_s T_i} < 0.75 \frac{T_i R}{E}, \qquad (IX.43)$$

then the porous particle can be considered as being isothermal.

The intraparticle transport effects, both isothermal and non-isothermal, have been analysed for a multitude of kinetic rate equations and particle geometries. It has been shown that the concentration gradients within the porous particle are usually much more serious than the temperature gradients. Hudgins [72] points out that intraparticle heat effects may not always be negligible in hydrogen-rich reaction systems. The classical experimental test as discussed in Section VII.15.3 to check for internal resistances in a porous particle is to measure the dependence of the reaction rate on the particle size. Intraparticle effects are absent if no dependence exists. In most cases a porous particle can be considered isothermal, but the absence of internal concentration gradients has to be proven experimentally or by calculation.

In gas–solid systems the *interparticle* gradients — between the external surface of the particle and the adjacent bulk fluid phase — may be more serious, because in general the effective thermal conductivity of the gas is much lower than that of the particle. For the interparticle situation the heat transfer resistances in general are more serious than the interparticle mass transfer effects; they may become important if reaction rates and reaction heats are high and flow rates are low. The usual experimental test for interparticle effects as outlined in Section VII.15.3 is to check the influence of the flow rate on the conversion while maintaining constant the space velocity or residence time in the reactor. This should be done over a wide range of flow rates and the conversion should be measured very accurately.

For the requirement that the conversion rate at the conditions of the particle surface should not differ by more than 5% of that under the conditions in the bulk of the reaction mixture flowing around the particle, Mears [73] derived the following criteria: if

$$\frac{\langle R_J \rangle d_p}{2 k_G \bar{c}_J} < \frac{0.15}{n}, \qquad (IX.44)$$

the mass transfer resistance over the gas film can be neglected, and if

$$\frac{(-\Delta H_r)_J \langle R_J \rangle d_p}{2\alpha \bar{T}} < 0.15 \frac{R\bar{T}}{E}, \tag{IX.45}$$

the interparticle heat transfer resistance can be neglected.

From the above it follows that in most practical situations a model, that takes into account only an intraparticle mass and an interparticle heat transfer resistance will give good results. However, in experimental laboratory reactors, which usually operate at low gas flow rates, this may not be true. Useful graphs are given by Suzuki and Smith [74] for computing the relative importance of the various transport resistances. Lee and Luss [77] studied the intraparticle temperature difference and Carberry [78] the relative importance of inter- and intraparticle gradients. However, as mentioned, for most practical cases the major thermal resistance is between the surface of the pellet and the ambient gas according to Luss [79], as has often been experimentally confirmed, by Kehoe and Butt [80] among others.

In the above criteria also the heat and mass transfer coefficients for the interparticle transport have to be known. These were amply discussed in Section VII.17. For gas–solid reactions we recommend the correlations from the school of Schlünder [75]. In a packed bed the void fraction is not constant over a cross-section, at the wall for spheres it even approaches $\varepsilon_w = 1.0$. Consequently also fluid velocities over a cross-section are not constant. The most extensive and accurate heat transfer correlation for fluid to particle transport has been presented by Martin [76]:

$$Nu_{\text{single particle}} = 2 + F \, Re^{0.5} Pr^{0.33}, \tag{IX.46}$$

in which $Re = u d_p / v$ and

$$F = 0.664 \sqrt{\left[1 + \left(\frac{0.0557 Re^{0.3} Pr^{0.67}}{1 + 2.44(Pr^{0.67} - 1)Re^{-0.1}}\right)^2\right]}$$

For packed beds Martin [76] found

$$Nu_{\text{packed bed}} = [1 + 1.5(1 - \varepsilon)] \, Nu_{\text{single particle}}, \tag{IX.47}$$

where Re now is $u d_p / v\varepsilon$, because the average fluid velocity now is u/ε. We see that in a packed bed the heat transfer rate is enhanced, e.g. for spheres and $\varepsilon = 0.4$, by a factor of 3.0 approximately. Based on the Chilton–Colburn analogy the same correlations hold for mass transfer if Nu and Pr are replaced by Sh and Sc respectively. The correlations were verified for $0.26 < \varepsilon < 0.935$, $0.6 < Pr$ (or Sc) < 10000 and for $Re \, Pr = Pe > 100$. In industrial practice always $Pe > 100$.

Example IX.2.b Check for intra- and interparticle resistances
A gas phase reaction is catalysed by a porous solid and takes place at 500°C. The global conversion rate has been determined at 240×10^{-6} kmoles key reactant converted/m³ reactor s and the reaction heat per kmole key reactant converted is

$(-\Delta H_r)_A = 700 \text{ MJ/kmol}$. The reaction is assumed to be pseudo first order with $E/R = 11\,600 \text{ K}$. Further data on the gas are $\mu = 3.5 \times 10^{-5} \text{ Ns/m}^2$, $\rho = 0.46 \text{ kg/m}^3$, $\lambda_G = 0.053 \text{ W/mK}$, $c_p = 1080 \text{ J/kgK}$ and the concentration of the reactant in the feed is $\bar{c}_o = 0.205 \times 10^{-3} \text{ kmol/m}^3$.

For the reactant the gas diffusion coefficient is $D_A = 3.9 \times 10^{-5} \text{ m}^2/\text{s}$ and the Knudsen diffusion coeficient $D_K = 0.42 \times 10^{-5} \text{ m}^2/\text{s}$. For the porous catalyst particles we have $d_p = 2.5 \times 10^{-3} \text{ m}$, $\lambda_s = 0.5 \text{ W/mK}$, $\varepsilon_p = 0.7$ and a tortuosity factor $\tau = 2$. The gas load on the reactor is $\Phi_m/S = 0.185 \text{ kg/m}^2 \text{ s}$. Check whether there are any transport resistances.

The effective internal diffusion coefficient in the catalyst pores D_i is given by the relation:

$$\frac{1}{D_i} = \frac{\tau}{\varepsilon_p}\left(\frac{1}{D_A} + \frac{1}{D_K}\right),$$

so that $D_i = 0.13 \times 10^{-5} \text{ m}^2/\text{s}$, and from equation (IX.47) we find that $\alpha_G = 83 \text{ W/m}^2\text{K}$ and $k_G = 0.13 \text{ m/s}$. Further $\Delta T_{ad} = 289 \text{ K}$, $\gamma = 15$, $\beta = 483 \times 10^{-6}$, $\gamma\beta = 7.2 \times 10^{-3}$, $Re = 13.2$, $Sc = 1.95$, $Pr = 1.40$ and $\bar{T} = 773 \text{ K}$. With these data we obtain the criteria:

intraparticle mass transport resistance: $\quad\quad 1.41 \not< 1.01 \quad\quad$ (cf. (IX.42))

intraparticle heat transport resistance: $0.68 \times 10^{-3} < 50 \times 10^{-3}$ (cf. (IX.43))

interparticle mass transport resistance: $11.3 \times 10^{-3} < 150 < 10^{-3}$ (cf. (IX.44))

interparticle heat transport resistance: $\quad 3.3 \times 10^{-3} < 10 \times 10^{-3}$ (cf. (IX.45))

These results lead us to the following conclusions. There are intraparticle mass transfer resistances, but the particles can be considered isothermal. Concentration gradients in the pores of the catalyst can be expected; the effectiveness factor must be <1. Fortunately there are no important concentration and temperature gradients over the gas film surrounding the catalyst particle. For the Thiele modulus we can derive $\phi = d_p(\langle R_A \rangle/\bar{c}_o D_i)^{0.5} = 2.4$. From figure VII.37 we can also read that there must be an intraparticle mass transfer resistance at this value of the Thiele modulus.

IX.2.2 Catalytic gas–solid reactors

Modelling of catalytic tubular reactors

This section stands somewhat apart from the previous one in which the non-uniform conditions were examined at the scale of the particles. The scale of scrutiny will now be enlarged, but it will still remain fairly small with respect to the smallest reactor dimension. The purpose is to discuss some effects of insufficient internal exchange of heat and matter which, particularly in solid catalysed gas reactions, may give rise to undesirable variations in temperature and concentration.

The catalytic tubular reactor (in general a pipe or a number of pipes in parallel, filled with solid pellets of catalyst) was mentioned in Section II.2 with respect to isothermal operation. The theory was developed further in Section VI.3, consideration being given to the heat effect of the reaction and to the external heat exchange. In that treatment, both the temperature and the composition of the reaction mixture were assumed to be uniform over the cross section of the reaction tube. This, however, depends very much on the possibilities of the transport of heat and matter perpendicular to the direction of the main flow. If these possibilities are limited, in the case of, for instance, an exothermic reaction in a cooled tubular reactor containing particles of catalyst, the temperature at the axis of the tube may become considerably higher than near the wall; consequently, the degree of conversion may be higher near the axis than near the wall. The greater part of this section will be devoted to the transverse mixing effects associated with a gas flow through packed beds and to their implications with respect to the design and operation of such reactors.

The internal exchange of heat presents no problem in the *fluidized bed reactor*. The mixing of solids generally is so good that no undesired 'hot spots' develop. A less satisfactory feature of this type of dispersion lies in the fact that, in most applications, part of the gas feed rises through the bed in the form of large bubbles or gas pockets, the contents of which may be insufficiently exchanged with the gas between the particles (see also Chapter VII and Example VII.17.1.). The resulting bypassing of reactant feed may severely limit the final conversion of the reaction.

The description of a *catalytic fixed bed reactor* may call for a very complex model because of heat and mass transport between the two phases and within each phase, as well as chemical reaction in one or more phases or at the phase boundary; further complications may arise from the difference in temperature dependency of the rates of physical transport and chemical reaction. A complete description of such a system can result in a very complex set of equations, in which especially the exponential relation between the temperature and the rate of reaction is a complicating factor. However, it is not always necessary to take all the phenomena into account and often a rather simple model can be used. In general one will choose the simplest possible model that still gives a sufficiently accurate description of the phenomena investigated. The description is never better than the assumptions made in deriving the model and the quality of the parameter values substituted in it, e.g. those of the chemical rate equation.

Most models for fixed bed reactors, usually a set of differential equations, must be solved numerically because they are non-linear. This is possible nowadays with high-speed computers, using advanced mathematical techniques [81], even for very complex models. To test the validity of the description by a model adopted, it is always necessary to compare results of calculations with measured data, which in general can only be obtained with much greater effort. Many mathematical fixed bed reactor models of varying complexity have been presented in the literature. Although the parameter space investigated is not always of practical interest, these analyses produce useful information on the conditions that must prevail in reaction systems to neglect the often secondary effects such as local mixing, resistance to mass and heat transport between phases and inside catalyst particles.

Froment [82, 83] introduced the classification given in Table IX.8 for steady state models. Ray [84] used the same classification for dynamic models. The models are divided into six classes: three types of pseudo-homogeneous models in which the reactor contents are considered as a continuum, and three types of heterogeneous models, in which variables like temperature and concentration differ in the catalyst and fluid phases. The disadvantage of this classification is that the differential equations of models of the same class are quite different in character, depending on the specific properties of the system considered, e.g. whether the reactor is cooled or operated adiabatically.

Table IX.8 Packed bed reactor models, according to Froment [82, 83]

A	Pseudo-homogeneous models $T_{fluid} = T_{solid}$ $c_{in\,fluid} = c_{in\,catalyst\,pores}$	B	Heterogeneous models $T_{fluid} \neq T_{solid}$ $c_{in\,fluid} \neq c_{in\,catalyst\,pores}$
A1	Plug flow one-dimensional	B1	External resistance only
A2	+Axial dispersion one-dimensional	B2	+External and internal resistance
A3	+Radial dispersion two-dimensional	B3	+Radial mixing

Another systematic approach was followed by Slinko [85], who divides a packed bed catalytic reactor into three levels, each level describing a sub-system inside the reactor. This approach was further elaborated by van Doesburg and de Jong [86], but now considering five different levels. The subsystems are defined as outlined in Table IX.9. Each of the levels is studied separately to assess the importance of the various possible physical or chemical phenomena.

The philosophy behind this scheme is to consider the reactor initially as a distributed parameter system by treating each level separately, although it does not imply that the description of the reactor obtained in this way is always a distributed parameter model. A distributed parameter system is a system where the parameter values for, e.g., the gas and the solid phase are different so that we have to describe the two discrete phases separately. If we can lump the properties of the two separate phases together into one single set of parameters, we speak of a lumped parameter system. From the discussion below it will become clear that often several simultaneous mechanisms can be described with one lumped parameter giving an overall description of the processes observed.

Level I contains the catalytically active surface and the molecules on and near this surface. The chemical reaction together with heat production or consumption for exothermic or endothermic systems, respectively, proceeds at the surface. Often reaction rate equations are based on the Langmuir–Hinshelwood adsorption theory and are expressed in terms of partial pressures of the

Table IX.9 Level scheme for model development of packed bed reactors, van Doesburg and de Jong [86]

Level I :	Separate stages of chemical change
Level II :	Transfer processes inside catalyst pellet Heat and/or Mass
Levell III:	Transfer processes in a film layer Heat and/or Mass
Level IV :	Transfer processes in a layer of catalyst Heat and/or Mass
Level V :	Interaction with the environment adiabatic or non-adiabatic

components. These rate equations have been obtained experimentally along the lines given in Chapter VII.

At *level II* the catalyst pellet can be considered as being quasi-homogeneous, since at any point in the pellet the temperature of the gas in the pores and the adjoining solid is equal. The heat of reaction generated or absorbed at the internal surface is distributed over adsorbed molecules, surface atoms or molecules of the catalyst, desorbing molecules and molecules in the gas phase. The large number of collisions guarantees a very good temperature equilibration. When the pellet may be regarded as a continuum relative to the concentration, the mass transport within the particle, mainly by diffusion, is described with an effective diffusivity, i.e. a combination of the molecular diffusion coefficient and the Knudsen coefficient, taking the pellet porosity and the tortuosity of the pores into account (see Example IX.2.b). Heat transport is characterized similarly by an effective thermal conductivity. Thus, a lumped description of a catalyst particle applies. Mass and heat balances for several pellet models are discussed in Chapter VII. When it follows from considerations at this level that the rate of mass transport in the catalyst pellet influences the overall process rate, an *effectiveness factor* can be used to lump level I and II as was outlined in Chapter VII.

At *level III* the system is extended to include the boundary layer around the catalyst pellet and special attention is now paid to heat and mass transfer processes through this layer. The external resistance to mass transport is usually lower than the corresponding internal resistance because the presence of pores reduces the surface through which reactants can diffuse into the particle. The tortuosity of the pores adds to the pore diffusion resistance. This does not imply that the internal rate of mass transfer usually controls the conversion rate. On the contrary, due to chemical enhancement of internal mass transfer by fast reactions, the controlling resistance is often located around the particles in the boundary layer. This is especially true for small particles as found in fluidized bed and riser reactors. When it follows from considerations at this level that the concentrations and temperature at the outer catalyst surface and in the adjoining gas phase do not differ, and when transport resistances within the particle are not present or lumped into an effectiveness factor, a pseudo-homogeneous model description is valid. In Chapter VII and Section IX.2.1 the various criteria were discussed whether phenomena at level II and level III can be lumped together.

Up to this level it makes no difference whether one uses a model based on differential equations or a so-called finite-stage or cell model, as proposed by Deans and Lapidus [87]. In the latter type of model, the voids between the particles are regarded as a structure of perfect mixers. In this book attention will be paid only to continuum models based on differential equations.

The number of mass transport mechanisms added at *level IV* is limited: convective mass transport, 'eddy' transport and bulk diffusion. The latter two are usually lumped together as dispersion in axial or radial direction. Heat transfer is the result of [75, 88]:

1. mechanisms independent of flow:
    1.1 thermal conduction through the solid particle
    1.2 thermal conduction through the contact point of two particles
    1.3 radiant heat transfer between the surfaces of two adjacent pellets (gas–solid system);
2. mechanisms depending on the fluid flow:
    2.1 thermal conduction through the fluid film near the contact surface of two pellets
    2.2 heat transfer by convection, solid–fluid–solid
    2.3 heat conduction within the fluid
    2.4 heat transfer by lateral mixing.

According to an experimental and theoretical study based on a one-dimensional pseudo-homogeneous model by Vortmeyer and Jahnel [89, 90] the contribution of radiation to the total heat flow is important at temperatures above 400°C. This is confirmed by others [91]. Below this temperature the various mechanisms of heat transport, except for heat transport by convection, are usually described by a lumped parameter, the effective thermal conductivity coefficient.

At *level V* the system is extended to interactions with the environment. Heat exchange is possible when the reactor is cooled (see Section VI.5). Cooled reactors require very complicated models when the radial temperature gradients and the resulting radial concentration gradients are so large that they cannot be neglected. A two-dimensional model should then be used and radial conductivity and heat transfer coefficients at the tube wall should be taken into account. The influence of the reactor wall may be important, even for an adiabatic reactor under transient conditions.

This is illustrated by Hoiberg *et al.* [92] who compared experimental data with calculations and found that the heat capacity of the reactor wall considerably influences reactor response. Eigenberger [93, 94] noted that even the response of a homogeneous tubular reactor to a step change in inlet temperature changes from fast and steep to slow and smooth when the heat capacity of the reactor wall is taken into account. Wall effects are particularly important in studies with pilot plant packed bed reactors, since the heat capacity of the wall may be of the same order of magnitude as that of the catalyst bed, or even higher. It thus appears that in real systems it is incorrect to neglect the heat capacity of the reactor wall. Nevertheless, the wall influence is assumed to be negligible in most theoretical studies. A similar conclusion applies to axial heat conduction along the reactor wall. Lunde [95] shows, experimentally and by simulation, that also heat conduction can be important.

In deriving a model for a catalytic gas reaction in a packed bed reactor we have to judge whether we can lump the various phenomena occurring at the different levels together in order to simplify the required model equations. But first we have to discuss the correlations that have been derived to describe the mixing phenomena in packed beds occurring at level IV.

Transverse mixing with flow through packed beds

As a fluid is passing through a bed of particles, its repeated lateral displacement combined with the mixing of volume elements belonging to different streamlines gives rise to a certain degree of mixing perpendicular to the main flow. The degree of mixing can be characterized by means of an apparent transverse dispersion coefficient D_t. As shown in Section IV.6 and figure IV.48 for a bed of spherical particles, Bo_t approaches a value of 10, which is reached when $\langle u \rangle d_p/v > 100$. Accordingly [96]:

$$\frac{\varepsilon D_t}{u d_p} = \frac{1}{10[1 + 19.4(d_p/d_t)^2]}, \tag{IX.48}$$

where u is the fluid velocity based on the empty cross section of the bed. It can be easily ascertained from equation (IX.48) that, for liquids, D_t in all practical cases greatly exceeds the molecular diffusivity D; with gases this is so only when Re_p is sufficiently large.

Heat can be transported perpendicularly to the main flow by the same mechanism if a transverse temperature gradient exists, resulting in a convective heat conductivity λ'_t. Apart from this transverse mixing effect, heat can be transported fairly well in a packed bed by conduction through the fluid from particle to particle, and, with gases, by thermal radiation between the particles. If, therefore, there is no flow, an apparent (isotropic) thermal conductivity of the bed, λ_o, already exists. Several methods have been published for estimating λ_o from the separate thermal conductivities of the particles and the fluid.

It is now assumed that λ_o and λ'_t act fairly independently. Accordingly, the total apparent transverse conductivity, λ_t, for a bed of spheroidal particles is given by

$$\lambda_t = \lambda_o + \lambda'_t. \tag{IX.49}$$

For this relation Schlünder [97] recommended for the convective conductivity term:

$$\lambda'_t = \frac{(\rho c_p)_G d_p u}{8[2 - (1 - 2d_p/d_t)^2]}, \tag{IX.50}$$

which is valid for $30 < Re_p Pr < 600$. The thermal conductivity λ_o of the bed under non-flow conditions is much more difficult to correlate reliably with the system properties. The Zehner–Bauer model derived by Schlünder et al. [75, 98, 99] most probably in the most accurate correlation. We will present here the less complicated, earlier correlation of Zehner and Schlünder [100]:

$$\frac{\lambda_o}{\lambda_G} = 0.67\varepsilon + (1 - \varepsilon)^{0.5} A \qquad \text{(IX.51.a)}$$

for $0.26 < \varepsilon_G < 0.93$, where

$$A = \frac{2}{\left(1 - \frac{\lambda_G B}{\lambda_S}\right)} \left[\frac{\left(1 - \frac{\lambda_G}{\lambda_S}\right) B}{\left(1 - \frac{\lambda_G B}{\lambda_S}\right)^2} \ln\left[\frac{\lambda_S}{B\lambda_G}\right] - \frac{B+1}{2} - \frac{B-1}{\left(1 - \frac{\lambda_G B}{\lambda_S}\right)} \right] \qquad \text{(IX.51.b)}$$

$$B = C\left(\frac{1-\varepsilon}{\varepsilon}\right)^{1.11}. \qquad \text{(IX.51.c)}$$

The constant C is 1.25 for spheres, 1.4 for crushed particles and 2.5 for cylinders and Rashig rings, and further λ_G and λ_S are the termal conductivities of the gas and the solid respectively.

Equation (IX.51) does not take into account the radiation in the bed at temperatures above 400°C. The equivalent heat conductivity for the energy transfer by radiation λ_R is given by

$$\lambda_R = \frac{0.23}{(2/\varepsilon_m) - 1} \left(\frac{T}{100}\right)^3 d_P \; \text{W/m K} \qquad \text{(IX.52)}$$

as originally derived by Damköhler [101]. Here ε_m is the emissivity of the radiating surface and T is expressed in K. Taking into account the radiation Zehner and Schlünder [100] gave the following correlation for the bed heat conductivity under non-flow conditions:

$$\frac{\lambda_o}{\lambda_G} = [1 - \sqrt{(1-\varepsilon)}]\left(1 + \frac{\lambda_R}{\lambda_G}\right) + (1-\varepsilon)^{0.5} A + \frac{(1-\varepsilon)^{0.5}}{\frac{\lambda_G}{\lambda_R} + \frac{\lambda_G}{\lambda_S}} \qquad \text{(IX.53)}$$

in which A is given by equation (IX.51.b) and (c). However, if the contribution of λ_o to λ_t is considerable, we recommend the use of the more extensive, more accurate Zehner–Bauer model [75, 98, 99].

The axial heat dispersion has hardly been investigated. Votruba et al. [192] derived the following correlation:

$$Bo_h^{-1} = \frac{ud_p}{\varepsilon\lambda_l} = \frac{\lambda_o/\lambda_G}{Re_p Pr} + \frac{14.5 \times 10^{-3}}{d_p\left(1 + \frac{C}{Re_p Pr}\right)}.$$

In this correlation λ_o is determined with equations (IX.51 or 53), d_p is expressed in m and C is generally between 0 and 5. Exact values of C can be found in the original publication; for $Re_p > 10$ the value of Bo_h is 3 to 6.

In principle, an additional heat transfer resistance should be taken into account for the transport of heat between the fluid stream and the wall containing the bed. Although the point velocity in the bed can be greater near the wall than its value averaged over the whole cross section, see, e.g., Schwarz and Smith [108] and Schlünder [75], the intensity of transverse mixing decreases as the wall is approached. A boundary layer exists at the wall itself; it has an average thickness determined by the velocity of flow and by the eddies produced by the particles adjacent to the wall. These effects can be conveniently combined into a wall heat transfer coefficient α_w. For packed beds Dixon and Cresswell [102] recommend for this wall heat transfer coefficient:

$$\frac{\alpha_w d_p}{\lambda_t} = 3.0 Re_p^{-0.25} \qquad \text{(IX.54)}$$

which was tested for the range $40 < Re_p < 6000$. Also Hennecke and Schlünder [103] presented a correlation, which is valid over a large range of conditions. Many surveys have been given in literature of the methods for estimation of effective transport properties in packed bed reactors, e.g. by Kulkarni and Doraiswamy [104] and by Gnielinski [105].

An estimate of the relative importance of the transverse conductivity and the wall heat transfer coefficient can be obtained by calculating the radial temperature distribution in a packed cylindrical tube under the assumption that

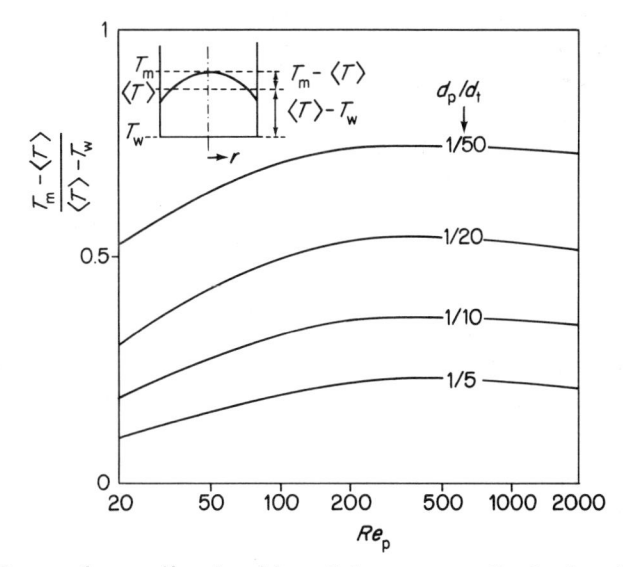

Fig. IX.23 Degree of non-uniformity of the radial temperature distribution with uniform heat production, from equations (IX.50), (IX.54) and (IX.55) with $Pr = 1$ and $\lambda_o/\lambda_G = 10$

the heat generated in it is distributed uniformly over the cross section. A parabolic temperature distribution results as shown in figure IX.23, top left. The ratio of $(T_m - \langle T \rangle)$, temperature in the centre minus average temperature and $(\langle T \rangle - T_w)$, average temperature minus wall temperature, is characteristic of the non-uniformity of the temperature distribution. With a uniform heat production it is found that

$$\frac{T_m - \langle T \rangle}{\langle T \rangle - T_w} = \left(1 + 8\frac{\lambda_t}{\alpha_w d_t}\right)^{-1}. \tag{IX.55}$$

This expression is equal to 1 when thermal conduction is entirely controlling ($\lambda_t/\alpha_w d_t \ll 1$), and it is equal to 0 when the wall heat transfer is the limiting mechanism ($\lambda_t/\alpha_w d_t \gg 1$, uniform temperature over the cross section).

Figure IX.23 shows that the non-uniformity of the temperature depends mainly on d_p/d_t and to a lesser extent on Re_p. It may be concluded that for $d_p/d_t > 0.2$ the temperature is nearly uniform over the cross section and that only α_w needs to be taken into account. Since equation (IX.54) is no longer valid in this region, α_w has to be derived from experimental results obtained under these conditions. The value of α_w can now be combined with the heat transfer coefficient between the wall and the cooling or heating medium to give the overall heat transfer coefficient U. The calculation of a tubular reactor then proceeds along the lines given in Section VI.3. If the influence of λ_t, and consequently that of D_t, cannot be neglected, a more detailed analysis of the temperature and concentrations distribution will be necessary.

The correlations described for the effective transport properties in packed beds have been verified experimentally for the range of $d_t/d_p > 4$. For the irregularities in the gas flow specially near the wall [106, 107] has been accounted for. With respect to the pressure drop the well-known Ergun equation [109] can be used. However, this relation will give erroneous results for $d_t/d_p < 15$; then a special procedure as e.g. outlined by Schlünder [75] has to be used taking into account the higher bed porosity near the wall.

Temperature and concentration distribution in a catalytic tubular reactor

Many reviews on calculation procedures and modelling of packed bed reactors have appeared in the literature, such as those by Finlayson [110], Karanth and Hughes [111] and Hlavacek [112]. The calculation of axial and radial temperature and concentration distributions requires the simultaneous solution of a number of balances to be derived for a small volume element in the packed bed reactor:

— a heat balance of the gas phase;
— a heat balance over the solid phase;
— component mass balances in the gas phase;
— component mass balances in the solid phase;
— the overall mass balance.

To this end the following reasonable assumptions may be made:

— there is cylindrical symmetry;
— the velocity profile is flat;
— D_1, D_t, λ_t and λ_1 are independent of the radial and axial position and of the temperature.

With respect to the last assumption we have to remember that λ_t may well depend on the temperature particularly at high temperatures because of the contribution of the thermal radiation to λ_o. We also have to remember that λ_t refers to the heat dispersion in the *gas phase*, in which the contribution of heat flow through the particles has been included (see equations (IX.51) and (IX.53)).

For level IV and the assumptions above we obtain for the steady state balances: for component J in the gas phase

$$\varepsilon D_t \left(\frac{\partial^2 c_J}{\partial r^2} + \frac{1}{r}\frac{\partial c_J}{\partial r} \right) - u\frac{\partial c_J}{\partial z} + \varepsilon D_1 \frac{\partial^2 c_J}{\partial z^2}$$
$$- k_G a_p (1 - \varepsilon)(c_J - c_{Ji}) = 0 \qquad \text{(IX.56)}$$

and in the solid phase

$$k_G a_p (1 - \varepsilon)(c_J - c_{Ji}) = \rho_B \sum_{}^{M} \langle R_S \rangle_J \ldots \qquad \text{(IX.57)}$$

The heat balance becomes for the gas phase

$$\varepsilon \lambda_t \left(\frac{\partial^2 \bar{T}}{\partial r^2} + \frac{1}{r}\frac{\partial \bar{T}}{\partial r} \right) - \rho c_p u \frac{\partial \bar{T}}{\partial r} + \varepsilon \lambda_1 \frac{\partial^2 \bar{T}}{\partial z^2} - \alpha a_p (1 - \varepsilon)(\bar{T} - T_i) = 0 \quad \text{(IX.58)}$$

and for the solid phase

$$\alpha a_p (1 - \varepsilon)(\bar{T} - T_i) = \rho_B \sum_{}^{M} (-\Delta H_r)_J \langle R_S \rangle_J. \qquad \text{(IX.59)}$$

The balances were made over a cylindrical ring as given in figure IX.24. In the equations the density of the reaction mixture flow has been taken constant, otherwise the term uc_J should have been differentiated; the meaning of some of the symbols is as follows: $1 - \varepsilon$ is the volume catalyst per unit of reactor volume, ε the bed porosity, ρ_B the bulk density of the catalyst (kg cat/m^3 reactor volume), $\langle R_S \rangle$ the conversion rate of J per kg catalyst at the conditions c_{Ji} and T_i including the effectiveness factor correction for level II phenomena as explained in Section VII.14, r the coordinate in the radial direction and M the number of independent reactions in which component J takes part.

The set of equations (IX.56)–(IX.59) together with their appropriate boundary conditions describe the *heterogeneous two-dimensional model* for a packed bed tubular reactor. This model comes very near to reality. For a reaction mixture with n components we would have to solve simultaneously $2n + 2$ partial differential equations and would need reliable correlations for some nine different parameters. It can be understood that in practice the design engineer will strive

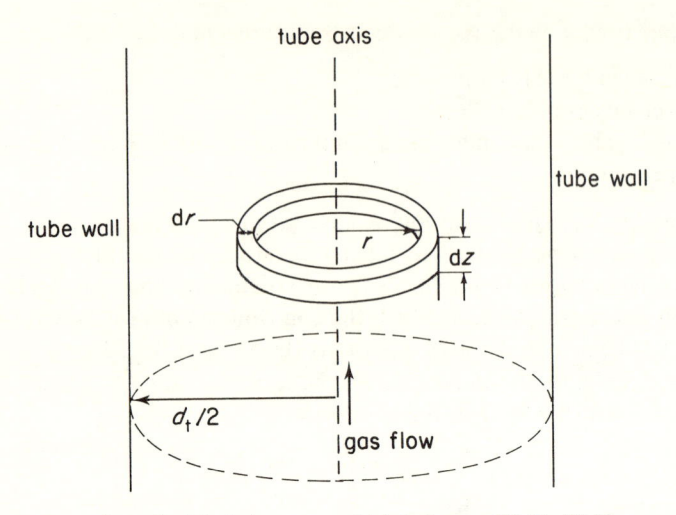

Fig. IX.24 The cylindrical ring over which balances IX.56–IX.59 are made

for further simplifications in order to reduce the computational effort. Fortunately this is very often possible in actual practice.

In Chapter VII and Section IX.2.1 criteria were discussed to judge whether we really have to use a heterogeneous two-dimensional model. Generally two additional simplifications are allowed:

(i) The axial mass dispersion and the axial thermal conductivity in the gas phase can be neglected, according to Young and Finlayson [113], if the following conditions are fulfilled:

for the axial mass dispersion

$$\frac{\langle R_S \rangle_{Jo} \rho_B L}{c_{Jo} u} \ll \frac{uL}{\varepsilon D_1} \tag{IX.60}$$

and for the axial heat dispersion

$$\frac{(-\Delta H_r)_J \rho_B L \langle R_S \rangle_{Jo}}{(\bar{T}_o - T_w)(\rho c_p)_G u} \ll \frac{(\rho c_p)_G uL}{\varepsilon \lambda_t}, \tag{IX.61}$$

in which o refers to the reactor inlet conditions. In practice for technical large scale packed bed reactors generally the reactor load u and inlet temperature $\bar{T}_o - T_w$ are chosen in such a way that the conditions (IX.60) and (IX.61) are fulfilled. This means that in technical packed bed reactors the axial dispersion can be neglected except for very short beds and extremely high conversion rates. Hlavacek and Hofmann [114] confirmed that the length of industrial fixed bed reactors eliminates the need for reactor models including axial dispersion. Therefore also the risks of multiple steady states due to axial dispersion are avoided in practice.

(ii) Applications of the theory developed in Sections VII.14 and IX.2.1 and experimental studies have demonstrated that under industrial conditions ($Re_p > 20$) and in the steady state the temperature difference over the boundary layer around the catalyst particles usually is not more than 0.5–5 K, so that in practice the temperature of the catalyst and the surrounding gas can be put equal as was suggested by Froment [115], among others. This often also holds in practice for the concentration difference over the boundary layer. We have to realize that catalytic tubular reactors are used only for relatively slow reactions. If a reaction were so fast in a tubular reactor that external mass transfer is conversion rate limiting, in practice the development engineer would switch to a fluidized bed or riser reactor with very small particles in order to avoid unnecessary rate limitations.

Assuming these two simplifications are allowed we get the *pseudo-homogeneous two-dimensional model*, see Figure IX.25 for the difference with respect to the previous model, for which the balances are:

the mass balance for component J

$$\varepsilon D_t \left(\frac{\partial^2 c_J}{\partial r^2} + \frac{1}{r}\frac{\partial c_J}{\partial r} \right) - u \frac{\partial c_J}{\partial r} - \rho_B \sum_{}^{M} \langle R_S \rangle_J = 0 \tag{IX.62}$$

and for the energy balance

$$\varepsilon \lambda_t \left(\frac{\partial^2 \overline{T}}{\partial r^2} + \frac{1}{r}\frac{\partial \overline{T}}{\partial r} \right) - \rho c_p u \frac{\partial \overline{T}}{\partial r} - \rho_B \sum_{}^{M} (-\Delta H_r)_J \langle R_S \rangle_J = 0. \tag{IX.63}$$

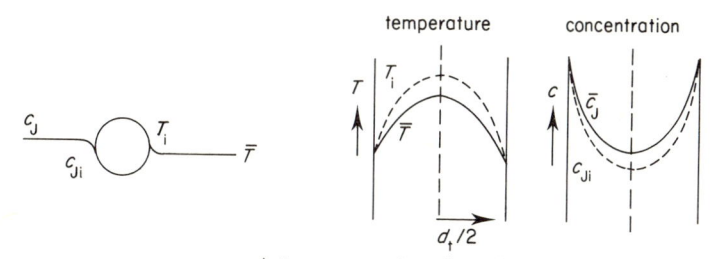

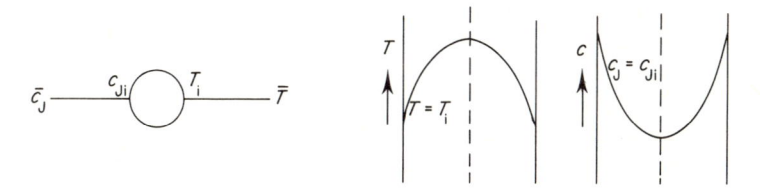

Fig. IX.25 Both two-dimensional models for cooled catalytic tubular reactors

Here again we add up over all the M independent reactions in which component J takes part. We know from Chapter VIII that the individual conversion rates for multiple reactions influence each other mutually: we refer to that chapter, where the appropriate rate expressions are elaborated. The boundary conditions usually applied are:

$$z = 0 \qquad c_J = c_{Jo} \qquad \bar{T} = \bar{T}_o \qquad \text{for} \quad 0 \leq r \leq d_t/2$$

$$r = 0 \qquad \frac{\partial c_J}{\partial r} = 0 \qquad \frac{\partial \bar{T}}{\partial r} = 0 \qquad \text{for} \quad 0 \leq z \leq L \qquad \text{(IX.64)}$$

$$r = d_t/2 \quad \frac{\partial c_J}{\partial r} = 0 \quad -\lambda_t \frac{\partial \bar{T}}{\partial r} = \alpha_w(\bar{T} - T_w) \quad \text{for} \quad 0 \leq z \leq L.$$

In short packed beds different boundary conditions (see also Section IV.1) have to be used, as was outlined by Wicke [116].

The relations in equations (IX.62)–(IX.64) can be made (partly) dimensionless by introducing $c_J = py_J/RT$ for ideal gases and by introducing a reference temperature T_o, T_w or T_c. We then obtain for the component mass balance

$$4\frac{\varepsilon D_t}{u d_p} \frac{d_p}{d_t} \frac{L}{d_t} \left(\frac{\partial^2 y_J}{\partial r^{*2}} + \frac{1}{r^*} \frac{\partial y_J}{\partial r^*} \right) - \frac{\partial y_J}{\partial Z} = \frac{L\rho_B RT}{up} \sum^{M} \langle R_S \rangle_J \qquad \text{(IX.65)}$$

and for the heat balance

$$4\frac{\varepsilon \lambda_t}{\rho c_p u d_p} \frac{d_p}{d_t} \frac{L}{d_t} \left(\frac{\partial^2 \vartheta}{\partial r^{*2}} + \frac{1}{r^*} \frac{\partial \vartheta}{\partial r^*} \right) - \frac{\partial \vartheta}{\partial Z}$$

$$= \frac{L\rho_B}{\rho c_p u T_c} \sum^{M} (-\Delta H_r)_J \langle R_S \rangle_J \qquad \text{(IX.66)}$$

if $r^* = 2r/d_t$, $Z = z/L$ and $\vartheta = \bar{T}/T_c$. Similar boundary conditions as in equation (IX.64) have to be applied including $\vartheta = \vartheta_o$ at $Z = 0$ and $-\partial\vartheta/\partial r^* = Bi(\vartheta - \vartheta_w)$ at $r^* = 1$. Equation (IX.65) still has the dimension kmol J/kmol mixture.

Depending on the reaction rate expressions there are various methods to make equations (IX.65) and (IX.66) completely dimensionless. This will be demonstrated in Example (IX.2.c). We prefer to do this on the basis of mass instead of on the basis of moles, because the number of moles is not necessarily constant over the reactor length, but the mass flow always is.

As already said, there is a radial flow, which is also partly induced by the radial temperature distribution, but these effects have already been accounted for in the experimentally determined values of Bo and Bo_h. The right-hand term in equation (IX.65) contains a kind of a Damköhler number ($k_c\tau$ for first order reactions) and a dimensionless reaction rate, e.g. $R_{SJ}/k_c\,c_{Jo}$; this same term is in equation (IX.66) multiplied by a dimensionless adiabatic temperature rise $\Delta T_{ad}/T_c$. The following dimensionless groups occur in the balances: Bodenstein numbers for radial mass and heat dispersion, the length to diameter ratio for the reactor tube, the tube to particle diameter ratio, a Damköhler number and an adiabatic temperature increase number. Moreover, in the boundary conditions the Biot number, $Bi = \alpha_w d_t / 2\lambda_t$ appears. Now, after all the assumptions made only three parameter values remain to be determined in this model: α_w, λ_t and D_t.

This two-dimensional pseudo-homogeneous model without axial dispersion is more suitable for design purposes, because with a good computer program concentration and temperature profiles can be computed relatively easily. To this end Hofmann [81] developed the FIBSAS program and has tested it extensively by comparing experiments with calculation results. The same program can also be used to evaluate parameter values by regression analysis from experiments, if in an experimental reactor radial temperature and concentration profiles have been measured: without reaction λ_t, α_w and d_t can be measured and with reaction effective kinetic parameters, activation energies and concentration dependences.

Hofmann [81] and also Chao et al. [117] found that under reaction conditions λ_t and α_w deviate significantly from those measured without reaction for heat transport in a packed bed. This is shown in figures IX.26: the difference is only caused by a difference in λ_0, the contribution under non-flow conditions to λ_t. Apparently there exists an interaction between this static thermal conductivity and the kinetic parameters.

Small changes in the kinetic parameters have a far larger impact on the reactor behaviour than considerable changes in the heat transport parameters, so that for

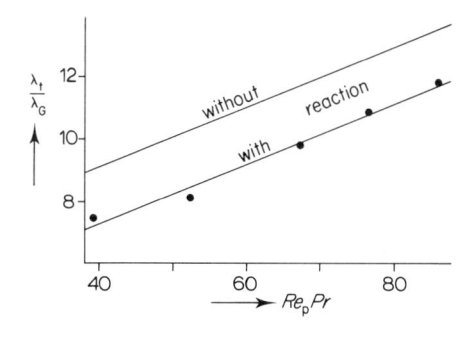

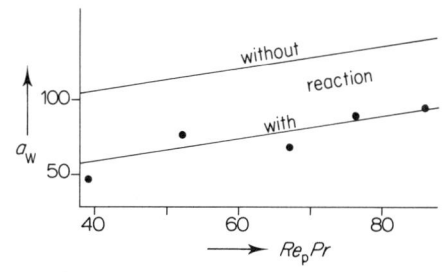

Fig. IX.26 Comparison of experimental and calculated values of λ_t and α_w in packed bed reactors under conditions with or without reaction. [From Hofmann [81]. Reproduced by permission of VDI-Verlag GmbH, Düsseldorf]

reactors it is of no value to look for more precise values and correlations of λ_t and α_w than those available nowadays. The largest uncertainty factor for industrial practice is the accidental character of the packing of the particles in tube bundle reactors, especially at small d_t/d_p ratios, which leads to changed porosities and a changed spread in local porosities after filling the tubes with fresh catalyst. Therefore the accuracy that can be reached in the prediction of the behaviour of packed bed reactors is limited by the poor predictability of the spread in the local porosity distribution of the packing. As shown in Section VII.17 this may greatly affect the reactor performance. This also means that the two-dimensional pseudo-homogeneous model is sufficiently accurate, as was first put forward by Richarz and Sgarlata-Lattmann [118]. Moreover, the model is rather insensitive to changes in D_t, but rather sensitive to changes in α_w and λ_t: mass is much more easily radially dispersed than heat, as was shown by Valstar et al. [119], Carberry [120] and Carberry and White [121], among others.

There remains the *one-dimensional pseudo-homogeneous model*, which has been discussed in Section VI.3. As has been said this model can be applied if $\lambda_t/\alpha_w d_t > 1$, see equation (IX.55). In the derivation of this equation a uniform heat production over the cross section was assumed. This is not valid, of course, and therefore Crider and Foss [122] suggested as a more realistic approximation to the overall heat transfer coefficient,

$$\frac{1}{U} = \frac{1}{\alpha_c} + \frac{1}{\alpha_w} + \frac{d_t}{12.3\lambda_t}, \qquad \text{(IX.67)}$$

in which α_c is the film heat transfer coefficient at the side of the cooling or heating medium. With this relation we can compare the radial temperature difference over the bed ΔT_t with the temperature difference in the fluid phase at the wall ΔT_w:

$$\frac{\Delta T_t}{\Delta T_w} = \frac{\alpha_w d_t}{12.3\lambda_t}.$$

If we wish to have only 25% of the resistance to heat transfer in the bed, this ratio should be 0.25 and the Bi number $0.25 \times 12.3 = 3.1$. For $d_t/d_p = 6$ we then find, from equation (IX.54), that in this case Re_p should be around 1200. This confirms that in packed tubes with a low value of d_t/d_p and a high gas load the one-dimensional plug flow model is adequate to describe the reactor behaviour.

Hlavacek [123] derived criteria for the adequacy of the one-dimensional model, which involve the knowledge or the one-dimensional T and ζ_A profiles, calculated with the value of U from equation (IX.67). Together with the profiles a function QR- the local ratio of the temperature derivatives of the heat production and heat withdrawal rate in the tube, which can be obtained from equation (VI.23), should be calculated at all axial points in the PFR:

$$QR = \frac{dH\ P\ R/dT}{dH\ W\ R/dT} = \frac{\Delta T_{ad} d\zeta_A/dT}{1 + \dfrac{4U}{\rho c_p d_t}(T - T_c)\dfrac{d\tau}{dT}}.$$

After comparison of the one-dimensional and two-dimensional pseudo-homogeneous models by computer calculations, Hlavacek distinguished three regions of adequacy of both models, depending also on the value of a dimensionless adiabatic temperature rise $\Delta T_{ad} = E\Delta T_{ad}/RT_o^2$:

— For $\Delta T_{ad} < 15$ and $QR \leq 1$ at all points in the reactor the one-dimensional model gives an excellent approximation; for $QR > 1$ the information obtained is adequate, although the two-dimensional model should be used.
— For $15 < \Delta T_{ad} < 50$ the one-dimensional approximation can be used only up to $QR = 1$.
— For $\Delta T_{ad} > 50$ the two-dimensional model must be used already when $QR > 0.5$.

We refer further to Section VI.3 and will give an example of the application of the pseudo-homogeneous two-dimensional model.

Example IX.2.c *Temperature and concentration distribution in an experimental catalytic reactor*

This Example was taken from a design problem worked out at the Institute of Chemical Technology, Technische Hochschule, Darmstadt, Germany. The results were kindly made available to us by Dr H Hofmann. It has been decided to build an experimental tubular reactor (diameter: 32 mm; length: 0.4 m) for the catalytic chlorination with HCl of benzene in the vapour phase (component A). The conversion rate is known as a function of temperature and of the degree of conversion of HCl (ζ_A) from measurements carried out in a fluidized bed with small particles of the catalyst to be used (see figure IX.27). The tubular reactor is to be filled with 2.5 mm particles of catalyst and it has been verified that under the reaction conditions mass transfer and diffusion inside these catalyst pellets do not significantly slow down the chemical conversion rate. The reactor will be cooled with paraffin oil of $T_c = 245°C$ at a fairly low flow rate. It is required to calculate the temperature and conversion distribution for the following operating conditions: feed composition as given in figure IX.27, a mass velocity of 0.314 kg/m² s and an inlet temperature $T_o = 245°C$.

The solution of this problem depends on a number of parameters which can be obtained by making the differential equations for the concentration and the temperature distribution dimensionless. When these are written down on a mass basis, and when the following new variables are introduced:

$\zeta_A = 1 - w_A/w_{Ao}$, $\vartheta = T/T_c$, $r^* = 2r/d_t$ and $Z = z/L$, equation (IX.62) can be written as

$$0 = \frac{4\varepsilon D_t L}{u d_t^2}\left(\frac{\partial^2 \zeta_A}{\partial r^{*2}} + \frac{1}{r^*}\frac{\partial \zeta_A}{\partial r^*}\right) - \frac{\partial \zeta_A}{\partial Z} - \frac{\rho_B L}{\rho u w_{Ao}} M_A R_{SA}$$

and equation (IX.63) as

$$0 = \frac{4\varepsilon \lambda_t L}{\rho c_p u d_t^2}\left[\frac{\partial \vartheta}{\partial r^{*2}} + \frac{1}{r^*}\frac{\partial \vartheta}{\partial r^*}\right] - \frac{\partial \vartheta}{\partial Z} + \frac{(-\Delta H_r)_A w_{Ao}}{T_c c_p M_A}\frac{\rho_B L}{\rho u w_{Ao}} M_A R_{SA}.$$

The boundary conditions (IX.64) can be easily transformed into the new variables, and the last condition of equation (IX.64) becomes

$$\frac{\partial \vartheta}{\partial r^*} = -\frac{Ud_t}{2\lambda_t}\vartheta \quad \text{at} \quad r^* = 1.$$

The following values of the parameters are obtained by calculation:

$$\frac{4\varepsilon D_t L}{u d_t^2} = 0.53, \quad \frac{4\varepsilon \lambda_t L}{\rho c_p u d_t^2} = 1.83,$$

$$\frac{\rho_B L}{\rho u w_{Ao}} = 3400 \text{ s} \left(= \frac{\text{kg of catalyst}}{\text{kg/s of HCl fed}}\right),$$

$$\frac{(-\Delta H_r)_A w_{Ao}}{c_p M_A} \equiv \Delta T_{ad} \simeq 122°C \quad \text{(increases somewhat with } T\text{)}$$

and $\Delta T_{ad}/T_c = 0.24$. Further $Ud_t/2\lambda_t = 0.46$ (in which the gas–wall heat resistance, α_w^{-1}, is 14% of the overall heat resistance, U^{-1}.

With these values of the parameters and with R_{SA} given by figure IX.27 the differential equations can be numerically solved; the steps taken in the z-direction are $\frac{1}{40}$ of the tube length L, and in the r-direction $\frac{1}{8}$ of the tube diameter d_t. The resulting temperature distribution is shown in figure (IX.28). Both the temperature at the axis and that near the wall show an appreciable increase (30 to

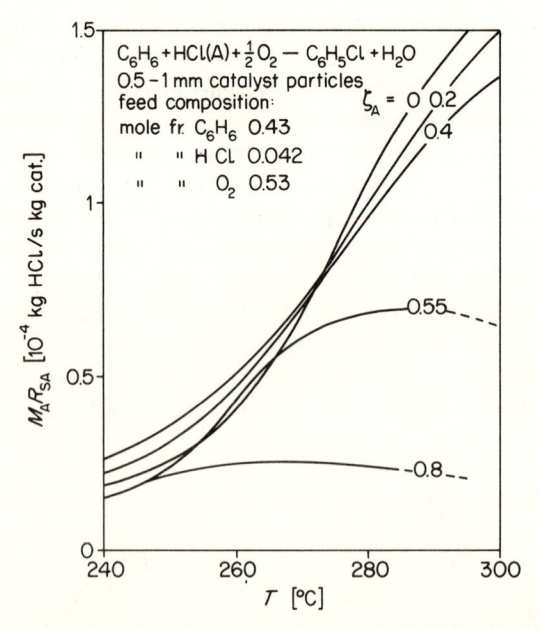

Fig. IX.27 Empirical data for the conversion rate of HCl (A) in the gas phase chlorination of benzene, obtained with small catalyst particles in a fluidized bed; Example IX.2.c

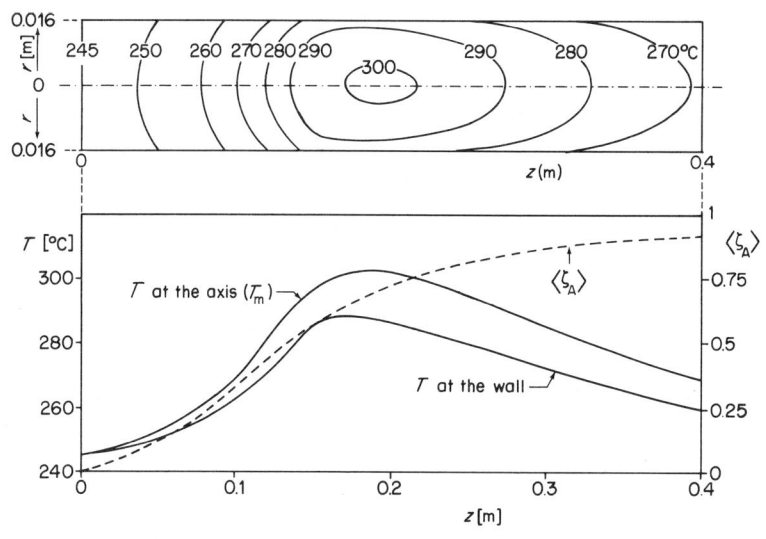

Fig. IX.28 Temperature and conversion distribution in an experimental catalytic reactor for the reaction indicated in figure IX.27; Example IX.2.c

40% of ΔT_{ad}), but the non-uniformity of the temperature over the cross section appears to be relatively small. This is due to the relatively low value of U which is determined mainly by the rather poor heat transfer coefficient between the cooling liquid and the outer surface of the tube. The degree of conversion averaged over the cross section $\langle \zeta_A \rangle$ as a function of z is also shown in figure IX.28. The radial variation of ζ_A is very small in this case. This is caused by the relatively small radial temperature variations and by the fact that, in the lower temperature range and below 50% conversion, the conversion rate is rather independent of ζ_A.

It may be concluded from these calculations that the reactor would not operate under sufficiently isothermal conditions if it were used for kinetic studies. A considerable improvement would be obtained if the outside heat transfer coefficient were greatly increased.

IX.2.3 The moving bed gas–solid reactor

Gas–solid reactions are the heart of many important industrial processes in metallurgy, combustion and gasification of coal, the cement production, the ceramics and refractories industry, etc. Much practical information is available for these processes but only in the last decades has good fundamental work been undertaken which gives an understanding of the reactors themselves and their behaviour. The computer is indispensable for transforming experimental results together with an adequate model into practical tools for the prediction of the behaviour of large scale reactors. For reactions between gases and solids the basic fundamentals specially on the scale of single particles are discussed in the book of

Szekely *et al.* [124], examples of complete industrial reactor studies can be found in recent publications, e.g. by Yoon *et al.* [125] and Amundson and Arri [126] on the Lurgi moving bed gasifiers, by Weimer and Clough [127] on a fluid bed gasifier and by Kam and Hughes [128] on the iron ore reduction in a moving bed reactor.

Three different basic types of gas–solid reactors can be distinguished:

— *moving bed reactors.* These are of the moving bed type such as shaft furnaces and multiple deck furnaces. Also the rotary kiln [129], which already approaches the fluid bed.
— *fluid bed reactors.* In these reactors one single bed or more beds in cascade can be applied, as also recirculation of the solids.
— *transport reactors.* In these reactors the solid particles are transported by the gas; examples are coal powder burners, coal gasifiers of the Shell or Texaco type, melt cyclones, riser reactors in cat crackers and so-called 'fast fluidized beds' in calcination and gasification.

The main distinction between these reactor types can be found in the residence time of the solids, which is in the range of hours to days, minutes to hours and seconds respectively. Gas residence times are always in the order of magnitude of seconds to even fractions of seconds in the transport reactors. The mixing behaviour of gases and solids can vary widely and, especially in fluid bed reactors, can become very complicated (see Example VII.17.1). Fluid beds behave differently from packed bed reactors; Re numbers are usually low due to the small particle sizes. The Nu and Sh numbers generally demonstrate a considerable spread; correlations can be found in Section VII.17.

A choice between the available reactor types is generally based on the reaction rate (the particle diameters of the available solids and the required solids residence time in the reactor). The particle diameter is generally chosen of such dimensions that ineffective particle zones are avoided. Therefore for fast reactions fluid bed and riser reactors are applied, while moving beds are used in practice for relatively slow reactions. This is the more true because fast reactions call for an excellent heat transfer characteristic, which can easily be realized specially in fluidized beds.

For the design and operation of a gas–solid reactor also the accurate control of the temperature of the solid particles is very important. A slight overheating of the particle can completely modify the texture and the properties of the required products. In practice the most often used methods for the temperature control of exothermic reactions are:

— cooling by cold reactant gases;
— cooling by cold solid reactants;
— recirculation of cooled products or inert heat carriers;
— cooling through the wall or through tubes in the bed;
— use of latent heat, e.g. the injection of water sprays;
— cooling by a simultaneous endothermic reaction.

Similar methods are used for heat supply to endothermic reactions.

The reaction between a gas and a solid takes place at the surface outside or inside in pores of the particles depending on the value of $k_G\delta/D_iE_A$ with $\delta = V/A$ (see Chapter VII). For heat and mass transfer rates the outer surface area of the particles is most important. This can be seen in figure IX.29, where for spherical particles the external area is plotted as a function of the particle diameter. However, in industrial practice particles are not often spherical; the usual rather irregular shape makes it hard to predict accurately the surface area available for reaction. To this end $\delta = V/A$ is used, as explained in Chapter VII. It must be noted that the surface area — and also other important conversion controlling parameters — can change with the course of the reaction. In particular, temperature changes can have important effects: the structure of the solid may change due to swelling, cracking, softening and sintering, pore formation, enlargement of pores or the progressive growth of a solid product layer.

The use of a model to describe the conversion rate of a particle depends on the particle behaviour during the reaction. For *non-porous particles* there can be distinguished, among others, the following:

— a shrinking particle, e.g. as with the combustion of some types of low ash content coals;
— a shrinking unreacted core. In this case the overall size remains unchanged but a porous product layer remains around the unreacted core, e.g. as with the reduction of metal oxides (see Example IX.2.d);
— the same, but now the particle size changes due to the difference of the volume of the solid before and after reaction.

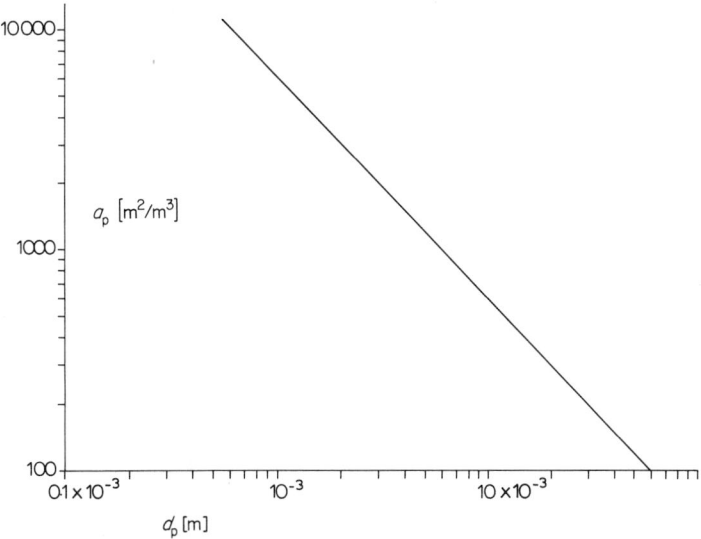

Fig. IX.29 Available external interfacial area of spherical solid particles as a function of the particle diameter

For *porous particles* different models have been developed with the following extreme cases:

— the homogeneous model. In this case the diffusive transport of the reacting gas inside the pores is so high relative to the chemical reaction rate that the conversion of the solid reactant occurs homogeneously all over the particle. In this case the reaction time is independent of the particle diameter, provided interparticle resistances are absent;
— the unreacted core model. In this case the gaseous reactant diffuses so slowly into the pores that the reaction takes place only at a sharp interface inside the particle. Now the reaction time is proportional to d_p if chemical reaction controls the overall rate, or to d_p^2 if back diffusion of gaseous products through the product layer does so (see Example VII.12.2.c).

Szekely and Evans [130] proposed the so-called grain model, which is intermediate between both extremes and which takes into account the fact that the pellet may often be composed of individual particles or grains (crystals), whereas the reactions take place on the surface of the grains. Another intermediate model is the local volumetric model of Groeveveld and van Swaaij [131]. In Chapters VII and VIII several of the models mentioned have been discussed (see Example VII.12.2.c for the shrinking unreacted core and Example VII.13 for the grain model).

None of the models currently available to describe the particle behaviour are yet completely satisfactory and new ones are still being developed [132]; moreover the different models and their characteristics over a widely varying regime have not yet been classified [133], especially with respect to their practical applicability. It is questionable, however, whether all the efforts for creating better models will give rewards in future, especially in view of the difficulty of predicting and controlling shape properties of solids and the ever-changing reactivity properties of natural raw materials, which have been excavated from our soil.

In mass and heat balances for gas–solid reactors new terms appear because of the solid flow entering and leaving the reactor. The usual assumptions made for moving bed reactors are:

— plug flow both in the gas and the solid phase;
— no radial mass gradients;
— often also isothermicity of the particles and constant heat capacities.

For a moving bed shaft reactor as in figure IX.30 and with the assumptions mentioned the heat balance for the solids is

$$\lambda_1 \frac{d^2 T_i}{dz^2} - \frac{4c_{pS}}{\pi d_t^2} \frac{d(\Phi_{mS} T_i)}{dz} - \alpha a_p(1-\varepsilon)(T_i - \bar{T})$$

$$- \frac{4\alpha_w}{d_t}(T_i - T_w) + \rho_B \sum_{J}^{M}(-\Delta H_r)_J \langle R_S \rangle_J = 0, \qquad (IX.68)$$

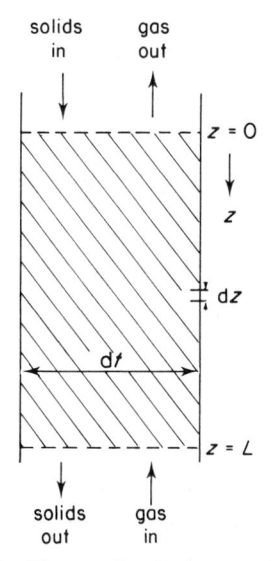

Fig. IX.30 The moving bed gas–solid reactor

the heat balance for the gas is

$$\frac{4c_{pG}}{\pi d_t^2} \frac{d(\Phi_{mG}\bar{T})}{dz} + \alpha a_p(1-\varepsilon)(T_i - \bar{T}) - \frac{4\alpha_w}{d_t}(\bar{T} - T_w) = 0, \quad \text{(IX.69)}$$

the component mass balance for the solid phase is

$$\frac{4}{\pi d_t^2} \frac{d(\Phi_{mS} w_{sJ})}{dz} = M_J \rho_B \sum^M \langle R_S \rangle_J \quad \text{(IX.70)}$$

and for the gas phase

$$\frac{4}{\pi d_t^2} \frac{d(\Phi_{mG} w_{GJ})}{dz} = M_J \rho_B \sum^M \langle R_S \rangle_J. \quad \text{(IX.71)}$$

Here T refers to the gas and T_i to the solids temperature, n to the total number of independent reactions and M to the number of independent reactions in which component J takes part. The usual boundary conditions are:

$$x = 0: T_i = T_{io}, \Phi_{mS} = \Phi_{mSo}, \quad w_B = w_{Bo}$$

$$x = L: \bar{T} = \bar{T}_L, \frac{dT_i}{dx} = 0, \quad \Phi_{mG} = \Phi_{mGL}, \quad w_A = w_{Ao}.$$
$$\text{(IX.72)}$$

The mathematical formulation of $\langle R_S \rangle_J$ strongly depends on the model that describes the gas–solid reaction. Also many numerical procedures for the solution of the above set of equations can be found in literature; as an example we refer to the work of Garza-Garza and Dudukovic [134]. If there are no heat losses through the wall, if there is no axial heat conduction and if the flows Φ_{mS} and Φ_{mG} are constant, we can see from equation (IX.69) that

$$T_i = \bar{T} - \frac{4c_{pS}\Phi_{mG}}{\pi d_t^2 \alpha a_p (1-\varepsilon)} \frac{d\bar{T}}{dz}.$$

After differentiation of this relation and substitution into equation (IX.68), we find

$$\left(\frac{4}{\pi d_t^2}\right)^2 \frac{c_{pG}\Phi_{mG} c_{pS}\Phi_{mS}}{\alpha a_p (1-\varepsilon)} \frac{d^2 \bar{T}}{dz^2} + 4 \frac{c_{pS}\Phi_{mS} - c_{pG}\Phi_{mG}}{\pi d_t^2} \frac{d\bar{T}}{dz}$$

$$+ \rho_B \sum_{J}^{M} (-\Delta H_r)_J \langle R_S \rangle_J = 0 \qquad \text{(IX.73)}$$

In this last relation of the first term represents the dispersion or back-mixing of heat caused by the countercurrent flow of gas and solids!

In the models for single particles the conversion rate is often expressed as the rate of change in size of the particle. This can sometimes cause difficulties in transforming to the scale of a volume element of the entire reactor. On the basis of the theory outlined in chapters VII and VIII we will outline the steps required from single particle behaviour to macroscopic reactor balances and will use the sharp interface shrinking unreacted core model. This model e.g. is often used for the iron ore reduction in a blast furnace of a steel plant. The reaction between iron oxide particles and reducing gas is assumed to take place at the sharp interface of porous iron surrounding the non-porous iron oxide core. The reaction is, e.g.,

$$3H_2 + Fe_2O_3 \rightarrow 3H_2O + 2Fe$$

or in symbols

$$\nu_A A + \nu_B B \rightarrow \nu\ P_P + \nu_Q Q,$$

in which A and P are gases and B and Q solids. The product layer is porous and the particle keeps its original diameter. In Example VII.12.2.c for this particle model we derived for the molar flow rate of A from the bulk of the gas towards the particle

$$\Phi_A = 4\pi r_p D_A \left(c_A - \frac{c_P}{K}\right)/A^*$$

and for the molar flux of A to the particle

$$J_A = \frac{D_A}{r_p}\left(c_A - \frac{c_P}{K}\right)/A^*$$

and for the conversion rate of A in moles of A consumed per unit time and unit reactor volume

$$R_A = \frac{3(1-\varepsilon)}{r_p} J_A.$$

In these relations A^* was given by

$$A^* = \frac{1}{Sh_B} + \left[\frac{D_B}{D_{iB}} + \frac{D_B}{KD_{iQ}}\right]\left(\frac{1-r^*}{r^*}\right) + \frac{D_B}{k''r_p}\frac{1}{(r^*)^2} + \frac{D_B}{KD_Q}\frac{1}{Sh_Q}.$$

These relations for Φ_B, J_B and R_B will now be used to formulate heat and material balances for the reactor. For the heat balance for one single, isothermal particle we can write, for the reaction scheme given above

$$\frac{4}{3}\pi r_p^3 c_{pS}\frac{d\rho_p T_i}{dt} = \frac{v_B}{v_A}4\pi r_p^2 J_A(-\Delta H_r)_A - 4\pi r_p^2 \alpha_w(T_i - \bar{T}).$$

After substitution of the relation for J_A, *assuming that* ρ_p remains approximately constant for this reaction and dividing by $4\pi r_p/3\lambda_G T_o$, in which T_o is the temperature of the particle at $t = 0$, we obtain:

$$\frac{d\vartheta_i}{d\tau'} = \left[\frac{6(-\Delta H_r)_A c_{Ao} D_A}{\lambda_G T_o A^*}\right]\left(\frac{c_A}{c_{Ao}} - \frac{c_P}{Kc_{Ao}}\right) - 3Nu_p(\vartheta_i - \vartheta).$$

In this expression $\vartheta = T/T_o$ and $\tau' = t/t'_c$ with the time constant $t'_c = 2r_p^2\rho_p c_{pS}/\lambda_G$. This equation has to be compared with equation (h) in Example VII.12.2.c. Further, we have

$$\frac{dr^*}{d\tau'} = \frac{1}{(r^*)^2}\left(\frac{c_A}{c_{Ao}} - \frac{c_P}{Kc_{Ao}}\right)/A^*$$

for the mass balance in the particle, in which $r^* = 2r_i/d_p$, $\tau = t/t_c$ and
$$t_c = c_{Bo}v_A r_p^2/v_B c_{Ao} D_A.$$

These particle mass and heat balances now have to be translated to the scale of a volume element of the entire shaft reactor of figure IX.30. The gases flow upward and enter with a temperature T_o. The solid flows downwards and enters at a temperature T_{io}. We will now make the following assumptions:

— both streams move in plug flow;
— the reactor operates adiabatically, there are no radial concentration and temperature gradients;
— there is constant pressure in the reactor and ideal gas behaviour;
— the voidage of the packing is constant;
— the particles are spherical and their density and diameter remain constant;
— the volumetric flow rates of the gas and solid streams remain constant.

Generally the particle density will change with the course of the reaction, but we may introduce an average density.

We will now consider a volume element of the reactor $\pi d_t^2\, dz/4$ and then find for the mass balance of the component A in the gas:

$$\frac{\pi}{4}d_t^2 u_G \varepsilon\, dc_A = J_A a_p(1-\varepsilon)\frac{\pi}{4}d_t^2\, dz = 3\frac{D_A(1-\varepsilon)}{r_p^2}\frac{p}{RT}\frac{y_A - y_P/K}{A^*}\frac{\pi}{4}d_t^2\, dz$$

and after introducing the dimensionless reactor height $Z = z/L$:

$$\frac{dy_A}{dZ} = \left[\frac{3(1-\varepsilon)D_A L}{\varepsilon u_G r_p^2}\right]\frac{y_A - y_P/K}{A^*}. \tag{IX.74}$$

For the solid reactant B we can write the mass balance as follows

$$d\left(\frac{\Phi_m}{\rho_s}c_B\right) = \frac{v_B}{v_{Ap}}J_A a_p(1-\varepsilon)d_t^2\, dz.$$

The concentration of B in the solid is $c_B = c_{Bo}(r^*)^3$, so that $dc_B = 3c_{Bo}(r^*)^2 dr^*$ and further $4\Phi_{mS}/\pi d_t^2 \rho_S(1-\varepsilon) = u_S$ is assumed constant, so that in dimensionless form we get

$$\frac{dr^*}{dZ} = \left[\frac{v_B D_A L p}{v_A r_p^2 u_S c_{Bo} RT}\right]\frac{1}{(r^*)^2}\frac{y_A - y_P/K}{A^*}. \tag{IX.75}$$

For the heat balance for the gas phase we have

$$u_G \varepsilon \rho_G c_{pG}\frac{\pi}{4}d_t^2\, dT = \alpha a_p(1-\varepsilon)(\bar{T} - T_i)\frac{\pi}{4}d_t^2\, dz$$

or in dimensionless form

$$\frac{d\vartheta}{dZ} = \left[\frac{3\alpha L(1-\varepsilon)}{r_p \rho_G c_{pG} u_G \varepsilon}\right](\vartheta - \vartheta_i). \tag{IX.76}$$

For the heat balance of the solids we find

$$(1-\varepsilon)u_S \rho_S c_{pS}\frac{\pi}{4}d_t^2\, dT_i = \alpha a_p(1-\varepsilon)(\bar{T} - T_i)\frac{\pi}{4}d_t^2\, dz$$

$$+ J_A(-\Delta H_r)_A a_p(1-\varepsilon)\frac{\pi}{4}d_t^2\, dz,$$

and if again we make this expression dimensionless,

$$\frac{d\vartheta_i}{dZ} = -\left[\frac{3\alpha L}{\rho_S c_{pS} u_S r_p}\right](\vartheta_i - \vartheta) + \left[\frac{3(-\Delta H_r)_A D_A L p}{\rho_S c_{pS} u_S r_p^2 T_o RT}\right]\frac{y_A - y_P/K}{A^*} \tag{IX.77}$$

and the usual boundary conditions are

$$Z = 0,\quad \vartheta_i = \vartheta_{io},\quad r^* = 1$$
$$Z = 1,\quad \vartheta = \vartheta_o,\quad y_A = y_{Ao}.$$

The dimensionless groups in equations (IX.74)–(IX.77) are all weak functions of the temperature; only A^* and K can strongly depend on the reactor temperatures. It is clear that the solution of the above set of simultaneous differential equations requires quite some mathematical effort with many iterations, but this is not difficult on modern computers. In the above equation r^*

is chosen as variable, but also easily the conversion of the solid reactant A can be taken as the variable:

$$\frac{d\zeta_B}{dZ} = -3\left[\frac{v_B D_A L p}{v_A r_p^2 u_S c_{Bo} RT}\right]\frac{y_A - y_P/K}{A^*}. \tag{IX.75.a}$$

In view of what has been said on the variation of reactivities of natural raw materials like ores or coal, in general only information on the behaviour of shaft moving bed reactors can be obtained. Fortunately a reduction of reactivity of the feed stock will be automatically compensated for by an increase in the average temperature level in the adiabatic shaft reactor, a well-known phenomenon in practice. This temperature level change is usually in the range of less than 50°C.

We will now present as an example the calculation of the temperature profiles in a semi-technical test reactor for the reduction of iron ore.

Example IX.2.d *Reduction of iron ore in a semi-technical reactor*
In a laboratory reactor with $d_t = 0.40$ m and $L = 2.0$ m iron ore pellets with $d_p = 0.01$ m are reduced by a mixture of 70 vol % N_2 and 30% vol H_2 at atmospheric pressure. Both gas and solid are fed at inlet temperatures of 800°C, $u_G = 2$ m/s and $u_S = 100 \times 10^{-6}$ m/s, the bed porosity is $\varepsilon = 0.4$. We will calculate the temperature profiles in this reactor and take the data from Ramachandran *et al.* [153].

The reaction $3H_2 + Fe_2O_3 \to 3H_2O + Fe$ is reversible and first order in H_2 and zero order in Fe_2O_3 with $k'' = 1.69 \exp(-4860/T)$ m/s, with $K = 38 \times 10^{-3} \exp(2800/T)$ and with the reaction heat $(\Delta H_r)_B = +70$ MJ/kmol Fe_2O_3 converted. Further data are, at 800°C:

$\rho_S = 4320$ kg/m³, $c_{pS} = 1100$ J/kg K and $c_{Bo} = 27$ kmol/m³

$\rho_G = 0.23$ kg/m³, $c_{pG} = 2080$ J/kg K and $\lambda_G = 2.6$ W/m K

$k_G = 1.82$ m/s, $\alpha = 2600$ W/m² K and $\mu_G = 21 \times 10^{-6}$ kg/ms

$D_A = 0.093$ m²/s, $D_{iA} = 300 \times 10^{-6}$ m²/s, $D_P = 0.035$ m²/s and

$D_{iP} = 110 \times 10^{-6}$ m²/s.

With these data we first determine the relative importance of the different resistances to mass transfer inside and outside a single particle at 800°C:

$$A^* = 0.033 + (310 + 1640)\frac{1-r^*}{r^*} + \frac{1020}{(r^*)^2} + 0.172.$$

Thus the resistances across the gas film surrounding the particle are negligible. Of course, at the start of the reaction ($r^* = 1$) pore diffusion is absent, but at a conversion of 90% ($r^* = 0.464$) the resistances are

$$A^* = 360 + 1890 + 4740,$$

so at a conversion $\zeta_B = 0.90$ and at 800°C of the total resistance 32% is originated by the pore diffusion. Now the differential equations become

$$\frac{dy_A}{dZ} = 188 \frac{y_A - y_P/K}{A^*}$$

$$\frac{dr^*}{dZ} = \frac{34\,800}{(r^*)^2} \frac{y_A - y_P/K}{A^*} \quad \text{or} \quad \frac{d\zeta_B}{dZ} = 104\,000 \frac{y_A - y_P/K}{A^*}$$

$$\frac{d\vartheta}{dz} = 4890(\vartheta - \vartheta_i)$$

$$\frac{d\vartheta_i}{dZ} = -6560(\vartheta_i - \vartheta) + 49.4 \times 10^6 \frac{y_A - y_P/K}{A^*},$$

and the boundary conditions are

$$Z = 0, \quad \vartheta_i = 1 \quad r^* = 1 \text{ or } \zeta_B = 0$$
$$Z = 1, \quad \vartheta = 1 \quad y_A = 0.30.$$

These four equations with their boundary conditions have been integrated numerically; the results are given in figure IX.31. We assume that all dimensionless groups are constant and independent of temperature except k'' and K containing groups, for which the relations above were used. We see that the solids temperature reaches a minimum due to the endothermicity of the reaction, as also does the gas temperature. Further in this test reactor the whole reactor consists of one single reduction zone and the reaction does not go to completion, as can be calculated from an overall heat balance.

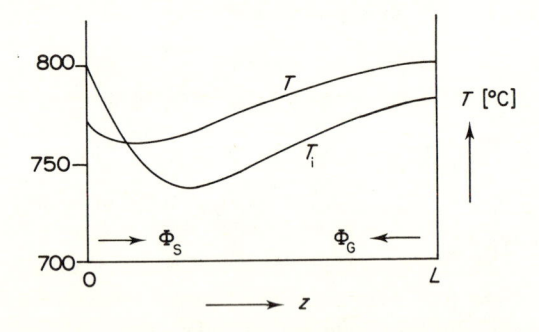

Fig. IX.31 Temperature profiles in a test reactor for the reduction of iron ore; Example IX.2.d

IX.2.4 Thermal stability and dynamic behaviour of gas–solid reactors

In principle all the phenomena of thermal stability and reactor dynamics, which were observed in single phase reactors and as discussed in Chapter VI, can also occur in heterogeneous gas–solid reactors: parametric sensitivity, multiplicity, hysteresis and limit cycles. These phenomena can now occur on different scales: the microscopic scale of a single particle and the macroscopic scale of the entire reactor and even including the heat exchange equipment in the reactor section (see Section VI.5 on autothermal reactors). Moreover due to the heterogeneity of the reacting system also new phenomena can be observed in gas–solid reactors such as 'wrong way behaviour' and 'creeping'. Pioneers in the field of the thermal and dynamic behaviour of gas–solid reactors were Amundson and Aris [136] and Wicke and Vortmeyer [137]. This field was reviewed by Luss [138] for single particles and by Endo et al. [139] for catalytic tubular reactors. Good surveys on stability and dynamics of gas–solid reactors in general were given by Gilles [140] and Eigenberger [141].

As already discussed in Section IX.2.1, multiplicity can occur for *single particles* or for single particle systems. This has been demonstrated for the case of the NH_3 combustion at a Pt gauze in Example IX.2.a. Luss [142] showed that multiplicity will exist for a single isothermal particle if

$$\frac{\gamma\beta}{1+\beta} > 8, \tag{IX.78}$$

where $\gamma = E/R\bar{T}$ and $\beta = (-\Delta H_r)_A k_G \bar{c}_A/\alpha_G \bar{T}$, and where $\bar{c}$ and $\bar{T}$ refer to the conditions in the bulk of the surrounding gas. The condition (IX.78) is also valid for an adiabatic fixed bed reactor, which is described by the one-dimensional two phase model with equal Péclet numbers for mass and heat and if $\bar{c}$ and $\bar{T}$ are replaced by the inlet conditions of the bed, c_o and T_o. The condition (IX.78) is not too severe: a study by van den Bosch and Luss [143] revealed that the industrially very important NH_3 and CH_3OH synthesis reactors operate in the multiplicity region. In extreme situations even intraparticle multiplicity may occur, if the effect of intraparticle heat transfer resistance on the local reaction rate is significant compared to the effect of intraparticle mass transfer resistance. In this case the acceleration of the reaction by a temperature increase inside the particle overcompensates the reduction of the reaction rate by the concentration decrease towards the particle centre [144, 145]. This was discussed in Section VII.14. Generally at high industrial flow rates the temperature differences between gas and particles are very low indeed, whereas heat conductivities of particles are large enough to prevent important intraparticle temperature gradients. Nevertheless, under transient conditions, e.g. when carbon deposited on catalyst particles is being burnt off, the temperature rise of the particles can be quite high, see Luss and Amundson [146] and Sohn [147], among others. Another exception in industrial practice is the NH_3 combustion, where the reaction has to be ignited to reach the upper stable operating state.

Multiplicity in *fixed bed reactors* — either cooled or adiabatic — can also occur due to axial dispersion of heat and/or mass. If multiplicity can take place, also hysteresis phenomena must be observed under special conditions. Another phenomenon is creeping, which is the movement of a very narrow reaction front upstream or downstream in a packed bed. Similar phenomena may lead to wrong way behaviour and runaway under dynamic conditions such as changing operating conditions — inlet concentrations or inlet temperatures — especially at start-up and shut-down of reactors. Instability in a packed bed reactor due to multiplicity of a group of particles most probably does not occur in industrial practice [139]. Limit cycles as discussed in Section VI.7 may also theoretically be expected in packed beds, as was demonstrated by Heinemann and Poore [148] for cooled tubular gas–solid reactors.

Fluidized beds generally are considered isothermal. However, also here single particles may have considerably higher temperatures than the average bed temperature, as has been observed by Aoyagi and Kunii [149] and Ross and Davidson [150] in coal combustion in a fluidized bed. Nevertheless it has never been observed that this would lead to an ignition of the entire bed: apparently local hot spots are dispersed rapidly by the strong agitation of the solids in the fluidized bed [151].

Parametric sensitivity, hot spots and runaway

Maximum allowable temperatures for exothermic reactions in gas–solid reactors are prescribed firstly because the temperature is not allowed to exceed a certain limit because of undesired side reactions and/or secondly because otherwise a temperature sensitive catalyst could become deactivated. To this end we discussed parametric sensitivity and temperature control in homogeneous tubular reactors in Section VI.3. In heterogeneous reactors the situation is more complicated because of the heat transfer resistance in the bed itself. Based on the assumptions of flat temperature profiles over a cross section, $T_o = T_w$ and only a temperature difference between solids and gas, Rajadhyaksha *et al.* [152] derived a method analogous to that of van Welsenaere and Froment (see Section VI.3), to calculate inlet concentrations for a cooled gas–solid reactor if a maximum allowable temperature has been prescribed. This method can be used for preliminary screening purposes.

McGreavy and Adderley [154, 155] derived a new runaway limit diagram to determine operating conditions to prevent particle instability, which could arise despite stable conditions in the surrounding gas phase. Their method is based on a heat balance for an isothermal particle and includes mass transport processes inside and outside the particle. Runaway lines in their diagrams depend on the Sherwood number and a modified Thiele modulus. Agnew and Potter [156] modified the Barkelew diagram for homogeneous reactors (see Section VI.3.1), into a similar diagram for heterogeneous reactors. They based their derivations on the cell model [87] and introduced two parameters to correct for the heterogeneity [157]:

$$M \simeq d_t/d_p \quad \text{for} \quad d_t/d_p > 8$$

and $St' = \alpha_w S/c_p \Phi_m$. In the runaway diagram of Agnew and Potter for $d_t/d_p \geq 8$,

$$\bar{a} \frac{N_c}{N_{ad}} = \frac{4\alpha'_w R T_w^2}{E d_t k_o \exp(-E/RT_w)(-\Delta H_r)_A c_{Ao}}$$

is plotted versus

$$N_{ad} = \frac{(-\Delta H_r)_A c_{Ao} E}{(\rho_G c_{pG})_o R T_w^2},$$

with $d_t St'/d_p$ as a parameter. Further parameters are defined as follows, if $M = d_t/d_p$:

$$\bar{a} = 0.5\left(\frac{4M}{2M-1} + \frac{M}{M-1}\right), \quad \alpha'_w = \frac{\alpha_w}{1 + Bi(1 - r_e^*)}$$

and

$$1 - r_e^* = \frac{1}{M} - \frac{2M-1}{8M(M-1)}.$$

Here $Bi = \alpha_w d_t/2\lambda_t$ can be determined with the correlations given in Section IX.2.2; α'_w is the wall heat transfer coefficient corrected for the cell model. In figure IX.32 the diagram for runaway determination is given for first order reactions and $T_o = T_w$. Agnew and Potter also give diagrams for half-order, second-order and product-inhibited reactions.

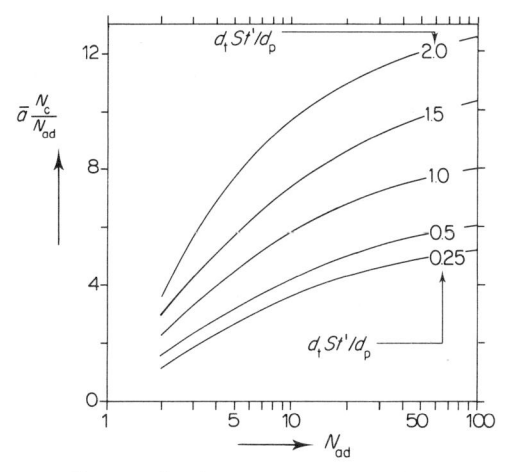

Fig. IX.32 Runaway diagram for heterogeneous tubular reactors and first order reactions. [From Agnew and Potter [156]. Reproduced by permission of Institution of Chemical Engineers]

Emig et al. [158] made experiments with a first order reaction — the synthesis of vinyl acetate — to test the different runaway criteria for cooled catalytic reactors. As an experimental runaway criterion they chose either $dT/dt \geq 3$ K/s or $T > 510$ K. In their case 510 K was the upper temperature limit for the stability of the catalyst they employed. Experimental observations on runaway were checked by calculating the temperature and concentration profiles with the FIBSAS program [81] for the pseudohomogeneous two-dimensional reactor model (see Section IX.2.2). They concluded that the Barkelew diagram gave good results, as also did the McGreavy and Adderley diagram. With the Agnew and Potter diagram results deviated somewhat; especially at high values of ϑ_{ad} the predicted values for runaway were no longer conservative. If accurate parameter values are available the FIBSAS program gives excellent results.

Recently Puszynski et al. [159] reviewed the idea that axial dispersion can be neglected in long tubular reactors, as we did in Section IX.2.2. They argued that this is no longer permissible near runaway conditions because near hot spots the temperature and concentration derivatives may become extremely steep and hence the terms

$$\frac{\partial^2 \vartheta}{\partial Z^2}/Pe_h \text{ and } \frac{\partial^2 \zeta}{\partial Z^2}/Pe$$

are of the type ∞/∞, so that they are no longer negligible. For a model reaction, the combustion of CO, they found that a rigorous model including the axial dispersion effects predicts hysteresis phenomena also in very long reactors with $Pe > 2000$. Ignition temperatures were equally well predicted by the plug flow model of Barkelew, the axial dispersion model of Hlavacek and Hofmann [114] and the rigorous model, but extinction temperatures not. They concluded that multiplicity and runaway are basically the same phenomena: the jump in the reactor exit conversion after a certain small change of the reactor inlet temperature manifests itself as an ignition temperature in the axial dispersion model and as a runaway in the plug flow model. In figure IX.33(b) temperature profiles are sketched for the upper stable operating points at points A and B in the conversion–reactor inlet temperature diagram IX.33(a) as they were calculated by the rigorous method. The profile B is more or less parabolic but the profile A is deformed because of overheating in the axis; this is caused by the difference in values for the Péclet numbers for heat and for mass dispersion. So in the parametric-sensitive regime very different profiles for steady state radial temperature can be expected in the cooled tubular reactor.

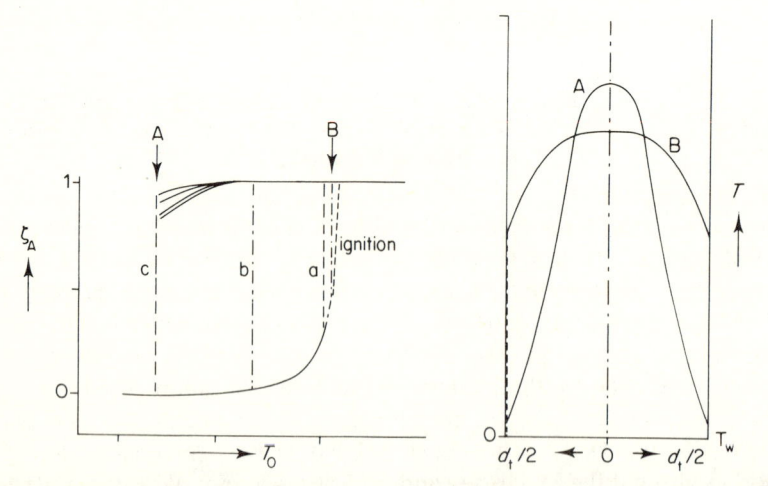

Fig. IX.33 Hysteresis in a cooled heterogeneous tubular reactor and radial temperature profiles. [From Puszynski et al. [159]. Reproduced by permission of Pergamon Press Ltd]. The ignition lines are line (a); the extinction lines are (a) for the plug flow model, (b) for the axial dispersion model and (c) for the rigorous model

Based on their evaluation with the rigorous model they [159] recommended for the design of industrial fixed bed catalytic reactors the following procedure:

(a) Solve the problem with the plug flow model and test for parametric sensitivity by Barkelew's method (Section VI.3). This should yield satisfactory results for the regime of low parametric sensitivity.
(b) For near parametric sensitive conditions near the ignition point calculate, using the axial dispersion model [114], whether hysteresis can occur. If there is no hysteresis, use the pseudohomogeneous two-dimensional model without the axial dispersion term (Section IX.2.2).
(c) If hysteresis can exhibit itself, calculate the extinction temperature by means of the more general rigorous model, which includes the terms for axial dispersion of heat and mass.

The rigorous model has not been analysed so far for other reactions in the region of near runaway.

Multiplicity in adiabatic beds

Multiplicity in adiabatic fixed bed reactors can occur due to the axial dispersion of mass and heat. This was analysed by Hlavacek and Hofmann [114] who concluded that this only happens for Péclet values below 200, so only in short beds. In the usual industrial reactors $Pe > 2000$, so that multiplicity is not to be expected. An exception is provided by the usually rather short first beds in adiabatic multibed reactors. More pronounced seems to be the axial dispersion of heat in adiabatic fixed bed reactors, which also causes multiplicity and hysteresis. This was demonstrated by, among others, Wicke et al. [160] and by Baiker et al. [161]; ignition and extinction phenomena have also been studied by, among others, Eigenberger [162] and Lübeck [163, 164] in adiabatic beds and by Vortmeyer and Stahl [165] in systems of separated catalyst layers.

From Baiker et al. [161] figure IX.34 is taken to demonstrate multiplicity and hysteresis in adiabatic fixed bed reactors. They used a reactor with $Pe \simeq 30$ and $L/d_p \simeq 200$. The concentration of reactant in the feed c_{Ao} was gradually increased

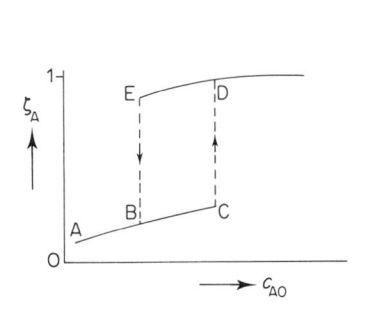

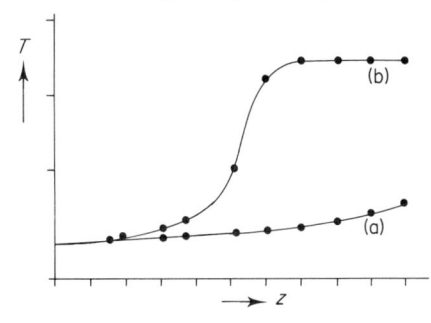

Fig. IX.34 Hysteresis in an adiabatic bed reactor and axial temperature profiles. Gas load and inlet temperatures are kept constant. [From Baiker et al. [161]. Reproduced by permission of Verlag Chemie GmbH]

from A to B to C in Figure IX.34(a). At a point very near to the ignition point C the axial temperature profile (a) in figure IX.34(b) was observed. After a slight increase in c_{Ao}, the conversion jumped to D in figure IX.34(a) and the steady state temperature profile in D changed to the curve (b) in figure IX.34(b). Reducing c_{Ao} again still kept the reactor at high conversions till point E in figure IX.34(a) was reached, where the conversion suddenly dropped to point B in figure IX.34(a): the reaction was extinguished. In all the experiments of figure IX.34 the reactor load u was kept constant.

Also small changes in load u or in reactor inlet temperature T_o can cause ignition or extinction. If the reactor operates at a low temperature level near or at point C in figure IX.34(a) a small reduction in flow rate leads to an acceleration of the reaction — due to the conversion increase at lower throughputs — at the end of the reactor: at the reactor outlet ignition starts. The strong back-mixing of heat by conduction and radiation moves the ignition front upstream until a new stable temperature profile at a high temperature and high conversion level has been established. Because of the hysteresis it can be understood that u has to be increased strongly in order to blow out the reaction again.

In large cooled industrial reactors, which operate at high reactor loads, ignition and extinction usually starts at the entrance of the reactor. Especially for partial oxidations, ignition is very unpleasant, because it leads to complete oxidation (combustion) of the reactant. By changing the operating conditions the reaction can be blown out and normal operation resumed again if nothing irreparable has happened, such as permanent reactor or catalyst damage due to the excessive reactor temperatures after ignition. Even reactor loads in every individual tube of a cooled multitubular reactor are difficult to achieve, so that one usually has to operate rather far away from the ignition point. Countercurrent cooling can also be dangerous, as explained in Section VI.7.2, and may even lead to limit cycles, although these limit cycles have hardly ever been observed in practice. For the calculation of ignition and extinction phenomena in adiabatic gas–solid reactors a two-phase model including axial dispersion terms has to be used.

Wrong way behaviour and creeping

Wrong way behaviour is the behaviour of a packed bed reactor which is contradictory to current thinking, e.g. a sudden decrease of the reactor feed temperature leads to a *temporary* temperature increase in the reactor bed. Some experimental observations of van Doesburg and de Jong [166] for the synthesis of methanol in an adiabatic bed reactor are given in figure IX.35. In figure IX.35(a) temperature profiles after a sudden increase of c_{Ao} are shown. In the first part of the reactor a new main reaction zone with complete conversion is formed. Due to the heat capacity of the catalyst packing, the reaction temperature can increase only belatedly, and the exit temperature transitorily decreases appreciably. Later the reaction zone expands until finally the new steady state temperature profile is reached. Figure IX.35(b) shows what happens if the feed

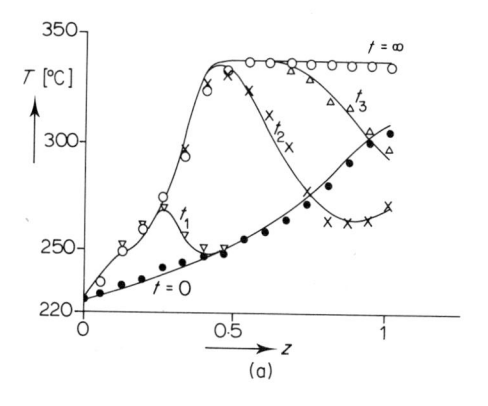

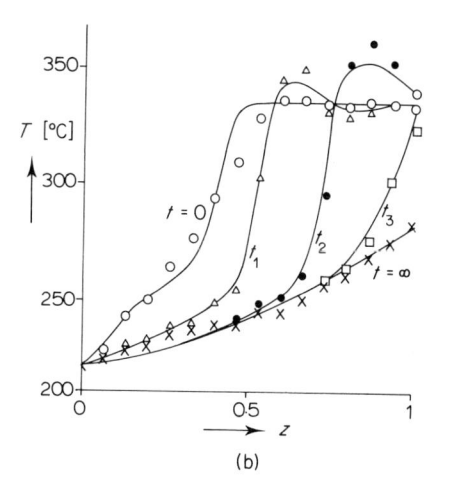

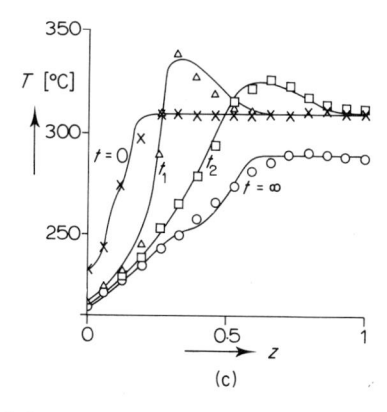

Fig. IX.35 Wrong way behaviour in an adiabatic bed reactor, according to van Doesburg and de Jong [166]. For explanation see text

concentration is reduced again: transitorily the reactor temperatures overshoot and the main reaction zone creeps towards the reactor outlet; the same happens with a sudden decrease of the reactor inlet temperature as in figure IX.35(c). This dynamic behaviour could be confirmed by van Doesburg and de Jong [106] by calculations with a pseudohomogeneous model without axial dispersion.

These phenomena are caused by thermal effects related to the movement of small main reaction zones — creeping ignited zones — as investigated already by Wicke and Vortmeyer [137]. In figure IX.36 creeping reaction zones as demonstrated by Padberg and Wicke [167] are given. At low values of u the reaction zone moves upstream towards the reactor inlet, at an intermediate value of u it remains in place and at high u values it creeps towards the reactor outlet. In the first case the cold particles upstream have to be heated; at a high u the reverse effect takes place. With an overall heat balance for a pseudostationary creeping reaction zone, Wicke and Vortmeyer derived

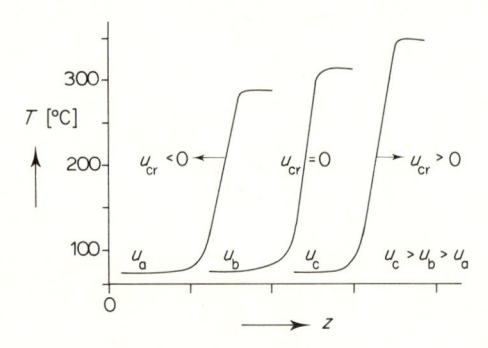

$$T_{max} - T_o = \Delta T_{ad} \frac{u\varepsilon\rho_G c_{pG}}{u\varepsilon\rho_G c_{pG} - (1-\varepsilon)u_{cr}\rho_S c_{pS}}. \tag{IX.79}$$

Fig. IX.36 Creeping reaction zones in an adiabatic bed reactor, according to Padberg and Wicke [167]

For a positive creeping velocity u_{cr} (in the downstream direction) the gas not only carries away the liberated reaction heat but also the heat stored in the particles, so that the temperature must increase to a value above the final adiabatic temperature increase. At

$$u_{cr} = \frac{\varepsilon\rho_G c_{pG}}{(1-\varepsilon)\rho_S c_{pS}}u,$$

the temperature increase must reach infinity according to equation (IX.79). This, of course, will not occur because u_{cr} is the result of a complex interdependence of heat and mass transport with chemical reaction. In deriving equation (IX.79) axial heat conduction was neglected, also cooling by radial heat losses through the wall has not been taken into account. See further Example IX.2.e.

Creeping reaction zones with ever-increasing maximum temperatures in the bed can occur in practice with deactivating catalysts, as shown by Butt et al. [168, 169]. At very low deactivation rates the maximum reaction temperature in the reactor decreases and the main reaction zone moves downstream [169]; however at high deactivation rates it moves towards the reactor inlet. Blaum [170] studied thermal catalyst deactivation and found temperature overshoots of more than 500 K! Gilles [171] made a theoretical analysis of these phenomena; Rhee et al. [172] and Elnashaie and Cresswell [173] analysed the creep velocities; Mehta et al. [174] derived expressions for predicting the maximum transient temperature rise in wrong way behaviour.

The most important conclusion is that a disturbance in operating conditions which moves the main reaction zone further downstream in the reactor — such as a decrease of c_{Ao}, T_o and catalyst activity for exothermic reactions and vice versa for endothermic reactions — liberates the heat stored in the packing and the reactor walls, so that temporarily the maximum reactor temperature can overshoot considerably. However, Endo et al. [139] questioned whether creeping as observed in laboratory reactors will ever occur in large industrial reactors.

For the study of dynamic behaviour in adiabatic reactors a one-dimensional heterogeneous or pseudohomogeneous model can be used. The basic equations for the heterogeneous model are the equations (IX.56)–(IX.59) without the radial dispersion terms but now also including terms for the accumulation of heat. For the solid phase the heat balance for one single reaction becomes:

$$(1-\varepsilon)\rho_S c_{pS} \frac{\partial T_i}{\partial t} = \lambda \frac{\partial^2 T_i}{\partial z^2}$$
$$+ \alpha a_p (1-\varepsilon)(T_i - \bar{T}) - \rho_S(1-\varepsilon)(-\Delta H_r)_A \langle R_S \rangle_A. \qquad \text{(IX.80)}$$

This equation can be made dimensionless by introducing $\tau = ut/\varepsilon L$, $Z = z/L$, $\vartheta = T/T_o$ and $\zeta_A = 1 - c_A/c_{Ao}$, if T_o and c_{Ao} are the gas inlet conditions for the adiabatic bed reactor. We then have

$$\frac{\partial \vartheta_i}{\partial \tau} = \frac{1}{Pe_{hS}} \frac{\partial^2 \vartheta_i}{\partial Z^2} + A(\vartheta - \vartheta_i) - B \langle R_S \rangle_A (\vartheta_i, \zeta_{Ai}). \qquad \text{(IX.80.a)}$$

Similarly we find for the heat balance in the gas phase,

$$\frac{\partial \vartheta}{\partial \tau} = \frac{1}{Pe_{hG}} \frac{\partial^2 \vartheta}{\partial Z^2} - \frac{\partial \vartheta}{\partial Z} - C(\vartheta - \vartheta_i) \qquad \text{(IX.81)}$$

for the component mass balance in the gas phase,

$$\frac{\partial \zeta_A}{\partial \tau} = -\frac{\partial \zeta_A}{\partial Z} + \frac{1}{Pe} \frac{\partial^2 \zeta_A}{\partial Z^2} - D(\zeta_A - \zeta_{Ai}) \qquad \text{(IX.82)}$$

and for the component mass balance in the solid phase,

$$\frac{\partial \zeta_{Ai}}{\partial \tau} = +D(\zeta_A - \zeta_{Ai}) - E \langle R_S \rangle_A (\vartheta_i, \zeta_{Ai}). \qquad \text{(IX.83)}$$

The significance of the constants in these equations is

$$Pe_{hS} = \frac{\rho_B c_{pS} uL}{\varepsilon \lambda_1}, \quad Pe_{hG} = \frac{\rho_G c_{pG} uL}{\lambda_1}, \quad Pe = \frac{uL}{\varepsilon D_1},$$

$$A = \frac{\alpha a_p \varepsilon L}{\rho_S c_{pS} u}, \quad B = \frac{(-\Delta H_r)_A \varepsilon L}{u c_{pS} T_o}, \quad C = \frac{\alpha a_p (1-\varepsilon) L}{\rho_G c_{pG} u},$$

$$D = \frac{k_G a_p (1-\varepsilon) L}{u} \quad \text{and} \quad E = \frac{\rho_S (1-\varepsilon) L}{u c_{Ao}}.$$

The dimension of B and E is kg solids per kmol A in the feed per second. For calculation purposes all these constants, of which A, C and D are dimensionless groups, are usually assumed independent of temperature; $\langle R_S \rangle_A$, of course, is very strongly dependent on the temperature. The boundary conditions normally used are

$$Z = 0: \quad \frac{\partial \zeta_A}{\partial Z} = Pe \zeta_A, \quad \frac{\partial \vartheta}{\partial Z} = Pe_{hG}(\vartheta - 1), \quad \frac{\partial \vartheta_i}{\partial Z} = Pe_{hS}(\vartheta_i - 1)$$

$$Z = 1: \quad \frac{\partial \vartheta}{\partial Z} = 0, \quad \frac{\partial \vartheta_i}{\partial Z} = 0. \tag{IX.84}$$

Sometimes other boundary conditions are taken (see Wicke [116], e.g. heat radiation of the catalyst at $Z = 0$ is taken into account. With this set of equations problems of dynamic behaviour of adiabatic bed reactors can be tackled; less accurate solutions can also be obtained if axial dispersion and/or temperature differences between solid and gas are neglected. An example where axial dispersion is neglected will be given in the next Example.

Example IX.2.e Decoking of a catalyst
It is required to regenerate a reforming catalyst by burning of the deposited coke. The reaction is $C + O_2 \rightarrow CO_2$ with $\Delta H_r = -394$ MJ/kmol. The regeneration gas is a mixture of air and nitrogen; the catalyst contains varying amounts of coke. It will be assumed that the conversion rate of the carbon deposit on the catalyst is completely limited by the interparticle mass transfer rate to the catalyst, which gives us the highest possible conversion rate. There is plug flow in the reactor and no axial dispersion of heat and mass. Further the reactor operates adiabatically, the coke is uniformly deposited on the catalyst and the regeneration gas enters the reactor at the initial bed temperature ($T_{oG} = T_{o\,bed}$). For this case equations (IX.80.a) to (IX.83) reduce to:

Eq. (IX.82): $\quad \dfrac{\partial \zeta_A}{\partial \tau} = -\dfrac{\partial \zeta_A}{\partial Z} + D(1 - \zeta_A)$

Eq. (IX.83): $\quad \dfrac{\partial \zeta_B}{\partial \tau} = +F(1 - \zeta_A)$

Eq. (IX.80.a): $\dfrac{\partial \vartheta_i}{\partial \tau} = -A(\vartheta_i - \vartheta) + G(1 - \zeta_A)$

Eq. (IX.81): $\dfrac{\partial \vartheta}{\partial \tau} = -\dfrac{\partial \vartheta}{\partial Z} - C(\vartheta - \vartheta_i)$.

The dimensionless group A, C and D were given before. The new dimensionless groups are

$$F = \dfrac{v_B k_G a_p L \varepsilon c_{Ao}}{v_A \rho_S u c'_{Bo}} \quad \text{and} \quad G = \dfrac{(-\Delta H_r)_A c_{Ao} k_G a_p \varepsilon L}{\rho_S c_{pS} u T_o}$$

and c'_{Bo} is expressed in kmoles C/kg catalyst. These equations can be solved for $t = 0$ and $0 < z < L$ with boundary conditions of $T_i = T_o$ and $c_A = 0$, and for $t > 0$ and $z = 0$ with boundary conditions of $c_A = c_{Ao}$ and $T = T_o$ and after introducing a new variable $\tau' - Z$ according to, e.g., Rhee et al. [175] or Boreskov et al. [176]. In our simplified case of no axial dispersion we find for the concentration profiles, which are independent of the temperature because the mass transfer resistance in the gas phase controls the following:

For $t < t_o$ and for oxygen,

$$\dfrac{c_A}{c_{Ao}} = \exp(-DZ);$$

and for the coke with $\tau' = t u_{\text{front}}/L$,

$$\dfrac{c'_B}{c'_{Bo}} = 1 - F(\tau' - Z)\exp(-DZ)$$

and for $t > t_o$ and $Z > Z_{\text{front}}$

$$\dfrac{c_A}{c_{Ao}} = \exp(-D(Z - Z_{\text{front}})) \quad \text{and} \quad \dfrac{c'_B}{c'_{Bo}} = 1 - \exp(-D(Z - Z_{\text{front}}))$$

and for $t > t_0$ and $Z < Z_{\text{front}}$:

$$\dfrac{c_A}{c_{Ao}} = 1 \quad \text{and} \quad \dfrac{c'_B}{c'_{Bo}} = 0.$$

In these relations t_o is the time required to burn off all the coke from the catalyst at the reactor entrance $Z = 0$:

$$\tau_o = \dfrac{u t_o}{\varepsilon L} = \dfrac{v_A c'_{Bo} \rho_S u}{v_B k_G a c_{Ao} \varepsilon L} = \dfrac{1}{F}$$

and Z_{front} is the location of the coke front. Some concentration profiles are given in Figure IX.37. There are three zones in the reactor: a regenerated zone free of coke, an oxidation zone and a zone downstream in the reactor, where the coke has not

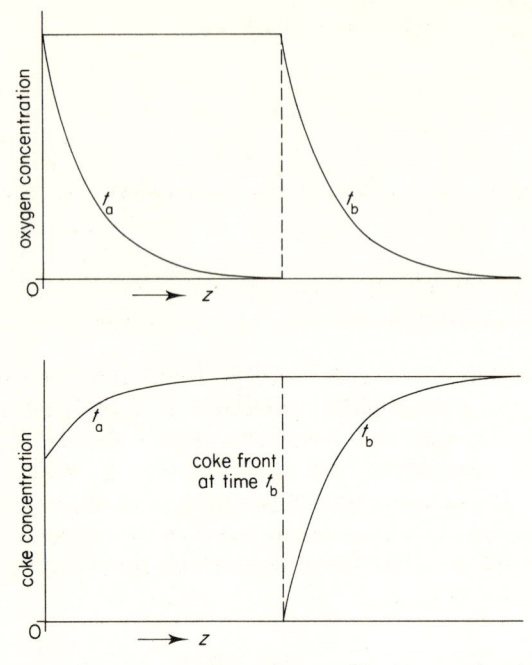

Fig. IX.37 Oxygen and carbon concentration profiles in a catalyst bed, which is being decoked

yet started to burn. After the initial period t_o the coke burns at a constant rate and the coke front moves at a constant velocity through the reactor, given by

$$u_{\text{front}} = \frac{uc_{Ao}}{\varepsilon c_{Ao} + \rho_S(1-\varepsilon)c'_{Bo}\dfrac{v_A}{v_B}} = \frac{F}{F+D}\frac{u}{\varepsilon},$$

where u is the gas load Φ_v/S. The location of the front in the reactor is the reactor entrance for $t < t_o$ and for $t > t_o$:

$$z_{\text{front}} = u_{\text{front}}(t - t_o)$$

The approximate length of the regeneration zone can be estimated for high Reynolds numbers if we use $Sh \simeq 1.90\,Re_p^{0.5}Sc^{0.33}$. We then have for spherical particles for the 'length of a reaction unit' as compared with the HTU concept:

$$HTU_G = \frac{u}{k_G a_p(1-\varepsilon)} = \frac{ud_p^2 v}{6Dv \times 1.90Re_p^{0.5}Sc^{0.33}} = 0.15d_p Re_p^{0.5}Sc^{0.67}.$$

For $Sc \simeq 1$ and say $Re_p = 400$ we then find for the HTU_G a value of $\simeq 3.5d_p$. So the oxygen is almost completely exhausted—99% converted—after a length of some 20 catalyst particles, see also page 617.

The heat balances can be solved analytically (Bessel functions) or numerically by computer. There are several dimensionless groups that determine the shape of the temperature profiles. As the gas passes through the bed causing the coke to burn it also carries the heat through the bed. The relative velocity of the coke front and the heat wave is determined by the ratio of the adiabatic temperature rises of the catalyst and of the gas:

$$\frac{\Delta T_{adS}}{\Delta T_{adG}} = \frac{(-\Delta H_r)_B c'_{Bo} \rho_G c_{pG}}{(-\Delta H_r)_A c_{Ao} c_{pS}} = \frac{v_A c'_{Bo} \rho_G c_{pG}}{v_B c_{Ao} c_{pS}} = \frac{AD}{CF}.$$

For $\Delta T_{adS}/\Delta T_{adG} > 1$ the heat is carried forward of the front; if <1 the heat wave follows the coke front. For $\Delta T_{adS}/\Delta T_{adG} = 1$ the heat remains concentrated near the front and T_{max} could rise to infinity — and the reactor go to eternity — if the reactor were long enough. Although the process is a transient one the temperature profiles rapidly approach a fixed shape, and once this temperature profile is attained, it moves through the bed at the same velocity as the coke front. For $\Delta T_{adS}/\Delta T_{adG} > 1$ the shape of the temperature profiles is also determined by the dimensionless group $\alpha/k_G \rho_G c_{pG} = C/D$. being the ratio of the heat dissipation rate to the reactant supply rate. This ratio based on the Chilton–Colburn analogy is equal to $(Sc/Pr)^{0.67}$, and is of the order of magnitude of one. Its influence on the shape of the heat wave is shown in figure IX.38.

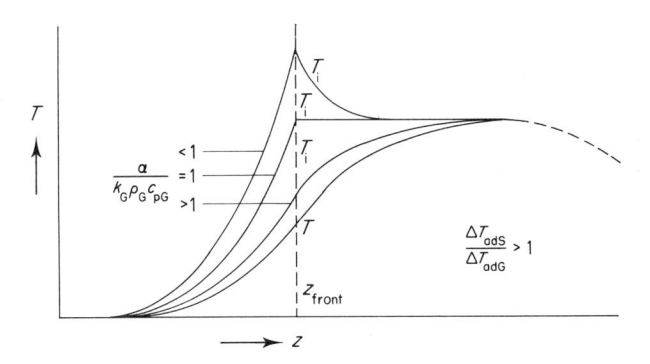

Fig. IX.38 The shape of the heat wave through the catalyst bed, which is being decoked

We will consider a reactor 8 m long and $T_o = 400°C$. The diameter of the spherical particles is $d_p = 5 \times 10^{-3}$ m and their density is $\rho_S = 1400 \text{ kg/m}^3$. The gas load to the reactor is $u = 1$ m/s and the coke content of the catalyst 2% by weight. We will consider different oxygen contents in the regeneration gas.

Further data of the system are for the gas: $\mu = 34 \times 10^{-6}$ kg/ms, $\rho = 0.50$ kg/m³, $c_p = 1020$ J/kg K, $D = 8.5 \times 10^{-5}$ m²/s, $Re_p = 190$, $Pr = 0.68$, $Nu = 7.9$, $\alpha = 77$ W/m² K, $k_G = 0.15$ m/s and $\varepsilon = 0.4$; for the solid $c'_{Bo} = 1.7 \times 10^{-3}$ kmol/kg, $\rho = 1400$ kg/m³, $c_p = 1050$ J/kg K and $1 - \varepsilon = 0.6$.

We first consider the case of an *oxygen content of 1 vol%*. Then $c_{Ao} = 0.18 \times 10^{-3}$ kmol/m³. In this case the induction time t_o before the coke front starts to move is $t_o = 73$ s, and the velocity of the coke front after t_o is $u_{front} = 126 \times 10^{-6}$ m/s, and the total regeneration time for a reactor of 8 m length is 63 470 s = 17.6 h. In this case $\Delta T_{adS}/\Delta T_{adG} = 4.6$. The temperature profile is given in figure IX.39 for a starting time of around 3 minutes. The dotted line indicates the maximum temperature that the particles will ever reach during the regeneration. We see the oxygen content is sufficient to increase the temperatures in the bed by 175°C. The coke front follows the heat wave.

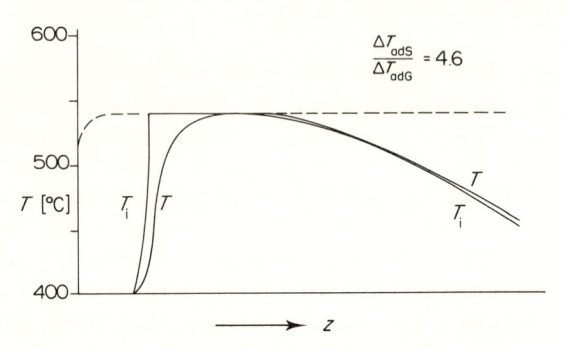

Fig. IX.39 The temperature profiles in a catalyst bed, which is being decoked. The coke front follows the heat wave

In figure IX.40 we use a regenerating *gas with 10 vol % of oxygen*. Catalyst temperatures become extremely high. Here $\Delta T_{adS}/\Delta T_{adG} = 0.5$. We see how the heat wave, which gets broader and broader, now follows the coke front. In figure IX.41 we have chosen the *oxygen content at 5 vol %* so that $\Delta T_{adS}/\Delta T_{adG} = 1$. Here the coke front and the heat wave have the same velocity and the particles get hotter and hotter because of the accumulation of heat near the coke front. In all these cases the temperature at the entrance of the bed never reaches the values found at the temperature peaks further downstreams.

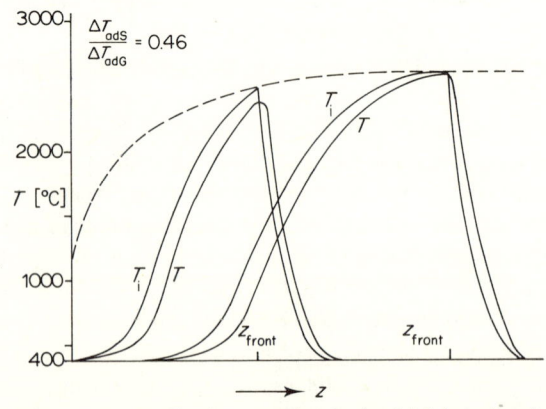

Fig. IX.40 The temperature profiles in a catalyst bed, which is being decoked. Now the heat wave follows the coke front

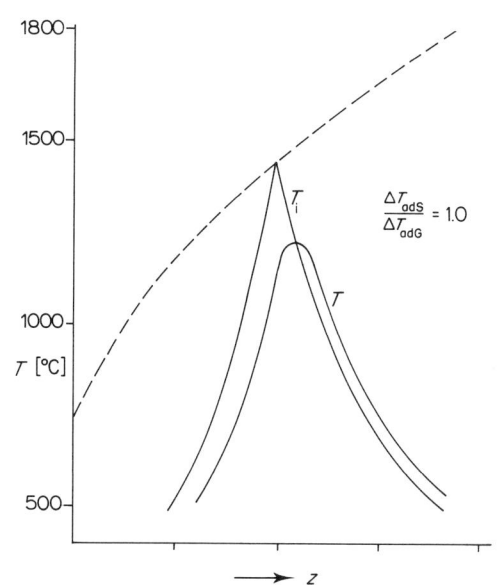

Fig. IX.41 Temperature profiles in a catalyst bed, which is being decoked. The coke front and the heat wave run at the same speed

We have to realize that we assumed mass transfer limitation for the conversion rate, so we calculated the highest possible temperatures. Our results further show that the conversion rate normally has no effect on the maximum temperature rise. The main determining factor is the relative velocity of the heat wave and the coke front as determined by the dimensionless group $\Delta T_{adS}/\Delta T_{adG}$. In order to prevent excessive temperatures during decoking it is therefore important to keep $\Delta T_{adS}/\Delta T_{adG}$ away from the value 1 and to keep ΔT_{adG} low, that is the oxygen concentration in the regeneration gas. Under our assumptions it must be kept below 1 vol %. We neglected the effects of axial dispersion and intraparticle diffusion and the temperature effect on the reaction rate as well. So in our model a sudden flame front passing through the reactor is impossible; this might occur if unreacted oxygen fills a portion of the reactor before an appreciable amount of reaction takes place.

IX.3 GAS–LIQUID–SOLID REACTORS

Gas–liquid–solid reactors are very important in industrial practice and occur in several different types, which can be divided into two categories: the particles are (a) stationary in a packed bed, or (b) are suspended in the liquid phase. The main characteristics of these reactor types are:

for *packed bed reactors* with usually rather large particles:

— the packed bed bubble column. Here the catalyst is packed in a vessel, which is filled with liquid; the gas bubbles through the liquid;

- the countercurrent trickle flow column. Here the gas is the continuous phase and the liquid trickles downward over the packing. This reactor type will exhibit flooding at too high loads;
- the cocurrent trickle flow column. Here because gas and liquid flow cocurrently, no flooding can occur. Depending on the gas flow rate, the liquid flow rate and their ratio, this reactor will operate in different flow regimes: spray, bubble, foaming or pulsing flow. Both gas and liquid are intensely mixed over a cross-section.

for *slurry reactors*, where the solid is suspended and the particle diameters are usually low:
- the slurry bubble column. Here the solids are carried away by the liquid flowing out of the reactor;
- the fluidized bed slurry reactor, where the particles largely remain in the reactor and the upper part of the liquid in the reactor behaves as a settler for the solids;
- the agitated gas–liquid slurry reactor. Here both gas and liquid are intensely mixed and the solids are carried away by the liquid.

In the packed bed reactors the solid phase is not necessarily a catalyst. Especially in cocurrent flow at high flow rates large interfacial areas are created. Therefore this reactor type is industrially used with inert packings to improve mass transfer rates in gas–liquid reactions (see Example IX.3). All the reactor types mentioned have been studied comprehensively and many data can be found in the literature, e.g. on two phase pressure drop, degree of wetting of the packing, hold-up of the phases in the reactor, axial and radial dispersion in the phases, interfacial areas and mass and heat transfer characteristics. We refer to the literature for data: several reviews have appeared recently in this field, for example by Satterfield [177] and by Hofmann [178] on packed bed reactors, by Goto *et al.* [179] on tricle bed oxidation reactors and by Gianetto *et al.* [180] on packed column hydrodynamics and wetting behaviour; whereas van Landeghem [11] and Kürten and Zehner [181] review both packed column and slurry reactors. Three phase slurry reactors were reviewed by Chaudhari and Ramachandran [182] and Deckwer and Alper [183], bubble column slurry reactors by Hammer [184] and bioreactors for aerobic processes by Schügerl [185]. Deckwer and Alper also gave recommendations for the optimal operation of three phase slurry reactors.

The main problem for these multiphase reactors is the formulation of the correct conversion rate expression: this was discussed in Chapters VII and VIII. Heat effects usually do not cause serious problems in gas–solid–liquid reactors. Not until recently has a start been made on the study and the correlation of effective heat dispersion coefficients λ_1 and λ_t and wall heat transfer coefficients α_w in packed bed reactors. For gas–liquid slurry reactors wall heat transfer coefficients have been correlated by Steiff *et al.* [186]. If we use the criterion (IX.45) of Mears for the absence of interparticle heat transfer resistance, we can quickly check that even for very rapid highly exothermic reactions in *slurry* reactors the temperature gradient over the liquid film will be negligible. The same holds for a possible intraparticle temperature gradient. This explains that at least over a cross-section the gas, liquid and solid phases are always considered to be isothermal. Only in *packed bed* multiphase reactors considerable temperature gradients occasionally may be expected. Some recent examples of temperature profile calculations in slurry reactors can be found in the studies of Deckwar *et al.* [187], of Shah and Parulekar [188] and Shah [189].

We will now discuss in an example temperature profiles along a cocurrent trickle bed reactor.

Example IX.3 *Sour gas absorption in a co-current trickle flow column*
The natural gas stream described in Example IX.1.c is now treated with MDEA in a cocurrent trickle bed absorber, packed with 1-inch Rashig rings. All data are the same as in Example IX.1.c. The inert packing is applied to obtain more interfacial area per unit reactor volume.

We will apply a cascade of cocurrent trickle bed reactors as shown in figure IX.42. In the reactors gas and liquid flow cocurrently downward over an inert packing, so that flooding cannot occur. In this way higher gas throughputs per unit cross-sectional area can be realized than in countercurrent operations. The liquid flow rate Φ_{vL}/S is set at 0.1 m/s. In cocurrent flow ultimately only one equilibrium stage can be attained in each reactor. This implies that for very deep H_2S removal several reactor beds are needed which have to be connected in a special way to provide overall countercurrent gas and liquid flows as shown in figure IX.42. This involves the installation of additional pumps. In the trickle bed calculation procedure each bed *i* is regarded as a series of ideally mixed reactor sections. As in the tray-to-tray calculations in Example IX.1.c, backmixing in the gas and liquid phases and pressure drop over the sections are neglected. By means of mass and heat balances over the cascade of reactors, which are essentially identical to those for the tray absorber, the temperature of and the total H_2S and CO_2 concentrations in the rich solution leaving reactor i are derived.

As far as the reactor balances for H_2S and CO_2 are concerned, the procedure is now slightly different. Flowing cocurrently in a reactor the H_2S gas concentration decreases and the unreacted H_2S liquid concentration increases due to the mass transfer. Ultimately the mass transfer driving force for H_2S is zero and only CO_2 is still absorbed at a relatively high rate. In order to avoid this for the case in which the selectivity is extremely unfavourable, the calculations are stopped at a preset positive absorption driving force for H_2S. This is realized, e.g. for reactor 1, by calculating the unreacted H_2S concentration in the liquid from the temperature and the total H_2S and CO_2 concentrations using the equilibrium model mentioned earlier in the publication of Blauwhof *et al.* [44]. The H_2S gas phase concentration, at which the reactor calculations are stopped, are obtained using an arbitrary but effective algorithm:

$$c_{H_2SG}^{1,n+1} = fc_{H_2SL}^{1,n+1} \text{ (moles/Nm}^3\text{)} \tag{a}$$

where f is a constant >1 which may have different values for each reactor. Condition (a) ensures a positive driving force for the H_2S absorption in each

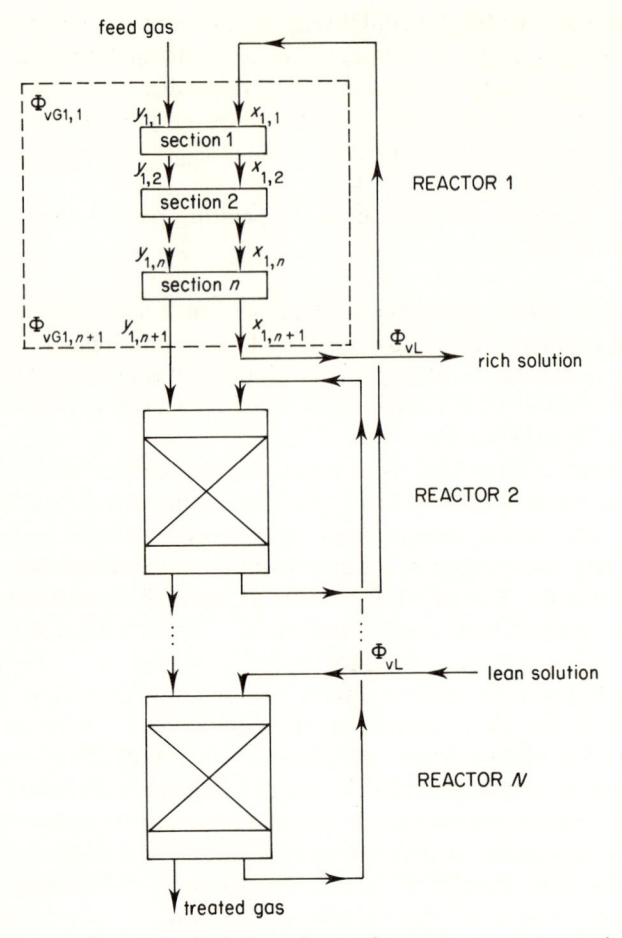

Fig. IX.42 Process flow and calculation scheme of a countercurrent cascade of cocurrent trickle bed reactors; Example IX.3

reactor section. An initial estimate of the CO_2 gas phase concentration leaving reactor 1, $c_{CO_2 G}^{1,n+1}$, now provides sufficient data to complete the overall mass and heat balances and the inlet data can be obtained. Next, a section to section calculation proceeds until the H_2S gas phase concentration falls below the value calculated by equation (a). If the calculated CO_2 gas concentration deviates more than 3% from the initial estimate the procedure is repeated using the calculated value as a new estimate. New reactors are added until the H_2S specification is met. If the CO_2 concentration in the treated gas deviates more than 5% from the initial estimate, the calculations are repeated, starting again with reactor 1. It will be understood that the choice of the factor f and the number of beds and their lengths in this cascaded reactor are interrelated. This requires an economic optimization study, which will not be presented here.

In the calculation procedure for the cascade of trickle-bed reactors the correlations of Fukushima and Kusaka [190, 191] for k_L and k_G in the pulsing flow regime are incorporated. For the interfacial area we have chosen a conservative value of $200\,m^2/m^3$ reactor volume for all trickle bed calculations. This value corresponds to the geometrical packing area of 1 inch Raschig rings. The mass transfer coefficients are converted to the operating pressures and temperatures by assuming a penetration theory dependency of k_L and k_G on the diffusivity:

$$k_{L/G\,at\,p,T} = k^*_{L/G} \frac{\sqrt{D_{at\,p,T}}}{D^*}\,(m/s), \qquad (b)$$

where D^* refers to the diffusion coefficient under the conditions at which the mass transfer coefficient relations of [190, 191] were obtained. The interfacial areas were assumed to be independent of pressure [43] and temperature.

The absorption characteristics of the cocurrent trickle bed absorbers are illustrated on the basis of gas concentration and temperature profiles, which are plotted as a function of the cumulative interfacial area in figures IX.43 and IX.44 respectively. Due to higher mass transfer coefficients and hence lower mass transfer resistances, the gas–liquid interfacial areas required to bring the gas on the H_2S specification in the trickle bed reactors are reduced to less than half of the area needed in the tray absorber (see figure IX.43). The higher k_G/k_L ratios realized in the trickle beds result in an improvement of the ratio of mass transfer resistances, which forms virtually the basis of the higher selectivity obtained in the cascade of trickle bed reactors (see Table IX.10). In figure IX.44 in each individual

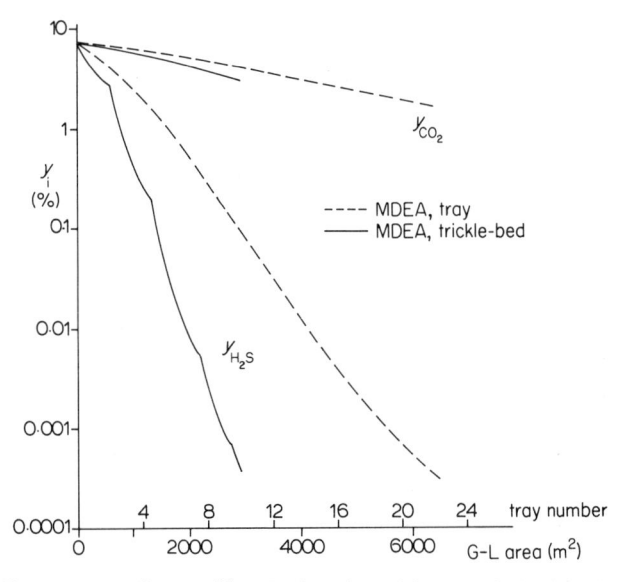

Fig. IX.43 Gas concentration profiles as a function of the cumulative interfacial area in a tray column reactor and a cascade of trickle bed reactors; Example IX.1.c and IX.3

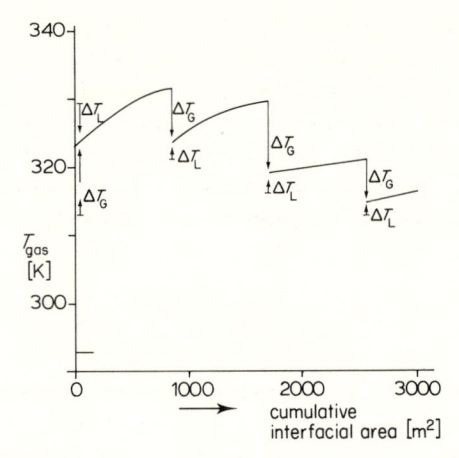

Fig. IX.44 Temperature profiles in a cascade of trickle bed reactors: Example IX.3

reactor bed both the liquid and the gas flow from the left to the right. We see how both are heated up in each bed due to the reaction heat liberated. The overall flow for the gas is also to the right and for the liquid to the left in figure IX.44. The temperature changes of the gas ΔT_G and of the liquid ΔT_L due to the mixing of gas and liquid streams at the bed entrances are indicated.

As explained by Blauwhof et al. [44] the selectivity of absorption is governed by the ratio k_G/k_L. Based on the correlation of Fukishima and Kusaka [190, 191] we have

$$k_G/k_L \simeq u_L^{-0.33} u_G^{0.15} d_p^{-0.33}.$$

The selectivity σ_P is expressed as the amount of H_2S absorbed divided by the total amount of $H_2S + CO_2$ absorbed. Since in our set-up the total amounts of liquid and gas pass through all reactors, u_L and u_G are directly proportional to each other by the overall mass balance equation (c) in Example IX.1.c, so that virtually only one really effective parameter remains:

$$k_G/k_L \simeq u_L^{-0.18} \quad \text{or} \quad \simeq u_G^{-0.18}.$$

According to this last relation a superficial velocity as low as possible is required in order to obtain a high k_G/k_L ratio and hence a high selectivity. This inevitably leads to larger trickle bed diameters, and hence to larger investments. An economic optimization is needed to determine the optimum design and operating conditions as far as the k_G/k_L ratio is concerned. The last degree of freedom mentioned above, the equilibrium approach factor f, hardly affects the selectivity of the operation, but controls largely the lay-out of the cascade. The optimum values of f for each reactor bed can be determined by an economical evaluation and will probably result in a set of beds of equal lengths.

The results for the tray column reactor and the countercurrent cascade of cocurrent trickle bed reactors are summarized in Table IX.10. We see that the

Table IX.10 Comparison of the dimensions and the performance of a tray column reactor and a trickle bed cascade reactor; Example IX.1.c and IX.3

Reactor type	tray	trickle
Absorber diameter (m)	2.95	1.62
Number of trays or sections	22	5
Absorber length (m)	19	17.5
Total interfacial area (m^2)	6120	2890
Selectivity, η (—)	0.568	0.637
Solvent flow Φ_L (m^3/s)	0.225	0.205
$y_{CO_2}^{out}$ (%)	1.68	3.01
$\Phi_{L\,trickle}/\Phi_{L\,tray}$		0.911
$\eta_{trickle}/\eta_{tray}$		1.121

cocurrent trickle bed reactors perform much better than the tray column reactor: they give smaller reactors, higher selectivities and lower liquid loads.

REFERENCES

1. Miyauchi, T. and Vermeulen, T.: *Ind. Eng. Chem.* (*Fund.*) **2**, 113 (1963)
2. Stemerding, S. and Zuiderweg, F. J.: *Chem. Eng.* **168,** CE 156 (May 1963)
3. Beek, W. J.: *Chem. Weekbl.* **58,** 37 (1962)
4. Mecklenburg, J. C. and Hartland, S.: *Int. Chem. Symp.* **26,** 115 (1967)
5. Mecklenburg, J. C. and Hartland, S.: *The Theory of Backmixing*, Wiley (London) 1975)
6. Rod, V.: *Coll. Czech. Chem. Comm.* **34,** 387 (1969)
7. Roes, A. W. M. and van Swaaij, W. P. M.: *Chem. Eng. J.* **17,** 81 (1979)
8. Roes, A. W. M. and van Swaaij, W. P. M.: *Chem. Eng. J.* **18,** 29 (1979)
9. Nagel, O., Hegner, B., and Kürten, H.: *Chem. Ing. Techn.* **50,** 934 (1978)
10. Charpentier, J. C.: *ACS Symp. Series* **72,** 223 (1978)
11. van Landeghem, H.: *Chem. Eng. Sci.* **35,** 1912 (1980)
12. Shah, Y. T., Stiegel, G. J., and Sharma, M. M.: *A.I.Ch.E. J.* **24,** (3) 369 (1978)
13. Astarita, G.: *Mass transfer with chemical reaction*, Elsevier, Amsterdam (1967)
14. Danckwerts, P. V.: *Gas–Liquid Reactions*, McGraw-Hill, New York (1970)
15. Danckwerts, P. V.: *Appl. Sci. Res.*, **A3,** 385 (1953)
16. Danckwerts, P. V.: *Chem. Eng. Sci.* **22,** 472 (1967)
17. Chiang, S. H. and Toor, H. L.: *A.I.Ch.E. J.* **10,** 398 (1964)
18. Mann, R. and Moyes, H.: *A.I.Ch.E. J.* **23,** 17 (1977)
19. Allan, J. C. and Mann, R.: *Chem. Eng. Sci.* **34,** 413 (1979)
20. Juvekar, V. A. and Sharma, M. M.: *Trans. Inst. Chem. Engrs.* **55,** 77 (1977)
21. Kaibel, G., Mayer, H. H., and Seid, B.: *Chem. Ing. Techn.* **50,** 586 (1978)
22. Nelson, P.: *Adv. Chem. Ser.* **109,** 237 (1972)
23. Geelen, H. and Wijffels, J. B.: *Proc. 3rd Chem. React. Eng. Symp.*, page 125, Pergamon, Oxford (1965)
24. Huang, D. T. J. and Varma, A.: *A.I.Ch.E. J.* **27**(1), 111 (1981)
25. Wöhler, F. and Steiner, R.: *Chem. Ing. Techn.* **42,** 481 (1970)
26. Barona, N.: *Hydrocarbon Proc.* **58** (7), 179 (1979)
27. Hanhart, J., Kramers, H., and Westerterp, K. R.: *Chem. Eng. Sci.* **18,** 503 (1963)
28. Ding, J. S. Y., Sharma, S., and Luss, D.: *Ind. Eng. Chem.* (*Fund.*) **13,** 76 (1974)
29. Hoffman, L. A., Sharma, S., and Luss, D.: *A.I.Ch.E. J.* **21,** 318 (1975)

30. Beskov, W. S., Charkova, T. V., and Novikov, E. A.: *Theor. Found. Chem. Techn.* **13**, 120 (1979) (in Russian)
31. Porter, K. E.: *Trans. Instn. Chem. Engrs.* **44**, T25 (1966)
32. Kishinevskii, M. K., Kormeho, T. S., and Popa, T. M.: *Theor. Found. Chem. Engng* **4**, 641 (1971) (in Russian)
33. Alper, E.: *Chem. Eng. Sci.* **23**, 2093 (1973)
34. van Krevelen, D. W. and Hoftijzer, P. J.: *Rec. Trav. Chim.* **67**, 563 (1948)
35. Calderbank, P. H.: *Trans. Instn. Chem. Engrs (London)* **36**, 443 (1958)
36. Westerterp, K. R., van Dierendonck, L. L., and de Kraa, J. A.: *Chem. Eng. Sci.* **18**, 157 (1963)
37. Reith, T.: *Brit. Chem. Eng.* **15**, 1559 (1970)
38. van Dierendonck, L. L.: Ph.D. thesis, Twente University of Technology, Neth. (1970)
39. van Dierendonck, L. L., Fortuin, J. M. H., and Vandenbos, D.: *Proc. 4th Eur. Symp. Chem. Reaction Eng.*, page 205, Brussels, Pergamon Press (1971)
40. Miller, D. H.: *AIChE J.* **20**, 445 (1974)
41. Nishikawa, M., Nakamura, M., Yagi, H., and Hashimoto, K.: *J. Chem. Eng. (Japan)* **14**(3) 219 (1981)
42. Teramoto, M., Tai, S., Nishii, K., and Teranishi, H.: *Chem. Eng. J.* **8**, 223 (1974)
43. Sridhar, T. and Potter, O. E.: *Ind. Eng. Chem. Fundam.* **19**, 21 (1980)
44. Blauwhof, P. J. M., Kamphuis, B., van Swaaij, W. P. M., and Westerterp, K. R.: *Chem. Eng. Fund* **1**(2) 1983)
45. Nonhebel, G.: *Gas Purification Processes for Air Pollution Control*, Newnes–Butterworth, London (1972)
46. Sharma, M. M. and Gupta, R. K.: *Trans. Instn. Chem. Engrs* **45**, T169 (1967)
47. Verma, S. L., and Delancey, G. B.: *A.I.Ch.E. J.* **21**, 96 (1975)
48. Beenackers, A. A. C. M. and van Swaaij, W. P. M.: *Chem. Eng. J.* **15**, 25, 39 (1978)
49. Schmitz, R. A. and Amundson, N. R.: *Chem. Eng. Sci.* **18**, 265, 391, 415, 447 (1963)
50. Luss, D. and Amundson, N. R.: *Chem. Eng. Sci.* **22**, 267 (1967)
51. Raghuram, S. and Shah, Y. T.: *Chem. Eng. J.* **13**, 81 (1977)
52. Raghuram, S., Shah, Y. T., and Tierney, J. W.: *Chem. Eng. J.* **17**, 63 (1979)
53. Huang, D. T. J. and Varma, A.: *A.I.Ch.E. J.* **27**, 481, 489 (1981)
54. Huang, D. T. J. and Varma, A.: *Chem. Eng. J.* **21**, 47 (1981)
55. Singh, C. P. P., Shah, Y. T., and Carr, N. L.: *Chem. Eng. J.* **23**, 101 (1982)
56. Hancock, M. D. and Kenney, C. N.: *Chem. Eng. Sci.* **32**, 629 (1977)
57. Shah, Y. T. and Sharma, M. M.: *Trans. Instn. Chem. Engrs.* **54**, 1 (1976)
58. Shah, Y. T., Juvekar, V. A., and Sharma, M. M.: *Chem. Eng. Sci.* **31**, 671 (1976)
59. Sharma, S., Hoffmann, L. A., and Juss, D.: *A.I.Ch.E. J.* **22**, 324 (1976)
60. Wicke, E.: *Z. Elektrochem.* **65**, 267 (1961)
61. Wicke, E.: *Chem. Eng. Sci.* **8**, 61 (1958)
62. Thoenes, D. and Kramers, H.: *Chem. Eng. Sci.* **8**, 271 (1958)
63. Roberts, D. and Gillespie, G. C.: *Adv. Chem. Series* **133**, 600, ACS, Washington (1974)
64. Zharov, D. V., Beskov, V. S., Kuzenkov, A. P., and Terteryan, A. M.: *Kin. i Kat.* **20**, 481 (1979)
65. Oele, A. P.: *Chem. Eng. Sci.* **8**, 146 (1958)
66. Kramers, H.: *Physica* **12**, 61 (1946)
67. Satterfield, C. N. and Cortez, D. H.: *Ind. Eng. Chem. Fund.* **9**, 613 (1970)
68. Shah, M. A.: Ph.D. thesis, Univ. Birmingham, UK (1970); Shah, M. A. and Roberts, D.: *ACS Symp. Series* **133**, 259, ACS, Washington (1974)
69. Mears, D. E.: *Ind. Eng. Chem. Proc. Des. Dev.* **10**, 541 (1971)
70. Petersen, E. E.: *Chemical Reaction Analysis*, Prentice-Hall, Englewood Cliffs (1972)
71. Anderson, J. B.: *Chem. Eng. Sci.* **18**, 147 (1963)
72. Hudgins, R. R.: *Chem. Eng. Sci.* **23**, 93 (1968)
73. Mears, D. E.: *J. Catal.* **20**, 127 (1971)
74. Suzuki, M. and Smith, J. M.: *A.I.Ch.E. J.* **16**, 882 (1970)

75. Schlünder, E. U.: *ACS Symp. Series* **72,** 110 (1978), ACS Washington
76. Martin, H.: *Chem. Eng. Sci.* **33,** 913 (1978)
77. Lee, J. C. M. and Luss, D.: *Ind. Eng. Chem. Fund.* **8,** 595 (1969)
78. Carberry, J. J.: *Ind. Eng. Chem. Fund* **14,** 129 (1975)
79. Luss, D.: *Proc. 6th Eur. Symp. Chem. React. Eng.*, Heidelberg, Dechema Frankfurt (1976)
80. Kehoe, J. P. G. and Butt, J. B.: *A.I.Ch.E. J.* **18,** 347 (1972)
81. Hofmann, H.: *VDI-Berichte* **349,** 119 (1979)
82. Froment, G. F.: *Adv. Chem. Ser.* **109,** 1, ACS, Washington (1972)
83. Froment, G. F.: *Chem. Ing. Techn.* **46,** 374 (1974)
84. Ray, W. H.: *Proc. 5th Eur. Symp. Chem. React. Eng. Amsterdam* review A8, Elsevier, New York (1972)
85. Slinko, M. G.: *Brit. Chem. Eng.* **16,** 363 (1971)
86. van Doesburg, H. and de Jong, W. A.: *Chem. Eng. Sci*, **31,** 45 (1976)
87. Deans, H. A. and Lapidus, L.: *A.I.Ch.E. J.* **6,** 656 (1960)
88. Balakrishnan, A. R. and Pei, D. C. T.: *Ind. Eng. Chem. Proc. Des. Dev.* **18,** 30 (1979)
89. Vortmeyer, D. and Jahnel, W.: *Chem. Eng. Sci.* **27,** 1485 (1972)
90. Vortmeyer, D. and Jahnel, W.: *Chem. Ing. Techn.* **46,** 11 (1974)
91. Berty, J. M., Bricker, J. H., Clark, S. W., Dean, R. D., and McGovern, T. J.: *Proc. 5th Eur. Symp. Chem. React. Eng. Amsterdam* B8-27, Elsevier, New York (1972)
92. Hoiberg, J. A., Lyche, B. C., and Foss, A. S.: *A.I.Ch.E. J.* **17,** 1434 (1971)
93. Eigenberger, G.: *Chem. Ing. Techn.* **46,** 11 (1974)
94. Eigenberger, G.: *Adv. Chem. Ser.* **133,** 477, ACS, Washington (1974)
95. Lunde, P. J.: *Ind. Eng. Chem. Proc. Des. Dev.* **13,** 226 (1974)
96. Fakien, R. W. and Smith, J. M.: *A.I.Ch.E. J.* **1,** 28 (1955)
97. Schlünder, E. U.: *Chem. Ing. Techn.* **38,** 967 (1966)
98. Zehner, P.: *VDI-Forschungsheft* **558,** (1973)
99. Bauer, R.: *VDI-Forschungsheft* **582** (1977)
100. Zehner, P. and Schlünder, E. U.: *Chem. Ing. Techn.* **42,** 933; **45,** 272 (1975)
101. Damköhler, G.: *Der Chemieingenieur*, Bd III, 1. Teil, Akademie Verlag, Leipzig (1937)
102. Dixon, A. G. and Cresswell, D. L.: *A.I.Ch.E. J.* **25,** 663 (1979)
103. Hennecke, F. W. and Schlünder, E. U.: *Chem. Ing. Tech.* **45,** 277 (1973)
104. Kulkarni, B. D. and Doraiswamy, L. K.: *Catal. Rev. Sci. Eng.* **22,** 431 (1980)
105. Gnielinski, V.: *Chem. Ing. Techn.* **52,** 228 (1980)
106. Lerou, J. J. and Froment, G. F.: *Chem. Eng. Sci.* **32,** 853 (1977)
107. De Wasch, A. P. and Froment, G. F.: *Chem. Eng. Sci.* **27,** 567 (1972)
108. Schwartz, C. E. and Smith, J. M.: *Ind. Eng. Chem.* **45,** 1209 (1953)
109. Ergun, S.: *Chem. Eng. Progr.* **48,** 93 (1952)
110. Finlayson, B. A.: *Cat. Rev. Sci. Eng.* **10,** 69 (1974)
111. Karanth, N. G. and Hughes, R.: *Cat. Rev. Sci. Eng.* **9,** 169 (1974)
112. Hlavacek, V., in *Chemical Reactor Theory: A Review*, Prentice-Hall, Englewoods Cliffs, N.J. (1977)
113. Young, L. C. and Finlayson, B. A.: *Ind. Eng. Chem. Fund* **12,** 412 (1973)
114. Hlavacek, V. and Hofmann, H.: *Chem. Eng. Sci.* **25,** 173, 181 (1970)
115. Froment, G. F.: *Chem. Ing. Techn.* **46,** 374 (1974)
116. Wicke, E.: *Chem. Ing. Techn.* **47,** 547 (1975)
117. Chao, R. E., Caban, R. A., and Irizarry, M. M.: *Can. J. Chem. Eng.* **51,** 67 (1973)
118. Richarz, W. and Sgarlata-Lattmann, M. A.: *Proc. 4th Eur. Symp. Chem. React. Eng. Brussels*, page 123 *Suppl. Chem. Eng. Sci.* (1971)
119. Valstar, J. M., Bik, J. D., and van den Berg, P. J.: *Trans. Instn. Chem. Engrs* **47,** 136 (1969)
120. Carberry, J. J.: *Ind. Eng. Chem.* **58**(10), 40 (1963)
121. Carberry, J. J. and White, D.: *Ind. Eng. Chem.* **61,** 27 (1969)

122. Crider, J. E. and Foss, A. S.: *A.I.Ch.E. J.* **11,** 1012 (1965)
123. Hlavacek, V.: *Ind. Eng. Chem.* **62**(7), 8 (1970)
124. Szekely, J., Evans, J. W., and Sohn, H. Y.: *Gas–Solid Reactions,* Academic Press, New York (1976)
125. Yoon, H., Wei, J., and Dunn, K. M.: *A.I.Ch.E. J.* **24,** 885 (1978)
126. Amundson, N. R. and Arri, L. E.: *A.I.Ch.E. J.* **24,** 87 (1978)
127. Weimer, A. W. and Clough, C. E.: *Chem. Eng. Sci.* **38,** 549 (1981)
128. Kam, E. K. T. and Hughes, R.: *Trans. IChem. E.* **59,** 196 (1981)
129. Helmrich, H. and Schügerl, K.: *Chem. Ing. Tech.* **51,** 771 (1979)
130. Szekely, J. and Evans, J. W.: *Chem. Eng. Sci.* **26,** 1901 (1971)
131. Groeneveld, M. J. and van Swaaij, W. P. M.: *Chem. Eng. Sci.* **35,** 307 (1980)
132. Lindner, B. and Simonsson, D.: *Chem. Eng. Sci.* **36,** 1519 (1981)
133. Kumisera, S., Nakagawa, J., Tone, S., and Otake, T.: *Chem. Eng. J. Japan* **14,** 190 (1981)
134. Garza-Garza, D. and Dudukovic, M. P.: *Computers & Chem. Eng.* **6**(2), 131 (1982)
135. Westerterp, K. R. and Crombeen, P. R. J. J.: *Chem. Eng. Sci.,* **38,** 1331 (1983)
136. Amundson, N. R. and Aris, R.: *Chem. Eng. Sci.* **7,** 121 (1958)
137. Wicke, E. and Vortmeyer, D.: *Ber. Bunsenges. Phys. Chem.* **63,** 145 (1959)
138. Luss, D.: in *Chemical Reactor Theory,* Prentice-Hall, Englewood Cliffs, USA (1977)
139. Endo, I., Furasawa, T., and Matsuyama, H.: *Catal. Rev. Sci. Eng.* **18**(2), 297 (1978)
140. Gilles, E. D.: *Chem. Ing. Techn.* **49,** 142 (1977)
141. Eigenberger, G.: *Chem. Ing. Techn.* **50,** 924 (1978)
142. Luss, D.: *6th Eur. Symp. Chem. React. Eng. Heidelberg,* VII-280, Dechama, Frankfurt (1976)
143. van den Bosch, B. and Luss, D.: *Chem. Eng. Sci.* **32,** 203 (1977)
144. Weisz, P. B. and Hicks, J. S.: *Chem. Eng. Sci.* **17,** 265 (1962)
145. Østergaard, K.: *Chem. Eng. Sci.* **18,** 259 (1963).
146. Luss, D. and Amundson, N. R.: *A.I.Ch.E. J.* **15**(2), 194 (1969)
147. Sohn, H. Y.: *A.I.Ch.E. J.* **19**(1), 191 (1973).
148. Heinemann, R. F. and Poore, A. B.: *Chem. Eng. Sci.* **36,** 1411 (1981)
149. Aoyagi, M. and Kunii, D.: *Chem. Eng. Commun.* **1,** 23 (1974)
150. Ross, I. B. and Davidson, J. F.: *Trans. I. Chem. E.* **60,** 108 (1982)
151. Jones, B. R. E. and Pyle, D. L.: *Chem. Eng. Sci.* **25,** 859 (1970)
152. Rajadhyaksha, R. A., Vasudeva, K., and Doraiswamy, L. K.: *Chem. Eng. Sci.* **30,** 1399 (1975)
153. Ramachandran, P. A., Zadeh, H. M., and Hughes, R.: *I.Chem.E. Symp. Series* **43,** 17-1 (1975)
154. McGreavy, C. and Adderley, C.: *Adv. Chem. Series* **133,** 519, ACS, Washington (1974)
155. McGreavy, C. and Adderley, C.: *Chem. Eng. Sci.* **28,** 577 (1973)
156. Agnew, J. B. and Potter, O. E.: *Trans. Instn. Chem. Engrs.* **44,** T216 (1966)
157. Olbrich, W., Agnew, J. B., and Potter, O. E.: *Trans. Instn. Chem. Engrs.* T207 (1966)
158. Emig, G., Hofmann, H., Hoffmann, U., and Fiand, U.: *Chem. Eng. Sci.* **35,** 249 (1980)
159. Puszynski, J., Snita, D., Hlavacek, V., and Hofmann, H.: *Chem. Eng. Sci.* **36,** 1605 (1981)
160. Wicke, E., Padberg, G., and Arens, H.: *Proc. 4th Eur. Symp. Chem. React. Eng. Brussels,* 425, Pergamon Press, Oxford (1971)
161. Baiker, A., Casanova, R., and Richarz, W.: *Germ. Chem. Eng.* **3,** 112 (1980)
162. Eigenberger, G.: *Chem. Eng. Sci.* **27,** 1909 (1972)
163. Lübeck, B.: *Chem. Eng. Sci.* **29,** 1320 (1974)
164. Lübeck, B.: *Chem. Eng. J.* **7,** 29 (1974)
165. Vortmeyer, D. and Stahl, R.: *Chem. Eng. Sci.* **36,** 1373 (1981)
166. van Doesburg, H. and de Jong, W. A.: *Chem. Eng. Sci.* **31,** 45, 53 (1976)
167. Padberg, G. and Wicke, E.: *Chem. Eng. Sci.* **22,** 1035 (1967)

168. Price, T. H. and Butt, J. B.: *Chem. Eng. Sci.* **32**, 393 (1977)
169. Weng, H. S., Eigenberger, G., and Butt, J. B.: *Chem. Eng. Sci.* **30**, 1341 (1975)
170. Blaum, E.: *Chem. Eng. Sci.* **29**, 2263 (1974)
171. Gilles, E. D.: *Regelungstechnik*, 191 (1977)
172. Rhee, H. K., Foley, D., and Amundson, N. R.: *Chem. Eng. Sci.* **28**, 607 (1973)
173. Elnashaie, S. S. E. and Cresswell, D. L.: *Chem. Eng. Sci.* **29**, 1889 (1974)
174. Mehta, P. S., Sams, W. N., and Luss, D.: *A.I.Ch.E. J.* **27**(2), 234 (1981)
175. Rhee, H. K., Lewis, R. P., and Amundson, N. R.: *Ind. Eng. Chem. Fund.* **13**, 317 (1974)
176. Boreskov, G. K., Matros, Y. S., and Kiselev, O. V.: *Kin. i Kat.* **20**, 773 (1979) (in Russian)
177. Satterfield, C. N.: *A.I.Ch.E. J.* **21**, 209 (1975)
178. Hofmann, H.: *Catal. Rev. Sci. Eng.* **17**(1), 71 (1978)
179. Goto, S., Levec, J., and Smith, J. M.: *Catal. Rev. Sci. Eng.* **15**(2), 187 (1977)
180. Gianetto, A., Baldi, G., Specchia, V., and Sicardi, S.: *A.I.Ch.E. J.* **24**, 1087 (1977)
181. Kürten, H. and Zehner, P.: *VDI Berichte* **349**, 157 (1979)
182. Chaudhari, R. V. and Ramachandran, P. A.: *A.I.Ch.E. J.* **26**, 177 (1980)
183. Deckwer, W. D. and Alper, E.: *Chem. Ing. Techn.* **52**, 219 (1980)
184. Hammer, H.: *VDI Berichte* **349**, 157 (1979)
185. Schügerl, K.: *Chem. Ing. Techn.* **12**, 951 (1980)
186. Steiff, A., Poggeman, R., and Weinspack, P. M.: *Chem. Ing. Techn.* **52**, 492 (1980)
187. Deckwer, W. D., Serpemen, Y., Ralek, M., and Schmidt, B.: *Ind. Eng. Chem. Proc. Des. Dev.* **21**, 231 (1982)
188. Shah, Y. T. and Parulekar, S. J.: *Chem. Eng. J.* **23**, 15 (1982)
189. Shah, Y. T.: *Reaction Engineering in Direct Coal Liquefaction*, Addison-Wesley, Reading, Mass. (1981)
190. Fukushima, S. and Kusaka, K.: *J. Chem. Eng. (Japan)* **10**, 461, 468 (1977)
191. Fukushima, S. and Kusaka, K.: *J. Chem. Eng. (Japan)* **11**, 241 (1978)
192. Votruba, J., Hlavacek, V., and Marek, M.: *Chem. Eng. Sci.* **27**, 1845 (1972)
193. Joshi, J. B.: *Trans. IChemE* **58**, 155 (1980)
194. Shah, Y. T., Kelkar, B. G., Godbole, S. P., and Deckwer, W. D.: *A.I.Ch.E. J.* **28**, 353 (1982)
195. Joshi, J. B., Pandit, A. B., and Sharma, M. M.: *Chem. Eng. Sci.* **37**, 813 (1982)

Chapter X

*The optimization of chemical reactors**

Optimization is the activity by which it is endeavoured to have a process executed under such conditions that an extreme value of a specific quantity is attained. All optimization procedures have the following steps in common:

1. Determination of the object of the optimization and definition of the optimization criterion, for example maximum profit or minimum reactor residence time. The object of the optimization has to be translated into some mathematical function: the objective function.
2. Development of the mathematical model of the process: the determination of the objective function and its correlation with the optimization variables.
3. Selection and determination of the optimization variables and their constraints, for example the reaction temperature as the variable together with a maximum allowable temperature as its constraint.
4. Selection of an effective search strategy for the calculation of the extreme(s) of the objective function, followed by the execution of the calculations.
5. Interpretation of the results and their translation into design and operating instructions.

The first three points belong to the task of the reactor engineer, selection of a search technique and the corresponding calculations will mostly be done by his mathematical colleague. However, the reactor engineer again has to judge and evaluate the results and translate them into useful design and operating information. The role of the reactor engineers is, therefore, centred around the steps 1, 2, 3 and 5 above. The development of a mathematical model for a chemical reactor was the subject of all previous chapters of this book; an adequate model to describe the reactor will be the starting point in this chapter.

The formulation of the objective function and the selection of optimization variables will be the subject of Section X.1. In Section X.2 we discuss some of the simpler optimization problems with reaction temperature as the main optimization variable. As the reactor engineer has to be somewhat familiar with

---

* H. J. Fontein, associate professor in computer applications to chemical engineering, contributed to this chapter.

the job of his mathematical colleague in order to be conversant with him, we will give a compendious introduction to the many optimization methods nowadays available and discuss the principles of a few spectacular methods in Section X.3. All along we will give some results obtained by optimization and their significance for design and operation of chemical reactors. For an introduction in more depth to optimization applied in chemical engineering we refer to the books of Ray and Szekely [1], Hoffmann and Hofmann [2] and Aris [3]. The book of Aris is exclusively dedicated to the application of dynamic programming to the optimization of chemical reactors.

X.1 THE OBJECT AND MEANS OF OPTIMIZATION

It is meaningless to discuss optimization unless the quantity of which the extreme is sought is properly specified. Since the production of chemicals is to be considered an economic activity, the quantity to be optimized is closely related to the economy of the plant; it will ultimately be its financial profit over the entire project life, which should be maximized. This means, among other things, that the plant unit should be so designed as to allow the desired amount of product to be obtained at the lowest possible manufacturing costs. Although the chemical reactor proper is frequently only a small part of the total equipment of a manufacturing unit, the reactor may considerably affect the overall economy of the plant. This will be illustrated below by means of the idealized economic diagrams introduced in Section II.8.

Let us assume that a material A is converted in an existing plant into a desired product P and by-products X and Y, and that the production rate of P is Φ_P at a plant yield η.

The plant yield is equal to the reactor yield η_{PL} if no recirculation of the raw material is applied; it is equal to the selectivity σ_{PL} of the reactor section if unconverted A is completely separated from the product stream and recycled to the reactor feed (see figure X.1). We assume that all separation efficiencies are

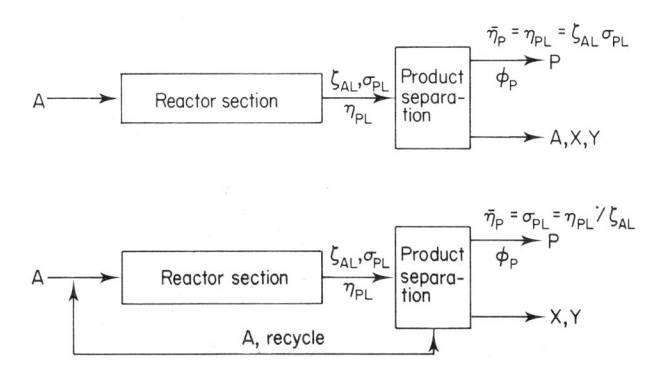

Fig. X.1 Plant yield $\bar{\eta}_P$ and reactor yield η_{PL}

100%. If the by-products have a negligible sales value, the gross profit I of the plant per unit time may be schematically represented by:

$$\dot{I} = C_P \Phi_P - \left[\frac{C_A \Phi_P}{\bar{\eta}_P} + C_{\text{var}} \Phi_P + \dot{C}_{\text{fix}} \right]. \tag{X.1}$$

C_A and C_P represent the total costs or net sales value of a unit amount of A and P, respectively; C_{var} are the variable costs per unit amount of P, and $\dot{C}_{\text{fix}}$ the fixed costs of the installation per unit time. The break-even point at which $\dot{I} = 0$ is obtained at a production rate $(\Phi_P)_{\text{nil}}$; the economic flexibility of the plant and the possibilities for making profit are improved as the ratio of the actual production rate and $(\Phi_P)_{\text{nil}}$ is increased.

Let the existing economic situation according to equation (X.1) be represented by figure X.2(O). This situation can, in principle, be improved by changes in the reactor section in a number of different ways:

(a) Yield increase by improved process conditions; the raw material costs

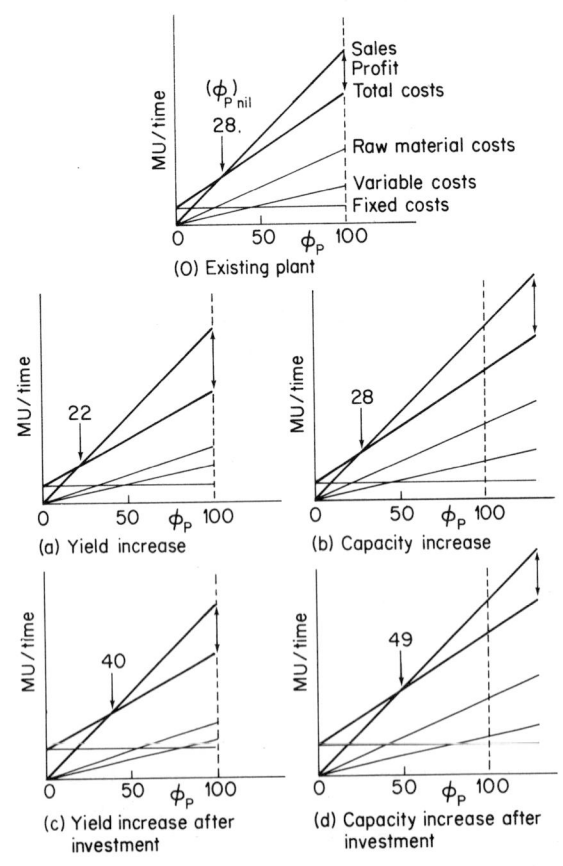

Fig. X.2 Several ways to improve the economy of a plant

diminish, and as a consequence the economic flexibility and the profit at the same production rate increases (figure X.2(a)).
(b) Capacity increases by improved process conditions; in this case the diagram (figure X.2(b)) is the same as figure X.2(a); the increased capacity results in greater flexibility and higher profit only if the increased production can be sold.
(c) As (a) after capital investment; it depends on the investment required for yield increase whether the profit is influenced favourably ($d\dot{C}_{fix}/d\eta < C_A \Phi_P/\eta^2$), or not (figure X.2(c)).
(d) As (b) after capital investment; it is seen from figure X.2(d) that the gross profit and the economic flexibility increase only if the production rate and sales volume rise considerably.
(e) Reduction of the variable costs; this has the same effect as (a).

Other combinations of measures taken and influence on profitability are possible, but will not be discussed.

It appears that the *plant capacity* and the *plant yield* influence the plant economy fairly independently, and also that the variable costs, in particular *heat economy* and *energy consumption*, can play an important part. Now, apart from the reactor section, the manufacturing unit will generally contain sections for feed preparation, product separation and product treatment, delivery, storage and waste treatment. Consequently, the extent to which the reactor itself influences the economy of the plant strongly depends on the costs connected with the reactor section in relation to the total costs.

The *cost function*, the right-hand term of equation (X.1), is still simplified, because we assumed the wasted by-products to have a negligible value. However, often they even have a negative value due to the corresponding separation and disposal costs. The fixed costs can be divided into two parts:

1. the truly fixed costs, for instance the personnel costs;
2. the fixed costs which depend on the investment in the plant, for example the maintenance, insurance and depreciation costs.

Once the investment has been made, so the plant has been built, the fixed costs no longer play a role in an optimization study, but in the evaluation and design stage, when the capacity of the plant and/or the investment still has to be decided upon, we have to take them into account by comparing the fixed costs for various alternatives. The total cost function now becomes

$$\dot{C}_{tot} = C_A \frac{\Phi_P}{\eta} + \frac{1-\eta}{\eta} C_{disp} \Phi_P + C'_{var} \Phi_P + \dot{C}_{Ifix} \left(\frac{\Phi_P}{\Phi_{Pr}}\right)^\alpha + \dot{C}'_{fix} \quad (X.2)$$

In this equation Φ_{Pr} is a reference production rate. The exponent α varies for different plants and pieces of equipment; often $\alpha = 0.6$ is taken according to the so-called Williams six-tenths rule. C'_{var} is the variable costs excluding the variable waste disposal costs C_{disp}, both per unit of P and unit of waste, $\dot{C}_{Ifix}$ is the investment related costs per unit time at a nominal production rate Φ_{Pr} and $\dot{C}'_{fix}$ is the truly fixed costs per unit time. If necessary, corrections for differences in molecular weight are included in the cost factors. In equation (X.2) it is assumed that the purification and disposal costs C_{disp} are equal for the waste products X and Y and the raw material A that is not recirculated. In the cost function (X.2) all costs except C'_{fix} are influenced by the reactor design and operation; the cost elements C_{var} and $\dot{C}_{Ifix}$ are influenced directly and/or indirectly only for the part which refers to the operating costs of and investments made in the reactor section. We will discuss this further in Example X.1.a.

X.1.1 The objective function

The only realistic objective of an optimization study is the *economic optimization* of a plant during the entire project life. The criterion used nowadays for measuring the profitability of a venture is either the *net present value NPV* or the *discounted cash flow rate of return DCFRR*. Both criteria are equivalent. The DCFRR is the interest rate at which $NPV = 0$. For economic acceptability of a venture the NPV must be positive for a preset minimum acceptable interest rate i_{min} or the DCFRR interest rate i_{DCFRR} should be higher than i_{min}. The objective function now is:

$$i_{DCFRR} = \max$$

or
$$NPV_{at\,i_{min}} = \max. \tag{X.3}$$

More information on these techniques for profitability studies can be found in the handbooks, e.g. [64, 65]. They require information on the cash flows, both positive, such as income from sales, and negative, like expenditures for the investment and in later years for the operation of the plant, taxes, etc. An estimate has to be made of sales prices, sales volumes, raw material prices and consumption, expenditures for personnel, energy, materials and equipment during all the years the project is in preparation, under construction or in operation. The NPV and DCFRR methods weigh cash flows in earlier years more heavily than those in later years by the technique of discounting to the present.

It can be intuitively felt that the formulation of the relation between all these factors and the properties of the chemical reactor is hardly possible. Once the plant has been built and is in *operation*, all the major investments have been made. Because it is no use to cry over spilt milk, it is now sufficient for optimization studies to minimize all actual and future expenses in an optimization study, without taking the investments made into account. Thus the economic optimization of an existing plant is simpler than of a plant in the design phase.

In view of the difficulties mentioned for economic optimization, therefore, mostly the second best solution of a *costs minimization* is taken for studies during the design phase, for instance by use of equation (X.2). There is a difference between cash flow and cost, mainly caused by the inclusion of a depreciation allowance in the costs. This depreciation allowance is not an expenditure, but an allowance for the investment expenditures made in previous years. It is a rather arbitrary cost, because depreciation periods and methods can vary widely. Cost minimization leads to the objective function:

$$\dot{C}_{tot} = \text{min.} \tag{X.4}$$

Due to the depreciation allowance both the investments made and the operating expenditures are included in $\dot{C}_{tot}$. Most often the costs are taken at the moment, when the plant runs at design capacity in some future year. Change of the economic climate and changes in price levels are mostly excluded.

Because of the difficulties in formulating the economic objective function often only a *technical optimization* is carried out. For chemical reactors this means that some technical property of the reactor is optimized, for example the residence time is minimized, the yield or the selectivity is maximized.

A relationship between the *technical* properties of the reaction section (such as capacity, yield, heat requirements) and the *economy* of the entire plant will, in general, be very difficult to formulate quantitatively. If, for example, a certain temperature level of the reactor section is surpassed, this may involve the need for an entirely different heating system or the use of different construction materials; this will result in a discontinuity in the relationship between total costs and operating temperature. Such considerations make the design of a plant which can operate near its economic optimum largely a matter of experience and sound judgment. At the same time, however, it is a wise policy to try to optimize parts of the installation, such as the reactor section, with reference to a more restricted criterion for optimization. In this way, the economic features of the entire plant can be studied for a number of combinations of plant sections which in themselves have been suitably optimized. In view of this procedure it is worthwhile considering optimization of chemical reactors in a more restricted sense. Accordingly, the rest of this chapter will be devoted to optimization with respect to capacity and/or yield of the reactor section, bearing in mind that its performance should be judged ultimately by standards of economy for the total plant and that technical optimization without regard for the economy of the entire plant is very dangerous.

Many basic data required for industrial optimization of a reactor are supplied by the chemical laboratory. Information should be available as to the cheapest raw materials, the preferred sequence of reaction steps in a complicated synthesis, the most suitable catalyst, the possible uses for by-products, etc. In evaluating the results of experiments related to these questions, the chemist is mainly guided by the obtainable yield of desired product. When a decision has been made on the particular reaction system to be used, more laboratory data are needed regarding the influence of the operating variables on the conversion rates of all reactions involved. So we have to be aware that many optimization decisions have already been taken in the laboratory, which require a thorough discussion with the manufacturing and economical specialists of the project team.

Hinger and Blenke [4, 5] found that a total of thirteen publications on the oxidation of SO_2 in a multi-bed adiabatic reactor with cold shot cooling with the cold feed in between the catalyst beds — a 'cold shot' converter or a 'quench' converter — gave thirteen different answers for the optimum conditions, because different objective functions and different optimization variables were chosen. Urbanek and Trela [73] established the same fact and compared the optimization studies with industrial practice. Le Goff [6] studied the optimization of the energy consumption alone in gas–solid reactors, including the energy consumed in the manufacture of the necessary equipment. Le Goff proved that fully turbulent conditions, which are necessary for obtaining plug flow, may lead to too high energy consumption during the entire project life. Then, specific flow rates should be reduced and reactor volume enlarged for minimum energy consumption.

X.I.2 The optimization variables

All variables that influence the design and operating performance of a chemical reactor can be chosen as optimization variables. Design variables are the reactor type, the reactor volume, the catalyst distribution among the catalyst beds, the number of tanks in a cascade, the heat exchange area in a cooled reactor, the reactor pressure, the feed distribution, the recycle of non-converted raw materials or intermediate products, etc. The reactor volume again is determined by the length and the diameter of the reactor. Also the size of catalyst pellets, the porosity and pore diameters can be chosen as optimization parameters. The concentration of reactants, the ratio or excess of reactant concentrations, the reactor inlet temperature, the cooling medium temperature and reactor temperature profiles are important.

All these variables are subject to *constraints* such as maximum allowable temperatures and pressures, which are mainly dictated by the construction of the reactor and the heat supply/removal system and the construction materials used. Experience has shown that many of these variables are of minor importance. The design engineer or plant operator has to select the most important variables. For example, in a cooled reactor the pitch, distance, diameter, number and length of the tubes can be chosen as variables, because they all influence the final reactor construction costs. In order to simplify the optimization study, it is better to choose only the tube diameter and the heat exchange area; and in an independent study the lowest construction costs can be determined for the selected values of d_t and A.

Hinger and Blenke [5] showed that the reactor temperature — and often also the pressure — is the most powerful optimization variable, that feed composition and velocities in the reactor are only of secondary importance and that geometric factors such as diameter and length of tubes, catalyst diameters and internal surface areas are only tertiary variables. We will therefore dedicate Section X.2 exclusively to the reactor temperature as a means of optimization.

X.1.3 Relation between technical and economic optima

Hinger and Blenke [4, 5] undertook a very extensive study on the relation between the choice of the objective function and the optimal reactor design. They considered a reactor section shown in figure X.3, in which a catalytic gas phase reaction is executed. The products are recovered in a column and the unconverted reactants are recompressed and recirculated to the reactor. Hinger and Blenke implicitly assumed the disposal costs to be negligible and used a pay-out time as the objective function. As a cost function they chose

$$\dot{C}_{RS} = \dot{C}_{RM} + \dot{C}_{fix} + \dot{C}_{var} + \dot{D}I_{RS}$$

in which $\dot{C}_{RS}$ is the total costs per unit of time in the reactor section of figure X.3, $\dot{C}_{RM}$ the raw material costs per unit of time, $\dot{C}_{fix} + \dot{C}_{var}$ the operating costs per unit of time of the reactor section excluding the raw material costs, $\dot{D}$ the depreciation rate (fraction/time) and I_{RS} the investment in the reactor section in MU. The pay-out time—their objective function—was defined as

$$\frac{I_{RS}}{\dot{I} - (\dot{C}_{RM} + \dot{C}_{fix} + \dot{C}_{var})},$$

in which $\dot{I}$ is the income from sales.

They studied three different gas phase reactions:
(a): $A + B \rightarrow P$
(b): $A + B \leftrightarrows P$
(c): $A + B \rightarrow P$ with $A + 2B \rightarrow X$ and $P + B \rightarrow X$
and optimized the costs of the reactor section only. Some of their results are shown in figures X.4 and X.5.

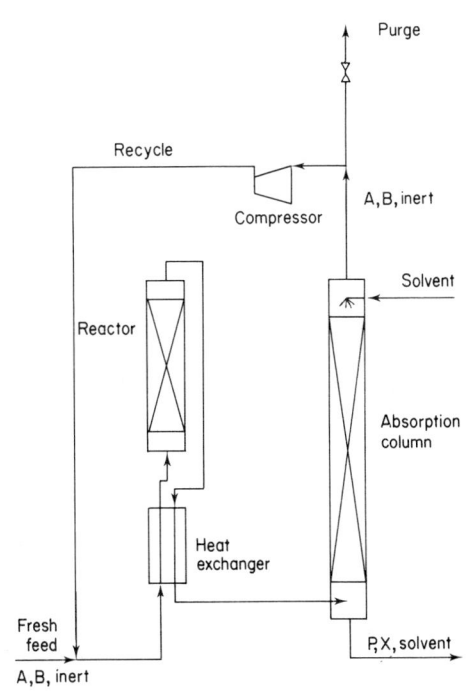

Fig. X.3 A reactor section of a plant as considered by Hinger and Blenke [4, 5]

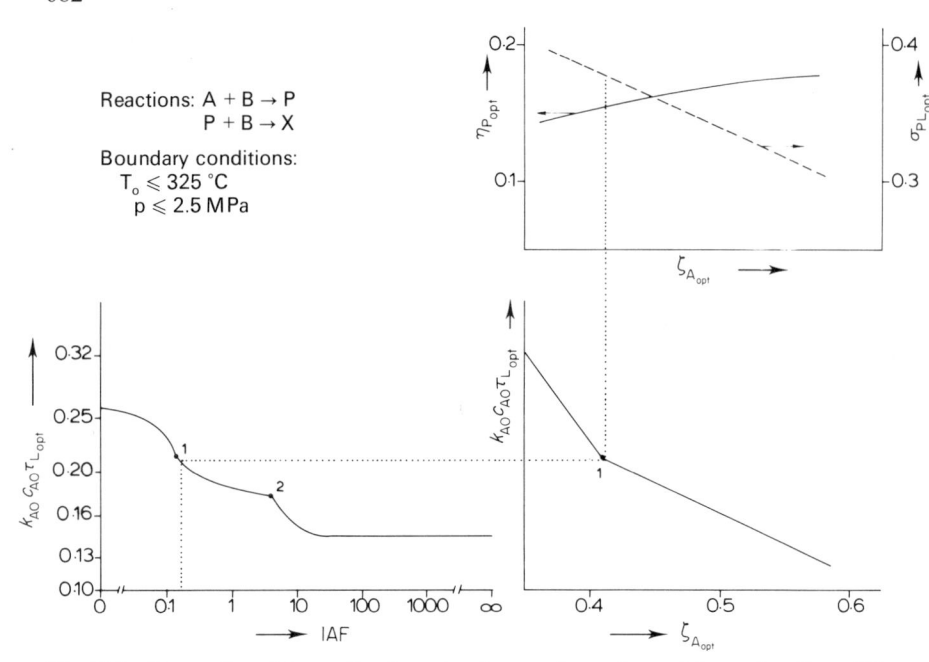

Fig. X.4 Properties of the optimal reactor for reaction scheme (c) as a function of the investment appreciation factor. Irregularities are caused by the pressure and temperature constraints [From Hinger and Blenke [4, 5]. Reproduced by permission of Verlag Chemie GmbH]

In these figures the optimal reactor is represented by $k_o c_{Ao} \tau_L$, where o refers to the reactor inlet temperature and inlet concentration. The reactor has been considered to operate under isothermal conditions. Lower values of the inlet temperatures and lower reactor pressures reduce the values of k_o and of the reactant concentration c_{Ao} in the gas phase respectively. Their reactor residence time is defined as L/u_o, the ratio of the reactor length to the linear gas velocity at the reactor inlet. This implies that a lower value of $k_o c_{Ao} \tau_L$ can be accompanied either by a smaller or a larger reactor volume, depending on the corresponding changes of the optimum inlet conditions. Despite this, their results demonstrate the large influence of economic parameters on the optimum value of $k_o c_{Ao} \tau_L$ for the reactor.

In figure X.4 the property $k_o c_{Ao} \tau_L$ of the optimal reactor is plotted versus the Investment Appreciation Factor IAF. This factor is defined by:

$$IAF = \frac{\dot{DI}_{RS}}{\dot{C}'_{fix} + \dot{C}_{var} + \dot{C}_{RM}}.$$

The kinks in the $k_o c_{Ao} \tau_L - IAF$ curve in figure X.4 are caused by the boundary conditions: at $IAF = 0.01$ the inlet temperature and at $IAF = 6.0$ the reactor pressure reach their upper boundaries.

The IAF is the ratio of the annual depreciation allowance of the reactor section

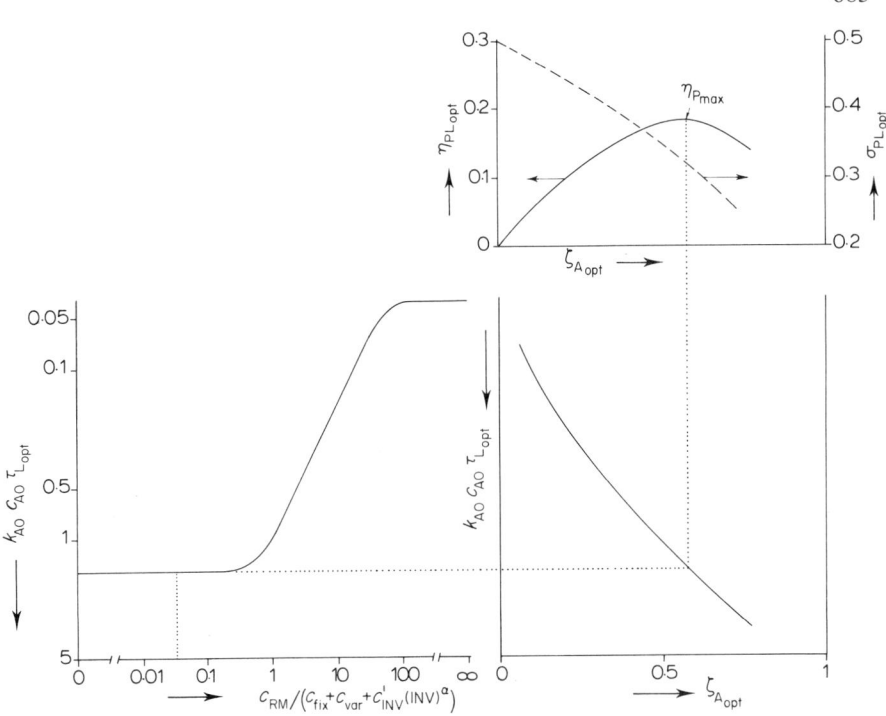

Fig. X.5 Properties of the optimal reactor for reaction scheme (c) as a function of the ratio of the raw materials costs to the operating costs [From Hinger and Blenke [4, 5]. Reproduced by permission of Verlag Chemie GmbH]

divided by the annual operating and raw materials expenses. We see from figure X.4 that the $k_o c_{Ao} \tau_L$ value of the optimal reactor decreases with increasing IAF. Following the dotted line in the same figure, we find the corresponding values of the conversion, yield and selectivity in the optimal reactor. A lower value of $k_o c_{Ao} \tau_L$ corresponds to a higher value of ζ_A: this means that the reactor becomes larger, the recycle rate lower and the sizes of the reactor product splitter, recycle compressor and heat exchanger smaller. A limit if reached for $IAF > 20$, where $\zeta_{AL} = 0.52$ and $\eta_{PL} = 0.17$ are the optimum values.

In figure X.5 the $k_o c_{Ao} \tau_L$ value of the optimum reactor is plotted as a function of the ratio of the raw materials costs to the operating costs. For low values of this ratio — cheap raw materials — the reactor should be run at maximum yield and for expensive raw materials the selectivity should be high. In the latter case the recycle rate is high with corresponding high investments in the splitter, compressor and heat exchanger. We see that the inlet temperature, pressure and residence time for the optimal reactor strongly depend on the IAF and on the ratio of raw material costs to operating costs including the depreciation charge. They studied many possible situations and many optimization variables and concluded for their special reactor section lay-out and reaction schemes:

I   For $IAF > 10$ or high investments and very low raw materials and operating

costs, their economic optimization criterion can be replaced by the more simple requirement that the investment in the reactor section should be minimized. This leads again to even more simple requirements:
(a) If the investment in the reactor and the column is lower than 10% of that in the compressor and heat exchanger, the yield for reactions (c) and the conversion for reactions (a) and (b) should be maximized.
(b) If the investment in the reactor is more than 50 times higher than that in the column, compressor and heat exchanger, then the volume of the reactor should be minimized.

II If $IAF < 0.1$, then the reactor section should be designed in such a way that the raw materials + operating costs are minimized. Here the two subcases are:
(a) If the raw material costs are more than 50 times higher than the operating costs, the selectivity should be maximized.
(b) If the raw material costs are less than 50% of the operating costs for reactions (c), the concentration of the desired product P in the reactor product should be maximized.

The conclusions of Hinger and Blenke strictly speaking only relate to their particular reactor section and reactions studied. It can be felt intuitively that they will have a more general validity.

From figure X.5 we see that for minimum total costs many optimal values of $k_o c_{Ao} \tau_L$ for the reactor exist: only in case the raw material costs far exceed the operating costs should a selectivity maximization be strived for; conversely if raw material costs are negligible a yield maximization coincides with the economic optimization. It will be clear that all technical optimization criteria leading to widely varying results in reactor designs only under very specific conditions will result in the economic optimal performance of the entire plant. As a consequence, every reactor optimization study should be preceded by a technical-economical evaluation of the entire plant in order to determine whether a more simplified technical optimization criterion can be used instead of maximizing the NPV or DCFRR. Regretfully, the extreme cases where, for example, only the reactor volume, energy consumption, yield or selectivity can be chosen as objective function seldom occur in practice except sometimes in the fine chemicals industry.

Example X.1.a *Influence of economics on optimal reactor design and operation*
Experimentally in a pilot plant the data given in figure X.6 were obtained for the reaction scheme A → P → X. A technical–economical evaluation has revealed that for this plant the investment costs will be so low that a minimization of the variable costs will fairly well represent the economic optimization of the plant. Separation efficiencies in the purification unit in the first instance can be assumed to be 100%. Moreover, to simplify the equations we assume $M_A = M_P = M_X$. The total *variable costs* amount to

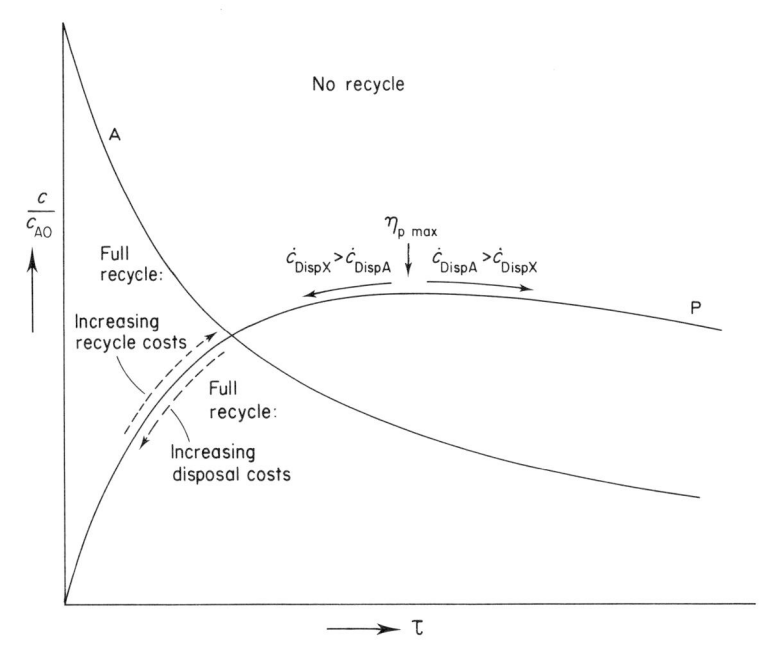

Fig. X.6 Influence of economics on reactor design and operation for consecutive reactions

$$\dot{C}_{tot} = + C_A(1 - R + R\zeta_A)\Phi_P/\sigma_P\zeta_A$$
$$+ C_{rec}R(1 - \zeta_A)\Phi_P/\sigma_P\zeta_A$$
$$+ C_{Xdisp}(1 - \sigma_P)\Phi_P/\sigma_P + C_{Adisp}(1 - R)(1 - \zeta_A)\Phi_P/\sigma_P\zeta_A \quad (a)$$

in which

$\dot{C}_{tot}$ = total variable costs [MU/time]
C_A = purchase price of A [MU/tonne A]
C_{rec} = recycle costs [MU/tonne A], mainly energy for splitter and pump
C_{Xdisp} = removal and disposal costs of X [MU/tonne X], mainly energy and chemicals in splitters and disposal unit
C_{Adisp} = as C_{Xdisp}, but for A [MU/tonne A]
R = recycle ratio
Φ_P = production of P [tonne/time], constant.

If we differentiate $\dot{C}_{tot}$ with respect to R, we find:

$$\frac{d\dot{C}_{tot}}{dR} = [C_{rec} - (C_A + C_{Adisp})](1 - \zeta_A)\Phi_P/\sigma_P\zeta_A.$$

The application of recycle makes sense only if the variable costs decrease, so in this case $d\dot{C}_{tot}/dR < 0$. So the condition for recycle application is $C_{rec} < (C_A + C_{Adisp})$. If this condition is fulfilled, costs are lowest at an extreme, because $d\dot{C}_{tot}/dR$ is independent of R. This extreme is $R = 1$, so full recycle should be applied. Similarly, it can be reasoned that for $C_{rec} \geq (C_A + C_{Adisp})$ we should choose the other extreme $R = 0$. We have an optimum if

$$d\dot{C}_{tot} = \frac{\partial \dot{C}_{tot}}{\partial \sigma_P} d\sigma_P + \frac{\partial \dot{C}_{tot}}{\partial \zeta_A} d\zeta_A = 0$$

or if

$$\frac{d\sigma_P}{d\zeta_A} = -\frac{\partial \dot{C}_{tot}/\partial \zeta_A}{\partial \dot{C}_{tot}/\partial \sigma_P}.$$

For a plant with *full recycle* ($R = 1$) we find for the optimality condition, after differentiation of equation (a),

$$\frac{d\sigma_P}{d\zeta_A} = -\frac{\sigma_P}{\zeta_A \left(1 + \dfrac{C_A + C_{Xdisp} - C_{rec}}{C_{rec}} \zeta_A\right)}$$

and for *no recycle* ($R = 0$) we find in a similar way that

$$\frac{d\sigma_P}{d\zeta_A} = -\frac{\sigma_P}{\zeta_A \left(1 + \dfrac{C_{Xdisp} - C_{Adisp}}{C_A + C_{Adisp}} \zeta_A\right)},$$

and for the technical optimum of a maximum yield ($d\eta_P/d\zeta_A = 0$) — because $\eta_P = \sigma_P \zeta_A$ and $d\eta_P/d\zeta_A = \sigma_P + \zeta_A d\sigma_P/d\zeta_A$ — the optimality condition is

$$\frac{d\sigma_P}{d\zeta_A} = -\frac{\sigma_P}{\zeta_A}.$$

Comparing the three optimality conditions we can conclude as follows:

— For $C_{rec} < C_A + C_{Adisp}$ full recycle should be employed. The economic optimum falls at shorter reactor residence times, because $C_A + C_{Xdisp} - C_{rec}$ is always positive. The higher the raw material costs C_A and removal and disposal costs C_{Xdisp} for X are, the lower the reactor conversion and the higher the reactor selectivity must be. We move farther away from the technical optimum η_{Pmax} in the direction of the origin. Vice versa, the more important the recycle costs, C_{rec}, we move towards an operation at the maximum yield point.

— For $C_{rec} \geq C_A + C_{Adisp}$ no recycle should be applied. In this case the reactor should be operated at its technical optimum only in case the removal and disposal costs for A and X are equal. If X is more expensive to remove and dispose of, we should operate before the maximum yield point and in case $C_{Xdisp} < C_{Adisp}$ beyond the maximum yield point as indicated in figure X.6.

Example X.1.b *Optimal reactant concentration in the reactor feed*
Rosas and Smith [7] studied a consecutive reaction of second order, where nicotinonitrile A is converted into the desired pharmaceutical nicotinamide P,

followed by the undesired formation of nicotinate X. Both reactions take place under the influence of alkaline hydrolysis (B = OH$^-$) according to

$$A + B \to P + B \to X$$

with $R_A = -k_P c_A c_B$
$R_P = k_P c_A c_B - k_X c_P c_B$
$R_X = k_X c_P c_B$
and $k_P = 2.70 \times 10^{10} \exp(-9260/T) \, m^3/kmol$
$k_X = 3.92 \times 10^6 \exp(-7500/T) \, m^3/kmol$.

As can be seen B acts in the first reaction as a catalyst and is consumed only in the second reaction. Reactant A cannot be recovered from the reactor product. Costs for raw material A are so high in comparison to all other costs that a maximum yield should be striven for. Product quality specifications lead to the requirements $\zeta_A \geq 0.99$ and $\eta_P \geq 0.95$. The reactions have to be carried out in a BR because of the long reaction times involved.

Activation energies indicate that the reaction has to be carried out at as high a temperature as technically and economically feasible. An optimum now requires choosing a correct initial value of c_B/c_A and a correct temperature so that product specifications are met and that B is not depleted by the second reaction before a yield $\eta_P \geq 0.95$ is reached.

Too much B initially present would cause the second reaction to continue, which implies a quality deterioration of the desired product.

Rosas and Smith [7] solved the mathematical problem by taking a new variable

$$\theta = \int_0^t c_B \, dt$$

so that $d\theta = c_B \, dt$. This variable transformation was originally proposed by French [66]. Writing the material balances now in the θ-domain, we find:

$$dc_A = -k_P c_A \, d\theta$$
$$dc_P = k_P c_A \, d\theta - k_X c_P \, d\theta$$
$$dc_X = dc_B = k_X c_P \, d\theta$$

These equations in the θ-domain are exactly equal to the equations for first order consecutive reactions in the t- or τ-domain for a BR or a PFR, as treated in Section III.4. The solutions are therefore exactly equal, but in the θ-domain. For isothermal operation the maximum yield is (see Section III.4.1)

$$\eta_{Pmax} = \frac{c_{Pmax}}{c_{Ao}} = \kappa^{\kappa/(1-\kappa)}$$

reached at $\zeta_A = 1 - e^{-k_P \theta_{max}}$

where

$$\theta_{max} = \frac{\ln \kappa}{(\kappa - 1)k_P},$$

in which

$$\kappa = \frac{k_X}{k_P} = 1.45 \times 10^{-4} \exp(1760/T).$$

The amount of B consumed equals the amount of X formed, given by

$$\frac{c_X}{c_{Ao}} = 1 - \frac{c_A}{c_{Ao}} - \frac{c_P}{c_{Ao}} = \zeta_A - \eta_P.$$

If we now feed an amount of B to the reactor so that the concentration of B in the maximum is almost zero, then the product deterioration stops. This leads to the condition for the optimal amount of B to be fed to the reactor, after elimination of θ:

$$\left(\frac{c_{Bo}}{c_{Ao}}\right)_{opt} = 1 - \kappa^{1/(1-\kappa)} - \kappa^{\kappa/(1-\kappa)}$$

The optimum reactor feed concentration is a function of the reaction temperature. Rosas and Smith [7] experimentally verified the derivations above. For reactor temperatures above 140°C product quality specifications could be met, a reaction temperature of 160°C was chosen finally for the design. In an isothermal reactor now the course of c_A, c_P and c_X can be calculated as a function of θ. The initial value c_{Bo} is known. With the relation $c_X = c_{Bo} - c_B$ and with $dt = d\theta/c_B$ the required reaction time can be calculated. The reactor can be overdesigned without penalty—or the reaction proceed longer than t_{max}—because after the depletion of B the undesired reaction stops, that is as soon as the maximum value of c_P is reached. At high temperatures the reactions are so fast that also a continuous PFR can be applied.

X.2 OPTIMIZATION BY MEANS OF TEMPERATURE

Once it has been established in what temperature range a given reaction system can be operated, it may be asked what will be the most favourable reaction temperature or temperature sequence with respect to capacity, final conversion and yield. For single irreversible reactions, the greatest conversion rate and hence the highest capacity are always achieved at the highest temperature which is economically and technically permissible. Particularly when great heat effects are involved, the arrangements for cooling or heating may be a much more determining factor in temperature selection than the reactor capacity. Capacity optimization may, however, play an important part in exothermic equilibrium reactions; a compromise must then be found between the requirements of a high degree of conversion (low temperature) and those of a high average conversion rate. The various optimization problems arising with this type of reaction will be dealt with below. In conclusion, some remarks will be made on the optimization of more complex reaction systems in which especially the reactor yield is involved.

X.2.1 The optimization of exothermic equilibrium reactions

With this type of reaction, the conversion rate passes through a maximum as a function of temperature when the composition of the reaction mixture is constant. At a low temperature this rate will be small, whereas a high temperature, T_e, will exist at which the mixture is in equilibrium and the conversion rate is zero. The temperature at which the conversion rate is a maximum will be somewhat lower than T_e and is determined by the condition

$$\frac{\partial R_J}{\partial T} = 0. \quad (X.5)$$

This condition can be further elaborated if the rate equation is known. Suppose we have the equilibrium reaction

$$A + B \rightleftharpoons P + Q$$

having a production rate of P defined by

$$R_P = k_1 c_A c_B - k_2 c_P c_Q,$$

with $k_1 = k_{1\infty} \exp(-E_1/RT)$, $k_2 = k_{2\infty} \exp(-E_2/RT)$ and $E_2 > E_1$. Applying equation (X.5) we find

$$\frac{\partial R_P}{\partial T} = c_A c_B \frac{dk_1}{dT} - c_P c_Q \frac{dk_2}{dT} = 0$$

or

$$\frac{c_P c_Q}{c_A c_B} = \left[\frac{k_1 E_1}{k_2 E_2}\right]_{T=T_{opt}} = \frac{k_{1\infty} E_1}{k_{2\infty} E_2} \exp\left[\frac{E_2 - E_1}{RT_{opt}}\right], \quad (X.6)$$

where T_{opt} is the temperature at which (X.5) applies. Since at equilibrium $R_P = 0$, we may also write

$$\frac{c_P c_Q}{c_A c_B} = \left[\frac{k_1}{k_2}\right]_{T=T_e} = \frac{k_{1\infty}}{k_{2\infty}} \exp\left[\frac{E_2 - E_1}{RT_e}\right].$$

The last two equations and the expressions for k_1 and k_2 yield

$$\frac{T_{opt}}{T_e} = \frac{1}{1 + \dfrac{RT_e}{E_2 - E_1} \ln \dfrac{E_2}{E_1}}. \quad (X.7)$$

When the difference between the absolute temperatures T_e and T_{opt} is relatively small, equation (X.7) may be approximated by

$$\frac{T_e - T_{opt}}{T_e} \simeq \frac{RT_e}{-\Delta H_r} \ln \frac{E_2}{E_1}.$$

Application of these equations is simple in the case of a continuous *tank reactor* with a minimum volume for a given product composition and throughput; T_e and hence T_{opt} and the maximum conversion rate can be readily calculated from the composition.

The optimum temperature profile in a *tubular reactor* can be found on the assumption that the criterion (X.5) and hence also equation (X.7) must be valid for any cross section of the reactor. Denbigh [8] postulated this condition in 1944 by intuitive reasoning; it can also be made acceptable in the following manner. The volume of a tubular reactor is proportional to the integral:

$$\int_0^{\zeta_{AL}} \frac{d\zeta_A}{R_J},$$

where R_J is a function of the initial composition, ζ_A and T for a given reaction system. For the optimum temperature policy, this integral should be a minimum, which leads to the condition

$$\int_0^{\zeta_{AL}} \frac{\partial R_J}{\partial T} \frac{d\zeta_A}{R_J^2} = 0. \tag{X.8}$$

This relation can be observed in many ways, and it is evident that one of these is to see that $\partial R_J/\partial T = 0$ (compare equation (X.6)) and the conversion rate has its maximum value as a function of temperature at any degree of conversion.

The procedure for calculating the optimum temperature profile then is as follows. First find T_e as a function of, e.g., the degree of conversion ζ_A (Curve 1 in figure X.7), and then T_{opt} by means of equation (X.7), Curve 2. Because of the known relation between T_{opt} and ζ_A, the conversion rate can be expressed in terms of ζ_A

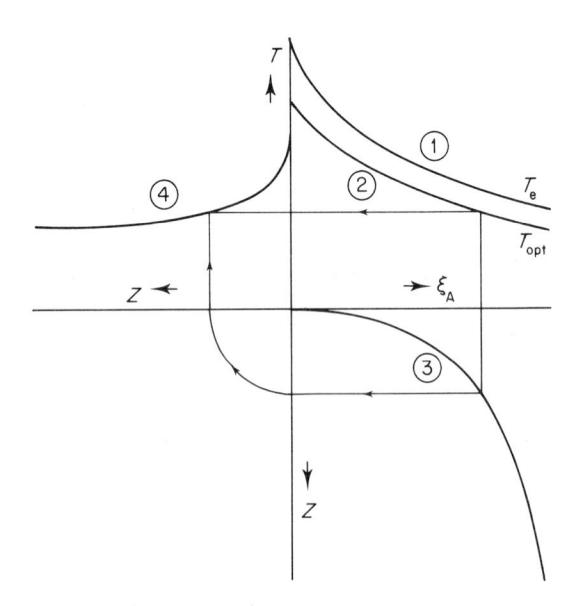

Fig. X.7 Principle of finding the optimum temperature profile for an exothermic equilibrium reaction in a PFR

only; hence, the material balance (e.g. equation (II.15)) can be integrated to yield ζ_A as a function of the distance z from the reactor inlet (Curve 3 in figure X.7). The optimum temperature as a function of z is obtained by combining Curves 2 and 3.

Only for very simple rate expressions is it possible to extend the above theory a little further with a general mathematical formulation. For example, in the reaction A → P, with $R_P = k_1 c_A - k_2 c_P$, $E_2 > E_1$, we may introduce the reference temperature T_R at which $k_1 = k_2 = k_R$ (also see Section VI.6). The ratio $k_1/k_R = \kappa$ then is a measure of the reaction temperature; the conversion rate can be written as:

$$R_P = k_R c_{A0} [\kappa(1 - \zeta_A) - \kappa^{E_2/E_1} \zeta_A]. \tag{X.9}$$

Application of condition (X.5) yields, for the dimensionless reaction rate constant at the optimum temperature,

$$\kappa_{opt} = \left[\frac{E_1(1 - \zeta_A)}{E_2 \zeta_A} \right]^{E_1/(E_2 - E_1)}. \tag{X.10}$$

By combining this result with the material balance for a *tank reactor*, it is possible, for example, to eliminate ζ_A and to find an expression for the minimum dimensionless residence time $(k_R \tau)_{min}$:

$$(k_R \tau)_{min} = \left[\left(\frac{E_2}{E_1} - 1 \right) \kappa_{opt}^{E_2/E_1} \right]^{-1}. \tag{X.11}$$

Similar calculations for a *tubular reactor* are more involved; they were carried out by Horn and Troltenier for first and second order equilibrium reactions [9]–[12]. Two of these papers [9], [12] contain tables which for various values of E_2/E_1 give the corresponding values of κ_{opt}, ζ_A and the dimensionless reactor length. Equation (X.10) indicates that the optimum conversion rate and temperature should be infinitely high at the inlet of the reactor ($\zeta_A = 0$); for practical reasons, however, there is always an upper temperature limit.

The theory of the optimum temperature profile was applied by various authors. Calderbank [13] calculated that, with an optimum temperature profile, SO_2 could be oxidized to a certain conversion level in a catalyst bed at a rate corresponding to a production of 54 tonnes H_2SO_4/tonne catalyst per day; with the same conversion level, this figure was 11.4 for an externally cooled catalyst bed and 3.4 for a two-stage adiabatic tubular reactor with intermediate cooling. In a paper on the same subject, Mars and van Krevelen [14] showed that the capacity of industrial reactors for the catalytic oxidation of SO_2 has been rising steadily. Applications to the synthesis of NH_3 were published by van Heerden [15] and by Annable [16].

The latter author compared the optimum temperature profile with the actual profile in an existing plant; they were sensibly different and the conversions were 22% and 19.2%, respectively. Consequently, it would be possible to increase the conversion considerably and at the same time to profit from the reduction of the recirculation of unconverted material as shown by Westerterp and Beek [17]. In order to find a solution for the technical realization of a temperature profile

Murase et al. [18] studied the optimum UA profile in a cooled tubular reactor for the NH_3 synthesis. As U is relatively invariant, this requires a continuous change of the number and/or the diameter of the tubes. Another optimization problem is the start-up procedure for an equilibrium reaction in an autothermal reaction system of the PFR + heat exchanger type in the shortest possible time without a considerable overshoot of the final reaction temperature, as was studied by Jackson [19]. In this case, the reaction has to be ignited with an outside heat source — an electric heater or a start-up furnace — in order to reach the upper stable operating point. In the first phase maximum heat is supplied and the bypass of the heat exchanger is shut. When the ignition starts, the heat supply is stopped and somewhat later the bypass should be completely opened to prevent the overshoot. Again after some time the bypass is gradually partially closed up to the final valve setting.

Although the above procedures make it possible to obtain a tubular reactor having the minimum volume for a given production rate and final conversion, it is questionable whether such a design will be attractive from the practical and economic points of view. It was shown by Westerterp [20] that it is technically not impossible to approximate the optimum temperature profile by suitable methods of heat exchange; the investment costs may, however, be considerable and, moreover, the heat economy may be poor. The feed has to be heated to a high temperature with high quality heat which is withdrawn along the reactor length by a much colder cooling medium. In many cases, therefore, alternative temperature policies have to be considered as well, involving, for instance, the isothermal and the adiabatic tubular reactor; these can also be optimized, as will be shown below.

For a given final degree of conversion, ζ_{AL}, the criterion for the highest capacity of an *isothermal tubular reactor* is given by equation (X.8), which is now used with the additional condition of isothermal operation. Equation (X.8) can be solved analytically with simple rate expressions. For example, with a first order equilibrium reaction with R_P defined by equation (X.9) the result is

$$-\ln[1 - (1 + \kappa')\zeta_{AL}] = \frac{(E_2/E_1 - 1)\kappa'(1 + \kappa')\zeta_{AL}}{\left(1 + \frac{E_2}{E_1}\kappa'\right)[1 - (1 + \kappa')\zeta_{AL}]},$$

where

$$\kappa' = \kappa_{opt}^{E_2/E_1 - 1}$$

and κ_{opt} is the value of k_1/k_R at which the reactor volume is a minimum. Figure X.8 shows κ_{opt} according to this result as a function of E_2/E_1 for $\zeta_{AL} = 0.9$. It may be concluded from this figure that the optimum temperature for isothermal operation is intermediate between the equilibrium temperature at $\zeta_A = 0.9$ and the temperature at which the conversion rate at the reactor outlet would be a maximum. This means that there is only one cross section at which equation (X.5) is valid, and consequently the reactor volume will be larger than that of a tubular reactor with the optimum temperature profile under otherwise identical

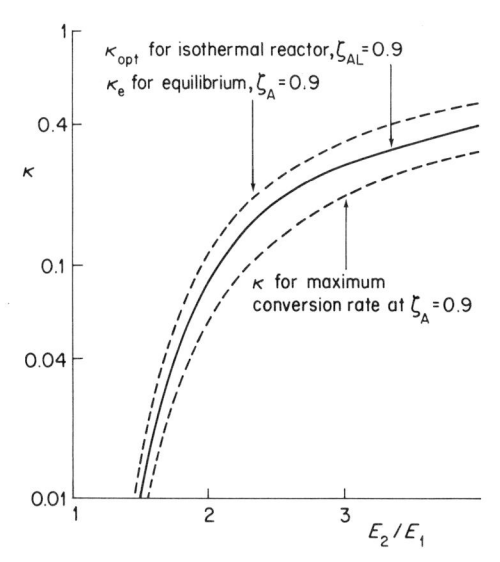

Fig. X.8 Dimensionless forward reaction rate constant κ for first order equilibrium reaction: its optimum value for an isothermal PFR lies between κ_e, at which $R = 0$, and κ for maximum conversion rate

circumstances. Horn [9, 21] calculated the ratio of these two volumes for first order and second order exothermic equilibrium reactions. Figure X.9 contains some of his results showing that this ratio is not excessively large at not too high degrees of conversion. Consequently, it will not be very serious if the optimum temperature profile is not adhered to, unless the reactor volume and/or the catalyst contained in it are extremely expensive.

It was already mentioned that the establishment of the optimum temperature profile may involve technical difficulties and expensive provisions; the same can be true for isothermal operation, although to a lesser extent. Particularly with gas reactions in a catalytic reactor, it is much simpler, from a technical point of view, to let the reaction proceed *adiabatically*. Now, in a single adiabatic reactor the temperature and the conversion are related by the expression

$$T = T_o + \Delta T_{ad} \zeta_A. \tag{X.12}$$

If a certain final conversion ζ_{AL} is desired, the temperature T_L at the end of the reactor should be lower than the equilibrium temperature T_e of the product mixture. Hence, the feed temperature T_o should be lower than T_e by an amount of the order of $\Delta T_{ad} \zeta_{AL}$. For many catalytic gas reactions this would lead to such low entrance temperatures that an excessively large reactor volume would be necessary or even the reaction would not start at all.

In practice, this difficulty is solved by the use of a *multi-stage tubular reactor* with intermediate cooling of the reaction mixture. Such a system is shown in figure X.10 together with the temperature as a function of the amount of catalyst passed by the reaction mixture. It may now be asked how, with a given final

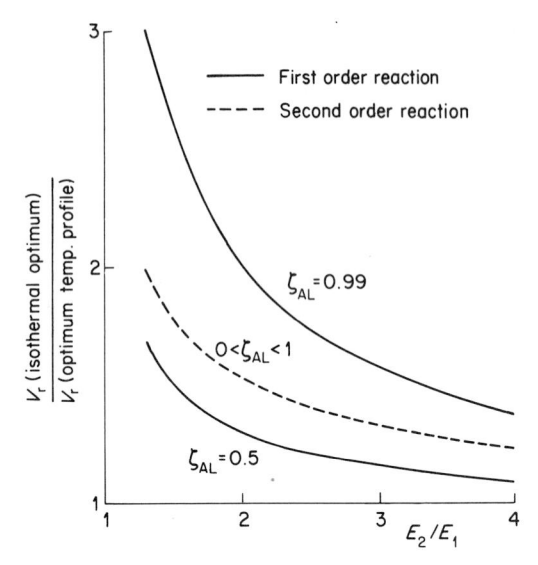

Fig. X.9 Volume ratio of tubular reactors with optimum uniform temperature and with optimum temperature profile; the curve for the second-order equilibrium reactions is for a stoichiometric feed and independent of the degree of conversion [From Horn [21]. Reproduced by permission]

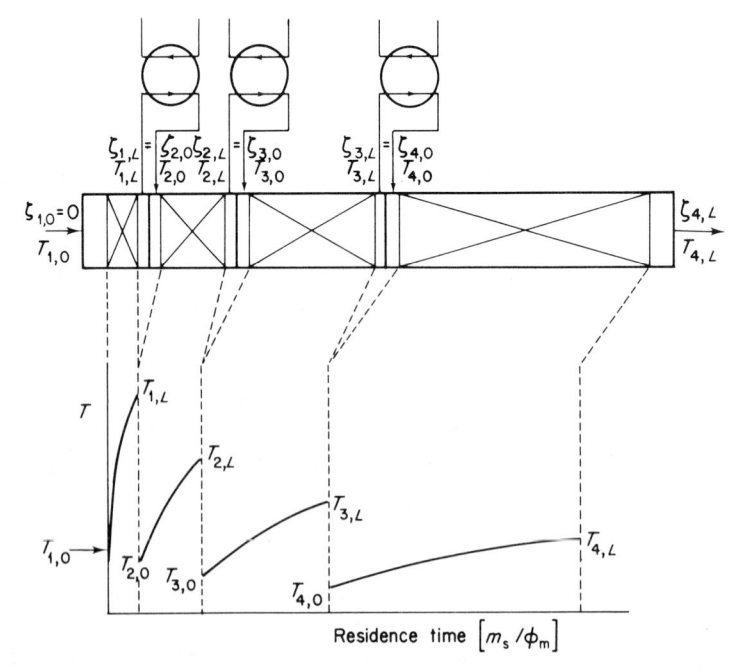

Fig. X.10 Multi-stage adiabatic tubular reactor with cooling between beds

degree of conversion, the reactor volume or catalyst mass should be distributed and what would be the best entrance temperature of each section for the capacity of the system to have a maximum value.

Horn and Küchler [23] provided the solution to a slightly different formulation of the same problem, i.e. that of determining the highest possible final conversion for a given number of sections and for a given entrance temperature of the first section. The conditions for this optimization appear to be:

(a) For each section (index n) equation (X.8) takes the form

$$\int_{\zeta_{n,0}}^{\zeta_{n,L}} \frac{\partial R}{\partial T} \frac{d\zeta_n}{R^2} = 0, \qquad (X.13)$$

where ζ_n and T are related by equation (X.12):

$$T_n - T_{n,o} = \Delta T_{ad}(\zeta_n - \zeta_{n,o});$$

(b) Between two sections $(n-1)$ and n:

$$R_{(n-1),L} = R_{n,o}; \qquad (X.14)$$

this condition relates $T_{n,o}$ to $T_{(n-1),L}$, since

$$\zeta_{(n-1),L} = \zeta_{n,o}.$$

For the sake of simplicity, the subindices referring to the components have been omitted; the quantity $M_J R_J$ is to be regarded as the conversion rate in mass units per unit mass of catalyst (R_s). The significance of the conditions above can best be illustrated by means of a temperature conversion diagram (figure X.11) of a given

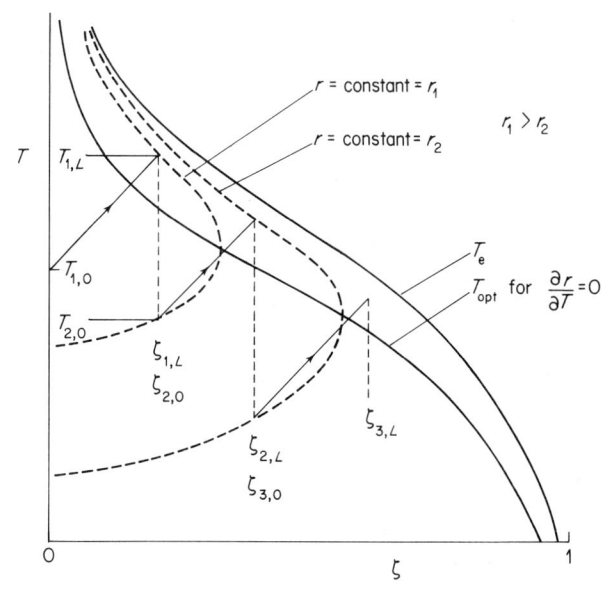

Fig. X.11 Illustration of the conditions (X.13) and (X.14) for the optimum design of a multi-stage adiabatic tubular reactor for exothermic equilibrium reactions

exothermic equilibrium reaction with a given feed composition. It contains a curve for T_e, the temperature at which the mixture would be in chemical equilibrium, and a curve for T_{opt} at which the conversion rate is at a maximum, both for the corresponding degree of conversion. Two lines of constant conversion rate have also been drawn. The feed enters the first stage at a temperature $T_{1,o}(\zeta_{1,o} = 0)$ and the reaction mixture is heated up proportionally to ζ in accordance with equation (X.12). When the T_1–ζ_1 line intersects with the curve for T_{opt}, the sign of $\partial R/\partial T$ changes from positive to negative. Condition (X.13) is fulfilled at the point with coordinates $(\zeta_{1,L}, T_{1,L})$; this point lies on the curve $R = R_1$, and equation (X.14) requires the reaction mixture now to be cooled to such a temperature $T_{2,o}$ that the mixture entering the second section has the same conversion rate R_1. It may occur that $T_{1,L}$ exceeds the maximum of the permissible temperature T^* so that only a corresponding degree of conversion $\zeta^* < \zeta_{1,L}$ can be reached in the first section. The first section then no longer has the minimum volume, since equation (X.13) will not apply any more. Horn [22] showed that in that case equation (X.14) should be replaced by

$$\Delta T_{ad} \int_0^{\zeta^*} \frac{\partial R}{\partial T} \frac{d\zeta_1}{R^2} + \frac{1}{R(\zeta^*)} = \frac{1}{R_{2,o}} \tag{X.15}$$

to find the proper entrance temperature to the second section.

This procedure is then repeated until the desired number of stages has been passed. The inlet temperature of the first stage, $T_{1,o}$, is an independent parameter which has to be varied in order to find the minimum amount of catalyst needed for a certain final degree of conversion, or the maximum degree of conversion obtainable with a given amount of catalyst in a given number of sections. A typical result of such calculations is shown in figure X.12. Starting from a given inlet temperature, $T_{1,o}$, the amount of catalyst is found with which a maximum degree of conversion, $\zeta_{N,L}$, is obtained. The greater the number of adiabatic sections, the more catalyst is needed for optimum operation and the corresponding increase in final conversion. The figure also shows that with a certain desired value of $\zeta_{N,L}$ the amount of catalyst can be drastically reduced by using more sections, provided the proper inlet temperature be observed. This is also clear from figure X.11 since the T–ζ zig-zag history of the reaction mixture grows closer to the curve for T_{opt} as the number of stages is increased. Moreover we see, that in a multiple-bed adiabatic reactor each successive bed is larger than the previous one and that the fourth bed in figure X.12 no longer adds much to the total conversion despite the very large amount of catalyst. For very expensive high pressure reactors a fourth bed is most probably not economically justified.

The above problem was also treated by Aris [25] in a more abstract manner, taking into account the economic implications. The technical optimization problem can be further complicated by considering the possibility of intermediate cooling by the addition of cold feed to the reaction mixture between sections. This reactor type is also called a 'cold shot' or a 'quench' converter. In this case, not only will the temperature drop between two sections but the degree of conversion

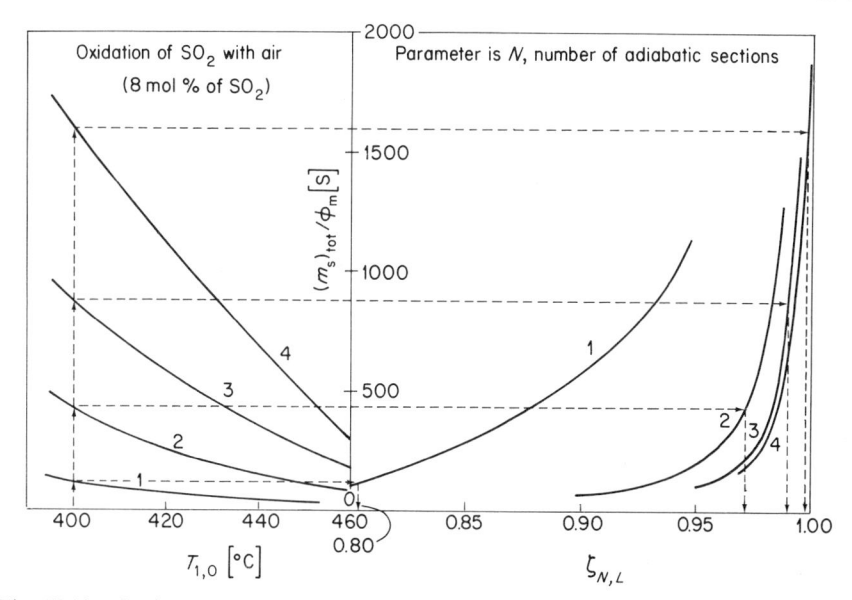

Fig. X.12 Optimum inlet temperature and minimum amount of catalyst for the oxidation of SO_2 in a multi-stage adiabatic reactor; calculated with Eklund conversion rate data [67] by Boreskov and Slinko [24]

will as well. Figure X.13 (path 2) shows this for the catalytic oxidation of SO_2, the optimization of which was treated by Boreskov and Slinko [24]. Schoenemann [26] worked out a similar problem for NH_3 synthesis. Bartholomé and Krabetz [27, 28] reported on the calculation of a multi-bed reactor for the equilibrium reaction $H_2O + CO \rightarrow CO_2 + H_2$. They discuss, for a multi-stage adiabatic reactor, the relation between the conversion and the inlet temperature, the influence of catalyst aging on conversion, and the possibility of increasing the conversion by distributing the catalyst over several beds with intermediate indirect cooling of the reaction mixture. It appears from their work that a deviation from the optimim distribution of catalyst in a two-bed reactor has only a small effect on the final degree of conversion, provided the inlet temperature has been properly adjusted.

It is seen that an optimum design can be made when the necessary information is available and reliable. Strictly speaking, such an optimum is then only valid for a given set of conditions regarding the feed composition and the reactor load. Since these conditions may vary in an operating installation, and particularly since the plant capacity in many cases must be increased or decreased over the rated capacity, it is important to examine the possibility of optimization under changed conditions. Two problems of this nature were reported on by Küchler [29]. His first example relates to an equilibrium reaction carried out in an isothermal tubular reactor. At a certain load the optimum temperature is 338°C and the degree of conversion 68%. If the reactor load is increased by a factor 2.5, the degree of conversion would be 45% at the same temperature. Under the new

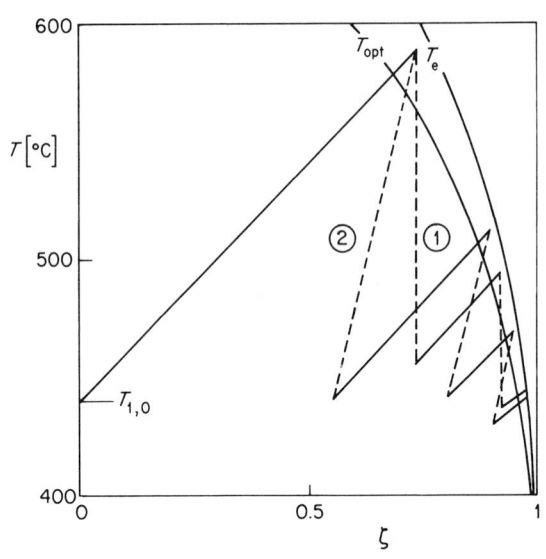

Fig. X.13 Temperature-conversion curve for the catalytic oxidation of SO_2 in a multi-bed adiabatic reactor: (1) three sections with intermediate indirect cooling; (2) four sections with intermediate supply of cold feed [Boreskov and Slinko [24]])

conditions, however, the optimum reactor temperature is 358°C and the corresponding maximum conversion 55%.

In the second example mentioned by Küchler, the same reaction is carried out in an adiabatic multi-stage reactor. Table X.1 shows the optimum conditions for two reactor loads differing by a factor of 2.7. It is seen that the optimum distribution of catalyst is only slightly different in the two cases; this is favourable since the distribution cannot be changed during plant operation. The various inlet temperatures, however, have to be drastically changed when the load is increased. The NH_3 synthesis reaction is the subject of many publications. Among others, Almasy et al. [30]–[32] studied optimization for minimum costs

Table X.1 Optimum conditions for equilibrium reaction in multi-bed reactor at different loads (Küchler [29])

	100% load		270% load	
	% of total cat. mass	Inlet temp. °C	% of total cat. mass	Inlet temp. °C
First bed	7.6	280	9.2	320
Second bed	14.9	279	17.5	311
Third bed	27.4	279	28.7	305
Fourth bed	50.1	277	44.6	300
Final degree of conversion	90.0%		75.2%	

and also the revamp of the reactor internals for maximum production in an existing plant, maintaining the original reactor shell. Gaines [33] studied the influence of operating variables in a cold shot converter with four adiabatic beds in an existing plant and determined how production could be maximized under varying operating conditions: he concluded that the inlet temperature to the fourth bed is the primary variable for production control. Burghardt and Patzek [34] made an excellent study of ammonia quench converters, which they optimized for maximum production with a maximum temperature constraint: their work will be discussed in more detail in Example X.2.a.

Lee and Aris [35] applied the method of dynamic programming to a cold shot converter for the SO_2 oxidation, also with a decaying catalyst activity [36]. For a gradually diminishing catalyst activity the inlet temperatures of the beds have to be adjusted (increased) in order to maintain at every instant the maximum production. This production then diminishes as operating time passes and is stopped as soon as a minimum acceptable production level is reached. Then the aged catalyst is replaced, the new charge activated and the reactor started up again. Mathematically much more difficult to solve is the problem of a dynamic optimum production. A maximum profit can be realized only if the influence of the process variables (such as temperature and concentration) on the change in time of the catalyst activity is also accounted for. The maximization of the profit during the whole production cycle, in which the dynamics of the changing parameter is taken into account, is called the dynamic optimization of a process. Meiring [37] solved this problem for a five bed cold shot converter for the methanol production from synthesis gas: the direct benefit in production volume was not impressive — only a few percent — but it resulted in the possibility of extending the catalyst life and therefore also the production cycle by about 40% owing to the dynamic optimum control. At the start of a cycle the dynamically optimum temperature level in the reactor, especially in the first beds, is lower than required for the instantaneous optimum; consequently, the production in the first beds is lower and in the last beds higher. The total production at the start of a cycle is lower than in the case of instantaneous optimum control. In the initial stage of the cycle the methanol production has been shifted to the last beds, so that the activity of the catalyst in the first beds is spared. This can be understood by looking at the temperature profiles, e.g. in figure X.10. Suppose that the total methanol production occurs in the first three beds of the reactor, so the reaction mixture flowing through the last beds has a high temperature: consequently, the catalyst in the last beds ages whereas the production in these beds is almost nil. Conversely, if the total production takes place in the last three beds, the reaction mixture flowing through the first beds has a low temperature: consequently, also the deactivation in the first beds is low. This means that in the first beds catalyst aging can be better prevented than in the last beds. In figure X.14 the production in each individual bed is shown for a 400 days' cycle according to Meiring [37].

Under certain conditions it is also practical to impose an *optimum pressure profile* on a reactor: Boreskov and Slinko [38] gave an interesting example of an industrial application, the endothermic equilibrium dehydrogenation of

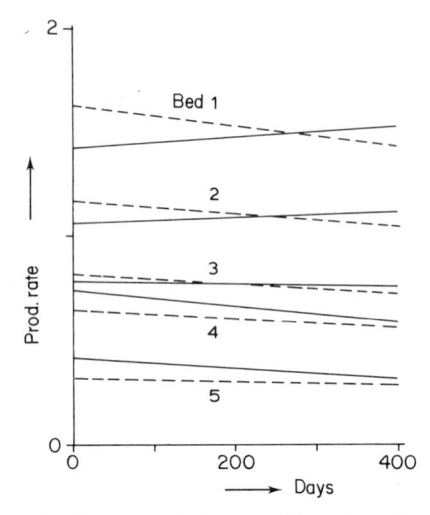

Fig. X.14 Optimum production rates during a methanol production cycle in a five beds cold shot converter with non-converted reactants recycle and aging catalyst: ——— dynamic optimum; .... instantaneous optimum [From Meiring [37]. Reproduced by permission of F. Meiring]

ethylbenzene to styrene. The equilibrium in this case shifts to the left due to the increase in the number of moles. So a decrease in pressure will move the equilibrium to the right. In practice this reaction is carried out with dilution by steam: due to the heat-carrying capacity of the steam the temperature drop of this endothermic reaction is reduced in an adiabatic bed. The steam addition also reduces the partial pressures. Boreskov and Slinko reported on a three stage adiabatic bed reactor with steam injection between the beds. Keeping the total amount of catalyst and of steam added constant, the total conversion could be increased from 38% in a single bed to 54% in a three bed reactor as shown in figure X.15.

For *adiabatic reactors* Aris [39] demonstrated that a *combination of reactor types*, such as a CISTR followed by a PFR, may give the *shortest total residence time*. This can be understood, because the CISTR will work at a much higher reactor temperature than the PFR at the start of the reaction and, therefore, at a much higher reaction rate if the reactor conditions are still sufficiently far from equilibrium. This is shown in figure X.16.a, where the reciprocal reaction rate in an adiabatic reactor for a given feed temperature is plotted as a function of the conversion of a key component in the reaction mixture. The reactor temperature in this case is, of course, uniquely related to the key reactant concentration and the feed temperature by $T = T_o + \Delta T_{ad}\zeta_A$. For equilibrium reactions with Langmuir–Hinshelwood kinetics depending on the influences of the adsorption terms in the denominator of the kinetic equation the R_A^{-1}–ζ_A curve for an adiabatic reaction can even exhibit a maximum followed by a minimum as shown in figure X.16.b. Then a sequence of PFR–CISTR–PFR has to be chosen for minimum total residence time. King [40] gave a follow-up by studying a cooled

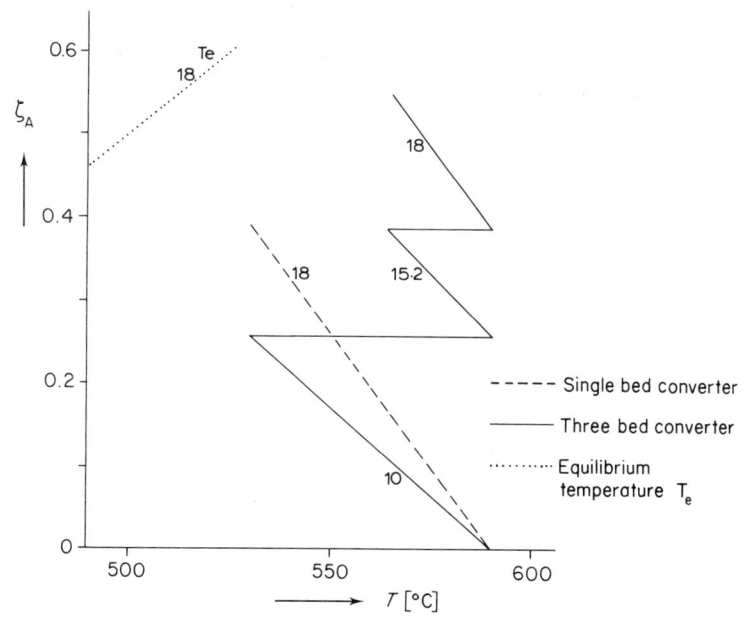

Fig. X.15 Conversion–temperature plot for single bed and three bed ethylbenzene dehydrogenation reactors. Total amount of catalyst is equal for both reactors, the parameter is the steam–ethylbenzene ratio [From Boreskov and Slinko [38]. Reproduced by permission of Institution of Chemical Engineers]

CISTR followed by an adiabatic PFR for exothermic equilibrium reactions: in this case for minimum total residence time the CISTR will become relatively larger and the PFR shorter than if the CISTR were not cooled.

Example X.2.a *Optimization of an ammonia cold shot converter with temperature constraints*

Burghardt and Patzek [34] made an extensive study of ammonia cold shot converters, in which the temperature was not allowed to exceed a certain maximum temperature T^* in order to protect the catalyst. We will follow their study in this Example and first derive the basic equations. The reactor layout is

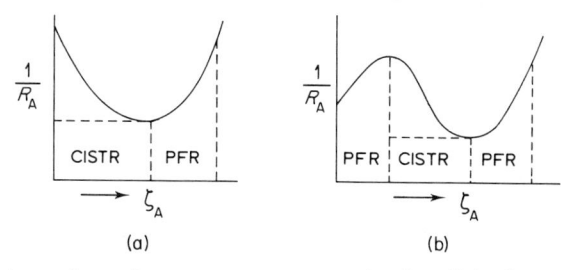

Fig. X.16 Reciprocal reaction rates versus conversion in adiabatic reactors: (a) simple order kinetics, (b) Langmuir–Hinshelwood kinetics and exothermic reactions, T and ζ_A are related by $T = T_o + \Delta T_{ad} \zeta_A$

shown in figure X.17 together with the relevant variables. The reactor feed contains a weight fraction w of nitrogen, of which a fraction $w\zeta_o$ is nitrogen already incorporated in NH_3 carried back to the reactor with the recycle stream, so that a fraction $w(1 - \zeta_o)$ is free nitrogen, which can be converted into NH_3 in the reactor.

A material balance for the free nitrogen at a mixing point between the beds leads to

$$(\phi_i - \phi_{i-1})\Phi_m w(1 - \zeta_o) + \phi_{i-1}\Phi_m w(1_i - \zeta_{L,i-1}) = \phi_i \Phi_m w(1_i - \zeta_{o,i}) \quad \text{(a)}$$

and a heat balance at a mixing point leads to

$$(\phi_i - \phi_{i-1})\Phi_m c_p T_o + \phi_{i-1}\Phi_m c_p T_{L,i-1} = \phi_i \Phi_m c_p T_{o,i}. \quad \text{(b)}$$

This leads to the two mixing equations, if c_p is assumed constant over the entire reactor:

$$\zeta_{o,i} - \zeta_o = \frac{\phi_{i-1}}{\phi_i}(\zeta_{L,i-1} - \zeta_o) \quad \text{(c)}$$

and

$$T_{o,i} - T_o = \frac{\phi_{i-1}}{\phi_i}(T_{L,i-1} - T_o). \quad \text{(d)}$$

Over an individual bed the relation $T_i = T_{o,i} + \Delta T_{ad}(\zeta_i - \zeta_{o,i})$ is valid. With this relation and with equations (c) and (d) we can derive a relation between the feed conditions and the initial conditions at the entrance of the ith bed:

$$T_{o,i} - \Delta T_{ad}\zeta_{o,i} = \frac{\phi_{i-1}}{\phi_i}(T_{L,i-1} - T_o) + T_o - \Delta T_{ad}\left[\frac{\phi_{i-1}}{\phi_i}(\zeta_{L,i-1} - \zeta_o) + \zeta_o\right].$$

Further,

$$T_{L,i-1} = T_{o,i-1} + \Delta T_{ad}(\zeta_{L,i-1} - \zeta_{oi-1}),$$

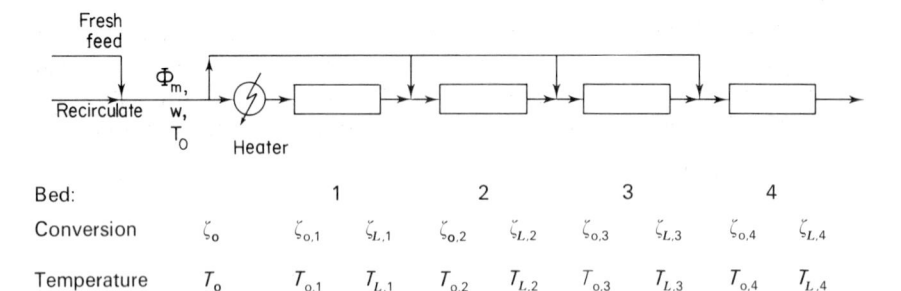

Bed:		1		2		3		4	
Conversion	ζ_o	$\zeta_{o,1}$	$\zeta_{L,1}$	$\zeta_{o,2}$	$\zeta_{L,2}$	$\zeta_{o,3}$	$\zeta_{L,3}$	$\zeta_{o,4}$	$\zeta_{L,4}$
Temperature	T_o	$T_{o,1}$	$T_{L,1}$	$T_{o,2}$	$T_{L,2}$	$T_{o,3}$	$T_{L,3}$	$T_{o,4}$	$T_{L,4}$

Fraction of the feed:
— flowing through reactor ϕ_1 ϕ_2 ϕ_3 1
— flowing through bypass $1 - \phi_1$ $1 - \phi_2$ $1 - \phi_3$ 0
— injected $\phi_2 - \phi_1$ $\phi_3 - \phi_2$ $1 - \phi_3$

Fig. X.17 Layout of the ammonia cold shot converter of Example X.2.a and the local values of the variables

so that we find:

$$(T_o - \Delta T_{ad}\zeta_o)_i = \frac{\phi_{i-1}}{\phi_i}(T_o - \Delta T_{ad}\zeta_o)_{i-1}$$

$$+ \left(1 - \frac{\phi_{i-1}}{\phi_i}\right)(T_o - \Delta T_{ad}\zeta_o). \quad (e)$$

If we now define $q_i = (T_o - \Delta T_{ad}\zeta_o)_i$, then we derive

$$q_i\phi_i = q_{i-1}\phi_{i-1} + q_o(\phi_i - \phi_{i-1})$$

$$q_{i-1}\phi_{i-1} = q_{i-2}\phi_{i-2} + q_o(\phi_{i-1} - \phi_{i-2})$$

$$\overline{\phantom{q_i\phi_i = q_{i-2}\phi_{i-2} + q_o(\phi_i - \phi_{i-2})}} +,$$

$$q_i\phi_i = q_{i-2}\phi_{i-2} + q_o(\phi_i - \phi_{i-2})$$

so that finally

$$q_i = \frac{\phi_1}{\phi_i}(q_1 - q_o) + q_o.$$

Because at the inlet of the first reactor $\zeta = \zeta_o$ and $T = T_o + \Delta T_{HE}$, where ΔT_{HE} = the temperature increase over the heater of the reaction mixture, we can finally write

$$(T_o - \Delta T_{ad}\zeta_o) = \frac{\phi_1}{\phi_i}\Delta T_{HE} + (T_o - \Delta T_{ad}\zeta_o). \quad (f)$$

The total mass of catalyst required in the ith bed is given by

$$m_{cat} = \phi_i\Phi_m w(1 - \zeta_{o,i})\int_{\zeta_{o,i}}^{\zeta_{L,i}} \frac{d\zeta}{R\left[\zeta, T_o + \frac{\phi_1}{\phi_i}\Delta T_{HE} + \Delta T_{ad}(\zeta_i - \zeta_o)\right]}. \quad (g)$$

The cold shot converter with N beds can now completely be described with $2N$ independent variables: $\zeta_{L,i}(i = 1, 2, \ldots N)$, $\phi_i(i = 1, 2, \ldots, N-1)$ and $T_{1,o} = T_o + \Delta T_{HE}$, the inlet temperature of the first bed. All other variables such as the temperatures and compositions of the reaction mixture in a cross section of a catalyst bed are unique functions of the set of independent variables. Now Burghardt and Patzek [34] investigated the conditions for an optimum cold shot converter in which, for a given total mass of catalyst,

$$\sum_{i=1}^{N} m_{cat i}/\Phi_m$$

the maximum conversion is obtained, whereas the reactor temperatures are not allowed to exceed the maximum temperature T^*. This is equivalent to minimizing $\sum_{i=1}^{N} m_{cat i}/\Phi_m$ for the constraint $T \leq T^*$. By introducing a penalty function they proved the conditions of Horn for optimality, equations (X.14) and

(X.15), to be correct. By partial differentiation they derived all the conditions for optimality.

The *calculation procedure* for the optimum cold shot converter is now as follows:

1. Assume an inlet temperature $T_{1,o} = T_o + \Delta T_{HE}$.
2. Calculate the mass of catalyst m_{cat}/Φ_m required by integrating Equation (g) and stop the calculation as soon as the condition of equation (X.13) is fulfilled or — if the maximum allowable temperature T^* is surpassed before the integral of equation (X.13) becomes zero — stop the calculation at

$$\zeta^* = \zeta_{L,i} = \frac{T^* - T_{o,i}}{\Delta T_{ad}} + \zeta_{o,i}.$$

3. The optimum inlet variables for the next bed can now be found from equation (X.14) or — if the maximum allowable temperature had been reached at the end of the previous bed — from equation (X.15). By eliminating ϕ_{i-1}/ϕ_i from the equations (c) and (d) we have a second condition:

$$\frac{\zeta_{o,i} - \zeta_o}{\zeta_{L,i-1} - \zeta_o} = \frac{T_{o,i} - T_o}{T_{L,i-1} - T_o}. \tag{h}$$

From equation (h) and equation (X.14) or (X.15) the two unknowns $\zeta_{o,i}$ and $T_{o,i}$ for the next bed can be found.

4. Repeat the steps 2 and 3 for all the following beds.
5. For the last bed stop the calculation as soon as the condition:

$$\Delta T_{ad} \sum_1^N i \int_{\zeta_{o,i}}^{\zeta_{L,i}} \frac{\partial R_i}{\partial T} \frac{d\zeta}{R_i^2} + \sum_1^N i \left(\frac{1}{R_{L,i}} - \frac{1}{R_{o,i+1}} \right) = 0 \tag{i}$$

is fulfilled, as was derived by Burghardt and Patzek [34].

6. Derive the optimum values of ϕ_i from the bed inlet conditions as calculated, for example with equation (f).
7. Repeat the calculations for different values of $T_{o,1}$ and find the locus of optimal N-bed cold shot converters as a function of the space velocity Φ_m/m_{cat}, where T^* is not exceeded.

Burghardt and Patzek worked this out for ammonia quench converters using two different kinetic equations due to Temkin–Pyzhev and to Benton. Results are given for a four bed quench converter in figure X.18. The maximum allowable temperature was $T^* = 520°C$. The feed contained 6.5% of ammonia, so $\zeta_o = 0.065$, and 10% inerts, argon and methane. The reactor pressure was 30 MPa and the feed temperature $T_o = 20°C$. From the results as shown in figure X.19 we can see that the difference between a four bed and a three bed converter is small, so that the incorporation of a fourth bed in such an expensive reactor seems hardly justified. For example the modern Kellogg cold shot ammonia converters for high capacity plants with daily outputs above 1000 tonnes have three beds only! In figure X.18 we have given results for two kinetic equations. With Benton

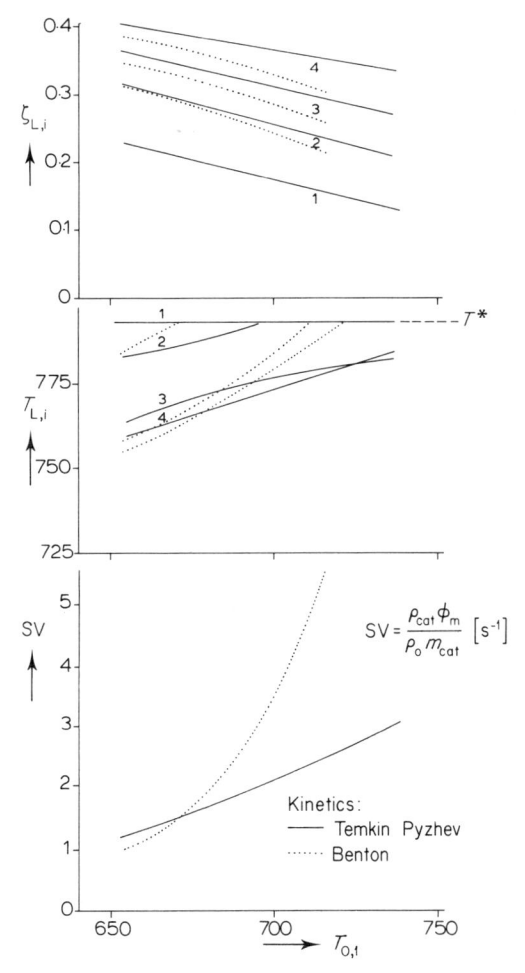

Fig. X.18 Outlet conversions and outlet temperatures of the individual beds and total catalyst requirements in an optimum four bed ammonia quench converter for Benton and Temkin–Pyzhev kinetics [From Burghardt and Patzek [34]. Reproduced by permission of A. Burghardt]

kinetics we find much higher catalyst requirements, slightly lower outlet conversions except in the first bed and higher bed inlet temperatures consequently. We once more see that the accuracy and reliability of the kinetic equations is a crucial factor in chemical reactor calculations!

X.2.2 Temperature optimization with complex reaction systems

The occurrence of one or more reactions by which reactants or desired products are turned into wasteful products greatly complicates the problems of technical optimization. The main reason for this is that more than one concentration

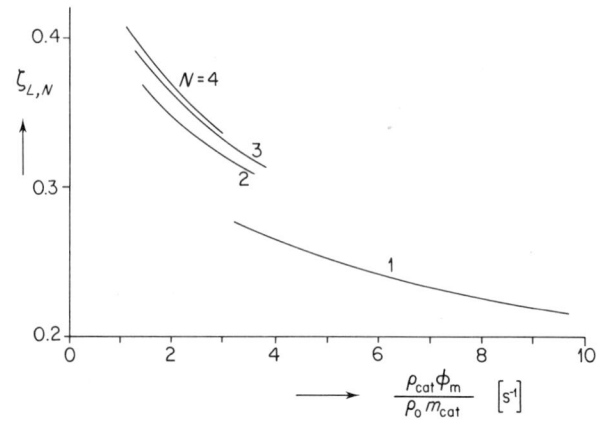

Fig. X.19 Optimum final conversions as a function of the amount of catalyst in an optimum N-bed ammonia quench converter for Temkin–Pyzhev kinetics. The parameter is N

variable is needed to specify the composition of the mixture and that several coupled material balance equations have to be handled. One of the mathematical techniques by which this can be done is the Lagrange multiplier technique; its elements will be outlined in Section X.3 with reference, among other things, to temperature optimization of complex reactions. For the time being, we shall deal only with the more evident aspects of this problem.

In Chapter III we used the concept of differential selectivity σ'_P indicating the ratio of the production rate of useful material and the conversion rate of reactant. This quantity depends not only on composition but also on temperature, and it will be clear that the sign of the quantity $\partial \sigma'_P / \partial T$ is closely related to the selection of the optimum temperature.

(a) In case $\partial \sigma'_P / \partial T > 0$, the desired reaction is favoured by a rise in temperature. Hence, the operating temperature should be selected as high as is compatible with other technical limitations; the highest possible yield is then obtained with parallel reactions and also with a system involving unwanted consecutive reactions provided the reaction mixture is discharged at the proper degree of conversion. At the same time, the high temperature ensures the highest possible capacity. As a consequence, this case does not generally present difficulties with respect to the optimum temperature.

(b) When $\partial \sigma'_P / \partial T < 0$, on the other hand, the yield is improved by lowering the temperature. This causes the capacity to be reduced, so that a compromise must frequently be made, the proper solution of which is again governed by the economy of the plant. For systems with two *parallel* reactions of equal order, Horn and Troltenier [9, 12] derived the condition for obtaining the maximum reactor capacity for a given yield. It appears that the temperature must rise with increasing conversion. This can be qualitatively understood. At a low degree of conversion the reactants still have a high concentration, and since it is important that most of them be converted to the desired product, a relatively low temperature is required. As the reactant concentration decreases, the differential selectivity is sacrificed in favour of a higher capacity and the temperature can be raised so as to enhance the conversion rate.

Bilous and Amundson [41], Horn [9] and Katz [42] investigated optimum temperature profiles for sets of two *consecutive* reactions where the intermediate product is the one desired. In this case a gain in capacity can in principle be obtained by letting the temperature decrease with increasing conversion. At a low degree of conversion, it is mainly the first of the two reactions that occurs, and a relatively high temperature is still permissible; when some of the intermediate product has formed, however, its decomposition must be slowed down by lowering the temperature. According to Horn, the increase in capacity obtainable by this method is at most 10–20% of the capacity of an isothermal tubular reactor giving the same yield. This means that the problem of maximum capacity hardly comes up in the case of consecutive reactions. The temperature level influences mainly the highest obtainable yield, and not much can be done to reduce the required reactor volume. Figure X.20 shows, for a system of two consecutive first order reactions in a tubular reactor with the optimum temperature profile, the dimensionless residence time and the dimensionless temperature at the outlet as a function of maximum yield. If the latter is to be increased from 0.9 to 0.95, T_L/T_R must be reduced from 0.75 to 0.71, and the reactor volume has to be increased accordingly by a factor of 3.6.

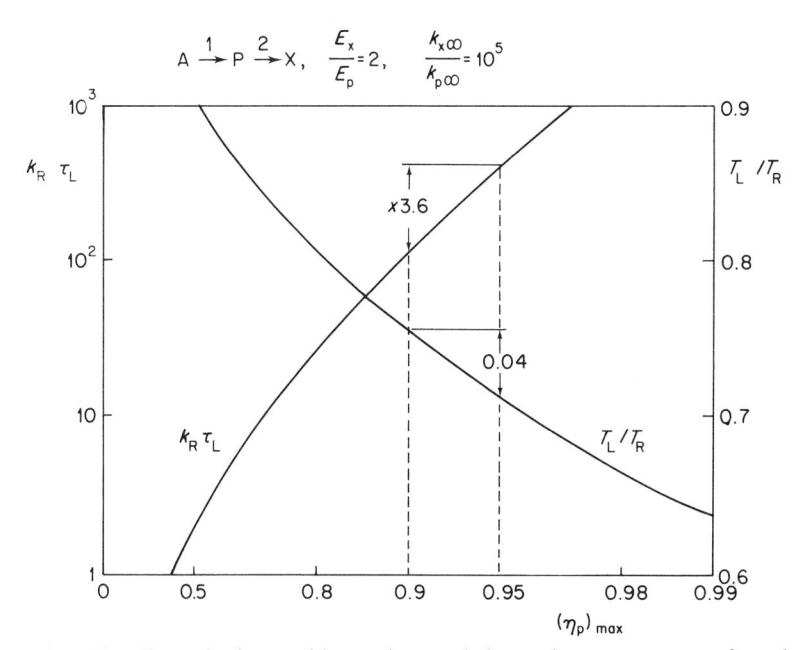

Fig. X.20 The dimensionless residence time and the outlet temperature of a tubular reactor with an optimum temperature profile for first order consecutive reactions, according to Horn [9]

Burghardt and Skrzypek [43] studied the optimum temperature profile for *combination reactions* of the type:

$$A + B \xrightarrow{k_P} P$$

$$P + B \xrightarrow{k_X} X$$

which maximizes the yield η_P in a PFR with a given residence time. The reactions are second order and there is an upper limit T^* and a lower limit $_*T$ for the reaction temperature, and further $E_P < E_X$.

For the consecutive reaction $A \to P \to X$ we would expect a falling temperature profile as explained above and for the parallel reactions $B \genfrac{}{}{0pt}{}{\to P}{\to X}$ we would expect a rising temperature profile. It can intuitively be felt that for our combination reaction the optimum temperature profile will be between the optimum profiles for consecutive and parallel reactions. Burghardt and Skrzypek proved this to be actually the case, depending on the values of E_P/E_X, $k_{P\infty}/k_{X\infty}$ and the total residence time they found the three different profile types shown in figure X.21. In this figure $\tau_{Lc} > \tau_{Lb} > \tau_{La}$.

(c) The value of σ_P' may pass through an extreme as a function of temperature. It is, of course, only of interest for temperature optimization if this extreme, defined by $\partial \sigma_P'/\partial T = 0$, is a maximum. This situation can arise with more wasteful reactions than one if one of these is relatively stimulated at low temperature and the other one at high temperature. Some remarks on the technique of yield optimization in such systems will be made in Section X.3. We shall treat below an example of this kind which has become classical.

Example X.2.b *Denbigh's problem*
Denbigh [44] considered a reaction system of the type:

$$A + B \xrightarrow{1} P \xrightarrow{3} Q$$
$$\downarrow_2 \quad \downarrow_4$$
$$X \quad\quad Y$$

where P is an intermediate product and Q the desired product. This could, for example, be a nitration of a hydrocarbon with P and Q as mono- and dinitro compounds, respectively, and X and Y as resinous by-products. In the case to be discussed, the production rate of X with respect to that of P from the reactants increases with rising temperature; at the same time, the production of Y is

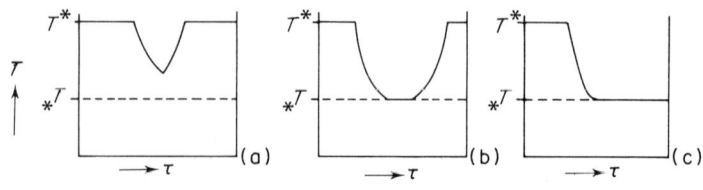

Fig. X.21 Optimum temperature profiles for a combination reaction, giving maximum yield (Burghardt and Skrzypek [43])

favoured at a low temperature. Consequently, the temperature should be low at a low degree of conversion, and high at a high degree of conversion.

Assuming all reactions to be of the first order and on the basis of a specific selection of the numerical values of the rate constants and their temperature dependence, Denbigh calculated the optimum temperatures for a cascade of two tank reactors. As an additional requirement it was postulated that the residence time in the first reactor entailed maximum concentration of P, and that the second reactor was so large as to occasion complete conversion of the reactants. The results of this calculation (figure X.22) indicate that, with two tank reactors operating at the same temperature, a maximum yield $\eta_Q = 0.25$ is obtained at 53°C. When the temperature of the second reactor, T_2, is kept constant and T_1 is varied, a yield maximum is found at which T_1 is considerably lower than T_2. It is seen from figure X.22 that the optimum value of T_1 is about 7°C and that T_2 should be as high as possible; in this example it is limited to 141°C. The corresponding yield has now increased to $\eta_Q = 0.53$.

Various other workers have taken up the same problem for further optimization studies. Horn and Troltenier [11] found that in a tubular reactor a maximum yield of $\eta_Q = 0.69$ if the temperature profile is optimized between the boundaries of 7 and 141°C. Using the method of dynamic programming (see Section X.3), Aris [45] verified Denbigh's calculations. He also calculated the optimum conditions for one, two and three tank reactors in cascade with an upper temperature limit of only 121°C and a restricted volume of the last reactor;

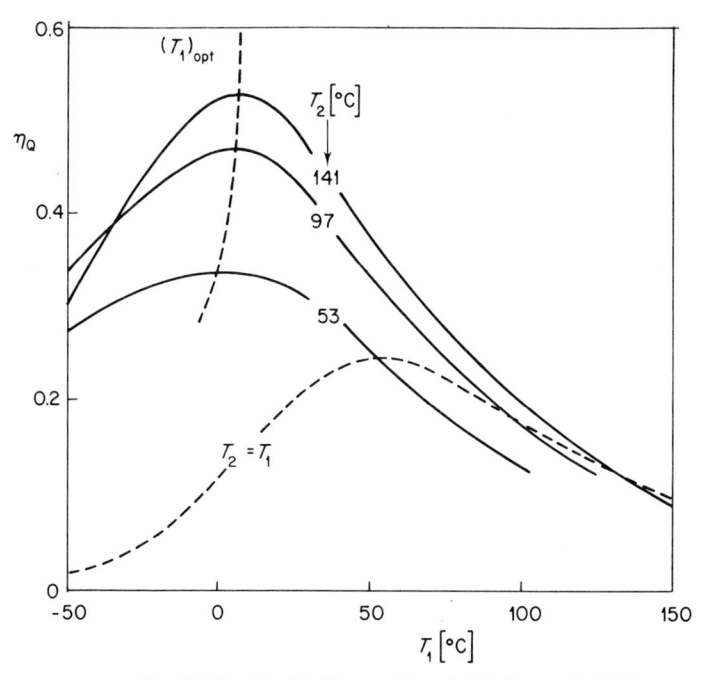

Fig. X.22 Denbigh's problem [44], Example X.2.b

Table X.2 Yields of Denbigh's reaction in a cascade, Aris [45] ($T \leq 121°C$, $k_1\tau \leq 2100$ for last reactor)

Number of tank reactors in cascade N	1	2	3
Optimum temperatures (°C)			
T_1	45	3	−13
T_2	—	121	15
T_3	—	—	121
Optimum volume ratios			
V_{r2}/V_{r1}	—	29	0.3
V_{r3}/V_{r1}	—	—	29
Yield of cascade η_{QN}	0.221	0.451	0.495

the results are shown in Table X.2. Storey [46] investigated the same problem using a different mathematical method (the numerical hill-climbing technique, Rosenbrock [47]). His optimum temperatures for a cascade of four tank reactors without volume and temperature restrictions are −15, 5, 130 and 300°C, respectively, with a final yield of $\eta_Q = 0.69$. Storey obtained reactor–volume ratios which are rather different from those found by Aris. The distribution of reaction volumes over different stages is probably much less critical than the temperature or the concentrations.

X.3 SOME MATHEMATICAL METHODS OF OPTIMIZATION

The general theory of finding an extreme value for a given function of a number of variables which are mutually dependent through a number of additional relations is beyond the scope of this book. When the number of variables is very small, say 1 or 2, the problems are reduced to finding an extreme of a curve or of a two-dimensional surface; these are solved by simple mathematical or graphical methods which can still be visualized. As the number of variables increases, thus also increasing the number of relations between them, it becomes necessary to use abstract methods, and the number of equations to be solved rises rapidly. Therefore, the actual solution of complex optimization problems is possible only with the aid of computers. Owing to the simultaneous development of these machines and of applied mathematics, powerful and practical methods have become available for solving complex optimization problems. In the following we shall discuss some of the most used methods. Although applications to reactor optimization will be considered, these methods are clearly quite general and by no means restricted to chemical reactors. An optimization problem is mathematically formulated as follows:

$$\text{minimize} \quad f(\bar{x}) \qquad \bar{x} = (x_1, x_2, \ldots, x_n) \qquad \text{(X.16)}$$

$$\text{for} \quad g_i(\bar{x}) = 0 \quad i = 1, 2, \ldots, m \quad (X.17)$$

$$\text{and} \quad g_i(\bar{x}) \leq 0 \quad i = m+1, m+2, \ldots, m+p. \quad (X.18)$$

In these equations $f(\bar{x})$ is the objective function; $g_i(\bar{x})$ are the equality and inequality constraints and $\bar{x}$, a vector, is the collection of variables, which describe the system. The difference $n - m$ is the number of degrees of freedom and should be at least 1, otherwise an optimization is not possible. We will explain the equations above with an example. Suppose we have to design a cylindrical storage tank of $300 \, \text{m}^3$ with a minimum contact area to the surrounding air. Space is limited so that the diameter d should not exceed 10 m and also the height h is limited to 12 m. The mathematical formulation is now

$$\text{Minimize} \quad f(\bar{x}) = \pi d h + \tfrac{1}{4}\pi d^2$$

$$\text{for} \quad g_1(\bar{x}) = \tfrac{1}{4}\pi d^2 h - 300 = 0$$

$$\text{and} \quad g_2(\bar{x}) = d - 10 \leq 0$$

$$g_3(\bar{x}) = h - 12 \leq 0.$$

In this case $\bar{x} = (h, d)$, $n = 2$, $m = 1$ and $p = 2$. The number of degrees of freedom is $n - m = 2 - 1 = 1$.

The usefulness of the results obtained for solving an optimization problem depends to a large extent on the care bestowed on formulating the objective function $f(\bar{x})$. The objective function can be a technical one or an economical one; the merits of each were discussed in the first section of this chapter. There are two kinds of constraint functions. Equality constraints (X.17) describe the physical laws pertinent to the problem: material and energy balances, transport equations, etc. Inequality constraints (X.18) are introduced to avoid undesirable or physically impossible solutions, such as a maximum reactor temperature to protect a catalyst or the construction material chosen. The optimum set of variables or the vector $\bar{x}^*$ that minimizes $f(\bar{x})$ also maximizes $-f(\bar{x})$, so that optimization problems, which require a maximization of the objective function, are included in the above formulation. A similar statement can be made about the inequality constraints (X.18).

Finding the optimal solution to a problem quite often involves a choice between alternatives, e.g. various reactor types, various raw materials or various methods of reactor cooling. The mathematical methods available for optimization cannot deal with this kind of variables. The best choice can only be made by comparing the results obtained from optimizations of the remaining variables for each of the alternatives.

Within the scope of this chapter only a compendious survey of the various available optimization methods can be given. Optimization methods are divided into *analytical* and *numerical methods*. Analytical methods are to be preferred because of the general character of the solution; application of them to reactor design, however, often involves unrealistic simplification of a complex problem. We will discuss geometric programming and the Lagrange multiplier techniques as important analytical optimization methods. The number of numerical methods that have been proposed in the literature may rightly be dubbed tremendous and even exasperating, at least for a tyro. Classification of numerical methods is possible in various ways: (1) according to the number of degrees of freedom — one-dimensional versus multi-dimensional methods; (2) whether derivatives are required or not and, if so, whether these are only first derivatives or also second derivatives; (3) whether the methods allow for constraints or not. The classification according to the number of degrees of freedom may be considered as the most systematically consistent one.

If the objective function has one degree of freedom, the result of the optimization will be a statement for the optimization variable: $x_1 < x^* < x_1 + \Delta x$; the range Δx can be chosen beforehand and is limited only by the accuracy of the computations. If, however, the number of degrees of freedom is higher than one, no such statement can be arrived at. In the latter case all methods have in common, that at each step a $\Delta \bar{x}$ is found which fulfills the condition that $f(\bar{x} + \Delta \bar{x})$ is an improvement on $f(\bar{x})$. The process is started at a certain set of values of the optimization variables $\bar{x}_o$ and is terminated if some breaking-off criterion is met, e.g. if the difference between $f(\bar{x} + \Delta \bar{x})$ and $f(\bar{x})$ is smaller than a given small number. There are no incontestable methods to ascertain whether the endpoint $\bar{x}_e$ of a multi-dimensional optimization is an acceptable approximation of $\bar{x}^*$; the best that can be done is comparing the $\bar{x}_e$ obtained for different starting values of $\bar{x}_o$. Equality constraints are in almost all cases included in the computation of the values of the objective function; inequality constraints can be taken care of in a number of ways, depending on the optimization method. A fairly general method is the penalty function: if a constraint is violated a penalty proportional to the amount of the transgression is added to the value of the objective function. The size of an optimization problem is an exponential function of the number of degrees of freedom. If the latter number is high, it is advisable to try to decompose the problem into subproblems. One of the best known methods is dynamic programming, which we will outline below. Also a numerical search technique will be described.

X.3.1 Geometric programming

Optimization methods based on differential calculus can become unwieldy and complex, involving a large amount of arithmetic and algebraic manipulation, so that it is easy to become confused and to make errors. To avoid this, geometric programming is an elegant method. Geometric programming can only be applied to objective functions that can be written as:

$$f(\bar{x}) = \sum_{j=1}^{t} C_j P_j(\bar{x}) \tag{X.19}$$

with

$$P_j(\bar{x}) = \prod_{i=1}^{n} x_i^{a_{ij}} \tag{X.20}$$

and
$$C_j > 0. \tag{X.21}$$

Here the number of terms in the objective function is t and the number of independent variables is n. From the equations follows, that geometric programming [48] is limited to cases where

- the objective function can be written as a sum of terms with index j, where each term is a product of powers of the independent variables with index i.
- the constants in each term are positive.

The last condition implies that geometric programming can be applied to cost minimization problems but not to profit maximization, where positive and negative constants are involved.

Now let $\bar{x}^*$ be an $\bar{x}$ that minimizes $f(\bar{x})$ and let

$$W_j = \frac{C_j P_j(\bar{x}^*)}{f(\bar{x}^*)}; \tag{X.22}$$

then

$$\sum_{j=1}^{t} W_j = 1, \tag{X.23}$$

and as can be mathematically proven [48],

$$\sum_{j=1}^{t} a_{ij} W_j = 0 \qquad i = 1, 2, \ldots, n. \tag{X.24}$$

Equations (X.23) and (X.24) constitute a set of $n + 1$ linear equations, from which $\bar{W}$ can be solved provided that $t = n + 1$. In that case the optimum value of the objective function $f(\bar{x}^*)$ follows immediately from

$$f(\bar{x}^*) = \prod_{j=1}^{t} \left(\frac{C_j}{W_j}\right)^{W_j} \tag{X.25}$$

If the value of $f(\bar{x}^*)$ has been obtained, the values of $\bar{x}^*$ follow from equation (X.22).

In case $t > n + 1$ a solution can still be obtained by maximizing a so-called 'dual' function $z(\bar{w})$, which because of equation (X.23) and (X.24) can be written as a function of only $(t - n - 1)$ of the variables W_j [48]. The theory of geometric programming has been extended to cover constrained objective functions [48] and is an elegant method, although its application is rather severely restricted by limitations on the shape of the objective function (X.19). We will give two Examples to elucidate its principles.

Example X.3.a *Minimum outer surface area of a cylindrical storage tank*
At the beginning of Section X.3 we formulated mathematically the optimization problem of minimizing the outer surface area of a cylindrical tank. We will find this minimum area by using geometric programming. First we introduce the

equality constraint $h = 4 \times 300/\pi d^2$ into the objective function. Our objective now is:

$$\text{Minimize } f(\bar{x}) = \frac{1200}{d} + \frac{\pi}{4} d^2. \tag{a}$$

This function contains two terms and one variable, so that $t = n + 1$. The first term represents the surface area of the wall, the second that of the roof of the tank. Conditions (X.23) and (X.24) now lead to

$$W_1 + W_2 = 1 \tag{b}$$

$$-W_1 + 2W_2 = 0 \tag{c}$$

The solution is: $W_1 = \frac{2}{3}$, $W_2 = \frac{1}{3}$.

Note that the application of condition (X.24) is very similar to the method used in dimensional analysis. Equation (X.25) now gives us directly the minimum value of the outer surface area:

$$f(\bar{x}^*) = \left(\frac{1200}{\frac{2}{3}}\right)^{2/3} \times \left(\frac{\pi}{4 \times \frac{1}{3}}\right)^{1/3} = 196.9 \, \text{m}^2.$$

Of this total surface area a fraction $W_1 = \frac{2}{3}$ is occupied by the wall and $W_2 = \frac{1}{3}$ by the roof. The value of d^* can be found from equation (X.22):

$$\frac{2}{3} = \frac{1200}{d^* \times 196.9} \Rightarrow d^* = 9.142 \, \text{m}.$$

The optimum height of the tank is then $h^* = 4.570$ m. The advantages of geometric programming over conventional differential calculus are now clear:

— the method is completely algebraic
— the values of the variables $\bar{W}$ give the fraction of the total area associated with each term
— the minimum area can be calculated directly without solving for d
— equations (b) and (c) can be solved easily and are much less complicated than the equation obtained by differentiation

Of course, this problem could have been solved by other methods. The choice often reduces to the selection of the technique that is easiest and offers additional insight into the problem. In this Example the designer could find the minimum area without solving for the optimum values of the system parameters. In this Example we see that the inequality constraints are not violated.

Example X.3.b *Minimum total volume in an isothermal cascade*
It is required to find the minimum total volume and the corresponding individual volumes of a cascade of N tank reactors for an isothermal first order reaction. If the reactors are numbered in the direction of the flow the material balance for the ith reactor gives

$$\frac{c_{A,i}}{c_{A,i-1}} = \frac{1}{1 + k\tau_i}. \tag{a}$$

The problem now is to minimize

$$k \sum_{i=1}^{N} \tau_i = \sum_{i=1}^{N} \frac{c_{A,i-1}}{c_{A,i}} - N \tag{b}$$

subject to the constraint

$$\frac{c_{Ao}}{c_{AN}} = \prod_{i=1}^{N} \frac{c_{A,i-1}}{c_{A,i}} = \frac{1}{1 - \zeta_{AN}}.$$

Because N is a known constant the problem can now be stated as

$$\text{Minimize } f(\bar{x}) = \sum_{i=1}^{N} \frac{c_{A,i-1}}{c_{A,i}}.$$

This constrained problem is easily changed into an unconstrained one by eliminating any of the $c_{A,i-1}/c_{A,i}$. We arbitrarily take $c_{A,N-1}/c_{A,N}$ and then find

$$\text{Minimize } f(\bar{x}) = \sum_{i=1}^{N-1} \frac{c_{A,i-1}}{c_{A,i}} + \frac{1}{(1 - \zeta_{AN}) \prod_{i=1}^{N-1} \frac{c_{A,i-1}}{c_{A,i}}}. \tag{c}$$

The variables are all $c_{A,i}$ except c_{Ao} and c_{AN}, so $n = N - 1$, and the number of terms in equation (c) is $t = N$, so that $t = n + 1$. We therefore can apply geometric programming. For condition (X.23) we find

$$W_1 + W_2 + W_3 + \cdots + W_{N-1} + W_N = 1$$

because there are N terms. From condition (X.24) we can derive for the different variables

$$\text{for} \quad c_{A,1} \quad : -W_1 + W_2 = 0$$
$$\text{for} \quad c_{A,2} \quad : -W_2 + W_3 = 0$$
$$\text{for} \quad c_{A,N-1} : -W_{N-1} + W_N = 0$$

From both conditions it follows that

$$W_1 = W_2 = \cdots W_N = \frac{1}{N}.$$

From equation (X.25) and (c) it follows that

$$f(\bar{x}^*) = \left[\left(\frac{1}{1/N}\right)^{1/N}\right]^{N-1} \left[\frac{1}{(1 - \zeta_{AN}) \times 1/N}\right]^{1/N}$$

$$= N \sqrt[N]{\left(\frac{1}{1 - \zeta_{AN}}\right)} = k \sum_{i=1}^{N} \tau_i^* + N.$$

For each individual tank $W_i = 1/N$, so that

$$k\tau_i^* = \frac{1}{N} k \sum_{i=1}^{N} \tau_i^* = 1/\sqrt[N]{(1 - \zeta_{AN})} - 1,$$

so the result is that all tanks have the same residence time. The objective function depends on the kind of reaction. For example, geometric programming cannot be used for consecutive reactions A → P → X to maximize the yield, because in this case the objective function would contain positive and negative terms, as can easily be verified.

X.3.2 The Lagrange multiplier technique

The Lagrange multiplier technique is a general mathematical method for finding the extreme value of a given function of a certain number of variables between which additional relations exist. The Lagrange function is defined as

$$L(\bar{x}, \lambda) = f(\bar{x}) + \sum_{i}^{m+p} \lambda_i g_i(\bar{x}) \qquad (X.26)$$

It can be proven that $f(\bar{x})$ is minimal for any $\bar{x}$ satisfying the following relations:

$$-\frac{\partial L}{\partial x_j} = 0, \qquad j = 1, 2, \ldots, n \qquad (X.27)$$

$$-\frac{\partial L}{\partial \lambda_i} = 0, \qquad i = 1, 2, \ldots, m \qquad (X.28)$$

$$\left.\begin{array}{ll} g_i < 0 & \text{and} \quad \lambda_i = 0 \\ \text{or} \quad g_i = 0 & \text{and} \quad \lambda_i > 0 \end{array}\right\} \quad i = m+1, m+2, \ldots, m+p. \qquad (X.29)$$

Here n is the number of variables, m the number of equations describing the process model and p the number of boundary conditions and constraints. It follows from the mathematical theory that the above is only true if some conditions are met. This being almost always the case in well-formulated problems the interested reader is referred to mathematical textbooks [1] and the work of Horn [9–12, 21–23, 49]. Horn (see, e.g., [9, 10, 49]) used this method extensively for the optimization of chemical reactors. In view of the great importance of the results obtained by him, we shall treat the theory in greater detail below, first for a cascade of tank reactors (i) and then for a tubular reactor (ii).

(i) *Cascade of N tank reactors.* We shall consider a reaction in which the composition of the reaction mixture is determined by the concentrations c_A and c_P of two species A and P. As usual, the conditions in tank reactor number n and its outlet will be indicated by the index n; they are determined by the composition c_{An} and c_{Pn}, by the average residence time, τ_n, and by the temperature T_n. It will be assumed that the inlet composition of the cascade, c_{A_0} and c_{P_0}, is fixed and not a variable of the problem; this assumption is not essential but merely a simplification.

The quantity to be optimized will generally be a function of the composition of the reaction mixture leaving the last tank reactor (conversion, yield) and of the total average residence time τ in the cascade (capacity):

$$M = M(c_{AN}, c_{PN}, \tau). \tag{X.30}$$

This function is meant to be representative of the merits of the plant performance and is the objective function. Now, in a system with given rate equations and with a given feed composition, c_{AN}, c_{PN} and τ are determined entirely by the values of τ_n and T_n in each of the given total of N reactors. Accordingly, M can, in principle, be written as a function of these 'primary variables',

$$M = M(\tau_1 \ldots \tau_N, T_1, \ldots, T_N). \tag{X.31}$$

However, the function M cannot be explicitly expressed in terms of these variables. The relation between c_{AN}, c_{PN} and τ and the variables τ_n and T_n (and c_{Ao} and c_{Po}) is governed by the following set of $(2N + 1)$ equations:

$$0 = c_{A(n-1)} - c_{An} + \tau_n R_{An}, \quad n = 1, \ldots, N, \tag{X.32}$$

$$0 = c_{P(n-1)} - c_{Pn} + \tau_n R_{Pn}, \quad n = 1, \ldots, N \tag{X.33}$$

and

$$0 = \sum_{n=1}^{N} \tau_n - \tau. \tag{X.34}$$

The Lagrange function now is

$$L = M(c_{AN}, c_{PN}, \tau) + \sum_{n=1}^{N} \lambda_{An}(c_{A(n-1)} - c_{An} + \tau_n R_{An})$$

$$+ \sum_{n=1}^{N} \lambda_{Pn}(c_{P(n-1)} - c_{Pn} + \tau_n R_{Pn}) + \lambda_\tau \left(\sum_{n=1}^{N} \tau_n - \tau \right). \tag{X.35}$$

All constraints are equality constraints so that the optimum is found by solving equations (X.27) and (X.28). If we apply equation (X.27), the following equations are derived:

$$\frac{\partial L}{\partial c_{An}} = \lambda_{A(n+1)} - \lambda_{An}\left(1 - \tau_n \frac{\partial R_{An}}{\partial c_A}\right) + \lambda_{Pn}\tau_n \frac{\partial R_{Pn}}{\partial c_A} = 0,$$

$$n = 1, \ldots, (N-1); \tag{X.36}$$

$$\frac{\partial L}{\partial c_{Pn}} = \lambda_{P(n+1)} - \lambda_{Pn}\left(1 - \tau_n \frac{\partial R_{Pn}}{\partial c_P}\right) + \lambda_{An}\tau_n \frac{\partial R_{An}}{\partial c_P} = 0,$$

$$n = 1, \ldots, (N-1); \tag{X.37}$$

$$\frac{\partial L}{\partial \tau_n} = +\lambda_{An}R_{An} + \lambda_{Pn}R_{Pn} + \lambda_\tau = 0, \quad n = 1, \ldots, N; \tag{X.38}$$

$$\frac{\partial L}{\partial T_n} = \left(+\lambda_{An}\frac{\partial R_{An}}{\partial T} + \lambda_{Pn}\frac{\partial R_{Pn}}{\partial T} \right)\tau_n = 0, \quad n = 1,\ldots,N; \qquad (X.39)$$

$$\frac{\partial L}{\partial c_{AN}} = \frac{\partial M}{\partial c_{AN}} - \lambda_{AN}\left(1 - \tau_N\frac{\partial R_{AN}}{\partial c_{AN}}\right) + \lambda_{PN}\tau_N\frac{\partial R_{PN}}{\partial c_{AN}} = 0, \qquad (X.40)$$

$$\frac{\partial L}{\partial c_{PN}} = \frac{\partial M}{\partial c_{PN}} - \lambda_{PN}\left(1 - \tau_N\frac{\partial R_{PN}}{\partial c_{PN}}\right) + \lambda_{AN}\tau_N\frac{\partial R_{AN}}{\partial c_{PN}} = 0, \qquad (X.41)$$

$$\frac{\partial L}{\partial \tau} = \frac{\partial M}{\partial \tau} + \lambda_\tau = 0. \qquad (X.42)$$

Equation (X.28) results in the relations (X.32)–(X.34). In this way a set of $6N + 2$ equations has been obtained, permitting the evaluation of the $6N + 2$ unknowns, being $c_{An}, c_{Pn}, \tau_n, \tau, T_n, \lambda_{An}, \lambda_{Pn}$ and λ_τ. It should be noted that the above equations are restricted in the sense that only two concentrations have been assumed to be needed for specifying the composition of the reaction mixture; the feed composition has been assumed constant, and intermediate feed or withdrawal has not been considered. Also, M has been assumed not to depend on the individual temperatures of the various reactors; however, if the objective function were influenced by the heat economy of the plant, a dependence on these temperatures would be possible. The same method as indicated above could be used, since there are no restrictions regarding the number of variables.

Only in very simple cases can the $(6N + 2)$ equations be solved analytically; when this is the case, the optimization problem may also be solved by more direct methods (see, e.g. Example X.3.c). In more complex problems, the simultaneous solution of $(6N + 2)$ equations containing $(6N + 2)$ unknowns is a typical task for a computer. The various routines available for this consist essentially of a sequence of successive approximations, so that numerous calculations have to be made. An elementary application of the above theory will be given below.

Example X.3.c *The maximum yield for first order consecutive reactions in one tank reactor*

It is required to find the condition for maximum yield of P for two consecutive first order reactions $A \to P \to X$ in a continuous tank reactor for the two following cases:

(a) Optimum load at a given reaction temperature;
(b) Optimum reaction temperature at a given load.

The feed contains neither P nor X.
We have:

$$R_A = -k_P c_A, \quad R_P = k_P c_A - k_X c_P,$$

while the quantity to be optimized, $M = c_{P1}/c_{Ao}$, is a function of c_{A1}, c_{P1} and τ. The equations to be solved are:

$$c_{Ao} - c_{A1} - k_P \tau c_{A1} = 0 \qquad \text{(cf. X.32)}$$

$$-c_{P1} + k_P\tau c_{A1} - k_X\tau c_{P1} = 0 \qquad \text{(cf. X.33)}$$

$$-\lambda_{A1}(1 + k_P\tau) + \lambda_{P1}k_X\tau = 0 \qquad \text{(cf. X.40)}$$

$$\frac{1}{c_{Ao}} - \lambda_{P1}(1 + k_X\tau) = 0 \qquad \text{(cf. X.41)}$$

$$-\lambda_{A1}k_Pc_{A1} + \lambda_{P1}(k_Pc_{A1} - k_Xc_{P1}) = 0 \qquad \text{(a)} \quad \text{(cf. X.38)}$$

$$-\lambda_{A1}c_{A1}\frac{dk_P}{dT} + \lambda_{P1}\left(c_{A1}\frac{dk_P}{dT} - c_{P1}\frac{dk_X}{dT}\right) = 0. \qquad \text{(b)} \quad \text{(cf. X.39)}$$

The last equation may be omitted in solving problem (a). The first two equations give

$$\frac{c_{A1}}{c_{Ao}} = \frac{1}{1 + k_P\tau}, \quad \frac{c_{P1}}{c_{Ao}} = \frac{k_P\tau}{(1 + k_P\tau)(1 + k_X\tau)},$$

and the next two

$$\lambda_{A1} = \frac{k_P\tau}{1 + k_P\tau}\lambda_{P1}, \quad \lambda_{P1} = \frac{1}{c_{Ao}(1 + k_X\tau)}.$$

After substitution of these expressions into equation (a), we find that the residence time for maximum yield is

$$\tau_{opt} = (k_Pk_X)^{-1/2}. \qquad \text{(c)}$$

In solving problem (b), equation (a) is not relevant since it was obtained by differentiation with respect to τ, which is assumed constant in this case. The expressions for c_{A1}, c_{P1}, λ_{A1} and λ_{P1} are now substituted into equation (b). The final result is that the maximum yield is produced at the temperature where

$$(1 + k_P\tau)\frac{dk_P}{dT} = k_P\tau(1 + k_X\tau)\frac{dk_P}{dT}. \qquad \text{(d)}$$

In this simple case, the results (c) and (d) could have been found directly by calculating the maximum value of c_{P1}/c_{Ao} as a function of τ and T since an expression for c_{P1}/c_{Ao} was available containing these two variables only. This example must, therefore, be regarded as an illustration of the method. It should be noted that the λs have a physical significance. In this case we have:

$$c_{Ao}\lambda_{A1} = c_{P1}/c_{Ao} = \eta_{P1} \qquad \text{(yield)}$$
$$c_{Ao}\lambda_{P1} = c_{P1}/(c_{Ao} - c_{A1}) = \sigma_{P1} \qquad \text{(selectivity)}$$

(*Example X.3.c ended*)

(ii) *Ideal tubular reactor.* The behaviour of a tubular reactor (without feed distribution or product removal) can be approached by taking a cascade of tank reactors of equal size and by letting the total number of reactors, N, go to infinity and the residence time of each reactor to zero. This property can be used in deriving the expressions for optimum conditions in a tubular reactor from the theory for the cascade. We shall again assume that the composition of the reaction mixture is determined by the two concentrations c_A and c_P and that the function to be optimized depends on the composition of the reaction mixture at the outlet, c_{AL} and c_{PL}, and on the total residence time τ_L:

$$M = M(c_{AL}, c_{PL}, \tau_L).$$

If the residence time τ is taken as a measure of the distance from the feed end of the reactor, the conversion equations (X.32) and (X.33) become (with ρ constant):

$$0 = -dc_A + R_A \, d\tau \tag{X.43}$$

$$0 = -dc_P + R_P \, d\tau. \tag{X.44}$$

The Lagrange function L defined by equation (X.35) now becomes

$$L = M(c_{AL}, c_{PL}, \tau_L) + \int_{\tau=0}^{\tau=\tau_L} \lambda_A(\tau)[dc_A - R_A \, d\tau]$$

$$+ \int_{\tau=0}^{\tau=\tau_L} \lambda_P(\tau)[dc_P - R_P \, d\tau]. \tag{X.45}$$

The Lagrange parameters λ_A and λ_P are now functions of τ. An extreme in L — and hence in M — is formed by equating the partial derivatives of L to zero with respect to c_A, c_P and T. The results are:

$$\frac{\partial L}{\partial c_A} = 0: \quad 0 = \frac{d\lambda_A}{d\tau} + \lambda_A \frac{\partial R_A}{\partial c_A} + \lambda_P \frac{\partial R_P}{\partial c_A} \tag{X.46}$$

$$\frac{\partial L}{\partial c_P} = 0: \quad 0 = \frac{d\lambda_P}{d\tau} + \lambda_A \frac{\partial R_A}{\partial c_P} + \lambda_P \frac{\partial R_P}{\partial c_P} \tag{X.47}$$

$$\frac{\partial L}{\partial T} = 0: \quad 0 = \lambda_A \frac{\partial R_A}{\partial T} + \lambda_P \frac{\partial R_P}{\partial T}. \tag{X.48}$$

Equations (X.46) and (X.47), which are analogous to equations (X.36) and (X.37), are obtained from (X.45) after the integrals $\int \lambda_A \, dc_A$ and $\int \lambda_P \, dc_P$ have been partially integrated. In addition to these three equations we have the boundary conditions:

$$\frac{\partial L}{\partial c_{AL}} = 0 = \frac{\partial M}{\partial c_{AL}} + \lambda_{AL} \tag{X.49}$$

$$\frac{\partial L}{\partial c_{PL}} = 0 = \frac{\partial M}{\partial c_{PL}} + \lambda_{PL} \tag{X.50}$$

$$\frac{\partial L}{\partial \tau_L} = 0 = \frac{\partial M}{\partial \tau_L} - \lambda_{AL} R_{AL} - \lambda_{PL} R_{PL}, \tag{X.51}$$

which come in the place of equations (X.40)–(X.42). From the five differential equations (X.43), (X.44), (X.46)–(X.48) with their boundary conditions (X.49)–(X.51), the relationship between τ, c_A, c_P, λ_A, λ_P and T can be found for which M will be extreme. Several methods of solution can be followed (Horn and Troltenier [11], Horn [21]), but computer aid will generally be necessary. It is possible to eliminate c_A, c_P and T from the five differential equations in the following manner:

$$(\text{X.46}) \times dc_A \rightarrow \quad \frac{d\lambda_A}{d\tau} dc_A + \lambda_A \frac{\partial R_A}{\partial c_A} dc_A + \lambda_P \frac{\partial R_P}{\partial c_A} dc_A = 0$$

$$(\text{X.47}) \times dc_P \rightarrow \quad \frac{d\lambda_P}{d\tau} dc_P + \lambda_A \frac{\partial R_A}{\partial c_P} dc_P + \lambda_P \frac{\partial R_P}{\partial c_P} dc_P = 0$$

$$(\text{X.48}) \times dT \rightarrow \quad \lambda_A \frac{\partial R_A}{\partial T} dT + \lambda_P \frac{\partial R_P}{\partial T} dT = 0$$

――――――――――――――――――――――――――― +

$$\frac{d\lambda_A}{d\tau} dc_A + \frac{d\lambda_P}{d\tau} dc_P + \lambda_A dR_A + \lambda_P dR_P = 0$$

When dc_A and dc_P are substituted by using equations (X.43) and (X.44), it is found that the following conditions must be met along the reactor:

$$d(\lambda_A R_A + \lambda_P R_P) = 0$$

or, with equation (X.51):

$$\lambda_A R_A + \lambda_P R_P = \frac{\partial M}{\partial \tau_L}. \qquad (\text{X.52})$$

If we are not interested in the capacity of the reactor but only in the yield, M will not be a function of τ_L and $\partial M/\partial \tau_L = 0$. λ_A and λ_P can then be eliminated from equations (X.48) and (X.52) to give

$$\frac{1}{R_A} \frac{\partial R_A}{\partial T} = \frac{1}{R_P} \frac{\partial R_P}{\partial T} \quad \text{or} \quad \frac{\partial}{\partial T}\left(\frac{R_P}{R_A}\right) = 0. \qquad (\text{X.53})$$

This condition simply states that the differential selectivity σ'_P should be a maximum at any cross section in the reactor. As already discussed in section X.2, it implies that the highest or lowest permissible temperature is the optimum temperature if there is no maximum in σ'_P. Interesting possibilities of temperature optimization exist if σ'_P has a maximum as a function of T within the permissible temperature range (see, e.g., Example X.3.d).

Another group of problems where the parameters λ can be fairly easily eliminated without the restriction $\partial M/\partial \tau_L = 0$ is encountered when the conversion rates are *linear functions* of the concentrations. In such case we can use the property that

$$c_A \frac{\partial R_A}{\partial c_A} + c_P \frac{\partial R_A}{\partial c_P} = R_A \qquad (\text{X.54})$$

if, e.g., R_A is linear in c_A and c_P.

The derivation of the optimum conditions will be shown below for the case where the two components A and P characterize the reaction mixture; it can be extended to more components. This is of some importance since the assumption of first order or linearized kinetics will in many cases provide a good approximation of practical circumstances and therefore a good estimate of optimum conditions.

If we write equations (X.46) and (X.47) in the form

$$0 = c_A \frac{d\lambda_A}{d\tau} + \lambda_A \frac{dc_A}{d\tau} - \lambda_A \frac{dc_A}{d\tau} + \lambda_A c_A \frac{\partial R_A}{\partial c_A} + \lambda_P c_A \frac{\partial R_P}{\partial c_A}$$

$$0 = c_P \frac{d\lambda_P}{d\tau} + \lambda_P \frac{dc_P}{d\tau} - \lambda_P \frac{dc_P}{d\tau} + \lambda_P c_P \frac{\partial R_A}{\partial c_P} + \lambda_P c_P \frac{\partial R_P}{\partial c_P}$$

and if we take the sum of these two equations, we obtain

$$0 = \frac{d}{d\tau}(\lambda_A c_A + \lambda_P c_P) + \lambda_A \left(c_A \frac{\partial R_A}{\partial c_A} + c_P \frac{\partial R_A}{\partial c_P} - \frac{dc_A}{d\tau} \right)$$

$$+ \lambda_P \left(c_A \frac{\partial R_P}{\partial c_A} + c_P \frac{\partial R_P}{\partial c_P} - \frac{dc_P}{d\tau} \right).$$

The last two terms in this equation are now zero because of the material balances (X.43) and (X.44) and of equation (X.54) for linear rate expressions. Accordingly

$$\frac{d}{d\tau}(\lambda_A c_A + \lambda_P c_P) = 0 \qquad (X.55)$$

and

$$\lambda_A c_A + \lambda_P c_P = \lambda_{AL} c_{AL} + \lambda_{PL} c_{PL}.$$

λ_A and λ_P can be calculated from equations (X.55) and (X.48). After substitution of these calculated values into (X.52) and using the values given by equations (X.49) and (X.50) for λ_{AL} and λ_{PL}, we find

$$\frac{R_A \frac{\partial R_P}{\partial T} - R_P \frac{\partial R_A}{\partial T}}{c_A \frac{\partial R_P}{\partial T} - c_P \frac{\partial R_A}{\partial T}} = \frac{-\frac{\partial M}{\partial \tau_L}}{c_{AL} \frac{\partial M}{\partial c_{AL}} + c_{PL} \frac{\partial M}{\partial c_{PL}}}. \qquad (X.56)$$

It is seen that this expression is reduced to equation (X.53) when the cost function does not contain the capacity (or τ_L) as a variable. The above theory was worked out in a more general form by Horn [21], who applied equation (X.56) in calculating optimum temperature profiles for various reactions. Analogous calculations were made by Bilous and Amundsun [41] for first and second order consecutive reactions in tubular reactors.

Example X.3.d *Temperature optimization of three parallel reactions*
It is required to find the optimum conditions in a tubular reactor for the three parallel first order reactions:

$$A \xrightarrow{1} P, \quad R_P = k_1 c_A = c_A k_{1\infty} \exp(-E_1/RT)$$
$$A \xrightarrow{2} X, \quad R_X = k_2 c_A = c_A k_{2\infty} \exp(-E_2/RT)$$
$$A \xrightarrow{3} Y, \quad R_Y = k_3 c_A = c_A k_{3\infty} \exp(-E_3/RT).$$

The activation energy E_1 lies in between the values of E_2 and E_3, so that the undesired side reactions 2 and 3 will dominate at low and high temperatures, respectively. The unconverted reactant A will be recovered and added to the feed. The by-products X and Y have no economic value. We shall require the gross profit per molar unit of P produced to be a maximum. The total gross profit of the plant per unit time will be supposed to be:

$$\Phi_v[c_{PL}C_P - (c_{Ao} - c_{AL})C_A] - V_r C_f - \Phi_v C_0$$

where

Φ_v = volumetric throughput,
C_P = sales value of one kmol of P,
C_A = cost of one kmol of A,
C_f = fixed costs per unit of reaction volume,
C_0 = variable costs per unit of throughput.

The profit per kmol of P then becomes

$$M(c_{AL}, c_{PL}, \tau_L) = C_P - \frac{c_{Ao} - c_{AL}}{c_{PL}} C_A - \frac{\tau_L}{c_{PL}} C_f - \frac{C_0}{c_{PL}}.$$

It is seen that, for a high value of M, the selectivity $[c_{PL}/(c_{Ao} - c_{AL})]$ and c_{PL} should be high and τ_L not too large. Selecting A and P as the characteristic components for which $R_A = -(k_1 + k_2 + k_3)c_A$ and $R_P = k_1 c_P$, we find for the left-hand side of equation (X.56),

$$\frac{k_1 - (k_1 + k_2 + k_3)\dfrac{dk_1}{d(k_1 + k_2 + k_3)}}{\dfrac{c_P}{c_A} + \dfrac{dk_1}{d(k_1 + k_2 + k_3)}}.$$

The right-hand side of equation (X.56) becomes, in view of the above cost function M,

$$\frac{C_f}{c_{Ao}C_A + \tau_L C_f + C_0} = C.$$

C happens to be independent of c_{AL} and c_{PL} in this case, but it varies with τ_L. Substitution of these expressions into equation (X.56) yields the following condition:

$$\frac{c_P}{c_A} C = k_1 - (k_1 + k_2 + k_3 + C) \frac{dk_1}{d(k_1 + k_2 + k_3)}. \qquad (a)$$

Figure X.23 shows the latter function of T for the given set of conversion rate constants indicated, with $C = 0$ and with a positive value of C. If the chemicals A and P are very expensive and the fixed costs relatively low, the reactor capacity may be a rather unimportant parameter, and consequently $C \simeq 0$. Equation (a) is then reduced to

$$\frac{d}{d\tau}\left(\frac{k_1}{k_1 + k_2 + k_3}\right) = 0, \qquad \text{(see equation (X.53))}$$

which means that an optimum temperature exists irrespective of the degree of conversion. Accordingly, the best procedure is to operate isothermally at the temperature which follows from the latter equation

$$T_{opt} = \frac{E_2 - E_3}{R \ln\left[\dfrac{k_{2\infty}(E_2 - E_1)}{k_{3\infty}(E_1 - E_3)}\right]}. \qquad (b)$$

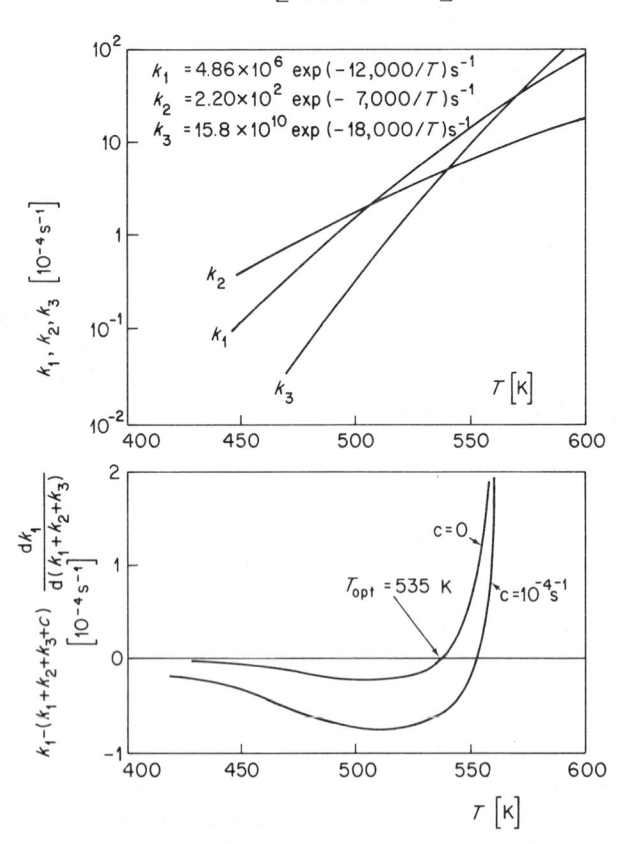

Fig. X.23 Yield optimization of three parallel first order reactions; Example X.3.d

For the ks indicated in figure X.23, we find $T_{opt} = 535$ K, in accordance with the intersection of the curve for $C = 0$ with the abscissa. It can be shown similarly that the maximum yield is obtained by isothermal operation for the reactions:

$$A \xrightarrow{1} P \underset{3}{\overset{2}{\rightleftarrows}} \begin{matrix} X \\ Y \end{matrix} \quad \text{and} \quad A \underset{2}{\overset{1}{\rightleftarrows}} P \xrightarrow{3} X,$$

provided E_1 lies between E_2 and E_3 and the reactor capacity is not a cost parameter (Horn and Troltenier [12]).

If the reactor capacity does play a part in the optimization problem, C has a positive value. Consequently, the curve representing the right-hand side of equation (a) is lowered (see figure X.23 for $C = 10^{-4}\,\text{s}^{-1}$). At the same time, the left-hand side of (a) has a positive value proportional to c_P/c_A. It can be seen from figure X.23 that a higher temperature is required in this case than is given by equation (b), and that the temperature should be slightly raised as c_P/c_A increases. It is intuitively understood that the temperature must be raised because it is only then that average reaction rates are faster and, consequently, the capacity is higher. In this optimization problem, the value of C cannot be fixed beforehand since it contains τ_L, which in turn is determined by the temperature profile in the reactor. Hence, the optimum temperature profile must be found by successive approximation, in each step of which the material balances have to be used for calculating the final conversion of A and the yield of P.

X.3.3 Numerical search routines

Numerical search routines are numerical procedures for finding the extreme of an objective function by successive approximation. They start from a complete solution of the reactor problem with assumed values of the primary variables. The corresponding value of the objective function will not be optimal. If it is now known how $f(\bar{x})$ will change when the primary variables are changed, it is possible to calculate improved conditions under which a new value of $f(\bar{x})$ nearer the optimum is found. If this procedure is repeated a sufficient number of times, the conditions for optimum performance can be approached as closely as is desired.

In applying this method to the general problem of a cascade of N tank reactors as stated in section X.3.2, we start, for example, on the assumption that all reactors have the same volume and operate at the same temperature T_n. The values of C_{An} and c_{Pn} can be calculated from the given inlet composition and from the material balances; the objective function $M(c_{AN}, c_{PN}, \tau)$ can therefore be evaluated for this non-optimal cascade. In order to know in what direction the primary variables have to be changed, the partial derivatives of M, i.e. of $M(\tau_1, \ldots, \tau_N, T_1, \ldots, T_N)$ (see equation X.31), have to be known under the conditions first assumed. To find these derivatives, the Lagrange method has to be used in the form

$$\frac{\partial M(\bar{x})}{\partial x_i} = \frac{\partial L}{\partial x_i} \quad \text{and} \quad \frac{\partial L}{\partial x_j} = 0.$$

This set of equations provide the values of λ_{An}, λ_{Pn} and λ_τ to be used for finding the values of $\partial M/\partial \tau_n$ and $\partial M/\partial T_n$ associated with the arbitrarily selected non-optimal cascade.

Now, if we suppose that the value of M moves closer to the extreme on increasing the assumed residence time τ_n, the best way to arrive at an improved value τ'_n from the residence time τ_n is

$$\tau'_n = \tau_n + \varepsilon_\tau \frac{\partial M}{\partial \tau_n}. \tag{X.57}$$

Similarly, an improved temperature, T_n, will be obtained from

$$T'_n = T_n + \varepsilon_T \frac{\partial M}{\partial T_n}. \tag{X.58}$$

These equations constitute the basis of the method of steepest ascent. Once a set of new τ'_n and T'_n values has been obtained, the whole procedure can be repeated a number of times until M does not change by more than a prescribed small amount.

The factors ε_τ and ε_T have to be selected in a judicious manner. If they are taken too small, a great many cycles may be needed to arrive at the final solution; if they are taken too large an overshoot of the optimum values of τ_n and T_n may occur. Many calculations obviously have to be carried out before optimum conditions are arrived at. However, the method appears to be very suitable for automatic computers. As an additional advantage, the machine can print out information concerning non-optimal conditions in the course of the computation. From these data, an idea can be obtained of the flexibility of the plant with regard to deviations from the best operating policy. It also is an easy matter to take into account additional complicating factors contained in the problem, such as temperature limits or limited tank sizes. In many chemical process situations the shape of the multi-dimensional surface of the objective function is not steep and the optimum is located at the highest or lowest point of a long, convex and flat plateau or a shallow, flat valley. The optimum point may not be more than a few percent larger or smaller than points far from the optimum. In cases such as this the gradient will be small and very precise calculations will be necessary for significant differences to be detected.

As said at the beginning of this section the number of search routines is exasperating: Horn and Klein [50] name a few. Several books have appeared on the subject of optimization methods. Some of these search routines do not require the computation of derivatives [51]. Usually, the chemical reactor engineer needs the help of a mathematical colleague in selecting the most suitable method and executing the computations. We will give an example of the application of a search routine as was published by Fontein and Groot Wassink [52].

Example X.3.e *Optimal design of an autothermic reactor*
Exothermic reactions that have to take place at high temperature levels can be brought to that level by installing a heat exchanger in which the reactor feed is

heated up by the hot reactor product. In principle all temperature levels can be reached — see Section VI.5 — provided the heat transfer area is large enough; in practice autothermal operation is restricted to reactions of high heat effects. Once autothermal operation has been decided upon many different line-ups can be applied and a choice has to be made. This choice is dependent on the conversion in the reactor. To determine the optimum conversion, in principle an optimization study of the entire plant is necessary. Fontein and Groot Wassink [52] made this choice on the basis of minimum annual costs for the reactor and heat exchangers as a function of the conversion, assuming that the rest of the plant is not affected by the reactor section. They studied three alternatives represented in figure X.24:

— external heat exchanger and reactor with internal cooling, the catalyst around the tubes;
— cold shot converter with three beds;
— three beds adiabatic reactor with external heat exchange with the reactor feed between the beds.

They studied the methanol synthesis from synthesis gas:

$$CO + 2H_2 \leftrightarrows CH_3OH.$$

Carbon dioxide was also added to the feed, taking advantage of the endothermic water-gas shift reaction to reduce the considerable heat of reaction. Kinetic data were taken for the high pressure methanol process as determined by Collina et al. [53] and if also CO_2 forms part of the reactor feed, by Cappelli et al. [54]. Reactor calculations were executed as outlined in Chapter VI and the heat exchanger calculations according to the method of Kays and London [52, 55, 56].

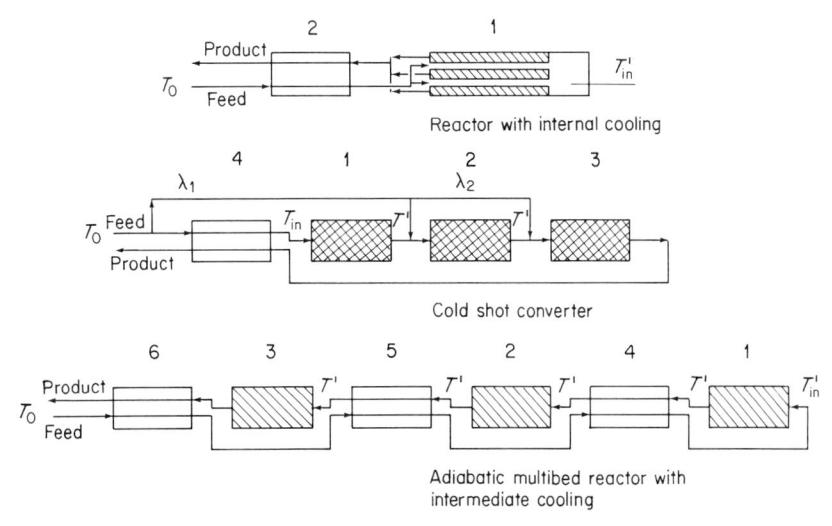

Fig. X.24 Three possibilities of autothermal operation: Example X.3.e

From an analysis of the degrees of freedom of each system the number of temperatures can be derived that can be fixed independently, when the feed rate and the conversion are known.

For the cold shot converter the distribution of the feed over the reactor beds follows as explained in Example X.2.a. By choosing values for the temperature marked with a prime in figure X.24, the reactors and heat exchangers can be calculated in the sequence indicated by the numbers. To protect the catalyst a maximum temperature $T^* = 380°C$ was introduced as a constraint. Shell diameters of the reactor and heat exchanger for the internally cooled reactor were taken as equal. Because tube diameters and the number of adiabatic beds are discontinuous variables, both were treated as constants.

It follows from the above that a multi-dimensional, numerical method has to be used to optimize the objective function. In this case a steepest descent method was used, which was based on the fact that at a point $\bar{x}$ the direction of greatest increase in the objective function is given by its gradient; the opposite direction is of course the direction of greatest decrease. Consequently the shortest path from a point $\bar{x}_0$ to the minimum $\bar{x}_e$ is the path that consists of a sequence of infinitesimal steps each in the direction opposite to the gradient, which is the dotted line in figure X.25, where a certain two-dimensional function is represented by a number of its contour lines, that is sets of points for which the function has a constant value. To determine a gradient $n + 1$ function evaluations are required so that this shortest path is not practicable. A more realistic approach is as follows: determine the gradient at the starting point $\bar{x}_0$ and take

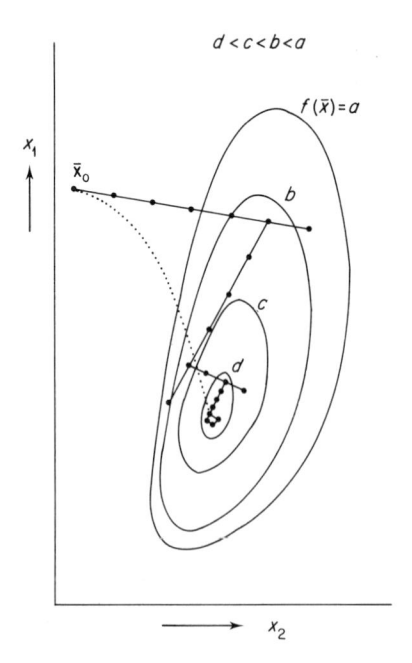

Fig. X.25 Optimization method of Example X.3.e

steps of finite length in the opposite direction till the value of the objective function decreases no more. At the point with the lowest value of the objective function the gradient is determined again and a new series of steps started. If the first step in a newly determined direction does not result in a decrease of the objective function, the step length is reduced. Computations are continued until the step has grown smaller than an *a priori* chosen small value. Constraints were taken care of by increasing the cost of the adiabatic bed reactor and the cost of the catalyst in the internally cooled reactor with a penalty function, when the maximum allowable temperature T^* was exceeded.

To test the reliability of the minima several optimizations were carried out starting from widely different values of the optimization variables. With the adiabatic reactors the convergence proved to be quite satisfactory. The internally cooled reactor required a reformulation of the optimization problem into a two-level hierarchical structure [52].

Results for a fixed annual methanol production are shown in figures X.26 and X.27. The reactor costs per kg methanol produced are highly dependent on the outlet conversion. These specific costs are minimal for the internally cooled reactor at an outlet conversion of 40%; for both adiabatic systems at much lower conversions. Remember that only the annual costs for the reactors and their heat exchangers were minimized; a lower conversion strongly influences the methanol separation and non-converted reactant recirculation costs, so that a definite selection of the reactor type can only be made by studying the complete installation. For two final conversion levels the dependence of the total annual reactor costs on the inlet temperature of the first reactor is shown; the minima are rather flat. If CO_2 is introduced into the reactor feed the same trends can be observed; the costs increase, however, due to the reaction rate decrease.

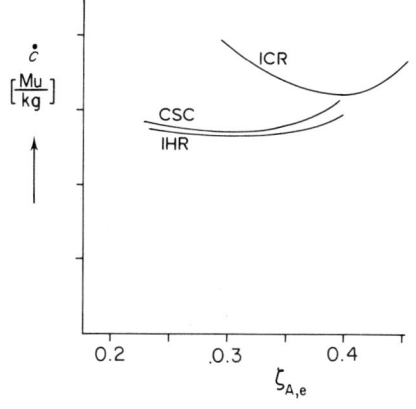

Fig. X.26 Specific costs — production costs per unit of methanol produced — depend on the reactor outlet conversion (Example X.3.e). ICR = internally cooled reactor, CSC = cold shot converter and IHR = adiabatic reactor with intermediate cooling

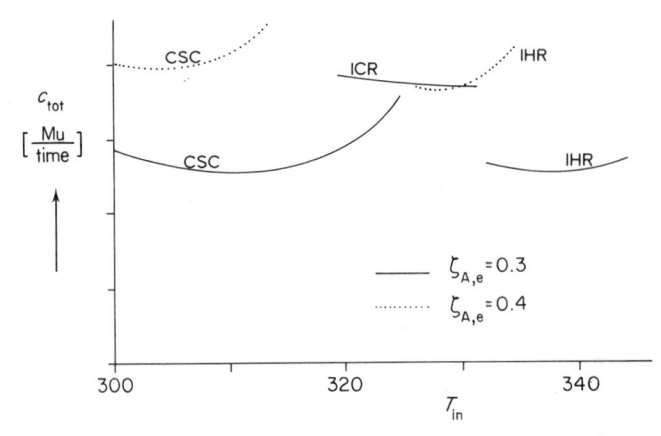

Fig. X.27 Dependence of the total reactor costs on the first reactor bed inlet temperature for two different conversion levels (Example X.3.e) and at constant production level

X.3.4 Dynamic programming

This method is mainly concerned with the optimum selection of a sequence of events. It is based on the 'principle of optimality' formulated by Bellman [57]: 'An optimal policy has the property that whatever the initial decisions are, the remaining decisions must constitute an optimal policy with regard to the state resulting from the first decision.' On the basis of this principle, an optimization problem involving the simultaneous selection of the values of a large number of parameters can be reduced to a sequence of decisions involving a smaller number of parameters. If this principle is applied to the optimization of a cascade of N tank reactors it may be said that, if the first n reactors have been optimized, the total system is optimal when the remaining cascade of $(N - n)$ reactors has been optimized with respect to the feed leaving the nth reactor. Reversing this argument, we may say that the remaining cascade can be optimized provided we know the feed conditions of the reactor number $(n + 1)$. Now, these conditions are not yet known as long as the optimization problem has not been solved; accordingly, they have to be treated as variables. The procedure now is to start with the last reactor, number N, and to find its optimum conditions as a function of the parameters of the feed entering it. Next, for each combination of these parameters the reactor number $(N - 1)$ can be optimized as a function of the feed conditions coming from number $(N - 2)$; hence, the optimum cascade of the last two tank reactors can be found as a function of its feed parameters, whereupon the procedure must be repeated with the next upstream reactor. If the flow of reaction mixture is characterized by more variables than one, the number of calculations to be made may become very great. According to Aris, who in a book [58, 59] discussed the applications to chemical reactors, the method is highly suitable for computers. As indicated by Storey [46], this procedure is limited to a true sequence (or, in the case of chemical reactors, to a cascade or an ideal tubular reactor); for reactor problems in which recirculation or feedback occurs, the method becomes quite cumbersome.

Interesting results have been obtained by means of dynamic programming. Grütter and Messikommer [60] showed that, with all *first order* reactions (including equilibrium, parallel and consecutive reactions) carried out *isothermally* in a cascade of tank reactors, maximum capacity is obtained when all reactors are of equal size. This is not so with a reaction order different from unity, but with isothermal second order reactions the capacity difference between optimum volume distribution and equal distribution appears to be small [61]. Therefore, practical as well as economic considerations indicate that all reactors in an isothermal cascade can best have the same volume. Aris applied the theory of dynamic programming to tubular reactors [58, 62] and also to multi-stage adiabatic reactors [25, 58].

The three examples given below are intended to illustrate the method outlined above. They relate to simple problems which can also be solved by more elementary methods in the first two examples.

Example X.3.f *The minimum reaction volume of an isothermal cascade*
It is required to find the minimum total volume and the corresponding individual volumes of a cascade of three tank reactors for a single first-order isothermal reaction [17]. The final degree of conversion is $\zeta_{A3} = 1 - c_{A3}/c_{Ao} = 0.9$. The material balances for the reactors numbered 1, 2 and 3 in the direction of flow give

$$\frac{c_{A1}}{c_{Ao}} = \frac{1}{1 + k\tau_1}, \quad \frac{c_{A2}}{c_{A1}} = \frac{1}{1 + k\tau_2} \quad \text{and} \quad \frac{c_{A3}}{c_{A2}} = \frac{1}{1 + k\tau_3}.$$

For the application of dynamic programming, we start with the third reactor, and we express $k\tau_3$ with the last equation in terms of the variable c_{A2}/c_{Ao}:

$$k\tau_3 = \frac{c_{A2}}{c_{A3}} - 1 = \frac{c_{A2}}{c_{Ao}} \frac{1}{0.1} - 1$$

This relationship is represented by the straight line in the lower half of figure X.28.a. Now the minimum value of $k(\tau_2 + \tau_3)$ has to be found as a function of c_{A1}/c_{Ao}. To this end we write the material balance of Reactor 2 in the form

$$k\tau_2 = \frac{c_{A1}}{c_{A2}} - 1 = \frac{c_{A1}}{c_{Ao}} \frac{c_{Ao}}{c_{A2}} - 1$$

This expression gives the curves shown in the top half of figure 28a with c_{A1}/c_{Ao} as a parameter. The minimum value of $k(\tau_2 + \tau_3)$ can then be found in the manner indicated, and the result is plotted as a function of c_{A1}/c_{Ao} in the lower half of figure X.28b. Finally, $k\tau_1$ is likewise plotted as a function of c_{A1}/c_{Ao}, and the minimum distance between the two curves in figure X.28b gives the minimum value of $k(\tau_1 + \tau_2 + \tau_3)$ with the corresponding value of c_{A1}/c_{Ao}.

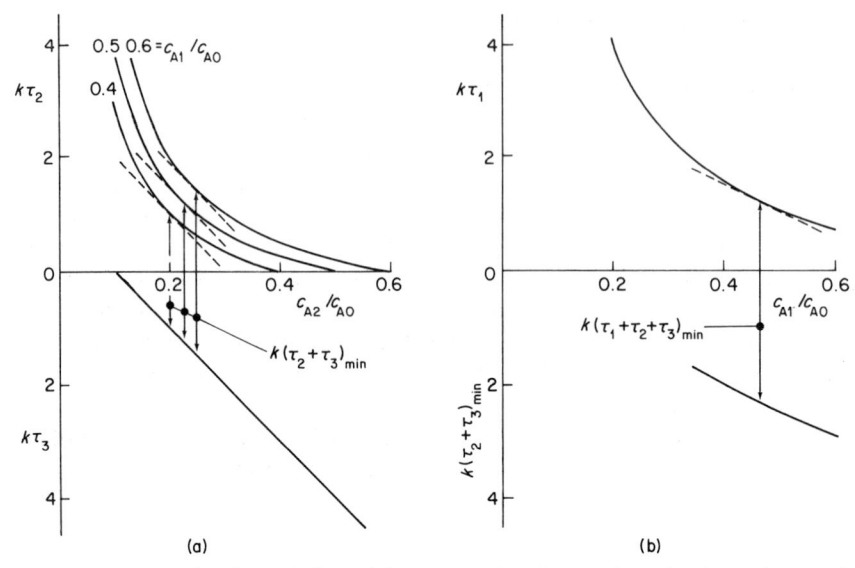

Fig. X.28 Determination of the minimum total volume of an isothermal cascade; Example X.3.f

It appears from this construction that:

$$k(\tau_1 + \tau_2 + \tau_3)_{\min} = 3.45$$
$$k(\tau_2 + \tau_3)_{\min} = 2.30$$
$$k(\tau_3)_{\min} = 1.15.$$

Hence:

$$\tau_1 = \tau_2 = \tau_3 \quad \text{and} \quad \frac{c_{A1}}{c_{Ao}} = \frac{c_{A2}}{c_{A1}} = \frac{c_{A3}}{c_{A2}} = 0.465.$$

This result has been obtained more quickly by geometric programming, see Section X.3.1. The above method may be useful with other rate expressions or with empirical conversion rates.

Example X.3.g *The minimum reaction volume of a non-isothermal cascade*
An equilibrium reaction with first order kinetics is carried out in a cascade of two tank reactors. The reaction is A → P, and the following relations apply:

$$c_A + c_P = c_{Ao}, \quad R_P = k_P c_A - k_X c_P$$

$$k_P = k_{P\infty} e^{-E_P/RT}, \quad k_X = k_{X\infty} e^{-E_X/RT}, \quad \frac{E_X}{E_P} = 2.$$

Find the minimum total reaction volume and the corresponding reactor temperatures when $c_{A2}/c_{Ao} = 0.1$.

We introduce as a measure of temperature $\kappa = k_P/k_R$, where k_R is the value of $k_P = k_X$ at the reference temperature T_R (also see Section VI.6 and equation (X.9)). The material balances of the two reactors can now be written in the form

$$k_R\tau_1 = \frac{1 - c_{A1}/c_{Ao}}{(\kappa + \kappa^2)c_{A1}/c_{Ao} - \kappa^2} \tag{a}$$

and

$$k_R\tau_2 = \frac{c_{A1}/c_{Ao} - c_{A2}/c_{Ao}}{(\kappa + \kappa^2)c_{A2}/c_{Ao} - \kappa^2}. \tag{b}$$

Since c_{A2}/c_{Ao} has been fixed at the value 0.1, $k_R\tau_2$ is, in principle, still a function of the two variables c_{A1}/c_{Ao} and κ. Now, one of the variables can be eliminated in this case because at each value of c_A/c_{Ao} there is an optimum temperature (corresponding to κ_{opt}) at which the conversion rate is at a maximum. For κ_{opt} we have here, from equation (X.10),

$$\kappa_{opt} = \frac{c_A/c_{Ao}}{2(1 - c_A/c_{Ao})}. \tag{c}$$

With this equation we find for the second tank, where $c_{A2}/c_{Ao} = 0.1$,

$$\kappa_{opt} = \frac{0.1}{2 \times 0.9} = \frac{1}{18},$$

and from equation (b),

$$(k_R\tau_2)_{opt} = 360\frac{c_{A1}}{c_{Ao}} - 36.$$

This relation is given by the straight line in figure X.29. The combination of equations (a) and (c) gives, for the first tank reactor,

$$(k_R\tau_1)_{opt} = \frac{4(1 - c_{A1}/c_{Ao})^2}{(c_{A1}/c_{Ao})^2}$$

which is the curve in figure X.29. In this graph, the sum of $(k_R\tau_1)_{opt}$ and $(k_R\tau_2)_{opt}$ has a minimum at $c_{A1}/c_{Ao} = 0.255$, from which the following conditions follow:

$$k_R\tau_1 = 34.4, \quad k_R\tau_2 = 55.9, \quad k_R(\tau_1 + \tau_2)_{min} = 90.2$$

$$\kappa_1 = 0.171, \quad \kappa_2 = 0.0556.$$

It is seen that the first reactor has a smaller volume than the second. It can now be asked whether the capacity will be much decreased if two tanks of equal volume are selected. It can be read from the graph (case b) that then $c_{A1}/c_{Ao} = 0.228$, with

$$k_R\tau_1 = 46.0, \quad k_R\tau_2 = 46.0, \quad k_R(\tau_1 + \tau_2) = 92.0,$$

$$\kappa_1 = 0.148, \quad \kappa_2 = 0.0556.$$

Only a very small increase in the total volume is found; the first tank is larger than in the optimal case and the degree of conversion is higher at a lower temperature.

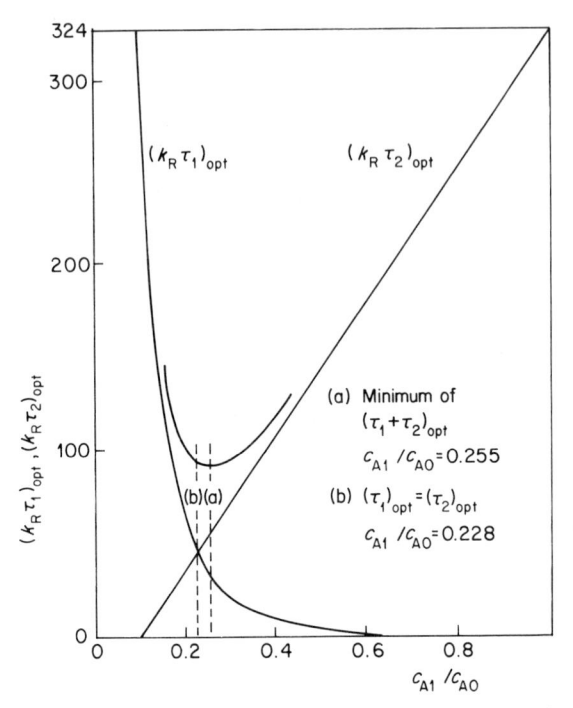

Fig. X.29 Finding the minimum volume of a cascade of two tank reactors for an equilibrium reaction; Example X.3.g

Here, too, it appears that the distribution of reactor volumes is not a very effective variable for optimization of the capacity. In this example, $k_R(\tau_1 + \tau_2) \leq 100$ (i.e., at the most 10% more than the minimum value) when τ_1/τ_2 varies from 1.7 to 0.2; when $\tau_1/\tau_2 = 0$ or ∞, we have only one tank reactor with $(k_R\tau)_{opt} = 324$.

In case this calculation has to be made for more than two tank reactors in a cascade, the beginning will be essentially the same, but the above conditions for the system of two tanks will have to be obtained as a function of feed composition. The same procedure can then be carried out for the third tank from the outlet, and so on.

Example X.3.h Cost minimization in a cascade
In a cascade of three tank reactors an equilibrium reaction is carried out of the type $A \rightleftharpoons P$ with no byproducts. The reactant A cannot be recovered from the reactor product. The reactions are first order with rate constants:

$$k_A = 3.58 \times 10^9 \exp(-12\,500/T)\,\text{s}^{-1}$$

$$k_{-A} = 3.19 \times 10^{16} \exp(-25\,000/T)\,\text{s}^{-1}.$$

The concentration of A in the reactor feed is $c_{Ao} = 1.0\,\text{kmol/m}^3$. We are required to choose the residence time τ_n and the temperature T_n in each reactor in such a way that the sum of the costs

- of the losses caused by non-converted reactant in the reactor product
- related to the investment in the reactors (depreciation, insurance, maintenance, etc.)

are minimized. The objective function will then become

$$f(\bar{x}) = C_A \Phi_v c_{A3} + C_v \Phi_v \sum_{n=1}^{3} V_n/\Phi_v,$$

in which C_A are the raw material costs per unit A [MU/kmol A], C_v the reactor costs per unit of reactor volume [MU/m³] and $f(\bar{x})$ are the total reaction costs per unit time [MU/time]. This objective function is an oversimplified cost function, which will be hardly ever encountered in practice, but will serve the purpose of this Example. After dividing by $\Phi_v C_A$ the objective function becomes

$$F = c_{A3} + \alpha \sum_{n=1}^{3} \tau_n. \tag{a}$$

Here α is the ratio between the fixed reactor costs and the raw material costs. For very expensive reactors and cheap raw materials α is high and α is low for very expensive A and cheap reactors. In this case the optimization variables are c_1, T_1, c_2, T_2, c_3 and T_3, so we have a 3 × 2-dimensional optimization problem. By the technique of dynamic programming this now can be reduced to four problems of the dimensions 2 and 1. The recursion equations follow from the mass balance:

$$c_{A,n-1} - c_{A,n} = \tau_n \{ k_{A,n} c_{A,n} - k_{-A,n}(1 - c_{A,n}) \} \tag{b}$$

The optimization scheme is now as follows:

$$F^* = \min_{\substack{c_{A1}, c_{A2}, c_{A3} \\ T_1, T_2, T_3}} \{ c_{A3} + \alpha(\tau_3 + \tau_2 + \tau_1) \}$$

$$= \min_{\substack{c_{A1}, c_{A2}, c_{A3} \\ T_2, T_3}} \{ c_{A3} + \alpha[\tau_3 + \tau_2 + \min_{T_1} \tau_1(c_{A1}, T_1)] \}$$

$$= \min_{\substack{c_{A1}, c_{A2}, c_{A3} \\ T_2, T_3}} \{ c_{A3} + \alpha[\tau_3 + \tau_2 + f_1^*(c_{A1})] \}$$

$$= \min_{\substack{c_{A2}, c_{A3}, T_3}} \{ c_{A3} + \alpha[\tau_3 + \min_{c_{A1}, T_2} (\tau_2(c_{A2}, c_{A1}, T_2) + f_1^*(c_{A1}))] \}$$

$$= \min_{\substack{c_{A2}, c_{A3}, T_3}} \{ c_{A3} + \alpha[\tau_3 + f_2^*(c_{A2})] \}$$

$$\vdots$$

$$= \min_{c_{A3}} \{ c_{A3} + \alpha f_3^*(c_{A3}) \}.$$

From the mass balance it follows that the condition for the optimum temperature $\partial \tau_n / \partial T_n = 0$ gives, for the optimum temperature in the nth tank,

$$T_n^* = \frac{E_{-A} - E_A}{R \ln\left[\dfrac{k_{-A\infty}}{k_{A\infty}} \dfrac{E_{-A}}{E_A} \dfrac{1 - c_{An}}{c_{An}}\right]}. \tag{c}$$

It is not possible to obtain an analytical expression for $c_{A,n-1}^*$ in a similar way. Therefore a numerical method has to be used to compute f_n^* for $n = 2, 3$. The computational scheme now could be as follows:

1. Choose a value of the integer p, $1/p$ being the step size.
2. Compute $f_1^*(c_{A1})$, using equations (b) and (c) for all values

$$c_{A1} = \frac{1}{p}, \frac{2}{p}, \ldots, \frac{p-1}{p}.$$

3. Now compute $f_2^*(c_{A2})$ for

$$c_{A2} = \frac{1}{p}, \frac{2}{p}, \ldots, \frac{p-2}{p}$$

as follows:
  3.1 Compute T_2^* using equation (c).
  3.2 Compute $(f_2)_j = \tau_2(c_{A2}, (c_{A1})_j, T_2^*) + f_1^*(c_{A1})_j$
  where $(c_{A1})_j = c_{A2} + \dfrac{j}{p}$
  for $j = 1, 2, \ldots$
  until $(f_2)_{j+1} > (f_2)_j$
  then $f_2^* = (f_2)_j$.

4. Compute $f_3^*(c_{A3})$ for $c_{A3} = \dfrac{1}{p}, \dfrac{2}{p}, \ldots, \dfrac{p-3}{p}$.

5. Compute $c_{A3} + \alpha f_3^*$ for $c_{A3} = \dfrac{1}{p}, \dfrac{2}{p}, \ldots, \dfrac{p-3}{p}$.

6. Find F^* and c_{A3}^* by inspection of the results obtained at step 5, find c_{A2}^* from the results obtained at step 4, and so on.

This computational scheme is based on the constraint $c_{A3} < c_{A2} < c_{A1} < c_{A0} = 1.0$, the unimodality of the objective function and on c_{An} being both a variable of state of f_n and an optimization or decision variable of f_{n+1}. A drawback of dynamic programming can be seen from the computational scheme. The accuracy of the results depends on the value of p, as can be found by varying p; for meaningful results p should be at least 1000. So 1000 values of $f_1^*(c_{A1})$ have to be stored in the computer memory and each individual value has to be used in step 3, where for $1000 - 1 = 999$ values of c_{A2} the value of c_{A1} is increased step by step till the optimum value of $f_2^*(c_{A2})$ is found. So a tremendous amount of information must be kept in memory. Several approaches have been suggested for getting round this, such as curve fitting the data to interpolation formulae. Other schemes are discussed by Aris [63].

Results obtained with a value of $p = 1000$ are given in Table X.3. If now the volumes of the three reactors are taken as equal, the number of degrees of freedom of the optimization problem is reduced to four: τ, T_1, T_2, T_3. Because of Equation (c) now a one-dimensional numerical optimization method can be used to find the best value of τ. Results are also given in Table X.3; again the difference in total residence time between optimally sized and equally sized reactors appears to be very small unless the raw material is relatively very expensive, as shown in Table X.4.

Total costs are only a small fraction higher for the cascade of equally sized tanks, whereas the average temperature level is somewhat lower, which will influence the heat withdrawal and feed preheating costs. The last costs, however, were not taken into account in this Example.

Table X.3 Optimal cascade of three tanks for an equilibrium reaction

α	τ_1	T_1	c_{A1}	τ_2	T_2	c_{A2}	τ_3	T_3	c_{A3}
Optimally sized tanks									
0.001	0.989	658	0.091	1.725	607	0.020	2.612	577	0.007
0.01	0.304	679	0.153	0.505	633	0.045	0.701	606	0.019
0.1	0.078	705	0.263	0.127	664	0.105	0.174	639	0.053
1	0.017	737	0.431	0.028	698	0.232	0.038	676	0.142
10	0.003	778	0.651	0.005	742	0.462	0.006	721	0.346
Equally sized tanks									
0.001	1.865	647	0.068	1.865	601	0.016	1.865	578	0.007
0.01	0.505	670	0.123	0.505	629	0.040	0.505	607	0.020
0.1	0.127	696	0.218	0.127	659	0.094	0.127	639	0.054
1.0	0.028	726	0.373	0.028	694	0.212	0.028	677	0.144
10	0.005	766	0.594	0.005	738	0.437	0.005	722	0.349

Table X.4 Optimal final values for the cascade of Example X.3.h

	Optimally sized tanks			Equally sized tanks		
α	c_{A3}^*	$\Sigma\tau^*$	F^*	c_{A3}^*	$\Sigma\tau^*$	F^*
0.001	0.007	5.326	0.0123	0.007	5.595	0.0126
0.01	0.019	1.509	0.0341	0.020	1.515	0.0348
0.1	0.053	0.379	0.0909	0.054	0.381	0.0923
1	0.142	0.084	0.2254	0.144	0.084	0.2278
10	0.346	0.014	0.4846	0.349	0.014	0.4879

X.3.5 Pontryagin's maximum principle

A general method of optimization of processes is the mathematical theory developed by Pontryagin [68]. It takes into consideration restrictions and the possibility that the optimum lies at the boundary of the allowed region of the parameters. If the process is defined by n parameters, the state of the process is defined by a point in an n-dimensional space. The course of the process is specified by the differential equation:

$$\frac{dx_i}{d\tau} = g_i(\bar{x}, \bar{u}), \qquad i = 1, 2, \ldots, n \qquad (X.59)$$

$$\bar{x}(\tau_0) = x_{i,o} \qquad \tau_0 \leq \tau \leq \tau_e, \qquad (X.60)$$

where the control vector $\bar{u}$ belongs to a class of admissible controls satisfying some definite equations. Among all admissible controls one, $u(\tau)$, must be found which ensures a maximum value of the chosen criterion of optimization — say $f(\bar{x}(\tau_e))$ — and satisfies the boundary conditions. According to the maximum principle the control vector should be chosen so that the Hamiltonian

$$H = \sum_{i=1}^{n} \lambda_i g_i(\bar{x}, \bar{u}) \qquad (X.61)$$

reaches its maximum value at any $\tau_0 \leq \tau \leq \tau_e$, where the vector $\bar{\lambda}(\lambda_1, \lambda_2, \ldots, \lambda_n)$ is defined by a set of differential equations

$$\frac{d\lambda_i}{d\tau} = -\frac{\partial H}{\partial x_i}, \qquad i = 1, 2, \ldots, n \qquad (X.62)$$

with the boundary conditions

$$\bar{\lambda}_i(\tau_e) = \frac{\partial f}{\partial x_i}(\bar{x}(\tau_e)). \qquad (X.63)$$

The theory of Pontryagin's maximum principle was first applied to chemical reactors by Boreskov and Slinko [69] and further elucidated by Aris [70]. Coward and Jackson [71] explored the difficulties in its application. The method has been widely used in the last decades; for example, Jackson [72] applied it to the mixing of different catalysts and the operation of reactors with decaying catalysts. We will explain the application of Pontryagin's maximum principle to first order consecutive reactions of the type A → P → X in a tubular reactor. At any place in the reactor the concentration of P depends on the temperature profile in the preceding part of the reactor. It is required to find the temperature profile $T(\tau)$ that maximizes c_P at the reactor outlet after the residence time τ_L. The point that distinguishes this problem from those discussed before is that the control or decision variable — the temperature — is now a function of another variable, residence time τ. We now have to maximize the function

$$f(\bar{x}) = \sum_{i=1}^{n} c_i x_i(\tau_L) = c_P(\tau_L)/c_{Ao}.$$

The process is governed by the equations

$$\frac{dc_P/c_{Ao}}{d\tau} = R_2(c_A/c_{Ao}, c_P/c_{Ao}, T) = k_P \frac{c_A}{c_{Ao}} - k_X \frac{c_P}{c_{Ao}}$$

and

$$\frac{dc_A/c_{Ao}}{d\tau} = R_1(c_A/c_{Ao}, c_P/c_{Ao}, T) = -k_P \frac{c_A}{c_{Ao}}.$$

(cf. X.59)

If $f(\bar{x})$ is to be maximized, then $T(\tau)$ should be chosen so that for each τ the Hamiltonian is maximized:

$$H = \sum_{i=0}^{n} \lambda_i g_i(\bar{x}, \bar{u}) = \lambda_1 R_1 + \lambda_2 R_2. \qquad \text{(cf. X.61)}$$

The components of the vector $\bar{x}$ are c_P/c_{Ao} and c_A/c_{Ao}. The differential equations governing the adjoint variables are:

$$\frac{d\lambda_1}{d\tau} = -\sum_{j=1}^{n} \frac{\partial R_j}{\partial x_1} \lambda_j = -\lambda_1 \frac{\partial R_1}{\partial x_1} - \lambda_2 \frac{\partial R_2}{\partial x_1} = k_P \lambda_1 - k_P \lambda_2$$

$$\frac{d\lambda_2}{d\tau} = -\lambda_1 \frac{\partial R_1}{\partial x_2} - \lambda_2 \frac{\partial R_2}{\partial x_2} = k_X \lambda_2,$$

(cf. X.62)

and the boundary conditions are:

$$\bar{\lambda}(\tau_L): \lambda_1(\tau_L) = 0 \quad \text{and} \quad \lambda_2(\tau_L) = 1 \qquad \text{(cf. X.63)}$$

$$\bar{x}(\tau_0): c_A/c_{Ao} = 1 \quad \text{and} \quad c_P/c_{Ao} = 0. \qquad \text{(cf. X.60)}$$

Furthermore it is required to keep the reactor temperature within the limits $T_* < T < T^*$. We see that now $\bar{\lambda}(\tau)$ is a vector of *place dependent* Lagrange multipliers.

A reasonably efficient procedure to solve these problems is given by the following algorithm:

1. Choose an initial temperature profile $T(\tau)$, $0 \le \tau \le \tau_L$.
2. Integrate 1. using the chosen $T(\tau)$ to find $c(\tau)/c_{Ao}$.
3. Integrate equation (X.62) backwards from τ_L to $\tau = 0$ using $T(\tau)$ and $c(\tau)$ and starting from boundary condition equation (X.63).
4. Compute a new temperature profile with

$$T(\tau)_{\text{new}} = T(\tau) + \varepsilon \frac{\partial H}{\partial T}.$$

The scalar ε is chosen in such a way that the objective function $f(\bar{c}(\tau_e))$ computed with $T(\tau)_{\text{new}}$ is maximal. To this end a secondary algorithm, not explained here, is used to find the optimal value of ε. If a temperature limit T_* or T^* is reached, this condition is imposed on $T(\tau)_{\text{new}}$.

5. Repeat steps (2)–(4) until convergence is reached.

Two elementary examples of Pontryagin's maximum principle are given below for optimal temperature programmes in batch reactors.

Example X.3.i *Optimum temperature profiles in several types of model reactors*
In continuous reactors consecutive reactions of the type $A \to P \to X$ take place. The first reaction is second order, the second is a first order reaction. Reactor temperatures are limited between 298 and 423 K; we will compare reactors with a total residence time of 3600 s and our aim is to maximize the yield of P.

Mathematically our problem can now be formulated for a PFR as:

$$\text{Minimize } f(\bar{x}) = -c_P/c_{A_0}$$
$$\text{for } dc_A/d\tau = -k_P c_A^2$$
$$dc_P/d\tau = k_P c_A^2 - k_X c_P$$

and $298 < T < 423$ K.

Further, we have

$$k_P = 0.833 \exp\left(-\frac{2516}{T}\right) \text{ m}^3/\text{kmol s}$$

$$k_X = 129 \exp\left(-\frac{5023}{T}\right) \text{ s}^{-1}.$$

From equation (X.61) it now follows that the Hamiltonian is given by

$$H = (\lambda_X - \lambda_P) k_P c_A^2 - \lambda_X k_X c_P,$$

and from equation (X.62) that the multipliers are given by

$$\frac{d\lambda_P}{d\tau} = 2(\lambda_P - \lambda_X) k_P c_A \qquad \lambda_P(\tau_e) = 0$$

$$\frac{d\lambda_X}{d\tau} = \lambda_X k_X \qquad \lambda_X(\tau_e) = 1;$$

the initial conditions are $c_A/c_{A_0} = 1$ and $c_P/c_{A_0} = 0$. The optimum temperature profile was determined with the algorithm described above; convergence was considered to be obtained if the improvement in $c_P/c_{A_0}(\tau_e)$ was less than 0.00003. To get some feel for the reproducibility two different initial temperature profiles were used, the isotherms at the upper and the lower boundaries. Table X.5 gives the results for the first, second and final iteration. Some conclusions can be drawn:

— the difference between the two optimum profiles is negligible.
— The isothermic reaction at the lowest possible temperature results in a very high selectivity at a low conversion; the reaction at the highest temperature gives a very poor selectivity at a high conversion. In both cases the yields are low. This is to be expected because $E_X > E_P$.

Table X.5 Results for the first, second and last iteration in an optimization study with Pontryagin's maximum principle (Example X.3.i)

Iteration	1			2			5		
Time, s	c_A/c_{Ao}	c_P/c_{Ao}	$T_{(K)}$	c_A/c_{Ao}	c_P/c_{Ao}	$T_{(K)}$	c_A/c_{Ao}	c_P/c_{Ao}	$T_{(K)}$
0	1.000	0.000	423	1.000	0.000	373	1.000	0.000	423
360	0.560	0.365	423	0.758	0.235	366	0.738	0.253	366
720	0.389	0.409	423	0.624	0.357	360	0.622	0.359	352
1080	0.298	0.375	423	0.538	0.429	355	0.547	0.425	348
1440	0.242	0.320	423	0.478	0.477	350	0.491	0.470	345
1800	0.203	0.266	423	0.434	0.509	346	0.448	0.503	343
2160	0.175	0.217	423	0.400	0.533	342	0.413	0.529	341
2520	0.154	0.176	423	0.374	0.551	337	0.384	0.549	340
2880	0.137	0.142	423	0.353	0.564	332	0.359	0.564	339
3240	0.124	0.115	423	0.336	0.575	326	0.338	0.576	338
3600	0.113	0.093	423	0.323	0.584	320	0.319	0.586	336
ζ_{AL}	0.887	—	—	0.677	—	—	0.681	—	—
η_{PL}	—	0.093	—	—	0.584	—	—	0.586	—
σ_{PL}	—	—	0.105	—	—	0.876	—	—	0.860

Iteration	1			2			8		
Time, s	c_A/c_{Ao}	c_P/c_{Ao}	$T_{(K)}$	c_A/c_{Ao}	c_P/c_{Ao}	$T_{(K)}$	c_A/c_{Ao}	c_P/c_{Ao}	$T_{(K)}$
0	1.000	0.000	298	1.000	0.000	351	1.000	0.000	423
360	0.939	0.061	298	0.813	0.184	351	0.743	0.249	364
720	0.885	0.115	298	0.686	0.305	350	0.625	0.357	353
1080	0.837	0.163	298	0.594	0.388	350	0.549	0.423	347
1440	0.794	0.205	298	0.524	0.447	350	0.493	0.469	345
1800	0.755	0.244	298	0.469	0.490	349	0.450	0.502	342
2160	0.720	0.279	298	0.424	0.521	349	0.415	0.528	341
2520	0.687	0.310	298	0.388	0.544	349	0.386	0.548	339
2880	0.658	0.339	298	0.357	0.561	349	0.361	0.564	338
3240	0.631	0.365	298	0.331	0.572	349	0.340	0.576	337
3600	0.606	0.389	298	0.309	0.581	348	0.321	0.586	336
ζ_{AL}	0.394	—	—	0.691	—	—	0.679	—	—
η_{PL}	—	0.389	—	—	0.581	—	—	0.586	—
σ_{PL}	—	—	0.987	—	—	0.841	—	—	0.863

— A falling temperature profile gives better results: at the start of the reaction c_P is still low, so the conversion of A is improved by high temperatures and the decomposition of P is still low. At higher c_P this decomposition is suppressed by lowering the temperature. The highest yield is obtained at a conversion level of 0.68 and a selectivity of 0.86.
— comparing the results obtained after two iterations with the final ones it is seen that ζ_A, η_P and σ_P are already very near to the optimum after two iterations; the temperature profiles, however, are quite different. This also leads to the statement that the costs of imposing the temperature profile

should be included in the objective function in order to get results of any practical value. The optimization methods based on Pontryagin's maximum principle also seem quite suitable to solve this economic optimization problem.

The results for a PFR will now be compared to those of a cascade of tank reactors. For the nth tank reactor the material balances are

$$c_{A,n-1} - c_{A,n} = k_P \tau_n c_{A,n}^2$$

$$c_{P,n-1} - c_{P,n} = -k_P \tau_n c_{A,n}^2 + k_X \tau_n c_{P,n}$$

or also

$$c_{A,n} = \frac{-1 + \sqrt{(1 + 4k_P \tau_n c_{A,n-1})}}{2k_P \tau_n}$$

$$c_{P,n} = \frac{c_{P,n-1} + k_P \tau_n c_{A,n}^2}{1 + k_X \tau_n}$$

The total residence time in the cascade is fixed at 3600 s. The number of decision variables for a cascade of N reactors is $2N - 1$: $N - 1$ residence times and N temperatures. We made three optimization studies:

— a cascade of 3 tanks
— a cascade of 3 equally sized tanks with $\tau_n = 1200$ s.
— a cascade of 10 equally sized tanks with $\tau_n = 360$ s.

To find the optimum conditions for this case another numerical optimization method was used. The results for the cascade of 3 tanks and for a cascade with equally sized tanks are not significantly different, as shown in Table X.6. A cascade of 10 tanks will approach the PFR more than that with 3 tanks only, as can be seen in Table X.7. The results have been summarized in Table X.8. We see

Table X.6   Optimum cascades of three tanks (Example X.3.i)

Cascade with optimal temperatures and residence times:

Reactor	1	2	3
Temperature (K)	352	347	341
Residence time (s)	1080	1120	1400
c_A/c_{Ao}	0.676	0.509	0.395
c_P/c_{Ao}	0.299	0.435	0.513

Cascade with equally sized vessels and optimum temperatures:

Reactor	1	2	3
Temperature (K)	350	345	343
Residence time (s)	1200	1200	1200
c_A/c_{Ao}	0.666	0.497	0.396
c_P/c_{Ao}	0.307	0.444	0.512

Table X.7 Conditions in a cascade of ten equally sized tanks (Example X.3.i)

Reactor	1	2	3	4	5	6	7	8	9	10
Temperature (K)	361	352	349	346	345	343	343	341	340	339
c_A/c_{Ao}	0.814	0.698	0.614	0.551	0.500	0.459	0.424	0.395	0.369	0.348
c_P/c_{Ao}	0.179	0.286	0.361	0.415	0.456	0.488	0.512	0.532	0.545	0.560

Table X.8 Summary of the results of Example X.3.i

Reactor type	Yield	Conversion	Selectivity
cascade, 3 tanks	0.512	0.604	0.848
cascade, 10 tanks	0.560	0.652	0.859
plug flow reactor	0.586	0.679	0.863

that in all optimal reactors the selectivity does not differ very much; the lower yields are mainly caused by lower conversions in a cascade. As it is easier to impose different temperatures on different tanks than a temperature profile on a tubular reactor, a cascade will probably be favoured, whereas economics will determine the number of tanks in it.

Example X.3.j *Parallel reactions in a batch reactor*
In a BR reactant A is converted into P, whereas also an undesired decomposition into X takes place. We will determine the optimum temperature programs for total reaction times of 3600 and 7200 s. Both reactions are first order and we want to maximize c_P/c_{Ao}. The material balances are

$$\frac{dc_A}{dt} = -(k_P + k_X)c_A$$

$$\frac{dc_P}{dt} = k_P c_A$$

with

$$k_P = 27.8 \times 10^3 \exp\left(-\frac{6000}{T}\right) \text{s}^{-1}$$

and

$$k_X = 0.556 \times 10^9 \exp\left(-\frac{10\,000}{T}\right) \text{s}^{-1}.$$

We wish to maximize $c_P(\tau_e)/c_{Ao}$ and find, for the Hamiltonian,

$$H = [(\lambda_X - \lambda_P)k_P - \lambda_P k_X]c_A$$

and, for the multipliers,

$$\frac{d\lambda_P}{dt} = (\lambda_P - \lambda_X)k_P + \lambda_P k_X \qquad \lambda_P(\tau_e) = 0$$

$$\frac{d\lambda_X}{dt} = 0 \qquad \lambda_X(\tau_e) = 1.$$

Moreover $298 < T < 398$ K.

The results are given in Table X.9.

We see that we have a rising temperature profile and a complete conversion. At the start of the reaction, the temperature is so low that the undesired side reaction is suppressed ($E_P < E_X$). Towards the end of the reaction the temperature has to be increased so that all A will be converted, although at a low selectivity. We see in the table that selectivity difference drops for the successive time intervals, but in the larger reactor with the lower average temperature the selectivity differences are on a higher level. For isothermal batch reactions with a total reaction time of 3600 s the results would be at 298 K: $\zeta_A = 0.172$ and $\sigma_P = 0.971$ and at 398 K: $\zeta_A = 1.000$ and $\sigma_P = 0.535$. We see that results are much improved by imposing a temperature program. (*Example X.3.j ended*)

In concluding this section it may be said that, thanks to the various mathematical techniques and the computing aids nowadays available, any optimization problem can be solved, provided the problem is realistic and properly stated. The difficulties of optimization lie mainly in providing the pertinent data and in an adequate construction of the objective function. This kind of information is quite specific for the process under consideration and for the plant environment. It cannot generally be obtained from *a priori* considerations. Carefully systematized data amassed from previous experience with similar processes and plants will have to guide improvements based on optimization.

Table X.9 Results of an optimization study applying Pontryagin's maximum principle (Example X.3.j)
(Here $\Delta\sigma_P = -\Delta(c_P/c_{Ao})/\Delta(c_A/c_{Ao})$)

Time (s)	c_A/c_{Ao}	c_P/c_{Ao}	$T_{(K)}$	$\Delta\sigma_P$	time (s)	c_A/c_{Ao}	c_P/c_{Ao}	$T_{(K)}$	$\Delta\sigma_P$
0	1.000	0.000	319	—	0	1.000	0.000	331	—
720	0.855	0.135	320	0.931	360	0.864	0.123	332	0.904
1440	0.718	0.261	323	0.927	720	0.735	0.238	333	0.891
2160	0.588	0.382	325	0.920	1080	0.608	0.350	336	0.882
2880	0.466	0.492	327	0.912	1440	0.488	0.455	338	0.875
3600	0.354	0.593	330	0.902	1800	0.378	0.550	341	0.864
4320	0.253	0.683	334	0.891	2160	0.277	0.636	345	0.851
5040	0.163	0.762	339	0.878	2520	0.184	0.713	350	0.828
5760	0.086	0.828	346	0.857	2880	0.103	0.778	357	0.802
6480	0.025	0.876	360	0.787	3240	0.034	0.828	371	0.725
7200	0.000	0.893	386	0.680	3600	0.001	0.848	397	0.606

REFERENCES

1. Ray, W. H. and Szekely, J.: *Process Optimization*, Wiley, New York (1973)
2. Hoffmann, U. and Hofmann, H.: *Einführung in die Optimierung*, Verlag Chemie, Weinheim (1971)
3. Aris, R.: *The Optimal Design of Chemical Reactors*, Academic Press, New York (1961)
4. Hinger, K. J. and Blenke, H.: *Chem. Ing. Techn.* **47**, 79 (1975)
5. Hinger, K. J. and Blenke, H.: *Chem. Ing. Techn.* **47**, 976 (1975)
6. Le Goff, P.: *Chem. Eng. Sci.* **35**, 2029 (1980)
7. Rosas, C. B. and Smith, G. B.: *Chem. Eng. Sci.* **35**, 330 (1980)
8. Denbigh, K. G.: *Trans. Far. Soc.* **40**, 352 (1944)
9. Horn, F.: Ph. D. Thesis, Techn. Univ. Vienna (1958)
10. Horn, F.: *Chem. Eng. Sci.* **14**, 77 (1961)
11. Horn, F. and Troltenier, U.: *Chem. Ing. Techn.* **32**, 382 (1960)
12. Horn, F. and Troltenier, U.: *Chem. Ing. Techn.* **33**, 413 (1961)
13. Calderbank, P. H.: *Chem. Eng. Progr.* **49**, 585 (1953)
14. Mars, P. and van Krevelen, D. W.: *Chem. Eng. Sci.* **3**, Spec. Suppl. 41 (1954)
15. van Heerden, C.: *Ind. Eng. Chem.* **45**, 1242 (1953)
16. Annable, D.: *Chem. Eng. Sci.* **1**, 145 (1952)
17. Westerterp, K. R. and Beek, W. J.: *De Ingenieur* **73**, Ch. 15 (1961)
18. Murase, A., Roberts, H. L., and Converse, A. O.: *Ind. Eng. Chem. Process Des. Dev.* **9**, 503 (1970)
19. Jackson, R.: *Chem. Eng. Sci.*, **21**, 241 (1966)
20. Westerterp, K. R.: *De Ingenieur* **73**, Ch. 69 (1961)
21. Horn, F.: *Z. Elektrochem.* **65**, 209 (1961)
22. Horn, F.: *Z. Elektrochem.* **65**, 295 (1961)
23. Horn, F. and Küchler, L.: *Chem. Ing. Techn.* **31**, 1 (1959)
24. Boreskov, G. K. and Slinko, M. G.: *Chem. Eng. Sci.* **14**, 259 (1961)
25. Aris, R.: *Chem. Eng. Sci.* **12**, 243 (1960)
26. Schoenemann, K.: *Génie Chim.* (Belge) **38**, 163 (1960)
27. Bartholomé, E. and Krabetz, R.: *Chem. Eng. Sci.* **65**, 224 (1961)
28. Bartholomé, E. and Krabetz, R.: *Chem. Eng. Sci. Suppl.*, 11 (1965)
29. Küchler, L.: *Chem. Eng. Sci.* **14**, 11 (1961)
30. Almsay, G. A., Hay, J. J., and Pallai, I. M.: *Brit. Chem. Eng.* **11** (3), 188 (1966)
31. Almasy, G. A., Jedlovski, P., and Pallai, I. M.: *Brit. Chem. Eng.* **12** (8), 1219 (1967)
32. Almasy, G. A., Veress, G., and Pallai, I. M.: *Chem. Eng. Sci.* **24**, 1387 (1969)
33. Gaines, L. D.: *Ind. Eng. Chem. Process Des. Dev.* **16**, 381 (1977)
34. Burghardt, A. and Patzek, T.: *Chem. Eng. Journal* **16**, 153 (1978)
35. Lee, K. Y. and Aris, R.: *Ind. Eng. Chem. Process Des. Dev.* **2**, 301 (1963)
36. Lee, K. Y. and Aris, R.: *Ind. Eng. Chem. Process Des. Dev.* **2** (4), 306 (1963)
37. Meiring, F.: Ph.D. thesis, Eindhoven Univ. Technology (1981)
38. Boreskov, G. K. and Slinko, M. G.: *Brit. Chem. Eng.* **10** (3) 170 (1965)
39. Aris, R.: *Can. J. Chem. Eng.* **39**, 87 (1962)
40. King, R. P.: *Chem. Eng. Sci.* **20**, 537 (1965)
41. Bilous, O. and Amundson, N. R.: *Chem. Eng. Sci.* **5**, 81, 115 (1956)
42. Katz, S.: *Ann. New York Ac. Sci.* **84**, 441 (1960)
43. Burghardt, A. and Skrzypek, J.: *Chem. Eng. Sci.* **29**, 1311 (1974)
44. Denbigh, K. G.: *Chem. Eng. Sci.* **8**, 125 (1958)
45. Aris, R.: *Chem. Eng. Sci.* **12**, 56 (1960)
46. Storey, C.: *Chem. Eng. Sci.* **17**, 45 (1962)
47. Rosenbrock, H. H.: *Computer J.* **3**(3), 175 (1960)
48. Duffin, R. J., Peterson, E. L., and Zener, C.: *Geometric Programming*, Wiley, New York (1967)
49. Horn, F.: *Chem. Eng. Sci.* **15**, 176 (1961)

50. Horn, F. and Klein, J.: *Adv. Chem. Ser.* **109,** 141, ACS, Washington (1972)
51. Wilde, D. J.: *Optimum Seeking Methods*, Prentice-Hall, Englewood Cliffs (1964)
52. Fontein, H. J. and Groot Wassink, J.: *Verfahrenstechn.* **8** (7), 200 (1974)
53. Collina, A., Ferrari, G. B., and Dente, M.: *Chimie et Ind., Génie chim.* **103,** 1751 (1970)
54. Capelli, A., Collina, A., and Dente, M.: *Ind. Eng. Chem. Proc. Des. Dev.* **11,** 184 (1972)
55. Kays, W. M. and London, A. L.: *Compact Heat Exchangers*, McGraw-Hill, New York (1964)
56. Fontein, H. J. and Groot Wassink, J.: *Eng. Process Econ.* **3,** 141 (1978)
57. Bellman, R.: *Dynamic Programming*, Oxford Univ. Press, London (1957)
58. Aris, R.: *The Optimal Design of Chemical Reactors*, Acad. Press, New York (1961)
59. Aris, R.: *Z. Elektrochem.* **65,** 229 (1961)
60. Grütter, W. F. and Messikommer, B. H.: *Helv. Chim. Acta* **44,** 285 (1961)
61. Grütter, W. F. and Messikommer, B. H.: *Helv. Chim. Acta* **43,** 2182 (1960)
62. Aris, R.: *Chem. Eng. Sci.* **13,** 18 (1960)
63. Aris, R.: *Discrete Dynamic Programming*, Blaisdell (1964)
64. Holland, F. A., Watson, F. A., and Wilkinson, J. K.: *Introduction to Process Economics*, Wiley, New York (1974)
65. Peters, M. S. and Timmerhaus, K. D.: *Plant Design and Economics for Chemical Engineers*, McGraw-Hill, New York (1980)
66. French, D. J.: *J. Am. Chem. Soc.* **72,** 1393 (1950)
67. Eklund, R. B.: *The Rate of Oxidation of* SO_2 *with a Commercial Vanadium Catalyst*, Almquist and Wiksell, Stockholm (1956)
68. Pontryagin, L. S., Boltyanskii, V. G., Gamkrelidze, R. V., and Mischenko, E. F.: *The Mathematical Theory of Optimal Processes*, Interscience, New York (1962)
69. Boreskov, G. K. and Slinko, M. G.: *Chem. Eng. Sci. Supplement*, 257 (1965)
70. Aris, R.: *Chem. Eng. Sci. Supplement* 265 (1965)
71. Coward, I. and Jackson, R.: *Chem. Eng. Sci.* **20,** 911 (1965)
72. Jackson, R.: *Ber. Bunsen Gesellschaft* **74**(2), 98 (1970)
73. Urbanek, A. and Trela, M.: *Catal. Rev. Sci. Eng.* **21**(1), 73 (1980)

Author Index

Aarsen, F. G., van den, 450
Abbi, Y. P., 225
Acetis, J., de, 453
Ackerman, G. H., 472
Adderley, C., 650
Adler, R. J., 198
Agnew, J. B., 650, 651
Aksel'rod, Yu. V., 519, 522
Alberda, G., 175
Allan, J. C., 578
Alper, E., 601, 664
Almasy, G., 288, 698
Ameno, T., 412, 527
Amundson, R. R., 293, 294, 339, 344, 443, 596, 640, 649, 657, 659, 707, 722
Anderson, J. B., 620
Annable, D., 691
Aoyagi, M., 650
Arai, F., 482
Arens, H., 653
Aris, R., 126, 153, 175, 191, 210, 221, 339, 383, 649, 675, 696, 699, 700, 709, 710, 730, 731, 738
Arri, L. E., 640
Asai, S., 410, 411, 416, 540, 558, 559, 561, 566
Asbjornsen, O. A., 246
Ashmore, P. G., 16
Astarita, G., 2, 521, 532, 533, 577
Austin, L. G., 430

Baans, C. M. E., 521
Baddour, R. F., 326
Baiker, A., 653
Bailey, J. E., 516
Baird, M. H. I., 216
Baker, C. G. J., 216
Balaceanu, J. C., 114
Balakrishnan, A. R., 626
Balasubramanian, S. N., 538

Baldi, G., 399, 487, 664
Baldyga, J., 256, 257
Barish, E. Z., 181
Barkelew, C. H., 210, 211, 295, 296, 298, 299
Baron, T., 210, 211
Barona, N., 596
Bartholomé, E., 697
Batchelor, G. K. J., 229, 251, 254
Bauer, R., 628
Beaverstock, M. C., 558, 559
Beckum, F. P. H., van, 531, 536
Beek, J., 247
Beek, W. J., 1, 4, 361, 362, 369, 371, 474, 479, 480, 489, 574, 691, 731
Beenackers, A. A. C. M., 454, 455, 476, 477, 487, 488, 514, 526–529, 531, 536, 540, 561, 563, 578
Belevi, H., 256
Bellman, R., 730
Bennett, C. O., 468
Benson, S. W., 16
Berg, H., van den, 557
Berg, P. J., van den, 636
Bergougnou, M. A., 216
Berty, J. M., 626
Beskov, W. S., 597, 617
Beutler, J. A., 419
Beveridge, G. S. G., 424
Bik, J. D., 636
Bilous, O., 293, 294, 344, 707, 722
Bird, R. B., 1, 3, 4, 361, 362, 400, 459
Bischoff, K. B., 126, 150, 198, 207, 213, 214, 391–393, 435
Blackmore, K., 511, 548
Blaum, E., 657
Blauwhoff, P. J. M., 668
Blenke, H., 679–684
Blok, J. R., 488
Boegborn, J. F., 146, 147, 148
Boelhouwer, C., 128

Bohle, W., 196, 483
Boltyanskii, V. G., 738
Bond, W. N., 479
Booth, F. B., 539
Boreskov, G. K., 659, 697–699, 701, 738
Bosch, B., van den, 649
Bosscher, J. K., 17
Boudart, M., 16
Bourne, J. R., 250, 254, 256, 257
Boyle, J., 473
Brian, P. L. T., 326, 402, 476, 558, 559
Bricker, J. H., 626
Bridgwater, J., 546, 547, 552–554
Brink, J. C., 479
Britton, M. J., 217
Brodkey, R. S., 256
Broeder, J. J., 209
Brothman, A., 181
Brounshtein, B. I., 412
Bruckhan, R., 216
Brunson, R. J., 399, 412
Bulicka, H., 466
Burg, C., 153
Burghardt, A., 699, 701, 704, 705, 708
Bush, S. F., 355
Butler, C., 335
Butt, J. B., 548, 657
Butt, M. A., 423, 429, 540, 541

Caban, R. A., 635
Calderbank, P. H., 255, 472, 487, 609, 691
Capelli, A., 727
Carberry, J. J., 139, 443, 468, 548, 621, 630
Carr, N. L., 597
Casanova, R., 653
Chan, M., 347
Chandalia, S. B., 539
Chao, R. E., 635
Charkova, T. V., 597
Charpentier, J. C., 192, 198, 216, 217, 461, 473, 475, 488, 489, 577, 596
Chaudhari, R. V., 519, 537, 540, 541, 664
Chen, B. H., 217
Chen, G. T., 338
Chen, M. S. K., 246
Chermin, H. A. G., 10, 114
Chiang, S. H., 577
Chiba, T., 483
Chin, M., 466
Cholette, A., 198
Christoffel, E. G., 469–471
Chuchvalec, H., 544, 545
Chung, S. F., 198, 215
Clark, S. W., 626

Clegg, G. T., 454, 455
Clements, W. C., 220, 223
Clough, C. E., 640
Cloutier, L., 198
Collina, A., 727
Converse, A. O., 692
Cornelisse, R., 526–529, 531, 536, 540
Corrsin, S., 229, 248, 251, 254–256
Cortez, D. H., 618
Coughanour, D. R., 521
Coursey, W. J., De, 411
Coussemant, F., 114
Coward, I., 738
Crandall, S., 521
Cresswell, D. L., 457, 629, 657
Crider, J. E., 636
Crombeen, P. R. J. J., 598
Croockewit, P., 211
Cutlip, M. B., 468
Curl, R. L., 248, 250

Damköhler, G., 2, 628
Dammers, W. R., 347, 348, 350
Danckwerts, P. V., 2, 17, 161, 200, 247, 370, 372, 375, 419, 454, 461, 463, 466, 473–476, 487, 521, 577
Darton, R. C., 484
Dautzenberg, F. M., 503
David, R., 247–249, 255, 258
Davidson, J. F., 482, 484, 650
Davis, H. S., 521
Dean, R. D., 626
Deans, H. A., 626, 650
Deckwer, W. D., 216, 593, 664, 665
Deemter, J. J., van, 173, 195, 196, 209, 215
Delancey, G. B., 578
Demaria, F., 335
Denbigh, K. G., 1, 86, 131, 133, 424, 690, 708–710
Dente, M., 727
Diboun, M., 216
Dierendonck, L. L., van, 472, 473, 489, 609
Dilman, V. V., 521
Ding, J. S. Y., 597
Dingman, J. C., 521
Dixon, A. G., 629
Doesburg, H., van, 624, 625, 654–656
Domask, W. G., 539
Doraiswamy, L. K., 519, 537, 538, 540, 541, 629, 650
Douglas, J. M., 288
Drinkenburg, A. A. H., 479, 480, 483, 488
Dubil, H., 349
Dudukovic, M. P., 644

Duffin, R. J., 713
Dun, C. L., 421
Dun, P. W., 17
Dunn, K. M., 640

Eagleton, L. C., 288
Eberhardt, E., 539
Effron, E., 217
Eguchi, W., 239
Eigenberger, G., 355, 626, 649, 653, 657
Eklund, R. B., 697
Eldridge, J. M., 57
Elnashaie, S. S. E., 657
Emery, J. P., 326
Emig, G., 651
Endo, I., 649, 650, 657
Ergun, S., 630
Eschard, F., 114
Esenwein, H., 488
Evans, J. W., 436, 640, 642

Fakien, R. W., 627
Fan, L. T., 175, 198, 207–211, 213, 246
Farina, I. H., 402, 423
Ferrari, G. B., 727
Fiand, U., 651
Finlayson, B. A., 630, 632
Fishbein, G. A., 412
Foley, D., 657
Fontaine, R. W., 175
Fontein, H. J., 674, 726, 727, 729
Fortuin, J. M. H., 347, 348, 350, 472, 489, 609
Foss, A. S., 626, 636
Frank-Kamenetski, D. A., 2, 438
Frederikson, A. G., 153
Freitas, E. R., 421
French, D. J., 678
Friend, L., 539
Froment, G. F., 126, 297, 299, 471, 624, 630, 633
Frost, A. A., 16
Fujine, M., 413, 415, 416, 418, 419, 511, 517, 519, 526, 527, 533, 560, 566
Fujita, S., 561
Fukuda, T., 216
Fukushima, S., 667, 668
Fulton, J. W., 489
Furasawa, T., 649, 650, 657
Furusaki, S., 482

Games, L. D., 699
Galloway, T. R., 481
Gamkredidze, R. V., 738

Gardiner, W. C., 16
Garnell, F. H., 520
Garner, F. H., 479
Garza-Garza, D., 644
Gaube, J., 349
Geelen, H., 421, 583
Gehlawat, J. K., 521
Gerrens, H., 239
Gestrich, W., 488
Gianetto, A., 487, 664
Gibilaro, L. G., 190, 191, 220, 221
Gilham, A. J., 461, 463, 466
Gilles, E. D., 346, 649, 657
Gillespie, B. M., 139
Gillespie, G. C., 617
Gioia, F., 520, 521, 532
Giraud, A., 114
Gnielinski, V., 629
Goar, B. G., 538
Godbole, S. P., 593
Godfrey, J. H., 461, 463
Goettler, L. A., 519, 521, 526–529, 531
Goff, P., Le, 680
Goldstein, R. F., 520, 538, 539
Golkeri, S. V., 516
Gomaa, H. G., 473
Goodenbour, J. W., 421
Goto, S., 664
Greassly, W. W., 239
Gregory, G. A., 487
Groeneveld, M. J., 642
Groot, J. H., de, 171, 177, 195, 215
Groot Wassink, J., 726, 727, 729
Grütter, W. F., 731
Gupta, R. K., 584
Gunn, D. J., 225, 521

Haderi, A. E., 339
Hahn, A. V., 539
Hammer, H., 664
Hammerton, D., 479
Hancock, M. D., 597
Hanhart, J., 171, 177, 217, 596
Hanson, C., 489
Harada, M., 239
Harriot, P., 175, 561, 571
Harrison, D., 482, 484
Hartland, S., 574
Hartog, H. J., den, 361, 371
Hashimoto, K., 561, 563, 566, 609
Hass, H. B., 121
Hasseltine, E. A., 402
Hatch, L. F., 539
Hatch, T. F., 537, 538

Hay, J. J., 698
Hedden, K., 441, 442
Heemskerk, A. H., 347, 348, 350
Heerden, C., van, 311, 323, 325, 335, 691
Hegner, B., 484, 576
Heinemann, R. F., 650
Hell, M., 557, 558
Hellin, M., 114
Helmrich, H., 640
Henderson, W. T., 421
Hendriksz, R. H., 407, 461
Hennecke, F. W., 629
Hicks, J. S., 455–457, 649
Higbie, R., 369, 370
Hikita, H., 410, 411, 416, 461, 463, 540, 558, 559, 561, 566
Himukashi, Y., 559
Hines, J. W., 217
Hinger, K. J., 679–684
Hinze, J. O., 216
Hlavacek, V., 457, 458, 628, 630, 632, 636, 652, 653
Hochman, J. M., 217
Hoelscher, H. E., 470
Hoffman, L. A., 596–598
Hoffmann, U., 651, 675
Hofmann, H., 346, 555–557, 623, 635, 651–653, 664, 675
Hoftijzer, P. J., 335, 401, 403, 407, 421, 561, 566, 601
Hoiberg, J. A., 626
Holland, F. A., 678
Honig, C. C., 211
Hoogendoorn, C. J., 192
Horak, J., 499, 544, 545
Horn, F., 691, 693–696, 706, 707, 709, 716, 721, 725, 727
Horn, F. J. M., 516
Hortavsu, O., 489
Hougen, O. A., 1, 4, 7–9, 11–13, 16, 263, 425
Hovorka, R. B., 198
Huang, C. J., 567
Huang, D. T. J., 592, 597
Hudgins, R. R., 620
Hughes, R., 630, 640, 647
Hugo, P., 67, 273–277, 347
Huibers, D. T. A., 128
Humphrey, A. E., 150, 153
Hurley, J. F., 402
Hutchinson, P., 126

Ihada, Y., 482
Imafaku, K., 216
Inoue, H., 548, 557, 561

Irizarry, M. M., 635
Ishikawa, H., 461, 463, 540

Jackson, R., 692, 738
Jahnel, W., 626
Jedlovski, P., 698
Jenney, T. M., 57, 58
Jhaveri, A. S., 461, 463, 487
Jirack, F., 499
Jockers, K., 539
Jones, B. R. E., 650
Jones, R. W., 58
Jong, W. A., de, 624, 625, 654–656
Joosten, G. E. H., 473
Jordan, D. G., 539
Joshi, J. B., 593, 596
Jungers, J. C., 114, 430
Juvekar, V. A., 411, 519, 537, 540, 541, 580, 597

Kaibel, G., 583
Kam, E. K. T., 640
Kamphuis, B., 668
Karanth, N. G., 630
Kataoka, M., 561
Kato, H., 482
Kato, Y., 216
Katz, S., 707
Kawazo, K., 470
Kays, W. M., 727
Kehoe, J. P. G., 621
Kelkar, B. G., 593
Kennedy, A. M., 461
Kenney, C. N., 597
Kilgannon, R. B. F., 455
Kim, S. D., 216
King, M. B., 487
King, R. P., 700
Kirk, R. E., 520, 538, 539
Kiselev, O. V., 659
Kishinevskii, M. K., 601
Klein, J., 727
Klein, J. P., 255, 256, 258
Klettner, H., 126, 158
Klinken, J., van, 503
Klinkenberg, A., 199
Kobayashi, H., 482, 483
Kobayashi, T., 413, 415, 416, 418, 419, 511, 517, 519, 526, 527, 532, 533, 560, 561, 566
Kobe, K. A., 539
Kobelt, D., 121
Koetsier, W. T., 146, 147, 148, 487
Koga, S., 153
Kohl, A. L., 521

Koide, K., 216
Kolmogoroff, A. N., 229, 251, 254
Komiyama, H., 548
Konczalla, M., 276
Kondo, M., 527
Kormeho, T. S., 601
Kothari, P. J., 466
Kozicki, F., 250, 254, 256, 257
Kraa, J. A., de, 473, 609
Krabetz, R., 697
Kramers, H., 171, 175, 177, 211, 217, 407, 461, 480, 596, 616, 618
Kraus, W., 488
Krause, F. E., 521
Krevelen, D. W., van, 2, 10, 19, 33, 114, 401, 403, 407, 421, 425, 521, 561, 566, 601, 691
Krishna, R., 485, 548
Kronig, R., 479
Kubicek, M., 457, 458
Kubota, H., 216
Küchler, L., 121, 122, 123, 695, 716, 697, 698
Kulkarni, B. D., 519, 541, 629
Kumazawa, H., 393, 419, 423, 429, 540, 541, 567
Kumisera, S., 642
Kunigita, E., 187
Kunii, D., 475, 481, 482, 650
Kuo, C. H., 516, 552, 567
Kuo, J. C. W., 126
Kürten, H., 487, 489, 576, 664
Kusaka, K., 667, 668
Kuzenkov, A. P., 617

Laidler, K., 16
Landau, J., 466, 473
Landeghem, H., van, 479, 488, 577, 593, 596, 664
Landsman, P., 211
Langbein, D., 121, 122, 123
Lapidus, L., 626, 650
Laurent, A., 461, 473, 475, 488, 489
Lauwerier, H. A., 209
Lee, J. C. M., 621
Lee, K. Y., 699
Lelli, U., 141
Leprince, P., 114
Lerou, J. J., 630
Levec, J., 664
Levenspiel, O., 41, 177, 191, 198, 207, 208, 213, 214, 217, 248, 249, 461, 463, 475, 482
Lewis, R. P., 659
Li, K. Y., 516
Lightfoot, E. N., 1, 3, 4, 361, 362, 399, 400, 459

Limido, G. E., 114
Lindner, B., 642
Linek, V., 473, 474, 476
Lips, J., 192
Lites, I. L., 521
Loffler, D. G., 567
Logeais, B. A., 326
London, A. L., 727
Long, R., 520
Longfield, J. E., 335
Lösch, J., 288
Lübeck, B., 653
Luckett, P., 511, 548
Lund, M., 141
Lunde, P. J., 627
Lundqvist, L. E., 557, 558
Luss, D., 126, 338, 482, 483, 516, 596–598, 621, 649, 657
Lyche, B. C., 626

Mah, R. S. H., 126
Mann, R., 411, 454, 455, 578
Manna, B. B., 217
Manogne, W. H., 17, 521
Marek, M., 628
Maria, K., de, 217
Marr, R., 217
Marrucci, G., 520
Mars, P., 691
Martin, H., 621
Maselkar, R. A., 473, 475, 488
Mason, D. R., 339
Matiatos, D. C., 476
Matros, Y. S., 659
Matsui, T., 561, 563
Matsuyama, H., 649, 650, 657
Mauser, H., 276
May, W. G., 195
Mayer, H. H., 583
McBee, E. T., 121
McGovern, T. J., 626
McGreavy, C., 457, 650
McKewan, W. M., 430
McMullin, R. B., 116
Mears, D. A., 449, 457, 620
Mecklenburg, J. C., 574
Mehta, K. L., 487, 488
Mehta, P. S., 657
Mehta, V. O., 488
Meîer, H., 539
Meiring, F., 699, 700
Menten, M. M. L., 149
Merchuk, J. C., 423
Messikommer, B. H., 107, 731

Methot, J. C., 256
Meyberg, W. H., 211
Mhaskar, R. D., 423, 541
Michaelis, L., 149
Michelson, M. L., 216, 222, 225
Midoux, N., 217
Miller, D. H., 609
Miller, R. S., 247
Miller, S. A., 539
Mischenko, E. F., 738
Miyamoto, K., 566
Modell, M., 1
Moergeli, U., 256, 257
Monod, J., 149
Moore, T. F., 521
Moo Young, M. B., 255, 487
Morooka, S., 482
Moser, F., 126, 158, 217
Moulijn, J. A., 213, 214
Moyes, H., 454, 455, 578
Muler, A., 327
Murakami, Y., 461, 463
Murase, A., 692
Muttzall, M. K., 1, 4, 362, 369
Myauchi, T., 482, 574

Nagasubramanian, K., 239
Nagata, S., 239, 561, 563, 566
Nagayasu, T., 561, 563
Nagel, O., 487, 489, 576
Nakagawa, J., 642
Nakamura, M., 609
Nakao, K., 561
Nanda, A. K., 466
Narsimkan, G., 424
Nauman, E. B., 190, 220, 248
Nauze, R. D., La, 484
Nawrocky, D. A., 519
Neher, G. M., 121
Nelson, P., 583
Nelson, P. A., 481
Nelson, W. L., 520
Newton, D. A., 479
Nijsing, R. A. T. O., 407, 461
Nishii, K., 609
Nishikawa, M., 609
Nonhebel, G., 584
Novikov, E. A., 597

Oblad, A. G., 539
Oden, E. C., 133, 134
Oele, A. P., 617, 619
Ohasi, Y., 470

Ohki, Y., 557
Olander, D. R., 419, 420
Olbrich, W., 650
Oliver, K. L., 539
Olney, R. B., 211
Onda, K., 413, 415, 416, 418, 419, 511, 517,
   519, 526, 527, 532, 533, 560, 566
Ostergaard, K., 216, 222, 225, 516, 649
Otake, T., 187, 561, 642
Othmer, D. F., 520, 538, 539
Ottengraf, S. P. P., 156, 157
Ouwerkerk, C., 524–531, 540

Pachovski, R. A., 566
Padberg, G., 653, 656
Pallai, I. M., 698
Pandit, A. B., 596
Pangarkar, V. G., 560, 561, 563
Papadopoulos, M. N., 421
Parulekar, S. J., 665
Pasquali, G., 141
Patzek, T., 699, 701, 704, 705
Pearson, L., 16, 17
Pei, D. C. T., 626
Pennell, A., 520
Perlmutter, D. D., 340
Perona, J. J., 284, 285
Peters, M. S., 5, 678
Petersen, E. E., 620
Peterson, E. L., 713
Pigford, R. L., 17, 364, 461, 519, 521,
   526–529, 531, 537, 538
Pikios, C. A., 338
Pinsent, B. R. W., 17
Piret, E. L., 57
Plasari, E., 247, 249, 255–258
Pniak, B. J., 146–148
Poggeman, R., 664
Pohorecki, R., 256, 257
Pontryagin, L. S., 739
Poore, A. B., 340, 650
Popa, T. M., 601
Porter, K. E., 361, 487, 601
Potter, O. E., 609, 650, 651, 667
Prater, C. D., 126
Prausnitz, J. M., 7, 10
Price, T. H., 657
Prochazka, J., 466
Pronk, K. M. A., 503
Prost, C., 461, 473, 475
Ptasinski, K. J., 339
Puszynski, J., 652, 653
Pyle, D. L., 650

Ragatz, R. A., 1, 4, 7–9, 11–13, 263
Raghuram, S., 596, 597
Rähse, W., 488, 557, 566
Rajadhyaksha, R. A., 650
Ralek, M., 665
Ramachandran, P. A., 519, 520, 524–527, 533, 538, 540, 558, 647, 664
Rao, M. A., 256
Rao, M. N., 48
Rao, Y. K., 430
Ray, W. H., 340, 624, 675, 716
Reid, R. C., 1, 7, 10
Reith, T., 472, 474, 489, 609
Rhee, H. K., 657, 659
Rice, R. G., 216
Richarz, W., 636, 653
Ridlehoover, G. A., 141
Riesenfeld, F. C., 521
Rietema, K., 156, 157, 256, 483
Rihani, D. N., 538
Ritchie, B. W., 251
Rivikind, V. Ya., 412
Roberts, D., 361, 461, 617
Roberts, G. W., 507–509
Roberts, H. L., 692
Robinson, C. W., 476
Rod, V., 423, 574
Rodiguin, N. M., 126
Rodiguina, E. N., 126
Roes, A. W. M., 221–224, 574
Roper, G. H., 537
Rosas, C. B., 686–688
Rosenbrock, H. H., 710
Rosensweig, R. E., 229, 251
Ross, I. B., 650
Roughton, F. J. W., 17
Rowe, P. N., 481
Rusinko, F., Jr., 430
Rys, P., 250, 254, 256, 257

Sada, E., 393, 412, 413, 415, 416, 418, 423, 429, 511, 517, 519, 526, 527, 533, 540, 541, 560, 566, 567
Sadek, S. E., 519
Saleem, A., 521
Sams, W. N., 657
Santiago, M., De, 402
Sater, V. E., 217
Sato, I., 567
Satterfield, C. N., 425, 488, 618, 664
Satyanarayana, M., 48
Schilson, R. E., 443
Schlünder, E. U., 213, 480, 482, 621, 626, 628–630

Schmalzer, D. K., 470
Schmidt, A., 538
Schmidt, B., 665
Schmidt, L. D., 567
Schmitz, R. A., 339, 340, 347, 596
Schoenemann, K., 556, 697
Schöön, N. H., 557, 558
Schügerl, K., 216, 640, 664
Schuler, J. C., 469, 471
Schuler, R., 521
Schwartz, C. E., 629
Schweyer, H. E., 5
Scott, D. S., 487
Scriven, L. E., 461
Seagrave, R. C., 141
Secor, R. M., 419
Seid, B., 583
Serpemen, Y., 665
Sgarlata-Lattmann, M. A., 636
Shah, M. A., 618
Shah, M. J., 322
Shah, Y. T., 216, 217, 423, 454, 489, 560, 561, 577, 593, 596, 597, 665
Sharma, M. M., 17, 216, 217, 423, 461, 463, 466, 467, 473–476, 487, 488, 519–521, 524–527, 533, 538, 540, 560, 561, 563, 577, 580, 584, 596, 597
Sharma, S., 596–598
Sherwood, T. K., 7, 10, 364
Shilimkan, R. V., 400
Shklyar, S. L., 519, 522
Sicardi, S., 664
Sideman, S., 489
Sie, S. T., 503
Sim, M. T., 411
Simonsson, D., 642
Singh, C. P. P., 597
Sinn, R., 487, 489
Sittig, M., 520, 539
Skrypek, J., 708
Slinko, M. G., 327, 624, 697–699, 701, 738
Smith, D. F., 42
Smith, G. B., 686–688
Smith, J. M., 7, 16, 136, 621, 627, 629, 664
Smith, T. G., 548, 567
Smith, W. K., 191
Snita, D., 652, 653
Sohn, H. Y., 436, 640, 649
Sokolinskii, Yu. A., 519, 522
Specchia, V., 399, 487, 664
Spiegel, M. R., 180, 223
Spieleman, L. A., 248, 249
Sridhar, T., 609, 667
Sridharan, K., 473–475

Stahl, R., 653
Steiff, A., 664
Steiner, R., 593, 596
Stemerding, S. F., 211, 409, 574
Stepanek, J. B., 400
Sterbis, E. E., 519
Stewart, W. E., 1, 3, 4, 361, 362, 400, 459
Stiegel, G. J., 216, 217, 577
Storey, C., 710, 730
Strand, C. P., 472
Strickland, H., 121
Sunagawa, T., 482
Suzuki, M., 470, 481, 621
Swaaij, W. P. M., van, 146, 147, 148, 173, 191, 192, 194, 195, 197, 198, 213, 215, 217, 221, 454, 455, 476, 477, 482, 483–485, 487, 488, 514, 526–529, 531, 536, 540, 561, 563, 574, 642, 668
Swain, C. D., 520
Swegler, E. W., 548
Szekely, J., 436, 552, 554, 640, 642, 675, 716

Tada, T., 566
Tagintsev, B. G., 521
Taglinger, L., 539
Tai, S., 609
Taitel, Y., 522
Takatsuka, T., 558, 559
Takeuchi, H., 532, 560
Tamir, A., 454, 522
Tanaka, K., 239
Tanaka, S., 216
Taweel, A. M., Al, 473
Taylor, G. I., 209–211
Taylor, M. D., 521
Teramoto, M., 561, 563, 566, 609
Terranishi, H., 609
Terteryan, A. M., 617
Teshima, H., 470
Thegze, V. B., 211
Thiele, E. W., 425
Thodos, G., 284, 285, 453
Thoenes, D., 480, 487, 616
Thomas, W. J., 511, 548
Thonon, C., 430
Tichacek, L. J., 210, 211
Tierney, J. W., 596, 597
Timmerhaus, K. D., 5, 678
Tobgy, A. H., 251
Tone, S., 642
Toor, H. L., 577
Towell, G. D., 472
Tram, K. E., 211
Trambouze, P. J., 131, 556

Trela, M., 680
Troltenier, U., 121, 691, 706, 709, 716, 721, 725
Truong, K. T., 256
Tsotsis, T. T., 339
Tsuchija, H. M., 153
Turner, J. C. R., 177, 208

Uppal, A., 340
Urbanek, A., 680

Vacek, V., 473, 474
Valstar, J. M., 636
Varma, A., 592, 597
Varshney, K. C., 487
Vasudeva, K., 650
Venderbos, D., 472, 489, 609
Venkatuswarlu, C., 48
Veress, G., 698
Verma, S. L., 578
Vermeulen, T., 574
Villermaux, J., 191, 192, 194, 195, 197, 198, 216, 217, 239, 245, 247–249, 251–258
Virkar, P. D., 454
Vivian, J. E., 476, 519
Voetter, H., 102, 107
Voncken, R. M., 49
Vortmeyer, D., 626, 649, 653
Votruba, J., 628
Vusse, J. G., van de, 102, 107, 135, 136, 198, 211, 488, 489, 557, 561, 584

Waddams, A. L., 520, 538, 539
Wagner, C., 311, 425
Walas, S. M., 16, 17
Walker, P. J., Jr., 430
Wall, R. J., 211
Wang, T. Y., 216
Wasch, A. P., de, 630
Waterman, H. I., 128
Watson, F. A., 678
Watson, K. M., 1, 4, 7–9, 11–13, 16, 263, 425
Weber, A. P. W., 58, 181
Weekman, V. W., Jr., 469, 470, 482, 483
Weeks, J. L., 516
Wehner, J. F., 200
Wei, J., 126, 640
Weimer, A. W., 640
Weinspack, P. M., 664
Weinstein, H., 245, 246, 258
Weisz, P. B., 449, 455–457, 548, 649
Wellek, R. M., 399, 412

Welsenaere, R. J., van, 297, 299
Wen, C. Y., 175, 198, 207–211, 213, 215, 482
Wender, L., 539
Weng, H. S., 657
Werther, J., 482, 484, 485
Wesselingh, J. A., 488, 489
Westerterp, K. R., 171, 177, 211, 217, 331–333, 336, 338, 473, 596, 598, 609, 668, 691, 692, 731
Wheeler, A., 496
White, D., 636
White, R. R., 217
Wicke, E., 33, 425, 441, 615, 634, 649, 653, 656, 658
Wieworowski, T. K., 538
Wijffels, J. B., 421, 503, 583
Wilde, D. J., 727
Wilhelm, R. H., 200
Wilke, C. R., 364, 476
Wilkinson, J. K., 678
Wilt, G. J., de, 488
Wirges, H. P., 347, 557, 566
Wöhler, F., 593, 596
Wojciechowski, B. W., 566
Wolfbauer, O., 126, 158

Wood, T., 17
Woodburn, E. T., 217

Yagi, H., 609
Yagi, T., 189
Yamadam, T., 239
Yang, C. C., 468
Yang, R. D., 153
Yarze, J. C., 539
Yokota, K., 532
Yoon, H., 640
Yoshida, F., 473
Yoshiharu, M., 473
Young, L. C., 632

Zadeh, H. M., 647
Zehner, P., 628, 664
Zeldowitch, J. B., 425
Zener, C., 713
Zharov, D. V., 617
Zoll, G., 216
Zoulalian, A., 245, 256
Zuiderweg, F. J., 195, 211, 215, 482, 485, 487, 574
Zwietering, T. N., 232, 239–242, 335

Subject Index

Absorption, of CO_2 and H_2S in amines, 529, 534–536
  of CO_2 and H_2S in hydroxide solutions, 531, 532
  of CO_2 and SO_2 in caustic solutions, 531
  of CO_2 in amines, 516–518
  of CO_2 in a packed column, 407–409
  of ethylene and oxygen in copper chloride solution, 533
  of HCl in ethylene glycol, the heat effect, 455
  of H_2S from natural gas in a tray column reactor, 584–589
  of H_2S from natural gas in a trickle flow reactor, 665–669
  of NH_3 in an acid solution, 366–367
  of phosgene and CO_2 in water, 532, 533
  with reaction, see Mass transfer with reaction
Activation energy, 16–19
  apparent, 23
Active sludge in waste water treatment, 155–158
Adiabatic temperature rise
  definition of, 267
  initial value of, 267–268
  true value of, 267
  ratio of, for gas and solid, 661–663
Absorption, test on occurrence of, 173
Agglomeration, see also Aggregates
  models for micromixing, 247–258
  random coalescence models in, 248
Aggregates, dimensions of in micromixing, 255, 257
  effective size of, 254, 255
  exchange with average tank contents (IEM model), 249
  life span of, in micromixing, 229
  micromixing time of, 247, 255–258
  model of eroding, 247

  random coalescence of, 248
Aggregation
  state of reacting fluid, definition, 229
Autocatalytic reactions, 22
  fermentation reactions as, 149
  in a tank reactor, 70–72
  in a recycle reactor, 68–72
  with mass transfer, 411, 412
Autothermal behaviour of a single particle, 613–615
Autothermal reactors, 310–331
  definition of, 310–311
  choice of operating conditions in an, 315–317
  dynamic behaviour of, 339–354
  extinction of, 311–317, 343–344
  hysteresis in, 316–317
  ignition of, 311–317, 343–344
  optimization of, 726–730
Axial dispersion, see also Check for absence
  in bubble columns, 216, 217, 592–596
  in gas–liquid reactors, 216–217, 592–596
  in fluidized beds, 214–215
  in packed beds, 213, 628–629, 632
  in Raschig columns, 409
  influence of, on conversion, 199–207
  influence of, on selectivity with consecutive reactions, 205–207
  of heat by countercurrent flow of gas and solids, 644
  reduction of, by spiralling tubes, 210
Axial dispersion coefficient, see Longitudinal dispersion coefficient
Axial dispersion models, see Dispersion models

Backmixing, see Non-ideal flow and Residence time distribution
Basket type mixed reactor, 467–468
Batch reactor, capacity increase of, 271

half value times in a, 40
isothermal for single reactions, 39–43
optimization of a, 744–745
policy of operation of adiabatic, 269–271
reaction time in a, 39
reactor volume for a, 41
Bifurcation line, definition of, 315–316
Biochemical reactors, *see also* Autocatalytic reactions
critical residence time in, 154
minimum residence time in, 150
multiplicity in, 152–153
wash-out in, 152–153
Blow-out of a reactor, 322–326
Bodenstein number, definition of, 207
for flow through an annulus with a rotating inner cylinder, 211
in fluidized beds, 214–215
in gas-liquid reactors, 216–217
in packed beds, 213
in rotating disc contactors, 211
in tubes, 209–211

Capacity of a reactor, *see* Residence time *and* Reactor capacity
Cascade of tank reactors, 54–61
calculation methods for a, 56–61
choice of temperature sequence in, 295
comparison of with a PFR, 55–56
conversion in, 199, 203
E diagrams and moments of, 183–185
heat transfer requirements in, 303–304
optimization of, 716–719, 731–737, 742–743
parallel reactions in, 99–101
relation between with dispersed plug flow, 189, 207
start-up of, 339
static stability in, 317–319
Catalyst deactivation, *see* Deactivation of catalysts
Catalytic gas-solid reactor, 622–639
Char, gasification of, 450–452
Check for absence in packed bed reactors of
axial heat dispersion, 632
axial mass dispersion, 632
Check for isothermicity of a gas-liquid reactor, 581–582
Check for resistance, *see also* Mass transfer, Heat effect *and* Diffusion limitation
gas phase for mass, 382
interparticle for heat, 621–622
interparticle for mass, 620–622

intraparticle for heat, 620–622
intraparticle for mass, 620–622
Chemical equilibrium, 10–14
Chemical kinetics, 15–24
and reaction mechanism, 20
biochemical kinetics, 149
influence of concentration on, 19–20
influence of temperature on, 15–19
lumping of, 146–148
measurement of, in a laminar jet reactor, 367–377
model reactors for investigation of, of multi-phase reactions, 458–471
Chemical production rate, 14
Chilton–Colburn analogy, 453
Chlorination of benzene, 116–120, 557, 637–639
Chlorination of toluene, heat effect in, 455, 579–580
Chlorination of xylose, 556, 557
Choice of a model for a gas–solid reactor, 624–627
Circulation reactor, *see* Recycle reactor
Closed-closed systems, 162
Closed-open systems, 162
Coal, *see* Combustion
Coalescence model, of micromixing, 248–251
relation of, to IEM model for micromixing, 251, 253
Cold shot converter, 322–323
optimization of a, 697–699, 701–705
Combination reactions, 124–148, 557–567
see also Consecutive parallel reactions
algebraic methods for, 135–148
interpretation of laboratory experiments with, 126–129
selection of reactor type for, 130–131, 138–142
Combustion, of ammonia, 617–619, 649
of coal, 441–442, 613–617
of coke deposits, 658–663
Competing reactions, *see* Parallel reactions
Complex reaction systems, optimization of, 705–710, 723–725, 740–744
Complex reactions, *see* Mass transfer with complex reactions *and* Multiple reactions
Concentration, flow averaged or mixed cup, 177
maximum allowable reactant, 299, 330–331
Concentration gradients, *see* Check for resistances

Consecutive reactions, 108–123
  and mass transfer, 541–567
  and reactor load changes, 115–116
  and recycle of products, 120–123
  and segregated tank reactors, 236
  enhancement factor for, 558, 561
  first order, in a CISTR, 113–114
  in a PFR, 109–113
  influence. of axial dispersion on the selectivity and yield for, 205–207
Consecutive parallel reactions with mass transfer, 557–567
Contact time, definition of, 370
Continuous ideally stirred tank reactor (CISTR), see also Tank reactor
  definition of, 227
  distribution of a, 117
Conversion, boundaries for micromixing, 239–243
  calculated from dispersion models, 199–207
  calculation for laminar flow in empty tubes, 208
  for maximum mixedness, 241–243
  for minimum mixedness, 239
  in a stirred tank with intermediate micromixing, 243–254
  influence of Pe on the, 201, 202
  prediction of, from the E diagram with first order kinetics, 172
Conversion rate, 15, 23
  see also Mass transfer with reaction
  and degree of conversion, 26–28
  and transport phenomena, 23–24
  distinction between and chemical kinetics, 15–16
  distinction between and chemical production rate, 15
  influence of the residence time distribution on the, 199–207
Convolution, of residence time distributions, 164, 165
  example of a calculation, 171
Cooling area, achievable in tanks, 277, 338
Criteria for catalytic tubular reactors for adequacy of
  one dimensional model, 636–637
  pseudo-homogeneous two dimensional model, 620–622, 652–653
Criteria for prevention of runaway
  in cooled packed bed tubular reactors, 650–653
  in cooled PFR, 297–299, 301–302
Creeping in gas–solid reactors, 656

Cross flow reactor, 101–108
  the idealized, 102–105
  types and yield, 107–108
CSSTR, continuous fully segregated stirred tank reactor
  definition of, 227, 228
CSTR, continuous partially segregated stirred tank reactor
  definition of, 228

Danckwerts model, 370, 371
  see also Penetration model
Deactivation of catalysts, 657, 658–663, 699, 700
  effect of, on the selectivity for multiple reactions, 503
Dead zones, identification of, by the E diagram, 172
  models including, 191, 197
Deconvolution of the E diagram, 171
Degree of conversion, 24–28
  definition of, 24
  relative, 25
  relation between concentrations and, 25–26
Degree of utilization, see also Effectiveness factor
  in a tank reactor, 388–390
Dehydrogenation of 2-butanol, 284–287
Dehydrogenation of ethylbenzene, 699–700
Delta function input, see Dirac delta function input
Denbigh's problem, 133, 708–710
Desired product, see Selectivity
Desorption with reaction, 420–424, 546–557
Desulphurization of gasoil, 146–148
Differential reactor, 467, 468
Differential selectivity, see Selectivity
Diffusion coefficient, effective internal, in porous particle, 425, 622
  Knudsen, 622
  measurement of, in pellets, 470
Diffusion in micromixing, 247, 257
Diffusion limitation, see also Mass transfer
Diffusion limitation, general criterion for the absence of, 448–449
Dirac delta function input, 174
Dispersed flow, see Plug flow with axial dispersion
Dispersion coefficient, definition of, 186
Dispersion model, calculation of conversion with, 199–207

combination models, 191–197
  for exchange with dead zones, 191, 197
  for fluid beds, 195
  for rotating disc contactor, 199
  for tanks in series, 183–185
  for trickle flow reactors, 197
  of a cascade of tank reactors with backmixing, 199
  of Adler and Hovorka, 199
  of Cholette and Cloutier, 198
  of Klinkenberg, 199
  of van Deemter, 195
  relation between the dispersed plugflow and tanks-in-series model, 189, 207
  the plugflow with axial, 185–191
  two regions models, 191
Distillation column as a reactor, 64–65, 421
  for consecutive reactions, 557
Distribution coefficient, definition of, 364
Distribution law for heat, 573
Dynamic behaviour, of gas–liquid reactors, 597
  of gas–solid reactors, 648–663
  of tank reactors, dynamic stability, 346–354
  of tank reactors, static stability, 317–319, 341–346
  see also Limit cycles
Dynamic programming, 730–737

Earliness of mixing, 229–232
  influence of, on conversion, 230, 231
  in a cascade, 230, 231, 242, 243
Economic balance, 5–6
Economic diagrams, 72–74
Economic optimization, criteria for, 678–679
E diagram, see also Residence time distribution
  definition of, 160
  calculation of variance from the reduced, 168
  conversion prediction from, 172
    approximated by Gaussean distribution function, 171
  for dispersed flow, 187–191
  for laminar flow in empty tubes, 207
  from the Laplace transform, 180
  of a cascade of tank reactors, 183–185
  of a CISTR, 179
  of a PFR, 178
  Laplace transform of the reduced, 180
  numerical example for the, 166–170
  reduced, 163
  for laminar flow in an empty tube, 208
  of a CISTR, 179
  of a PFR, 178
  relation of the with the F diagram, 165
Effectiveness factor, definition of, 384
  with intraparticle heat effects, 456
  with mass transfer and reaction in series, 427
  general definition of, 392
  with first order equilibrium reaction in a pellet, 440–442
  with Langmuir–Hinshelwood kinetics, 435
  with parallel reactions, 507–509
Emulsion, micromixing in an, 229
  polymerization, 237
Endothermic reactions in a CISTR, 309–310
Energy balance, 4–5
  and work done, 263
  in flow processes, 261–263
  in non-flow processes, 261–266
Enhancement in micromixing, 244, 245, 249, 253
Enhancement factor, see also Mass transfer with reaction
  definition of, 375
  for homogeneous irreversible reactions of order n, m, 410–411
  for instantaneous equilibrium reactions with desorption, 423
  for instantaneous reactions, 398–400
  for instantaneous parallel reactions, 512
  for Langmuir–Hinshelwood kinetics, 435
  for mass transfer with reversible reactions, 416–420
  in mass transfer of two reactants, 537–541
  with consecutive reactions, 558, 561
  with parallel reactions, 512, 522, 528, 531
  general equation for, for a reaction of the order (1, 1), 402
  negative value of, 535–536
Enthalpy, influence of temperature on, 7
Equilibrium constant, relation between and Gibbs free energy, 10
Eroding aggregates, mixing model of, 247, 255
  time of erosion of, 255
Erf function, 369, 370
Esterification, reactor for, 421

Exothermic reactions, heat effects in mass transfer with, 452–458
Experimental reactors, see Chemical kinetics
Explosion, 271–273
Extinction of reactors, see Hysteresis and Autothermal reactors
Extractive reactions, 546–557

Fanning friction factor, influence of, on Bodenstein number, 210
Fast reaction, definition of, 380, 396
  definition of for mass transfer and reaction in series, 428
F diagram, definition of, 160
  calculation of variance from the, 169
  for dispersed flow, 189
  for turbulent flow in tubes, 212
  numerical example of, 166–170
  relation of with E diagram, 165
Feed distribution, see Cross flow reactor and Cold shot convertor
Fermentation reactions, see Autocatalytic reactions and Biochemical reactors
Fick's law, 362
Film model, see also Mass transfer with reaction
  description of, 367–368
Film thickness, see also Penetration depth
  definition of, 368
  for stagnant bodies, 379, 383
  for heat transfer, 453
First order reaction, see also Mass transfer with
  conversion prediction from E diagram for, 203–205
  in a cascade of tank reactors, 199
  in dispersed flow, 199–202
Fixed bed reactor, as a model reactor, 467, 468
Fluid–fluid reactions, see Mass transfer with reaction and Selectivity
Fluidized bed reactor, axial dispersion in, 214
  Bodenstein number in, 215
  bubble growth in, 484
  heat balance for, 623
  model for the design of, 482–488
  two phase model for, 195
Fluid–solid reactions, see Mass transfer with reaction and Selectivity
Fourier transformation, application of, in residence time distribution measurements, 222–225

Fractional yield, see Selectivity
Furfural synthesis, 556, 557

Gasification of char, 450–452
Gas–liquid reactors, agitated tank reactors as, 596–612
  bubble column reactors as, 590–596
  characteristic data for, 479
  check for isothermicity of, 581–582
  column reactors as, 580–589
  heat effects in, 575–612
  types of, 575–577
Gas–liquid–solid reactors, 663–669
  packed bed reactors, 663–669
  slurry reactors, 664
Gas phase resistance, criterion for, 382
  with reaction in liquid, 382–383, 403–406, 407–409
  with reaction in porous pellet, 438, 451
Gas–solid reactors, see also Fluidized bed reactor, Mass transfer with reaction, Packed bed, Riser, Selectivity and Modelling
  choice of reactor type for, 640–642
  heat effects in, 612–663
  models for, 642
  length of reaction zone in, 615–619, 660
  moving bed, 639–648
  thermal behaviour of a single particle, 613–622
Gaussian distribution curve as an approximation of the E diagram, 183, 185
Geometric programming, 712–716
Grain model, 450–452

Hatta number, see also Reaction modulus
  definition of for first order reactions, 374
  definition of for reactions of order 1,1, 395
  definition of for reactions of order n, m, 410
  general definition of, 391
  physical meaning of, 378, 379
Heat effect, and desired reaction temperature, 260
  at fluid–fluid interface, 454–455
  in BR, 264–273
  in CISTR, 302–310
  in mass transfer with reaction, 452–458
  in PFR, 281–302
  in SBR, 273–281
  intraparticle, 455–458

Heat production rate (HPR), definitions of, 304–307
  in a CISTR, 305–310, 327–339
  in agitated gas–liquid reactors, 601, 603–609
Heat of reaction, 6–10
  and heats of combustion, 6
  and heats of formation, 6
  calculation of the, 8–10
Heat transfer coefficient, *see also* Overall, Wall, Nusselt
  correlations for packed beds, 481
Heat withdrawal rate (HWR), definitions of, 305–307
  in a CISTR, 305–310, 327–339
  in agitated gas–liquid reactors with a vaporizing solvent, 601–609
Heaviside step function input, 174, 175
Henry's law, 364
Heterogeneous reaction, 357
Heterogeneous reactors, *see* Multiphase reactors
Higbie model, description of, 369, 370
  see also Penetration model
Hinterland ratio, definition of, 386
Holding time, 163
  as opposed to mean residence time, 173
Hold-up, dynamic, 192
  in trickle flow reactors, 216
  influence of on Bodenstein number, 217
  static, 192
Homogeneous reaction, 357
Hot spot temperature, 293
Hydrogenation, 566
Hyperbolic functions, 378
Hysteresis, 649, 652–654
  see also Autothermal reactor

Ideal tank reactor (CISTR), 49–54
  residence time in a, 49
IEM model, definition of, 250
  application of the in conversion calculations, 251–259
  in unmixed and premixed feedstocks, 251–253
  relation of, to random coalescence model, 251, 253
Ignition of reactors, *see* Hysteresis *and* Autothermal reactors
Intermediate micromixing in continuous stirred tank reactors
  experiments, 254–258
  theories, 243–254

Instantaneous fractional yields, *see* Selectivity
Instantaneous reaction, definition of, 394, 395
  criterion for, 398
  enhancement factor for, 398–401
Integral reactor, 467, 468
Intensity of dispersion, *see* Péclet number *and* Bodenstein number

Kinetics, *see* Chemical kinetics

Lagrange multiplier technique, 716–725
Laminar flow, Bodenstein number for, 209, 210
  compared with dispersed plug flow, 209
  conversion equation for, 208
  E diagram for, in empty tube, 208
  influence of molecular diffusion on the RTD for, 209
  Péclet number for, 209
Laminar jet reactor, 376, 377, 459–461
Langmuir absorption coefficient, 430
Langmuir–Hinshelwood kinetics, influence of on mass transfer, 430, 435
Laplace transformation, application of in residence time distribution measurements, 180, 185, 189–190, 219–220, 222
Length of reaction zone, in gas–solid reactors, 615–619, 660
Lewis number, definition of the, 453
  use of, 615–619
Life expectation of fluid elements, 240–242
Limit cycles, 346–354
  in gas–liquid reactors, 597
Liquid–fluid reactions, *see* Mass transfer with reaction *and* Selectivity
Liquid–solid reactions, *see* Mass transfer with reaction *and* Selectivity
Liquid–solid reactors, *see* Packed bed, Fluidized bed, Slurry reactors, *etc.*
Locus of maxima curve, 297–298
Longitudinal dispersion coefficient, 186
  for fluidized beds, 214
  for gas–liquid reactors, 216–217
  for laminar flow, 209, 210
  for packed beds, 213, 214
Longitudinal mixing, causes of, 182
Lumping of kinetics, 146–148, 158

Macro fluid, 160, 161
  see also Segregation
  definition of, 229

Macro kinetics, 3
Macro mixing, definition of, 228, 229
Macropores, 548
Mass transfer, of two reactants, 536–541
  with independent parallel reactions, 497–503
  with parallel reactions, 518–536
  with a single reaction, 357–490
  with autocatalytic reactions, 411–412
  with complex irreversible homogeneous reactions, 390–393
  with consecutive reactions, 541–567
  with dependent parallel reactions, 503–518
  with equilibrium reactions in a porous particle, 436–443
  with heterogeneous reactions, 424–448
  with homogeneous bimolecular reactions, 393–412
  with homogeneous irreversible first order reactions, 371–390
  with Langmuir–Hinshelwood kinetics, 393, 435
  with Michaelis–Menten kinetics, 567
  with multiple reactions, 495–567
  with parallel reactions, 503–520, 531
  with reaction, general material balance for, 362–363
    heat effects in, 442, 452–458
    and desorption, 420–423
    in porous solids, 433–448
    in series, 426–432
    in two phases, 423–424
  with reversible homogeneous reactions, 412–420
  with reversible heterogeneous reactions, 436–443, 445–448
  with reversible parallel reactions, 516, 533
  with zero order reaction, the boundary layer problem in, 511
Mass transfer coefficient, definition of, 363, 364
  for particles, 479–483
  from bubble to dense phase in a fluid bed, 482
  in laminar jets or wetted wall columns, 459
  in penetration models, 370–371
  in stirred cell reactors, 459
  in the film model, 368
  measurement of, 471–478
Mass transfer limitation, general criterion for the absence of, 448–449

Mass transfer resistances, 364–366
  combination of, 364, 382–383, 403–406, 407–409, 438, 451
Mass transfer unit, definition of, 193
Material balance, 3–4
  summary of for model reactors, 51
Maximum allowable temperature, 274–276, 290, 300–302, 319–320, 331–339
Maximum mixedness, definition of, 240, 241
  conversion for, 241–243
Maximum principle of Pontryagin, 738–744
Mean residence time, see Residence time, mean
Michaelis–Menten kinetics with mass transfer, 567
Microfluid, 160, 161
  definition of, 229
Micro kinetics, 2
Micromixing, agglomeration models for, 247
  boundaries for mixers and plugflow reactor, 232–237
  coalescence model of, 248, 251
  dimensions of aggregates in, 255, 257
  enhancement in, 244, 245, 249, 253
  experimental results on, in stirred tanks, 254–259
  formal models of, 244
  ideal, definition of, 227
  IEM model of, 250
  intermediate, definition of, 228
    boundaries for reactors with arbitrary RTD's, 239–243
    in multiphase systems, 243, 244
    in stirred vessels, 227, 232–239, 251, 256, 257
  models for intermediate in tank reactors, 244–250
  nature of, 228–232
  role of diffusion in, 247, 257
  states of in a tank reactor, 227
  time scales of, 247, 250
  typical cases where is important, 228
Micropores, 548
Minimum mixedness, definition of, 239
  conversion for, 239
Mixed cup concentration, 177
Mixed flow reactor, see Continuous ideally stirred tank reactor
Mixed reactor, see Continuous ideally stirred tank reactor and Tank reactor

Mixing, earliness of, 247
  identification of, 172
  models for, 182
Modelling of catalytic tubular reactors, 622–627
  heterogeneous two-dimensional model, 631–633
  pseudo-homogeneous one-dimensional model, 48–49, 284–287, 636–637
  pseudo-homogeneous two-dimensional model, 633–635, 637–639
  temperature and concentration distribution in a cooled reactor, 630–639
Moments of the E function, 162
  first moment, 163
  for a cascade of tank reactors, 184–185
  for dispersed flow, 188–191
  from Laplace transformations, 180
  general definition, 162
  higher moments, 169
  second moment, 163
  zeroth moment, 161
Moments of the reduced E function, 163
Multibed adiabatic reactors, with cold feed injection in between the beds, *see* Cold shot converter
  optimization of bed inlet temperatures in, 693–699, 726–730
Multiphase reactors, expressions for mass transfer rates in, 357–471, 495–567
Multiphase systems, micromixing in, 243, 244
Multiple reactions, *see also* Mass transfer *and* Selectivity
  mass transfer with, 495–567
    and product distribution, 87–93
    and the reaction path, 87–93
    and yield or selectivity, 84–93
    in cooled tubular reactors, 339
  types of, 22
Multiplicity, definition of, 312
  condition for no, in a CISTR, 328
  in adiabatic beds, 653–654
  in agitated gas–liquid reactors, 597–598, 607–612
  in biochemical reactors, 152–153
  in bubble column gas–liquid reactors, 592
  in gas–solid reactions, 613–617
  in packed bed reactors, 648–650
Multistage reactors, *see* Cascade *and* Multibed

Nernst model, *see* Film model

Nernst's law, 364
Nicotinamide production, 686–688
Non-ideal flow, 160–226
  see also Dispersion model
  multiparameter models for, 191–199
  in fluidized beds, 195, 214, 215
  in packed beds, 212–214
  in pipe flow, 207–212
  in rotating disc contactors, 199, 221
  in stirred tanks, 198
  in trickle flow reactors, 197, 216–217
  in tube reactors, 185–191
  modelling, 160–259
  tailing, 172, 192
  with dead zones, 172
  with short cut flows, 172
Number of transfer units, 367
Numerical search routines, 725–730
Nusselt number, for single particles, 621
  in packed beds, 481, 621

Open-closed systems, 162
Open-open systems, 162
Operating point, lower and upper stable in a CISTR, 312, 314, 322, 325, 330
  stable in an autothermal reactor, 312–317, 334
  unstable in an autothermal reactor, 312–317, 334
Optimization of reactors, constraints, 680, 696, 704, 708–710, 711
  dynamic programming for, 730–737
  economical, 675–679, 726–730, 734–737
  geometric programming for, 712–716
  Lagrange multiplier technique for, 716–725
  mathematical formulation of the problem of, 710–712
  mathematical methods for, 710–746
  numerical search routines for, 725–730
  objective function for, 678–680
  objectives of, 675–680
    by choice of reactant concentration in reactor feed, 142–146, 686–688
    by means of pressure, 699–701
    by means of temperature, 688–705
    for complex reaction systems, 705–710, 723–725, 740–744
    for exothermic equilibrium reactions, 742, 743
    for maximum capacity, 689–705, 714–716, 731–734
    for maximum yield, 705–710, 718, 719, 740–744

for minimum reactor section costs, 726–730, 734–737
in adiabatic bed reactors, 693
in adiabatic multibed reactors, 693–699, 726–730
in batch reactors, 743, 744
in cold shot converters, 697–699, 701–705
in isothermal tubular reactors, 692
in tank reactors, 689–690, 732–734, 742, 743
in tubular reactors, 690–692, 726–730
with deactivating catalyst, 699, 700
optimization variables in, 680
Pontryagin's maximum principle in, 738–744
procedure of, 674
relation between technical and economic optima in, 680–690
technical, 677, 679
Optimum residence time, for consecutive reactions, 111, 114
Overall fractional yield, see Selectivity
Overall heat transfer coefficient, based on coolant inlet temperature, 268
for cooled packed bed tubular reactors, 636
true, 268
Oxamide, synthesis, 592–596
Oxidation, of hydrocarbons in gas–liquid reactors, 598–612
of naphthalene, 335–339
of SO_2, 679–680, 691, 697–699
Oxochlorination, 533

Packed beds, axial and radial dispersion in, 212–214
mass transfer coefficients in, 479–481, 621
versus risers, 433–445
Parallel reactions, effectiveness factor for, 507–509
enhancement factor for, 512, 522, 528, 531
in segregated tank reactors, 235, 236
of differing reaction order and selectivity, 93–95
of equal reaction order and selectivity, 93–95
with mass transfer, 497–536
with two reactants, of differing reaction order and selectivity, 101–108
Parametric sensitivity, controlling parameters for, 291
and choice of coolant temperature, 290
and choice of tube diameter, 290
in cooled PFR, 289–302
in gas–solid reactors, 648–653
Particles, reaction in, see Mass transfer with reaction
Péclet number, definition of, 186
influence of on conversion, 201, 202
methods for measuring of, 221–225
for laminar flow, 209
in packed beds, 213
related to cascade of tank reactors model, 189, 207
physical meaning of, 187
Penetration depth, for heat and mass, 577–579
from penetration model, 374
Penetration model, see also Mass transfer with reaction
of Danckwerts, 370–371
of Higbie, 369–370
Penetration time, see Contact time
Pipe flow, dispersion in, 207–212
Piston flow, see Plug flow
Plant economics, 5, 29
influence of reactor design on, 675–677, 684–686
influence of yield or selectivity on, 675–677, 684–686
and maximum production rate, 79–82
and reactor design, 77–82
and reactor economics, 6
and reactor selectivity, 29–31
and reactor yield, 31–32
Plug flow reactor (PFR), adiabatic, 287–289
adiabatic with heat exchange, 320–326
comparison of, with a BR, 44
cooled, 281–302
countercurrent flow of coolant in, 354–355
definition of, 43
isothermal laboratory, 283
optimization of, 719–730, 740–742
with internal heat exchange, 323–326
residence time distribution in, 177
residence time in, 44
Plug flow with axial dispersion, 185–191
boundary conditions for, 187, 190, 191
conversion in compared to an ideal plug flow, 202
E diagrams of, 187–191
compared with cascade of tank reactors model, 189, 207

prediction of conversion with, 199–207
Point thermodynamical properties, 264–265
Polymerization reactions, influence of micromixing on, in a suspension, 229, 237
  kinetics of, 19
  in a cascade, 59–61
Pontryagin's maximum principle, 738–744
Pore diffusion, *see* Mass transfer with reaction
Pore mouth blocking, effect of, on the selectivity with multiple reactions, 503
Power jet reactor, 489
Power jet tube reactor, 489
Product distribution, *see* Selectivity *and* Yield, 87–93
Pseudo-homogeneous model, *see* Modelling
Pulse input, *see* Dirac delta function input

Radial dispersion, *see* Transverse dispersion
Rate controlling step, *see also* Mass transfer with reaction
  examples of the concept of, 382, 431, 439, 448
Reaction, influence of mass transfer on, *see* Mass transfer with reaction
  in porous solids, 433–448
Reaction and dispersion, 199–207
Reaction heat, feedback of, 310–311
  use of, 310–326
Reaction modulus, *see also* Hatta number
  definition of, for first order reactions, 379
    for reactions of the order (1,1), 395
    for reactions of the order $(n,0)$, 392
    for reactions of the order (n,m), 410
    with heat effects, 456
  general definition of, 391
  physical meaning of, 378, 379
    as a function of observables, 449
    for first order equilibrium reactions, 438
Reaction order, definition of, 22
  shifting, 22, 23
Reaction path, 87–93
Reaction types, 20–22, 496
Reaction velocity constant, 16
  dimensions of the, 23
  for surface reactions, 425, 426, 434, 435
Reactor, correct choice of a realistic model for multiphase reactors, 574–575
Reactor capacity, definition of, 38

Reactor holding time *see* Holding time
Reactor load, definition of, 38
Reactor operation, batch versus continuous, 72–74
  PFR versus CISTR, 74–77
Reactor types, classification of, 32–36
Recirculation reactor, *see* Recycle reactor
Recycle of non-converted reactants, influence of, on plant economics, 29–32, 78–79, 675, 684–686
Recycle reactor, 68–72
  as a model reactor, 181, 182, 467–468
Reduction of iron ore, 644–648
Reference temperature, definition of, 332
Residence time, mean as opposed to holding time, 173
  mean for a cascade, 165
  mean from the E diagram, 163, 167
  mean from the F diagram, 165
  measurement methods for the mean, 217–225
Residence time distribution, 160–226
  see also E diagram
  boundary conditions for, 162
  influence of, on conversion, 160–226
    in the CISTR, 177
    in the gas in agitated gas–liquid reactors, 596
    in the PFR, 177
  of laminar flow in an empty tube, 208
Residence time distribution measurement, illustration of various methods for, 218–225
  injection and detection methods for, 176, 177
  input functions for, 173–176
  requirements of tracers for, 175
  transfer functions for, in open–open systems, 221
Resistances, *see also* Mass transfer *and* Checks
  examples of combinations of, 382, 431, 439, 448
Reversible reactions, influence of mass transfer on, 412–420, 436–443, 445–448
  in a CISTR, 307–309
Riser reactor, 640
  as a model reactor, 467–468
  versus packed bed reactor, 443–445
Rotating disc contactor, Bodenstein number in, 211
  dispersion models for, 199

RTD, *see* Residence time distribution *and E* diagram
Runaway, *see also* Criteria for prevention
  definition of, 291
  in cooled PFR, 291–302

Sauter mean diameter, 385
Second order reaction in dispersed plug flow, 202
Segregated fluid, 160, 161
Segregated reactors, conversion in of the mixed vessel type, 233–235
  conversion in, with an arbitrary RTD, 239
  polymerization reactions in, 237
  selectivity in, with consecutive reactions, 236
  selectivity in, with parallel reactions, 235, 236
Segregation, complete in mixers, 233
  in suspensions, 229
Selectivity, definition of, 29, 85
  definition of differential, 86
  definition of differential ratio, 86
  definition of ratio, 86
  definition of types I, II and III, 496
  effect of cyclic changing of concentrations on, 516
  effect of dispersion on, 202
  effect of intraparticle heat effects on, 516
  effect of pore mouth blocking on, 503
  effect of stripping of unstable products on, 546–557
  influence of axial dispersion on, 205–207
  influence of mass transfer on, 495–567
  overall, 85
    and reaction temperature in a CISTR, 331–339
    in mass transfer with consecutive reactions, 541–567
    in mass transfer with parallel reactions, 497–536
    in multiphase systems of, type I, 497–503
    type II, 503–518
    type III, 541–567
Semibatch reactor (SBR), 61–68, 273–281
  relative conversion in a, 66
  versus batch reactor, 273–281
Series reactions, *see* Consecutive reactions
Sharp interface, shrinking unreacted core model, 644–648
Shift conversion, 697
Shrinking unreacted core model, 445–448

Short-cut flow, identification of, by *E* diagrams, 172
Shuttle kinetics, 516–518
Side by side reactions, *see* Parallel reactions
Single pellet diffusion reactor, 469–471
Single reactions, types of, 20–22
Slow reaction, definition of, 381, 395
  for mass transfer and reaction in series, 428
Slurry reactors, description of mass transfer with reaction in, 428–432
  as a model reactor, 467–468
Solvent evaporation, in gas–liquid reactors, 598–612
  influence on conversion in gas–liquid reactor, 605–606
  influence on enhancement factor, 604–605
  influence on HWR and HPR, 603–606
Solubility of gases in liquids, influence of temperature on, 577–580, 602
Space velocity, definition of, 45
Specific contact area, correlations for, 385, 484, 487–490
  measurement of, 472–478
Stagnant film model, 367–368
Standard deviation of a distribution, 163
Stanton number, modified, 306
Stimulus–response techniques, 176
Stirred cell reactor, 459–463
Stirred tank reactor, *see* Tank reactor
Stripping, *see also* Mass transfer with reaction
  of reaction product, 420–423, 546–557
Successive reactions, *see* Consecutive reactions
Sulphonation, heat effect in, 454, 455
Surface reaction rate constant, conversion of, to volumetric rate constant, 434, 435
  definition of, 425, 426
Surface renewal model, *see* Penetration model
Sweetening, *see* Absorption of CO_2 and H_2S
Synthesis of a reactor model, 572–575
Synthesis of ammonia, 8–13, 267, 325–326, 649, 691, 692, 697–699, 701–705
Synthesis of maleic acid mono-ester, 276–281
Synthesis of methanol, 649, 699, 726–730

Tailing in *E* functions, 172, 192
Tanks in series model, *see* Dispersion

model *and* Cascade
Tank reactors, *see also* Continuous ideally stirred tank reactors
   circulation time in, 50
   conversion with segregated flow in, 232, 333
   determination of safe operating conditions for, 327–331, 335–354
   dispersion models for, 198
   dynamic behaviour of, 339–354
   equations for mass transfer with first-order reaction in, 385–390
   intermediate micromixing in, 243–258
   mass transfer and consecutive reactions in, 552–557
   mixing time in, 50, 52
   modelling of non ideal, 198
   sensitivity to operating conditions, 313–317, 348–354
   start-up of, 52–54
   with a microfluid, *see* CISTR
Temperature, choice of reaction, 260
   influence on transport properties, 577–579, 602
Temperature gradients, criterion for the occurrence of, in pellets, 443
   see also Check for resistances
   in boundary layers, 452–455, 578–582, 620–622
   in pellets, 455–458
Temperature rise at the interface, 452–458
Thermal conductivity of a packed bed, *see* Transverse dispersion
Thermal stability of gas-solid reactors, 648–663
Thiele modulus, *see* Reaction modulus *and* Hatta number
Tortuosity factor, 622
Tracer inputs, 174–176
Trajectories, temperature conversion, 290–298
Transverse dispersion in packed beds
   dispersion coefficient for heat, 627–628, 635
   dispersion coefficient for mass, 214, 627
   importance of, 629–630
   influence of tube to particle diameter ratio, 630
Transport reactors, 640
Trickle flow reactors, axial dispersion in, 216–217
   dispersion model for, 197

Trifurcation point, definition of, 315–316
Tubes, axial dispersion in empty, 207–212
Tubular reactor, *see* Plug flow reactor
Two-film model, 382, 383

Uniqueness of the steady state in a CISTR, 329–331
Unit step function, *see* Heaviside step function input

Van de Vusse kinetics, 135–142
Van Deemter model, 195
Van Krevelen–Hoftijzer approximation method, 400–403
Van Krevelen–Hoftijzer plot, 403
   for equilibrium reactions, 418–419
   for reactions of the order (n,m), 411
Variance of a distribution, 163
   calculation of the, by short cut methods, 169–170
   for a cascade, 165, 184
   from E diagrams, 163, 167
   from F diagrams, 166, 169
Vector, definition of, 711
Venturi scrubber, *see* Gas–liquid reactors, 489

Wall heat transfer coefficient in packed beds, 629, 635
Waste water treatment, 149, 154–158
   sludge recycle in, 155–158
Wetted wall column, 459–461, 463–466
Wrong way behaviour in gas–solid reactors, 654–656

Yield, *see also Selectivity*
   definition of, 29, 85
   improvement of the, by stripping of product, 552–557
   influence of axial dispersion on the, for consecutive reactions, 205–207
   maximum, for consecutive reactions, 111, 114
   optimization of the, 705–710, 718–719, 723–725, 740–744
   optimum in a cascade, 131–135
   and reaction temperature in a CISTR, 331–339

Zero-order reaction boundary problem, 511

DATE DUE